PRAISE FOR
The Kennedy Men

"Finely written. . . . Full of considered analysis. . . . He subtly makes the intriguing connections among every aspect of these men's lives. . . . It surely will be viewed as one of the most valuable accounts of a family that vowed to help shape its country, then did." —*Newark Star-Ledger*

"[Leamer] once again proves himself to be by far the most thorough and the strongest narrative chronicler of the Kennedy family. . . . His research is thorough, but it is his elegant writing and gift for narrative that makes *The Kennedy Men* such a thorough delight." —*Palm Beach Post*

"Evenhanded. . . . The writing is always tight and often graceful. . . . Worthy of being placed on the same shelf with Doris Kearns Goodwin's *The Fitzgeralds and the Kennedys.*" —*Library Journal*

"Engaging and fast-moving. . . . A stirring narrative." —*Publishers Weekly*

"Vivid. . . . Leamer's writing is impressive throughout. . . . He sets up the themes that will be the motivation and ruination of Kennedy men."

—*Booklist*

"[Leamer] portrays the triumphs and tragedies of the male side of the family melodrama with appropriate flair." —*New York Review of Books*

"Laurence Leamer has the virtue of a lucid, compelling narrative style, a feeling for history, and a wonderful eye for the telling detail. In this companion to his vivid portrait of the lives of the Kennedy women, he sees the whole tapestry of the Kennedy men with their magnificent obsessions, and at the start of a new century he reproduces the Kennedy family saga with a mix of compassion, accuracy, intelligence, and great human insight—unputdownable!"

—Nigel Hamilton, author of *JFK: Reckless Youth*

"Laurence Leamer is a great writer and storyteller, but he is also a careful, balanced, and evenhanded historian. He spent a significant amount of time at the Kennedy Library researching primary sources. For example, in his account of the Cuban Missile Crisis, he actually listened to the ExComm tapes rather than depend on published transcripts. This is an honest account of a remarkable, perhaps unique, American political family whose influence will undoubtedly continue well into the twenty-first century."

—Dr. Sheldon M. Stern, historian,
John F. Kennedy Library, 1977–1999

"*The Kennedy Men* presents my great-grandfathers, uncles, and cousins as they were. Their story is America's story. Jack Kennedy loved the written word and had a deep appreciation for history. In *The Kennedy Men* finally we have a book Jack Kennedy would have enjoyed."

—Kerry McCarthy, Joseph and Rose Kennedy's
grandniece and a Kennedy family historian

"I don't see how anyone can read *The Kennedy Men* and ever see the Kennedy family the same way again. This book is full of extraordinary revelations written in a compelling style. It may read like a novel, but this is serious history written by a masterful storyteller. Here, finally, is the Kennedy story in all its richness, drama, and nuance."

—Dan Moldea, author of *The Killing of Robert F. Kennedy:
An Investigation of Motive, Means, and Opportunity*

Arthur Grace

About the Author

Laurence Leamer is the author of nine books, including the *New York Times* bestseller *The Kennedy Women.* He has written for numerous publications, including *Harper's, Playboy, The New Republic, New York, Washingtonian,* and *The New York Times Magazine.* He lives in Washington, D.C., and Palm Beach, Florida.

ALSO BY LAURENCE LEAMER

Three Chords and the Truth: Hope and Heartbreak and Changing Fortunes in Nashville

The Kennedy Women: The Saga of an American Family

King of the Night: The Life of Johnny Carson

As Time Goes By: The Life of Ingrid Bergman

Make-Believe: The Story of Nancy and Ronald Reagan

Ascent: The Spiritual and Physical Quest of Willi Unsoeld

Assignment: A Novel

Playing for Keeps: In Washington

The Paper Revolutionaries: The Rise of the Underground Press

The KENNEDY MEN

1901–1963
The Laws of the Father

LAURENCE LEAMER

Perennial
An Imprint of HarperCollins*Publishers*

A hardcover edition of this book was published in 2001 by William Morrow, an imprint of HarperCollins Publishers.

HarperCollins books may be purchased for educational, business, or sales promotional use. For information please write: Special Markets Department, HarperCollins Publishers Inc., 10 East 53rd Street, New York, NY 10022.

First Perennial edition published 2002.

Designed by Claire Naylon Vaccaro

The Library of Congress has catalogued the hardcover edition as follows:

Leamer, Laurence.
The Kennedy men : 1901–1963 / Laurence Leamer.
p. cm.
Includes bibliographical references and index.
ISBN 0-688-16315-7
1. Kennedy, Joseph P. (Joseph Patrick), 1888–1969. 2. Kennedy, Joseph P. (Joseph Patrick), 1888–1969—Family. 3. Kennedy, John F. (John Fitzgerald), 1917–1963. 4. Kennedy, family. 5. Politicians—United States—Biography. 6. Men—United States—Biography. I. Title.

E748.K376 L43 2001
973.9'092—dc21
[B] 2001031689

ISBN 0-06-050288-6 (pbk.)

09 ❖/RRD 10 9 8 7 6 5 4

TO MY FATHER

LAURENCE E. LEAMER

AND

IN MEMORY OF MY FRIEND

DR. STEPHEN A. COLE (1940–2000)

Contents

Book One

Book Two

Book Three

Book One

1

A True Man

Twelve-year-old Joseph Patrick Kennedy may have been dressed like a young gentleman, but he walked with the bold strut of an Irish tough full of the lore of the streets. As he hurried down Webster Street, his blue eyes exuded a hungry intensity for whatever life might offer. He was taller than most boys his age and had reddish hair and an abruptness to his features that left him just short of handsome. His strong, willful face had already lost whatever boyish innocence it once held.

Joe had been brought up on the island enclave of East Boston and knew the streets and byways with perfect acumen. Today, for the first time, he would be traveling without his family to the proud city across the bay. He would be passing through streets full of uncertainty, confronting strange new people. It was a prospect that would have filled many youths with apprehension, but nothing in Joe's demeanor suggested that he was worried about the adventure.

Joe's mother, Mary Augusta Hickey Kennedy, had arranged for her only son to deliver hats from a prestigious shop to the great ladies of Boston. Before Joe set off on his delivery in the summer in 1901, Mary Augusta looked at her son with what the family called "Hickey eyes." They were piercing, dismissive eyes that with a mere glance could stop a vulgarity in midsentence or send a supplicant reeling backward in shame. Joe's mother admonished him to behave impeccably and to refer to himself as the proper "Joseph," not the vulgar "Joe."

Joe rushed off down the street from the Kennedys' two-story home located in the best residential area. From up here on the highest elevation on the island, Joe could look down far below where passenger ships glided into the harbor packed with immigrants. Driven from their land by the great potato famine, between 1846 and 1849 nearly one hundred thousand Irish immigrants had arrived on Boston's pristine shores. Among them were Joe's grandparents. Patrick Kennedy had disembarked in 1849 on these very streets, where he and his bride, Bridget Murphy, set up residence in a tiny apartment.

After only nine years in East Boston, Joe's grandfather died. He left his thirty-seven-year-old widow with four children under the age of eight and an estate of seventy-five dollars. Bridget worked first as a servant but eventually found a job in a small variety store only a few blocks from where Joe now walked. In what was a difficult accomplishment for an immigrant widow, Bridget managed to buy the store.

Joe found his way to the hat shop and stepped up into the horse-driven wagon. As the driver guided the horse through the streets, the air was full of the stench of horse manure, the foul odor of rendering plants, the fumes of the steamers, the acrid malodor of the New England Pottery Company, and the smells of the Atlantic Steel Works.

The carriage rolled toward the mainland ferry, passing numerous taverns, dark havens that marked their presence by small signs. If Joe's father, Patrick Joseph "P. J." Kennedy, had set a symbol of his success on his mantelpiece, it would have been a humble glass of beer. As a youth, P. J. worked a short while as a stevedore. Then P. J.'s mother had grubstaked her only son to open a pub. As for her daughters, Bridget followed the pattern of her people and her time. She sent one daughter off to work in the jute mills and settled for another to become a shirtmaker, while she did everything for her son.

P. J. drank only enough so that he would not appear a parsimonious sort, his shot glass filled not with whiskey but with beer. In P. J.'s tavern, as in most others in East Boston, the talk was usually of politics. P. J. carefully built his clientele, expanded into the wholesale liquor business, and entered politics as a state legislator. Favors were the mortar of P. J.'s career, and he built his career one brick at a time.

By the time Joe was born, P. J. was the Democratic ward boss for East Boston, one of the most powerful political figures in the city. With his husky figure and handlebar mustache, P. J. appeared the perfect rendering of an Irish-American politician. Every evening the petitioners arrived at the house on Webster Street, bewildered new arrivals clutching legal notices, unemployed workers looking for a city job, and widows about to be evicted.

As the carriage turned onto Meridian Square and the ferry landing, it passed the Columbia Trust Company, an imposing four-story brick and iron building. Joe's father was a founder of this new bank, one of the many businesses in which he was involved. The *East Boston Argus-Advocate*, in a rare moment of candor, described P. J. as "slick as grease." Slick as grease he was, and slick as grease he had to be to climb out of the prison of poverty and accumulate a fortune, all without ever moving from East Boston.

When a husband died, P. J. was there with his condolences, but he was also there to buy the widow's house at a good price. He and his business associates bought extensive real estate and other businesses in East Boston, usually keeping their interest quiet. P. J. used his political power as a lever to push him into all kinds of deals, including a major position in the liquor wholesale business, an industry that he helped oversee in the state legislature.

No matter how well off he became, P. J. never flaunted his wealth. Though he sailed a yacht in the harbor in the summer and wintered in Florida, he still rode the trolley and tipped his hat to the ladies.

P. J. was a shrewd, practical man who endowed his son with his own deep insight into all the machinations of human beings. He was a man, however, who had none of his wife's overweening ambition, a man who saw East Boston as world enough for himself and for his son.

Everywhere Joe cast his eyes, he spied new arrivals from the ports of Europe and heard the rancorous clamor of peddlers. The horse-drawn wagon brushed past Russian and other Eastern European Jews selling goods from pushcarts and gesticulating Italians hawking sausages and vegetables where thirty years before Irish widows had begged passersby for a coin or two. These new immigrants, especially the Jews, were an exotic, threatening element to Irish-Americans. They were pouring into East Boston, crammed into triple-decker houses and tenements. There would soon be enough of them to become the largest Jewish community in New England, and by the time they founded their synagogues and opened kosher markets, the second-generation Irish-Americans were pulling out. Joe's father could have left too, but this was his political bailiwick, and he was not giving it up to these new arrivals.

The carriage waited in line before moving onto the ferry that sailed between East Boston and Boston proper on the mainland. The pedestrians hurried onto the boat, paying their one-penny fares and passing through the turnstile to share the ride with a polyglot cross section of Irish, Italians, Jews, businessmen, shrouded widows, and youths. Teamsters quieted their teams of horses while peddlers shoved their pushcarts aboard.

The ferry, like the island community itself, was an inelegant, practical affair, a low-bottomed vessel with a long smokestack set amidships, belching a spume of black smoke into the air. As the ferry approached Battery Wharf, Joe saw the commerce of Boston in all its diversity. Joe's immigrant grandfather Patrick had been a cooper, a barrel-maker. Wooden barrels full of foodstuffs and sundry items rested everywhere along the wharves, sitting on horse-drawn wagons or standing dockside waiting to be lifted onto the ships.

Joe's wagon rolled off the ferry and moved slowly through the clogged streets of the North End. Here, where almost a century and a half ago Paul Revere had created his elegant silver pieces, immigrants sat in sweatshops sewing pants and shirts, often for more hours than the day had light. The North End was a foul, fetid area where more than twenty-five different nationalities lived in uneasy proximity. Over twenty-five thousand people were jammed together there as tightly as in any city in the world except Calcutta.

As much as Joe's mother might have wanted him to discard much that marked him as Irish-American, that heritage was his free pass through these dangerous streets. The Irish were diminishing in numbers, but they still controlled the waterfront, and at night, if an Italian or a Jew dared trespass in these precincts, he might leave with a broken nose or a bleeding face. The ethnic groups struggled against each other, the Irish against the Jews, the Jews against the Italians, the Italians against the Greeks. Stick to your kind. That was the basic axiom of life.

Beyond these sad streets lay the commercial areas of downtown Boston. These stores drew their clientele from all over the city. Highborn ladies in expensive finery stepped gingerly through the crowded streets. Chattering shop girls hurried back from their break. Workingmen with wages in their pocket shopped for Sunday suits, and unemployed men wandered aimlessly along.

The wagon moved up these teeming roads, finally coming upon an open space that exploded with light and the appearance of freedom. Here lay Boston Common and the Boston Public Garden. The Common, founded in 1634, is a massive version of the public parks found in towns across New England, emblematic of the democratic ideals of the region. The gallows once stood on the Common, and until 1830 Bostonians reserved the right to graze their cattle there. The formal, elegant Public Garden, founded in 1839, fit in with the aristocratic ideals of nineteenth-century Boston and the Protestant elite that controlled it. Along the pathways even the weeping willows and beeches seemed as properly garbed as the Bostonians who strolled past the swan boats.

The elite walked sedately among the flowerbeds and statuary, but there was another world as well, a boy's world full of danger. Outsiders like Joe knew that they had to be wary. In winter, Boston Common became a field of battle. The Irish boys made their way up from the North End to take on the highborn Protestant boys in epic snowball fights. The Irish toughs were a dark and terrible legend, merciless in their attacks against the young blue bloods who asserted their own young manhood and held their ground against assaults on their turf.

Joe was not much of a fighter himself, and he had not come here today to confront any of the boys who lived near Boston Common. His carriage moved on toward the townhouses and mansions spread along Commonwealth Avenue in the new Back Bay area and along Beacon Street. It was a world so different from the one below that it was as if life itself should have a different name here. So in a way it did. Here on these broad avenues the Protestant elite ruled over the space and grandeur of the city, over its elegance and art, and applied fine-sounding old English names to their streets and apartments. Clarendon. Exeter. Somerset. The names resonated like fine old crystal.

When a servant answered the door, Joe politely stated his business and waited for the lady of the house to come to the door to try on her new hat. The elite ladies believed that they had nostrils of such refinement that they could catch the scent of an Irishman before they even saw him. They had as their guides not only their own servants but also half a century of magazine caricatures by Thomas Nast and others portraying Irish-Americans as quasi-apes, as looming, salivating simian wretches.

Joe's face at first glance showed nothing of what the ladies considered the crude excess of an Irishman's features. The matrons could try on their hats with the pleasing knowledge that their bonnets had been touched not by a rough Irish hand but by the fine fingers of a young man who could have been their own son.

These ladies were "New England Brahmins," a term coined by one of their kind, Oliver Wendell Holmes. The hereditary aristocracy of the region fancied itself much like the Hindu religious caste: a natural elite, sanctified by an all-knowing God and a just social order. The Brahmin was, as Holmes wrote, "simply an Americanized Englishman. As the Englishman is the physical bully of the world, so the Bostonian is the aesthetic and intellectual bully of America." Bullies they were, protecting their sacred precincts from loathsome pretenders who dared to dress themselves in the language and lingo that was not theirs, attempting to pass as one of their betters. The Brahmins had an almost perfect self-confidence. Nothing, no momentary fall in their economic well-being, no peasant races disembarking on their land, could move them off their high ground.

The Boston upper class was largely without irony and had a blessed ability to forget what should best be forgotten. They tended not to focus on a past in which many of their ancestors had made fortunes in a three-sided trade that had slaves as one of its sides, or a present in which their coffers were enriched by the cheap labor of the immigrants they largely despised. They were proud that at their Somerset Club on Beacon Street no member would think of engaging in the disgusting practice of doing business in their social bastion, blissfully forgetting that they were such a close-knit elite that they could easily do their dealing elsewhere.

These were a people of restraint who at times mistook manners for morals. The flowering of a distinctive, dominant New England literature and culture was largely over. The blossoms had fallen, leaving the thorns of reaction and regression in a people who had turned from history to genealogy, from literature to antiquarianism. The Brahmins were facing the melancholy mathematics of democracy: one foul Irish immigrant vote was worth as much as that of the most refined Brahmin gentleman. The Irish politicians would soon have the votes to take over the Brahmins' city, and there was little the Protestant elite could do to prevent it.

The Protestant upper class, however, did not simply slink into the night, carrying away the burden of their culture and their past. They were astute businessmen. They sat atop vast wealth that they continued to amass, dominating the economic life of New England. In their leisure hours they asserted themselves where a man's vote did not matter, over the cultural and philanthropic life of the city.

As diminished as this world might soon be, it was still the summit toward which Mary Augusta pointed her son. Joe went from one imposing home to the next, learning the lesson of this exercise: these Brahmins were a royalty to whose company he could dare aspire. There was nothing about him—not his name, not his features, not his manner—that marked him as someone apart from these rich ladies and their elegant homes. Nothing seemed to prevent him from living the life they led.

When Joe arrived back in the house in East Boston, he was in his mother's universe. Mary Augusta was the monarch with absolute sovereignty over her small kingdom. She was five feet seven inches tall, towering above most of the women of her era, with a posture so straight that it seemed to add even more inches to her height. Mary Augusta was her own greatest creation, having reinvented herself as an aristocratic lady. No one who saw her walking to church with stately grace would ever imagine that her father had been a laborer. Even when she was a young woman and her father had risen to the point where he listed his occupation as engineer, the Hickeys were still not well off enough to live in anything but a rented house.

Mary Augusta was twenty-nine years old in 1887, approaching spinsterhood, when she spied P. J. walking past her kitchen window and set her cap for him. She became a splendid wife, no less so because she was so aware of her virtues. A woman of deep faith, she had been educated by the nuns for her role as wife and mother.

Mary Augusta loved her two daughters, Mary Loretta and Margaret Louise, but Joe was the measure of all things. Joe, not his younger sisters, would go out into the world. Mary Augusta taught Joe that there was no horizon on which he could not set his eyes. His sisters could be coddled and spoiled, for if they married well and properly, they might spend their lifetimes coddled and spoiled. As for Joe, his mother did not so much give him love as the promise of love. She spooned out her affection to Joe like a tonic that had to be taken in only the smallest of doses.

Mary Augusta was so concerned with the impression Joe would make on the world that when he was born on September 6, 1888, she insisted that he be named Joseph Patrick, not Patrick, after his father and his grandfather. Patrick was the most common Irish name, and she would not have her only surviving son forever marked by his immigrant forebears. Mary Augusta was trying to bring Joe up as her little Catholic gentleman, all frills and fanciness, but her son had never fully gone along. For his first formal photographs, she had Joe photographed in a long dress with a bow around his neck. Even then Joe stared out at the camera with firm unyielding eyes and a clenched fist.

Mary Augusta's regimen as a mother was to teach her first and only surviving son the merciless rituals of civility. For Joe's mother, the relentless pursuit of civility was not a trivial matter. In the radical egalitarianism of America, people learned to mimic the manners of those whose company they sought. The most vulgar and ill bred could affect the manners of their betters for a time, but eventually the mask of civility would fall. These ersatz ladies and pseudo-gentlemen often exposed themselves at dinner by choosing the knife as their favored implement for eating rather than the fork. Eating with a fork became such a symbol of civility that its teaching was laid out in Joe's parochial school curriculum ("shall eat with a fork, rather than a knife; shall take small mouthfuls of food and masticate quietly").

Young Joe lived largely in a female world, and from that world he took many of his ideas of womanhood. He had his mother as his guide and goad, a constable of civility. He had his two younger sisters, who constantly deferred to him. The Irish servant girls treated the young master as a royal being. He was the center around whom all things revolved, a condition that he took as the natural order of things. Joe saw even the Catholic Church largely through the eyes of women, particularly the nuns who taught him at parochial school.

Joe listened to the moral axioms proffered by the nuns and followed his mother's detailed course in manners, but he chafed at all the restrictions she put on him. Mary Augusta represented a secret danger to her only son. Her idea of civility, of culture, was a seductive call that risked closeting him away so that he might never become a true man. Her house was a sanctuary of rectitude and security, but it was on the streets below, unprotected by his mother's sheltering skirts, that Joe had to journey to become a man.

Joe was on two journeys: one toward civility along a pathway led by his mother, the other a struggle toward true manhood. President Theodore "Teddy" Roosevelt feared that an insipid, feminized culture was castrating men, robbing them of their vitality. Roosevelt saw each nation engaged in its own struggle for survival, a fight in which only a nation of true men might survive. "Any nation that cannot fight is not worth its salt, no matter how cultivated and refined it may be, and the very fact that it can fight often obviates the necessity of fighting," Roosevelt asserted. "It is just so with a boy."

G. Stanley Hall, the most prominent psychologist of the age, taught that boys replicate the evolutionary process of civilization, from savagery to barbarism to refinement. It is a process, he lectured, that each boy has to go through to become a true man. A teenage boy who was "a perfect gentleman has something the matter with him.

"An able-bodied young man, who can not fight physically, can hardly have a high and true sense of honor, and is generally a milk-sop, a lady-boy, or a sneak," wrote Hall, a Harvard Ph.D. and the president of Clark University. "He lacks virility, his masculinity does not ring true, his honesty can not be sound to the core." Risk was risk and danger was danger, and Hall did not shrink from the implications: "Better even an occasional nose dented by a fist, a broken bone, a rapier-scarred face, or even sometimes the sacrifice of the life of one of our best academic youth than stagnation, general cynicism and censoriousness, bodily and psychic cowardice, and moral corruption, if this indeed be, as it sometimes is, its real alternative."

Out on the streets of East Boston, the games were sometimes rough. Boys had their noses bitten, ears half torn off, groins kicked, heads stomped on, and lips split. Boys yelped and screamed, exhorted and cursed. A boy asked no quarter and gave none. A boy who whined was a sissy, a dandified, effete mommy's boy.

Danger was everywhere. Joe often hitched a ride on the back of one of the long coal pungs that moved laboriously from the docks to the ferry. These horse-drawn wagons were so long that the driver rarely could strike his unwanted passenger, as often happened when boys jumped on a carriage. Joe still might have fallen off and had his legs crushed by one of the wheels. There was danger even when he stayed home. Joe had been injured playing with a toy pistol; so had one of his friends, and the boy died of blood poisoning. The dead boy's brother invited Joe to go sailing with him. It was the first of the month, the day on which young Joe always took confession, so he said no, and the boy upset the boat and drowned. Danger might be omnipresent, but Joe stepped around it like a puddle of water in his path.

Life in the streets, though, was not all danger and risk. Young Joe loved the rituals of patriotism. He always attended the parades to watch the drum corps and the Civil War veterans and the bands marching proudly by. One Memorial Day he got together all his friends in uniforms, and they marched in the parade, falling out long before its end. Back at the house, he orchestrated a flag-raising ceremony with all the neighborhood children present, and his own sister Loretta swathed herself in a flag and wore a glorious "Columbia" crown.

Joe's father and mother could easily have given their precious son an allowance large enough that he never would have had to bother getting a foul taste of the workaday world of America. They did not do so, however, and were proud that young Joe went out hustling jobs. He ran errands at P. J.'s bank. He hawked newspapers on the street corner. He lit stoves on the Sabbath for Orthodox Jews.

One summer Joe got together with a friend, Ronan Grady, to raise pigeons, which many in East Boston considered a delicacy. Ronan had the coop and the pigeons, but Joe didn't fancy himself as a pigeon farmer, feeding the birds expensive food, cleaning the coops, and waiting months for them to fatten. Instead, he and Ronan regularly picked out two of the most likely pigeons, secreted them under their shirts, and took them to Boston Common. There they released the birds. By the time the boys got home, their pigeons had already returned, bringing amorous partners with them. The boys sold the birds and split the profits. Joe was beginning to learn that there was nothing worth more than a good idea and a better angle.

The nuns might educate his sisters, but for Mary Augusta's son, only the finest secular education would do. Joe set out in September 1901 to take the ferry to attend seventh grade at Boston Latin School on the corner of Dartmouth Street and Warren Avenue. He was entering what was probably the finest public school in America. Alumni included Samuel Adams, one of the fathers of the American Revolution; Ralph Waldo Emerson, the author; Charles Eliot, the president of Harvard; and George Santayana, one of the university's most distinguished professors.

The youths who surrounded Joe in the Boston Latin classrooms did not bear the great names of the old city. The upper-class Protestants thought of Boston Latin as *their* school, but they had largely given it up rather than have their sons' sweet souls soured by sitting next to the likes of a Joseph P.

Kennedy. They had made their hegira to the undefiled premises of a group of new private prep schools scattered across New England, where their sons would sit only among their own kind. That left only a few of the poorer Brahmin brethren to sit in classrooms full of immigrant sons and grandsons.

The boys who dominated the academic life of the school were almost all Jews. Joe was not one of those thrusting his hand up, waving it for attention. He was not a good student. His grades were pathetic, including Cs in elementary and advanced Greek and his second year of elementary French; Ds in English, elementary history, elementary Latin, elementary algebra, and geometry; and Fs in his first year of elementary French, elementary physics, and advanced Latin.

These grades did not temper Joe's ego. He looked disdainfully at the humorless, relentless, merciless struggles of grade grubbers. For him, the glory of these years lay elsewhere, especially on the athletic field. On the baseball diamond all of his natural aggressiveness played out. He slid with spikes up, argued with umpires whose casual ineptness appalled him, and batted each time as if the game depended on him stroking the ball over the fence. For Joe, there was purity in this world that he found nowhere else. Years later he would lovingly remember the details of each school game, reliving the glory of those long past days.

Joe befriended one of the best athletes at Boston Latin, Walter Elcock. Not only was Elcock captain of the football team, but he was sure to be chosen captain of the baseball team as well. Joe took his friend to steak dinners that he would not have been served at home and talked him into stepping aside so that Joe would be named captain. Thus two years in a row, Joe was captain in name and deed.

Joe learned of the profound dangers of sex, not of its pleasure. There was no purity in the world of sex, especially not in the Ireland from which Joe's ancestors had emigrated. In the name of God, the peasant priests drove the sexes apart, patrolling the Irish countryside in search of couples so foolish as to seek out a dalliance. Men married late and reluctantly, seeking another farmworker as much as a wife. Then, and only then, did they partake in the short, brutish business that was sex and prove their manhood nightly by continuing to lift a few with friends at the village tavern.

In the Boston of Joe's childhood, a gentleman did not talk about sex. As for children, when they talked of "the dirty place" or "the dirty parts," it was clear of what and where they spoke. They might disguise the words, but they could not disguise the acts. Hall recalled that, growing up in the small town of Worthington, Massachusetts, youths experimented with "homosexuality, exhibitionism, fellatio, onanism, relations with animals, and almost every form of perversion."

The list of sexual experiments in East Boston may have been smaller, but life for an adolescent was presumably not radically different. No one of any honor and decency spoke of such matters, and Joe most likely stayed clear of such behavior in word and deed. Joe's own family displayed attitudes suggesting that they considered sex a matter largely peripheral to the serious business of life. As a young man, his father was too busy succeeding to squander his time in momentary flirtations. He had not married until he was nearly thirty, only then starting his family of four children. Joe's own uncle, John Hickey, a doctor, and his aunt Catherine had never married, and indeed, they lived together—another fine example of how the bothersome business of sex could be exorcised.

Joe was six feet tall, far above the average height for his generation, and his height advertised his virtue and manhood. He was a hardy, athletic, outgoing youth. He had an interest in the opposite sex, but it appeared to be confined to the narrow parameters of civilized life. In the summer of 1907, Joe met a petite young woman in Orchard Beach, Maine, where his family spent part of each summer. He had met the vivacious sixteen-year-old a decade before on the same beach, but he did not remember her. Her name was Rose Fitzgerald, and she had all of the virtues that his mother had taught him to hold dear in a woman. She was a deeply religious Catholic. She was a cultivated woman who could play the piano. She was a far better student than Joe. And not least among her merits, she was the beloved favorite daughter of the mayor of Boston, John "Honey Fitz" Fitzgerald.

Honey Fitz was everything Joe's mother had taught her son to deplore: a professional Irish-American full of self-conscious blarney. This man sang "Sweet Adeline" at the hint of an invitation and cried Irish green tears on cue. He was the kind of Irish-American politician the Brahmins hated, a mountebank who, on a congressman's salary, had built a mansion in Dorchester and jig-danced away from anyone who tried to investigate him. He was an embarrassment to those Irish-Americans who were attempting to brush the straw of Ireland off their clothes. He was nonetheless a man of such immense native sagacity that he had served three terms in Congress from Boston's North End while living in Concord, a full sixteen miles from his district and the constituents whom he vowed he loved so much. Honey Fitz was deeply possessive about power, publicity, and his beloved daughter Rose. He would choose her suitors, and he was not about to see her wooed by Joseph P. Kennedy.

Joe began a romance with Rose that was both innocent and clandestine. The couple met in the Boston Public Library after Joe's baseball games and wherever they could manage a few delicious moments together. For Joe, the risks were penny-ante, a momentary embarrassment. For Rose, however, this mild dalliance was an adventure of high order. The religious guides were full of terrifying warnings of the fate in store for the young woman who did not protect her virginity as life itself ("He is shut out from the Kingdom of God. His portion will be the worm that never dieth, the fire that is never quenched. O Christian maiden, tremble before this awful sin!").

Rose had wanted to attend secular Wellesley College. Instead, at the insistence of Bishop William O'Connell, Rose's father enrolled his daughter at the Convent of the Sacred Heart in Boston. Then a year later, in the spring of 1908, when his almost eighteen-year-old daughter expressed her intention of marrying Joe, he sent Rose and her younger sister, Agnes, off to Europe to the Convent of the Sacred Heart at Blumenthal, Holland. There she would be cloistered away from Joe and his pernicious influence.

The Joe from whom Rose was sailing away had become an ideal specimen of young manhood. At nineteen, he was half a foot taller than most of the men of his father's generation, and his striking face was perfectly groomed, his reddish hair impeccably brushed, his eyes a brilliant blue. He had a proud military bearing. His manners were military too, given more to abruptness than graciousness, but among his peers at Boston Latin he was a popular student.

The fact that Joe had taken an extra year to graduate from Boston Latin hardly diminished his popularity. He was colonel of the Cadet Corps, president of his class, and a baseball player of legendary repute, just having won the city trophy for the best batter with the awesome average of .667. If there was one disconcerting note, it was the prediction in the 1908 Boston Latin School yearbook that Joe would earn his fortune "in a very roundabout way."

2

Gentlemen and Cads

It would have taken a keen observer to realize that the arresting young man striding along with such confidence through Harvard Yard in September 1908 had no business being there. Joe's academic record at Boston Latin School was so abysmal that he had repeated his last year and was a year older than most of the other freshman. On his first day at Harvard, twenty-year-old Joe was as much a child of special privilege as some of the denizens of the Boston elite who had little to recommend them to the Ivy League school but old money and old names.

Even if Joe had had a good academic record, the son of a leading Boston Catholic politician belonged at Boston College, the proud new Jesuit institution that sat on Chestnut Hill, where Bishop O'Connell intended that it would look down upon Harvard and its secular world. But for a young man who aspired to the pinnacle of American life, the most prestigious university in America beckoned irresistibly. No university has ever dominated the intellectual, social, and athletic life of an American city the way Harvard dominated Boston in the early years of the twentieth century. Joe was not walking this day along mere bricks and stone but on a noble path toward everything that he aspired to be: a civilized gentleman fulfilling his mother's dreams, a celebrated athlete, a brave true man of Harvard.

On the same day that Joe matriculated alongside other public school graduates, the prep school youths arrived from their ghettos of privilege. Most of them set up housekeeping on the celebrated Gold Coast, the row of private dormitories on or near Mount Auburn Street. They brought their carriages, cars, and servants, and they greeted each other with casual familiarity. They fit into their college life as comfortably as they would have checked into a first-class stateroom on a transatlantic crossing. They were now "Gold Coast men," and so they would be known during their years at Harvard.

As these young men settled in, reminiscing about summers full of European travel, sailing, tennis, or western sojourns, Joe found more meager accommodations on campus. In doing so, he defined himself as a "Yard man." Yard men were in steerage on a ship they did not know, among passengers they had never met, on a journey they had never taken. The dorms were frequently foul places, with underground toilets and so few showers that many men went unwashed. The filthy windows let in little light, and at night the students studied by flickering gas jets. "Compared to any respectable house or hotel they are all vile," wrote one undergraduate critic in the *Harvard Advocate.*

Joe did not want to be relegated to a tedious life among his dormmates but sought to stand among the privileged men of the Gold Coast. He imitated their dress, manners, and social attitudes. He had a rare gift of social mimicry and a constant wariness among his social betters never to betray his past.

Imitation is not always the sincerest form of flattery; sometimes it disguises its opposite. Joe was no unctuous wanna-be, but a young man whose pride matched that of the most arrogant of the Brahmins, a pride that he hid at times under a cloak of deference. Joe had a profound desire to be accepted in their world, but he could not openly admit to such a goal. Social ambition is the one human aspiration that dares not speak its name; to be caught at it is to fail.

Joe perfectly mimicked the attitude of the Gold Coast men toward their academic work. To many of them, it was little more than a tedious aside to the real business of college life—election to their private club. A club was the only proper place for a gentleman to eat and socialize. The winnowing process began in the sophomore year when upperclassmen chose members for the Institute of 1770.

"A hundred or so of the class are devoured by the Institute and carefully told that there are two kinds of men at Harvard—'gentlemen' and 'cads,' " wrote Paul Mariett in the *Harvard Illustrated* in May 1911. "The Institute contains the gentlemen." The first seventy or so chosen became members of DKE, or the Dickey. They in turn joined "waiting clubs" out of which the new members of the final ten clubs would be chosen, the most prestigious being Porcellian, followed by AD and Fly.

Joe found it impossible to get to know many of the Gold Coast men. They kept to themselves and their clubs. Their motto was "Three Cs and a D, and keep out of the newspapers." "Our friendships are made in our rooms, with men who appreciate a good cigar much more than a Greek pun," one of them wrote, dismissing the tedious world beyond Mount Auburn Street.

For a young man who aspired to great wealth, it was natural that Joe gravitated toward the Harvard upper crust. Everywhere Joe looked, he saw irrefutable evidence that money and class were the same. The names of over half the millionaires in Boston were listed in the Social Register. About two-thirds of the Bostonians who were officers and trustees of major American businesses came from the upper class. They sent their sons to a Harvard that those young men largely dominated.

By the time Joe entered Harvard, he was disgorging anything that might mark his Irish immigrant heritage. He did not drink, sidestepping one stereotype: the bulbous, blustering, belligerent Irish drunk. He had been born and brought up in an East Boston known as an immigrant enclave. During his Harvard years, the family moved to the prestigious seaside suburb of Winthrop.

Joe could change his accent, dress, and home address. He could not change the fact that his grandmother had been a servant, as had most of the Irish immigrant women of the famine generation, and that his ancestors had been peasants. His grandmother's name, Bridget, had been so ubiquitous that the Brahmin ladies referred to female servants as "their Bridgets," and the now-debased name had largely disappeared with the next generation.

For the most part Joe's professors felt nothing but contempt for the immigrant onslaught that they believed had so besmirched the pristine reaches of their Boston. One of them, Barrett Wendell, reflected that almost everyone of his class had contemplated suicide because of the immigrants.

Joe was not one to query his teachers and challenge their ideas or expose his background by rubbing against the wrong kind of ideas or people. As at Boston Latin School, he was no student. Joe took no pleasure in the bounty of courses set before him. In his freshman year he barely managed a "gentleman's C." That was prime evidence that he had not been infected by the contagion of academe, losing his manhood by sitting too long in class and library. The very mediocrity of his grades suggested that the professors and their arid pedantries had not produced what Teddy Roosevelt called another "over-civilized man, who has lost the great fighting, masterful virtues."

The Brahmin world that Joe wanted so desperately to enter was more than a charade of social rituals and endless disdain for the vulgar masses. Harvard gentlemen bravely shed their blood in their country's wars. For Joe and the other students, the martyred dead were not simply names on monuments that they breezed by on their way to class. There were Harvard men still living who had fought in the Civil War and were the living testament to noble acts. When the students entered Memorial Hall, dedicated to the memory of the Civil War dead, they doffed their hats; those who neglected this modest gesture of respect were greeted by the sound of hundreds of students drumming silverware on their water tumblers.

Harvard took itself seriously as an incubator of courage, considering its classrooms and playing fields as the highest training grounds for a true manhood that would have its final test on the fields of battle. Even William James, the celebrated Harvard psychology and philosophy professor, thought that war was the natural arena for young men to prove themselves. "Our ancestors have bred pugnacity into our bone and marrow, and thousands of years of peace won't breed it out of us," he wrote in a famous essay attempting to create some "moral equivalent to war." "So far as the central essence of this feeling goes, no healthy minded person, it seems to me, can help to some degree partaking of it. Militarism is the great preserver of our ideals and of hardiness, and human life with no use for hardihood would be contemptible."

The Brahmin elite had more modest sacrifices to make in peacetime too. Joe could hardly aspire to be part of the Brahmin world without realizing the extent to which the New England Protestant elite, more than any other group in America, developed the idea that great charity is the natural concomitant of great wealth. If Joe was ever to stand with the elite, he too must be seen as a man of beneficence. America was not a land in which a life of the senses, be it of Epicurean indulgence or of cultural pursuit, was a worthy goal. The vigorous pursuit of wealth was a manly endeavor, but the mere spending of that wealth an heir's pallid pleasure.

Joe, first of all, had to be accepted among the elite. Three-quarters of the former prep school students made the final clubs, and except for an occasional athlete, almost none of the public school men gained entry. As a public school graduate and an Irish-American Catholic, Joe had no chance of making a final club unless he made his name widely and positively known on campus. Joe was not the populist sort who enjoyed mixing with a broad range of humanity, but he understood the value of getting his name out there. He joined the finance committee of the Freshman Smoker, the one event where most of the class socialized together. He also was one of the fifteen ushers for the class dinner, a position that involved signing up twenty fellow freshmen for his table. Both of these positions involved raising money, typically the least desired and thus the most accessible entry point to an organization.

Athletics was the arena for attracting attention to one's name, and here the Gold Coast men confronted Joe. As much as Joe loved sports, these prep school youths had gone to schools where athletics was far more important than at Boston Latin and other public schools. Prep school graduates were generally bigger, taller, heavier, and stronger than the public school men. In 1912 the *Harvard Crimson* noted that a study by Dr. D. A. Sergent of the Harvard Hemenway Gymnasium of roughly one thousand freshmen concluded that "the private school men were in every way superior physically to public school men."

The prep schools dominated the varsity sports. They made up nearly the entire first-string Harvard football team; in one typical year, 1911, all but one of the twenty-three football letters went to prep school men. So did most of the crew and track letters. Only baseball, a sport that many thought plebeian and not quite gentlemanly, had a more balanced mix of public and private school alumni. Even here, eight of the fifteen lettermen came from prep schools.

The Harvard obsession with masculine sport was in part a class struggle, a way to maintain supremacy in a rude and brutal new industrial world. This idea that scholarship and athletics were natural halves of an education resonated deeply within Joe. The veritable machine of legend, the field of manhood, was the Harvard football team. As much as Joe desired athletic fame, however, and as large and strong as he was, he simply did not like the bruising physical demands of the game.

The Harvard man's spirit was unleashed on the football field in a torrent during the annual Harvard-Yale game. In Joe's sophomore year the undefeated Harvard Crimson lost to the undefeated, unscored-upon Yale Blue, with Vice President James Sherman in the Cambridge stands. An observer, a former Army coach, pronounced the line play "the most magnificent sight I ever saw. Every lineman's face was dripping with blood."

The nearest Joe could safely get to the hallowed playing field was to have prominent football players as two of his closest friends. Robert Fisher was a guard of such ability that he became an all-American. Fisher had the prep school credentials of a year at Andover, even if he had gone there on a scholarship. He was so poor that he had to commute from Dorchester. Joe invited the star athlete to live with him free in his room at Perkins Hall, one of the Harvard dorms. In one neat move, at no extra cost to himself, he had defined himself as a generous friend while attaching himself to the man who would become one of the Harvard sports heroes of his time. Tom Campbell, his other new friend, was a star halfback and a graduate of Worcester Academy. Campbell's primary social demerit was that he was a Catholic, and he did not bring Joe the cachet gained from walking home to his dorm with his roommate, the celebrated Bob Fisher.

During his freshman baseball practice Joe made his third close friend at Harvard, Robert Sturgis Potter. Here Joe had found a student who struck every social high note. Potter came from an old Philadelphia family. He was a graduate of St. Mark's, the crème de la crème of prep schools, and lived in Randolph Hall, one of the most desired residences on the Gold Coast.

Joe was not a cynical arriviste who befriended these men only because they might advance him. He enjoyed them, and it marked not simply his social ambition but his confidence that he dared to reach out to make such friends. In Potter and Fisher, Joe had made brilliant friendships, for Fisher was class president their sophomore year, with Potter succeeding him their junior year.

Although Joe did not fancy himself a football player, he knew that on the baseball field he could show that he had the true stuff. Joe's name might be carried far beyond the reaches of the playing field. Spectators came to watch by the thousands. Not only other Harvard men but also the public considered these young men heroes as much as a later generation would celebrate professional athletes.

"Important fall baseball practice will commence Monday," the *Harvard Crimson* announced on the front page on October 1, 1908. "Every man who is eligible for the University nine and cannot be of service to the football team is requested to report for this practice."

Joe had already taken the full measure of the other freshman players. That first day, walking over to take batting practice, Joe said to his friend Arthur Kelly: "We're the two best damn ballplayers on the team!"

Joe became the first-string first baseman, and one of the outstanding players on the freshman team. One of the best batters, he was also flawless on the field. The team lost only one game all season and tied another in which his absence due to a knee injury was noted. At the end of the season the *Harvard Crimson* described Joe as the most likely prospect to move up to a starting position on the varsity. "It is expected that this year's successful Freshman team will also contribute valuable material," the paper noted. "J. P. Kennedy is a likely man for first-base."

That summer Joe was riding on a bus when the driver mentioned that the vehicle was for sale. This gave Joe the idea of buying the contraption and turning it into a tourist bus. He lined up Joe Donovan, a fellow Harvard man, as his partner. They painted the wagon a glorious cream and blue and with neat lettering along the sides christened the bus "The Mayflower." In a story crafted by Horatio Alger, the two young Irish-American entrepreneurs would have set off along Boston's hallowed streets picking up passengers. The reality was that most of the prospective customers arrived at South Station, and another company had the right to park there.

Joe was not one to nibble around the edges of this new business. He went to see Mayor Fitzgerald. As much as the politician disliked his daughter's boyfriend, Fitzgerald could at least appreciate Joe's initiative. Within a few days the man at South Station learned that his buses could no longer park there, and the new possessor of the coveted space maneuvered the Mayflower into the vacated spot. For a short while Joe worked as the guide, wearing a black and white cap and shouting through the megaphone. Such plebeian endeavors, however, were not for him, especially when he could cheaply hire others. For three summers, Joe and his friend ran the service, netting more than five thousand dollars, an enormous sum for summer labors. Joe had seen again that initiative alone was a fool's parlay and fairness a loser's gambit. No matter how good your idea, it was equally who you knew and how you used them that mattered.

On the baseball diamond Joe had defined himself as one of the better-known members of his class, a student athlete who had every prospect of becoming one of the stars of the Harvard team. In the classroom he had established himself as proudly mediocre, displaying contempt for scholarly endeavor worthy of the elite. His friendships with Fisher and Potter had elevated him into the company of the most revered men of Harvard. He had not let matters rest there but had gotten his name on various committees, further advertisements for himself.

Joe had done everything to ensure his election to one of the esteemed private clubs. The selection process for the ten final clubs was arbitrary in its means, and final in its judgments. The members of the Institute of 1770, the first step in the process, were chosen ten at a time, a social plebiscite brutal in its finality. Joe's confidence and shrewd social maneuvering paid off. In the fall of his sophomore year, Joe and his three friends were chosen together by the inner club, the Dickey. For Joe his selection may have appeared inevitable, but it was a measure of the gauntlet he had passed that in a typical year, out of the 116 men chosen for the Institute of 1770, 112 were Gold Coast men, and only 4 roomed elsewhere, including 2 Harvard lettermen.

If Joe had been brashly overconfident in assuming that his election to the Institute of 1770 was a foregone conclusion, he had every right to think he would now be chosen as a member of one of the ten final clubs. After all, he was a Dickey, already in an honored special circle. It was not a question of whether Joe would make a club, but which club would choose him. That was a matter of ample debate among his friends as the day finally arrived. Would it be Porcellian? Or perhaps the gentlemen from AD would cherish his company? Then again, what of Fly or Spee, or, God forbid, one of the lesser clubs—Phoenix, DU, or Iroquois?

On a gray day in the late winter, Joe and Fisher were waiting in their room at Holyoke for the expected knock. As they paced anxiously, the clubmen spread out across the campus with the cherished invitations in their hands, knocking at the doors along the Gold Coast, though occasionally entering the dormitories in Harvard Yard. After handing out the precious letter, they took their newest member with them and moved on to the next person on their list. As the afternoon wore on, the groups grew larger—three, then five, and finally ten new members, along with the older clubmen, off in their world, beyond and above the public universe of Harvard. It was a moment of euphoria, leading finally to the clubhouse and inevitably to an evening of dinner, drinks, cigars, and manly conversation.

The tap on Joe's door came, as he knew it would. It was the gentlemen of Digammas, but they were coming for Fisher, not for Joe. And so he sat in the room and waited while outside the singing and the shouts grew louder. And he waited and he waited, and as the short winter sun descended across the Cambridge sky, he was still waiting.

For Joe to have traveled so far up this road only to be turned back meant that he had been ostracized, pointedly thrown back into the common ground of the greater Harvard. He was as marked now as if he had been made a Porcellian man.

Being passed over for club membership was only part of what Joe considered the terrible unfairness of that year. That spring Joe did not move up to a starting position on the Harvard varsity but sat on the bench watching players whom he believed he could outperform. As frustrating as it was to sit and watch, Joe's identity at Harvard was based largely on being a varsity ballplayer, and he could not possibly leave the team. He knew that he still could be chosen for one of the final clubs. He was on eternal probation, his conduct monitored, his gestures noted. He continued his sagacious assault on the social world by joining the Junior Dance committee, which included a Frothingham, a Lowell, and his dear friend Potter. The event was held in the Union, which was festooned with potted palms, laurels, Japanese lanterns, and frosted bulbs. "Everywhere was to be seen evidence of the activity and good taste of the committee in charge," the *Harvard Crimson* noted.

Joe's Brahmin friends had assumed that his Irish uncouthness would eventually work its way to the surface, but on the committee he had appeared an arbiter of good taste. Joe's only disappointment was that Rose had been unable to be his date. She had returned to Boston after her year studying at the Convent of the Sacred Heart in Blumenthal, Holland, and another year at the Sacred Heart boarding school at Manhattanville in New York City, an exquisite rendering of Catholic womanhood. For all his social ambition, Joe had no intention of marrying anyone but a Catholic, and there was no Catholic woman in all of Boston like Rose Fitzgerald. Her mother had retreated into her private life, and Rose had largely become her father's hostess. Mayor Fitzgerald ruled over his family with a narrow severity that he did not attempt over Boston; he insisted that Rose accompany him for a vacation in Palm Beach, Florida, rather than attend the Junior Dance.

Joe had kept up his romance with Rose over the years and over the distance, and despite her father's attempts to dangle other suitable young men before his beloved eldest daughter. Honey Fitz might not like Joe, but his daughter's innocence was safe with Joe. There was no greater taboo than to touch the sweet maiden whom he hoped might one day be his wife.

The ideal of manhood at Harvard had nothing to do with sex. Passion was for the football field, not the boudoir. Teddy Roosevelt was proud that his mother had taught him that he should be as pure as the woman he married. Joe's generation of educated young men learned not only that there were good women and bad women, but also that the bad women were either prostitutes or amoral seductresses who preyed on vulnerable young men.

If Joe should partake of their fatal charms or chance on an unknown working-class woman at a public dance hall, he risked horrors beyond measure. Even these seemingly innocent young girls might already be infected with syphilis that had progressed to the stage when the victim has "a peculiar kind of sore throat with white mucous patches [and] . . . the moisture from the lips is as venomous as the poison of a rattlesnake."

Such warnings were enough to keep many young men in the library and the club, preferring Quaker oats to wild oats. Joe, however, was not to be dissuaded. He headed to Boston, where he attended musicals and squired around young actresses and showgirls. This dangerous world lay far beneath the carefully prescribed pathways of society and morals. Here he was, as his friend Arthur Goldsmith remembered him, "a ladies' man." On one occasion he and Goldsmith went out with a couple of charmers from the chorus line of *The Pink Lady*. Joe was arm in arm with a young lady, whirling her around a roller rink, when he came upon Rose skating by herself. "He talked himself out of that one," Goldsmith said, impressed at Joe's ability to excuse the inexcusable.

Joe had learned the sweetest lesson of all. Down this road lay not disease and death but pleasure. He could be the good layman at church on Sunday, the good gentleman around Harvard Square during the week, the honorable escort to Rose on special evenings, and still take his trips down into the demimonde of pleasure.

During Joe's junior year Harvard hired its first professional baseball coach, Dr. Frank Sexton, a former major league baseball player. "Any men who through indolence or carelessness handicap Dr. Sexton in his initial efforts as coach deserve the strongest condemnation," the *Harvard Crimson* editorialized. Harvard was tired of losing, especially to Yale. The college was bringing in a whole new breed of highly paid coaches. Sexton was supposed to win. He could hardly afford to indulge in gentlemanly inclusiveness; he cut the squad in two, spending most of his time with the A Division while relegating Joe and the rest of the flotsam to the B Division.

In later years Joe attempted to rationalize the failure of his Harvard baseball career by saying that he had thrown out his arm against Navy, sacrificing his physical well-being for his beloved Crimson nine. This was simply not the case. The four times that Joe did get into a game, he did as well as the regular first baseman.

Joe loved baseball with rare passion, and as painful as it must have been to be so dismissed, he had no choice but to sit on the bench game after game, week after week. Few would remember how rarely he had played, but everyone would know if he quit. Dr. E. H. Nichols, the baseball and football team physician, told the *Harvard Crimson*: "No year and no season seems to go by that I hear applied to one or more athletes, the term 'quitter,' which is quite the most contemptuous and derogatory term that one college boy can apply to another, and which implies a lack of physical courage."

As Joe watched his teammates out on the playing field, he was observing not simply a game of baseball but all the tensions between the traditional ideals of fair play and sportsmanship and modern competitiveness. It was the tension between the Brahmin world Joe sought to enter and the means he was willing to use to get there, means that helped destroy the very world he thought so enviable.

The ideal Harvard man, in the words of Charles Eliot, the just retired president of Harvard, was a gentleman "carrying in his face his character so plainly to be seen there by the most casual observer, that nobody ever makes to him a dishonorable proposal." He was a man like Eliot, himself, who believed that pitchers on the baseball diamond should not resort to such despicably low cunning as throwing a curveball. On the football field, Eliot thought, backs should take the ball and charge into the toughest part of the line, never taking the cowardly expedient of attempting an end run. In Eliot's world, there were no end runs, just charges straight up the middle.

Those of Eliot's academic generation were appalled by the spectacle of baseball. They despised this game played as much by professional teams as college amateurs, the vulgar masses cheering on the players in raucous discord, the gambling on the games, and the drumbeat of hype in the sooty tabloids. The purists believed that the players should be quiet on the field and stop such sharp practices as chatting up their pitcher or yelling at the batter. As for the spectators, a sportsman should sit quietly, applauding at every example of good play by either team.

The student fans liked to sit together in the bleachers, cheering their cherished Crimson on with organized shouts and songs, a scene that the *Harvard Crimson* declared "a rather hysterical and often unfair attempt to compel victory, rather than a recognition of good playing." These noisy, disreputable students appalled the good academics of Harvard. "Baseball is on trial as a game for gentlemen," Dean D. R. Briggs stated with all the authority of his high Harvard office behind him. "If it is the duty of patriotic students to make all the noise they can while the visiting pitcher is facing their representatives . . . if baseball must, as the Yale Alumni Weekly puts it, 'degenerate into vocal competitions on the part of the players, or into efforts to rattle the opposing pitchers on the part of the grandstands,' the sooner we have done with the game the better."

At the beginning of the season the entire Harvard varsity team had been given Harvard *H* sweaters. Joe could wear the black sweater with its crimson *H* as he walked through Harvard Yard. He knew, however, that unless he played in one of the two games against Yale, he would have to return his sweater with its emblem of honor. As the season went by the prospect of winning his letter grew dimmer and dimmer. Sexton's team was winning, and even when Harvard was far ahead, the new coach rarely substituted, and when he did, he usually would not reach far enough down on his roster to put Joe in the game.

Sexton was so dismissive of Joe's abilities that he did not even choose him to travel to New Haven for the first Harvard-Yale game. The Crimson team decimated the Yalies eight to two, led in part by Joe's friend Bob Potter, who hit a home run and slid home after another hit. Despite Harvard's insurmountable lead, Coach Sexton did not make a single substitution.

Three days later the Yalies traveled up to Cambridge for the second game of the series. It was Class Day at Harvard, a sparkling June afternoon and a splendid setting for the twelve thousand spectators who jammed into Harvard Stadium, even perching on the edges of the field and standing up on the roof of the stadium. Joe's friend and social mentor, Bob Fisher, was one of the two cheerleaders, directing shouts from enthusiastic ranks of alumni and students. When N. P. Hallowell marched in behind the band with his white bearded face held high, carrying his class banner bearing the numerals *1861*, he received a thunderous ovation. The honored classes of 1901 and 1908 marched in too and took their seats behind home plate. So did the captains of many previous Crimson baseball teams.

These alumni were not here in their suits and ties and bowler hats to watch a mere baseball game. These Harvard gentlemen had come to celebrate the Harvard ideal of sportsmanship. They had also come to watch Harvard win, and they were the driving force pushing Harvard to do what it had to do to be competitive, while paradoxically enough staying true to the sentimental myths of their sweetly remembered college years. They could be merciless toward a losing team—indeed, in the *Harvard Alumni Bulletin* they had called this year's successful contingent "rather crude material"—while threatening to boycott games if the players did not stop the incessant, vulgar, ungentlemanly chattering on the field.

Rose had come too, with Mayor Fitzgerald, to watch her beau perform manly feats. For Joe it was an exquisite setting in which to display his athletic heroism in front of these thousands of Bostonians and others, his sweetheart, and her father, an avid baseball fan. He sat there in the dugout, however, as the Harvard nine fulfilled its mandate on this glorious afternoon, taking an early lead. It was unthinkable that he would play even a moment of this contest.

The game stayed close, and if Coach Sexton had made no substitutions in the first match, he surely was not about to make them now. The score was four to one when Yale came to bat in the ninth inning, and the coach sent the starting team out on the field for the last time. The first batter grounded out, but the next Yalie stroked a single into left field. A pinch hitter attempted to move the batter on, but he hit into an easy out.

Two outs. One more out. The Yalies were a scrappy bunch who had not been defeated twice by Harvard in eight years. The Harvard fans knew that it might not be over yet. Then, as the fans waited for Charles McLaughlin to pitch to what they hoped would be the final batter, the coach called time and put Joe in at first base. Only then did McLaughlin pitch. The batter hit to the shortstop, who threw the ball to Joe for the final out.

While the rest of the team converged on the pitcher's mound, where they enveloped McLaughlin in their hugs, Joe walked silently off the field, clutching the winning ball. McLaughlin finally separated himself from his teammates and ran over to Joe to claim the coveted ball that was doubly his due, as the winning pitcher and as captain. Joe stuffed the ball into his own back pocket and walked on.

Joe had thought that the baseball diamond was a sacred field of play, removed from all the sordid duplicities of the outside world. He had attempted to live in that world, but the coach had dismissed him from the A squad like a day laborer. The winds of a new age were blowing through Harvard Stadium, even if most of the alumni could not feel them.

The plaudits won on this field could be cashed in out there in the world, and those who did not know that were fools. Sexton was paid far more than any professor. He was paid to win, and win he had. McLaughlin, the hero of the team, was about to cash in too, by using the fame and honor that he had won on the field to help him get into the new movie business. A few days before this final game, Joe's father's friends had gone to the senior captain and told him that if he wanted those movie theater licenses in Boston, then Joe better get into that final game against Yale. And so Joe had won his letter, and he had not cared how.

To a previous generation, Joe's Harvard *H* would have been a badge of shame, not of honor, but he did not see it that way in the world in which he was living. Rose did not see it that way either, wearing the rose-colored glasses that were part of a woman's natural wardrobe. "My father and I saw him the day he won his 'H,' " she recalled, "when he made the winning play against Yale."

Joe had his letter, and now he left the team for good. His excuse was that he had agreed to coach the freshman team, but Sexton probably had had quite enough of Joseph P. Kennedy. Although the first-string first baseman was graduating, the *Harvard Crimson* did not even mention Joe as a prospect to take over the position.

That fall, Joe was one of only 36 Harvard men among the 2,262 undergraduates who had the right and honor to wear the Crimson sweater with the black letter. He was chosen now for one of the lesser of the final clubs, Delta Kappa Epsilon, in part probably because of his Harvard *H*. He had achieved his great goal, but he had achieved it so late that he had little time to savor the pleasures of life as a clubman. He entered Delta Kappa Epsilon by the front door now with the men of the Gold Coast. He had achieved what was socially impossible for the Catholic son of an Irish-American politician, and by rights he should have savored his triumph. Yet he was angry that he had been snubbed and had not received all that was his due. He had no interest in sitting around in his club endlessly socializing. He had had whatever he thought he could wring out of his Harvard experience, and he petitioned to graduate at midyear. The university turned him down, presumably because of his academic record; during his four years at Harvard he received not a single A, four Bs, nineteen Cs, and ten Ds, including one in social ethics.

As he strolled through Harvard Yard, a stalwart senior and the very model of a Harvard man, who knew or cared how Joe Kennedy had won his letter or how he had made his final club? He had entered Harvard hoping to be like the men of the old Harvard world. He was leaving a man of the raw new century.

3

Manly Pursuits

Joe stood next to Rose in Cardinal O'Connell's private chapel repeating his marital vows. This tender ceremony was a victory not only of love and devotion but also of power and cunning. Joe had wrested Rose from her father, Honey Fitz, in a struggle that had lasted most of a decade and was not yet over even on this radiant October day in 1914.

For Honey Fitz, as for Joe, life was a matter of battles between men. Joe could see through Honey Fitz as if he were a pane of glass. There was a hard nub of petty jealousy in the former mayor that exhibited itself to any man who might stand above him, even in the private kingdom of his own home. Joe was not a man to bow in deference to his new father-in-law; Joe was a threatening figure to Honey Fitz, especially since the mayor had left office and no longer wore the mantle of power. Joe, for his part, had an acute understanding of Honey Fitz's vulnerabilities and took pleasure in probing them.

With neither a fortune nor a great name to bestow on Rose, Joe had set out upon graduation to establish himself in the business world. He first took a job as an assistant bank examiner for the state. It was not a grand position, but it was an enviable one in which to learn how banking truly worked and just how close to the brink of illegality a bank could go without falling off the edge. "If you're going to get money, you have to find where it is," Joe told his cousin Joe Kane.

Joe had been at his new position only a little more than a year when his father came to him and said that a major Boston bank, the First Ward National, was trying to take over Columbia Trust, the bank P. J. had co-founded. The keys to holding on to the bank were the shares held by the estate of the late John H. Sullivan, another founder of the bank. With the barest hint from Joe, Uncle James and Aunt Catherine Hickey came up with the money that, when added to that of his father and several others, saved the bank from a takeover.

In partial repayment for Joe's acuteness, the twenty-five-year-old was named president of Columbia Trust, supposedly becoming the youngest bank president in America. He was now a man of such position and promise that he could rightfully claim Rose Fitzgerald's hand. For Rose, it was an exquisite dream finally realized. "I had read all these books about [how] your heart should rule your head," she remembered decades later. "I was very romantic and there were no two ways about it."

As the newlyweds walked out of the chapel into the bountiful sun of an Indian summer day, they waited on the steps while photographers took pictures. The reporters and photographers were not there for Joe. They would feature the couple on the front page of the next day's Boston newspapers only because Joe Kennedy was marrying the former mayor's daughter, the most eligible Catholic woman in Boston. Even in this moment Honey Fitz took his own pound of publicity, standing there with Rose while Joe huddled in the background.

The wedding party drove over to the Fitzgerald mansion in Dorchester for a wedding breakfast for seventy-five guests. When Rose had made her debut, her father postponed the city council meeting so that his colleagues might attend, and he also invited the governor-elect and other high officials, as well as hundreds of celebrants young and old. For his daughter's wedding to Joe, he was putting on only the minimally acceptable celebration, a reception at his home primarily for family members.

It was just as well that so few outsiders were present, for in the midst of the party Joe and Honey Fitz began yelling at one another. Rose was so distraught that she took off her wedding ring and put it on the mantel. By the time the couple left for their honeymoon at the Greenbriar Hotel in White Sulphur Springs, West Virginia, however, her ring was back on her hand, and her smile was back on her face, and they departed as if it had been a perfect day.

When the newlyweds returned from their two-week-long honeymoon, Rose was probably pregnant. As flattering as that was to Joe's sense of manhood, it also meant that if he followed accepted practice he could have no further sexual relationships with his bride until after the birth of their child. If sex outside of marriage was unthinkable and unspeakable, sex inside of marriage had to be carefully rationed. Some experts believed that there was an unfortunate tendency among frisky young married couples to indulge in sex gluttonously every night. Before long, warned Winfield Scott Hall, the leading sexual hygienist, the poor wife "is in a condition of neurasthenia, or nervous prostration, and the husband is consciously depleted in his powers, and his business efficiency noticeably decreased." Those who indulged too frequently, or in what was considered an unnatural manner, suffered, as Dr. A. T. Reinhold warned, "the most desperate cases of paralysis and epilepsy."

Some experts believed that ideally sex should take place once a month, or among the passionate few, even twice a month. The reason for having sex only once a month had nothing to do with religious mandates that sex was for procreation, not pleasure, but with medical and psychological factors. Although the wife's condition was duly and necessarily noted, the primary concern was the man. In this mechanistic concept of manhood, sperm was the precious elixir of life that, when unnecessarily spilled, dissipated the man, who thus abused himself and the future of his race.

Joe was a man of immense vitality who sought in his marriage to have an active sex life. Instead, almost from their wedding day, sex was a matter of tension between the couple. Rose was pregnant so often during the first decade of their marriage that if the couple followed the conservative regimen, he would have had sex with her hardly more than half a dozen times a year.

Joe, like his wife, was a man of profound discretion when discussing personal family matters, but Rose's sexual reticence was so upsetting to him that he talked about it among friends. "Now listen, Rosie, this idea of yours that there is no romance outside of procreation is simply wrong," he lectured Rose in front of their neighbors, the Greenes. "It was not part of our contract at the altar, the priest never said that, and the books don't argue that. And if you don't open your mind in this, I'm going to tell the priest on you."

Joe moved his bride to a new house in the suburb of Brookline. The town, one of the wealthiest communities in America, was home to many of the most affluent members of the Protestant elite, their mansions set in cul-de-sacs or shrouded in trees, far back from the vulgar streets. Joe had come out to Brookline in search not so much of fresh suburban air but of a good address signaling that he was an American success pure and simple, no rude immigrant, no hyphenated Irish-American. His neighbors on Beals Street were not old Yankees but for the most part people like himself whose forebears had arrived more recently in steerage.

Joe and Rose were at the beginning of a migration of middle-class Irish-Americans and largely German Jews to Brookline. He had headed out to the western suburb to get away from the taint of his immigrant past, but Brookline was a veritable colony of the Irish: at 11 percent, they made up the largest foreign-born element by far. For the most part these were not bankers and businessmen, though, but cooks, maids, firemen, cops, plumbers, and sanitation workers, a caste of workers and servants.

Joe's new home, not far from the trolley into Boston, was a small three-story structure that the newlyweds filled with what they considered all the accoutrements of civility. Part of that was the maid and later the nurse who lived in tiny rooms on the third floor, where they shared the same bathroom and were so close to Joe and Rose that the couple always had an audience. Rose had the maid/cook wear a black uniform at dinner and serve her banker husband on fine china.

Each evening Joe returned home to what was in many respects a miniaturized upper-crust world, or more accurately, Joe and Rose's imitation of what they thought that world was. There was the same self-conscious civility between the two of them as between their parents, no vulgar Irish excess, no loud arguments within hearing of the maid. They practiced the decorum that never faltered and wore the masks that never fell.

When Rose was about to give birth, Joe took his wife to a rented summerhouse on the ocean at Hull. To minister to her, he brought a special nurse, a maid, and two doctors, notably Dr. Frederick Good, who became the family's pediatrician. There, on July 25, 1915, Rose gave birth to their first child, Joseph Patrick Kennedy Jr., a squalling healthy son who weighed in at a formidable ten pounds.

Joe was proud of this son who bore his name, but he kept a distance from the tedious business of raising the baby, handing him off to nurses or his mother. Joe was acting no differently than most men of his generation, rendering unto women what was theirs. He went out into the world, leaving the home early each morning in his Model-T Ford, arriving home at night when the baby was asleep or resting and Rose was tired from her motherly exertions.

Joe kept that same distance on May 29, 1917, when Rose gave birth to John Fitzgerald Kennedy upstairs in the bedroom of the Brookline home. Dr. Good arrived to deliver the Kennedys' second son as he had the first, but this was not quite the event that Joe Jr.'s birth had been. Grandfather Fitzgerald did not stand below on Beals Street trumpeting to all who would hear that this son would one day become president of the United States, as he had done two years before on the beach at Hull when Joe Jr. was born.

As Joe saw it, there had never quite been a son like his firstborn, a rousing, exuberant child who charged on into life. He loved his second son too, but little Jack seemed a lesser sort in almost everything—smaller, slower, weaker, and greater only in his susceptibility to disease.

For most of the men of Joe's generation, World War I was the defining event of their young lives. In the summer of 1915, after the German sinking of the passenger liner *Lusitania,* 1,200 American men, one-third of them Harvard graduates, had descended upon Plattsburg, New York, at their own expense for four weeks of military training under General Leonard Wood, a Harvard man. The following summer, with government money, about 16,000 men attended twelve camps similar to Plattsburg across America.

While these men prepared for war, Joe and three of his Harvard friends, Tom Campbell, Bob Fisher, and Bob Potter, gathered together on the Friday evening before the long pre–Fourth of July weekend at Joe's parents' waterfront home in Winthrop. As they reminisced, across the Atlantic the Battle of the Somme had begun. Of the 110,000 British troops who moved out of their trenches toward the German lines on an exquisite spring morning, 60,000 would fall, either killed or wounded.

On Saturday morning the Boston papers were full of details of the immense human slaughter juxtaposed against idealistic paeans to their sacrifices. Joe's three Harvard friends read the accounts as true men, seeing the bloody field as the plain of honor where a man belonged and celebrating the nobility of these Englishmen and their selfless sacrifice. Joe sat silent.

Fisher and Potter were the models of the Harvard gentlemen in whose company Joe had sought to be included. The two men had been his mentors in learning the nuances of the Brahmin world, and they were mouthing the rhetoric of the true man's credo that was so much a part of Harvard life. It was unthinkable for Joe to stand up to his friends. To do so was to stand up to the very world to which he had so long aspired. Joe nonetheless spoke out, deriding what he considered his friends' mindless idealism. He told them, as Rose remembered, that their "whole attitude was strange and incomprehensible to him." His friends, as Joe saw it, were not innocent in their mindless jingoism. He told them that "by accepting the idea of the grandeur of the struggle, they themselves were contributing to the momentum of a senseless war, certain to ruin the victors as well as the vanquished."

Joe had monumental insight into the dark and senseless part of human nature. He had his father's cunning sense of the political world and looked directly at all the twisted aspirations, avarice, and endless folly of humankind. He had his mother's profound sense of social aspiration and her haughtiness toward most of humankind. He was a shrewd and subtle judge of humanity, knowing just who might be useful to him and how and why, and whom he could use before they used him.

Joe was a man of what he later called "natural cynicism," which to him was a dysphemism for the highest realism. His natural cynicism was a philosophical stance that he believed he shared with a tiny elite of men who looked down on humankind with bemusement and disdain. As he saw it, most human beings could never reach for long beyond their basest instincts.

Joe's fellow Harvard men contributed as much as any group in America to what he considered "the momentum of a senseless war." By the time war was declared the following April, America had the nucleus of a college-educated officer corps, in which Harvard men were disproportionately represented. If Joe's friends were naive, they shared their naivete with millions of their compatriots. Joe's three close friends all honored their ideals by serving in the war.

In all, 11,319 Harvard men served in World War I, and 371 died in the service of their country. Of Joe's class alone, the majority of graduates either enlisted or were drafted, and 7 did not return. Some marched off to war because they were patriots; some marched off seeking fame in the cannon's roar, and some marched off hardly knowing why. Those, like Joe's three close friends, who set off as idealists returned as men to whom irony was often the highest political emotion. There had been several Harvards when Joe was there, but for many of the men of his generation there were now only two: those who served and those who did not.

Joe was not one of those who publicly opposed America's entry into the war, a cause that might have subjected him to public rebuke and hurt his bank's business. He was a man of natural cynicism after all, and it would have been futile to stand up to the folly of his fellowman. While some of Joe's classmates had connived to get into battle, joining the French Foreign Legion or the Canadian forces, Joe connived to get out.

As soon as war was declared and his friends had donned their khakis, twenty-eight-year-old Joe wangled a fifteen-thousand-dollar-a-year position as the assistant manager of the Bethlehem Steel Fore River Shipyard in Quincy, Massachusetts. In peacetime Joe would never have been considered for such a seat, but all across America executives and managers were exchanging their suits for officers' uniforms. Joe was willing to help build warships to be used by men who might die in a war in which he did not believe.

When Joe arrived at his new office in October 1917, he knew nothing about administering a shipyard. One of the first things he learned was that the craft union had negotiated a new wage scale with his predecessor to go into effect with the next pay cycle. Instead of honoring the agreement or alerting the union that he wanted to renegotiate, he simply kept the old pay rate.

The workers, many of them immigrants, including a large number of English and Scottish skilled workers, were not that different from the kind of men who had nightly sat in his father's home seeking advice and aid. If Joe had observed that scene with any acumen, he would have realized that you could push men like this, and you could push them some more, but if you broke their trust in you, dishonored your word, then they had a boundless fury.

When the workers got their pay envelopes and saw that they had been deceived, about five thousand of them went on strike. That was no trivial action on their part, for the District Exemption Board of Quincy immediately took away their draft exemptions. "The strikers, in refusing to work in the shipyard . . . become automatically eligible for the trenches," one district board member said, the intimidation hardly veiled.

Within days of his arrival Joe had managed to create a crisis of potentially immense magnitude. The ships in the dry docks were crucial to the war effort, and if the strike spread, its costs would soon become incalculable. No one knew this as intimately as did Assistant Secretary of the Navy Franklin D. Roosevelt. The aristocratic Roosevelt, a Harvard graduate, class of 1903, had never had workingmen sitting around in the evenings in the living room of his family estate in Hyde Park, New York. Yet Roosevelt had an instinctive awareness of how to deal with these workers. He sent a telegram in which he flattered the men by telling them that there was "probably no one plant in the country whose continuous operation is more important to the success of this country in the war, than that at Fore River." He spoke to their devotion to country by appealing to their "patriotic duty." Most important, Roosevelt agreed to honor the new pay scale.

In all of this Roosevelt demonstrated the incipient awareness of a great politician who understood that the essence of democratic politics is empathy. A leader must first understand what the other person wants. Only then can he act. Joe, for his part, saw life as a brutal Darwinian struggle in which men of will and power imposed themselves on the mediocre, the passive, and the slow-witted.

The men went back to work, and Joe left his position after scarcely a month. By any measure but his own, Joe was disgraced, shuttled aside into a lesser post at the new Squantum yard, handling all the company stores. For another man, this defeat would have been a painful moment of self-awareness in which he would have taken stock of his own excesses and mistakes. Joe, however, apparently walked away from this defeat incapable of or unwilling to admit to his culpability, having learned little but that even in the business world he operated in a democracy where the weak majority could gain ascendancy over the strong few.

Joe could not turn to Rose to talk of this failure. Rose was only a woman, and as *The Catholic Encyclopedia* expressed it, "The female sex is in some respects inferior to the male sex, both as regards body and soul." Women were seen as incapable of a man's high seriousness, and for Joe to turn to his wife for counsel would have been both unmanly and unseemly.

Rose had no inkling of the emotional price her husband was paying. To her, Joe was a heroic figure who worked terribly long hours for the war effort and suffered from an ulcer for his relentless endeavors. He was nervous and high-strung, and Rose worried about his health. Rose did not think that the debacle in the shipyards would have been reason enough for a nervous ulcer; to think such thoughts would have broken the covenant between them. Joe, for his part, could not and would not see that Rose was probably often depressed, a word and an emotion that were simply not allowed. She was married to a prominent and honored Catholic gentleman. She was a formidable woman in Catholic society, president of the Ace of Clubs, a leading Catholic women's club, with her own social life and prestige. Her children were looked after by nurses and maids. She had a life that most women would have thought close to perfection.

Despite his new position, Joe's draft board had the audacity to try to call him up. He pointed out how indispensable he was to the war effort. When the board thought otherwise, Joe's boss went all the way to Washington to see that his young associate did not have to serve.

It was a good time to be an ambitious young man in America away from the stench of the trenches. While one million Americans served in the armed forces, the economy was roaring ahead, and a man could make it in a way he never could before. Joe had only a small office, but he was the turnstile through which all the goods had to pass, and he made the most of it. He had a fine salary, bonuses, the right to run the canteen for his own profit, important new contacts, and the knowledge that Bethlehem Steel was a stock that a smart man had better get into.

To Joe, all the prissy rules and moral guidelines of Harvard did not apply to him and his life. In July 1919, he joined Hayden, Stone and Company as a stockbroker. His employer, Galen Stone, made much of his money, not by the tedious route of collecting customer commissions but by employing insider information to drive a stock up or down. The technique, while then technically legal, preyed on the avarice and ignorance of the average stock buyer, an approach that fit perfectly with Joe's view of human beings. He became as adept at this game as his employer. "Tommy, it's so easy to make money in the market we'd better get in before they pass a law against it," he told one friend, Tom Campbell. On one stock alone, Pond Coal Company, in which Stone was chairman of the board, Joe made close to seven hundred thousand dollars on an investment of only twenty-four thousand dollars. It was a high-stakes game in which Joe had thrice lost everything, he told his friend Oscar Haussermann, but he had come back and was again on top of the game.

Joe was a competitor in everything, and to some degree his home life was not measuring up. He could not complain that Rose was anything less than the woman he had married: profoundly religious, Catholic-educated, and socially conservative. But she was not growing into the new age, a time when women could snap wisecracks as quickly as men, roll their stockings on the dance floor, smoke cigarettes, and vote.

Joe might not have wanted such a woman for his wife, but Rose remained a Catholic provincial. She might talk of culture, but he was the one who truly loved classical music and looked forward to their nights at the symphony.

Joe Jr. was his father's namesake in every way, a healthy, vibrant four-year-old, but the other children were not quite measuring up. Jack was a scrawny, whining, sickly tyke, and the newest, Rose's namesake Rose Marie, or Rosemary as she was called, born in September 1918, was painfully slow in every part of her life.

In January 1920, Joe came home from his new office on Milk Street in downtown Boston and found that his pregnant wife had returned to her father's home in Dorchester. It was unspeakable and unthinkable that Rose should leave him, an insult to his manhood, to his children, to his family, and a full measure of the silent pain his wife was suffering. Joe was a good provider and a good husband by all the measures that mattered in the world in which he lived. He could not fathom the idea of divorce, which would sever him and Rose forever from the sacraments of the Church, making him such an outcast from Catholic society that his economic future would be compromised. He had no choice, however, but to wait his wife out, and wait he did for three full weeks.

In the end it was his father-in-law, a man Joe considered in part a mountebank, who told his daughter to return to her husband. In Honey Fitz's world, appearances were reality. As mayor, he had orchestrated photos of himself as a devoted family man, the public image so different from the reality, in which Honey Fitz honored his love of home by rarely being there. So there would be, if necessary, a similar portrait of Rose and Joe and their family. Honey Fitz would have no divorces or separations to stain his family name, no disgrace brought on by his favorite daughter.

Rose's father did not inveigle his son-in-law to share more of his life with his wife, to attempt to understand her despair, or even perhaps not to arrive home with the scent of chorus girls on his lapel. He saw Joe's role first of all as a provider, and if there was any failure, it was that he had not provided well enough. "If you need more help in the household, then get it," he told Rose. "If you need a bigger house, ask for it. If you need more private time for yourself, take it. There isn't anything you can't do once you set your mind on it. So go now, Rosie, go back where you belong."

Joe was relieved that Rose was back, but he was as drawn as ever to women whose laughter rang freely in the night. He had a suggestively intimate style in his letters to young women. "I don't know how close you will be obliged to stick to your boss tonight," he wrote Vera Murray, the executive secretary of a theatrical producer in August 1921. "I know how close you would have to be if I were your boss."

A month later Joe wrote Arthur Houghton, a theatrical manager and friend: "I hope you have all the good-looking girls in your company looking forward, with anticipation, to meeting the high Irish of Boston, because I have a gang around me that must be fed on wild meat lately, as they are so bad. As for me, I have too many troubles around to both[er] with such things at the present time. Everything may be better however, when you arrive."

Rose returned to Joe having made a silent compact that she would build her life around her God, her children, and the acquisition of material goods. She was hardly back when little Jack came down with scarlet fever, his body covered with red spots. It was a fierce, highly contagious disease whose very mention made mothers shiver and lock their doors. Rose was of little help, for she had just given birth at Beals Street to their second daughter, Kathleen, in February 1920. Like her sister and brother, the baby risked being infected by the disease. For the first time in their marriage Joe was thrust into the center of his children's lives.

Joe knew that to save Jack, and perhaps to save all his children, he had to find a hospital bed for his son. The Brookline Hospital had no contagion ward, and his son was not eligible to enter the special children's ward at Boston Hospital. There were 125 beds in the ward, and more than 600 Boston children sick with scarlet fever. The illness fell equally on the poor and the rich, the children of the North End and the children of Back Bay. There was a terrible triage at work in the choice between those who would enter the hospital, and probably live, and those who would not and might well die or pass on the disease to their siblings.

Power and influence could mean life and death, and Joe saw that he had not enough of either to save his son. His father-in-law, though, still had the power to see to it that little Jack got a bed that should have gone to a child living in Boston.

Joe was a man who thought he could solve any problem, but he felt now a parent's helplessness in the face of illness. Watching little Jack in his sterile white room at Boston Hospital touched Joe in places in the heart that he had not known he had. Jack was a likable lad, with his good humor and gentle warmth, and a son who did not cry at having been taken out of his home and placed among strangers. Joe went to church and prayed to God promising that if his son lived he would give half his wealth to the church. And when

Jack lived, his father wrote out a check for $3,700 to the Guild of St. Apollonia. It was a noble gesture, but given Joe's earnings, it would seem that he was a calculating negotiator even when he was dealing with God.

Joe believed in the saving grace of money, but he now had two more reasons to seek great wealth and power. His wife had returned to him after having been told that her marital salvation lay in more servants and a bigger house. And his son had lived only because of his father-in-law's power.

Joe was already making hundreds of thousands of dollars, but now, with the enactment of Prohibition in January 1920, he saw an opportunity to make even more. His own father had made his way in life largely through the liquor business—first with his own tavern, later with a wholesale liquor business—and the political clout of the industry. An Irishman's pub was his Somerset Club, and for every worker who tumbled out drunk, ten others had a drink or two and went home to their families.

When the good ladies of the Woman's Christian Temperance Union campaigned for the Eighteenth Amendment to enact Prohibition, men like Joe's father saw it as a malicious attack on them and their ways. With Prohibition, it was the Italian winemakers of northern California who saw their vineyards turn to weed, the German beermakers of St. Louis whose breweries were shut down, and the Irish tavern keepers whose doors were shut forever.

An observer might even have viewed the Volstead Act instituting Prohibition as a regressive piece of legislation, targeted at the poor and the foreign. Since the law allowed citizens to stock up beforehand, the forward-looking men of New York's Yale Club put away enough for nearly a decade and a half of good drinking. For the well-to-do, there was always a supply at their clubs and homes. But the saloons of the poor were shut and dry.

The promise of bootlegging beckoned to the quick and the daring, but not to Harvard men. Crime was often a poor man's capital, the quickest and surest way out of the ghetto. Joe, however lived far from the cusp of poverty and was within reach of the refined upper-class world where he sought to live. But Joe saw an opportunity. That he saw it and acted upon it is one of the most extraordinary facts of Joe's entire life. His father had for decades been a major wholesaler and importer of liquor, and it was probably through him that Joe had contacts in Canada and England. Testimony suggests that Joe had the best of it, for the most part delivering the merchandise offshore to bootleggers who brought the liquor into the United States. In 1926 the Canadian Customs Commission looked into liquor export taxes that were not being paid and found the name "Joseph Kennedy" on many documents. Although the commission never definitively linked the name to the young

Boston businessman, there is no other Joseph Kennedy whose name has been prominently linked to bootlegging.

Joe did not enter this illegal business as some desperate expedient. He took no more risks than he did with some of his early gambles on Wall Street. If anything, liquor was another part of his portfolio; he was a businessman spreading his risks. Joe kept such distance from the business that his name was never formally linked to bootlegging. Cartha DeLoach, the former deputy director of the FBI, recalls that "there was a great deal of suspicion concerning his being possibly involved in smuggling in the early days. But as it is in America, you overcome these things."

Joe chose his partners as cannily as he chose his part of the business, one of them being Thomas McGinty, an Irish-American known as "the King of Ohio Bootleggers." "Joe brought the liquor to the middle of Lake Erie, and the boys picked it up," recalled McGinty's daughter, Patty McGinty Gallagher.

McGinty had been a flyweight boxing champion, and he was a man armed with Irish blarney as well as a steel fist. He went on to become the leading Irish presence in the Jewish-run Cleveland syndicate. He founded the famous Mounds Club outside Cleveland, ran racetracks, managed the casino at the Hotel Nacional in Havana, and became a hidden owner of the mob-controlled Desert Inn in Las Vegas, all the time maintaining a relationship with the Kennedy family. His daughter Patty, in the bar in her home in Palm Beach, Florida, has a picture of her father and a smiling Jack Kennedy taken in Havana in the 1950s.

Another witness is Benedict Fitzgerald, an attorney who not only knew the Kennedys intimately but also represented Owen Madden, a gangster who controlled leading nightclubs in New York City. Fitzgerald says that Joe was involved in bootlegging deals with Madden, a view confirmed by Q. Byrum Hurst, another of Madden's attorneys.

Madden, though English-born, was of Irish blood, a son of Erin brought up in Hell's Kitchen, the Manhattan slum that bred little but disease, despair, and crime. Madden spent a term in Sing Sing, the upstate New York prison, for a killing he vowed he did not commit. Madden was a dapper dresser whose manner may have been appropriated by his friend George Raft in his screen portrayals of gangsters.

Madden's pathway was greased with ample payoffs to cops and Tammany Hall that allowed him to be one of the three biggest importers of bootleg booze along the East Coast. He moved from criminal deals to entertainment, backing his lover Mae West in her Broadway debut and running the celebrated Cotton Club, the Harlem boîte. He moved into the world of sports, managing the career of Primo Carnera, the giant heavyweight boxer. In Madden's world there was a seamless journey from crime to politics to sports to entertainment, and Joe was now embarking on that journey. It was a world in which politicians acted like criminals, and criminals such as Madden acted like politicians or sports heroes. In this world a man like Joe Kennedy was just another player.

Madden was forced out of New York and took up residence in Hot Springs, Arkansas, where he reputedly controlled much of the gambling and vice. Even then, Fitzgerald says, Joe maintained a business relationship with Madden. "I know that Joe gave him some finances while he was down there," Fitzgerald recalls.

According to former Florida Senator George Smathers, "everybody knew" that Joe sold liquor. "But he wasn't a bootlegger in the sense that he ran across state lines and carried illegal whiskey."

Yet another credible witness to Joe's activities is Zel Davis, a former Florida state's attorney, whose uncle was the leading bootlegger in Palm Beach County. "Joe was having scotch sent into the Bahamas from England," Davis says. "He was doing business with Roland Simonette, who became the first premier of the Bahamas in the late sixties. When Simonette was young, he owned shipyards and had boats. He ran one boat from Nassau to New York and New England. Joe also had connections with Palm Beach County bootleggers. That boat came from West End in Nassau to Palm Beach. They built warehouses to hold it."

The most intriguing tale of all about Joe's alleged illegal involvements was given in 1994 by ninety-three-year-old John Kohlert in a videotaped oral history interview conducted shortly before he died. As a young man in the twenties, Kohlert worked his way through college tuning pianos, including those in speakeasies and at the Cicero, Illinois, home of Al Capone. Kohlert, a Czech immigrant, knew little about Capone's reputation as a vicious mobster and was delighted that Capone offered him work. "When I worked on his player piano, he said, 'How would you like to stay for spaghetti?' I tell you, I never had a spaghetti supper such as I had at Al Capone's place. And he said, 'Well, we're going to have company.' I said, 'That's fine.' The company he had was Joe Kennedy. He introduced me to him. He said, 'This is Joe Kennedy from Boston, and we have a little business deal to make at supper. I hope you don't mind.' I said, 'Hell, no, I don't mind.' So while we had spaghetti, they made a deal.

"Al Capone owned a distillery in Canada. I'm not sure, but anyway, it was quite a prominent distillery, that made whiskey. And so they made a big deal of Capone selling whiskey and Kennedy selling Irish whiskey to Capone. And they made a deal to exchange it on Lake Michigan off Mackinac Island. That's where the Kennedy ships and the Capone ships were going to make the exchange of the two whiskeys that they agreed to there at that spaghetti dinner."

An obscure dying man in a short aside in a lengthy oral history has little cause to invent such an extraordinary tale, and little time to create such details. Patty McGinty Gallagher and Zel Davis have no reason to invent a family history so profoundly different from their own lives. Benedict Fitzgerald has no reason to exaggerate his extravagant life. There remains a maddeningly anecdotal character to this evidence and the allegations that have been made in other books, but the sheer magnitude of the recollections is more important than the veracity of the individual stories.

Joe's grandson Christopher Kennedy suggests that it would have been impossible for his grandfather to have had such far-flung dealings, doubly so without leaving any record. "About ten years ago a Wharton MBA with a degree in accounting who had audited several of the East Coast's wealthiest families conducted an audit of Joe Kennedy," Christopher Kennedy says. "He transcribed to the dollar every source of income and use of cash, and every dollar is accounted for."

That result strongly suggests, however, that Joe understood one of the fundamental rules of illegal dealing. The dealer does not get caught holding the goods. Joe occupied the enviable end of the business, away from the police and the riffraff, away from the parceling up and the danger. It is probably testimony to the sheer acumen of Joe Kennedy that no one has come up with any hard physical evidence linking him to bootlegging, but the circumstantial evidence strongly suggests that Joe was a financier and supplier of illegal liquor.

Joe agreed to supply the liquor for his tenth Harvard reunion in June 1922. During his college days he would no more have been associated with an illegal activity than he would have danced an Irish jig in Harvard Yard. Joe had wanted desperately to be associated with these Harvard gentlemen, but he now no longer thought that it was enough merely to pattern his life on theirs. He understood their quiet disdain for men like him, and yet he was willing to provide their whiskey, fulfilling every stereotype of the Irishman.

Joe was no retail bootlegger with his stash of liquor, no matter what some of his classmates might have thought. He got his liquor through his father's associate, charging his class $302 for twenty-six and a half gallons of liquor that he did not drink. Some of these Harvard graduates had ancestors who had arrived on the *Mayflower,* and Joe had the liquor delivered in a boat that landed in Plymouth where the Pilgrims had disembarked.

Joe caroused with his Harvard mates, dressed in a sparkling white shirt, a bow tie, and the inevitable crimson sweater. For Joe, this was a uniform for a summer's weekend. For his old friends Bob Fisher and Tom Campbell, this was practically their daily wardrobe. Fisher had given up his career in merchandising to become the Harvard football coach. Campbell had joined his old football teammate as Harvard's freshman coach and also served as assistant graduate treasurer of the Harvard Athletic Association.

Only his other old friend, Bob Potter, had the kind of life that Joe could fully admire. He was vice president of the National Shawmut Bank, lived in a townhouse in Back Bay, and lunched at the Somerset Club, within whose portals Joe was not welcome. Some of these fellow graduates were cutting a fancy picture, but they were coming to old Joe now to hit him up for loans at Columbia Trust. Even Potter, two years later, asked to borrow thirty-five hundred dollars. "It's all right to give it to him if you can get collateral accepted in the savings department." Joe wrote the bank officer with a dismissive tone. "Otherwise, just tell him that you have not got the money."

Joe, for his part, had long since duly noted the hypocrisy of the Brahmin class. Were the good gentlemen who drank bootleg whiskey cut of a finer moral cloth than Joe, who sold it to them? With all their fine talk of law and honor, these Harvard men drank the liquor brought to them by a new class of criminals. He could stand and talk with these self-satisfied burghers, listening to their tales of business, banking, children, golf, and tennis. He was an amusing raconteur, enjoying the casual camaraderie. He appeared no different from his classmates, though he traveled on a stage wider than any that they traversed and dealt with men like Tommy McGinty and Owen Madden as well as with cardinals and magnates. He was making his way into the shadowy centers of power in America, places his Harvard friends and professors did not know and would never understand.

4

"Two Young 'Micks' Who Need Discipline"

Gee, *you're* a great mother to go away and leave your children alone."

Jack may have been frail in body, but there was a daring quality to a five-year-old boy who would so confront his mother. It was April 1923, and Jack was upset that Rose was going off to California with her sister for an extended trip of four or five weeks, leaving him and his three siblings behind. It was a risky business to speak out, for if he displeased his mother, she might bring out her infamous wooden coat hanger. To Rose, a coat hanger was not a biblical "rod of correction" with which she beat goodness and wisdom into her offspring. Hers was a more scientific endeavor. The coat hanger was her little tuning fork that she judiciously applied until her children were singing the song she wanted them to sing. She never lost control, never struck them out of sheer anger, but hit them with what she considered the proper, healthy dosage of pain.

This time Rose did not bring out the coat hanger but she did note her son's sarcastic outburst in her diary. The next day, as the children were playing on the porch, she said good-bye and drove down the elm-lined street. Realizing that she had forgotten something, she turned back and was relieved to see that little Jack and his siblings were playing, mindless of the fact that their mother had gone.

Rose may have set off on her trip in guiltless bliss, but Jack had struck at an unpalatable truth. As a youth, he would tell his closest friend, that he had cried every time his mother left on one of her endless trips, until he realized that his tears not only did no good but irritated Rose and caused her to pull away emotionally even more from her second son. He realized that he had "better take it in stride." The terrible threat was that if he did not buck up, Rose would return physically to the home in Brookline, but she might not return emotionally. As an adult, Jack reflected that his father was "a more distant figure" than his mother. Rose, however, although more often physically present, was "still a little removed which I think is the only way to survive when you have nine children."

Jack rationalized the psychological distance that Rose kept from him and his brothers and sisters. It was in part the same distance that Rose's mother and Joe's mother had kept from their children, a distance common in Irish-American homes. Love was a sweet cake given out only on special occasions, and then only in modest nibbles.

Rose took her children to see Bunker Hill and Plymouth Rock. She did not allow Jack and Joe Jr. to romp over the sites but insisted that they stand there while she inundated them with details and anecdotes worthy of a tour guide. Upon returning to the house in Brookline, she quizzed them over lunch, pushing them to remember and recite. When she took her children to Boston's Franklin Park Zoo, the boys and their sisters did not simply stand and watch the lions prowling back and forth. Rose stepped forward and lectured her children on the Christian martyrs who had been eaten by lions such as these.

Most women of Rose's generation might have attempted to follow the child-care experts, but the burden of raising children was such that they often had no time to follow this "scientific" imperative. Rose, however, knew that her one legacy would be her children. She was a literalist who took the ideas of the modern scientific experts and applied them as if they were absolute dictums.

Regularity was the key to scientific child-rearing, and Rose treated it as a moral imperative that played into all of her natural instincts for order and discipline. The authoritative guidebook, *The Care and Feeding of Children* by Dr. L. Emmett Holt, instructed mothers that babies had to be fed and put to sleep "at exactly the same time every day and evening." They were to be fed for no longer than twenty minutes, and if they started falling asleep, they were to be kept alert by "gentle shaking" until they finished feeding.

Dr. Holt warned against displaying too much affection, hugging a crying child, or indulging in mere emotion. The expert considered plumpness the mark not of a healthy bonny baby but of a lethargic, indulged child who, he believed, might be eating "twice as much food as is proper." Rose had a mandate to take her own obsession with weight and apply it to her children so they would not have a peasantlike girth. She apportioned food to them like medicine, holding off on helpings to Joe Jr. and Rosemary and adding food to Jack's plate.

Rose monitored everything that might touch her children's lives in the home like a guard patting down visitors for contraband. She could not have her children polluted by improper reading material, and she purchased approved books for them at the Women's Exchange. She was appalled when her mother bought a popular book, *Billy Whiskers*, and gave it to Jack. "I wouldn't have allowed it in the house except that my mother had given it to him," Rose recalled. Rose found the brash colors of the book offensive. It was not the pictures that made the series Jack's favorite but brash, bold Billy himself. Billy would not be tethered or harnessed. He ate whatever he wanted to eat, from popcorn to cakes, and butted anyone who got in his way. For Jack, it was surely an exquisite dream of a book about a goat that lived free, as a boy could not.

Rose wanted to be a perfect young mother, for perfection was the only passing grade. The experts warned her that her children's failures were her failures. "Lack of precision in the mother is responsible for most of the failures we see," wrote Mrs. Burton Chance in *The Care of the Child*. Experts tried to turn the nursery into a pristine conservatory where children were carefully nurtured and monitored like exotic flowers. If these flowers did not grow tall and splendid, the fault lay in the gardener, and a terrible fault it was, a shame visible to all.

Rose had a strapping son in Joe Jr., an advertisement for her virtues as a mother. Alas, she had a sickly weak child in Jack. In her namesake Rosemary, she slowly realized, she had a child who was mentally retarded—a moron, in the scientific nomenclature of the time. It was a devastating realization to both her and Joe, and it forever changed the family.

Rose had wanted to be the mother of mothers. By the merciless standards of science, she had probably created, through slovenliness, excess, or oversight, these two children who demonstrated her failures. Whatever guilt she may have felt was covered by an elaborate brocade of optimism. She had a relentlessly upbeat spirit and belief. She pushed her half-sick children out to play, denied her own pain and doubt, and smiled, if sometimes through clenched teeth.

When Joe Jr. and Jack and their siblings needed a touch of unbridled love and a hug, they turned to their stout Irish nanny, Kikoo Convoy, a woman untutored in modern child-rearing techniques. As much as the children adored Kikoo, as they grew older they could see that in their world such unregulated emotion was a servant's indulgence.

Rose often dressed her two sons in identical sailor suits or other clothes that marked them as if they were twins. They needed no such costuming to mark them as brothers in blood and destiny. Joe Jr. had been an enormous baby, anointed not simply by his parents but by life itself. Teachers told Rose and Joe how brilliant their firstborn was and showed them Joe Jr.'s IQ tests that confirmed his high intelligence, so much so that he was chosen for a study of gifted children. Rose could not help but think that there was a direct linkage between intelligence and virtue, and it seemed unthinkable that her mischievous second son could be as smart, or even smarter, than his brother.

"I didn't think you could have two in one family," she said later. His father had much the same opinion as his mother. "He told me once that he didn't think Jack would get very far and he indicated he wasn't very bright," recalled Henry Luce, the publishing magnate, of a conversation a few years later.

Jack was younger, smaller, weaker, and lesser in everything but spirit. He was second-born and bore all the markings of his diminished rank. "The mood of the second-born is comparable to the envy of the dispossessed with the prevailing feeling of having been slighted," wrote Alfred Adler, the psychologist. "His goal may be placed so high that he will suffer from it for the rest of his life, and his inner harmony be destroyed in consequence. This was well expressed by a little boy of four, who cried out weeping, 'I am so unhappy because I can never be as old as my brother.' "

The jealous competitiveness was a Darwinian struggle in which Jack never yielded but rarely won. Jack would admit decades later that Joe Jr. had been "rather heavy on me on occasions. Physically we used to have some fights that, of course, he always won. . . . I was somewhat brighter than he was, but I would say he was physically ahead of me." It was a telling mark of their competitiveness that Jack would only grudgingly admit that Joe Jr., two years his senior, "was physically ahead." Jack had few memories of his childhood, but one of them was a bicycle race between the two boys. "We ran a race against each other around the block and hit head-on," Jack recalled. "You know, we started in the opposite direction, came round this way, and I really tore this all up and got, I guess, twenty-eight stitches and he emerged unscathed."

Here was poignant testimony to the drama of birth order. The Kennedys took this typical sibling struggle and ratcheted it up to create much of the essential psychological drama of the family. Joe and Rose made their eldest son a little father with authority over his brothers and sisters. They were not so foolish as ever to say that Joe Jr. was their favored son, but they did everything to show it. At the dinner table, when Joe quizzed his children on current events or history, he went always first to Joe Jr., who as often as not replied in words that could have been his father's. Under the table the two boys kicked each other, careful not to be detected by their mother. Their father took the competition to higher and higher levels. "Remember that Jack is practicing at the piano each day an hour and studying from one-half to three-quarters of an hour on his books so that he is really spending more time than you," Joe wrote Joe Jr. in July 1926.

When Rose looked back on these years, she said that she had been so preoccupied with Rosemary's difficulties that she felt guilty that she had neglected Jack. Rose recalled: "When his sister was born after him, it was such a shock, and I was frustrated and confused as to what I should do with her or where I could send her or where I could get advice about her, that I did spend a lot of time going to different places or having her tutored or having her physically examined or mentally examined, and I thought he might have felt neglected." She did not worry that she may have neglected her other daughters, Kathleen and Eunice, born in July 1921, who presumably did not need the mothering that a son needed.

Little Jack lay in his bed in a dressing gown reading books, each illness and each condition taking more days away from the rugged fields of play and the struggle for manhood. The specter of homosexuality—a moral disease, a betrayal of masculinity—was omnipresent. What could be more horrifying than if sickly little Jack ended up as one of *them*, as a "man of broad hips and mincing gait, who vocalizes like a lady and articulates like a chatterbox, who likes to sew and knit, to ornament his clothing and decorate his face"? Dr. Joseph Collins wrote in his best-selling *The Doctor Looks at Life and Love* that "they are the most to be pitied of all of nature's misfits. . . . They are constantly between the devil and the deep sea; tormented by desires that will neither be subdued nor sublimated and unable to obtain even vicarious appeasement."

Rose was herself a carrier of the vice of femininity and did not want to make her sickly child into a mommy's boy. Rose inspected Joe Jr. and Jack before they left to walk to Edward Devotion Elementary School, where they were as neat and proper as the outfits they wore. After school they changed into clothes that made them look, in Rose's phrase, "like roughnecks." Once clad in these coarse garments, they were like unfettered young beasts, so wild that at least one of the neighbors' boys, Robert Bunshaft, was not allowed to play with them. They went traipsing down to the stores at Coolidge Corner. When they saw a sign outside a restaurant, NO DOGS ALLOWED IN THIS RESTAURANT, they scribbled in the word *HOT* before *DOGS*, then headed off after admiring their witticism. Two days later they returned to the shopping area and stole false mustaches from a store. On other occasions they walked into the public library in Brookline and disrupted the quiet, as if it were their right to carouse through the stacks, yelping and joking, while their nanny sat reading a book. Then they were out on the streets of adventure again, where Joe Jr. led his brother through the alleys and the back streets, a boy's jungle, once getting caught up on a neighbor's roof. "The boys have a new song about the Bed Bugs and the Cooties," Rose noted in her diary in February 1923. "Also a club where they initiate new members by sticking pins into them."

For the sedate neighbors, it was bizarre, inexplicable, and unseemly. The Kennedys, after all, had moved into a new large house on the corner of Naples and Abbottsford Road, had a Rolls-Royce with an English chauffeur, and were the wealthiest family in this largely middle-class area. And yet their two sons, beatific altar boys at St. Aidan's on Sunday, gentle scholars at school, became rude ruffians on the street.

No one considered Joe responsible for the children's problems. Joe was living at a time when even the word "Father" was suffering what many considered a diminution into the words "Dad" or "Daddy," terms that his own children used while, like most of their contemporaries, continuing to refer to Rose as "Mother." In the twenties "Dad" was the butt of jokes and cartoons, the neutered "Pop" who had nothing more to offer his children than his wages. Poor Pop. He was not wanted in the nursery, and he was widely rumored to be inadequate in the bedroom.

"It has often been said that American husbands are the best providers and the poorest lovers in the world," wrote Dr. Joseph Collins in 1923. "They frankly admit the first charge and tacitly the second." The image of the father was such that in 1924 the *New York Times* ran an article about the almost stillborn Father's Day and its official flower, the dandelion, chosen "because the more it is trampled on the better it grows."

Many fathers wanted to get into the nursery and into the center of their children's lives but feared that they would be emasculated. Some of them were willing even to help with housework if they could do so without being sissified. They sought a companionate relationship with their children, one in which, as Chester T. Crowell wrote in the *American Mercury* in October 1924, "once children are accepted as associates rather than duties, living with them becomes a lot of fun."

Joe was radically different from the diminished, companionate fathers of his time. He had his own ideas about fatherhood, and in carrying them out he became the most important political father of twentieth-century America. Rose referred to him once as "the architect of our family," and an architect he truly was as he set in place the master plan.

Joe had no knowledge of the nursery. He did not care much about fathering when his children were mere infants; he was preoccupied with his business pursuits. One winter day he was pulling two-year-old Joe Jr. on his sled while carrying on a conversation with Eddie Moore, a lifelong associate. The toddler fell off into the snow while his father continued walking, pulling the empty sled. As for his daughters, Joe loved them deeply, but he had no interest in raising them, whatever their ages. As he saw it, that was best left to his wife and to the nuns. If his daughters were properly protected, they would not struggle into womanhood but would largely assume it, taking on the natural and narrow virtues of their sex.

For his sons, however, once they reached a certain age, he was ready to lead them on the arduous journey into true manhood on which he had been thwarted. His sons would not be prissy inheritors. He would give them the wealth that the Brahmins assumed was synonymous with virtue, and he would push them up that great road. He would develop his sons as a happy meld of gentleman and athlete, democrat and aristocrat.

With his sons not only dressing like roughnecks after school but acting like ones, Joe saw that their rambunctious masculinity needed to be channeled into the proper venues. For the first time since Jack's scarlet fever, Joe assumed dominance over his sons' lives, a dominance that in a sense he never relinquished. His daughters could go to public and parochial schools, but his sons had to enter the elite Protestant social world where the progeny of their families entered. "The old man wanted them to mix with money," recalled Tom Finneran, one of their classmates at Edward Devotion. "That's why they went to private school."

Joe sent the two boys to the exclusive Noble and Greenough School to match wits and fists with the likes of Storrows, Bundys, Littles, and Coolidges. Generations of refinement had not honed the natural boyish viciousness out of the students, and they took exquisite pleasure in testing the mettle of the two Kennedy interlopers, probably the only two Catholics in the school. The upper-crust boys hurled the word "Irish" at the two new arrivals as if it were the crudest invective.

Joe wanted his sons to compete manfully on the fields of sport. Jack was team captain in a baseball game played in May 1926. Joe would have been there, but he was in New York. So he cabled his "Good Luck" to eight-year-old "Captain Jack Kennedy." Even without such encouragement, young Jack knew that on the baseball diamond and the football turf he and his brothers were to show their mettle and spirit as Kennedys.

One of the biggest of the boys, John Clark Jones III, tormented Joe Jr. after school until young Kennedy ran scurrying down the street to take up sanctuary in St. Aidan's. "This is a shrine!" Joe Jr. yelled. "You can't touch me here." A Catholic church was a dark mystery to the Brahmin boys. They would no more have chased Joe Jr. in there than risk bad luck by chasing him under a ladder. The Kennedy boys used their sacred preserve again and again. "When they'd shoot a snowball in our snowball fights, they'd duck over to the church and get inside the building there for protection," recalled Holton Wood, a fellow student.

As much as Joe Jr. and Jack fought each other, they stood together against this foreign world. On the playground Joe Jr. would challenge other boys to a fight, the bigger and older the better. Jack would stand by, betting on his big brother, with marbles as the currency. As the days went by, Jack's little bag of marbles grew larger and larger. To boys like Augustus Soule Jr., Joe Jr. was the perfect archetype of the despicable Irish thug that he had been told about—"very pugnacious, very irritable, very combative." Joe Jr. was a strange Irish bully, however, a fourth-generation American picking on boys who were bigger and older than he was.

When the headmaster sold the school to a developer, Joe was one of seven members of a committee to found the new Dexter School. For the first time in his life, he was on an equal footing with the other private-school fathers. His sons had been the opening wedge, but he was an equal only because he had the money to contribute substantially to the building fund. The parents whispered about Joe and passed on rumors about the illegal ways in which he was making his fortune. But his dollars gave off no foul odor, and they rationalized that even though they took his money, the Kennedys would never grace their homes socially.

Joe did not have to tell his sons that they were different, but he wanted that memory stamped on them indelibly. He was often gone from Brookline, but he pointedly took them out for football practice the first time in September 1926. To Joe, the football field was still the plain of honor where manhood was forged. He had recoiled at the struggles on that field himself and turned back toward more sedate sports, but he wanted his sons to prove their mettle, as he had not.

"Well, coach, you're going to have quite a problem, because here are two young 'micks' who need discipline," Joe told Willard Rice, the new coach, a Harvard man. "Mrs. Kennedy and I will give you carte blanche for any disciplinary measures that you need to take to get them into line." "Mick" was a term of derision. By using it, Joe was raising the stakes, which, on this field, were already high enough. The boys were the only two Catholics on a team that wore imitation Harvard uniforms. Their teammates did not need added incentive to tackle and block these papist encroachers until they ran whining off the field. The coach inspired laggards by kicking them in the behind, a practice that Joe and Rose considered "a wonderful idea."

The Dexter boys thought of football as their sport. They had every reason to believe that someday they would wear these same colors in Harvard Stadium, as did their big brothers and older cousins and friends. There they would compete against young men like themselves. What the youths discovered was something that many people would learn over the years: the Kennedys bore up to pain as if it were some paltry diversion that they refused to recognize.

The Kennedy boys were tough beyond measure. As a back, Joe Jr.'s greatest attribute was his crushing aggressiveness. Pounding into his opponents, he preferred to run over them rather than to finesse his way around the defenders. Joe Jr. treated the football field as a testing ground where the crack of tackles and blocks was so vicious that all but the most heedless and fearless were driven to the sidelines. Jack, a wiry quarterback who led the Dexter eleven to victory, was made of more subtle stuff.

Joe Jr. was even more his father's son on the baseball diamond. He would argue with the umpire, an unthinkable affront to the sportsmanlike ideal of Dexter. He scurried after every fly ball within reach, shouting the other outfielders aside. When he pitched, he threw the ball with such force that he nearly knocked the catcher backward. Only Jack could catch his brother, although tears welled into his eyes as the ball burrowed into his glove and he sometimes dropped the ball, to the glowering disdain of Joe Jr.

At Dexter, as for the rest of their lives, each brother had not simply friends but admirers who championed them and put down the other brother. Those who cared for Joe Jr. celebrated his exuberant extroversion, the way he grasped for every ball and fought every fight, seeing him as a model of boyhood. Those who cared for scrawny Jack considered him a deeper sort and thought that extroverts shouted their insensitivity to the world. But both boys played with intensity, not understanding those who sauntered after fly balls or blocked halfheartedly in football games.

The boys would have found these afternoons on the playing fields of Dexter perfect if only their father had been there. He was gone sometimes for months, but when he returned, it was as if a firestorm of life had descended into the Kennedy living room. Joe purchased a controlling interest in FBO, a film company, and had gone to the West Coast to become a Hollywood magnate. "At that time he was the only Christian in the movie picture business," Rose recalled. "And I believe the story was that the Jews would get this Irishman. But on the contrary, he had banking experience while the other people had not."

FBO made cheap films, its greatest asset being Tom Mix, a cowboy star beloved by boys across America. On one of his visits to Brookline, Joe returned with Tom Mix outfits, an awesome prize for Joe Jr. and Jack. When he arranged for films to be shown at Dexter School, it was a special occasion that none of the other fathers could provide.

In Hollywood Joe learned one of the profound lessons of his life. He had always believed in family as a manifestation of his will. He saw in Hollywood how the Jews were able to maintain control of this important new industry.

As much as they competed with and berated their competitors, they stood together against the Gentile world. That was their strongest weapon.

Joe would build his family to stand together too, against all who might challenge them. "As long as they stick together that is the important thing," Rose recalled as the essence of her husband's thinking. "That is what he believed and expected."

When Joe was home, he always accompanied Rose to St. Aidan's for mass on Sunday. The Kennedys cut fine figures as they entered the Tudor church. Joe had taken away from Harvard a sense that clothes were the way a man advertised himself, telling the world about his class, his aspirations, and his confidence. He was an impeccable dresser—as concerned about his clothes as was his wife about hers—and he walked with a sprightly, self-confident step. His fellow parishioners could easily see that Joe Kennedy was a man of the world, with a pretty, youthful wife and a handsome family. The Kennedys did not mix much with their Brookline neighbors, barely nodding a greeting before sitting down.

As they kneeled in their pew, the Kennedys were worshiping at different churches. Rose went to mass practically every day. With a faith both deep and true, she was making obeisance to a God who gave her solace and peace. Joe sat with much the same reverential cast to his face, the very picture of the perfect Catholic layman. Yet he slept with actresses and chorus girls, manipulated stocks so as to exploit widows, pensioners, and other less shrewd manipulators, and set up bootlegging deals with mobsters from Cleveland to Palm Beach.

Joe did not rustle nervously in the pew, worrying that a just God might smite him down for using the Church as a shield for his sins. He had ample reason to believe that there were two churches: the one ministered to by simple parish priests like Father John T. Creagh, here this Sunday, and the sophisticated, worldly church of power and substance, led in Boston by Cardinal William O'Connell. The cardinal was as much a player in the world of power as Mayor James Michael Curley or Senator Henry Cabot Lodge. Cardinal O'Connell had officiated at Joe and Rose's wedding and blessed their lives. His nephew, Monsignor James O'Connell, may have been present that day too, for he then lived with his uncle in the official residence.

Monsignor O'Connell had a certain disability for a priest—a wife in New York City. So did Father David J. Toomey, editor of the *Pilot*, the publication read by good Catholics in Boston. When Father Toomey was secretly excommunicated, he claimed that the cardinal's nephew had bought his uncle's silence by threatening to expose the cardinal himself for embezzling from the archdiocese as well as his "sexual affection for *men*." By apparently lying to Pope Benedict XV about his nephew, the cardinal defused the controversy and returned to Boston to inveigh against immoral movies and sin in general. The cardinal was not one to look too deeply into the cellars of Joe Kennedy's life, for he might find his own skeletons hidden away there in the darkness. Beyond that, Joe was not only a generous contributor to his church but, in *Photoplay* magazine's authoritative words, "the screen's . . . leading family man."

For Joe, the Catholic Church was like the Democratic Party—an institution that he was born into and that he used as he saw fit—but he had no more deep faith in one than the other. A great family, as Joe defined the term, was a wealthy family, and he was now a millionaire several times over. Wealth, however, could be either the rich sustenance out of which accomplishment grew for generations or an overrich banquet that left those who feasted on it satiated and weak. Joe had seen both aspects of wealth.

Joe's father-in-law, Honey Fitz, had feared that his sons might outdo him. If anything, Joe feared the opposite. He had an astute understanding of the psychology of money. One key to the success of the great families lay in institutionalizing money in irrevocable trust funds. Thus, no one generation could squander the family's assets, and each member could know that he would go into life spared the tedious necessity of scrambling for a basic living.

Early in 1926, Joe institutionalized his belief in family when he established the first of a series of trust funds. The trust agreement was an artful document, for it created a family wealth that could go on for generations. Until they were thirty-five years old, the trustees had discretion over the percentage of the income they would give the Kennedy offspring. The Kennedy daughters received the same share as their brothers, but with his low opinion of their financial acumen and assumption that women were naturally profligate, Joe added a typical "spendthrift clause for female beneficiaries."

Joe told the financier Bernard Baruch, who shared his cynicism about human nature, that the trust fund would allow his children to "spit in his eye." It was not that at all. Joe's belief in family was in some ways an immortality wish, a way of living on through his sons and his sons' sons. A trust fund was as much a part of that vision for his sons' lives as private school education and athletic competition, as well as a part of his vision of what he thought a true man should be and have and do.

5

Moving On

In September 1927, the Kennedy chauffeur drove the family from Brookline to South Station to take a train to their new home in New York. Joe had a gift for mythic self-creation that was as American as the curveball. He could not admit that he was moving to New York largely because it was a more convenient place for him. He had to create a moral drama. He was fond of saying later that he had left so that his children would not have to suffer from the anti-Catholic, anti-Irish ambience of Boston.

"I felt it was no place to bring up Irish Catholic children," Joe said. "I didn't want them to go through what I had to go through when I was growing up." His sons knew that among their schoolmates the term "Irish" was not a term of honor, but they surely would have been bewildered if they had been told they were leaving their friends and their neighborhood to save them from the horrors of prejudice.

The Kennedys were leaving a Boston in which the Protestant upper class dominated banking and the law, but the city itself, by the mid-1920s, was largely run by Irish-Americans like James Michael Curley, who was mayor during most of this period. Joe, moreover, was moving to a suburb of New York that was even more an enclave of the Protestant elite than Brookline.

On one occasion two decades later he did admit why he had left Boston, though he made it seem as if he had been a poor man driven out of the city because he could not get a job. "It is not a pleasant thing for a young man born and reared and educated in Boston to have to pull up his stakes and seek opportunity elsewhere," he said. "I know, for I had to do it."

Joe fancied not only that he had fought against outrageous prejudice and was driven in hunger out of Boston but that he had struggled upward from the pits of poverty. That too was a common American belief: if you were not born rich, then the next best thing was to have been born poor and to have pulled your way up with nary a helping hand. As the train pulled out of South Station, however, Joe was leaving behind not poverty but most of the inconvenient witnesses to his past.

Those most irritated at his attempt to wrap his past in the sackcloth of poverty were his maternal aunts and uncles. They were proud of their family's achievements and thought that Joe was diminishing them and their lives. His Aunt Catherine had so much believed in young Joe that she had lent him much of her life savings, when he was trying to save Columbia Trust Company from a takeover bid, without so much as asking for a promissory note.

It was not a woman's place to know much about banking and finance. When, several years later her check bounced at the Columbia Trust Company, she had gotten dressed and gone down to the bank to find out why it had made such an embarrassing mistake. Joe had never replenished her account, and indeed he would never do so. Nor did he return money to his other relatives. Catherine and her brother and the other Hickey relatives were nothing more than reminders of a past he wanted to forget. As he left Boston that day, he was leaving unpaid debts. He was a sagacious judge of character, as ready to take advantage of the nobility of a relative as the ignobility of a stranger. He knew that these debts would never be called, that his actions would never be known beyond his uncles and aunts, who would bear a silent shame.

Joe was used to traveling endlessly. Rose's roots were far deeper, and she would never grow them as deeply again, away from her family, from her identity, from her own separate status as the mayor's daughter. She now had seven children, including her third daughter, Eunice, born July 10, 1921, her fourth daughter, Patricia, born May 6, 1924, and her third son, Robert Francis, born November 20, 1925. She was pregnant now with her eighth child, and as the train rolled onward she was traveling each mile farther away from the security of Dr. Good, who had delivered her babies and in whose care she planned to return to give birth to her newest child. Rose, even on this day, kept studious notes of her children's lives. She wrote that six-year-old Eunice was suffering from stomach problems. Another mother might have scribbled down that perhaps her sensitive daughter was upset by the move. That was a guilty suggestion that Rose would never consciously admit, or her daughter dare to speak.

Rose was proud that her children were so wonderfully resilient, and the seven young Kennedys soon settled into life in the gracious thirteen-room house that Joe had rented in Riverdale, just outside of New York City, overlooking the expansive Hudson River. Rose was heavy with child and alone in a new community in which she knew no one, but she understood that this was all women's business, something with which Joe must not be distracted. Joe left his family there and headed off into his other life.

Joe considered life a banquet at which he could feast on whatever and whomever he pleased. He had taken a liking to the sun and sociability of Palm Beach when Rose introduced him to the Florida resort a few years before. Now, in January 1928, instead of spending time in the frigid North with his pregnant wife, he was down in Palm Beach, setting off on a daring new romantic adventure.

Joe's choice for his newest dalliance was twenty-eight-year-old Gloria Swanson, a celebrated Hollywood star. Greta Garbo was more classically beautiful, Mae West was more voluptuous, but no other contemporary actress had Gloria's mysterious, exotic aura. She was a thrice-married egotist, now saddled to a fawning gentleman whose most notable feature was his name, the Marquis de la Falaise de la Coudraye. Gloria fancied herself a woman of her time, not a silly flapper but a spirited, independent, passionate woman who thought that eroticism was her natural due. While Rose perused her Catholic missives, Gloria read such up-to-the-minute works as *Sex and the Love Light* and *The Art of Love*.

Seduction comes in many forms, and Joe began by offering Gloria's husband a position at Pathé Studios in Paris, far away from his wife. Each day he huddled with Gloria or worked alone, straightening out her tangled finances. Evening after evening he escorted the couple to the most splendid of parties and balls, always deferring to the marquis. Then one afternoon, when he was convinced that the apple would fall from the tree on its own accord, he had his associate Eddie Moore take the grateful marquis deep-sea fishing. While the nobleman was far out on the Atlantic, Joe knocked on the door to Gloria's suite at the Royal Poinciana Hotel.

Joe left no memoir of the events of that afternoon, and we have only Gloria's autobiography to tell us what transpired. As she recalled, he stood in the doorway, a perfect study of the Palm Beach bon vivant in his white flannel pants, sweater, and two-colored shoes. "He moved so quickly that his mouth was on mine before either of us could speak," the actress recalled. "With one hand he held the back of my head, with the other he stroked my body and pulled at my kimono. He kept insisting in a drawn-out moan, 'No longer, no longer, now.' He was like a roped horse, rough, arduous, racing to be free. After a hasty climax he lay beside me, stroking my hair. Apart from his guilty, passionate mutterings, he had still said nothing cogent."

Joe's mutterings may have been passionate, but they were surely not guilty. Millions of American men watched Gloria on the screen, but she was a siren they caressed only in their fantasies. Joe had seen those movies too, and he had moved to this front with the same calculation and cunning that he used on Wall Street. Joe was not a man who liked risk, be it in war, business, or romance. He was, however, in love with Gloria, or at least in love with the idea of Gloria, and love was always a danger. He was passionately attracted to this daring, sensual, perfumed being so different from staid and proper Rose, from whose mouth came axioms and homilies and to whom sex was largely one of the obligatory rituals of marriage.

Joe considered giving birth a wife's work. He saw no reason to be with Rose to observe the messy, painful business. So his pregnant wife traveled without him to Boston. She would have preferred to deliver at her home, but instead she went to St. Margaret's Hospital, where she gave birth to Jean Ann, February 20, 1928, with Dr. Good and his team at her side.

As Rose lay in bed, she received a thick sheath of congratulatory telegrams, letters, and flowers, including an especially stunning arrangement from Gloria Swanson. In adultery the act itself is only the start of the duplicities. Joe sent a diamond bracelet to Boston for his wife. He arranged for Rose to have catered food from the Ritz. "Well, he felt sorry for me being in the hospital," Rose asserted. While Rose spent a month by herself in Boston recuperating, Eddie Moore and his wife, Mary, watched over the children in Riverdale. Eddie was a full partner in Joe's deceptions, and his presence in the house was another deceit.

Joe and Rose's marriage may have appeared little more than an elaborate, exquisitely rendered masquerade. They spoke carefully chosen words to each other, rarely stepping into territory that might bring pain or exposure. The Kennedys did not hide in their charade but actively solicited accolades for their wondrous model family. Joe was as proud of Rose as a mother and proper wife as Rose was proud of Joe as a father and proper husband.

Rose and Joe's children were their mutual business, and when they were around them, there was usually some other agenda at work, some life lesson being imparted. They kept so much of their emotional lives from each other that Rose and Joe's life together always had elements of a performance. So did their children's lives. They were often performing for their parents, mouthing the script they were supposed to speak. When they got older and Rose sent them round-robin letters, she mentioned the children one after another, holding them up to scrutiny, judging them on a report card whose standards she alone knew.

Joe rented a house in the center of Beverly Hills on Rodeo Drive, a few minutes from Gloria's home on Crescent Drive. Hollywood had only enhanced Joe's belief that there were always two worlds: the facade, whether the celluloid screen, the speech on the political platform, or the price of a public stock offering, and behind it, a truth known to the wary few. He had arrived in Hollywood celebrated by Will Hays, the industry censor, as the man who would clean up Hollywood and bring back films that the whole family could watch without shame or embarrassment.

His creative contribution turned out to be a series of low-budget films that suggested that morality and mediocrity were blood brothers. As for his personal conduct, in Hollywood hypocrisy was elevated to the level of philosophy, and no one found it unseemly that the celebrated family man was carrying on an assignation with a married star.

Joe's relationship with Gloria was not simply an erotic diversion. He descended on her life and took it over, from the scripts that she read and the details of her financial statement to the minutiae of her social life. A woman who had so easily betrayed her husband might betray him as well, and several years later the next occupant of Gloria's dressing room found a bugging microphone embedded in the ceiling, presumably placed there by Joe. Joe did little to hide his paramour from his children, even inviting Gloria to his homes in Hyannis Port and Bronxville. "Would you please get me a picture of Miss Swanson with her name on it," Kathleen wrote her father in January 1930. "How is little Gloria?" Kathleen asked her father three months later, inquiring about Gloria's daughter.

Joe's largest gift to Gloria, or so it seemed, was to star her in his most expensive production, *Queen Kelly*, directed by the celebrated Erich von Stroheim. Joe was not one to let love get in the way of commerce. He had written Gloria's contract so that although he split the profits on the film with her, if the film lost money she had to pay any losses by herself. In this instance, life truly was in the details, for the film was an unreleasable debacle, and the actress was saddled with enormous debts.

Joe tried in a modest way to balance his various lives. He made periodic visits to Rose and the children and traveled to Boston in the spring of 1929 when his seventy-one-year-old father lay ill in the hospital. His mother had died, at sixty-five, of cancer six years before. Joe made his obligatory appearance at his father's sickbed. P. J. seemed to be recovering, and Joe hurried off again on the train heading back to Los Angeles. The old man died soon after Joe left, and his son did not return for the funeral. Joe loved his father. His absence was probably not a sign of disregard but more likely indicative of his inability to stare into the implacable face of death. He was a man who avoided blood and pain, and he probably could not confront the terrible finality of his father's death.

If Joe had been there at the Church of St. John, he would have seen his own thirteen-year-old son, Joe Jr., greeting the mourners with solemnity and grace, standing where his father should have been. He would have heard Joe Jr. described as Honey Fitz's natural heir and seen his firstborn son take his first step toward manhood. He would have seen mourners as varied as life itself, from the powerful to the powerless, from men of wealth and position to modest men whose only tie to P. J. was that once he had helped them.

P. J. had asked that with his death all the IOUs that he had amassed over the years be burned. His two daughters, Loretta and Margaret, followed their father's mandate, burning notes totaling at least $50,000. Beyond that, P. J. left an estate that one of Joe's closest associates, James Landis, estimated as between $200,000 and $300,000, half to his two daughters and half to his son.

Joe's affair with Gloria had begun with what the actress called "passionate mutterings," and it continued with duplicity within duplicity. Joe was the daring director of the romance, with life as his great stage. He brought the actress home to meet Rose and the children. He invited her to travel with him and Rose on a European trip. Through all this Rose played the humble hausfrau and Gloria played the star, and hardly an honest word passed between them.

Gloria had a friend whose face had been scarred by broken glass in an automobile accident. The actress feared that fans pressing against her car might break the glass and destroy her face and career. Rose was sent forward into the crowds around the vehicle. "Who are you?" they shouted as she got in the car. "What are you doing? We want to see Gloria."

Joe was perfectly willing to send his wife into the yearning, starstruck crowds to protect his mistress. Joe knew that no matter what he did, Rose would act as if she did not see how she was being treated. She had her children and her faith, her honored name and great houses. That was what had been rendered unto her. What transpired beyond that was not her life.

"The story got around that Joe and Gloria had gone on a trip to Europe together," Rose recalled decades later. "A long story started until they said her child was named after him. The boy Joseph had been named after her father. He was four or five when she [Gloria] first laid eyes on Joe Kennedy. But the story had got around that he was Joe's son."

Gloria wrote in her autobiography that at one point Cardinal O'Connell implored her to end the affair. The Church had supposedly turned down Joe's request to be allowed to marry the actress, and now it was time for them to sever their relationship. That was Gloria's story. But it is doubtful that Joe would have gone to the Church to ask that he, Joseph P. Kennedy, honored Catholic layman and father of eight children, be allowed to become Gloria's fourth husband.

Gloria was an actress, and she used all her skills in her autobiography, as she did in her decade-long affair with Joe Kennedy. She continued to visit him even as the affair slowly dissipated. Joe was left not with sweet memories of nights of bliss, but with an object lesson in the danger of passions, not only for himself but also for his sons. "Forty is a dangerous age," he reflected to Harvey Klemmer, an aide. "Look out, boy. Don't get in trouble. When I was forty, I went overboard for a certain lady in Hollywood of which you may have heard. It ruined my business. It ruined my health, and it damn near ruined my marriage."

After a year in their rental house in Riverdale, Joe bought a splendid estate in the exclusive Westchester County community of Bronxville. When he was home, Joe joined the other men in this town of sixty-five hundred residents, commuting to New York City each morning by train, leaving Rose and their children behind in the brick Georgian mansion.

Alice Cahill Bastian, the nurse, recalled that Joe "was always the first one up in the morning in the household. Shortly thereafter, the sound of childish voices came from his room. The little ones had quietly crept down the hallway to enjoy a romp and reading the funnies with Daddy. This early morning visit was a highlight of their day and his. As he kissed them good-bye when leaving for his office or on a trip, he was never too hurried to ask them about school or their plans for the day. Oftentimes the younger ones would accompany him to the train."

Joe was the more demonstratively affectionate parent. His children responded with a love that was far more deeply emotive than what they felt for Rose. They learned quickly not to lie to their father. He tolerated none of the casual dissembling he called "applesauce." Nor did he want any "monkey business" at home or school. Though their mother's hand disciplined them, they feared their father's displeasure far more deeply. His slightest reprimand stung far greater than a coat hanger applied to their backside. Joe, however, like Rose, was often gone. "Daddy did not come home last night," Kathleen wrote her mother in Palm Beach in February 1932. "We do not know when he is coming."

Gossip was one of the town's primary products, and a favored subject was the mysterious Kennedys living in the most expensive property in town. "They were considered nouveau riche," recalled David Wilson, who went to school with the Kennedy boys. There were no Jews and few Catholics in Bronxville, but it was neither the Kennedys' faith nor their ethnic background that had set the town buzzing. It was Joe's sexual indiscretions.

Bronxville fancied itself a sedate, churchgoing, conservative place, but there were dalliances galore among the affluent couples. It did not matter what one did as long as one did it quietly. Joe was so public in his womanizing that even many of the children in town knew about his assignations. He was so disrespectful of the shallow sanctimoniousness of the village that he brought women to the Gramatan Hotel. Some of the teenagers in town were so brazen as to chase after Joe, hoping for a glimpse of his current companion.

Joe's most flagrant adventures did not begin until after Rose gave birth to Edward Moore "Teddy" Kennedy, their ninth and final child on February 22, 1932. "People said, 'Why do you want to have nine children [if] you have had eight?' " Rose recalled. "You are over forty years old, and you will be all tired out, and you will lose your figure and looks. Why do you want to pay any attention to those priests? So I got rather indignant and made up my mind that neither Teddy or I were going to suffer and [we] were going to be independent and make it in a superior fashion. We weren't going to have anybody feel sorry for us."

Rose was forty-one years old, and it was a difficult, tiring labor that left her exhausted and seriously ill. She was in bed for over a month. Rose had ample reason to believe that if she became pregnant again she might die in labor. Since she was not going to violate the church's mandate against birth control, she saw that her only choice was no longer to have sexual relations with her husband. Joe had needed no excuse for his affairs, but now there was an unspoken agreement that he could live as he wanted to live.

On the first day that ten-year-old Jack attended Riverdale Country Day School, he came running out late for the sparkling Reo bus, clutching a piece of toast in his hand, his shoes in the other hand, his tie askew. Jack had a preternatural instinct to take whatever was best. Instead of moving toward the back of the bus, he plunked himself down in the seat directly behind the driver. The seat had been claimed already as the prized property of one of his new classmates, Manuel Angulo. The boy cherished the seat in part because whoever sat there got first dibs in the afternoon when the driver stopped for the Good Humor ice cream truck. Manuel was no more going to argue his case with Jack than Jack was going to listen to his belligerent seatmate, and within a few minutes the boys started pounding each other, rolling into the aisle, setting off screams and shouts of encouragement. The bus driver pulled over to the roadside and separated the two boys. "After that we became good friends," Angulo recalled.

Joe Jr. and Jack both had an intense, nervous hunger for all the minutiae of a boy's life, treating touch football games as epic contests and passionately competing for even something like a bus seat as if the struggle were for life itself. Their friends relished their time at the Kennedy house, for everything was intensified there.

When Rose was home, she was not like many of their mothers, who were interested in bridge games and garden clubs. Rose was a mother who directed her children's every endeavor. She had ample time for her children in part because the ladies of Bronxville had largely ostracized her. She was blackballed from the Bronxville Women's Club, an outcast not because of her faith but because of her husband's infidelity. It was also unthinkable that the Kennedys would be admitted to the Bronxville Country Club.

Jack and Joe Jr. were among the most popular boys at Riverdale. Joe Jr. was the dominant brother, pushing his little brother into the background. "Perhaps Joe Jr. was kind of spoiled," Angulo recalled, "but then again, that was because he was the apple of his father's eye. I wouldn't say that Jack was spoiled."

Joe Jr. was ahead of his class in everything, including his interest in girls. Jack was so shy around girls that when it came time for him to go to dancing class he would hide in the bathroom. He was a handsome lad, but when girls started calling him on the phone, he could not even bring himself to speak.

Girls weren't everything. One fall afternoon Jack was sitting in the upper field with his football-playing mates, waiting for his turn to take the field. "What are you going to do when you grow up?" one of the boys asked. "I want to be a doctor," one boy said. "I want to be an engineer," a second asserted. "I'm going to be president of the United States," Jack said matter-of-factly.

Little Bobby would never have expressed such bold dreams, or if he had, his words would have been lost in his big brothers' boastful shouts. His own mother often scarcely paid attention to him. She ruefully admitted years later that by the time the seventh of her nine children was born, even she was dragged down by the relentless routine of mothering.

"As a mother, yes, I did get a bit tired after fifteen years of telling the same bedtime stories, celebrating the same holidays," Rose confessed. She was traveling more now, including twice-yearly sojourns to Paris to be fitted for the latest styles, and much of the time she foisted Bobby off on governesses and nannies. She might have felt differently if Bobby had been a brilliant boy, but lost in the middle of the family, he seemed to have nothing singular about him.

His big brother Jack could not remember Bobby until he was three and a half. Jack's first significant memory is perfectly emblematic of Bobby's upbringing. There stood this tiny tyke on the deck of a yawl in Nantucket Sound, jumping again and again into the rough waters, teaching himself how to swim, while Joe Jr. watched from another boat. Joe Jr. might have stopped him, lecturing him on the dangers, but short of saving him from drowning, Joe Jr. let his little brother continue.

Bobby could dive into that sea a thousand times, but he would never be able to knife into the water with the ease and elegance of his big brothers. Bobby scrambled on in life, fearful that he would be dismissed as unworthy and that when he arrived at the table of life the food would be gone and the guests departed. Once, when he was only four years old, he came careening down to the dining room, terrified that he would be late, and ran into a glass door, severely cutting his face. Another day he was playing in the toolshed when he dropped a rusty radiator on his foot, breaking his second toe. The pain was excruciating. Most boys would have burst into tears, dragging their injured foot behind them as they stumbled toward the house. Bobby sat there grimacing in pain, not even removing his shoe. A half hour later he finally took off his shoe: his sock was soaked in blood, and he had to be taken to the hospital.

In the high stakes of inheritance, Bobby seemed to have drawn the worst card. Unlike his brothers, he wasn't a handsome child whose presence could charm the uncharmable. Bobby was scrawny and small, always struggling to keep up, running along double time while his brothers forged ahead with long strides. As a boy, he had soft, gentle features that suggested he had best stay away from the tough playing fields of manhood. His mouth was often pursed in a wry expression, suggesting nothing if not bewilderment. Some called him shy, but it was a strange shyness, for he would suddenly burst out like a cuckoo clock on the hour, making a few loud noises, before shutting himself up again. He was not especially smart either. Not only was he not a top student, but he showed none of Jack's flashes of brilliance that excused his otherwise mediocre school record. Joe Jr. was one of the top students in his class at Riverdale, while in sixth grade Jack won a commencement prize for best composition.

"Bobby looked on both [of his parents] as if they were saints," reflected Kirk LeMoyne "Lem" Billings, Jack's closest friend. "He could see no defects in either. They did not reciprocate. They did not return his love." Joe and Rose surely loved their son, but he seemed at times an interloper, an observer to the dramatic comings and goings of his big brothers, half ignored as his parents doted over little Teddy. Bobby was the most emotionally vulnerable of the boys. "Bobby got along better with his mother than with his father," his younger sister Jean reflected. "Jack, the reverse. His father was often very rough on Bobby; his mother would console him, 'You're my favorite,' half jokingly." Rose may have told Bobby on occasion that he was her favorite, but she most likely did so because he was not.

Life for Bobby was a foreign language that he spoke only haltingly, stumbling over the syntax, his accent showing that this was not a tongue that came naturally to him. He struggled to stay up with the world his father taught him must be his. As a boy in Bronxville, he signed up for a paper route, an endeavor that Joe thought admirable training for a youth. Admirable training it was, but primarily for the Kennedys' chauffeur, who delivered the papers each morning. "I put an end to that," his mother recalled. "Bobby said he was so busy with his schoolwork."

Bobby had the deepest faith of any of his brothers. He appeared so devout at St. Joseph's Church in Bronxville that one of the nuns, Sister M. Ambrose, thought that he "might have a religious vocation." "Bishop Bernard from the Bahama Islands used to be given permission by Father McCann to collect at the Masses at least once a year," Sister Ambrose recalled in a letter to Bobby years later. "After an appeal at the nine o'clock mass, you went home, got your bank, and gave the contents to the Bishop."

The Kennedy boys had set their roots in Bronxville far deeper than their parents ever would or could. Rose wanted her sons to go to an elite Catholic school or, if Joe spurned that idea, to public school, "where they'd meet the grocer's son and the plumber's son as well as the minister's son and the banker's son." In Bronxville there were few plumber's and grocer's sons in the classrooms, but the progressive school system was among the best in America.

Joe, however, had no use for what he considered the tedious malarkey of American democracy. He had received his education in American social mores at Harvard. He was not going to have his sons go through life bearing the stigma of public school. Rose argued with her husband, but this was not a debate that she had any business joining or the possibility of winning.

In the fall of 1929, fourteen-year-old Joe Jr. left his classmates at Riverdale to head off to Choate, one of the half-dozen or so top prep schools in America. "He is a rare youth and you will be most fortunate to have him," Frank S. Hackett, the Riverdale headmaster, wrote in support of Joe Jr.'s admission. "I regret exceedingly that this family are contemplating any change."

Jack stayed at home another year attending Riverdale before being sent to Canterbury Prep, a Catholic school in New Milford, Connecticut. Since his early years of sickness, he had been a healthy, vibrant boy in Bronxville. Away from home, thirteen-year-old Jack began to suffer from a myriad of minor maladies. He joined the football team, but being small and weak, he was knocked up and down the field. "My nose my leg and other parts of my anatomy have been risked around so much that it is beginning to be funny," he wrote Bobby.

By now Jack was fluent in the emotional language of the Kennedys. He knew he could never complain too loudly that he missed his mother or his friends, or that he was full of wistful homesickness. He might allude to such things, but sickness was the only avenue of emotional release left to him. His frequent letters home were full, not of the joys and vicissitudes of a boy's life, but of problems with his health. He wrote his father that at mass he "began to get sick dizzy and weak. I just about fainted and everything began to get black." He said he was only saved from collapsing onto the floor by an alert proctor, who held him up.

Jack could not tolerate the suggestion that he might be weaker than his brother, and he pointedly reminded his parents: "Joe fainted twice in church so I guess I will live." On another occasion he wrote about problems he was having with his eyes: "I see things blury [*sic*] even at a distance of ten feet," he told his parents. "I can't see much color through that eye either." He wrote that he was losing weight and that he was "pretty tired."

Such letters would have sent most parents hurrying to the school to succor their sickly son. Neither Joe nor Rose apparently ever visited Jack at the school. They wrote frequent letters. They kept in contact with the school administration. But they treated the school year as a sentence that Jack had to serve out on his own and whose life lessons would be diminished by their presence.

Jack was already imbued with the Kennedy attitude that individuals outside of the family were largely interchangeable. That winter of 1930/31, Jack wrote his mother complaining about the suit that she sent him and detailing problems with his eyes. Only then did he mention what another boy would have considered the most important news. Jack had been out sledding with some of the other youths. The hill was steep and slick with ice. Jack estimated that the sleds careened down the pathway at close to forty miles an hour. He was traveling so fast that he did not make a turn and sailed into a ditch. Jack set out again, and as he wrote his parents, "I smashed into a sled that was lying on the ground and I saw the other boy, Brooks, lying on the ground holding his stomach." Jack went on in clinical detail describing the youth's "grayish color" and speculated that the injured boy was probably operated upon. "He had internal injuries and I liked him a lot," Jack concluded, describing the student as if he were deceased.

The friend had been carried off life's playing field, and though he presumably recovered, Jack moved on. In her reply Rose might have tried to teach her son that without a sense of responsibility for his own actions, he would never be a true adult. She did not even address the matter, however, or ask about the boy's well-being. If she did, the letter has been lost.

Jack's grades were as sickly as his health, and his teachers knew him more as a patient than as a pupil. Joe finally caught on to the message Jack was sending him and allowed his son to come down to Palm Beach for a vacation. "I hope my marks go up because I guess that is the best way to say thanks for the trip," Jack wrote his father, fully understanding that Joe considered life a matter of exchanges. Jack was not able to repay that debt, for as soon as he returned to Canterbury he was stricken with stomach pains. The surgeon who was flown down to attend him pronounced that Jack had appendicitis and needed an immediate operation. Jack never talked about the fear he surely must have felt clutching his stomach in terrible pain, then being carried off to Danbury, Connecticut, where he was operated upon, alone and isolated. He did not return to Canterbury that year but was taken to Bronxville, where Rose monitored his recovery, making him study so he would not lose his academic year.

Jack and his siblings looked forward that summer, as they did every summer, to their sojourn at the Kennedys' house in the hamlet of Hyannis Port, Massachusetts. Joe had rented the white clapboard house for three years before purchasing it in 1928. The oceanfront house, set on two and a half acres of land, had plenty of room for tennis courts, a pool, and an expanse of grass for football games.

Hyannis Port became as close to a spiritual home as the Kennedys would ever have. These Cape Cod summers were not vacations, filled with the natural lassitude of hot, humid days. Hyannis Port was the school in which more than anywhere else Joe and Rose created the emotional ethos of the young generation of Kennedy men.

Joe believed that every moment of life had to be squeezed of its juices until only dry pulp remained. In the morning he was the first to get up and go for his hourlong horse ride. After breakfast he took his place on a deck outside his upstairs bedroom window, where he could survey his domain. He could not abide seeing his children sitting around, even for a moment. They moved from tennis to swimming to football to sailing, sometimes led by a full-time sports instructor. Joe had taken the playing fields of Harvard and brought them to Hyannis Port, out to the tennis court and out on Vineyard Sound, anywhere his sons might challenge each other and the lesser sons of other vacationing families.

When the boys played touch football, their friends soon learned that "touch" meant something different to the Kennedys than it did to others. It was the Kennedys' field and the Kennedys' football, and they usually claimed quarterback as their natural due. They had their own rules, often changing the parameters of the field on each play. They threw every pass as if it were the last play of the game and they needed a touchdown to win.

In the summer of 1937, Joe Jr. took Teddy out in his sailboat for his first race. Five-year-old Teddy was a natural sailor, and he and his big brother had the sails up just as the starting gun sounded. "Pull in the jib," Joe Jr. shouted as the boat cruised ahead. "Pull in the jib." Teddy looked around as if looking for some implement with "JIB" written on it in big letters. As the other boats drew farther ahead, Joe Jr. jumped up and grabbed the jib. Then he took Teddy by the pants and threw him into the sea. As Teddy felt the cold water and the stark fear of the moment, Joe Jr. grabbed him by the shirt, lifted him up, and dumped him on the deck like a fish. After the race, in which they came in second, Joe Jr. warned Teddy not to talk about the incident but to keep it eternally between them.

Joe was the master of competitiveness, and he doubtless would have found his eldest son's action only mildly excessive. Joe set the example in part by taking on his own sons in the sports in which he knew that he could beat them. He was a strong golfer and could easily defeat his sons. He was as much a strategist on the tennis court as in the boardroom, and he played a shrewd game with his sons, driving the ball back and forth, and then neatly dropping it away from their most desperate feint.

Each time he played them, however, it got harder to win. Then finally, one summer day when he was in his early forties, he struggled harder than ever against Joe Jr. Joe came off the court that day a victor, but he never played his sons again, preferring to give up tennis rather than lose to them.

Joe turned the luncheon and dinner table into another playing field, quizzing his sons about events in the world. He might ask a perfunctory question or two of one of their friends, or even squander a moment on one of his daughters, but his sons were his pedagogical target. When he was not there, Rose continued the questioning, often reading from a prepared list, relentlessly asking her queries in her tiny, grating voice.

Some of the Kennedys' friends dreaded sitting there, having to respond to questions they could sometimes hardly understand, much less answer. Harry Fowler, one of the friends, believed that for a boy summer was vacation time. "My lord, this is a nice summer afternoon," he thought as he sat there with the family at lunch. "What in the hell is Mrs. Kennedy doing anyway?"

Rose and Joe were attempting to imbue every aspect of their sons' lives with a fever-pitched competitive intensity. Joe was proud of saying that second best was not good enough. Winning was everything. "Don't come in second or third," he admonished his children. "That doesn't count—but win."

To Joe, life itself outside this pristine precinct was nothing more than an extension of it, an epic competition that went to the daring and the determined. "He always trusted experience as the greatest creator of character," his daughter Eunice said.

Joe offered his sons the opportunity to gain a fine education. He bought them status and entrée to the heights of society. But for them to be the kind of men he wanted them to be, they needed to pass through a crucible of experience. He did not flinch when Joe Jr. and Jack took their boats out in the highest of seas. He understood that it was dangerous, but he wanted his sons one day to sail bravely out into the storms of life.

Little Teddy, the last-born, was as much a part of this drama as Joe Jr., the firstborn, even if his big brothers treated him at times like a puppy that they could either ignore, coddle, or tease until it barked. Teddy was the last of nine children and the fourth of four sons, and in that one fact lay much of the drama of his life. "All children can be de-throned, but never the youngest," wrote Alfred Adler.

> He has no followers but many pacemakers. He is always the baby of the family. . . . In every fairy tale the youngest child surpasses all his brothers and sisters. . . . And yet the second largest property of problem children comes from among the youngest, because all the family spoils them. A spoiled child can never be independent. Sometimes a youngest child will not admit to any single ambition, but this is because he wishes to excel in everything, be unlimited and unique. Sometimes a youngest child may suffer from extreme inferiority feelings; everyone in the environment is older, stronger, and more experienced.

One of Teddy's few early memories is of the day he walked home alone from kindergarten in Bronxville. "I wasn't supposed to—and I remember getting spanked," he recalled. "Someone was supposed to pick me up, and I walked home, so the person who was supposed to pick me up didn't know where I was and it caused needless worry. I think she [Rose] used a hairbrush to spank me."

If his older brothers and sisters had walked home at the age of six, their conduct would probably have merited nothing more than a rebuke. But since the kidnapping and murder of Charles and Anne Morrow Lindbergh's baby within a few days of Teddy's birth, Rose looked at the world beyond her walls with a new wariness. Up until then she had seen the Kennedys' wealth, privilege, and celebrity as a perfect thing. What had Lindbergh's fame been, however, but a beacon attracting tragedy?

When *Life* featured the Kennedys and their wealth, Rose rebuked Henry Luce, the publisher, for needlessly pointing kidnappers toward her children. Teddy was brought up to mimic the intrepid lives of his big brothers as they moved out fearlessly into the world. Yet he was also taught to be suspicious of strangers, and what was the world he saw beyond family but an endless sea of strangers?

Jack treated his past like a prosecutor's brief, remembering every rebuke, every unfairness, but Teddy had a different kind of memory, selecting what was sweet and good from the remnants of the past. Teddy did not remember the mother who was gone for weeks at a time, but the mother who was there.

"Probably three times a week she'd read us a Peter Rabbit or a Thornton Burgess story," Teddy recalled. "Jean and I would go up to her room, and she'd read that. At the end of that I'd go down and go to bed, and in a few minutes she'd come down and kiss me goodnight."

With his impeccable business instincts, Joe had gotten out of the stock market before black Tuesday in October 1929. While some of his less prescient neighbors were relegated to living in mansions without electricity and pawning their family valuables, Joe did not suffer during the Great Depression. He profited from it as one of the leading bears, selling the market short from a desk at Halle and Stieglitz on Madison Avenue and Fifty-second Street.

Despite the millions of dollars that Joe continued to earn, he feared that the malaise was so severe that it might drive the four million unemployed and the two million vagrants into the streets, fomenting a revolution or anarchy that would take away all that he had earned. He was the darkest of Cassandras, apprehensive that he might end up as penniless as the forlorn men peddling apples on the street corners. "I am not ashamed to record that in those days I felt and said I would be willing to part with half of what I had if I could be sure of keeping, under law and order, the other half," Joe reflected years later. "Then it seemed that I should be able to hold nothing for the protection of my family."

Joe decided to back New York Governor Franklin D. Roosevelt for the presidency in 1932. He did so largely out of what he considered self-interest, not out of concern for the commonweal. He did not care much that Roosevelt might have the compassion, ideas, and intelligence to stem the tide of hollow-eyed, desperate wanderers filling the migrant camps of California or to keep good farmers from being driven bankrupt off their family land. Joe cared that Roosevelt would help him maintain for his family what was *his*. "I wanted him [Roosevelt] in the White House for my own security and the security of our kids, and I was ready to do anything to help elect him."

Joe was honest in his proud avowal of expediency, refusing to cover it with the tinsel of idealism. Self-interest is a thin parchment on which to swear fidelity to a cause or a person. Almost from the day he signed on to the campaign he began to extract his pound of benefit. "I doubt that Joe Kennedy felt like tugging his forelock to anybody on God's earth, and I don't think he ever did," said Frank Waldrop, the late editor of the *Washington Times Herald* and a friend of Joe. "But I also don't think he was under any illusions as to what Roosevelt was up to when he was dealing with him."

Joe fancied himself a man of ideas, but he had seen at Harvard that no matter how people pretended otherwise, money was always part of the admission price. He not only contributed $50,000 himself to the Roosevelt campaign but solicited funds from other wealthy men. Joe borrowed a private plane from a speculator, William Danforth, and crisscrossed New England raising money. Joe knew how to fill the most coddled heart with fear, and he took money from Republicans as well as Democrats.

Some of the Republicans insisted on anonymity, thinking it best to keep their beneficence quiet until after the election. Joe happily obliged, sending the checks in with his own name proudly displayed. Joe flew out to meet with the press magnate William Randolph Hearst and to attempt to woo him away from his candidate, Speaker of the House John Nance Garner.

At the Democratic convention in Chicago Stadium in June 1932, FDR fell short of the two-thirds majority then necessary for nomination. Although he was the convention's clear choice, FDR's supporters knew that if they did not win soon, some delegates might start looking elsewhere for a candidate. At this point Joe made one of the crucial phone calls to Hearst, pushing him to ask his first choice, Garner, to release his delegates. Hearst did so, and on the fourth ballot FDR finally won the nomination.

Afterward Roosevelt left for a leisurely cruise, fishing from a yawl along the coast of New England. Behind sailed a yacht full of advisers, would-be advisers, and putative advisers, meeting the candidate in port each evening for strategy sessions. Proximity is the first law of politics, and Joe made sure that he was there each evening, whispering his wisdom into the candidate's ear. He had to whisper with special intensity, for many of FDR's advisers despised Joe. They thought that in any true tribunal he would have been in the first ranks of the greedy speculators and manipulators blamed for helping to bring on the Great Depression.

On one of those days when Joe's competitors for Roosevelt's ear were isolated out on the yacht, Joe flew into New York City for a meeting with Roy Howard, the Scripps-Howard publisher and a prominent backer of one of the defeated candidates, Newton Baker of Ohio. Here was an opportunity for Joe to promote FDR's candidacy with one of the most powerful publishers in America, but if he did so, he chose a curious means.

"He, himself, is quite frank in his lack of confidence in Roosevelt, as evidenced by his statement to me yesterday that he intends to keep constant contact with Roosevelt during the cruise on which the latter embarked yesterday," Howard wrote Baker afterward. "Kennedy expects to fly to whatever port Roosevelt is in for the night, to be present at the evening conferences, because he knew if he were not present the other men—notably Louis Howe and Jim Farley—would 'unmake' Roosevelt's mind on some of the points which Kennedy had made it up for Roosevelt."

The Scripps-Howard chain was a rival of the Hearst newspapers, and Howard worried that Hearst might have an unseemly influence on a new Democratic administration. Joe assured the publisher that "Roosevelt was under no obligation whatsoever to Hearst and had not communicated with him personally either by telephone, or letter. . . . He protested . . . that Roosevelt is under no obligation to Hearst because Hearst is motivated not by any desire to nominate Roosevelt."

Despite his passionate avowals, Joe afterward solicited Hearst for a $25,000 campaign contribution. He wrote the press magnate a thank-you note that could scarcely be misunderstood: "You may rest assured, and this I want to say in order to go on record, that whenever your interests in this administration are not served well, my interest has ceased."

Joe could not seem to understand that what he thought of as candor others took as duplicity. In the minds of important men like Baker and Howard, Joe was a man not to be trusted. To them, he appeared a mercenary of conscience, with his mind and ideas for sale to the highest bidder.

Joe had a bigger game in mind than men like Baker and Howard could even imagine. This election and presidency was only a prologue. As Roosevelt's cruise continued, Joe stayed onboard the yacht for a while, sharing a cabin with Eddie Dowling, an actor and producer. On Cape Cod, Joe left the boat for good, leaving his bed to Eddie Moore, his lifelong associate. Moore was in an especially expansive mood, in such proximity to the man who would probably be the next president of the United States. He confided in Dowling about his employer. "If we live long enough and he is spared, the first Irish Catholic in the White House will be one of this man's sons," Moore told the incredulous Dowling. Irishmen were always full of their reveries, and Dowling marked it off as just another "pipe dream."

When Joe acted in a generous manner, it often turned out to be simply a loan that was expected to be repaid with interest. One of those who learned this lesson was the president's son, James "Jimmy" Roosevelt.

Jimmy was something that none of Joe's sons would ever be, a sad inheritor endlessly trading on his father's name and power. Joe helped Jimmy win some of the Ford Motor Company's business for Jimmy's insurance company. Joe then invited Jimmy to go along with him to England in September 1933, where he hoped to acquire liquor distributorships before the end of Prohibition.

Jimmy's presence next to Joe at business meetings signaled to the British business leaders that if they wanted to please the new administration, here was a good way to do it. Joe had no interest in risk, leaving that sad concept to entrepreneurs, plungers, and business school professors. He preferred certainty, and the exclusive Dewar's Scotch, Gordon's gin, Ron Rico rum, and Haig & Haig Scotch import licenses were the very definition of certainty. Once that was locked up for his new company, Somerset Importers, Joe finagled a "medical" license to bring in large shipments of Scotch to be sitting in warehouses on December 5, 1933, when liquor could be sold legally again in America.

Joe expected a cabinet post for his contributions to FDR's victory and in recognition of the acumen that he thought he would bring to the new administration. Instead, he was frozen out, the only trifle thrown his way being possible membership in the American delegation at the London Economic Conference, which would be writing a reciprocal trade agreement with Latin American nations. James Warburg, Roosevelt's chief representative, scotched that possibility. "I didn't want anything to do with a delegation selected to pay off political debts," he recalled, "for one thing because I thought it would be by definition an incompetent delegation." A few days later, on April 8, 1933, Warburg had lunch with Harrison Williams, a young administration official, who provided even more devastating insight into Joseph P. Kennedy. "Found out all about Kennedy from him," Warburg noted in his diary, "a completely irresponsible speculator who has been spreading malicious stories about the President. . . . I think Kennedy was probably spreading these tales because he hadn't got his payoff for his $50,000 campaign contribution."

In June 1934, Roosevelt made the "completely irresponsible speculator" chairman of the new Securities and Exchange Commission (SEC). Joe's critics scoffed at the idea that he would end the ruinous practices that he and his kind had for so long employed to such profit and harm.

Joe knew that he had to make the sort of authentic reforms that would satisfy his critics among the New Dealers. Yet he could not so overregulate the securities industry that the golden god of greed would never return to Wall Street. Joe was a decisive man, and in hardly more than a year he did magnificent service, creating many of the laws and rules that have formed the underpinnings of Wall Street ever since his time and helped to keep the American economy on an ever-rising trajectory.

Joe had the confidence—and this became an essential part of the way the Kennedys worked—to bring in the smartest, most ambitious people he could find and let them work unfettered. In this instance, he brought in such brilliant minds as two Yale law professors, William O. Douglas and Thurman Arnold, both of whom would make a firm mark on history.

Joe was not one to accept all the humbug of the New Deal by living in a narrow townhouse in Georgetown. Instead, he rented Marwood, a magnificent twenty-five-room estate in Maryland. He happened to be in town on the final Sunday in June 1935, a time by which most of his critics had been stilled. In the middle of the afternoon, Roosevelt telephoned to ask if he could come out to Marwood for dinner that evening. At seven the president and his small group arrived, including Missy Le Hand, his assistant, and the irrepressible Tommy Corcoran, carrying his accordion.

Joe had no use for the inane dinners and endless cocktail parties that clogged up life in Washington and that he usually managed to avoid. He preferred to stay at home and have people come to him for what he considered a truly good time. That did not mean liveried butlers and four kinds of forks, but sheer fun. That was what Joe liked, and that was what the president liked as well. The evening progressed with ample supplies of mint juleps and the freshest political gossip, a hearty dinner, and a new Hollywood movie, *Ginger*, shown on a screen set up on the lawn.

Then Corcoran took out his accordion, and it was time for song mixed liberally with joshing. The president was a man who could set aside the burdens of office with a song and a joke, that deep melodic laughter like a benediction. Joe felt comfortable enough with Roosevelt to trade him quip for quip, joke for joke.

The president was full of a sailor's yarns, tales of Northeast fishermen told in dialect, and of the ineptness of Roosevelt's own sailing companions at Harvard. Joe was not a sailor himself and had never been part of that world, but his sons were, and he matched the president story for story.

Each year Roosevelt went off on the Astors' yacht, another outing to which Joe, even now, would not have been invited. "Your taste in dumb cruise mates doesn't seem to have changed very much," Joe told the president.

Roosevelt laughed loudly and asked Corcoran to play "Tim Toolan," a favorite Irish ballad about a boy who makes good in the Irish business of politics. Roosevelt knew the words, and when it came to the refrain, he sang out loud and pure:

The majority was more
Than it h'd ever been b'fore
And our hero h'd carried the day!

It had been a wondrous evening, and Joe's guests did not leave until well after midnight. By the way Washington looked at things, Joe, given full imprimatur as a friend and counselor to the most powerful man in America, had arrived.

6

"Most Likely to Succeed"

In the fall of 1930, thirteen-year-old Jack went off to Choate with his big brother. Rose did not accompany her sickly son that day. Nor did Jack's mother ever visit the school in Jack's five years there. She nonetheless inundated the headmaster and his wife with letters. She had all kinds of advice and suggestions, but always from a distance. She did not let her son know how much she loved him, how conversant she was with every last detail of his education, how great a concern she had over the myriad of illnesses that he faced.

Joe also wrote often to the headmaster about both his sons but rarely came up to Connecticut to see them. For years he had been off in New York or Washington or Los Angeles, seeing his children primarily during vacations and on weekends. He nonetheless dominated his sons as assuredly as if he had been present from morning until night.

Joe's sons were extensions of him, young men bearing his name who would go out into the world to capture cities that he had seen only from the outer walls, to climb mountains that he had only gazed upon. Choate was a splendid crucible on which to forge the kind of men his sons were to be. There were few Jewish students, who had so dominated Boston Latin when Joe had gone there; the Choate application asked specifically whether the youth was "in any part Hebraic," a question that helped to keep that particular contagion to a minimum. Nor would his sons' unformed minds be contaminated by foul radicalism. A perfect indication of Choate sentiments was that the students voted each year for the "most conservative" student but not for the "most liberal." That shameful category stayed largely where it belonged: outside the gated precincts of the school.

That first evening at Choate, Jack was invited down to the home of the headmaster, George St. John, and his wife, for an evening of ice cream, singing, and what Mrs. St. John called "friendliness," as if it were another category of dessert. St. John was an austere, pointedly serious gentleman who ruled over the Georgian-style buildings and playing fields of the Connecticut campus with autocratic zeal.

The headmaster had an admirable concern for the individual lives of the boys at the Episcopalian-founded school, treating each one as a work of art in the making. He was occupied with his charges with a depth and intimacy that some of their own parents did not show. He was generous in spirit and fair as long as his authority was not questioned and youths did not imagine that they might construct another world beyond his purview. "I can't live in a school where the waters are troubled," he admitted.

St. John fancied that he treated all the boys the same, but the headmaster showed a special benevolence toward those, like the Kennedy boys, whose parents were willing to further their advance with added contributions. There was a certain irony to this. One of the teachers, Harold Taylor, avows that the headmaster had a distaste for Catholic upstarts like Joe Kennedy who dared foul his beloved Protestant school with his papist sons. In seeking added assistance from Joe, St. John was not so foolish as to indulge in the same honesty and forthrightness that he taught his minions to make the basis of their own lives. Better a little disingenuousness. Soon after Jack arrived, the headmaster wrote a letter to the former Hollywood mogul asking that "some one send to me information which one so ignorant as I can rely on" concerning a new sound motion picture projector. That was a mite too subtle, and so the headmaster appended a postscript: "It may be some time before I can raise money enough, but at any rate I want to be ready to go when I find it possible."

Joe did not miss the message, and he sent off an expensive sound system to the school, ingratiating himself with the headmaster and presumably helping to ingratiate young Jack and Joe Jr. with their schoolmates as well.

"We'll try to show our appreciation, our sheer gratitude, in every way we know," he wrote Joe. "I'm keeping close to Jack." That was the headmaster's intention, but Jack was hardly the kind of youth to accept the St. Johns' invitation to "feel free to drop in at our house at any time—even without a special invitation."

That first evening St. John talked to Jack admiringly of his big brother. St. John wrote Rose afterward that Joe Jr. was "one of the 'big boys' of the School, on whom we are going to depend." Joe Jr. had already been at Choate for two years. Now Jack faced the prospect of living in a house in which his brother had spread his belongings from room to room, leaving little space for another Kennedy.

Joe Jr. was a paradigm of what St. John envisioned a Choate boy to be. His time at the school formed a perfect rising arc, from a modest, difficult beginning to a noble finish. He had arrived in the fall of 1929 and had suffered through the hazing of the sixth-formers. He had been so foolhardy as to arrive back on a Sunday evening three hours late from Thanksgiving vacation. That gave St. John ample opportunity for a character lesson: he ordered the boy to be kept at school an added period over Christmas.

In his class work Joe Jr. had been equally foolhardy. As the headmaster wrote his father, Joe Jr. was "too easily satisfied and does not go that second mile that would make him a real student. Joe is still somewhat superficially childish. We like Joe so much that we want his best and Joe himself really wants to give it to us."

Joe Jr. stopped fighting against the yoke of discipline and brought his grades way up. At the beginning he had been no better an athlete than a student, just another second-string guard on the junior varsity. He began to run and to train on his own, and he worked his way up to a starting position on the Choate varsity.

Joe Jr. also became one of the most admired young men at the school, not just by the headmaster and the teachers but also by his own peers, and especially by the younger boys of Jack's age. He was a model to which they aspired, an exuberant, masculine straight shooter who had a smile and a word even for the newest of the boys.

Joe Jr. would have no part of hazing underclassmen but indulged in a full share of innocent wickedness. He delighted in filling his housemaster's shoes with sand, short-sheeting his bed, and then shaking his head mournfully as he commiserated with the teacher about whoever would do such a dastardly prank.

Optimistic. Conservative. Ambitious. Athletic. Friendly. Loyal. Intelligent. Exuberant. Joe Jr. was everything Joe wanted in a son. In comparison to Jack, it seemed at times as if Joe had passed on to his firstborn son everything of value. Joe Jr. was that most rare of spirits, a natural extrovert. The center of the room was Joe Jr.'s natural resting place, and he strode into any setting with that energetic manner.

Jack had the most tortured, complex feelings toward his big brother. That admixture of love, jealousy, anger, and competitiveness had jelled into the seminal relationship in young Jack's life. Jack, for all his sensitivity, could not see that Joe Jr. probably felt threatened by his brother and saw Jack's potential far better than Jack saw it himself.

Joe wanted his sons to be loyal as brothers, and much of the time they acted with deep affection toward each other. Yet there remained an undercurrent of rivalry and endless competitiveness. Though Joe was sure that Joe

Jr. would win, he watched like a promoter who owned both boxers. He created a family climate in which Joe Jr. could berate his little brother and Jack could fight back with words as much as he did with his clenched fists.

"When Joe came home he was telling me how strong he was and how tough," Jack wrote his father in Palm Beach while he was still at Canterbury. "The first thing he did was to show me how tough he was to get sick so that he could not have any thanksgiving [*sic*] dinner. Manly youth." Jack was the brother for whom the infirmary was a second home, and the sight of his sick brother was sweet vengeance. "He was then going to show me how to Indian wrestle. I then through [*sic*] him over on his neck."

As Jack saw it, Joe Jr. had nothing to teach him, and he had colleagues in his campaign against his big brother. "Did the sixth-formers lick him? Oh Man he was all blisters, they almost paddled the life out of him. He was roughhousing in the hall a sixth-former caught him he led him in and all the sixth formers had a swat or two. What I wouldn't have given to be a sixth former."

At Choate, Jack was for the most part too sly, too ferretlike, to be caught and bullied by the upperclassmen, but what a ransom he would have paid to have been there among those beating "the life out of" his brother.

Jack appeared shy, but his was not that sort of shyness that plagues a person in public places, taking immense energy to accomplish what for others is nothing but the routine of daily social life. Jack was not so much shy as reserved, keeping a distance from everyone and everything. Part of this manner was his way of building his own niche apart from his brother.

And partly it was the result of his illnesses, of lying in bed watching adults scurry around him trying to hide their knowledge of his maladies. Although the Kennedys projected an image of a family of radiant good health, only young Joe Jr. seemed immune to disease. Rosemary was slow. Eunice was plagued with illnesses. Kathleen had asthma. None of them, however, suffered from the scourge of afflictions that affected Jack.

Jack had scarcely settled into his routine at the prep school before he was in the infirmary. Mrs. St. John wrote Rose that Jack had what "seemed to have been the beginnings of a little cold." Rose might have surmised that Jack's condition was potentially more serious or the infirmary would have been overflowing with sniffling students, but the two women spoke in a genteel code language, always minimizing, always downplaying. Life would be just fine if everyone would simply say it was fine.

In early January, Jack was back in the familiar confines of the infirmary with "a mild cold in his head." Mrs. St. John wrote Rose that Jack had arrived with "a lavender bathrobe and lavender and green pajamas" and appeared to be "settling in for a pleasant stay," as if he were setting off on a Caribbean cruise.

Rose was consumed with her son's condition. Yet even after he had been in bed for two weeks, it was unthinkable that she would drive the sixty-five miles to Wallingford to see her sick son. Instead, she wrote letters telling the school that three years before Jack had had "mumps, and the doctor thought it was probably a cold." She suggested that he be given a teaspoon of Kepler's Malt and Cod Liver Oil after every meal. Jack's mother was assured that not only would he be forced to take a daily dosage in the infirmary, but that the tonic would continue back in his dormitory room. It was hardly a routine to please a boy who would have to be paraded in front of his peers after each meal to receive his medicine, a regimen that would continue until Jack was "full of pop."

The school treated Jack like a fragile seedling that had to be sheltered lest he be torn away in the storm of life. They were about to let him return to his room when the weather turned cold, wet, and unpleasant. So he was kept a bit longer in the infirmary. Whether owing to the mysteriously regenerative benefits of cod liver oil or simply the "glorious sunshine" that had finally graced the Connecticut winter, Jack was allowed to return to his room and to the dining hall, where the masters attempted to fill their 117-pound charge with salads and vegetables and in the afternoons to get him to down glasses of eggnog.

Rose telephoned Mrs. St. John asking that Jack be pushed to "finish well this term so that he will not have to do any summer work." Jack was back in the infirmary in April with a mysterious "swelling" and urine that "was not entirely normal." Despite Rose's entreaties, Jack had to return in August for the summer session.

The following academic year Jack was plagued with a whole new range of illnesses. He had problems with weak knees. He had bad arches that required special built-up shoes, a condition that alone merited ten letters from Rose to the administration. He was in the infirmary with possible "pink eye" and on another occasion with a high temperature and "a little grippy cold."

Scrawny, frail Jack was not the sort of youth who went out for football, not at Choate, and not anywhere else. But as a Kennedy son, Jack had to go out for the team on which his brother starred. Earl Leinbach, one of the junior coaches, was a severe disciplinarian who egged his charges on by chasing after them with a paddle and striking them full force on the buttocks. Jack's most distinguished contribution to the team was managing to avoid the coach's paddle as he swerved away from the strokes. In the end Jack was so unhealthy that he had to leave the dream and the honor that he had sought on the football field.

Jack had gone out for football and for two years fought with tenacity, but he was simply too small and weak. In the end, as he wrote his father, the closest he could get to the Choate football team was to be a cheerleader.

Golf was scarcely a worthy alternative to the struggles of the gridiron, but even here Jack feared that he was not up to the mark. "The golf is going good," he wrote his father, "and I have a slight chance for the team because it is rather bad this year."

Jack needed to find a separate sphere where he could stand tall and apart, not always in the shadow of his brother, whose light blocked out the younger Kennedy's accomplishments. He found those spheres largely in collaboration with Kirk LeMoyne "Lem" Billings, the classmate who became his closest friend and conspirator in the games of youth. Lem was a six-foot, 175-pound, gawky, bespectacled son of a socially prominent Pittsburgh physician, with a sense of humor almost as darkly ironic as Jack's. What linked the two youths most profoundly were their older brothers. Jack and Lem both had brothers who carried the family name to heights the two younger boys could hardly hope to attain.

Frederic Tremaine Billings Jr., like Joe Jr., bore his father's name, and he too carried his father's values into the world. At Choate his brother was, as Lem only grudgingly admitted years later, "rather outstanding." He was president of his class, editor in chief of the yearbook, chairman of the student council, and, like Joe Jr., winner of the Choate Prize, the highest honor.

At Princeton, Frederic was chairman of the student council, Phi Beta Kappa, captain of the football team, an all-American honorable mention for football, and winner of the Pyne Prize, the highest of honors. Then he went on to England as a Rhodes scholar and became, like his father, a doctor.

Lem's father had set forth the rules and expected both sons to run onto the same field of play. "My father did try very hard to have me line up as well as my brother in every area and was very disappointed that I didn't," Lem reflected. "I tried very hard."

The friendship between Lem and Jack was unlike anything either one had experienced in the past. Like most adolescent friendships, theirs was based on a commonality of experience that they took to be life itself. They spoke in their own shorthand.

They both had pimples and were struggling with their sexuality. They built their own lives away from the prying eyes of others, including most of the other students. "I think that people who knew him [Jack] liked him very much," Billings recalled. "I think others possibly didn't because he had a sharp tongue and could make fun of people very easily if he didn't think they lived up to what he felt they should. I wouldn't say he was overly popular."

Jack and Lem were constantly together at Choate and elsewhere, and apart they corresponded regularly. In his scores of letters Jack never even mentioned Rose and hardly named Joe Jr. As for his father, he was the "old man," a figure who was there primarily to berate his son when he got close enough, an austere, prickly presence who had to be gotten around.

Teenage boys often belittle each other in boisterous, rancorous put-downs, the only kind of display of manly affection with which they feel comfortable. Jack was merciless in his attacks on Lem, beating him down with a constant stream of criticisms. "Dear Unattractive," he began one letter, a salutation he could have used every time he wrote his friend. Lem was the foil for all Jack's insecurities, and whatever weight these attacks took off Jack's own self-doubt, they surely added to Lem's.

Jack was capable of generous acts of solicitude for his closest friends, but he inevitably tossed the gesture out with caustic disdain. As Jack saw it, Lem was forever the second best, and second only because there were just two of them.

Joe Jr. graduated from Choate in the spring of 1933, but his shadow remained, hovering over Jack. Joe Jr. had won the Choate Prize as the student who best combined scholarship and athletics, the exemplar of what a Choate graduate should be and what his father envisioned. His name was engraved on the bronze Harvard Football trophy, and he was celebrated in the *Boston Globe* as "a very popular hero."

Joe Jr. might have headed off to Harvard in the fall, but his father had a different idea in mind for his firstborn son. Joe decided to send Joe Jr. to England to study. He could have sent him to Oxford or Cambridge, where he would have settled in among the kind of privileged young men whom Joe considered the Kennedys' natural company. Instead, he enrolled Joe at the London School of Economics, a fervidly intellectual atmosphere full of Socialists and others who fancied themselves on the cutting edge of economic and political theory.

Joe had carefully evaluated the school and what Joe Jr. might learn there. He wanted his son to be able to cope with ideas about socialism and communism, all the supposedly most advanced thinking. Beyond that, he wanted his son to learn about the workings of capital and wealth. He told Rose: "These boys, when they get a little older and have a little money, I [want] them to know the whatnots of keeping that."

Joe Jr. was not an intellectual, and his wit, while genuine, was narrow. To some, Joe Jr. seemed narrowly and ignorantly conservative. At the London

School of Economics he was dim-witted compared to some of the other students, notably three brilliant Jewish Socialists from London's East End. In one third-year seminar that he sat in on, the three students and Professor Harold Laski batted ideas back and forth so quickly that poor Joe Jr. could not even follow.

Afterward, instead of leaving sheepishly with his head down, or arrogantly dismissing the discussions as pedantic foolishness, Joe Jr. showed up at Laski's office and asked the professor to explain to him what he had not understood. It would have seemed preordained that Laski, an acerbic, deeply opinionated man of the Left, would see in young Joe Kennedy prima facie evidence for why capitalism was doomed. Laski realized, however, that Joe Jr. had the stuff in him to realize that his classmates knew much that he did not know and were worthy men with whom one day he hoped he might be able to argue on a more equal level.

Laski saw that young Joe had character and an incomparable zest for life, qualities that were not a matter of right or left, brilliance or dullness. "He had set his heart on a political career," Laski recalled. "He has often sat in my study and submitted with that smile that was pure magic to relentless teasing about his determination to be nothing less than President of the United States."

Following his year in London, Joe Jr. set off that summer with his friend Aubrey Whitelaw for a trip around Europe. Joe Jr. had a blessed quality of joyful self-assurance that drew people to him. Before they left England, Joe Jr. purchased auto insurance in London even though he was too young. Not every youth on his first European sojourn wandered the Continent behind the wheel of a Chrysler convertible. Joe Jr. was a young man that summer who, as Whitelaw recalled, laughed with detached amusement at a dubious Italian guide who had stolen his wristwatch, made caustic remarks at each sighting of Mussolini's public portraits, and came close to fisticuffs in Munich with a Nazi who wanted Joe Jr. and Aubrey to "Heil Hitler."

Although Joe Jr. might not have wanted to salute Hitler himself, he was impressed by aspects of Nazism. After all, he had been brought up to believe in power, order, and discipline, and to Joe Jr.'s way of thinking, Hitler was carrying out those principles on a national scale. In a letter to his father, he expressed what he clearly intended as a sophisticated, detached view of the situation, but in fact his observations were primitive, passionately engaged, and dangerously naive. He thought that the Germans had good reasons to dislike the Jews. He accepted the Nazi propaganda that the Jews dominated the Weimar Republic as "the heads of all big business, in law, etc. It is all to their credit for them to get so far, but their methods had been quite unscrupulous." Joe Jr. felt sorry that "noted professors, scientists, artists, etc.

so should have to suffer," but he was sympathetic to the Nazis' dilemma. He concluded that it was impossible to sort the good Jews from the bad, and that the only reasonable answer was to throw them all out of Germany. He did not favor excessive private violence, but as he lectured his approving father, "in every revolution you have to expect some bloodshed."

Once the blood had been washed from the streets, Hitler could begin wholeheartedly with such progressive measures as his eugenics campaign, carrying out his new law that would sterilize those so richly deserving of the measure. "I don't know how the Church feels about it," Joe Jr. wrote, though he surely could have guessed, "but it will do away with many of the disgusting specimens of men who inhabit this earth." If Joe Jr. had thought a bit more about the matter, he might well have realized that one of those disgusting specimens would presumably be his own mildly retarded sister, Rosemary.

Joe Jr. was not simply observing Nazism as his father's surrogate, or as an abstract student of politics, but as a young man searching for ideas that he might one day implement in America. Joe Jr. intended to be president, and he was already planning his cabinet, telling his friend Aubrey that he would name him to a newly created post of secretary of Public Education. "We were going to make that office much more important than a Secretary of Defense," Whitelaw recalled in a letter to Jack, "in accordance with Plato's admonition that a government should be judged not by its Secretary of War but by its Secretary of Education."

Joe Jr. was an impressionable young man full of pretensions to political wisdom that he did not have. As proud as Joe Sr. was of his son's observations, he gently cautioned Joe Jr. that Hitler might have gone "far beyond his necessary requirements in his attitude toward the Jews." Joe Jr. returned later that summer to Hyannis Port bubbling not with quotations from *Mein Kampf* but with all the ideas he had learned from Laski and his colleagues.

"Joe came back about 3 days ago and is a communist," Jack wrote Lem. "Some shit, eh." Rose was appalled at the heretical ideas that Joe Jr. was spouting, ideas that he never would have learned if he had stayed in America. "Joe, if you feel like that why don't you just give away your boat and live like all these other people," she told him slyly.

"Oh, Mother, just giving away one boat would not make any difference," he replied. Rose was even more appalled when Joe Jr. appeared to be making his first convert. "Joe seems to understand the situation better than Dad," Jack told his mother. That was enough to send Rose hurrying off to her husband to warn of the incipient revolutionaries in their own house. Joe was not worried. "I don't care what the hell they think about me," he told Rose bluntly. "I will get along all right if they stick together."

Joe Jr.'s departure from Choate did not change Jack's behavior at all. He was still the merry prankster standing apart from the rules and rigors of the school and refusing to wear any harness of responsibility. Sloppy in dress and manner, he walked on the borderline of disdain, daring his masters to attempt to pull him back.

John J. Maher, the football coach and housemaster, had applied the switch of discipline to the backs of scores of recalcitrant scholars. Maher wrote the headmaster: "At first his [Jack's] attitude was: 'You're the master and I'm a lively young fellow with a nimble brain and a bag full of tricks. You will spoil my fun if I let you, so here I go—catch me if you can.' " After two years even Maher essentially gave up. "I'm afraid it would be almost foolishly optimistic to expect anything but the most mediocre from Jack," he wrote St. John.

"There is actually very little except physical violence that I haven't tried," Jack's French teacher reported. After a rare visit to the school, Joe left with a devastating sense that his second son was a spoiled, petulant young man, just the sort who might one day dishonor the Kennedy name. "I can't tell you how unhappy I felt in seeing him and talking with him," Joe wrote St. John in November 1933. "He seems to lack entirely a sense of responsibility. The happy-go-lucky manner with a degree of indifference does not portend well for his future development."

Jack wasn't concerned about St. John but about his father. Joe had taught him that no matter what he did, if he came to his father and told the truth, everything would be fine. Joe absolved all. He forgave all. He made what was bad good and what was broken whole. The curtains of his confessional were the boundaries of the family, and every word and every deed stayed within those precincts.

The truth can be as manipulative as a lie, and Jack learned to finesse his father with candor. "I thought I would write you right away as LeMoyne and I have been talking about how poorly we have done this quarter," he wrote Joe in early December 1934. "I really feel, now that I think it over, that I have been bluffing myself about how much real work I have been doing." That was all true, but Jack's was the calculated candor of a defense attorney who outlines his client's worst crimes before the prosecuting attorney does it. Jack had confessed before the quarter was even over, and well before his father would see his son's grades. No matter what punishment he faced from his father in Palm Beach, the letter would soften the blows.

Jack had gauged his father perfectly. "I got a great satisfaction out of your letter," Joe replied. Jack's father was a busy man, one of the wealthiest, most powerful men in America, yet he answered immediately. Joe was full of the most wearisome cynicism toward everything in life except for his sons. They were the repositories of his ideals and aspirations. He was doing whatever he could so that on the day when his sons' wealth and privilege were weighed against their achievements, an honest assayer would say that the Kennedy men's lives more than balanced the scales.

This was not a rhetorical shield behind which he pushed his sons ahead on their crude accession. This was his profound belief, and at this point he pushed it with principle and tact and nuance. "I would be lacking even as a friend if I did not urge you to take advantage of the qualities you have," he wrote Jack in early December 1934. "It is very difficult to make up fundamentals that you have neglected when you were very young and that is why I am always urging you to do the best you can. I am not expecting too much and I will not be disappointed if you don't turn out to be a real genius, but I think you can be a really worthwhile citizen with good judgment and good understanding."

Jack did not heed his father's heartfelt words. He read the *New York Times* every day and had a nearly encyclopedic knowledge of current events, but he sat daydreaming in class, responding listlessly to his teachers' queries, barely getting by academically.

As a father, Joe was facing one of the conundrums of great wealth. How did a rich man raise sons with purpose and a sense of destiny and responsibility? He thought he had been doing just that in part by sending Joe Jr. and Jack to Choate, but he saw in young Jack all the wages of his failure. Joe's fear was that wealth and all the mindless ease that came with it had distorted his son's values.

Joe was proud that he had earned money as a boy and that he had adhered so tightly to adulthood's harsh laws. It appalled him, as he wrote a Choate administrator, that he and Rose had "possibly contributed as much as anyone in spoiling him [Jack] by having secretaries and maids following him to see that he does what he should do, and he places too little confidence on his own reliance."

Jack was a young man of terrible carelessness. He was careless about his clothes, careless about appointments, and careless about money. Worse yet, he was intellectually careless. He was living through the Great Depression, with millions unemployed and the roads and rails full of hollow-eyed wanderers, and yet he knew and felt nothing of what his compatriots were suffering. "I have no memory of the Depression," he told journalist Hugh Sidey years later. "We lived better than ever. We had bigger houses, more servants. We had more money, more flexibility, more power than ever before." Jack was a rich young man who saw wealth not as something that had to be earned or maintained, or that others might try to wrest away, but as something that was simply his, as much a part of him as his feet and his fingers.

Ralph "Rip" Horton, Jack's friend and classmate, was just as scrawny and malnutritioned-looking as Jack, and just as rich. He had, moreover, certain perks of wealth that Jack didn't have but aspired to, such as a card granting entry to the more elegant New York clubs and a casual familiarity with the café society of the metropolis. The youths dressed up in ersatz adulthood on their forays into the Manhattan nightlife, acting with the nonchalance of the regular habitués.

Jack was getting over his shyness toward the opposite sex and slowly developing a lustful, cynical self-assurance. To Jack and many of his classmates, one of the major allures of attending Choate was the inspired presence of the wife of one of the masters. "Queenie," as she was known by the boys, was a voluptuous young woman, and perfectly aware of the dry-mouthed attention she received when she strolled into the dining hall. The highest honor that several of the boys dreamed of was not the Choate Prize that Joe Jr. had just won but the bedding of Queenie, a fantasy so unthinkable, so daring, so impossibly sweet that it could scarcely be whispered about.

One of Jack's classmates, Larry Baker, has the distinct recollection that Jack bragged that he had that honor. That claim was doubtless one of Jack's first forays into fiction, for what rankled him, as it did several others, was that often Queenie invited Maurice Shea, the Choate quarterback, to her house for tea. Jack was so irritated by the sheer unfairness of this that he turned to a song he had written:

Maury Shea, Maury Shea
Drinking tea every day
Maury Shea, what's your appeal?
Queenie, we want a new deal.

Queenie had never had a song written to her before, at least not by one of her husband's students, and one day she asked Jack and his friends to sing it to her. "We damn near died," Horton recalled. "Why we weren't thrown out of school, I'll never know."

During spring break of his junior year Jack and a group of students drove down to Palm Beach in three cars. The youths roared southward, the cars playing tag with each other, roaring along the macadam at seventy miles an hour. In North Carolina they were stopped for speeding. Jack pleaded poverty to the small-town judge, displaying his empty pockets. The magistrate cut their fines, and off they roared again. Jack and his friends did not suffer the mild discomfort of stopping at anonymous motels. They lived in their private world of privilege. They knew people all along the route, and they stopped at several estates.

At Sea Island a tollbooth blocked their way. One of the boys picked up a fire ax that happened to be lying there, and they roared past, waving the ax at the tollbooth keeper. Later in the day the young men drove Larry Baker's new Model A convertible into the ocean, a mild diversion that upset only the car's owner. Larry already had one souvenir from his friendship with Jack—a dark front tooth that had been deadened when Jack shoved him against the stucco wall of their house at Choate.

In Palm Beach, after a meal at the Kennedy home, Jack insisted that the group make a visit to the Cuban Tearoom, a brothel in West Palm Beach. Larry waited in the car while the others went inside. "They made fun of me," Baker recalls. "The kids went deep-sea fishing the next day. They teased me because I hadn't gone to the brothel, and decided to make me seasick."

Baker was far from the only detractor among Jack's classmates. One of his former friends accused Jack of telling the dean that he, the friend, and another student had a motorbike hidden in the countryside, a motorbike that Jack had ridden. Others felt that Jack was nothing but a spoiled snob, tooling off to Miss Porter's School to pick up his date in his father's chauffeur-driven Rolls-Royce and limiting his friendships to those of social or athletic status.

For Jack, the illnesses continued. Over Christmas 1933, he had another relapse. His condition was now so "interesting" that it merited a discussion before the American Medical Association. After his recovery, Joe wrote Joe Jr.: "It is only one of the few recoveries of a condition bordering on leuchemia [*sic*], and it was the general impression of the doctors that his chances were about five out of one hundred that he ever could have lived." If the youth's father wrote his eldest son such a terrifying letter, surely Jack knew his prognosis as well. He could hardly help knowing, for when it was thought that he had leukemia and would die, back at Choate the headmaster led a prayer for Jack in the chapel.

Jack's fragile health was also shameful, not only because it singled him out as a weak figure, but because of the sheer embarrassment of some of the diseases with which he was afflicted. When he returned from Palm Beach after the Christmas holiday, he came down with a terrible case of hives, covering his entire body, and was taken to the hospital in New Haven. "Well, you know, Jack, the doctors are simply delighted to have the trouble come out to the surface instead of staying inside," Eddie Moore, his father's associate, told him.

"Gee!" Jack exclaimed. "The doctors must be having a happy day today!"

When the hives began to disappear, Jack wished the doctors would stop poking at him and let him get back to his life. "If this had happened fifty years ago, they would just say, 'Well, the boy has had a case of hives, but now he's all over it,' " he told Clara St. John, the headmaster's wife. "Now they've got to take my blood count every little while and keep me here until they correspond to what the doctors think it ought to be."

Jack wanted to get out of the hospital, but he was still so mysteriously weak that the doctors continued their tests. By the summer Jack had more vague symptoms, and his father decided to send him to the famous Mayo Clinic in Rochester, Minnesota, for a battery of tests. It was a situation that cried out for his mother's presence. Rose's visit, however, would have been an omen of death or a sign of intolerable weakness. Instead, Eddie Moore, a surrogate parent, accompanied Jack.

Jack was a fiercely intelligent young man who looked with wry bemusement at his life. Never was there a moment of self-pity in Jack. Never did he muse aloud about why the God his mother worshiped daily should have plagued him with these constant illnesses. Never did he ask the heavens why he couldn't have a disease that he might defeat instead of these inexplicable conditions that the doctors never seemed able to diagnose or resolve.

In a series of letters to his friend Lem, Jack sought to turn his weeks at the Mayo Clinic into a roguish adventure. His closest friend was staying at the Kennedy house in Hyannis Port, partaking of all the summer revelry that to Jack was the sweetest part of the year. Lem was living *his* life. Only occasionally did Jack even allude to the terrible uncertainty of his plight. "The reason I'm here is that they may have to cut out my stomach!!!" he scribbled on the side of one letter. He began another letter by exclaiming: "God what a beating I'm taking. I've lost 8 pounds and still going down." He did not go on bemoaning his condition however, but wrote how he had gone to the movies and found himself sitting next to a couple. "You've never smelt anything so vile as that girl," he wrote. "She stank. I mentioned to Ed to move over. He moved over and then I moved. Well the girl looked at me and then whispered something to the fellow. He took his arm down and stood up. I, nothing daunted, stared back. The girl grabbed his arm and he sat down. It was a lucky thing for him because he was only about 6'3" and I could have heaved him right into the aisle on his ass."

Jack was lying in a hospital bed at St. Mary's Hospital in Rochester, where, as he wrote Lem, "I've got something wrong with my intestines—in other words I shit blood." He thought he might have piles. He was obsessed with his adolescent plague of pimples. The doctors performed procedures that seemed like base assaults on his dignity. Jack turned the tables on his tormentors by employing the only weapon he had: a dark sense of humor,

> Yesterday I went through the most harassing experience of my life. First they gave me 5 enemas until I was white as snow inside. Then they put me on a thing like a barber chair. Instead of sitting in the chair I kneeled on something that resembles the foot rest with my head where the seat is. They took my pants down!! Then they tipped the chair over. Then surrounded by nurses the doctor first stuck his finger up my ass. I just blushed because you know how it is. He wiggled it suggestively and I rolled 'em in the aisles by saying "You have a good motion." He withdrew his finger. And then, the shit stuck an iron tube 12 inches long and 1 inch in diameter up my ass. They had a flashlight inside it and they looked around. Then they blew a lot of air in me to pump up my bowels. I was certainly feeling great as I know you would having a lot of strangers looking up my ass-hole. Of course when the pretty nurses did it I was given a cheap thrill.

Jack's bed symbolized not only sickness but also sex. He told Lem that he had managed to masturbate only twice, and his "penis looked as if it had been run through a wringer." His pajamas were dirty and "stiff from sweat." He "feels kind of horny," especially after reading a dirty book. All day long nurses entered his room. They were "very tantalizing and I'm really the pet of the hospital . . . and let me tell you nurses are almost as dirty as you, you filthy minded shit." He boasted to Lem that one of the nurses "wanted to know if I would give her a workout," but the nurse did not return later to his room.

The boastful boyish bravado and self-conscious obscenity are so overwhelming that it is easy to forget, as surely Jack wanted to forget, that he was a seventeen-year-old youth half a continent away from his family, lying in a hospital bed with doctors and nurses poking away at his body, seeking to understand the mysterious affliction that tortured him and held him back from the life he wanted so much to live. He was sick and he was hurting and he was full of pain that he could not even fully admit to himself, and never to Lem and the rest of the world.

Jack was at the Mayo Clinic while his closest friend sat in his house, sailing and swimming, enjoying the marvelous days of summer. He wrote Lem about "dirty minded nurses" and confrontations in a movie theater as if, even in this worst of times, he was having daring experiences that Lem would never have. Jack had a wondrously inventive mind and, like most adolescents, was full of braggadocio about his supposed sexual encounters.

Jack was probably inventing these stories, and doing so with detail, imagination, and nuance. He was affirming his own vitality, even if it was in the guise of a fictional creation. There was a heroic quality in the magnitude of Jack's denial and the daring invention of this picaresque life. He could have given in to his illness and accepted an invalid's pale life. Instead, Jack had learned a denial technique when dealing with the implacable realities of life, and it proved to be irresistible. He had made a mask that he could put on whenever he needed it, or if he chose to, he could put it on and never take it off again.

As the perpetual observer of his own life, Jack lived in two worlds. It was this extraordinary duality that set him apart not simply from Joe Jr. and his other siblings but from everyone else as well. Young Jack was in a situation that would have filled most adults with fear, but there was none of that in him. Here in this hospital room his sense of irony grew hard and became the preeminent means by which he looked at the world.

Jack left the hospital that summer with no firm diagnosis of his condition, and that fall he returned to Choate. In chapel the headmaster stood before the students and in his Olympian manner talked about the small minority of mindless troublemakers who were destroying the peace of his beloved Choate. St. John said that these miscreants represented no more than 5 percent of the student body, but that they were a slow poison. If he ever could determine their names, he would root them out quickly and send them home disgraced. They were muckers, nothing but muckers, mucking up Choate.

As Jack and his friends listened, they were delighted at St. John's coinage. Muckers! That was exactly what they were. Muckers! The boys decided to turn their casual fraternity into a small secret society. To formalize matters, they trooped into Wallingford and had special golden shovel emblems made for themselves at a jewelry shop.

The thirteen members of the Muckers Club met every evening between dinner and chapel in Jack's room. Theirs was the most benign of conspiracies, their worst crimes being such misdemeanors as stealing out for milk shakes and playing their radios after hours. They planned at Spring Festivities to bring shovels into the dance on which they would pull their dates. Then they would all go outside and be photographed with the shovels next to a manure pile.

When St. John heard of this mysterious group and their nefarious scheme for Spring Festivities, he reacted with unrelieved fury. He was a man whom humor had bypassed. He simply could not abide such a challenge from the likes of Jack Kennedy. Mucker. The very name was a confrontation. Jack was Mucker Number One. He and his friends had created a little revolutionary nest within the sacred reaches of Choate, a nest that had to be eradicated. He called the boys into his office and raged at them, threatening to throw them out, ruining their academic records and their chances of attending an Ivy League college.

The matter was so serious that the headmaster wired Joe at the SEC in Washington: "Will you please make every possible effort to come to Choate Saturday or Sunday for a conference with Jack and us which we think a necessity."

Joe arrived that Sunday afternoon and went immediately to the headmaster's study, seating himself in a leather chair next to St. John's desk. The headmaster saw this as an occasion to teach this Catholic upstart an unforgettable lesson by heaving his troublesome son out the front gates, even though other faculty members, including Harold Taylor, had implored St. John to let poor Jack finish out his term.

Another father would have pleaded with the headmaster to allow his son to graduate. Joe did not do that. He was a brilliant businessman in part because he figured out his opponent's motivation and tried to play to that. In this instance, St. John was angry because Jack and his friends dared to attack his authority by creating their own separate world. Joe flattered the headmaster by berating his own son as much as St. John himself ever did.

While this was going on. Jack was back in his room prowling back and forth, dreading the reaction his father would have. He was finally shown into the office where his father and the headmaster sat. He had a look of doom on his youthful countenance. While St. John took a phone call, Joe turned to his son and half whispered: "My God, my son, you sure didn't inherit your father's directness or his reputation for using bad language. If that crazy Muckers Club had been mine, you can be sure it wouldn't have started with an M!"

That was enough to relax Jack, but as soon as St. John returned, Joe went back to berating his son, the two men punching verbally at Jack one after the other. "We reduced Jack's conceit, if it was conceit, and childishness to considerable sorrow," St. John remembered decades later. "And we said just what we thought, held nothing back, and Mr. Kennedy was supporting the school completely." As the two Kennedys left, St. John felt richly vindicated, even though he had not thrown Jack or the other youths out of school.

Joe had acted with magnificent panache. He had come up to Choate that weekend to see to it that Jack graduated. That way his son would go to a top college and have the kind of life his father intended him to have. He had whispered to Jack in St. John's office, not because he was winking at the headmaster's authority, but to put his son at ease.

Joe was enraged at his son's behavior. But nothing, no conduct, no crime, nothing, could diminish Joe's aspirations for his son's future or lessen his pride in family blood. Joe was always there for Jack. No matter what the crisis, what the time, what the cost, what the pain, Joe was there. For Jack, that was the lesson of his father's weekend visit and of the Muckers' revolt.

In the aftermath the headmaster arranged for Jack to speak with Dr. Prescott Lecky, a Columbia University psychologist. The analyst could see that Jack was "definitely in a trap, psychologically speaking. He has established a reputation in the family for thoughtlessness, sloppiness, and inefficiency, and he feels entirely at home in the role." Jack had his own deep insights into his psyche. He understood the psychological bounds that constricted him, but the truth did not set him free. "A good deal of his trouble is due to comparison with an older brother," the psychologist noted. "He remarked, 'My brother is the efficient one in the family. I am the boy that doesn't get things done. If my brother were not so efficient, it would be easier for me to be efficient. He does it so much better than I do.' Jack is apparently avoiding comparison and withdraws from the race, so to speak, in order to convince himself that he is not trying."

Rose, a woman of remorseless optimism, wrote in her memoirs that "this silly episode of the Muckers may well have been a turning point in his life." He did pull up his grades a bit those last months, but for all four years of high school at Choate he did not receive one honor grade and finished 65th in a class of 110. His SAT scores for college entrance were respectable (verbal 627, math 467), but hardly the marks of a brilliant young man. Choate's "General Estimate" of Jack for his Princeton application was both fair and devastating. "Jack has rather superior mental ability without the deep interest in his studies or the mature viewpoint that demands of him his best effort all the time. He can be relied upon to do enough to pass."

One of the first occasions in which the "new" Jack made his appearance was during the very Spring Festivities that the Muckers had planned to disrupt. Each year the evening began with a performance of Gilbert and Sullivan. It might have been expected that a much-chastened Jack would have been in the front row, applauding vigorously. But Jack and his friends could not abide sitting there. Instead, Jack and his date, Olive Cawley, and Lem and his date, Ruth "Pussy" Walker, drove off with Porter "Pete" Caesar, who had graduated the year before and was now back from Princeton, driving a spiffy roadster. Leaving Choate that evening would have been grounds for expulsion by itself, but the crime was dramatically escalated when they stopped at a roadhouse for some beer.

On their drive back to campus, the group realized it was being followed by proctors whose mandate was to catch such miscreants as Lem and Jack. Caesar sped away, roaring out into the countryside, trying desperately to shake off their tormentors. He turned into a farmyard and cut off the headlights, while Jack, Lem, and Olive ran to secrete themselves in the barn. Jack and Lem in tails and Olive in a ball gown hunkered there among hay and animals. When the proctors arrived, they found Caesar and Lem's date making out in the car. The proctors had seen more figures in the car than these supposed smoochers, and they sat in their car waiting. Caesar tore out of the farmyard and led the Choate teachers on a new chase, finally shaking them.

"Finally, Pete Caesar came back and drove into the barnyard very fast; no other car was in sight," Lem recalled. "Olive rushed out to the car and lay on the floor at Pussy's feet with a coat thrown over her. I dashed to the back—Pete opened the trunk and I jumped in. He closed me in. We didn't dare call for Jack—so we left him behind. . . . It was only a mile to the school. . . . He [Caesar] delivered Olive and myself to the dance and we danced into the crowd together despite the fact that Olive had lost one heel from her shoe and looked a pretty mess. About a half an hour later, I'm glad to report, Jack showed up."

Jack had one other parting gift for dear old Choate. As one of the last acts of the year, the seniors voted for yearbook honors in a number of categories. The highest honor of all was "Most Likely to Succeed," given inevitably to a serious, studious young man who exemplified all the Choate ideals. Jack was as likely to win in that category as to be asked to give a sermon in chapel. Jack's friends, however, waged a vigorous, whimsical campaign, and he not only won the signal honor but won it overwhelmingly. It was a final poke in the eye for Choate as he drove out the gates for the last time.

7

The Harvard Game

While other Harvard freshmen lit up cigarettes, trying to look grown up, nineteen-year-old Joe Jr. smoked a cigar, puffing on it, as his friend said, "like an old man of the world." Like his father before him, Joe Jr. arrived at Harvard a year older than most of his classmates, and he seemed even more mature. He was six feet tall and 175 pounds, a genial, expansive figure who strutted across Harvard Yard in the fall of 1934 as if he knew that he belonged in the steps that he trod. He may have been four generations from the wheat fields of County Wexford, but his blood was Irish and he had a handsome, full-blown Irish face, with wide cheeks and perfect white teeth owing to his mother's obsession with orthodontia.

Joe Jr. was used to servants, helpmates, and advisers, and his father's stipend of $125 a month was enough to enable him to live far better than most of his contemporaries. He was not one of those who thought the tutoring services of Harvard Square an unfair, cynical business that destroyed much of what a college education was supposed to be. He availed himself so much of the tutors that at least one of his contemporaries believed he used them for every one of his classes. He wasn't about to clean up after himself either. He had his own valet, George Taylor, a black man who passed out business cards proudly announcing that he was a "gentleman's gentleman."

Joe Jr. was the kind of Harvard man his father had aspired to be, but without his father's overweening social calculation and youthful self-consciousness. Joe Jr. knew that the men of the Gold Coast looked down upon his kind, but it didn't bother him as it had his father. If many of his friends were Catholics, it was simply that he enjoyed their company. He had met his closest friend, Timothy "Ted" Reardon, a fellow Catholic, one day after freshman football practice when Joe Jr. asked him to stay after and throw the ball around.

Joe wanted sons of matchless physical courage, privileged sons whose money and education had not tempered their toughness. Joe Jr. was fearless, taking pure joy in the crunch of a hard tackle or a savage block, in either football or rugby. His picture was featured in *Life* in a story on a rugby game against Cambridge University; he stood on the sidelines, his head bandaged, a wounded gladiator ready to go back into the fray. He broke his arm playing football in the spring of 1935, but that accident hardly slowed him.

Joe Jr. did not leave his intrepid spirit behind when he left the football field. One winter night he was walking near Harvard Square when he heard a woman's anguished screams. When he turned and saw that a man was beating her, he could have called for the police or admonished the attacker to stop. Instead, Joe Jr. charged the man and started flailing away. The police arrived to find two men fighting in the street. Joe Jr. ended up spending the night in jail, a rare setting for a Harvard student.

On another occasion, Joe Jr. told his friends how he had jumped into the Charles River to save a drowning man. To some of his friends, it seemed a boastful Irish yarn, and they started calling him "the Life-saver," their words touched with more derision than admiration. But if Joe Jr. was one to garnish his stories with hyperbole, he was not known as a liar, and if he said he saved the man, he probably had.

Joe Jr. was quicker of temper than of mind, always ready to rush into a scrap or to charge into the toughest part of the line. His friends soon learned that he could turn on them in righteous anger as quickly as he could on some lowlife tough striking his girlfriend on a dark Cambridge street. Unlike his father, Joe Jr. brimmed with pride at his Irish-American heritage and took umbrage at those who dared to demean it. He clenched his fist at those who spoke with disdainful hauteur at the corruption of Irish politicians.

Reardon learned that he could not proffer even a gentle jibe. When a group sat discussing politics in Joe Jr.'s room in Stoughton Hall, everyone was in agreement that corruption was a problem in city politics. "Sure, these city officials are all the same," Reardon said, taking it from generalities to specifics. "Now take even John F. . . ."

"Get out this room!" Joe Jr. shouted even before Reardon had the name "Fitzgerald" out of his mouth. As Reardon hurried out the door, Joe Jr. attempted to give him a last good-bye with a kick in the posterior.

Reardon was the son of a barrel-maker who had lost it all during the Depression. He made the daily jaunt to Harvard from his modest home in Somerville. The young man had everything Joe Jr. had except for money, panache, and boundless self-confidence. In the end he settled into a position of subservience. He played the archetypal role of a Kennedy "friend," one not that much different from the relationship Joe's father had during his Harvard days with Robert Fisher, or that between Jack and Lem Billings. "I always thought that Joe's father was putting Ted Reardon through college, paying for him," recalled Robert Purdy, one of their housemates. "Ted just waited on Joe like he was a valet. He took care of his car, took him home every night."

Joe Jr. was good at everything, from catching a pass thirty yards downfield to waltzing a debutante across the dance floor, from debating the latest New Deal policy with assurance and knowledge to carousing over beers with rowdy friends down at the pub. He did not care about grubbing grades, but the implication was clear that Joe Jr. could have bested the greasy grinds if he had wanted to and had chosen to take time for such silliness.

Joe Jr. might have spent his weekends at deb parties and postdeb gatherings, squiring around the most eligible and beautiful young Catholic women. But the games that he enjoyed playing were not on those dance cards, however, and good Catholic that he was, Joe Jr. went elsewhere for his pleasures. When it came to women, he had the natural predatory instincts of his father but the good sense to play his games of momentary romance primarily among women who had played before: actresses, showgirls, and café society women of certain means and uncertain morals.

Often Joe Jr. left Cambridge on Friday to spend his weekends in New York. When he returned, he had stories to tell of a sophisticated world that would have spurned most of his college mates. To him, women were short-term lovers, not long-term friends. They were perfumed beings put on earth for the pleasure of men like him.

Joe Jr. treated romance as a game that had few limits. Not until Robert Purdy married his college sweetheart did he learn that his friend had made an unsuccessful pass at his fiancée. "My impression of Joe was that he would never date the same girl twice," recalled Purdy. "He was rough on girls, I gather. He was not apparently the kind of guy that nice girls liked to date."

Joe Jr. had his claque of admirers. During freshman year he chaired the committee for the annual "smoker" and became a minor college legend by obtaining the services of the famous Rudy Vallee and his orchestra. He served on the student council and the Winthrop House Committee, including the chairmanship his senior year. He was a member of Hasty Pudding and Pi Eta but did not make one of the exclusive final clubs. This did not appear to rankle him in the least. He was a proud Catholic American, a member of the St. Paul Catholic Club. He espoused a passionate but manly Catholicism in which he prayed on his knees but winked at the sins of the flesh and of human corruption.

In the years since his father's graduation, Harvard had attempted to democratize itself in part by forcing freshmen to live in the Harvard Yard dormitories and then creating seven houses for upper-class students. One of them, Winthrop House, was a sprawling red-brick structure by the Charles River. The resident tutors included such young instructors as B. F. Skinner, who would become the father of behavioral science, and John Kenneth Galbraith, the wry, arrogantly self-assured liberal economist.

Galbraith had the disconcertingly dangerous idea that liberalism was not only an abstract philosophical position but a guide to daily life. He looked askance at the way Winthrop House chose its residents, attempting to replicate the narrowly snobbish club world that the new house system was supposedly trying to end. The housemaster, Ronald M. Ferry, presented Galbraith with a ruled sheet featuring the cryptic letters: St. G. Ex, E & A, O.P, H.S., and X. Thus was displayed the social order, exactly as it had been in Joe Kennedy's days.

At the top stood St. Paul's, St. Mark's, Groton and Middlesex, Exeter, and Andover, other private schools, public high schools, and then the all too common Xs, the Jews. Galbraith wanted to choose an outstanding student, Theodore White, but he was told that the quota of Xs was already filled, and so the man who would become one of the seminal journalists of his age went elsewhere. If he had been accepted, White would have found himself living on a floor with all the other Xs. Joe Jr., as a Choate graduate, was not up to the level of St. G. Ex., but he was far above H.S. and X, and he was readily chosen.

Many rich men's sons would have spent their days at Harvard primarily with the sons and daughters of wealth. Joe Jr. enjoyed few things as much as heading off to Suffolk Downs with Joe Timilty, the rotund Boston police commissioner, sitting there among touts and pols. One day he took the incomparable Ethel Merman to the track with him, shepherding her as if she were some postdeb from Wellesley, not one of the stars of the musical stage. On another occasion he wangled a date with the beautiful young star Katharine Hepburn, who arrived accompanied by her mother.

Back in Cambridge, Joe Jr. was a gregarious young man who traveled with an entourage of friends and burst into Winthrop House full of energy and exuberant greetings as if to announce that life itself had arrived. Despite his temper, Joe Jr. was a man of myriad courtesies. Josephine Fulton, one of the managers of the house, answered the phone for Joe Jr. and often took calls from his grandfather, Honey Fitz. Then she would get in her car and go fetch him, most likely from a coffee shop where he hung out with his friends.

He thanked her with boxes of chocolates, a small courtesy with which most young men could scarcely be bothered. When he went to visit his beloved Grandpa Fitzgerald at the Bellevue Hotel where he resided, Joe Jr. always stopped to say hello to the elderly housemaid.

Joe expected that Jack would follow his big brother to Harvard, but Jack said that he intended to join his friends Lem Billings and Rip Horton at Princeton. "You didn't go to school because your friends are going there," Joe Kennedy recalled telling his son. "That's not the real reason, is it, Jack?" His son had had two blessed years away from his brother and all the tedious, inevitable comparisons. Doubtless Jack was looking forward to building his own identity away from Joe Jr.'s transcendent presence.

If Joe sometimes treated his sons like marionettes, he held the strings so loosely in his hands that they could not always feel the pull on their backs. He did not push Jack to join his older brother at his alma mater. He insisted, however, that before he went off to college he follow further along Joe Jr.'s pathway and study under Harold Laski at the London School of Economics.

Another young man would have been ecstatic to be sailing off to Europe in September 1935 on the *Normandie.* In Jack's letters to Lem, however, it is as if he had been marooned, cast off into someone else's life. Jack's letters read like the ribald mutterings of a precocious fourteen-year-old, not the Ivy League–bound eighteen-year-old son of Joseph P. Kennedy. He might have deferred to his father at the captain's table, but with Lem Jack described his father as a humorless pain in the posterior who nagged his second son about feeding his pimples with rich desserts, monitoring his conduct like a German nanny. Jack was so appallingly thin, only 135 pounds on his six-foot frame, that his face was like a series of stark images. Look at it one moment—the great shock of hair, the deep intense eyes, the aquiline nose, the pearly teeth—and he appeared a handsome young man with matinee idol looks. Look at his face again—the thin cheeks, the eyes deep set, the face so long—and one thought only of ill health. Look yet again, and there was a strange feline quality to the face, nearly androgynous, a youth too sweet for the rigors of Kennedy manhood.

Outside of his father's purview, Jack had a little adventure on the crossing that would have appalled Joe. "I have had a very strange experience," he wrote Lem. "There is a fat Frenchie aboard who is a 'homo.' He has had me to his cabin more than once and is trying to bed me." Strangest of all was that Jack was intrigued enough by the man's entreaties that he ventured back to his cabin. There was a preening, narcissistic quality in young Jack, and he clearly found it flattering that a man would be so attracted to him.

Jack had hardly arrived at Claridge's in London when he was stricken again with some mysterious malady and sent off to a hospital for tests. He wanted Lem to know that Jack Kennedy was not some self-pitying wretch, but an adventurer of such daring that he could turn even a hospital bed into a playground for his manhood. Only Jack would think of naming his penis "J J Maher," after his much-hated nemesis at Choate. "Today was most embarrassing as one doctor came in just after I had woken up and was reclining with a semi on due to the cold weather. His plan was to stick his finger under my pickle and have me cough. His plan was quickly changed however when he drew back the covers and there was 'J J Maher' quivering with life. As the nurses were 3 deep around the bed, I was rather nonplussed for a time."

Jack's visitors included "a very good looking blond whom Dad seems to know, about 24 who is a divorcee. . . . She is going to St. Moritz with me at Christmas but I have not as yet laid her." Once again Jack's room stirred with sex, not sickness, the borderline between fantasy and reality perhaps not even visible to Jack. There were the nurses, always the nurses, "very sexy and the night nurse is continually trying to goose me so I have always to be on my guard."

Jack knew that not only Lem but also Rip and their friends would read his letters. There was nothing of the antiseptic smell of a hospital on these pages, nothing of fear, nothing of prayers or bewilderment, nothing of the sheer unfairness of his plight. His friends would laugh about Jack and "J J Maher" and wish him back among them, when life would be just a little more brilliant.

His friends got their wish, for Jack's father decided that his son would be better back in the States studying at Princeton. Lem and Rip shared their modest fifth-floor quarters with their former Choate classmate. As Jack lumbered up those stairs, he had a strange patina to him. Bud Wynne, a friend, remembered him "turning yellow, a yellowish-brown tan, almost as if he had been sunbathing." He was hardly at Princeton for six weeks before he left, victim of this undiagnosed, mysterious illness. To the doctors who examined him, he had become, in that most ominous of phrases, an "interesting case."

After the new year of 1936, Jack was back in the familiar surroundings of a hospital room at Peter Bent Brigham Hospital in Boston. From his bed, Jack wrote jumbled, scribbled missives to his friend Lem. Once again it was suspected that he had leukemia. By now his friend was used to the blend of scatology, sexual boasts, medical descriptions, and putdowns in Jack's letters, but there was a fever pitch of intensity to them now, the sexual imagery even uglier, the assaults on poor Lem more merciless, the sexual boasts larger.

Jack's penis was his nasty, irreverent friend, always ready to perk up at the most morbid of moments. He had scarcely arrived at the hospital when he suffered what he called "the most harrowing experience of my storm-tossed career." The doctors put a rubber tube up his nose and into his stomach and poured alcohol into him. "I have this thing up my nose for 2 hours and they just took it out (don't be dirty Kirk) and now I have a 'head-on' and a 'hard on' as . . . a beautiful nurse came in and rubbed my *whole* body."

Jack might have been lying in bed, poked at by a team of doctors, but even here he saw himself as a vibrant sexual being, unlike Lem and his other wimpy friends. "I don't know why you and Rip are so unpopular with girls," he wrote. "You're certainly not ugly looking exactly. I guess they're [*sic*] is just something about you that makes girls dislike you on sight."

He was ready to instruct his pitiful friends in the art of sexual conquest. "I'm writing Rip against taking Nancy Williams," he told Lem about a planned weekend. "She is an ugly bitch and we're not going to load the place with your women." As for Lem, at least he could pay to have sex. "It seems to me, you prick, that if you can afford a week fuck fest with you paying Joe and I and Caesar $5.00 each for every fuck and paying the doctor bills for sif [syphilis] which you are certain to pick up in this nigger place."

Jack slashed at Lem with one blow after the next, attacking his friend's sexuality and his relative poverty, his two greatest vulnerabilities. Lem either ignored Jack's thrusts or attempted his own minor feints, but he was largely defenseless.

When Jack could get out of the hospital in the evenings or on weekends, he spent much of his time looking for women. He did not share his big brother's belief that a man—a gentleman, that is—did not attempt to seduce women of a certain quality, manner, or faith. Jack was nothing if he was not fair. He considered all women fair game. He discussed women with other young men like a woodsman sharing tips on how to set his steel traps.

One of his friends, Pete Ramney, had gone out with one of Jack's dates. He boldly asked Ramney how far he had gotten with her. When Pete told him he had reached "second base," Jack was incensed. Ramney had gotten further than he had. "The next time I take her out she is going to be presented with a great hunk of raw beef," he wrote Lem, "if you know what I mean, although I doubt it."

Here, as at Choate, Jack's proud sexual boasts may have been written with a novelist's imagination and read with guileless credulity by a Lem Billings who was far more sexually insecure than his friend. Jack's close college friend and roommate James Rousmanière, is one of the skeptics. "I think he was making it up," Rousmanière asserts. "That was the masculine ethic. And I think he made up three-quarters of it. And I don't hold it against him."

As a young man, Jack wanted an audience for his sexuality. He was at his most aggressive when there were other men around whom he wanted to impress even more than he did the woman of the moment. She was interchangeable; they were not.

These letters then should be read in part as Jack's vision of what he could be or would be. He created a Don Juan image of himself long before he became one in reality. In sexual conquests, there are few things as helpful as a bad reputation, and in the end Jack became the sexual being he thought he wanted to be.

Jack was "getting rather fed up with the meat here, if you know what I mean," he wrote Lem, though he knew his friend didn't know. "They haven't found anything as yet except that I have *leukemia* and agranulocytosis. Took a peak [*sic*] at my chart yesterday and could see that they were mentally measuring me for a coffin."

But two sentences were quite enough on the dark subject of a disease that probably meant his death. "Eat drink and make [out], as tomorrow or next week we attend my funeral," he continued. "I think the Rockefeller Institute may take my case. Flash! Got the hottest neck ever out of Hansen Saturday night. She is pretty good so am looking forward to bigger and better ones."

The doctors tinkered away a few more weeks after they decided that Jack did not have leukemia. Many patients would have been infuriated that the doctors could make such a mistake, but for Jack it was just another pratfall visited upon him by men in white. Finally, after almost two months, he left the hospital at the end of February 1936 knowing as little about his malady as when he arrived. At least he was free to sun himself in Palm Beach and then to spend a few weeks at the Jay Six Cattle Ranch in Benson, Arizona. While there he applied for admission as a transfer student to Harvard College. He gave his father's birthplace as Winthrop, a more socially acceptable address than the déclassé East Boston where he was in fact born.

Jack preened in the sun and the desert air. His narcissism grew out of not only his immense vanity but also a natural obsession with his star-crossed health. "If you could see what a thing of beauty my body has become with the open air, riding horses and Mexicans, you would stuff such adjectives as unattractive when you are speaking of my body right where they belong," he wrote Lem from Arizona. He seemed to be flirting with his friend, taunting him with his alluring physicality. And yet there was a hint of self-mockery in his narcissism, for as much as Jack bragged about his beautiful body, something always went wrong. "I woke with a hacking cough," he noted in another letter. "It will be the fucking last straw if I come down with TB in addition to a good load of clap in this my health cure." That latter disease was a decided possibility, for he claimed he had ventured south of the border in search of his favorite pleasure. "I ended up in this 2-bit hoar-house [*sic*] and they say that one guy in 5 years has gotten away *without* just the biggest juiciest load of claps . . . so boys your roomie is carrying on . . . upholding the motto of 'always get your piece of arse in the most unhealthy place that can be found.' "

In the fall of 1936, Jack joined Joe Jr. at Harvard, where his father had thought he belonged in the first place. That he was willing to transfer suggested that Jack was moving closer to accepting the family destiny that his father had laid out for him. Joe Jr. was a young man of rare popularity, a football player and an exemplary Harvard man, and at first Jack had to nestle into what was his brother's world.

"Dr. Wild, I want you to know I'm not bright like my brother Joe," he told Payson S. Wild, an assistant professor of government and the acting master of Winthrop House. Jack placed his feet in his brother's large footsteps, even sleeping often on a couch in Joe Jr.'s suite at Winthrop House. If Jack had been too small, weak, and vulnerable for football at Choate, then he was too small, weak, and vulnerable for football at Harvard. He was a Kennedy, though, and he had to play football. He had gained weight, but even at 160 pounds on his tall, still-weakened frame, he remained a gnat that runners swatted away as they dashed down the field. He was quickly demoted to the freshman "B" team, which was beaten by the likes of a prep school such as Exeter.

Jack found a different and more familiar way to stand out among his football teammates. He invited five of them down to Hyannis Port, enticing them with the prospect of dates that his father's aide, Eddie Moore, had set up for them. "Four of us had dates and one guy got fucked 3 times, another guy 3 times (the girl a virgin!) + myself twice," Jack wrote Lem. The former virgin had the considerable audacity to write her date "a very sickening letter letting [him know] how much she loved him etc + as he didn't use a safer he is very worried. One guy is up at the doctor's seeing if he has a dose + and I feel none too secure myself. We are going down next week for a return performance, I think."

Few Harvard freshmen could provide such splendid weekends for their classmates, and they doubtless spread tales of their wonderful new friend, a swordsman who understood the true meaning of hospitality. When the coaches heard about the famous weekend in Hyannis Port, Jack found himself not only berated but also demoted even further to the third team. He might not ever win his crimson "H," but there were other kinds of honors. "I am now known as 'Play-boy,' " he wrote Lem. It was a label that would stick with him the rest of his life.

"I swear I don't think he ever made love to a girl, told her how wonderful she was, how sweet she was," reflected Rip Horton. "I just don't think he ever did that. I don't know but I don't think so. I don't think he was sentimental. I don't think he was ever dependent on the companionship of a girl. He always felt they were a useful thing to have when you wanted them, but when you didn't want them, put them back."

Jack further enhanced his reputation as "Play-boy" Kennedy by chairing the Freshman Smoker Committee. Jack not only snared Gertrude Niesen, a stunning singing star, but also had the considerable audacity to think that the singer might put on two performances that evening in Cambridge, one before his classmates, and a second for him alone. He wanted to fly down to New York to escort her back to Boston but lost the coin toss to his classmate Hunt Hamill. Jack was there, however, at the airport to greet Niesen and, with his friends, spent several hours with her before the performance.

Joe Jr. had made a triumph of his own freshman smoker, and he considered Niesen his droit du seigneur. "Get lost, Baby Brother," he told Jack abruptly. "I'll take over." That was a phrase that Jack had heard more than once, but after all, it was Miss Niesen's choice. "I was very young," she recalled decades later. "And it was very exciting, very flattering, very wonderful. Joe was a *terribly* good-looking guy. He was much better looking than Jack at that particular time. If I'd been a little older and really understood what was going on—"

Jack had a wondrously wry sense of humor. With women he could be playfully piquant and flirtatious but rarely crossed over the border into rudeness or vulgarity. His wit cascaded out of him. Niesen matched him rejoinder for rejoinder. Her performance that evening was a triumph for the singer herself, but equally so for Jack. "Gertrude Niesen was just enjoying the hell out of it, and Jack Kennedy was joking with her the whole time," Rousmanière recalled. "Suffice it to say that Jack was not there at the end. Hunt Hamill was there at the end with Gertrude Niesen. But it was the culminating social event of the year, and Kennedy was the chairman of the committee. That was his first political success."

In his sophomore year, Jack moved into a Winthrop House in which his brother was one of the leading figures. Like his father before him, Jack's overwhelming concern that year was not his class work but his admission to a proper club. The clubs had stayed as they were since his father's day, places where a gentleman ate and drank and socialized. Those with social aspirations still roundly desired membership. The clubmen were invited to all the fancy debutante events in Boston—dances at the Somerset, the Ritz, the Women's Republican Club, and the Brookline Country Club.

Despite the unchanging rituals of the final clubs, life outside of them was moving away from the purview of the good gentlemen of the clubs. Jack's profound identification with the club world, and his calculated and diligent quest for membership, set him far apart from an emerging new Harvard that sought to break down not only the walls within Harvard but also the walls that enclosed it.

Jack's approach was not that much different from his father's. He surrounded himself with a mix of the socially prominent leavened with star athletes. The blue-blooded Rousmanière was one of his close friends in Winthrop House. Jack had also befriended Torbert "Torby" Macdonald, a football star, who became his roommate their sophomore year. When the ten clubs started sending out their invitations to their socials, where the clubmen had an opportunity to assess possible new members, Jack and Torby were rarely asked. Half a century later Rousmanière could say simply: "I guess he [Macdonald] wanted to be part of it, but Torby would have been a hard sell," leaving the implications unsaid.

Not only was Torby a Catholic, but his father was a high school football coach. As the clubmen saw it, Jack was a hard sell too, with his ethnic background, his dubious faith, his questionable father, and the fact that not all playboys were gentlemen. In his favor were his wit and charm. He was a devotee of New York clubs and the high life, and undeniably wealthy. In sum, he was a plausible candidate, but just barely. Rousmanière and two of his eminently acceptable friends, Peter E. Pratt and William C. Coleman Jr., agreed that they would go together with Jack as a package. "It was obvious that only a couple of clubs were going to accept Jack Kennedy," Rousmanière recalled. "And we, the three of us, said okay, we'll play it out. In the end the Spee Club became the one place that seemed to be acceptable. So that's what happened."

Jack was a clubman of the first order, spending most of his free time at the handsome ivy-covered building at 76 Mount Auburn Street and taking most of his meals there. There was no one-way mirror at Spee, as there was at the Porcellian, from which the clubmen could stare unwatched at the outside world, but Jack and his friends looked on from afar at much of the Harvard world. The intellectual climate of Harvard in the thirties had been immensely broadened since Joe's days, both by the greater diversity of the students and faculty and by the dangerous world that lay outside the open gates of the college. Of course, there were still such undergraduate endeavors goldfish gulping, reported on the front page of the *Harvard Crimson*, or a kissing contest, but these activities were nothing but the spring silliness of any college generation. The students were at times confronted with a diversity of ideas as wide as that in the outer world. The faculty included Granville Hicks, a 1923 graduate and a Communist, as well as Earnest A. Hooton, an anthropology professor and eugenicist who believed that robbers could be discerned by such distinctive marks as "attached ear lobes, heavy beards, and diffused pigment in the iris." New Dealers shuttled down to Washington, trading in their professional gowns for the mantles of power. In Cambridge, students and faculty alike debated Roosevelt's reforms. A few Communist undergraduates met regularly, while on the right-wing extreme a group calling itself Yankee-American Action held discussions at the university.

At Winthrop House, Galbraith was intensely interested in the debates of his time and surrounded himself with students of like mind. The young professor made a quick judgment that Jack was not serious, but an amusing dilettante, whose main elective was the opposite sex. Though it might have been pleasurable being around Jack, pleasure was a dangerous business for an ambitious young scholar, hardly a commodity of value in his academic world. "One did not cultivate such students," Galbraith reflected.

Even some of Jack's fellow students kept a wary distance from him, lest they too suffer some dreary consequence. One of them, Blair Clark, watched in appalled fascination as Jack hustled waitresses from Dorchester in and out of Winthrop House. Clark was convinced that sooner or later Jack would find himself expelled.

"Jack devoted himself to personal enjoyment, social matters, but Joe Jr. sought out members of the faculty, sought me out, particularly perhaps, and was enormously interested in world affairs, much more so than Jack," Galbraith recalled. "Jack had a social agenda to pursue. Joe Jr. had a much stronger scholarly bent."

Joe Jr. may have impressed his professors with his high seriousness, but he was hardly the kind of student to bury himself in the bowels of Widener Library. Even Jack was impressed with his brother, the swordsman. "Did you see about Joe in Winchell's column—Quote—'Boston Romance—J. P. Kennedy Jr., The Wall Street's lad, and Helen Buck of the Boston Back-Bay bunch, are keeping warm,'" Jack wrote Lem. "Fucking gold-fish is the way I would describe it."

For Joe Jr., a football weekend was exactly that, and after the Harvard-Princeton game he and his friend Tom Bilodeau jumped the team train in New York and headed out for an evening in the most elegant boîte of the city. Sherman Billingsley, the owner of the Stork Club, practically invented café society, that spirited mix of the wealthy amiables of both sexes, stunning young women, stars, politicians, and the transitory celebrities of the gossip columns. To gauge one's place in this new society, all a person had to do was show up at the Stork Club without a reservation.

Joe Jr. did not have to wait behind the velvet rope but was immediately shown to a good table. As he scanned the room, his eyes tended to focus on the most beautiful women in the club. There at a ringside table sat a gorgeous young woman, and next to her none other than John F. Kennedy. Joe Jr. thought of the Stork Club as his club, and he was hardly amused that his brother should have usurped what he considered rightfully his. Joe Jr. went immediately to a pay phone and paged Jack. As Jack threaded his way back through the tables to take the call, his brother hurried to ringside and, with a dialogue as original as it was duplicitous, talked the young woman into leaving with the two Harvard football players before Jack returned. Later that evening, when the two young men arrived at the Bronxville house after escorting the woman to her apartment, Jack was already there, angry enough to want to fight his big brother.

This was hardly the only time the two brothers came close to drawing blood. Joe Jr. was usually the instigator. He could couch even his best advice with such cavalier dismissal that Jack was bound to do just the opposite. On the football field in the fall of 1936, Joe Jr. jogged over from the varsity to where Jack was practicing with the other freshmen. As Torby listened in, Joe Jr. said: "Jack, if you want my opinion, you'd be better off forgetting about football. You just don't weigh enough and you're going to get hurt."

Torby could see his friend's face flash with anger at Joe Jr.'s arrogant dismissal of his athletic prospects. "Come off it, Joe," Torby exclaimed, interjecting himself between the brothers. "Jack doesn't need any looking after." Instead of echoing his friend's sentiments, Jack turned on Torby: "Mind your own business! Keep out of it! I'm talking to Joe, not you."

Joe Jr. was right. Jack should not have been out on the football field. In a scrimmage the next year Jack would hurt a spinal disc so severely that not only would his third-rate football career be over, but he would suffer from back problems the rest of his life. That was Jack's legacy from the Kennedys' obsession with football—an often-debilitating pain that plagued him and at times nearly crippled him. Here on the field of play where he was supposed to have won the laurels of true manhood, he would hurt himself in such a deep and largely unperceivable way that much of his life became a struggle to pretend that he walked as other men did, and to prove that his spirit would never be crippled.

Torby was not the only friend to learn that as disputatious as the Kennedy brothers might seem, their father had taught them to stand arms linked against the world. Despite this commonality, both young Kennedys stood at a distance from everyone else around them, and apart from each other as well. "I suppose I knew Joe [Jr.] as well as anyone, and yet I sometimes wonder whether I ever really knew him," Jack reflected a decade later. "He had always a slight detachment from things around him—a wall of reserve which few people ever succeeded in penetrating." Jack could as easily have been speaking of himself as his big brother.

Joe Jr. and Jack were not like so many young men who defined their own manhood by thrusting off from their father's vision of the world. Joe Jr. largely affirmed his father's world as his own, while Jack steered warily around Joe's mammoth presence in search of his own identity. Both sons, however, not only tolerated their father's sexual mores but largely adopted them and were endlessly amused by them.

When Joe Jr. and Jack went to Hyannis Port, they were not surprised to find their father seated in the front row of the private movie theater next to his latest blonde "secretary" while their mother stood in the background, a shadowy, silent presence. They found it, if anything, droll that their father cast his lecherous eye on any young woman, even their own dates. Nor did they find it irregular that their parents had separate apartments in New York.

As the Kennedys saw it, they were not a jaded family in which appearance was the only reality and hypocrisy passed as a code of honor. Joe Jr., Jack, and all their brothers and sisters considered that they had a great, loyal, caring family. In their own minds, they could integrate all these contradictions where some outsiders saw only falsehood. They could see too that what passed as family among many of their peers was pallid kinship. That was why their friends loved being around them, for the Kennedys lived with passionate intensity and with reverential attention to all the formalities of family. Joe might parade his mistresses in front of his wife, but at the dinner table his sons stood up when their mother entered and left the room. Joe Jr. and Jack respected their mother, but their love for their father was deeper, more emotional, and full of complexity.

When Joe arrived at Winthrop House in the fall of 1937 to give a guest lecture, Joe Jr. and Jack entered the room just as their father was about to be introduced. They stood in the back waiting to greet their father after his talk was over. Joe, however, called his sons forward, and in front of their friends the two young Kennedys kissed their father.

Joe treated the students to a dramatic, intimate narrative of the life of power in Washington. By then he was justly celebrated as the man who in only fourteen months as SEC chairman had shored up the foundations of capitalism, paring away the mechanisms that had allowed men like himself to make fortunes that risked destroying the very system that had so richly benefited them. Joe had gone on in 1937 to a ten-month stint as chairman of the U.S. Maritime Commission, where he had for the most part further distinguished himself.

As candid as he was, Joe did not let the Winthrop men in on what he considered the inner secrets of power. Joe was a man who believed that life did not just happen, but that the inspired few created a world that the mediocre masses accepted as truth. Part of that creation was the image that Joe created for his family. He was hungry for praise for himself and for his sons. The Kennedy men lived in the glow of publicity in part because Joe importuned those who held the spotlight to shine it brightly on them and their achievements.

Joe received an honorary doctor of laws degree from Oglethorpe University in Atlanta, Georgia, a prize that the unwary would have thought came to him unbidden. The reality was that in 1937, his friend Bernard Baruch, the financier, donated $62,500 to Oglethorpe, an institution that had a reputation largely as a diploma mill. A few weeks later Baruch received a letter from a school official saying that he would recommend Joe for an honorary degree. That same day Baruch wrote Joe asking him which honorary doctorate degree he preferred. "I never heard of any but Laws but you're the chairman," Joe wrote back.

Two years later Arthur Krock, the *New York Times* Washington bureau chief and columnist, attempted unsuccessfully to obtain another honorary degree for Joe from Krock's alma mater, Princeton. Krock was one of the most powerful journalists in America, and yet he moonlighted as a hapless flack for the Kennedys, compromising his integrity. He wrote Joe's 1936 campaign tract, *Why I'm for Roosevelt*, wrote glowing columns about the man and his sons, and was available for editorial services and shameless shilling for any family member.

Joe Jr.'s Harvard friend Charles Houghton was startled to see that any event, even a sailboat race, that involved Jack or Joe Jr. received enormous publicity, and he thought he had puzzled out one reason why. "The old man hired Arthur Krock for twenty-five thousand dollars to keep the Kennedy name in the papers," Houghton said. "The old man had a blue print. . . . Arthur Krock was supposed to publicize any member of the Kennedy family to make them look good. It's the first time I ever heard about PR. This was back in 1937–38 when twenty-five thousand dollars was a lot of money." Krock denied that Joe ever paid for his services. Whether paid or not, he was one of the first of many journalists to put his craft to work advancing the Kennedys' careers more than he advanced the truth about them.

Thanks to men like Krock, Joe had managed to become one of the most celebrated members of Roosevelt's administration. He was also one of its most disloyal. Joe did most of his bellyaching not to Roosevelt but to conservative journalists such as Krock, who heard his complaints not as treacherous whining but as a principled assault on New Deal doctrines.

Roosevelt realized that in Joe he had a subordinate who approved of neither his fiscal nor social policies, a man who in less troubled times perhaps would have been sacked. The president, however, was holding together a disparate and troubled coalition. Joe was no more disloyal than many Democratic Catholics who listened on the radio to the vicious anti–New Deal jeremiads of Father Charles Coughlin as if they were Sunday sermons. Joe was friendly enough with the demagogic priest that he wrote Coughlin in August 1936 thanking him for calling Joe "a shining star among the dim 'knights' of the present administration's activities."

During the interlude between his two positions in Washington, Joe had hired himself out to various troubled corporations for princely sums. He was brilliant at ferreting out excess, fraud, and corporate squalor and singularly unsympathetic to the mediocre inheritors who often sat at the heads of America's major corporations.

In a stint at troubled Paramount Studios, Joe prepared a harshly candid portrait of a company in which directors who were the architects of "dissension and division in management policies" still sat on the board. He condemned the "cumulative effects of a chain of incompetent, unbusinesslike and wasteful practices," and he backed up his critique with figures as firm as his words.

Joe received $50,000 in return, plus another $24,000 for his assistants, but he apparently thought that Paramount itself would have been fairer compensation. "Roosevelt is worried about anti-Semitism in America and one of the causes is all the movie companies being owned by Jews," he told the board of directors, including Edwin Weisl, a senior partner in the Wall Street firm of Simpson, Thatcher & Bartlett. Weisl recalled the incident to his son, Edwin Weisl Jr., who, like his father, was a leading Democratic operative.

"What are we supposed to do?" a board member replied, painfully aware of the wages of fear rising across the world.

"The company ought to be owned by a non-Jew, and Roosevelt says you should sell it to me."

Weisl had excellent political connections, and before he advised the board to sell, he wanted to confirm everything with Harry Hopkins, one of Roosevelt's closest aides. "Hey, Chief," Hopkins asked Roosevelt as Weisl stood near by. "What's this about Joe Kennedy and Paramount?"

"What's Paramount?" the president queried.

"Have you heard enough?" Hopkins replied, and the two men left the president alone. Weisl went back to the board to tell them that they should not sell Paramount to Joe but pay him off with money and thanks and be rid of him.

Joe returned to Cambridge in November 1937 for the Harvard-Yale game, in which Joe Jr. would have his last chance to win a letter. His son had had a sad, unfulfilling career on the Harvard team, condemned to perpetual life on the bench.

He had practiced as hard as any of the varsity players and was never known to shirk an exercise or cut practice. Joe Jr.'s tenure on the team, however, had been plagued by injuries. In his freshman year he had broken his arm. The next year he had suffered such a serious knee injury that in his junior year his leg was operated upon. His father, despite his quasi-religious belief in football, feared that his eldest son was going to end up with a gimpy leg or worse, an honor that he would wear longer than a Harvard "H." "You should think very seriously whether it is worthwhile or not," Joe wrote his son.

Joe Jr. took his admonitions from the book of life he had learned from his father in the summers at Hyannis Port. A man did not give up or give in. He had lost weight, and he was far down on the list of ends on the team. Yet he went out for his final year, another dispirited season largely spent watching his teammates from the sidelines.

For his sheer devotion to the team, Joe Jr. deserved to be shuttled in for at least one play to win his Harvard "H." Joe Jr. had no doubt but that he would indeed see action, for it was the honorable Crimson tradition to let all the seniors in for at least one play during their final Yale game.

His father could have told his son that as far as he was concerned, that tradition had ended a quarter-century ago on the baseball diamond. Since then his father had come to believe that fate was another word for pathetic fatalism. On the Friday evening before the game, as the backfield coach, Al McCoy, sat in his office perusing the list of seniors who were likely to play, Coach Dick Harlow received a call from a person identifying himself as a friend of Joe Kennedy. "He wanted to know if Joe [Jr.] was going to get his letter tomorrow because he wanted to tell the father," Harlow fumed as he hung up the phone. "Well, nobody's going to high pressure me!" Later that evening a man who identified himself as Joe Kennedy attempted unsuccessfully to reach the coach to put in a further plug for his son.

Saturday afternoon several games were played before the fifty-seven thousand fans as the damp wind blew fiercely across the Charles River. For some, the least interesting of the games was the one played on the frigid field. There the Yale and Crimson elevens struggled up and down the turf until, in the last quarter, with six minutes remaining, Harvard broke the tie with a touchdown.

For the seniors, the game that truly mattered centered on the coach's decision to let them play in their last competition and win their letter. With that final Harvard touchdown, they were sure their moment had arrived. Up until then only seven substitutes had seen play, and the sideline was full of restive, nervous athletes ready for their moment of glory. The new coach had been brought in from Western Maryland, where he had a stellar record. Harlow's mandate was to win. It was one of the ironies of this day that of all men it should have been that champion of winning at all costs, Joseph P. Kennedy, who glowered down on the coach, wanting his own son to play, winning be damned.

Harlow looked up and called for a new end to enter the game. If power and history meant what Joe thought they did, the coach would have bypassed Green, Jameson, and Winter and called out "Kennedy." Instead, Harlow barked out, "Jameson." The coach had made his last substitution, and when the final gun went off, there was as much sadness as joy on the long Harvard bench.

While Joe Jr. was lost in the surging crowds, his father made his way to the center of the field, where he greeted Coach Harlow with words that did not make their way into Monday's *Harvard Crimson.* The Harvard paper did mention the names of seven worthy seniors who did not play that day. To add a modicum more of indignity to Joe Jr.'s final game, his name was not even listed.

8

Mr. Ambassador

When Joe was named the new ambassador to the Court of St. James's, the announcement was greeted with wide approval. Joe was not what Americans perceived as the stereotypical diplomat, a pinstriped, lisping, top-hatted fop, but a straight-shooting, straight-talking American whom the British could not bamboozle. He was not going to be taken in by the highfalutin' tomfoolery that had supposedly seduced previous ambassadors and turned them into hapless agents for the British establishment.

Before setting off from New York, Joe planned a dramatic gesture against those elitists who thought themselves better than a third-generation American. The most privileged young ladies of America came to England each year to be presented at court, a practice that Joe decided to end as soon as he arrived in London. Joe would keep the honor intact for Americans resident in London. Thus the debuts of *his* daughters could be occasions uncluttered by a hundred fluttering American young women. "That neat little scheme you cooked up, before you left . . . to kick our eager, fair and panting young American debutantes in their tender, silk covered little fannies, certainly rang the bell," the journalist Frank Kent wrote Joe. "A more subtle and delightful piece of democratic demagoguery was never devised."

The forty-nine-year-old ambassador arrived in London on March 1, 1938, to assume one of the most crucial assignments in American diplomatic history. Joe had not even presented his credentials when he was confronted with the darkening dilemma of Europe and the transcendent question of what America's role should be in the growing conflict. That month German armies marched into Vienna and Adolf Hitler pronounced the Anschluss, the uniting of the two countries. From the Austrian capital, Hitler cast his predatory eye on Czechoslovakia. In Germany itself, those Hitler considered his enemies—Jews, Communists, pacifists, and democrats—were being herded into camps with names such as Dachau and Buchenwald. In Spain, the Fascist forces of Francisco Franco, aided by their German and Italian allies, moved forward in Catalonia, driving the Republican armies back.

Joe's first speech was the traditional one given by each new American ambassador to London's Pilgrim Club, a prestigious gathering of leaders in business and politics as well as ranking diplomats. It was a fitting venue for a modest address by an ambassador new both to diplomacy and to Britain. Joe wanted to say something substantial, however, "thereby breaking a precedent of many years' standing," as he wrote in his diary. The new ambassador imagined himself a fearless man who would serve up healthy platters of unparsed reality to an audience unused to such simple fare. He drafted a preachy discourse that sought to push American foreign policy up the road of isolationism, away from Britain and the struggle against Nazism. Joe put a far higher value on candor than it deserved, for truths and policies changed, and it was a fool's game proudly to state the obvious, rubbing British noses in the face of American policies that offended many of them. Joe was full of arrogant self-assurance and his idée fixe that America had to stay out of the sordid, dangerous, deadly strife of Europe.

As much as Joe despised and feared communism, he shared with Marxists the belief that economics was the bedrock reality beneath politics. People were a largely mediocre, self-serving lot whose most important organ was not their head but their stomach. "An unemployed man with a hungry family is the same fellow, whether the swastika or some other flag floats over his head," he wrote Kent. Joe made the same sort of cynical assertions in his draft speech. "I think it is not too much to say that the great bulk of the people is not now convinced that any common interest exists between them and any other country."

When Joe sent his proposed address back to the State Department, Secretary of State Cordell Hull needed to use a full measure of his diplomatic skills to get his newest ambassador to cut the most offensive passages without taking his changes as a rebuke. As tactful as Hull tried to be, he played his trump card, ending his lengthy telegram: "I have shown this to the President and he heartily approves."

Joe's agenda, as he wrote Bernard Baruch, was to "reassure my friends and critics alike that I have not as yet been taken into the British camp." Since he had just arrived in London, it is hard to see how he could have already become a fifth columnist, seeking to seduce an innocent America into a marriage of inconvenience with a declining, troubled Britain. As Joe looked out on the cordial gathering of much of the British establishment, he wrote

Baruch that he found it "difficult to let them have the unpalatable truth I had to offer." But brave man that he thought himself to be, he overcame his reluctance and signaled to the British that the newly arrived ambassador had already fully made up his mind about the crucial issue of the day. He was startled, though he should not have been, that "parts of it fell flat."

Joe was not a man who liked to dawdle. He had no time or patience for silly posturing and the daily inanities of the diplomatic world. Diplomacy, however, is a game largely of tiny victories, of nuanced rituals in which everyone is a player, both friends and enemies. He seemed hardly to understand that an ambassador is called a "diplomat" for a reason. The refined manners and cautious language of the diplomatic circles are not silly affectations but the very procedures that allow friends to stay friends, enemies to sup with one another, and belligerents to enter into civilized discourse.

Joe's candor was a gift that he passed out promiscuously. In June he had his first meeting with Herbert von Dirksen, the German ambassador. The Nazi official reported afterward that Joe had expressed his full understanding of the plight of the misunderstood Nazis. The American ambassador had even attempted to advise him on ways to minimize the unfortunate image that much of the world had of the Nazis. The Nazi ambassador said that Joe told him that what was harmful to the Nazis was not that they wanted to get rid of the Jews, "but rather the loud clamor with which we accompanied this purpose. He himself understood our Jewish policy completely; he was from Boston and there, in one golf club, and in other clubs, no Jews had been admitted for the past . . . years." The Nazi diplomat quoted Kennedy as saying that the great things Hitler had done for Germany impressed him. Von Dirksen added that Kennedy believed "there was no widespread anti-German feeling in the United States outside the East Coast where most of America's 3,500,000 Jews were living."

Years later Joe denied that he had ever made such a statement, calling it "complete poppycock." But the words resonated with the harsh tenor of his candor. Joe was not pro-Nazi, but he rationalized the unrationalizable, turning his eyes away from the worst of Hitler's excesses.

Joe's anti-Semitism—and there is no more benign word for his beliefs—was common among those of his class and background. The hard center of anti-Semitism in America lay not among the poor but among the rich, perpetuated not by the assaults of street thugs but by genteel whispers. Hoffman Nickerson, in his 1930 book *The American Rich,* celebrated the fact that the wealthy class had raised barriers so that the Jew lost "his hope of concealing his separateness in order to rise to power within non-Jewish societies, half unseen by those among whom he moves." The more Joe and his family had risen in society, the more they observed the wages of anti-Semitism.

Joe's life paralleled the social exclusion of Jews from elite American life, all done without any clamor at all. In 1922, President A. Lawrence Lowell of Harvard gave a graduation address at Harvard in which he proposed quotas limiting Jewish students. In Bronxville, the Kennedys lived in a community proud of not having Jewish residents. In Palm Beach, Jews were not welcome at the premier hotels and were excluded from membership in the most desirable clubs.

Joe believed himself a man of the immutable world of power, part of the elite world. He considered it only natural to point out that America had its "Jewish problem" too. And it was natural that the Germans believed that there were millions of Americans like Joseph P. Kennedy who understood what the Führer was doing. After all, a nation that had systematically excluded Jews from the haunts of its social elite could certainly understand why another nation felt that it had to exclude them from its very precincts.

From his early days in London, Joe was obsessed with the Jews and their perilous propensity to yell out when they were struck. As he saw it, they were fully capable of manipulating America into a war to save their lives and possessions. Joe feared war not for himself or for his nation as much as for his precious sons. His boys might die or suffer in a conflict that he considered none of his country's business.

When Joe's aide Harvey Klemmer returned from a trip to Germany, he told the ambassador of the horrendous things he had seen as he wandered through the streets. Nazi storm troopers molesting Jews in the streets, painting swastikas on windows, and trashing the merchandise in Jewish-owned stores. Joe listened to this newest witness to the Fascist savagery. Then, as Klemmer recalled, he turned to his aide and said, "Well, they brought it on themselves."

Joe was certain of his own judgment, and deeply suspicious of Roosevelt. The new ambassador began sending a series of essaylike letters to a select group of influential friends in the United States, each one marked "Private and Confidential," as if the letter had only one recipient. It was a foolish, needlessly provocative thing to be doing. He was supposed to be the eyes and ears of his government. Yet in one of these often-weekly missives he wrote that he had "been to no great pains thus far in reporting to the State Department the various bits of information and gossip which have come my way, because they don't mean anything as far as we are concerned."

Joe considered himself smarter than most of those around him, able to read the self-serving motives that propelled society along. That was his most dangerous illusion, for there was often a transparency to his manipulations that made even those he called his friends suspicious of him. His associates may have been flattered to receive these candid memos. But they were less flattered when they realized that they were simply names on a list. Kent, his irony perfectly in place, wrote another recipient, Baruch: "Just had another of Ambassador Kennedy's syndicated 'Private and Confidential' letters."

One of those not on the list who was less than amused by Joe's machinations was Roosevelt. The president was contemplating running for an unprecedented third term in 1940 and was inordinately sensitive to any Democrat who would dare to think of challenging him. Joe's letters seemed an attempt to curry favor with some of the most important opinion makers in America. "Will Kennedy Run for President?" asked *Liberty* magazine in May 1938, a question that many were beginning to ask, no one more seriously than Roosevelt himself.

Roosevelt was worried in part because Joe had told him that America would have to "come to some form of fascism here." Joe believed that only an authoritarian government would be able to contain social unrest, hold down the grasping masses, and build a strong economy. It was a foolish thing to tell Roosevelt and only exacerbated the president's growing distaste for his new ambassador. Roosevelt could have recalled Joe, but that would have brought him back to America with two full years to create mischief before the presidential election. Joe probably would have been able to pry away several million Catholics from the unwieldy New Deal coalition. Roosevelt decided that he would cauterize this little wound before it became serious.

Steve Early, the president's press secretary, called in Walter Trohan, a reporter for the *Chicago Tribune*. The *Tribune* was the most consistently anti–New Deal major paper in America, and it was a measure of the subtlety of FDR's chicanery that Trohan should be the vehicle that he chose. Trohan was Joe's supposed friend, but he said that he was ready to "write a story against any New Dealer."

"The boss thought you would," Early replied. "Joe wants to run for president and is dealing behind the boss's back." The press secretary tossed a bunch of letters toward the reporter, a collection of the "Private and Confidential" letters Joe had written Krock. The *New York Times* reporter had sent the correspondence to the White House as evidence of his patron's perceptive thinking.

"The guy's working both sides of the street," Early said.

Joe was in the United States for an official visit when he was confronted with the *Tribune*'s front-page headline—"Kennedy's 1940 Ambition Open Roosevelt Rift." Trohan wrote that "the chilling shadow of 1940 has fallen across the friendship of President Roosevelt and his two-fisted troubleshooter, Joseph Patrick Kennedy." The story said that Joe had "besought a prominent Washington correspondent to direct his presidential boom from

London." For Joe's career, the most ominous words were that inside the administration he was being called "the soul of selfishness" in words "crisp with oaths."

Joe was in Washington on June 25, 1938, when he learned of the story that had run two days earlier. "It was a true Irish anger that swept me," Joe noted in his diary, the first time in years he had acknowledged that he might have some of his forebears' more dubious qualities. Joe talked to Roosevelt, who did a superb job of pretending innocence. Joe was proud that he did not "mince words," but berating the president was an indulgence of the worst sort. An enemy will do you ten times the harm that a friend will do you good, and President Roosevelt was the worst enemy of all. For all Joe's proud bluster and Early's "half hearted denial," the new ambassador to the Court of St. James's realized "that something had happened."

Roosevelt held his diverse New Deal coalition together in part with the glue of manipulativeness. He nodded approvingly to one subordinate's ideas, then nodded approvingly to another who suggested quite the opposite. Joe represented many of the twenty-one million Roman Catholic Americans, a crucial if uncertain part of Roosevelt's coalition. Roosevelt needed Joe, even if he neither liked nor trusted him. He needed him to pull back fully within the parameters of the administration, and to do in London what he had been sent there to do. It was a message as clear in its delivery as it was impossible for a man of Joe's character to carry out.

As disdainful as Joe was of the often tedious work of diplomacy, he reveled in the more pleasurable rituals of life at the Court of St. James's. In the evenings, the palatial thirty-six-room ambassadorial residence at 15 Prince's Gate often resounded with gay laughter and spirited discussions as members of the British elite took their measure of this irrepressible, irreverent new ambassador; many of them were no more interested in taking on the Nazis than he was. Rose was usually there beside him in a splendid Parisian gown.

Joe introduced his happy brood and had his older children sitting at the dinner table, imbibing the sophisticated conversation. Kathleen was making her debut that season, and she was full of exuberance and wit, her laughter cascading across any gathering. She was a triumphant success, pursued by several of the most desirable young men in the kingdom. Eunice, Pat, and Jean were studying at the Sacred Heart Convent at Roehampton, where many of their classmates were from the titled families of Europe.

Bobby and Teddy attended the Gibbs Preparatory Boy's School on Sloane Street, within walking distance of the ambassadorial residence. Bobby took the school's rigid discipline in stride, but Teddy took umbrage at being whacked on the fingers with a ruler for some silly infraction. Rosemary was sent the farthest away, to a Montessori school in Hertfordshire.

Luella Hennessey, an ebullient, warmhearted nurse, had come over from Boston to watch over the Kennedy children. She saw that Joe was extraordinarily busy with his diplomatic duties while Rose had a more leisurely existence, her days lightly punctuated by social events and fittings. And yet Joe told her: "Any problems you have with the children, Luella, you are to bring them to me, not to my wife." Joe dominated his children's lives with an often mysterious presence. He investigated any young man so bold as to date one of his daughters and even went so far as to check out the men whom Luella was dating.

Joe applied true diplomatic skills to his own sex life, practicing a discretion that had eluded him in America. He arrived at a Wimbledon tennis match with a young blonde on his arm, while his daughter Kathleen attended the match watched over by a chaperone. But for the most part Joe kept his affairs quiet. He was enamored enough of British life to have forsaken actresses and showgirls for sophisticated, upper-class British ladies.

These aristocratic ladies were at what the British upper class considered the age of dalliance—old enough to have successfully bred an heir or two, young enough to have their looks and desire fully intact, bored enough to welcome occasional diversions. As he approached his fiftieth birthday, Joe was a powerful, virile man who quickly adapted to the more subtle forms of sexual conquest practiced in London.

Occasionally Joe took his aide Harvey Klemmer aside and bragged to the younger man of his latest conquests. This at first startled Klemmer. After all, Joe insisted that Klemmer always include in the speeches he wrote for the ambassador a few paragraphs about the ambassador's wonderful family, his loyal, loving wife, and their nine precocious children. But Klemmer found himself mesmerized by Joe's detailed accounts, especially when he started dropping one famous name after another. "His name was connected to various women all the way up to the top," Klemmer recalled. "Once he said the queen was one of the greatest women in the world. He wanted even that left to speculation, when there was absolutely nothing."

Early in their stay Rose and Joe spent one wondrously memorable weekend at Windsor Castle as the guests of King George VI and Queen Elizabeth. Rose recalled later in her autobiography that after the master of the household had shown them to their rooms in the tower of the castle and the servant bearing crystal glasses of sherry had left, Joe turned to his wife and said: "Rose, this is a helluva long way from East Boston." As full of momentary awe as he may have been, Joe was not the kind of man ever to admit to being impressed by mere surroundings or acquaintances. He did not spend this weekend in fawning obeisance but over dinner expressed his isolationist views to Queen Elizabeth in a blunt American idiom. Joe was nothing if not bold in word and manners. When the queen lost her napkin and said that it was gone, he noted in his diary that he had spied it "sticking up and retrieved it."

That weekend Joe met sixty-nine-year-old Prime Minister Neville Chamberlain for the first time. The dour, cryptic, uncommunicative Chamberlain was in some ways as curious a choice as prime minister as Joe was as ambassador. Both had achieved success in business. In their lifestyles however, Chamberlain, the ascetic, had little in common with Joe Kennedy, the libertine.

The two men did hold the same pessimistic vision of the world situation and had the same dark foreboding about the cost of confronting the growing menace of Nazi Germany. From their first meeting they had a special relationship. The State Department had deep doubts about Joe, but it was a major coup for the ambassador to strike up such an immediate and deep rapport with the prime minister.

All during the summer of 1938 the Nazi propaganda machine blared out its tales of the poor Sudetenland Germans cut off from their beloved fatherland in a Czechoslovakia that repressed them. Czechoslovakia was a polyglot affair, an artificial construct, but it was a democracy whose sovereignty France had pledged to support. Hitler screamed that he would wait no longer before moving into Czechoslovakia, and the timid democracies of the West shivered in fear that they would be driven into a war they did not want and might not win.

In August, Joe prepared a speech to be given in Aberdeen, Scotland. He intended to say that "for the life of me I cannot see anything involved which would be remotely considered worth shedding blood for." There was Joe's entire political philosophy laid out in a single sentence. He couldn't understand why anybody would want to go to war to save anybody but himself. To save himself and his kind, he was all for a triage in which those outside the protective circle would have to fend for themselves. He was willing to announce to the Nazi wolf that the shepherd was not guarding the sheep and he could move on his prey at will.

Roosevelt was dismayed when he saw the speech. "The young man needs his wrists slapped rather hard," he told Secretary of the Treasury Henry Morgenthau Jr., a curious choice of words since Joe was only six years younger than the president. Roosevelt had come to believe that the defeatist group in England had taken in Joe. So, Roosevelt believed, had Chamberlain. Morgenthau wrote in his diary: "The president called him [Chamberlain] 'slippery' and added, with some bitterness, that he was 'interested in peace at any price if he could get away with it and save his face.'"

That was a searing judgment. But it was Chamberlain whose nation was poised to go to war, not Roosevelt's, and given the temper of America and the strictures of the Neutrality Act, if Chamberlain did go to war, all that the president would be able to offer was a toast to the intrepid British.

In early September, Joe called Rose where she was vacationing on the French Riviera and told her that she had to return to London immediately. Hitler was tired of waiting and was about to send his armies into Czechoslovakia to claim the Sudetenland as his own. The acrid smell of war was in the air, and Joe was afraid, not for himself so much as for Rose and his sons and daughters. He was, as he wrote a friend, "trying to keep up my contacts so we would know what really was going on before it actually happened so that we would not be caught unprepared contemplating the possibility of the bombing of London with eight children as prospective victims." Joe personalized politics, seeing every event in terms of his own fortune and family. That gave immediacy and passion to his every move and now, as war appeared imminent, an increasingly desperate urgency.

Lord Halifax, the British foreign secretary, asked Joe how America would react if Hitler ran over Czechoslovakia. "I don't have the slightest idea," Joe replied, "except that we want to keep out of war." This led the British diplomat to ask why *his* country should defend all the ideals and values of democracy by itself. "The British made the Czechoslovak incident part of their business . . . and where we should be involved, the American people failed to see."

Joe paraded before the British his friend Charles Lindbergh, with his terrifying tales of the awesome arsenal of the Luftwaffe, the German air force, the strongest in the world. As Chamberlain attempted to negotiate with Hitler, Joe seemed incapable of realizing that language itself—a word, a phrase, a subtle interpretation—had become the very stuff of life and death, peace and war. He gave an interview with the Hearst papers in which he said that Americans must "not lose our heads."

Roosevelt was aghast at the idea of Joe spouting off once again and told Hull that he would have to take his ambassador to task. In the end he thought better of risking a confrontation. Instead, he wrote Joe a letter in the grammar of political Washington, in which words meant the opposite of what they said: "I know what difficult days you are going through and I can assure you that it is not *much* easier at this end!"

The president was attempting to support morally those who stood up against the menace of Hitler while not making promises of aid he could not keep and in proffering would only give solace and strength to his enemies and set back even further what he considered the common cause. In London he needed a loyal ambassador with a subtle mind and a nuanced character, an ambassador who worked for Roosevelt, not for himself.

Chamberlain flew to meet Hitler once again and returned with empty hands and a heavy heart. Rose noted in her diary: "Everyone unutterably shocked and depressed." Those who had taken Hitler's measure may have been saddened and disheartened, but they were not shocked that the man had acted as they knew he would act.

Joe was not only morose and depressed but full of self-pity. He felt no comparable sympathy, however, for the several hundred thousand opponents of Hitler living in the Sudetenland, whose risk was somewhat more imminent than his own. "I'm feeling very blue myself today," he wrote Krock, "because I am starting to think about sending Rose and the children back to America and stay here *alone* for how long God only knows. Maybe never see them again."

Chamberlain made a final dramatic trip to Munich, where he reached a settlement with Hitler. The Anglo-German agreement, the Munich Pact, was not so much a compromise as a capitulation, giving Hitler the Sudetenland in return for a promise that he would stop there and respect what was left of Czechoslovakia. With this agreement, Chamberlain flew back to England and announced, "I believe it is peace for our time." The church bells rang out, peeling a joyful timbre. The streets were full of Londoners shouting their approval.

Joe fought his way through the crowds to the embassy, as exuberant and exhilarated as any of the celebrants. Later he happened to meet Jan Masaryk, the despondent Czech minister to London. The Czech diplomat knew that all Chamberlain had done was to throw a sacrificial lamb to the Nazi beast. Once Hitler had digested his meal, he would move on to the rest of Czechoslovakia. "I hope this doesn't mean they are going to cut us up and sell us out," Joe recalled the Czech leader saying. "Isn't it wonderful?" Joe exclaimed, as Masaryk remembered. "Now I can get to Palm Beach after all."

Joe was a truthmonger who shouted his truths so loudly that all could hear them. Three weeks after Munich on October 19, 1938, Joe spoke at the Navy League's Trafalgar Day dinner. He told his audience that it was foolish to emphasize the difference between democracies and dictatorship when, "after all, we have to live together in the same world whether we like it or not."

This was the same little drum Joe had been beating on since he arrived in England, but it had begun to sound hollow and thin. The audience in that room was full of navy officers, some of whom had fought against the Germans in World War I. The larger audience was hardly more receptive. America was slowly waking up to the stark menace of Hitler, and Joe's words set off a firestorm of criticism across America. Joe had been used to laudatory press treatment; now, as he wrote later, he was "hardly prepared, despite years in public office, for the viciousness of this onslaught."

Joe had his own fervent supporters, including Jack, who from Harvard wrote him a comforting letter: "The Navy Day speech while it seemed to be unpopular with the Jews, etc. was considered to be very good by everyone who wasn't bitterly anti-Fascist."

Jack was echoing his father's sentiment that the tedious, troublesome Jews had the audacity to object when their brethren in Germany were either shipped off to camps or stripped of their belongings and thrown out of their native land. As both Kennedys saw it, the Jews could not seem to understand that there were geopolitical considerations far more important than the survival of some of their people.

In mid-December 1938, Joe returned to the United States for a vacation and consultations. In Washington, as Joe recalled in his unpublished memoirs, Roosevelt told him that "he would be a bitter isolationist, help with arms and money, and later, depending on the state of affairs, get into the fight." Joe was utterly opposed to such a plan, and by rights he should have resigned or Roosevelt should have fired him. Instead, the two men chatted amiably in a discussion that was nothing if not duplicitous.

"I have never made a public statement criticizing you," Joe told the president. "As to what I say privately, you know perfectly well that I will never say anything that I have not said to you face to face. But you know the way I feel about you and I won't be any good to you unless we are on good terms."

Roosevelt knew indeed what Joe felt about him, but this was not the moment for candor. "Don't worry," Roosevelt said. "I know how you feel—and as for those cracks at you, some people just like to make trouble."

The president knew that Joe was one of those who liked to make trouble, but he preferred having him an ocean away rather than committing political mischief at home, rallying the forces of isolationism around his banner. As for Joe, he returned to London in February 1939 a member of isolationist ranks that had once been full of many honorable, if misguided, men and women. Now, after Hitler's troops had marched into the rest of Czechoslovakia in March, even Chamberlain himself, the personification of isolationism, realized his policy had failed. In reluctantly signing a pact to come to the aid of an invaded Poland, he drew a line in the parched earth of Central Europe from which he could not withdraw.

Out of the debacle of the failed policy came a new opening for Joe. He knew what Roosevelt was thinking, and he had unique credibility as Chamberlain's friend. He could have subtly insinuated to Chamberlain that his island nation would not be alone in fighting Hitler. Aid might not come in the quantities sought, and the soldiers might not arrive as soon as they were needed, but in due course they would probably come.

Joe could not have said such things boldly, for the repercussions for Roosevelt and his race for a third term would have been devastating. But he could have gently pressed his president's agenda. He could have listened to the voices of Britain, truly listened, and gauged the moral fiber of a people and passed on that word to Washington.

The only voices that Joe listened to, however, were the upper-class accents that entered his salon at 15 Prince's Gate. He was disdainful of the privileged young men who came to the ambassadorial residence as his daughters' escorts and his sons' friends, and he took them as an honest measure of British strength. They listened to his loud contempt for them and their nation, and they put the old man on, telling him that they would never fight. And Joe took them at their word.

On the evening Joe Jr. arrived in London after graduating from Harvard in June 1938, he headed out to a favorite haunt of the sophisticated upper class, the 400 Club. He was the ambassador's son, and he sat that evening at a table with a Turkish pasha, an Argentine polo player, a Dutch baron and his beauteous wife, and the daughter of American humorist Will Rogers. Night after night during those summer months Joe Jr. was either there at ringside or off with some of the most attractive women in the kingdom.

Joe Jr. could have gone on for months spending his evenings as one of the playboys of the city, putting in a few desultory hours during the day as his father's aide. He was, however, a young man who would have grown bored measuring out his days in champagne bottles. He had always been a child of fortune, and he looked out on the troubled seas of Europe like a sailor who takes his craft out only in high seas.

Joe Jr. set out for Paris, where for two months he worked as an attaché to William C. Bullitt, the American ambassador. From there he traveled with a diplomatic passport to Prague, Warsaw, Leningrad, Copenhagen, and Berlin. Like his father, he believed that strength created, if not its own morality, then its own imperative. "Germany is still bustling," he wrote a Harvard classmate. "They are really a marvelous people and it is going to be an awful tough time to keep them from getting what they want."

After his journey across Europe, Joe Jr. arrived in Saint Moritz for a Christmas vacation with his family. His father was in the United States, but if Joe had been there he would have seen a son who was the exquisite perfection of what he thought a man should be.

Joe Jr. had hardly arrived when he was arm in arm with Megan Taylor, the beautiful eighteen-year-old world figure-skating champion. They made the most stunning of couples as they skated together across the rink. Despite his romance, Joe Jr. cherished his time with his younger siblings. Unlike his mother, he was demonstrative in showing his love, running up the stairs to greet the six-year-old he called "Teddy Boy," spinning the child in his arms. Ted, whose eyes grew bright as stars, worshiped his big brother.

Joe Jr. took to the winter sports of Saint Moritz as if they were no more difficult to learn than checkers. He jumped on a one-man bobsled and, zooming down the run at seventy-five miles an hour, came close to setting a world record. On the ski slopes there were no records to challenge, but if there had been a trophy for speed and daring, he would have been a finalist.

On one spectacular spill he fell on his arm, cutting his skin. He skied the rest of the way down and called the family nurse: "Luella, I need a Band-Aid." Luella took one look at the injury and sent Joe Jr. off on a sleigh to the hospital, where he was treated for a broken arm. In her years as a nurse, Luella had seen how pain that drove one man to scream would cause nary a whimper in another, but she had rarely seen a person who seemed entirely immune to pain, who almost enjoyed it, like a tonic. Joe Jr. laughed at the injury, as if he had nicked himself shaving. Sure enough, he was immediately out skating again with Megan Taylor.

His younger brothers applauded Joe Jr. by imitating him, with results that were only slightly less harsh. Bobby sprained his ankle on the nursery slope, while Teddy suffered a wrenched knee. Little Teddy enlivened his enforced convalescence by playing with matches in his room at the Palace Hotel before coming down to the lobby. "Eddie Moore came in and found the whole wastebasket on fire," Ted recalled. "My father being outraged, I think I got another spanking for that."

Teddy had his defender in Luella. "Joe Jr. used to tease Teddy terribly," the nurse recalled. "It was all in fun, but he was the only one who did it. If he was teasing Teddy or saying, 'Why don't you do this,' or, 'You're too old to act that way,' I'd say, 'Now don't be teasing. He's my Edward.' Teddy was my favorite, a happy little fellow. I just loved Teddy."

For Joe Jr., his vacation in Saint Moritz had been a glorious winter adventure, but it was also inevitable that a young man of his passion, political ambition, and daring would want to be a witness to the Spanish Civil War. As Joe Jr. headed south, he was an anomaly among the young men of America and Europe who had gone to Spain. Most of them had arrived to fight with or support the loyalist Republican troops. Among them were a number of the greatest writers of the time. Ernest Hemingway. George Orwell. André Malraux. Stephen Spender. W. H. Auden. Langston Hughes.

Unlike these men, Joe Jr. was naturally sympathetic to the reactionary General Franco and his German and Italian Fascist allies. In church Joe Jr. had heard the priests talking in menacing detail of the godless hordes who burned churches and killed priests and nuns. His own father had protested when the president had contemplated lifting America's embargo on arms to the belligerents. At Harvard, Joe Jr. wrote his thesis on "Intervention in Spain," a document that perhaps was destroyed so that the Kennedys would not have to explain Joe Jr.'s enthusiastic support for Franco. As Catholics, not only Joe Jr. but his whole family saw Mussolini and Franco as bulwarks against atheistic communism.

In early 1939, Franco and his allies were in the ascendancy. Hemingway had already left Madrid to write his great novel *For Whom the Bell Tolls*. George Orwell was gone too, after fighting in the trenches and writing his classic book *Homage to Catalonia*, documenting the deceit of the Communists and their role in destroying Republican idealism.

Joe Jr. was not a rightist ideologue who reveled in the Republican defeat. He was a man who felt as much as he thought. Even before he entered Spain, he had visited a camp in France along the border where members of the International Brigade were interned. They had come from all over the world to fight for the Loyalist cause. He was shocked at the conditions in which they lived, and he listened to their tales with empathy.

"The last few weeks, before they got across into France, were terrible," Joe Jr. noted. "Without arms and with no idea just where they were, as they retreated they expected at any moment to find themselves surrounded by Franco's men. . . . Most of them were Communists, whose families at home were being supported by the various trade unions."

Joe Jr. understood the universal language of valor. He saw that these men were brave and true. He was sorry that their adventure had ended in the squalor of a camp, where they were reduced to squabbling over crusts of bread. He traveled into Spain, arriving in Barcelona in February only a week after its fall to Franco. The area around the harbor was a charred ruin, but gleaming steel Italian and German warships sat proudly in the port.

From Barcelona, Joe Jr. sailed on a British destroyer to Valencia, one of the last strongholds of the Loyalists. Joe Jr. had what Orwell would have called a soldier's vision of the war; that is, he was no jingoistic cheerleader viewing the destruction as God's revenge on a heathen people. German bombs had leveled the area around the port, but back a short way from the rubble children played as children always play. "It made me sick," Joe Jr. wrote in a letter published in the *Atlantic Monthly*. "The sound of those little kids' voices."

Joe Jr. was not only an American but a Kennedy; his father was an enemy of the Loyalists. Joe Jr. had turned in his diplomatic passport for ordinary documents, but he was peculiarly vulnerable. He was an uninvited guest in a society that stood on the edge of panic and paranoia over what measure of blood Franco would consider his just revenge.

The story of the civil war lay not among the orange fields of Valencia, however, but in Madrid, held by the Republicans. It was here that the war had begun and it was here that it would end. Joe Jr. talked his way onto a military bus traveling to the besieged city, rationing out cigarettes to his new friends on the bus as it bounced northward along ravaged roads. Joe Jr. didn't speak much Spanish, but his smile and his backslapping good cheer served him well, and doubtless his new comrades thought that Joe Jr. was one of them.

The Madrid that Joe Jr. entered was a beggar city; its boulevards had been stripped of trees, chopped down for firewood, and its sidewalks were full of shuffling, hungry, bewildered people. Stray dogs and cats had long since been eaten, and the meat on sale in the Plaza Mayor was rat. Foreign journalists and most other observers had fled the city as well, and Joe Jr. was a unique witness to the last days of the Spanish Republic.

He staked out a room at the abandoned American embassy and set out to imbibe the full flavor of the beleaguered city. He could hear Franco's artillery sounding in the distance. Down the street lay savage fighting between the Communists and their erstwhile allies, the Socialists and anarchists. Thus sounded the death knell for the republic: the soldiers who had fought years together against an implacable foe were now dispensing the last ounces of their resolve against each other.

Joe Jr. was perhaps the only nonresident American in Madrid at the end of the Spanish Civil War. He had proved his bravery just by being there, but he had an even more daring idea. He wanted to find the Franco underground. He was told that he should go to a certain address, 19 Castelló Street. There he saw affixed to the door an American flag and a certificate stating that the house was a diplomatic enclave.

Joe Jr. met Antonio Garrigues y Díaz Canabate and his American wife, Helen Anne. And he saw basement quarters where the couple sheltered nuns and others. They were not passive recipients of Catholic refugees; they were running operations in a city that verged on anarchy.

Joe Jr. asked to join them. He squeezed into Garrigues's small car with three others, and they headed out into the Madrid streets. The Spaniard carried several sets of documents with which he played his own peculiar form of Russian roulette. When the vehicle was stopped, he had to decide whether the soldiers were Communists or Republican Loyalists. If he chose wrong, he and his passengers might be arrested or summarily executed. At a roadblock at Calle del Esse, Garrigues decided to show the soldiers his Red Cross credentials. The soldiers ordered the men out of the car and lined them up against a wall. Joe Jr. took out his passport, a document that, shown to the wrong soldiers, would have been greeted with the firing squad. This time the soldiers shrugged and waved the group on.

Joe Jr. was there in March, when the city fell, not in some great battle, or with Franco marching proudly and dramatically into Madrid, but in a strangely desultory manner. Joe Jr. saw a car rushing through the city, the Nationalist flag waving from its window, then a second car, a truck full of gesticulating young men shouting, "Franco is coming!" The once-banned scarlet and gold Nationalist colors were everywhere, hanging in windows, worn on scarves, draped in restaurants. "We were touched by the expressiveness in their voices and the look in their eyes," he wrote his father, "by occasional sad faces, by a woman in black holding two children with a bitter look on her face."

Joe Jr. flew to London to greet a father full of immense pride at his son's daring and resolve. While his son was in Spain, Joe had read some of Joe Jr.'s letters to Chamberlain at a dinner one evening. Joe had not wanted to inflict too much of this on the prime minister, but Chamberlain had asked him to read on, and so he had.

Joe wanted not only for his sons to experience life in all its rich manifestations, but for their adventures to be trumpeted in newspapers and magazines, spreading the glory of the family name. That was the approach to everything they did. Win a sailing race. Triumph in a football game. Travel to war-torn Spain. Then make sure that your achievement is widely known and richly celebrated.

Joe Jr. had written a letter to the *Atlantic Monthly* about Valencia for which he received twenty-five dollars, but his ambitions were greater now. He had had an experience that he could weave into a book, and he began by writing a series of six articles.

For weeks Joe Jr. had gone through events in Spain that should have been burned into his consciousness. Yet there was a distant quality to much of the writing, as if he were remembering events of twenty years before. A passage would be full of vividly observed detail, followed by pages with no more verve than a legal brief. What was missing was a sense of politics, without which the events he had seen were largely meaningless, and his account merely a travelogue of adventure.

Joe Jr. was a highly opinionated young man. Whatever the issue, he had his stance, strongly argued and strongly felt. He also had political ambitions, and he may well have felt that he should position himself as an honest observer and keep his political views to himself, confining them to private letters and memos to his father. In doing so, however, he stood back from his own sense of truth and wrote a paltry outline of what he had seen and felt, leaving out what he believed. In the end, none of these articles saw publication, and Joe Jr.'s journey was celebrated more within his own family than in the greater world.

Joe thought that his sons were never too young to learn to be the public men he wanted them to be. Joe's sons felt the need not simply to emulate their father but to defend him. Even young Bobby got into the fray. In November 1938, Walter Lippmann wrote a thoughtful column rightfully criticizing Joe as one of those "amateur and temporary diplomats [who] take their speeches very seriously. Ambassadors of this type soon tend to become each a little state department with a little foreign policy of their own." Twelve-year-old Bobby wrote a tedious rebuttal that was little more than a regurgitation of some of his father's more extreme views. There was a slovenly quality to the letter, with myriad typos, including a misspelling of Lippmann's name.

Although Bobby would presumably have found it offensive if others had suggested that his father believed as he did simply because he was an Irish-American Catholic, he was perfectly willing to condemn Lippmann's writing as the rationalizations of a Jew seeking to protect "his" people, and not his nation. Bobby dismissed Lippmann's thoughtful, reasoned critique as nothing more than "the natural Jewish reaction."

For one so young, Bobby had a deeply offensive arrogance, much like his father's. He lectured Jews that they had better accept the realities of making accommodations with Hitler. "I know this is extremely hard for the Jewish community in the US to sto mach, [*sic*]," he wrote with all the wisdom of his years, "nbut [*sic*] they should see by now that the fcourse [*sic*] which they have followed the last few years has brought them nothing but additional hard ship [*sic*]."

In April 1939, thirteen-year-old Bobby was invited to be one of a group of children laying a stone at the Clubland Temple of Youth in Camberwell. Joe, the subtle mastermind of his sons' journeys into manhood, had the event scripted with all the detail of a major diplomatic meeting. As ambassador, Joe had been asked to preside, but this was his son's first important public appearance, and he wanted the spotlight to shine unabashedly only on Bobby.

The press attaché notified news organizations that the ambassador's son would be making a short speech. James Seymour, the ambassador's aide, prepared elaborate notes for the event, including Bobby's little speech ("All the temples I've read about in history books are very old . . . but this 'Temple of Youth' is awfully young"). On the appointed evening, Bobby pulled a crumpled, penciled note out of his pocket and read what appears to have been his own remarks, not Seymour's version of what a boy scarcely a teenager should say ("Many years from now, when we are very old, this Temple of Youth will still be standing to bring happiness to many English children").

Young Bobby already understood that part of his role in life was to take the work of others and by his own subtle infusions make it his own. He was a boyish mixture of shyness and confidence. When the seventeen-year-old Japanese representative found herself next to the eleven-year-old daughter of the Chinese ambassador, there was deadly quiet, and the event risked becoming not a celebration of the commonality of children but a minor diplomatic incident between children of the two warring Asian nations. "I wonder shall we all be here to see Her Majesty perform the opening on May 20?" Bobby asked, his innocent query giving the two young women a neutral matter to discuss.

Even six-year-old Teddy realized that he was not just a boy. His parents told him that wherever he went he had to remember that he was the son of the ambassador to the Court of St. James's. That was a burden his older brothers had not had. The little boy was living two lives, his own and his father's glorified idea of him. He came home from school one day and asked his mother for permission to punch a classmate. "Why?" Rose asked, believing that in the pantheon of virtues, civility stood below only godliness. "Well, he's been hitting me every day, and you tell me I can't get into fights because Dad is the ambassador." After a family discussion, Ted was told that this once he could strike back at his tormentor.

In what would prove to be the last summer of peace, Joe Jr. headed out once again across Europe, a footloose, privileged witness to his times. He was frequently an insightful, prophetic observer. In Germany he saw that the people were largely united behind Hitler and that "there is only one thing that the

Germans understand and that is force. All attempts at conciliation are taken as signs of weakness, and furthermore are used as propaganda by the Germans to convince the smaller countries that the English won't fight."

Joe Jr. may have been his father's son, but in these dispatches he did not simply parrot his father's isolationist views. He had a young man's rage at the hypocrisies of the world, and he pointed them out in passionate detail. Europe was full of ethnic loathing. Hitler was the primary architect of malevolence, but there were others building their bonfires of hate across the ancient landscape. This world might suck America into all its malice and complexities.

Joe Jr. raged against Roosevelt's hypocrisy, egging Britain on from a safe distance. He was not, however, the cocksure, often arrogant young man of two years before. He was even willing to consider a policy unthinkable to the ambassador to the Court of St. James's. Joe Jr. concluded one dispatch: "I had always thought that we should stay out of war and that being a rich nation we can live by ourselves . . . but if we can't . . . then I think we should have a real policy in Europe entirely fitting for the greatest power in the world rather than a half-hearted, namby pamby policy skipping one way then to the other so no one knows what will happen if their [*sic*] is a war."

While Joe Jr. spent much of his time journeying around Europe, Jack was back at Harvard. Exemplifying what can be the terrible carelessness of wealth, he believed that there was always someone else to pick up for him. In his suite at Winthrop House, he had the disconcerting habit of dropping his clothes in the middle of the floor. One evening as he hurriedly dressed to go out, he threw his pants and shirt into a pile in the middle of the room. His roommate, Torby Macdonald, looked at the heap and declared their room had the distinct appearance of a rummage sale. "Don't get sanctimonious," Jack snapped back. "Whose stuff do you think I'm throwing mine *on top of?*" That may have been true, but before long George Taylor, Jack's black, self-styled "gentleman's gentleman," would be dropping by to hang and press Gentleman Jack's clothes while Torby's would stay where they fell.

At times Jack treated the law like a petty hindrance that should not bother someone who carried the name Kennedy. Jack wrote Lem that he had a "rather unpleasant contact with a woman in a car who was such a shit that I gave her a lot of shit." Euphemism is often a liar's cloak, and Jack admitted to his friend that the woman had written the Registry of Motor Vehicles complaining that "I had leered at her after bumping her four or five times, which story has some truth although I didn't know I was leering. . . . Anyways they got me in and are sore at me."

Jack had apparently become so angry that he had rammed into this woman's car a number of times. When he was called to account for his deplorable act, he lied, telling the officers that he'd "loaned my car out that night to some students." Jack had given the police the name of one of these students—none other than his friend Lem. Now Lem was supposed to take the blame, saying "you're sorry and realize you should not have done it." Confession was good for the soul, even if it was a lie masquerading as honesty, as long as Jack did not have to take responsibility.

It remained a wondrous time to be rich, as long as you kept your eyes high, away from the unseemly sight of the poor and the hungry. Jack was a Spee Club man, and his friends were either rich and wellborn or stellar athletes. He did not sit at dinner at his club or around Winthrop House bemoaning the fate of the poor. Not once in any of his letters did he ever mention the dreadful consequences of the Great Depression.

Jack traveled from one watering hole of the wealthy to the next. At the wedding of his classmate Ben Smith in Lake Forest, Illinois, Jack left a faucet on all night in the home where he was staying. By morning the plaster had fallen down from the ceilings, and the newly decorated home was a shambles. Jack admitted his culpability with such self-deprecating charm that the hosts couldn't possibly dislike him, and so they turned their wrath on a friend whose only crime was being there.

Women were one of his primary divertissements, if they were pretty. If they were not pretty, he simply ignored them, or occasionally ridiculed them. During Easter 1938 Jack and Lem were down in Florida, where they heard that the Oxford Meat Market was having a picnic for the servants in Palm Beach. The two young men thought that it might be good pickings there, that pretty working-class girls would be happy to spend an evening with young gentlemen.

Jack's eyes fell on one pretty Irish lass. Jack cut her away from the rest, but though she agreed to go out with him, she insisted that her friend Rachel come along. Rachel was enormous, at least 250 pounds, dressed in a sailor suit. Jack, always ready for amusement, said that he would arrange a date for Rachel. He then called his snobbish Princeton friend Sandy Osborn and told him that he had a date for him. "Well, we went to pick up the girl at the corner of such-and-such street, and Sandy was so eager and excited: and suddenly the lights hit her standing on the corner," Lem recalled. "It was one of the funniest experiences I've ever had in my life!"

Jack was also careless in his vision of the world. In the summer of 1937, he traveled through Europe with Lem, keeping a diary of his experience. It was an exceptional time to be journeying from the border of Spain to Berlin, but for the most part he wrote like a snooty prep school boy criticizing nations as if they were bad restaurants.

To young Jack, ethnicity was more important than ideology. In Saint Jean de Luz, he talked to refugees driven out of Spain by Franco's minions. He noted in his diary: "Story of father starved kept in prison without food for a week brought in piece of meat, ate it—then saw his son's body with piece of meat cut out of it."

Jack did not look for reasons in the politics of his time but in the Spanish national character. "In the afternoon went to a bull-fight," he scribbled in his notebook. "Very interesting but very cruel, especially when the bull gored the horse. Believe all the atrocity stories now as these southerners, such as these French and Spanish, are happiest at scenes of cruelty."

In Germany, although he did not express pro-Nazi sentiments, Jack was impressed with the quality of life. "All the towns are very attractive showing that the Nordic races certainly seem superior to the Latins." Near the end of his journey, he concluded: "Fascism is the thing for Germany and Italy, Communism for Russia and democracy for America and England."

On his trip to Europe in 1937 Jack had been sailing along when, in London at the end of the summer, he developed a bewildering case of hives that four different doctors looked at before the problem disappeared as mysteriously as it had begun. Jack's health was a veritable dictionary of illnesses, and years later Lem joked to a friend that if he wrote Jack's biography, he would title it "John F. Kennedy, A Medical History."

With his weak stomach, Jack was fortunate that he could eat at the Spee and be served a special diet. The club even had an ice cream–making machine to produce the one food that he could easily digest and that might put some meat on his bones. He went out for the Harvard swim team, but in March 1938, he entered New England Baptist Hospital with an intestinal infection that knocked out any chance of a swimming letter. Once again he lay in a hospital bed, trying to rise out of his pain and get back into the world.

By the middle of June, Jack was back in the hospital, nagged by weight loss and continuing intestinal problems. In the fall, doctors wanted him back in the hospital for more tests. He wrote Lem, "I've been in rotten shape since I've got back and seem to be back-sliding." The following spring he was so sick that he had to agree to go to the Mayo Clinic for tests. It was simply endless.

Jack made a point of lying out in the sun at Palm Beach or Hyannis Port getting a tan, making sure his face always had a robust sheen. He was not going to be pitied or excused. No one was going to push him aside from the races of youth. His few friends who knew the truth realized that Jack's greatest creation was the illusion of health. If he talked about his condition, and that was rarely enough, he did so only to make a joke about it. "Jack didn't discuss it," Rousmanière recalled. "We never made it a part of our lives what his health was. He was deceptive and his health was never an issue."

Joe had taught his sons that nothing was more valuable in life than time, for life itself was no more than a blink of an eye. To be sickly was to live half a life, and half a life was not life enough. Jack willed himself to live his life as something he was not and never would be—a healthy man.

Jack kept his poor health secret all his life. If sometimes denial is another name for a lie, at other times it is the mask of courage, and friends such as Lem and Rousmanière believed that was the mask their friend was wearing. Later in his life, Jack would lie about his health for political reasons, but now he pretended that he was something he was not so that he could live as he thought a man must live.

During his last two years at Harvard, Jack had become the student that his father and older brother never were. Jack burrowed into books not to avoid the world but to enter it more fully. There was a dramatic immediacy to many of the classes, especially in government and modern history. The professors themselves often stood on different sides of the issues of war and peace. Education was no longer an abstraction. Education, at least part of it, was about life and death, a struggle in which these young men might soon be involved.

In the spring of his junior year, Jack received a leave of absence to go to Europe and prepare his senior honors' thesis on England's failure to prepare for war. During these seven months in 1939, Jack traveled throughout Europe and Palestine, sending his father a series of detailed accounts of his journey, most of which have been lost. He wrote Lem as well without having to play the diplomat.

In these letters to Lem there is a curious dichotomy between the fledgling public man and the spoiled, petulant, narcissistic youth. To that latter Jack Kennedy, life was a dress parade. Jack wrote Lem how he "went to court in my knee breeches and made my gracious bow before the king and queen and was really quite a figure," while he was not "sporting around in my morning coat, my 'Anthony Eden' black homburg and white gardenia." He was, as always, deeply sensitive to anything that smelled of the effete. He read in the *Choate News* that a lone bat had terrified students, zooming around Mem House, hiding until an intrepid housemaid attacked the animal with a mop. "My God, they all sound like fairies, both faculty and student body," he wrote Lem.

To Jack, sex remained the preferred avenue of manly adventure, his sexual safaris taking him from the brothels of West Palm Beach to the refined enclaves of the demimonde, where at times the behavior did not differ that much from that of the whores of South Florida, only the price.

No small part of the adventure was the imminent possibility of carrying home an unwanted souvenir. In London he saw a woman who had lived with the Duke of Kent and had among her baubles of romance "terrific diamond bracelets" from the duke, a "big ruby" given to her by a Nepalese prince, and a cigarette case "engraved with Snow White lying down with spread legs, and the seven dwarfs cocks in hands waiting to screw her—very charming." Jack wrote Lem: "I don't know what she thinks she is going to get out of me but we'll see. Meanwhile very interesting as am seeing life."

Jack was seeing life too on what he called "the serious side." His father got him introductions to ambassadors and other high officials. But it was his self-confident manner and astuteness that led them to spend time with him that they probably would not have granted to a less politically sophisticated young man. "The whole thing is damn interesting," he wrote Lem, "and if this letter wasn't going on a German boat and if they weren't opening mail—could tell you some interesting stuff."

Some politicians, even in their youth, live their lives as if they are biographies, measuring each word for posterity, writing letters with their ideas and feelings self-censored, all sanitized. Jack, however, wrote in an emotive stream, going in a few sentences from a rude put-down of his contemporaries to a telling insight into the world in which he lived, and then back to more gossip.

Again and again Jack told Lem that it was all "very interesting," and so it was. He traveled to the open city of Danzig, where swastikas were in full array and he "talked with the Nazi heads and all the counsels up there." He realized that Poland would never give up Danzig to Hitler. Nor was Germany about to back off. "What Germany will do if she decides to go to war," he wrote Lem on May 1, 1939, "will try and put Poland in the position of being aggressor—and then go to work."

Jack's diplomatic reportage was perfectly prescient, but he had no interest in talking to the unwashed, the unlearned, and the unstylish. "All of the young people own estates of around 1,000,000 acres with 10,000 or so peasants," he wrote Lem, as if young peasants were some degraded form of humankind unworthy of being called "young people." If Lem came to visit, he promised that they would visit an estate rented by the Biddles "with around 12,000 on it who tip their hats with one hand and push forward their daughter with the other."

While twenty-year-old Jack was in London, he read *The Young Mel-*

bourne, a new book by David Cecil. Jack was at an age when a young man of a literate bent reads books not as abstract fodder but as a guide to conduct. Young men of Jack's generation were reading Hemingway's *For Whom the Bell Tolls* and learning of love and courage, or John Steinbeck's despairing *Of Mice and Men*. Their favorite book would never have been a literary biography of a Whig aristocrat who at a young age "envince[d] that capacity for compromising genially with circumstances," a capacity that the young rarely consider a virtue.

In this Whig world that Jack so admired, daring was for the battlefield, not for the art that hung on one's walls, the literature that one admired, or the politics that one espoused. It was a life of appreciation, not creation, and perfunctory religiosity, not deep spirituality. It was above all a man's world of conventional excess—drinking and gambling and wenching. "The ideal was the Renaissance ideal of the whole man, whose aspiration it is to make the most of every advantage, intellectual and sensual, that life has to offer," the author wrote in what could have been an adage for Jack's own life.

Jack was a young man of intellectual precociousness. Many of the ideas in the book resonated within him and passed as wisdom. Young Melbourne was "a skeptic in thought, in practice a hedonist," a description that fit Jack's own vision of himself. This Whig hedonism was not vulgar sexual indulgence, but a life in which pleasures, sensual and otherwise, were aesthetic treats. Passion was in the pursuit. Loyalty was for friends, not for spouses or lovers. Moral perfection was left where it belonged, in heaven.

"Nature had meant for him that rare phenomenon, a philosophical observer of mankind," Cecil wrote. That sentence too would surely have resonated with Jack. He was in ways a spectator to his own life. There seemed nothing, no romance deep enough, no danger intense enough, no thought profound enough, to pull him fully outside the circle of himself.

There was the conundrum, not only for young Melbourne but also for young Jack Kennedy. Is a man who sees the world as it is—with all its compromises, dishonesty, self-interest, and impediments to change—less able to make a deep mark than another man who is blithely unaware of his own limitations and those of the world? Was it the lot of a man who understood the sheer futility of most human effort—the illusions and vanity that drove men to seek fame and power—to stand on the sidelines, his head tilted ironically, merely observing the life that ran before him?

9

"It's the End of the World, the End of Everything"

Joe believed that England would lose in a war with Germany. Everything he did and said followed from that conviction. If he was correct, he was not a treacherous defeatist whose sentiments prolonged the period before America would begin to support its greatest ally with its immense military arsenal.

No, he was a daring prophet and patriot who, once war seemed imminent, sought to extract whatever he could from the dying island empire, holding the British up to repatriate hard currency for Hollywood films, hanging tough in barter deals with the British, and promoting those British leaders who sought peace at almost any price. He had gone into politics to save his own fortune, or a good part of it. If his liquor business benefited from the diminished tariffs he was now helping to negotiate, that was only a fitting compensation for his many contributions as a public servant.

As for the Jews, it pained Joe that they mocked him so when he considered himself practically their best friend in government. He believed that he shared with Chamberlain the attitude that if the world exploded in war, the troublesome Jews would share a good deal of the blame. In his thinking, he had nonetheless worked on several fronts to get some of them out of Germany and accepted in one of the many countries that shunned them. One of the new ambassador's first tasks was to deal with the Evian conference in France. It was called a "refugee" conference, but that was a euphemism that hid the fact that 90 percent of the refugees were Jewish. The delegates focused on what to do with German "refugees," but the problem was far broader. The Poles and the Romanians wanted to be rid of their "refugees" too. The British feared that too liberal a solution in Germany would inspire several other European nations to deport their "refugees."

Joe tried to work out a solution to allow Jews to buy their way out of Germany and into other countries. Joe's proposal pleased his eldest son. "The baby is tossed right into the laps of the people themselves for the real concern now is money," Joe Jr. wrote in his diary on November 21, 1938. "If the Jews come through and especially the other people this problem can be cleared up. Dad doesn't think they will put up the necessary money, but he thinks he has done his part, and the rest is up to the others."

London was full of rumors that Joe was making a new fortune selling out the British by trading on inside information. There is no evidence of such dealings in the British Foreign Office's extensive files, and Joe may well not have been taking advantage of his position. But he was a man of such monumental cynicism, who glorified the rudest reaches of self-interest, that those around him often thought the worst of him.

Joe was not cynical, however, about his children. No father cared more for the future of his sons than he did, and when he spoke out of that feeling, he touched what was deep and true. In May 1939, he traveled to Liverpool University to receive an honorary degree. There, as young British faces looked up at him, he spoke of his own children and their futures. "I have a couple of boys, and two or three daughters, who think that they know what's wrong with the world," he told his British audience. "They are quite outspoken in their opinion of the way we old folks have been doing things. I shouldn't want them to know it but I must admit, just between us, that I can't blame them. . . . The important thing to remember is that the majority of our difficulties are man-made. . . . They are the result of human carelessness, human short-sightedness, human greed."

Joe's relationship with Roosevelt was by now so full of duplicity that endless scrubbing would not have washed away all the lies. In July, Joe's friend and journalistic promoter Arthur Krock wrote a column titled "Why Ambassador Kennedy Is Not Coming Home," in which he said that the president's militant young New Dealers had set out to destroy the ambassador with a series of lies. It was whispered that he had gone to bed with the British appeasers and was badmouthing the president. "None of these statements was true," Krock said. "But they were sedulously circulated for quite a while."

Roosevelt knew the source of this missile, but he was too shrewd to signal that he had spotted his attacker. Instead, in a letter to Joe he unloosed his venom on Krock. "He is, after all, only a social parasite whose surface support can be won by entertainment and flattery, but who in his heart is a cynic who has never felt warm affection for anybody—man or woman." Although he supposedly was attacking Krock, the president clearly had another target in mind.

In August 1939, the Soviets and Nazis signed a pact, making allies of the two great totalitarian powers. While Nazi troops surreptitiously massed at the Polish border, the American ambassador was off in southern France on what he considered a much-needed vacation. From Cannes, Joe wrote a letter to the president, replying in part to Roosevelt's letter. If ever there was a time for the ambassador to step back from all the tedious minutiae of the diplomatic life and impart all the precious insight he had gained during his tenure in London, that time was here. But he had no deep thoughts, no telling perspicacity, and no warning call of the events to follow.

Joe wrote from Cannes that "the chief thing I have noticed in the South of France, on the part of caddies, waiters and residents, is a very strong anti-Semitic feeling. Beyond that . . . I can contribute nothing to an understanding of the international state of affairs."

Caddies and waiters did not pop off with anti-Semitic cracks unless they felt the remarks would be well received. That aside, it was a curious state of affairs indeed when the ambassador to the Court of St. James's believed that his sole insight into European affairs was the mindless gossip that he picked up around the golf course and the dinner table.

As he swam at Hotel du Cap Eden Roc and dined on the Riviera, Joe was full of pathetic self-pity. "About my position in England my only thought was to wonder whether my experience and knowledge were not being completely wasted," he continued in his letter to the president. He nonetheless assured FDR that he would stay where he was asked to stay. He told Roosevelt that as a boy his father had "taught me two principles: gratitude and loyalty. He said 90% of people seem to forget favors and kindnesses done them. Of the second principle, he said, no matter how you may fail in ability you can make it up by being unfailingly loyal to your friends. I have tried to live up to those two principles and, to you personally, I owe a debt on both counts."

Joe's father had indeed acted on those principles, and he was a revered soul. Joe had honored his father's ideals by mouthing the words and acting the opposite. He had no deep gratitude toward the president, but a mounting anger at the man's refusal to heed his despairing messages. As for loyalty, he had talked so loudly behind the president's back that Roosevelt had long since heard his rude imprecations and taken his subordinate's measure. Roosevelt knew, however, that at this crucial time he had to keep Joe at his post; back home in the United States he would be an articulate critic and a natural leader of American isolationists.

During those long August days, Joe had opportunity to spend hours with his family, who had all convened at the villa Domaine de Ranguin near Cannes. His sons were the measure of the man, and he took pleasure in twenty-four-year-old Joe Jr., with his stories of the fall of Spain, and twenty-two-year-old Jack, a Harvard senior. At thirteen, Bobby was still too young to join in the spirited discussions on war and peace, and little Teddy was still little more than a family mascot.

During the day Rose took the family to Hotel du Cap Eden Roc to swim. High above the water stood a cliff that only the most daring made their diving platform. Joe Jr. and Jack dove off again and again while the rest of the Kennedys watched in awed admiration.

Then the brothers called upon seven-year-old Teddy to join them in diving thirty feet into the water. If Joe Jr. and Jack believed that Teddy would rush fearlessly up to that high perch, willing to do anything to impress them, they did not know their little brother. He walked up the stone steps, but he did it leaden of feet and spirit, heavy with fear. Most mothers would have risen and scolded their grown sons for putting their little brother up to such a foolhardy dare, but Rose sat quietly, believing this exercise a part of manly education.

Like his big brothers, Teddy dove into the water again and again. Years afterward he could remember the day as he did few things in his childhood. "I think it was Joe [Jr.] for the most part who encouraged me to jump off the rock," he recalled. "I could barely swim at that point. I'd jump in the water and grab hold of them. I did this several times. I think I was pretty scared."

If Rose even saw the fear that Teddy felt, she would have considered it something that had to be purged. She did see the bonding of her sons, and the trust that little Teddy showed in his brothers. "Joe [Jr.] took Teddy to the high cliff and Teddy dove down with confidence that Joe would be there to catch him when he went down or at least rescue him when he dove down from this tremendous height," she said. "I used to watch them with that relationship between the brothers."

On September 1, 1939, Germany invaded Poland with all the fires of hell. World War II had begun. Two days later Joe sat at 10 Downing Street with Chamberlain reading the speech that the prime minister was about to deliver to Parliament declaring war. Joe read the haunting, searingly honest words: "Everything that I have worked for, everything that I have hoped for, everything that I have believed in during my public life has crashed in ruins." Joe was an emotional man, and he cared for Chamberlain as a man and as a vehicle for ideas that were as much his own as the prime minister's. And as he sat there reading the words, his eyes filled with tears.

Back at the embassy Joe put in a call to Roosevelt, even though it was 4:00 A.M. in Washington. Joe began by telling the president the substance of

Chamberlain's address. Roosevelt had known that it would come to this one day, and if he was solemn at the thought of what faced him, he was full of resolve.

Joe, however, was almost hysterical. He was a man without hope. He portrayed a desperate, impotent Britain that soon would be overrun. In their fight against fascism, the democracies would come to resemble the very dictatorships that they fought. The world economies would crumble. Those with full stomachs now would go hungry, and those who were hungry now would starve. Joe told the president that the shadows of a new dark age were falling on Europe. Everywhere there would be chaos.

Roosevelt had incalculable weight on his shoulders, and matters more urgent to attend to than listening to the endless laments of a man who despised him. But listen he did, trying to calm an ambassador who should have been trying to calm the president. "It's the end of the world," Joe moaned, "the end of everything." For Joe, the world was his family, and his overwhelming fear was that he would lose his sons.

Joe had no intention of attending Chamberlain's speech, but Rose, Joe Jr., Jack, and Kathleen hurried out into the streets to Parliament to hear the prime minister's historic address. Joe's children did not share their father's brooding morbidity but presented a frolicsome air to a photographer who captured the impeccably dressed contingent.

They had been a gay presence in the house at 15 Prince's Gate. For months Joe Jr. and Jack had debated the issues of peace and war vigorously with their father. Kathleen had been full of her tales of life among the privileged, and the other children were almost as outspoken. Now they were leaving England, departing after a final weekend in the country.

America was neutral, but the flames of war burned without reason or selection. On September 3, a German torpedo sunk the *Athenia*, a British liner whose fourteen hundred passengers included three hundred Americans. Joe asked Jack to travel to Glasgow, Scotland, with his aide, Eddie Moore, to meet and comfort the surviving Americans.

It was no easy audience Jack faced that day. "You can't trust the goddamn German navy!" a survivor shouted. "We won't go home without a convoy!" Jack comforted the Americans with the savvy of a seasoned diplomat. He listened to their wrath and sorrow and fear. He felt that not only should a ship be sent directly to Glasgow to pick up the survivors, but also that it should be sent back to America as part of a convoy. "The natural shock of the people would make the trip to America alone unbearable because of the feeling that they will have that the United States exposed them to this unnecessarily," he noted in a memo. After serving as his father's surrogate, Jack flew back to

America from Foynes, Ireland, on the *Dixie-Clipper* to begin his senior year at Harvard.

After his family returned to America, Joe was alone, and for a man who cherished his family as he did, solitude was like a fog that rolled over his life and stayed there, covering everything with gray. The palette on which Joe painted his picture of the world had only dark colors on it now. His judgment ultimately was based on what political judgment always comes down to, not issues, but men themselves, their strengths and resolution, and he found the British a doomed race. His judgments were sweeping in magnitude. He told the president in a letter that "there are signs of decay, if not decadence, here, both in men and institutions . . . democracy as we now conceive it in the United States will not exist in France and England after the war, regardless of which side wins or loses."

Soon after the invasion of Poland, Joe had lunch with Winston Churchill. For years Churchill had been sounding the alarm against the rising forces of fascism, but his warnings had gone unheeded. His champions considered him eloquent and truthful, while his detractors condemned him as a verbose, exaggerating, drunken has-been. Now, with Hitler's invasion of Poland, Churchill had become a prophet with honor, and the likely wartime prime minister.

There was an eerie emptiness in the skies. The war against Britain had not yet begun in earnest. Churchill knew that one day the "phony war" would end and that he might look up and see the sky full of German planes flying to bomb English towns. "If the Germans bomb us into subjection," Churchill remarked to Joe, "one of their terms will certainly be that we hand them over our fleet. If we hand it over, their superiority over you becomes overwhelming and then your troubles will begin."

Joe considered Churchill's remark tantamount to blackmail. The old mountebank was daring Washington to be so stupid as not to allow Britain to buy war goods in America. As Joe saw it, Churchill was as bad as the Jews in his determination to suck America into the conflict.

"Maybe I do him an injustice," Joe wrote in his diary, though he was sure that he was right, "but I just don't trust him. He always impressed me that he was willing to blow up the American Embassy and say it was the Germans if it would get the United States in."

Hitler had possibly burned down the Reichstag in Berlin and blamed the disaster on the Communists. If Churchill was willing to do the same, then the world that Joe prophesied had already arrived. There was no difference between fascism and democracy. Joe did not see that, as desperately as Churchill wanted America to enter the war, there were moral limits on what he would do. He pleaded. He seduced. He manipulated. He threatened. But Churchill would simply not commit an evil act of his own, stage his own Reichstag fire, in order to seduce his erstwhile ally into entering the war.

Joe recalled that it was the day after that luncheon that the first of a series of secret letters from Roosevelt to Churchill arrived by diplomatic pouch at the embassy, the beginning of a historic correspondence. Joe resented the exchange, seeing it rightly as an attempt to bypass him. He had failed Roosevelt as his honest prolocutor, and the president was seeking out another channel.

It was unprecedented for the American president to be sending secret letters to the new first lord of the Admiralty, the most vociferous foe of Chamberlain's policies. It would have been judged foolhardy and provocative if the stakes had been any less. But Roosevelt and Churchill saw themselves and their nations allied in a great and noble struggle against Nazism, and they endeavored to push forward the tacit alliance in every way they could. The odds were terribly against them, as Joe had reminded the president ad nauseam. The two leaders, however, were not desperate gamblers throwing their last chip on the roulette wheel, but men who believed that, if they lost, civilization lost and life as the world had known it would be gone forever.

At its highest level, the level at which Churchill and Roosevelt operated, politics is not the art of the possible, but an alchemy that transforms the impossible into the possible. That is the political alchemy that the two leaders were practicing, alchemy that cynical, sullen Joe could not possibly grasp.

In March 1940, Joe led Sumner Welles, the undersecretary of State, to Churchill's office at the Admiralty for an afternoon meeting. Churchill sat in a large chair in front of the fireplace, reading a paper and smoking one of his famous cigars with a drink at his side. Joe noted these details in his unpublished memoirs as if they were proof of the man's decadence.

During Welles's visit, Joe stopped the British officials cold when they started spouting what he considered sentimental garbage about their noble purpose. "For Christ's sakes," he exclaimed, "stop trying to make this a holy war, because no one will believe you; you're fighting for your life as an Empire, and that's good enough." Joe, however, had not reckoned with Churchill, who rose up and spoke in words charged with passion. He described the Nazis as a "monster born of hatred and of fear," but to Joe such eloquence was at best little more than gaudy gift-wrapping. That Churchill used it on his two auditors suggested his further chicanery; he did not see the two American diplomats as worthy of simpler words. Joe did not see that eloquence was one of Churchill's strongest weapons, one he used wherever and whenever he thought it might forward his cause.

"What we have lost on balance is not significant," Churchill said, looking at a chart of naval losses. Welles suggested the obvious, that the war had not truly begun and that when it did the losses would be enormous and the world economy would suffer horrendously.

"I am not sure of that," Churchill said. "The last war did not bring about conditions of that type. In fact, the standard of living all over the world has been bettered since then."

Joe had had quite enough of this dissembling British politician who masqueraded as a statesman. "That is taking a short view of it," he interrupted, giving Churchill a dose of what he considered unvarnished, ineloquent truth. "The well that the water has been taken out of has become drier and drier. This war is the climax of your process of 'raising standards.' To keep on doing it, even to maintain the standards that have been reached, seems to have made it necessary now to go out and tap other people's wells. That kind of an economic structure is nothing to brag about."

Churchill went on as if Joe had never spoken, rising again to eloquent heights as he declared that Nazi Germany had to be destroyed once and forever. The gulf between the two men was as wide as the ocean that separated their countries. If Joe listened, he did not hear, and if he heard, he did not believe.

Later, when Churchill had heard more tales of Joe's defeatist diatribes, he said: "Supposing, as I do not for one moment suppose, that Mr. Kennedy were correct in his tragic utterance, then I for one would willingly lay down my life in combat rather than, in fear of defeat, surrender to the menaces of these most sinister men." Joe, for his part, had the cowardice of his lack of conviction. He was unwilling to have himself or his sons die in what he considered a foolish, futile fight.

Back at Harvard in the fall, Jack worked diligently on his honors' thesis. His whole intellectual life these last two years had prepared him for analyzing why England had woken up so belatedly to the Nazi threat. In exploring the subject, he risked confronting his own father and the views that Joe had professed.

No one watched Jack's progress with more interest than Joe Jr., who was now studying at Harvard Law School. "Jack rushed madly around the last week with his thesis and finally with the aid of five stenographers the last day got it [in] under the wire," Joe Jr. wrote his father.

The 151-page essay was the most sustained intellectual effort of Jack's life. As he saw it, the fault for Britain's belated rearmament lay largely not in its leaders but in the nature of democracy. The naturally self-indulgent British democracy could not be awakened by a few shouts but only by the thunderous barrage of cannon fire. Jack shared his father's profound pessimism about democracy, believing that only a totalitarian state could defeat a totalitarian foe.

"The [British] nation had failed to realize that if it hoped to successfully compete with a dictatorship on an even plane [*sic*], it would have to renounce temporarily its democratic privileges," Jack wrote. "It meant actual totalitarianism, because after all, the essence of a totalitarian state is that the national purpose will not permit group interest to interfere with its fulfillment."

Jack had some diversions from his academic work that left him with another health problem that would plague him the rest of his life. That college year he suffered from a condition described as "nonspecific urethritis," which was treated at the Lahey Clinic with "local urethral treatment and sulfonamides." The disease was a swelling and inflammation of the urethra, the duct that drains the bladder. "Nonspecific" is the medical way of saying that the exact cause may not be known, though the condition was usually the result of unprotected sexual intercourse.

The doctors most likely would have made a different notation in Jack's medical record if the condition had derived from something other than sex. It was, after all, an age of sexual euphemism. Harvard's course on general hygiene, mandatory until 1935 and known by the cognoscenti as "Smut I," covered the hygiene primarily of what many called "the dirty parts." Men who ventured downtown to indulge with prostitutes were "resort[ing] to less polite alternatives."

Jack had good reason to cover up his disease. It was widely rumored—inaccurately, it turned out—that anyone so daring as to apply for a Wassermann syphilis test was abruptly segregated from the rest of his college mates. The Harvard Hygiene Department estimated that far less than 1 percent of the student body was infected with venereal disease each year, suggesting that, in his sexual conduct, Jack was an aberration.

Jack's illness did not prevent him from finishing his senior thesis. His essay had all the markings of a student term paper, from its twisted syntax to its pages peppered with typos, but it was a provocative analysis, and it won him a cum laude citation, the lowest of the three honor grades. When Arthur Krock read the manuscript, he pronounced that it could be published as a book and offered to help. Jack's father not only concurred but sent his son a thoughtful seven-page letter suggesting that he might have "gone too far in absolving the leaders of the National Government [Chamberlain and his cabinet] from responsibility for the state in which England found herself at Munich."

Although Joe did not say it so directly, if Jack criticized Chamberlain he would in effect be criticizing his own father. Joe didn't care. It was a mark of Joe's love for Jack that he wanted his son to have his own ideas even if they conflicted with his father's, as long as those ideas would lead him to a powerful, privileged position in American life. As it stood, however, Jack was hardly suggesting that Chamberlain and men like him had helped create a mentality in which the British people were comatose from fear or apprehension at the awesome power of the Nazi war machine.

Jack had the kind of help in preparing his book manuscript that few authors receive. His Harvard friend Blair Clark, whose journalistic skills were such that he was the editor of the *Harvard Crimson*, says that he and Jack worked together rewriting two of the chapters. Arthur Krock applied his skilled editorial hand to the manuscript. Joe bought hundreds of copies of his son's book and was largely responsible for creating the impression that *Why England Slept* was a popular success. He wrote Churchill that "it is already a best-seller in the nonfiction group," though in fact it made no best-seller lists. In the end the book sold about twelve thousand copies, a worthy total for a serious, often turgid political book by a twenty-three-year-old first-time author, but hardly the best-seller of legend. As Joe saw the world, image was everything, and he had created this idea that his son was an immensely successful young author.

At Harvard in the spring of his senior year, Jack learned that one of his oldest friends, Bill Sweatt, had died of a mysterious ailment while on a trip to South America. Jack and Bill had been boys together in the winter in Palm Beach. Both had gone off to Choate. Bill was the kind of stellar youth admired not only by the headmaster and his toadies but even by self-conscious miscreants like Jack and his friends in the notorious Muckers Club.

Jack had known nothing of death except what he read about it in books and poems. He was stunned by the arbitrariness and randomness of his friend's demise. Why Bill? Why now? Unlike Jack, who sorted out human beings like a barrel of mainly rotten fruit, Bill had seemed to get along with everyone. On one occasion, when Bill had spoken negatively of a fellow student, Jack's father had commented: "Well, if Bill Sweatt doesn't like him, there is something wrong with him."

On his Spee Club stationery Jack wrote a letter unlike anything he had written before. In seeking to comfort his friend's mother, his words were gracious, restrained, deep, and truthful. All his life Jack had disdained what he considered the silly rituals of a gentleman's life, but now he heralded young Bill as a model of what a gentleman should be.

Jack's attitude toward the rituals of upper-class manhood may have changed during his time in London, after reading about Melbourne, or simply because he had grown more comfortable with himself. Jack expressed his ideals of how a man of his class should behave, an ideal very far from how he had lived most of his two decades of life.

> At Choate, I remember he came in 2nd in the class votes as the most gentlemanly and those who knew him voted for him with the feeling that the word "gentleman" in its fullest and broadest sense—in the sense of having great consideration for other people and what they felt—that that best described Bill. I think we all gained something from knowing him and for that reason, I think he probably gave and got far more out of his life than many others who lived to a greater age.

The house at Prince's Gate was shrouded in blackout curtains no darker than Joe's own mood. Much of the time he spent at St. Leonard's, the huge estate in the countryside that he had taken in the event of the anticipated German bombing of London. He was losing weight. His hair was turning gray. His stomach was so sensitive that he had a special diet at the Mayfair restaurants where he often supped alone. For a while he was taking belladonna to get to sleep. He was no longer an honored guest at the great houses of London. He was alone and he was depressed, and that bleak mood fell over everything he thought and said and wrote.

Joe believed that those who had so recently wooed him ostracized him now because he continued to shout the truth. This banishment was severe not simply for what he said but how he said it.

In March of 1940, Raimund von Hofmannsthal, a socially prominent Austrian writer, sent a memo to Clare Boothe Luce titled "Joseph Kennedy and Diplomatic Corps." The Luce publications portrayed the ambassador to the Court of St. James's as a dedicated, accomplished public servant, but in private memos a different Joseph P. Kennedy emerged. Von Hofmannsthal pointed out that Joe was far from the only neutralist ambassador in London. The writer suggested that it was not Joe's views but the way he lived that so offended the British. He spent so long away in the United States, sent his family home, and lived in the British countryside proving "that he lacks the solidarity towards English which is expected of an ambassador."

"People particularly resent the fact that he has been a popular figure in London and court life, that they did overlook his childish Prairie County, Ohio, mannerisms and now feel that if they had been more severe with him from the beginning, he would not have let them down," Von Hofmannsthal wrote. "When he had the king and queen to dinner, he had had the awesome nerve to have photographers there so that the evening would be well publicized, a rudeness to the extreme." Joe's worst faux pas, as von Hofmannsthal related it, was at the last Court Ball, where Joe, the Connecticut Yankee at King Arthur's Court, had walked blithely up to Queen Elizabeth and asked her to dance. "Actually the incident was only known to a few people and to this day has not got into the press," von Hofmannsthal noted.

"His behavior as ambassador was outrageous," said Henry Luce, the *Time* publisher. "It was outrageous because he said that England was bound to get beaten. Oh, he had a lot of courage. But there he was sitting there in the middle of the blitz phoning to me on the open transatlantic phone saying the jig was up for England. You just don't do that kind of thing. The British never forgave him for that."

Joe was roundly detested not because of his views, which were not that unknown in Britain and not uncommon in America, but because of the way he professed them. He seemed to take pleasure in his proclamations of doom. His defeatism was dangerous to the British because it was like an infectious disease that he was attempting to spread. They saw him as peppering the air with germs of despair that could become a plague. London was a city of spies, and the British Foreign Office was deluged with reports of Joe's comments on the weakness of British manhood and the impossibility that the island nation could stand up to the onslaught of German steel and might.

Some observers attributed Joe's stream of comments to ineptness, cowardice, or sheer bullheaded Americanism. Victor Perowne of the British Foreign Office astutely concluded "that Mr. Kennedy's perpetual 'spilling' of these views is not out of naiveté, but very much on purpose and the effect on . . . our interests is regrettable." The fear was not that Joe's words would affect British morale. Instead, the British Foreign Office worried that Joe's words would do the most damage in the smaller neutral countries in Europe and in the United States.

In February 1940, while he was in America, Joe sent a telegram to the American embassy in London asking staff to RUSH PACIFIST LITERATURE. To the British, who intercepted the message, it was a further example of what seemed treacherous behavior. In reality, Joe was only trying to help Jack research his Harvard thesis, but the climate of duplicity was such that the British suspected the worst.

The Foreign Office had information that Joe was "quite unpopular with his own staff and the American press correspondents here," whose distaste for him was so immense that "they wax indignant at the mere mention of his name." Joe's belief that the war would radically diminish Britain's place in the world and bring an end to the British Empire was prophetic. It was nonetheless a businessman's view of the world, in which nothing mattered but the tallies of economic power. He was, as T. North Whitehead wrote in a Foreign Office note, "playing off his own lot in his stupid private conversations and uncalled for remarks to the press."

Joe's obsession with economics as the fundamental bedrock of foreign policy was, if anything, accentuated by the onslaught of war. America's greatest problem with the British, he told Jay Pierrepont Moffat, the European bureau chief at the State Department, was that they might liquidate their American securities by dumping them on the market. So that they would not "treat us rough," Joe said, America had to "keep them dangling."

"It is always difficult for an ambassador to get below the surface of London politics and London society," Harold Nicolson wrote in *The Spectator* on March 8, 1940, as Joe returned to London. "If he could go, heavily disguised, to Leicester sometimes, and sometimes to Glasgow, he would realize that Great Britain, although a difficult proposition, is also extremely tough."

Even in the salons of Mayfair and the country homes of lords and ladies that he still did frequent, Joe heard passionate, determined voices. But he considered mere determination a fool's capital. Morale did not matter, only matériel. After meeting with Joe, British Foreign Secretary Lord Halifax wrote that, "as he [the American ambassador] saw it, the winning of the war had little to do with changes of Government, or accusations or complacency or lack of drive, it was simply a question of whether one had enough aeroplanes."

Everywhere Joe looked he saw evidence of the truth of his views. The seemingly invincible armies of the Third Reich moved inexorably forward into Denmark and across Belgium and Holland, through France, to the English Channel. And now the sound of the German planes droned high above London as they dropped their bombs. Spreading death and destruction below, the German pilots looked down on what they called "an ocean of flame."

The British intercepted a number of triple-priority letters from Joe to the secretary of State, missives that were devoid of even a hint of awareness that the British were fighting implacable evil. As Joe saw it, the British could not possibly hold out alone; unless the United States intervened, they were finished. Joe believed that although the Conservative government, headed by the new prime minister, Winston Churchill, might have been the putative leaders, the Socialists were "running the government." On the surface, the British may have been standing up to the Nazi air raids, but there was an "undercurrent of ill-will."

If Roosevelt had depended on Joe for his knowledge of Britain, he would never have proposed "lend-lease"—essentially lending American ships and planes to a hostage island on the verge of defeat or starvation. It was true, as Joe wrote Washington, that there were poor people bemoaning their fate, grumbling, "How can we be worse off than we are today? After all Hitler gives his people security." But for every bloke in a factory bitching about his fate, there were ten stalwarts working double shifts, cleaning up rubble, shaking their fists at the steel that rained from the sky.

It was true, as Joe told the secretary of State, that if things got tougher some members of the upper class might decide they had had quite enough of Mr. Churchill's war. But for every blue-blooded defeatist there were ten young men flying their Supermarine Spitfires and Hawker Hurricanes against the Nazis in the skies above England.

Joe had not listened to the voices of Britain. Nor had he looked deeply into English faces. Joe observed the worst in men, and he considered it their true value. He heard doubts, and he thought it was defeatism. He sensed fear and mistook it for cowardice. He called himself a realist, but he mocked what was heroic and noble and selfless.

Joe's emotional state tainted everything he said and did. He told a story of a little church in the town of Horta, which was holding a vesper service. Across the street stood a crowd of British men. The men said that there was no point going across the street to church. A man could no longer believe in God. They were disillusioned. They were without hope.

Those men may have stood there that evening, but it was Joe who had no faith. In his world of endless grays, Joe could not see evil, and thus he could not see good either. "I am depressed beyond words," he wrote Krock. Joe was a man who might well have been diagnosed as clinically depressed. On Joe Jr.'s twenty-fifth birthday, he wrote his son a sad letter saying that, though he felt he would survive, he had confidence that if necessary, Joe Jr. could "run the show."

On August 1, Roosevelt called his ambassador to the Court of St. James's. Afterward Joe dictated his recollection of the conversation. "I wanted to ring you up about the situation that has arisen here so that you could get the dope straight from me and not from somebody else," Joe remembered the president saying. "The sub-committee of the Democratic Committee desire you to come home and run the Democratic campaign this year, but the State Department is very much against your leaving England."

"Well, that's all very good," Joe remembered replying, "but nevertheless I am not at all satisfied with what I am doing and I will look at it for another month and then see what my plans are. I am damn sorry for your sake that you had to be a candidate, but I am glad for the country."

Nary an honest word had passed between the two men. The president preferred to keep his troublemaker away from America until after the election, while Joe feared what a third term for FDR would mean. Joe believed that Roosevelt, Churchill, the Jews, and their allies would manipulate America into approaching Armageddon. He had begun to press his case not only with words but also with what could only be perceived as threats.

"For the United States to come in and sign a blank check for all the difficulties that are faced here is a responsibility that only God could shoulder unless the American public knows what the real conditions of this battle are," he wrote Welles on September 11, 1940, leaving unsaid who he considered the best-qualified person to inform the Americans. "I can quite appreciate their desire to help this country fight this battle but they should have a very clear notion for what the responsibility will entail for the American people to take up a struggle that looks rather hopeful on the surface but is definitely bad underneath."

Joe found ample time to write his children from London. He took special care in his letters to little Teddy. Joe was aware of the petty dissembling of his youngest son. He was not going to confront his son, but told him, "I certainly don't get all of those letters you keep telling me you write to me."

Little Teddy had been shuttled around from school to school, and from home to home. He did not have the strong hand of his father pointing him down the pathway that all the Kennedy sons must tread. Instead, he had these letters from Joe in which his father placed himself at the epicenter of danger.

"I am sure, of course, you wouldn't be scared, but if you heard all those guns firing every night and the bombs bursting you might get a little fidgety," Joe wrote. "It is really terrible to think about, and all those poor women and children and homeless people down in the East End of London all seeing their places destroyed."

All eight-year-olds are literalists. Poor Teddy couldn't know how much his father was exaggerating, and that Joe was safely in the country, belittled by Londoners as a coward. He must surely have feared that his father might never return. "I know you will be glad to hear that all these little English boys your age are standing up to this bombing in great great shape. They are all training to be great sports."

Life was a merciless competition, and even here Teddy was being compared to others; he too was supposed to be a "great sport." His father concluded: "Well, old boy, write me some letters and I want you to know that I miss seeing you a lot, for after all, you are my pal, aren't you?" Teddy wasn't his pal at all, for Joe was never a pal to his sons, never a comrade.

In his time alone, Joe had apparently sought solace in the arms of Clare Boothe Luce, the brilliantly acerbic playwright and journalist. Clare combined the coquettish skills of a courtesan with an ambition for power and influence the equal of any man's. That she and Henry, her husband, no longer had a sex life together hardly appeared to shake their marriage, for the games of power that the couple enjoyed playing took place outside the boudoir.

When Joe met her in Paris in April 1940, using the excuse that he had traveled across the Channel to visit a sick Eddie Moore, he spent much of his time with Clare at the Ritz Hotel. Clare was a bold writer who rarely employed euphemisms, but this time she noted coyly in her diary that Joe had been "in bedroom all morning."

Clare was everything Rose was not: a daring, passionate woman who after dinner stayed with the men in the salons of power instead of demurely rising and taking her coffee and brandy with the ladies in another room. Her pillow talk was not simply the cooing words of love but a bold dialogue on great events.

Clare was a woman of hard reactionary views who trumped even Joe in her disdain for the lower classes and the Jews and what she considered the mongrel races of the world. She shared with Joe the belief that America had better quickly rearm, turning itself into a fortress that would be impregnable to the onslaughts of war.

She had a gift for self-dramatization that was a journalist's common failing, but here Joe bested her. He wrote her on October 1, 1940: "Yesterday a Messerschmitt just missed the house as it crashed. I could see the pilot's face, his head lolling over one side . . . headed straight for the ground. . . . I imagine it will take a long time to get the drone of German motors out of my ears after I get back; and not to hear gunfire nine or ten hours a night will make me rather lonesome for the battlefront. When somebody asks me what I did in the second war I'll say I lived in London, and that's a damn sight worse than anything else I can think of, unless it is Dunkerque."

Beneath the boyish bravado was a shameless quality to Joe's dissembling. Most of the time he was living well outside London, and he had suffered nothing compared to those who nightly weathered the Nazi bombs.

With Clare and Henry, Joe had begun what Roosevelt could only believe was a treacherous dialogue with his enemies. The ambassador was in frequent contact with the Luces, who were backing the Republican candidate, Wendell Willkie, in his race against Roosevelt. Willkie campaigned on the theme that Roosevelt was lying in his pledge to keep America out of war. If he was reelected for a third term, American sons would soon be dying on foreign shores.

Joe agreed largely with this thesis and proffered the possibility that he would return to the United States to endorse Willkie, a gesture that, as he wired Luce, would produce "25 million Catholic votes," enough "to throw Roosevelt out."

Joe may have been exaggerating the numbers, but his was not an idle boast. Many of his fellow American Catholic voters were reluctant tenants in the house of the New Deal, and as the election neared they seemed to be moving away in droves. Joe was the most powerful Catholic in the administration, and if he left the New Deal in dramatic fashion, the road behind him would be full of Catholics leaving the Roosevelt camp.

Felix Frankfurter, a Supreme Court justice and one of the president's closest advisers, saw that Joe could be a force of great and crucial good if he would return to America and give an impassioned radio address confirming his support for Roosevelt. That would not only stop the flow of Catholics away from Roosevelt but also bring enough of them back to ensure a third term for Roosevelt.

In the United States his own son, Joe Jr., was playing an active role as a youthful isolationist. At Harvard Law School, Joe Jr. became his father's proud surrogate, one of the leaders of the Harvard Committee Against Military Intervention. Some of his opponents considered "isolationist" little more than another name for cowardly expediency, but men of principle espoused this cause as well, and Joe Jr. was no coward physically or intellectually.

Joe Jr. took on the internationalists on their turf, talking at Temple Ohabei Shalom in Brookline among the very people whom his father accused of willfully manipulating America into war. He debated Harvard professors in whose classes he had only recently studied, giving them not one iota of deference in his attacks on their positions.

Joe Jr. was a more vociferous isolationist even than his father. He opposed U.S. lend-lease aid to Britain, calling it a prelude to the sending of men and the inevitable entry of America into the war. "I urge you to consider . . . that convoys mean war. I support all aid to England but we must not throw away our greatest asset, our hemispherical position. . . . We will only sacrifice everything by going in," he said.

Joe Jr. took the politically daring step of supporting James Farley as the Democratic candidate for president, in opposition to Roosevelt's third term. Joe Jr. had gone as a delegate to the Democratic National Convention in

1940, and even knowing that Farley could not win, Joe Jr. insisted on voting for him on the first ballot.

Joe Jr. was speaking out against intervention, but his was a voice that could hardly be amplified. Joe Sr., for his part, knew that the moment had arrived when he might stand at the epicenter of history. Henry Luce wired him that he should return to the United States to tell the truth about Roosevelt's war plans, the "ordinary antiquated rules" of loyalty and protocol be damned. Clare Luce sent another cable: WHEN YOU LAND TELL THE PRESS AND THE PEOPLE THE TRUTH AS YOU HAVE ALWAYS TOLD IT TO ME.

In one moment Joe would pay back all the rebuke, the humiliation, the misunderstanding that he felt he had suffered. Roosevelt had word of Joe's plans, and he raised the ante by refusing Joe's request to return to the United States for consultation. Joe decided that if he could not come back in person, he would send his words back instead. He wrote a devastating article that he called "an indictment of President Roosevelt's administration for having talked a lot and done very little," and he vowed that if he was not called home, he would publish his article the week before the election.

Joe had simply had it with London. In early October, he told Lord Halifax that he intended to give up his ambassadorship, going out with a bang, not a whimper. Joe confided to the British diplomat that he had sent his article attacking Roosevelt to the United States, where it was scheduled for publication just before the presidential election.

Roosevelt understood that in a close election the whole future of his administration and his alliance with Churchill and the British might stand or fall on the actions of one man, and he reluctantly agreed to call Joe back to Washington. Roosevelt prepared for the meeting with all the staging of a great director. The newspapers were full of rumors about Joe resigning or coming out for Willkie. Knowing Joe's brash propensity for mirthless candor, Roosevelt knew that he might well make some intemperate remarks to the scribes who waited at the airport in New York.

Roosevelt stipulated that when Joe's plane landed, he was "not to make any statement to the press on your way over nor when you arrive in New York until you and I have had a chance to agree upon what should be said. Please come straight through to Washington on your arrival." That would give Joe no time to meet with the president's opponents, and no opportunity to stoke the fires of his fury even higher. The president insisted that Rose be invited as well, a brilliant and crucial part of his strategy.

Rose's years as an ambassador's wife had been the happiest moments of her public life. As the plane flew south to the capital, Rose pleaded with her husband not to resign. "The president sent you, a Roman Catholic, as ambassador to London, which probably no other president would have done," she argued. "You would write yourself down as an ingrate in the view of many people if you resign now." Rose went on to tell Joe that he risked hurting not only himself but also his own sons and their political futures. If Joe's remarks won the election for Willkie, Joe would have his moment of revenge, but the retribution would be meted out not on him but on his sons and their careers. As long as the Democrats held forth their banner, Kennedy would be a traitor's name.

Joe listened to his wife's counsel and afterward admitted that she had "softened" him up. He did not care about what he considered the dubious honor of the ambassadorship any longer; most likely what rang deepest and truest to him were his sons' possible fates. By his actions in London he was hoping to save their lives, but he was not doing so in order to destroy their futures.

Roosevelt had alerted his secretary, Grace Tully, to "be sure and butter up Joe when you see him" before she showed him into the private quarters. Joe, who was planning a bold, merciless confrontation with the president, found Roosevelt seated over a cocktail shaker mixing drinks for his close friend Senator James F. Byrnes of South Carolina and Mrs. Byrnes.

Over a Sunday dinner of scrambled eggs and sausages, Joe was at his civil best, relating tales of life in beleaguered London. "I've got a great idea, Joe!" Byrnes said, as if a glowing lightbulb had appeared above his head. "Why don't you make a radio speech on the lines of what you have said here tonight and urge the president's reelection?"

At the level that Roosevelt was playing politics, and for the stakes that were on the table, not a moment of this evening was unscripted. Joe was not for a moment taken in by Byrnes "acting as though a wonderful idea had just struck him." Roosevelt, for his part, surely knew that Joe was not fooled and realized also that his angry ambassador would not dare call the president's bluff. They were like two men sitting across a game board from each other, but while Joe was playing checkers, Roosevelt was playing a master's game of chess.

Joe did not respond to the senator's "great idea" but sat and fumed quietly. Roosevelt had already felt Joe's anger in other meetings, and he had staged this modest dinner in part so that with other guests present, his ambassador would not dare show his venom. Despite the staging, Joe was not about to spend the evening in chitchat and meaningless pleasantries.

"Since it doesn't seem possible for me to see the president alone, I guess I'll just have to say what I am going to say in front of everybody," Kennedy said suddenly. As Joe went through the litany of abuses he felt he had suffered, Rose noticed that Roosevelt's eyes snapped nervously, the only sign of emotion the president allowed himself.

"I am damn sore at the way I have been treated," Joe went on, like a prosecutor making his final arguments. Joe was not so bold as to attack the president, whom he considered the architect of his abuse. Instead, Joe trashed State Department officials, such as Sumner Welles, who had bypassed their ambassador, humiliating him. Welles and his subordinates had only been the honest messengers of Roosevelt's policy, but Joe berated them in fiery assault.

Roosevelt was not interested in speaking the truth now, but only in placating this enraged and dangerous man. So the president started attacking the State Department with ferocity even greater than Joe's. After the election the president would have "a real housecleaning," throwing out these officials who had so wronged the ambassador to the Court of St. James's. Joe would suffer no longer. There were many words of untruth in Roosevelt's harangue, but his words did what they were supposed to do. They calmed Joe down and made him and Roosevelt momentary allies against their common foe.

Joe was an angry, cynical man who still might lash out in vitriol against Roosevelt and his third term. His greatest vulnerability lay not in the many things he hated but in the few things he loved. The man loved his sons, and it was of his sons that Roosevelt spoke now.

"I stand in awe of your relationship with your children," Roosevelt said in his first words of the evening that rang with some semblance of truth. "For a man as busy as you are, it is a rare achievement. And I for one will do all I can to help you if your boys should ever run for political office."

A promise is sometimes only another name for a threat. According to Roosevelt's son James, the president went on to say that if Joe decided to throw his lot in with Willkie now and abandon the president, he would be an outcast and his sons' political careers would be destroyed before they began.

Years later Joe told Clare Luce that Roosevelt offered him an irresistible deal that evening: if Joe would endorse Roosevelt in 1940, "then he would support my son Joe for governor of Massachusetts in 1942." Even if Roosevelt did not make such an explicit offer, the implications were clear that if Joe cared about his sons' futures, he had best be quiet. A year and a half later Jack declared in a conversation that "his father's greatest mistake was not talking enough; that he stopped too quickly and was accused of being an appeaser. He stated that the reason his father stopped talking and didn't go on and present his side of the question fully was due to the fact that he believed it might hurt his two sons later in politics."

The evening had not changed Joe's belief that Roosevelt was slowly manipulating the country into war. Joe could have walked out of the White

House that evening, flown back to New York to meet the Luces, and cast his lot with Willkie. If the Republican won, Joe would be the man who had dared to stand up and say what had to be said.

Roosevelt had just promised that he would get rid of Joe's enemies, and if momentarily Joe had believed him, upon reflection he was too shrewd to think that the president would change. But his love for his sons outweighed even his own ambition for power and position. He was not going to give them a tainted name as their inheritance or hobble them in the race of life.

What he was about to do was as noble and selfless as anything he would do in his life. His sons were not unaware of the sacrifice their father made that evening. Afterward Jack discussed his father with his friend Torby Macdonald. Torby wrote Jack that few people realized that "self-success" was more important than "worldly success," and that Jack's father was one of the few who had done so by "putting family over Ambassadorship."

"All right," Joe finally said that night, giving in to Roosevelt's request that he make a radio speech. "I will. But I will pay for it myself, show it to nobody in advance, and say what I wish."

Joe spoke Tuesday evening on CBS Radio like the bearer of truth from the bloody fields of war. For months he had been saying privately that Roosevelt was leading America into war; now he stated that "such a charge is false."

Joe knew that many of those listening across the nation had heard of his disagreements with Roosevelt, and he did not deny them but asked how many employees agreed completely with their employer. "In my years of service for the government, both at home and abroad, I have sought to have honest judgment as my goal," he said. "After all, I have a great stake in this country. My wife and I have given nine hostages to fortune. Our children and your children are more important than anything else in the world."

Joe had touched the deepest chord within his own life, a chord that resonated in the lives of most Americans. His speech was a triumph, lauded in the press and applauded by Democratic politicians. In the end Roosevelt won in an electoral landslide, and Joe's speech was not the seminal event that it would have seemed in a close election. Joe, however, in one great public moment, had proved his fealty to Roosevelt. He had good reason to believe that his sons would be rewarded for their father's loyalty.

Roosevelt saluted him, but the Luces and the Willkie forces considered Joe a betrayer. "There was that radio address when everyone thought he was going to come out for Willie," Henry Luce recalled. "We thought he was and Clare tried to get hold of him and he wouldn't answer the phone. We were appalled when he came out for FDR as the man to keep us out of war."

On Saturday after the election, Joe's secretary showed Louis Lyons of the *Boston Globe* and two reporters from the *St. Louis Post-Dispatch* into his suite at the Ritz-Carlton Hotel in Boston. Joe was in suspenders, eating apple pie, having a casual morning on a casual day.

In London, Joe had become used to calling the press into his office and spewing out his most intemperate attack on his enemies, knowing that the scribes would cut his remarks down to fit into the narrow confines of acceptable discourse.

Joe believed his remarks this morning were not to be printed, but it was a mad gamble to talk as he did to journalists whom he did not know intimately, betraying all the glorious rhetoric of the radio speech he had just given. He said nothing that he had not said a hundred times before, that "democracy is all finished in England" and "it may be here" as well. He prophesied that if America entered the war, "everything we hold dear would be gone."

Joe had learned little in London about the necessary parameters of diplomacy, and he took unseemly pleasure in speaking the unspeakable. He laced his diatribe with wildly inappropriate comments on everyone from Queen Elizabeth ("more brains than the Cabinet") to Eleanor Roosevelt ("She bothered us more on our jobs in Washington to take care of the poor little nobodies who hadn't any influence than all the rest of the people down there altogether").

The story that appeared the next day, November 10, 1940, in the *Boston Globe* finished off whatever effectiveness Joe still had in London and destroyed whatever residue of trust Roosevelt might still have had in his ambassador to the Court of St. James's. Joe submitted a letter of resignation and retreated to Palm Beach.

He did not sit in Florida playing and replaying the scenes of his life in London, trying to pick out where he had gone wrong. Devoid of self-criticism, Joe was even more convinced of the correctness of his views, and more determined to struggle against what he considered his and the nation's implacable enemy, the internationalists who were wooing America toward the crimson fields of war.

10

Child of Fortune, Child of Fate

Joe had tied his two eldest sons to himself so tightly and with so many entanglements that to be men they had to sever the bonds to break free. Yet in doing so they risked falling into an abyss. Success was the tightest knot that bound them. Few fathers did as much for their sons as Joe did for his, and few fathers demanded more of his children.

Joe inscribed their names on a contract that they had not seen. He gave them privilege, opportunity, and wealth. In return, they would have to be among the great men of their time. They could not have inheritors' pale lives, measuring out their days in games and chitchat at their clubs, envied for their lives of pleasure and revered for nothing except their good fortune.

Joe had marked off the high road to manhood that his sons must travel. Joe Jr. marched proudly up the middle, far ahead of his younger brother. When Joe Jr. looked back and saw Jack coming up behind him, it irritated him. Jack's thesis "seemed to represent a lot of work but did not prove anything," Joe Jr. jealously wrote his father when he first read it. Despite what his brother might say, Jack kept walking ahead, occasionally wandering off onto strange byways, but always returning to continue up the same arduous pathway.

Jack, not Joe Jr., had written *Why England Slept,* winning the first major laurels of adult life. Yet he did not have the imposing dignity and presence that Joe Jr. could don like evening clothes and shuck just as easily. Jack was twenty-three years old, though he looked far younger. He had the same sloppy nonchalance about his wardrobe that he had about the world. His boyish insouciance may have been irresistible to women but hardly marked him as a future leader.

At times Jack talked about becoming a journalist. He could have parlayed his book and its stellar reviews into a position that would have been the envy of his friends at the *Harvard Crimson.* He thought about law, but that was a tedious regimen at best, and he could not face the prospect, especially with his star-crossed health. Instead, after graduating from Harvard, he decided in the fall to go out to northern California to study at Stanford University. He would be a nondegree student, able to pick his way through whatever courses took his fancy.

In late September 1940, Jack took a small apartment at The Cottage, a modest complex popular among graduate students. "Still can't get used to the Co-eds," he wrote Lem, "but am . . . taking it very slow as do not want to be known as the beast of the East." That aspect of Stanford had its unique appeal to the handsome heir. It was a tonic watching the book-toting, chattering coeds hurry across the quad in their obligatory silk stockings.

Jack may have been unformed in other parts of his life, but he had established his adult sex life. His bad back required him to sleep on a bed board. That was an ideal excuse to have women do what he wanted them to do.

"Because of his back he preferred making love with the girl on top," recalled Susan Imhoff, one of the first coeds to make a visit to Jack's room at The Cottage. "He found it more stimulating to have the girl do all the work. I remember he didn't enjoy cuddling after making love, but he did like to talk and had a wonderful sense of humor—he loved to laugh."

Jack had a taste for gloriously attractive, smart women, but as eagerly as they entered into their affairs with him, they usually left with a sense of disquiet. They may have had other affairs that did not end well, but there was something deeply unsettling about Jack. He swept down on women, wooing them with his charm and wit, and then flew off again, never having been touched, leaving only a whiff of emotion.

Jack was no crude predator who lured women to his room, but a sly, sophisticated gamesman who seduced by seeming not to seduce. Once a woman succumbed, he quickly and efficiently disposed of the matter. As much as he might pretend otherwise, the fact that a woman slept with him showed that she was no better than all the others. "I'm not interested—once I get a woman," he told his Stanford friend Henry James. "I like the conquest. That's the challenge. I like the contest between male and female—that's what I like. It's the chase I like—not the kill!"

Jack pursued one coed, Harriet "Flip" Price, more than any of the others. Harriet was beautiful, intelligent, wellborn, and athletic. For all his desire to score yet another conquest, Jack was not the sort to promise eternal fidelity or to vow that when he looked into Harriet's eyes he heard wedding bells. They laughed and joked together and had sweet good times riding around in Jack's new Buick convertible.

As much as Jack wooed her, and as much as she felt herself "wildly in love," Harriet would not sleep with him. Virginity was part of her capital. She was not going to exchange it for anything less than marriage. Beneath it all, they shared the belief that marriage was too serious a business to leave solely to the whims of romance. "I think Jack knew what he was doing all the time," Harriet recalled. "And I think he knew exactly what kind of woman he wanted to marry, and did exactly what he set out to do." And so, indeed, did Harriet.

Jack hid from Harriet how troubled he was about his health, though he could hardly disguise the fact that after an hour of driving his back hurt so much that he had to stop for a while. Jack was chosen high in the draft in October 1940. He was fully aware of the irony of what a war it would be if it were fought with the likes of him. "This draft has caused me a lot of concern," he wrote Lem in November. "They will never take me into the army, and yet if I don't it will look quite bad. I may be able to work out some sort of thing." He understood that his very manhood was at stake, and he could not sit on the curb waving a flag while other young men paraded off to war.

Jack often did not show up for class and rarely participated in discussions, and by that measure he appeared little more than a silly dilettante and playboy. Harriet saw that beneath the veneer of frivolity and merciless wit Jack was deeply concerned with the world around him. He struggled to forge his own ideas, and in doing so he struggled for an identity apart from and beyond his father.

Joe was a fierce and powerful force who gave no quarter. "When I hear these mental midgets [in the United States] talking about my desire for appeasement and being critical of it, my blood fairly boils," Joe wrote Jack in September, as if to say that if his son veered from Joe's truth he too would shrink to nothing in his father's estimation. "What is this war going to prove?"

Jack was at an age when most men have resolved their feelings about their parents, but he talked to Harriet endlessly about Joe. "He talked about his father's infidelities," she recalled. "I think his father was a tremendous influence. I don't think there's any question about that, but not all to the good! It seemed to me that his father's obvious rather low opinion of his wife and the way he treated her, that some of that rubbed off on Jack. He wasn't mean or anything about his mother, but I think that denigration, that came from the father rubbed off on the son. And that's where all the womanizing and everything came from!"

Jack's relationship with his father was changing, evolving into a far more complicated bond than what Joe had with his other sons. Jack no longer simply mimicked his father's behavior and ideas. After Joe's self-destructive candor with the *Boston Globe,* Jack began work on a document suggesting how his father should reply to his critics. Joe pressured him as he would any subordinate. WHEN WILL OUTLINE ON THAT APPEASEMENT ARTICLE BE READY REGARDS JOSEPH KENNEDY, he cabled Jack from Palm Beach early in December.

The nine-page, single-spaced letter was the first truly political document Jack ever wrote. His father considered the British a weak, shuffling, defeatist people who would be stomped into the earth by the Nazi jackboots. Jack wanted Joe to say: "I have seen the English stand with their backs to the wall and not whimper. I have seen the grim determination."

Jack told his father that he would have to temper his candor and camouflage his bitter truth. "I don't mean that you should change your ideas or be all things to all men," he told Joe, "but I do mean that you should express your views in such a way that it will be difficult to indict you as an appeaser unless they indict themselves as war mongers."

For the first time Joe treated Jack as an intellectual equal, and his son responded with an astutely calculated defense of the ambassador's tenure. In Jack's memo there is nothing of deference; the document simply reflects two men dealing with a problem. Jack had some sound ideas for his father, but truth was no more than an occasional visitor to these pages, welcome only when it would burnish Joe's image. Jack had articulately presented his father's case, but in the end Joe decided not to defend himself in such a dramatic manner.

When Jack flew out of San Francisco after his semester at Stanford, he sat in the United Airlines plane to Los Angeles writing a note to his father, further elucidating his views. A good politician learns to treat policies the way a good doctor prescribes medicine, always aware that its side effects may outweigh its benefits. Jack was opposed to American entry into the war, though he sensed that it would come. He feared that the isolationist movement had led to the diminution of the aid America was giving to a beleaguered Britain. "The danger of our not giving Britain enough aid, of not getting Congress and the country stirred up sufficiently to give England the aid she needs now—is to me just as great as the danger of our getting into war now—as it is much more likely."

Jack sensed what might happen if Germany defeated Britain. He envisioned America "alone in a strained and hostile world," spending "enormous sums for defense yearly" with the electorate asking "why we were so stupid as not to have given Britain all possible aid."

As Jack flew away, Harriet missed her beau enormously. Who else among her acquaintances could change in the space of a few minutes from a cheerful, witty young man whose greatest attribute was his charm to an adult pondering the dark question of his time, and then back again to his gay, seemingly carefree self? He seemed to express so much less than he felt, and to feel so much less than he thought.

To Jack, being emotionally vulnerable was like being bound in silk so fine that it was hardly noticeable until he sought to pull away. He could not abide the idea that he might have exposed himself to Harriet by something he said or felt or wrote. "As I remember some of your most interesting mornings would be the mornings you spent reading your letters from your beaus in the East to a squad which you gathered down at the restaurant," he wrote her on October 9, 1941. "Do me a favor and don't read this."

The couple exchanged letters for a number of months as the romance faded away like the images in old photographs. In one note Harriet told Jack that she was almost killed in an automobile accident; thrown out of her car after hitting a tree, she was miraculously unhurt. "But as you say 'That's the way it goes,' " she wrote.

That was one of Jack's favorite phrases. His father believed that life was something that a man could create in his own image. Jack believed that fate was the god to whom one showed obeisance, not by prayers, but by shrugs.

Shortly after reading Jack's ardent missive about the dangers of simple-minded isolationism, Joe decided to testify before Congress in favor of aid to Britain. Joe agreed with Jack now that the best way to keep out of war was to build up the American defense and send ships full of armaments to England from what Roosevelt called "the arsenal of democracy."

As for Joe Jr., he had taken the isolationist torch from his father. He was now one of the most vocal leaders of the Harvard movement. While his father was presenting a very different message before the House Foreign Affairs Committee, Joe Jr. was back in Boston at the Foreign Policy Association fervently arguing that the United States should not send convoys of food and weapons or it would find itself standing next to Britain in the front lines.

Joe Jr. made one unusual point for a man so passionate in his isolationist beliefs. He said that if most Americans decided they wanted to go to war, he was willing to go too. Joe Jr. hoped to be president. He could feel the winds of war blowing in across the Atlantic.

If Joe Jr. held his position too long and too firmly, he risked seeing his political future swept away, a minor casualty in all the carnage of war. He saw himself as a proud patriot, not as a pacifist or an ironic bystander like his younger brother. "I think in that Jack is not doing anything," Joe Jr. wrote his father early in 1941, "and with your stand on the war, that people will wonder what the devil I am doing back at school with everyone else working for national defense."

In June 1941, months before many student internationalists who despised Joe Jr.'s views even thought of military service, Joe Jr. enlisted in the Naval Aviation Cadet Program. His physical examination at the Chelsea Naval Hospital showed him to be in splendid condition, standing five-feet-eleven, weighing in at 175 pounds, and bearing not a single glitch on his health. Joe had greased things as best he could for his namesake, writing his son that he had set up the navy physical on "a personal favor basis" and then arranging for him to see Admiral Chester Nimitz in Washington.

Not every cadet arrived at Squantum Naval Air Facility that summer escorted by such a notable as Joe Timilty, the Boston police commissioner, driving his official car. Nor had most cadets taken private flying lessons, as Joe Jr. had done thanks to a family friend, Benedict Fitzgerald. He nonetheless underwent the same rigors as the other cadets, facing a gauntlet that washed out half the would-be pilots. He ran double-time between classes and studied the arcane minutiae of navigation. He sat in a double cockpit with instructors who liked nothing better than to wash out another incompetent wretch who, if not for their good judgment, would one day have lost a good plane, killing himself in the bargain.

For the Fourth of July weekend, Joe Jr. returned to Hyannis Port. The Kennedys were a family of special occasions, and no holiday was so brilliantly memorialized as the nation's birthday. "Year after year, they were the highlights of our summers on Cape Cod," Teddy recalled nearly four decades later. "Even today, I can see the gaily decorated porch, the long wooden table piled high with boiled potatoes and green peas and its centerpiece of fresh salmon, which Dad had brought in from New Hampshire or Maine, or even Newfoundland if he'd heard that the best salmon were running there."

The table was as heavily laden with food as ever, and the laughter as deep, but over the gathering hung Joe's terrible fear that if war came there would never be a family gathering like this again. Joe Jr. and Jack, along with their father, dominated the conversations. Kathleen, Eunice, Pat, and Jean adored their brothers. They were honored enough just being there without injecting themselves often into the manly conversation. Rosemary sat there too, slightly reticent, a gentle presence, never entering into the quick-witted repartee. Bobby and Teddy were observers of their big brothers and the great world about which they reported like scouts back from reconnaissance.

The major event of the weekend was the Fourth of July sailboat race. As they always did, the whole family went down to the pier, where they piled into a motor launch to watch Jack or Joe Jr. lead the pack of sailboats to victory.

For the most part Jack and Joe Jr. ignored their baby brother, but this afternoon Jack motioned to little Teddy to join him in his beloved sloop, *Victura*, to serve as his crew. As splendid a moment as it was for Teddy, he must surely have dreaded that Jack might do what Joe Jr. had done four years before, dumping him in the water in frustrated competitive zeal.

If victory had been what mattered today, Teddy might have found himself tossed into the drink like unwanted ballast. But on this day, unlike so many others, something else mattered more than coming in first. "We lost, but I admired Jack all the more, because he should have blamed me and didn't," Teddy recalled. "Winning was important, he said, but loving sailing was even more important."

Jack and Joe Jr. left after the weekend, but Bobby and Teddy stayed at Hyannis Port for the summer. They were only boys, but even they could sense how much their world was changing. Their father was selling their house in Bronxville, the only home they had ever known. From now on they would live itinerant, if privileged, childhoods, shuttled between vacation homes and boarding schools.

Bobby had started out at St. Paul's, but Rose decided that the Episcopalian school was more interested in proselytizing an untrue faith than educating her seventh child. She transferred him to Portsmouth Priory, where she believed the Benedictines would educate him in true Catholic principles. Rose's letters to her son were as preachy as anything Bobby heard in chapel ("Remember, too, that it is a reflection on my brains as the boys in the family are supposed to get their intellect from their mother, and certainly I do not expect my own little pet to let me down"). He tried to make his name an emblem of success, working hard to get decent grades and make the football team. He was as mediocre at one as the other and ended up as the manager of the hockey team.

Bobby was always struggling to keep up and to achieve the honors that his brothers won with such grace and ease. Bobby wanted to swim as fast as Joe Jr. and Jack, to throw the football as far, and to stand as high in his class, but he had neither the brawn of his biggest brother nor the brains of either one. He had another weakness for a boy who sought to compete in his brothers' manly world—a passionate religious faith that tempered everything he sought to do.

One day at Portsmouth, Bobby was studying for his final Latin exam when his closest friend, Pierce Kearney, came rushing into the room. In his hand he held a smudged mimeographed paper that had been plucked out of the housemaster's wastebasket. "Look, Bob, what we've got," Pierce said, pushing the document toward him. "It's tomorrow's exam!"

Bobby began to tremble, in Kearney's recollection, "shaking like a leaf."

In the end Bobby scribbled some desultory notes and passed the paper back to his friend. For Bobby, quivering as he nervously perused the questions, this moment was a precursor to the great moral dilemmas of his life.

Bobby wanted to be what he thought his brothers were and what his father told him he must be, but he did not have what he considered their great and noble gifts. He struggled harder than they ever did, but even that wasn't enough to lead him to the head of the crowd. Everywhere he looked along the marathon of his life he saw moral shortcuts, hidden routes that might lead him to the front of the pack.

This was tempting, but Bobby was not simply moral, but moralistic. He had to be able to justify his actions in the name of goodness. On this occasion Bobby didn't take down all the questions. He could probably rationalize that he was doing something different from the boys who cribbed every answer, that his actions were a cut above those of the cheats and the dissemblers.

Rose sent Teddy to join his brother at Portsmouth in May 1941. He arrived in short pants, a pudgy, freckled-faced Little Lord Fauntleroy led up the driveway by his impeccably attired father. Teddy was at a double disadvantage: he was several years younger than the other boys, and he had been unceremoniously placed at the school most of the way through the year. That was the story of Teddy's childhood, shuttled from school to school, ten in all, adhering to his parents' schedules, rarely staying long enough to make real friends. "That was hard to take," Teddy reflected. "I can't remember all those schools. I mean, at that age, you just go with the punches. Finally I got through schools where I spent some time learning and trying to find out where the dormitory is and the gym."

Teddy complained to Bobby that the bigger boys were picking on him. Bobby was a scrapper, but he had a small boy's shrewdness. He knew enough not to attempt to take on his brother's cause, at least not now. "You'll just have to look out for yourself," he told Teddy, advice that went against everything his father had taught his boys about family.

Teddy was still recovering from a serious illness, and he found it hard to look out for himself. "I had been at a boarding school in Riverdale," Teddy recalled.

> And when I went there, I got whooping cough and pneumonia, and I was very, very sick. I was in the hospital for about four or five weeks and missed a central part of the year. Mother just canceled all of her plans, and just she and I went up to Cape Cod and spent the better part of three and a half months up there. I almost died. And then they

> said I would need a good deal of rest and attention. So when I first went up there, I was in bed, and it was really sort of the two of us. Someone came in and cleaned up. And Mother would cook breakfast and lunch and dinner and read all afternoon. And then as I got a little stronger we went out for walks, short walks, longer walks, together. And then I had a relapse up there actually, which she blamed herself for. And so we stayed up there. And so my father came up periodically. But she stayed there the whole time.

Rose had never nurtured Teddy's older siblings as long and as deeply as she nurtured her Teddy. He was her last child, and she lavished her emotions on him. Her youngest son returned her dedication in kind, with a deep, simple, abiding love. He never berated her, as Jack had, for her lack of physical affection. She had succored him, and he cherished such sweet memories of his early years as Rose each evening coming to his room where "she'd read a Peter Rabbit story and sort of act out the little characters and the voice. And then we'd say prayers and I'd go to bed."

Teddy's brothers and sisters loved their summers at Hyannis Port, but for him the place meant even more. It was a symbol of his mother's love and his father's devotion. "The house for us was home," Teddy said. "I don't remember in my lifetime ever having strangers to dinner in our house, in my lifetime. Never . . . I don't ever remember them having a party at the house. Very rarely they would have someone for lunch. Very rarely, so that house, that home, was there."

In the summer at Hyannis Port, Bobby spent more time with his younger brother than many boys would have spent. Nine-year-old Teddy's closest friendship was not with fifteen-year-old Bobby, however, but with his ten-year-old cousin, Joseph "Joey" Gargan. Rose's sister, Mary Agnes, had died four years before, leaving Joey and his two sisters motherless. Even with their own large family, Rose and Joe invited the young Gargans each summer to Hyannis Port. As generous as his uncle and aunt were to Joey, there was always a quid pro quo with the Kennedys. Although nothing was said formally, Joey was deputized to watch over little Teddy, to be his guardian and his mentor and his friend. Joey was loyal and vigilant, steering his little friend away from conduct that would have brought a firm rebuke from Teddy's father. For both youths, summers in Hyannis Port resonated with all the glorious pleasures of boyhood and some of its dangers.

"Teddy went outside the harbor [in his sailboat] one day, and they got caught in the fog and they got very nervous and hysterical," Rose recalled. "Some of them started to weep, and after that he understood how important it was not to go out of the harbor when there might be fog."

For Teddy, these radiant times were the source of many of his ideals of family. Hyannis Port was the private preserve of family and home to which he would always return. Joe was no longer flying off to Washington or New York. He rarely left this house except for his early morning horseback ride.

Rose could no longer sail off to Europe on her shopping trips, and she too was omnipresent. "I saw a man and a woman in a relationship which was one of love and affection and respect for one another," Joey recalled. "I never saw two people who spent that much time together. If I spent as much time with my wife as Joe spent with Rose, she'd be asking me to take a vacation."

That was the vision of family love that Teddy carried within, never to be bleached out by the fierce sun of life. Hyannis Port was the home where he had been nestled in his mother's arms and had sat boldly at the table beside his brothers and sisters. These grounds were hallowed, alive with memories. His grandfather's songs carried on the wind, and Jack was out there sailing *Victura,* out there somewhere beyond the last horizon.

While his big brother learned about flying, Jack was off traveling around Latin America. It had not been the easiest of trips. Jack wrote his Harvard friend, Camman "Cam" Newberry: "I don't know about the army, my back was snapped several times."

The armed forces were not so bereft of fit young men that they would happily take an emaciated-looking recruit of such dubious health that he could not even get life insurance.

Jack had struggled to play football when he was a sickly wisp of a youth, and now he could not possibly stand on the sidelines as the leaders of his generation donned uniforms to enter the ultimate field of play. His whole conception of manhood was at stake, as well as everything his father had taught him about how he must live. Rightly, then, it was to his father that he went to seek help in getting into the navy, which would probably have rejected him as easily and quickly as it had accepted his brother.

Joe feared war for his country and himself, but mostly he feared it for his sons. He had ominous premonitions of what war would bring. Although he rarely mentioned his sons in his dark prophecies, he feared that their lives might be part of the terrible wreckage of war. He could have attempted to talk Jack out of his plans to enlist. Joe, however, had wanted sons who were what he considered true men, and now he had them—brave sons who would run toward the cannon's bright fire.

Joe contacted his former naval attaché in London, Captain Alan Kirk, the director of Naval Operations of Intelligence in Washington. Kirk wrote Captain C. W. Carr at Chelsea Naval Hospital: "The boy has taken the attitude that he does not wish his father's position used in any way as a lever to secure him preferment. This is an excellent point of view but, nevertheless, it has occurred to me he might be helped in one way—vis., his physical condition." The malleable doctor gave Jack a perfunctory enough examination that he passed.

Kirk was not finished with favors. In October, Jack was appointed an ensign under the chief of Naval Operations of Intelligence in Washington. Joe Jr., who had just been sworn in as an aviation cadet in the U.S. Naval Reserve, had decidedly mixed feelings about his kid brother donning the navy blue.

After all, Joe Jr. had paid his dues in weeks of rigorous training before he could affix the gold anchors that he was now privileged to wear. Jack had done nothing but pass a fake physical, and now, without a day of military training, he was an ensign, outranking his big brother. But as he told a friend, he was also worried about Jack's troubled back and felt that his father should have exerted his influence to keep his younger brother out of uniform, not to get him in.

When Jack arrived in Washington in September, he was not the first Kennedy living in the city. His favorite sister, Kathleen, was working as a secretary at the *Washington Times Herald*, while Rosemary was living at a convent. Rosemary was so voluptuous, sweetly spoken, and demure that during the summers at Hyannis Port, Jack and Joe Jr. had to ward off young men. All their lives they had been good brothers watching out for their sister, protecting her. Rosemary had the mental age of a fifth- or sixth-grader. She could not keep up with the quick banter around the family dinner table, but to most people she looked like just another young Kennedy.

From the time she was a little girl and they realized she was "slow," Joe and Rose had brought up their firstborn daughter as much like their other children as possible. She had special teachers at the convent schools where they sent her, but they tried to keep her life integrated with those of the rest of their children. In London, she had made her debut and been presented to the king and queen; after spending days learning the elaborate curtsy that other young women picked up in a few hours, she had awkwardly tripped before the monarchs.

Teddy knew nothing of the difficulties his big sister was facing. He only knew that good Rosemary was his gentle friend. She was not rushing out on dates or off with her friends like his other big sisters. She was there, ready to talk to him and play. To him, she was a dream of what an older sister should be. "I just had the feeling of a sweet older sister . . . who was enormously cheerful, affectionate, loving perhaps even more so than some of the older ones," Teddy reflected. "She always seemed to have more time, and was always more available."

Rosemary could have handled a menial job, but in 1941 there was no place for her to go. In recent months, she had begun to suffer from terrible mood swings. She had uncontrollable outbursts, her arms flailing and her voice rising to a pitch of anger. In the convent school in Washington the nuns were having a difficult time managing her. She sneaked out at night and returned in the early morning hours, her clothes bedraggled. The nuns feared that she was picking up men and might become pregnant or diseased.

Joe felt that he had to do something. "Mr. Kennedy was so afraid of her getting in trouble or of being kidnapped," recalls Luella Hennessey, the family nurse. "It would be better for her not to be exposed to the general public in case she ran away. It would be better to almost 'close the case.' Then there wouldn't be any more trouble."

The newspapers and magazines were full of stories about a dramatic new surgical technique, the prefrontal leucotomy. The operation, known in the United States as a "lobotomy," cut away the prefrontal lobes of the brain and changed the emotional life of the patient forever. In 1941, fewer than five hundred lobotomies had been performed throughout the world. Even its proponents considered the operation experimental.

In the United States, the medical kings of lobotomy were Dr. Walter Freeman, a neurologist, and James Watts, a surgeon, who had their practice in Washington, not far from Rosemary's convent. Joe liked to cut away at a problem, and then move on. Thus, it was perhaps understandable why Joe took Rosemary to Freeman and Watts's office. What is less understandable is why he was not immediately shown the door.

Dr. Freeman, a showboating self-promoter, had repeatedly stated that he performed the operation only when all other approaches had failed. The doctors operated on tragically sick individuals, the long-term depressed, lifelong alcoholics, and hopeless schizophrenics. They had performed only one of their eighty operations on a patient younger than Rosemary, and never on an individual with mental retardation.

It may be that Freeman was drawn to the idea of operating on the daughter of one of the most famous men in America, thereby raising his stature dramatically. It may be that Joe pushed the doctors to do what they knew they should not have done. In any event, Joe gave the go-ahead without getting his wife's approval, surely knowing what Rose's opinion would have been.

On the morning of the operation, Rosemary was wheeled into the operating room fully cognizant of her surroundings. The doctors gave her Novocain, a local anesthetic, and with her fully awake, Dr. Watts drilled two small holes into her cranium. While Dr. Watts employed a tiny spatula to dig out the white matter of the frontal lobe, Dr. Freeman kept up a conversation with Rosemary.

Dr. Freeman was a charming, outgoing man, and he was good at his job, which was to keep the patient talking, getting her to sing if possible. Rosemary viewed the world around her with trust. The more Rosemary cooperated with the doctor, the more she talked, and the more she sang, the more Dr. Watts cut.

When Rosemary finally grew quiet, the surgeon knew that he had cut enough. Dr. Watts put in sutures, and she was wheeled back out of the operating room. When she woke up, the doctors found that Rosemary had talked too much and sung too long and that Dr. Watts had cut too deep. She was like an infant, capable of speaking only a few words, staring out into a world she did not know or understand.

Rosemary loved and trusted her father. She had been isolated, shut off away from her family in a convent. She had good reasons to suffer from what Watts called "agitated depression," but that was woefully little reason for the radical operation that destroyed everything but her soul.

There was one other person in the operating room that morning, the nurse who worked for the two doctors. The nurse was so horrified by what she saw happening that she left nursing and never returned to the profession. There were other operations, so many others, but it was Rosemary she mostly remembered.

Over half a century later it was a horror that still kept coming into her consciousness. She might be sitting in a restaurant with her daughter talking about her childhood, and it would come back to her again, and she would talk about the guilt and sadness that never fully left her.

The nurse was not the only person haunted by Rosemary's operation. The lobotomy is the emotional divide in the history of the Kennedy family, an event of transcendent psychological importance. This was the first family tragedy. Unlike all the subsequent deaths and accidents, no mark of patriotism, heroism, daring, or even dread circumstance could be attached to this act.

In this family where all the important events of the day were discussed over the dinner table, surely it was time to confront Joe with what he had done, to have it out, to discuss, to cry, to ask God's mercy and forgiveness, and then go on. But it did not happen. And it is here that the Kennedy pattern of denial is implanted in the psyches of the children. The truth becomes a form of betrayal. The mumbling inarticulateness with which many of them discuss personal history has its beginnings here.

Rosemary was shipped away to a private sanatorium, and for years her siblings simply did not talk about her anymore. Her mother wrote letters to her children in which her name was never mentioned. Rosemary's name was excised from the family and its history and its aspirations. For little Teddy, already uprooted, shuttled around schools, his own sister, whose sweet gentleness had been one of his few constants, was gone, as far as he knew forever, unmentioned and unmentionable.

There had been no more vociferous opponent of American entrée into war than Joe Jr., but now that he wore the navy blue he turned all his energy toward becoming a pilot. At the Jacksonville Naval Air Station, Joe Jr. continued his training. He got up at 5:00 A.M. cursing the bugle and did his jumping jacks as the sun rose. He had no gentleman's gentleman to pick up after him now, and he received an unhealthy quota of demerits during inspections.

On one occasion, his wastebasket did not pass muster, and he was handed four demerits and five hours marching. His mother considered this punishment so outrageous that she had to be talked out of writing "the Navy Department and the War Board and some Commanders—and maybe Mr. Roosevelt himself."

For all his attributes as a cadet, Joe Jr. was not a natural pilot with the intuitive sensibility in which a man and a plane mesh together as one. In the evenings, he often played a strong game of blackjack or bridge, gleefully sweeping up the winnings.

Joe Jr. was a popular cadet who, after some shrewd maneuvering, won election as the president of the Cadet Club. He had little time to chat with his new friends at the club, but it was not all a merciless grind. "I am sorry Gunther's storage department sent Joe's morning coat instead of his tails down to Jacksonville, but if he had wired back in time, they could have sent his tails by air," Rose wrote the family.

Joe Jr.'s mother would have been impressed by one aspect of her eldest son's life, and that was his faith. Joe Jr. was far more religious than Jack. As a boy, Joe Jr. had met the princes of the church in his own house, and he was comfortable around priests in a way Jack would never be. He was the last one in line for confession on Fridays, but he was always there. The base chaplain, Father Maurice S. Sheehy, heard what passed as sins among these cadets, the carnality, the lies, the blasphemies, and in the end there stood Joe Jr. This moment to the priest was one of the blessings of faith.

"There is nothing comparable to the beauty of the soul of a young man who opens the window of that soul to a priest in confessional, and who reveals, in so doing, the passion for what is honest and decent and good,"

Father Sheehy recalled. "The faith in many generations of Kennedys and Fitzgeralds was revealed in the humility of this favorite son of fortune whenever he went to Confession."

Young Joe Jr. was a child of fortune. Life's blessing shone on him. He was a talisman of good cheer.

Unlike his brother, Jack was in no hurry to thrust himself up in the front lines to be ready when war began. In Washington he put in his hours in the Navy Department on Constitution Avenue, but he took far more interest in his evenings and weekends and his pursuit of a different sort of prey. Washington was full of young single women, and Jack doubtless would have had scores of sexual adventures if one day soon after his arrival his sister Kathleen had not introduced him to Inga Arvad, her colleague on the *Washington Times-Herald.* "He's coming to Washington," Kathleen exulted. "I'm going to give a party at the F Street Club, you will just love him!" And so she did.

Inga's ripe, sensual beauty would have been enough to attract Jack for a short-lived affair, but it was not her blonde looks that led him into the deepest relationship of his young life. At the age of twenty-eight, Inga had already lived half a dozen lives, shedding her identities like last year's fashion.

There was a mystery about her that Jack could hope to fathom, but in the end he was the one whose mysteries were unraveled, not hers. Of all the women he had ever known, Inga was the one who saw him, with her shimmering blue eyes, not only largely as he was but as what he might become.

Jack knew immediately that this woman had a history. Most of her past he knew nothing about, and much of it he never would know. Inga called herself an "adventuress." She had a courtesan's subtle style and cunning. Inga had been born into a well-to-do Danish family in 1913, or so she said. Her father died when she was only four, and her mother seems to have seen her beautiful daughter as a vehicle for her own advancement. Inga told Jack that she had been such a natural actress that "the Royal Theatre in Copenhagen declared that I probably would be a second Pavlova."

Instead, at the age of fifteen, with her thirty-six-inch bust, eighteen-inch waist, and thirty-five-inch hips encased in a pink Empire-style dress, she won the Miss Denmark contest. In Paris for the Miss Europe contest, sixteen-year-old Inga met a young Egyptian student and diplomat with whom she eloped. The man was rich largely in debts, and Inga employed her theatrical skills fending off creditors. She traveled with her husband to Cairo and Alexandria, where she left him and returned to Denmark.

In 1935 Inga met Paul Fejos, a film director nearly twice her age. Fejos starred her in a film shot in the Norwegian fjords. Disillusioned with film-making, Inga traveled to Berlin. Despite her lack of experience in journalism, she arrived in the German capital with credentials from *Berlingske Tidene*, the leading Danish paper.

Beauty is its own calling card, and Inga quickly gained access to the Nazi elite that any journalist would have envied. She interviewed Joseph Goebbels, Heinrich Himmler, Hermann Göring, and Hitler himself. She attended the 1936 Berlin Olympics and on one occasion sat in Hitler's box. The Führer gave Inga a signed photo of himself in a silver and red leather frame with the peculiar inscription: "To an indefinite Fr(au)." Hitler generally did not hand out such special photos to any but his intimates, and by any measure, it was extraordinary that Hitler gave one to Inga.

Inga married Fejos in 1936, and after leaving Germany in circumstances as mysterious as her arrival, she and her second husband journeyed to the East Indies to do some anthropological explorations. There, in remote villages, Inga said that she was worshipped as a goddess, and a primitive statue was built to replicate her blonde beauty. She left, however, because this life left her with hardly more fulfillment than the life of an actress. She arrived in New York with her mother in February 1940.

In the fall, Inga entered the Columbia University School of Journalism. She had a way of arousing both jealousy and suspicion in people, especially women. In November, one of her classmates wrote a letter to the FBI after spending an evening at Inga's apartment, where "the conversation slid into a discussion of the large number of Jews in the class, and the danger of civil war in this country. We left very late, dazed by her charms, but with the uncomfortable feeling that we had been somehow threatened." Her accuser said that, although she had no evidence, she believed that Inga had been set up at the school for the purpose of "influencing morale in this country for the benefit of the German government."

Inga was doing little more than reiterating opinions she had casually picked up on the social circuit in Berlin and elsewhere. Her beauty was a magical amulet that she could use only among men. At the journalism school she may have had detractors among her classmates, but she nonetheless developed useful friendships with several professors. At the same time, although she was still married to Paul Fejos, she was carrying on an affair with Nils Blok, a Dane of artistic bent who was working for the Danish consulate. Her mother found her daughter's behavior reprehensible and regularly berated Inga, whom she had brought up to be something more than an adulteress. She berated Inga's lover as well, writing in her diary: "[I] lost control of myself when I saw him making love to Inga. He undoubtedly has Inga's permission."

Inga moved to Washington upon graduation, in part to get away from her mother's ceaseless hectoring. She had met Arthur Krock at Columbia, where the *New York Times* columnist led her to believe that he was "a skirt chaser." Krock used his formidable position to advance Inga as a candidate for a job at the *Washington Times-Herald.*

"I've got another one for you," Krock told Frank Waldrop, the editor. "What are you, our staff procurer?" Waldrop remembered replying. Inga decided to display her abilities as a reporter by doing an interview with Axel Wenner-Gren, her husband's employer, one of the world's wealthiest men and a suspected Nazi spy. As she talked to the multimillionaire, she was under FBI surveillance.

Waldrop assigned Inga to write a benign, chatty column profiling the powerful and intriguing, the best possible entrée to the highest ranks of the capital. Washington was a city of powerful men who fancied pretty women as one of the natural accoutrements of power. Inga had scores of admirers. In those early months in Washington, she was awakened every morning at seven-thirty by a call from Bernard Baruch, the legendary seventy-one-year-old financier. "He can help me a lot," she wrote her mother, "but it won't do any good for him to be so much in love."

Inga knew that in all likelihood her future depended on powerful older men like Baruch and Krock, but she was drawn now into a liaison with Jack. Inga saw the deep complexities of the man. She saw how he could manipulate most people as easily as she could beguile men.

When they walked into parties, he lit up what he called his "BP" (big personality), charming his way across the room. And when he left, he dismissed everyone he had met. "What a drag!" he exclaimed. "What a bore!" Jack didn't have friends. He had supplicants, lackeys, men like Torby Macdonald, whom Jack on occasion derided as little better than a stupid oaf.

When Jack arrived at Inga's modest apartment, he whipped off his clothes, took a shower, and pranced around the living room in a towel. They made love when he wanted to make love. It did not matter if Inga had just dressed herself for a fancy party, if Jack was ready for sex, Inga had to accommodate him. "We've got ten minutes," he told her. "Let's go."

At times Jack might have treated Inga with the same casual disregard that he did other women, but he was venturing into a deep emotional jungle where he had never gone before. Betty Coxe Spalding, Kathleen's roommate, found it a strange, disconcerting relationship. "I think he was terribly dependent on her," she recalled. "I thought, 'God, it's sort of . . . she's motherly to him.' I don't know that he had too much physical affection from his mother, more from his father probably than their mother."

John White, a journalist on the paper, talked to Inga about Jack and sensed her discomfort at how dependent Jack was becoming on her. She was used to older, powerful men who sheltered her from the world; now she was harboring Jack in her arms.

"She said that he began to come apart," recalled White. "She said that Jack's attitude was, if the girl wouldn't go to bed, that was all right. But if she goes to bed, it was under his terms that she gets out and goes and that is it. No in between. No affection and no lasting relationship. But he wanted to hang on to Inga. It was a little embarrassing to have this uniformly successful man suddenly become groping."

It had been unseasonably warm that December Sunday that would forever after be known as Pearl Harbor Day. The next morning everyone went back to his or her same job, but now that the war had begun, everything had changed.

At the *Washington Times-Herald,* one of the leading isolationist papers in America, a reporter on the paper felt that Inga might be a German spy. This was no longer a jealous suspicion but a matter of such significance as to merit an immediate meeting at the FBI, in which Inga called the allegation absurd.

Inga did not know that she already had been under surveillance and had her own file at the FBI. The agency went ahead with a full-scale investigation. Agents broke into Inga's apartment, photographing letters and other documents. They purloined her mother's diary and letters. They wiretapped her phones and began around-the-clock observation.

The cold blasts of paranoia played across Washington that January, and everywhere the agents looked they found strange inconsistencies. An attaché at the Danish legation reported that he had never heard of Inga's family in the circles in which she claimed acquaintance. And wasn't it strange that Inga spoke such perfect English when she had arrived in America less than two years before? She said she didn't know German, yet sources reported that she would occasionally utter German expressions. At one point the FBI inventoried her possessions and discovered the Hitler photo. A Nazi spy may not have been likely to be traveling with a signed photo of Hitler, but the FBI would have been foolhardy and derelict not to have thoroughly investigated Inga. It was all beginning to add up, or so it seemed. The agent in charge noted that the case had "more possibilities than anything else I have seen in a long time."

Jack and Inga had been lovers probably no more than three weeks when their little dollhouse of romance began to come tumbling down. Jack was a navy intelligence officer bedding down with a suspected Nazi spy. If he had been a man of narrow political ambition, he would have fled from Inga. He knew that his lover was no Mata Hari, but in her fatal embrace might lie the end of his career as a naval officer as well as any future in politics.

Instead, Jack took what he considered precautions, silly little gestures like corralling a friend, John White, to accompany the couple on outings pretending that he was Inga's date. Jack may have been an intelligence officer, but he did not notice the extensive government surveillance or the presence of two sets of detectives watching him and Inga—not only the FBI but also private detectives hired by Inga's husband.

On January 12, 1942, Walter Winchell, the gossip columnist, carried an item blind only to the unsighted and the uninformed. The cognoscenti woke up that morning to read: "One of ex-Ambassador's Kennedy's eligible sons is the target of a Washington gal columnist's affections. So much so she has consulted her barrister about divorcing her exploring groom. Pa Kennedy no like."

"Pa Kennedy" not only did not like this situation but cringed at the potential consequences of this foolish affair. He had seen how close his affair with Gloria Swanson had come to destroying his marriage, and he could not understand why his second son couldn't follow his father's lead and keep his dalliances dalliances. Joe had received a telephone call from Inga's jealous husband, an enraged, bitter man who was not going to go shuffling away into the dusk. Joe had long ago prophesied that war would mean the end of democracy, and he could surely see that if Jack did not end his affair with Inga, he might be swept away in the hysteria of a warring nation. Within twenty-four hours after the Winchell column, Jack found himself reassigned to the navy base at Charleston, South Carolina.

Jack spent his last three nights in Washington in Inga's apartment. He could have gorged himself on her womanly charms before heading off satiated, ready to move on and to forget. Inga, though, was not just another fitful affair, and his very first weekend in Charleston she took the train down to South Carolina to spend more time alone with her lover.

Inga saw into Jack as no one else had—the light and the darkness, the glory and the doom, the nascent idealism and the desperate cynicism. She was lifetimes older. She talked and wrote to him as if he were all youth and future and hope.

"He is full of enthusiasm and expectations, eager to make his life a success," Inga wrote him, as if they were exploring a separate person together. "He wants the fame, the money—and what rarely goes with fame—happiness. . . . There is determination in his green Irish eyes. He has two backbones: His own and his father's."

Inga saw the depth of Jack's ambition. He was not his brother's pale shadow, but a man who weighed the cost of aspiration with an appraiser's cautious eye. Inga spied two pathways before Jack. She called one direction "the West," though it was not a direction but many directions, not a mere road but endless space.

Out this way a man lived as he wanted to live. Here there was room for a woman like Inga, whose past disappeared in the openness, room for babies, laughter, and adventures that had no beginning and no end. Then there was that tortured road to power, twisting through the salons of Georgetown and the corridors of influence, a narrow, lonely road that ended at the White House.

Inga believed that her Jack could journey up both roads, and she implored him to keep both dreams alive. "You are going away," she wrote him. "And more important than returning with your handsome body intact . . . come back with the wish both to be a White-House-man and wanting the ranch—somewhere out west."

As much as Jack might want to journey up both roads, Inga knew that most successful men chose only one road. "Put a match to the smoldering ambition, and you will go like wild fire," she implored him. "It is all against the ranch out west, but it is the unequalled highway to the White House. And if you can find something you really believe in, then my dear you caught the biggest fish in the ocean. You can pull it aboard, but don't rush in, there is still time. Nonsense? Maybe badly expressed, but it is right, perfect and powerful like Young Kennedy in person." She saw that if he could only reach out and find the ideals that gleamed out there on the horizon, then he might indeed have both power and principle, great accomplishment and noble ends.

Inga was daring Jack to break out of his prison of intellect, forgoing his dependence solely on calculation and rationality. "Maybe your gravest mistake, handsome . . . is that you admire brains more than heart," she wrote him, "but then that is necessary to arrive in the end so that Jack could prove that he was indeed 'a man of the future.'"

When Jack and Inga met or talked on the telephone, he savagely rebuked the world in which he found himself. He wrote Inga condemning the excesses of "stinking New Dealism." His stint in Washington had convinced him that the city was a political brothel. He despaired at the boastful headline screaming victory above a story that "stinks of defeat."

Jack believed, like his father, that the war "may call for us to be regimented to the point that make the Nazis look like starry-eyed individualists." He had a Puritan's rage at the triviality of politics and the pathetic self-interest of people chasing pensions, not Nazis. He despaired even at himself and people like him who were thinking gloomy thoughts and "writing gloomy letters" instead of fighting the war that had to be fought.

One of the persistent themes of Jack's life was the natural lethargy of democracy. He believed that most citizens, if they had heard Paul Revere's clarion call in the night, would merely have turned over and gone back to sleep. It took fires in the night lapping against the very foundation of their homes to wake up the citizens of a democracy and send them out into the street to attack the forces that would destroy them.

Jack had seen it in England, and now he saw the same thing again in Washington and Charleston. In the nation's capital, the politicians and bureaucrats and reporters trafficked in the trivial while, as he wrote Lem, "all around us are examples of inefficiency that may lick us—Nero had better move over as there are a lot of fiddlers to join him."

As he sat in Charleston perusing the newspapers, he felt that "Washington is begginning [*sic*] to look more and more like the Cuban Tea Room on a Saturday night with the Madame out." Jack, like his father, looked down with despairing eyes on the world in which he lived. He, however, could also look up and proclaim, "The reason we're not witnessing a true tragedy is that we can do something that the Greeks couldn't, we can prevent the gloomy ending." Here then was the potential greatness of young Kennedy. Jack looked at humankind in all its frailties, weaknesses, and self-interests, shirking from none of the darkness, but then looked up and saw what might be.

At the same time Jack looked up from his cynical disregard for the female sex and conferred on Inga a passion that was as much emotional as physical. If love is emotional idealism, Jack bestowed that on her, in deed if not in word. She constantly showered her lover with verbal roses: "Honey," "Darling," "Honeysuckle," "Honey Child Wilder." She called him all these names and more. She proclaimed to Jack, "I love you." He said nothing in return but sidled silently away from such professions. He may not have been demonstrative, but he told Inga that he had talked to the Church, presumably about the possibility of marrying a twice-divorced woman.

For all the hours that the FBI recorded their conversations over the telephone or in hotel rooms, not once did Jack mention Rosemary and her terrible fate. He did not talk of his own physical pain and how he hid that from the world. He did not explore the complicated relationships with his father, mother, or Joe Jr. This was the deepest love affair of Jack's twenty-four years, but even in bed with Inga in their most intimate moments, he kept a distance from her, harboring his deep concerns like keys to an inner life that no one would ever see.

As tightly as Jack guarded his own psyche, he was astute about human beings and their motivations. He may well have sensed that within the layers of Inga's complexity lay ample room for carnal duplicities. During the first weekend in February that she took the train down to Charleston, the FBI reported that she and Jack "engag[ed] in sexual intercourse on numerous occasions." When she returned to Washington, she invited Nils, her former lover, to Washington; he would come only if she would "go to bed with him." Inga told him that though she didn't "want to sleep with a dozen men at one time," she would "be with him."

When Inga talked about her fear that she might be pregnant and her desire to get an immediate annulment, Jack remained silent, the loudest possible message. Jack's father called and talked to his son about the affair. Inga blamed Jack's father for injecting himself into their pure, perfect love.

She wrote him later: "If I were but 18 summers, I would fight like a tigress for her young, in order to get you and keep you." She was not only four years older and twice married, but she was carrying on affairs with two different men. Joe was right in prophesying disaster if Jack did not back away from Inga. Whatever he said to his son, whether it was the logic of his words, the sheer emotional brutality of his revelations, or the strength of his threats, it was enough for Jack.

Jack obtained permission to fly to Washington on February 28; there he spent one last night with Inga and told her what he felt he had to tell her. "I may as well admit that since that famous Sunday evening I have been totally dead inside," she wrote him afterward. She went out to Reno to obtain a divorce, returning afterward to Washington.

Jack's heart had led him into an emotional jungle where he had never ventured before. Inga may have betrayed him, but Jack's heart had been a betrayer as well, leading him into this dangerous, uncertain world. He had fought his way out again, and he stood now in a world of clarity and distance. In her pain and emptiness, Inga admonished her lover: "If you feel anything beautiful in your life—I am not talking about me—then don't hesitate to say so, don't hesitate to make the little bird sing."

However much Jack mourned his loss, that suffering was probably nothing compared to his physical agony. He was racked with pain in his spine. His stomach was knotted up in cramps. Even Jack, for all his willful denial, could not pretend that all was well. His only choice was to return to his familiar haunts at the Mayo and Lahey clinics to consult specialists who so far had done little to help him.

On his way to Rochester and Boston, Jack flew down to Palm Beach to spend a weekend with his family and his houseguest, Henry James. Joe had just endeavored to help extricate Jack from what his father considered his romantic debacle. These two days should have been a sentimental respite.

Joe, however, was in a savage mood, and when Jack showed up late for lunch, his father exploded. "For God's sake!" he yelled as the family quivered at every syllable. "Can't you even get on time to meals? How do you expect to get anything done in your life if you can't even arrive on time!"

"Sorry, Dad," Jack responded, and blamed his tardiness on his friend Henry. Joe prided himself on his punctuality. It irked him that his second son could not even adhere to this basic tenet of manners. Jack's rudeness bordered on the insulting, but that was not what irritated Joe so much.

As soon as he had heard about the Japanese attack on Pearl Harbor, Joe telegraphed President Roosevelt offering his services. He had received no immediate response, and that only darkened his deep melancholy about the future. At fifty-three, he was a vital, energetic man who could have contributed much to the war effort. The president finally responded with the suggestion that the former chairman of the U.S. Maritime Commission "could be of real service in stepping up the great increase in our shipbuilding." He also had an offer to develop the civil defense program in the eastern United States. Joe turned down the proposals that were suggested to him, however, and told the president that he would "just be a hindrance to the program."

Joe had not been offered a cabinet-level position, but he had been offered what not all men receive: a second chance. He could have exorcised most of the bad feeling that his ambassadorship had engendered by pitching in and doing what he was asked. He was a man of such overweening pride, however, that he refused. He might merely have been trolling for a better offer, but a message from J. Edgar Hoover that arrived at the White House on April 20 probably ended that possibility.

The FBI director addressed the confidential, hand-delivered letter not to Roosevelt personally but to his secretary, General Edwin M. Watson. This procedure gave the president the option of saying that he had not even seen the letter, for it involved his profligate son, Jimmy, in a matter that, if true, could have sent him to prison. The accuser, described by Hoover as a person of "unknown reliability," charged that before the war Joe Kennedy and Jimmy Roosevelt had bribed then–Postmaster General James Farley to seek lower liquor tariffs.

Even if the accusation was untrue, it could be enormously embarrassing, and it may well have been the reason that FDR offered Joe no more opportunities. Joe still could have found valuable work to do, but he did nothing. As vitally energetic as he was, Joe largely sat out the war overseeing his investments. His only contributions to the war effort were his complaints and criticisms, and the gift of his sons.

Joe performed what he considered his patriotic duty in 1943 by becoming a "special service contact" for the FBI, passing on whatever information he thought Hoover might find useful. "In the moving picture industry he has many Jewish friends who he believes would furnish him, upon request, with any information in their possession pertaining to Communist infiltration into the industry," noted the FBI's Boston field office. He also proved himself useful to the FBI Hyannis agent in a case involving the shipbuilding industry.

Pride was the prison in which Joe lived, such willful, narrow pride that he refused to take a position that would place him beneath his exalted vision of himself. "When I saw Mr. Roosevelt, I was of the opinion that he intended to use me in the shipping situation, but the radicals and certain elements in the New Deal hollared [*sic*] so loud that I was not even considered," Joe wrote Lord Beaverbrook, the press magnate. As always, the error lay not in him but in others. He found it discouraging that while "there is a great undercurrent of dissatisfaction with the appointment of so many Jews in high places in Washington," the president "had failed to appoint an Irish Catholic or a Catholic to an important war position since 1940."

Jack left his father that weekend and flew north to begin medical treatment. He had become a medical kaleidoscope. Each doctor who looked at him saw something different and prescribed his own unique remedy. The doctors at Mayo and Lahey at least agreed that Jack needed a back operation, but the navy doctors said no, such a dramatic solution was uncalled for, and Jack returned to where he began—a quasi-invalid pretending to be a stalwart navy man.

Jack should have been relegated to a desk job for the duration of the war, but instead he applied to and was accepted for midshipmen's school. "He has become disgusted with the desk jobs and all the Jews," Joe wrote Joe Jr. on June 20, 1942, "and as an awful lot of the fellows that he knows are in active service, and particularly with you in the fleet service, he feels that at least he ought to be trying to do something. I quite understand his position, but I know his stomach and his back are real deterrents—but we'll see what we can do."

In July, on his way to the Chicago training facility, Jack stopped in Washington, where he saw Inga. He wanted to come to her apartment, but she preferred to keep her former lover just that. After chatting with Jack, she telephoned a friend and told her that the poor Jack she knew looked like a "limping monkey from behind. He can't walk at all. That's ridiculous, sending him off to sea duty."

In the first months of the war the allies had suffered a series of humiliating defeats, from the capture of Singapore to the fall of the Philippines. The

American public had little to celebrate but the saga of the PT boats and their skippers. These cowboys of the seas were a daring, dauntless lot, darting in and out of combat in their eighty-foot wooden boats. It was just the image Americans had of themselves—quick, smart, intrepid, and inventive.

The skippers, or many of them, came from upper-class backgrounds, having learned to sail as youths on sailboats or family yachts. The world of the PT boats was like that of Teddy Roosevelt's Rough Riders, gentlemen officers and roustabout sailors united in a gallant quest. For the men of Jack's father's generation, war was supposed to be an arena of heroism. This was the world that Joe had brought up his sons to believe still existed, a world where one day they would trade in their medals for their country's highest political honors.

While Jack was going through the ten-week officer training program at Northwestern University, one of the war's earliest heroes, John D. Bulkeley, arrived at the training facility to recruit officers to captain PT boats in the Pacific Theater. The young lieutenant commander's exploits had just been chronicled in a bestseller, *They Were Expendable,* and he had even had a ticker tape parade down Broadway.

Bulkeley had won his hero's appellation by rescuing General Douglas MacArthur from the besieged beaches of Bataan and carrying him through 560 miles of enemy waters to safety. Now he told his enthralled audiences that all he needed were five hundred PT boats, and he and his colleagues could almost singlehandedly defeat the Japanese navy.

It was an irresistible idea. Brave Americans would skim across the seas with speed and daring, slashing at their dull-witted, sneaky, evil enemy, and then zigzag away, each day closer to victory. They would vanquish their foe, not with steel and fire and might and blood but with wit and daring.

Joe wanted to make sure that his son was among the chosen few. He might have attempted to dissuade Jack, for he, more than anyone, knew of Jack's back problems and the punishment he would take in an eighty-foot armed vessel hurtling away from the enemy, planing along the top of the water. Joe, however, played this high-stakes game as if he had a marked deck and knew that he could not lose. In New York, he went to see Bulkeley and promoted Jack's candidacy, suggesting that it would be good for his son's future political career.

Even without his father's efforts, Jack probably would have been chosen for the Melville Motor Torpedo Boat School in Rhode Island. He was a superb sailor and had all the attributes of a PT skipper except one: good health. In this deception lay a moral conundrum of immense importance. Was it a willful, selfish, dangerous act for this man to attempt to brace himself on a deck where only an officer in good health could stand firmly? By doing so, was he risking not only himself but also his men in an impetuous attempt to prove himself a true man? Or did Jack have a spiritual fortitude that would help him stand up to trials of combat that might send a more physically fit officer crumbling to the deck?

In the Quonset huts at Melville, Jack was cheek by jowl with men of backgrounds as diverse as the country itself. His old friend and new mate at Melville, Torby Macdonald, had wryly commented on the spectacle of his snooty friend among the unwashed and the unlettered. "I swear to God Jack I thought I'd die of exhaustion from laughing—I'd laugh going to bed and wake up at it having dreamt of you in camp with Moe Sidelburg and Joe Louis as bunkmates."

Now, however, when Jack popped off with an anti-Semitic aside, accusing the Jews of sneaking out of danger's way by "going into the Quartermaster Corps," he not only had Torby presumably nodding his agreement—and perhaps adding a few more ugly insinuations—but a Jewish classmate, Fred Rosen, standing there to confront him. Rosen knew how to count, an ability that Torby would most likely have attributed to race. Rosen added up the number of Catholic and Jewish volunteers at Melville ready to risk their lives. When he showed Jack his tabulation, Jack had little choice but to apologize.

After graduating from Melville, Jack served as an instructor for five weeks before he received orders in January 1943 to take a convoy of four of the squadron's boats to Jacksonville, Florida. It was a tedious journey, the only heartening aspect being that the farther south they traveled the warmer it became.

On the third day, one of the boats ran aground. Jack wanted to throw a towline to the beached boat and yank it out, but as he attempted to do so, not only did his own boat run aground, but the towline managed to snarl itself around the propeller. Someone was going to have to dive into the frigid waters and free the rope. By any measure, including health, seniority, or simple logic, Jack was the last person who should have been designated to strip down and jump overboard.

"It was wintertime, and it was goddamn miserable in those boats in the wintertime," recalled Holton Wood, who attended Melville with Jack. "Because if you were going at any kind of speed, you got cut by this cold wind and spray and so forth. And he ran aground, and it was cold. It was sufficiently aground so that they couldn't back it off. If you hit rocks, you could grind up the propellers.

"And so he decided that somebody had to go over the side. Now, he could have designated anybody else, but he did not. He went over himself. He could have been badly injured. He went under the boat to see what the bottom was like. And he then came up, and as he was getting up on the slippery, icy deck, he slipped and fell and injured his back again."

Jack was successful, but the following day, when the boats stopped in Morehead City, North Carolina, he was so sick that he ended up in a sick bay with what was diagnosed as "gastroenteritis acute." Jack had performed an act as brave as it was foolhardy, and the boats had to sail the rest of the way to Florida without him. Jack's conduct had not answered the question of whether he should have been there, but he had displayed one of his psychological paradoxes.

Jack's health was so bad, and his denial so extreme, that he was attempting to will himself not only into some semblance of normality but also into a veritable superman. He had such a need to test his manhood and his courage that he telephoned Senator David Walsh for help in getting reassigned to a PT-boat squadron in the South Pacific.

11

A Brothers' War

On the troop ship sailing eastward, Jack told a new friend, James Reed, about his favorite book, *Pilgrim's Way* by John Buchan. The book tells the story of the generation of upper-class British who fought and died in World War I. Buchan knew the dread realities of the trenches, but the stench of rude death does not hang over the book. Young men die, but there is perfect beauty in their deaths. "For the chosen few . . . there is no disillusionment, they march on in life with a boyish grace," Buchan writes of one fallen hero. "Death to him was less a setting forth than a returning," he writes of another.

There is no sentimentalist like a cynic in his moments of vulnerability, and as Jack sailed toward combat, *Pilgrim's Way* was among his bibles of life. Jack was skittish when it came to blood. So were Buchan and Teddy Roosevelt and other philosophers of true manhood.

When Jack arrived in the New Hebrides, there were no philosophers present to explain the unexplainable, no eloquent memoirists to take away death's sting. He saw life and death and combat not through another's lens but through his own clear eyes. In April 1943, Jack was ferried on a small transport ship to Guadalcanal. The young men crowded on the deck knew that death might face them out there over the horizon, but few of them realized as intimately as Jack did that death resided within their own bodies. He had already cheated mortality even before he saw the enemy and was carrying into combat a body that in some ways was already wounded.

All his life Jack had been a rich boy among rich boys, but now he was among young men who ridiculed his strange accent and his seemingly affected ways. These men figured they knew life the way a rich boy never could, and he became the butt of their jokes. One of them, Ted Guthrie, a poor boy from the hills of South Carolina, joined in the merriment. "We made fun and called you a sissy," he wrote Jack later. "This due to the fact you were a rich man's son."

Then they sailed into the midst of an air attack on other boats and the joking ended. Jack's boat was so loaded with fuel oil and bombs that it risked exploding in instant conflagration. Those on board were no longer rich and poor, northerner and southerner, but men who knew that they might soon die together. "I was only sixteen years old and scared to death," Guthrie remembered. "Our ship had just been straddled by bombs and our gun tub was knee deep in water. I wanted to run but gained strength by the courage shown by Mr. Kennedy."

Jack figured the captain would hightail it out of the battle zone. Instead, in a pause in the attack, Jack's boat sailed over to pick up a Japanese pilot who had parachuted into the water. "He suddenly threw aside his life-jacket + pulled out a revolver and fired two shots at our bridge," Jack wrote Lem. "I had been praising the Lord + passing the ammunition right alongside—but that slowed me a bit—the thought of him sitting in the water—battling an entire ship. We returned the fire with everything we had—the water boiled around him—but everyone was too surprised to shoot straight. Finally an old soldier standing next to me—picked up his rifle—fired once—and blew the top of his head off."

Jack was startled to discover that what Americans considered mindless fanaticism was the common code of the Japanese officers. In *Why England Slept,* he had described how hard it is for democracies to come together in peacetime on difficult issues. Now he observed why democracies find it so hard to fight a war with the narrow will and focus of totalitarian regimes. The Americans had the finest planes, the newest ships, and equipment beyond measure. Their weakness was human lassitude. He was appalled by the "lackadaisical way [the swabs] handle the unloading of the ships," as if they were Stateside with nothing more at stake than filling a warehouse, not the lives of men. "Don't let any one sell the idea that everyone out here is hustling with the old American energy," he wrote his parents in May. "They may be ready to give their blood but not their sweat, if they can help it, and usually they fix it so they can help it."

Back at Northwestern, Jack had heard John D. Bulkeley tell the awestruck midshipmen that a mere five hundred PT boats would defeat the Japanese in the naval war. Out in the South Pacific, life was a little different. On the Blackett Strait, the men had much to fear from their own planes shooting at them, and many of them died from shelling by their comrades. Jack had a sense of life stripped of all the literary pretense, all the philosophical garnishes, all the propaganda and cant.

Jack willed himself to health, dismissing his back pains with a twist of wit. He was not as good at disguising his problem, however, as he thought. Leonard "Lenny" Thom, his executive officer, wrote home to Kate, his fiancée, that Jack Kennedy, the new skipper of PT-109, was half sick but pretended he was healthy. Lenny, who had played football at Ohio State, knew what it meant to play hurt, and he admired the cantankerous disregard the man showed for his own well-being, refusing to sign in for sick bay.

Just as Jack believed he could will himself to good health, so too he believed that he could will himself to live. Back in the States in April 1942, he had spent a weekend at George Mead's family plantation in South Carolina. George, an heir to the Mead paper fortune, had enlisted early and was already a marine corps officer. George had been afraid that his fear of dying might make a coward of him. That spring weekend Chuck Spalding had cheered up glum George with Jack's philosophy: if you thought you were going to live, you'd live. It was that simple. George just had to get his head in the right place. George would go on to fight at Guadalcanal. He would be no coward, but he would die in those jungles, not that far from where Jack stood now. George Mead was the first of Jack's friends to go, shot in the head with a Japanese bullet.

Jack figured that death had her own timetable and could take those who huddled in fear as easily as those who sailed intrepidly to meet her. Joe Jr. was talking about flying in the Pacific, but Jack told his parents that "he will want to be back the day after he arrives, if he runs true to the form of every one of us." As for Bobby, if he enlisted, the family was foolish to think that they could fix it so that he would be out of harm's way. "He ought to do what he wants," Jack lectured them. "You can't estimate risks, some cooks are in more danger out there than a lot of flyers."

Jack and his fellow PT-boat captains were supposedly dashing cowboys of the sea who came sweeping down on the Japanese barges and destroyers, firing torpedoes and sweeping out again before the enemy could fire with accuracy at the small dark boats. That was the beau ideal, but as Jack was learning, this war had little place for romantic adventure but endless room for dark comedy and mishaps. He had come roaring back to base one day, trying to beat the other boats to be first in line to be refilled. He had gone charging into the dock, damaging his boat and earning himself the nickname "Crash Kennedy."

On the night of August 1, 1943, Jack's boat, PT-109, sailed out into Blackett Strait along with fourteen other torpedo boats to attempt to intercept the "Tokyo Express"—the Japanese vessels attempting to supply their forces. The U.S. Navy's big ships had made life improbable and short for the Japanese destroyers and cruisers. The enemy had turned to self-propelled, low-bottomed, armed barges. These were much more difficult targets, and at night they moved up the strait in droves.

On a moonless evening like this one, it was so dark that for Jack and the others it was if they were sailing through an underground cavern. Lookouts wearied of their watch. They looked everywhere and nowhere. It was so dark that men at times saw visions in the blackness. It was difficult to tell foe from friend, land from ship.

A strong searchlight punctured the darkness. Jack thought that a Japanese shore battery had latched on to his boat, and he guided the boat in a twisting pathway until he was once again in the embrace of darkness. The Japanese rarely had searchlights onshore, and the light probably came from a Japanese boat, not a shore battery. PT-109 should have attacked, not roared away from this ersatz island, but Jack hadn't been briefed as well as he should have been, and he was playing a game knowing only some of the rules.

In the blackness the phosphorescence stirred up by the torpedo boat's propeller left a shimmering white trail that risked exposing its position. Jack ordered the boat throttled down so that it would operate on only the center engine. As PT-109 idled in the black water, the men saw a ship looming toward them. They assumed it was one of the other PT boats and would soon veer off. The vessel continued its course, bearing down on them.

A Japanese destroyer pressed toward them only a hundred yards away. Jack frantically turned the wheel, but working on one engine, the PT boat responded with languid indifference. The destroyer cut through the PT boat as if it were no more than foam floating on the surface and moved on into the darkness. As Jack fell to the deck, he thought to himself that this was death, that this was what it felt like.

Around the wreckage, gasoline burned furiously. The sea was lit up like a gigantic searchlight sweeping down on the survivors, pinpointing them in the blackness. Luck is always a matter of perspective. Although two of Jack's men had died, the wind blew the flaming sea away from the eleven survivors.

The front half of the PT boat sat in the water perfectly intact, as if the Japanese ship had performed surgery, neatly amputating the back half of the ship. It was all bizarre and inexplicable. Some of the men paddled in the water half giddy after inhaling gasoline fumes. They tried to make sense of the senseless. The wrecked boat, which to George "Barney" Ross looked like a fishing bobber, was the only mooring. One by one they swam over and laid themselves out on it, the only sound the water slowly seeping into the watertight passages. Pat McMahon was burned and William Johnston was sick, they were all in a state of semishock, and they knew they could not stay on the wrecked boat much longer.

Jack said that he was not the commander, that they were equals in this situation, but the men asked him to lead. And he led them now, not because he outranked them, but because they wanted him to lead, and he was willing to think and to do and to plan and to dare. By midmorning the remnant of PT-109 appeared as if it might soon sink into the dark blue waters. Jack decided the men would have to swim to a small island that they could see three or four miles away. Floating in the water was a plank that they could all cling to as they paddled along, but McMahon was too far gone for the exertion. So Jack took a tie from McMahon's "Mae West" life jacket, put it like a bit in his teeth, and pulled the stricken sailor along with him as he swam. After five hours, the group reached their precious refuge. The island turned out to be not much bigger than a football field, its only virtue that there were no Japanese soldiers.

Jack and the men huddled on the ground hoping that the Japanese would not spot them. Jack had one wounded man on his hands, one sailor with the look of terror in his eyes, and several others in middling shape at best. If he had decided that they should just stay there on the tiny island, like shipwrecked sailors on a raft, no one would have thought that he had done less than his duty. As Jack lay there exhausted, he called upon much that he had learned in his twenty-six years. Jack's father had taught his sons that the world was not a place where history was done unto them, but that they had the right and the will and the power to help shape the events of their time. Jack had listened to his father, but for most of his adult life, he had stood on the sidelines, a passive, astute observer of those events.

Jack decided that in the evening someone should swim out into Fergusson Passage to flash a lantern at a passing PT boat. It was in some ways a crazed idea, but it was the only idea they had, and Jack said that he should be the one to make the attempt. Ross, who had had the exquisite bad luck to hitch a ride on PT-109, thought that it was "either courageous or foolish," though perhaps it was a bit of both. Jack had seen how often Americans shot by mistake at their own comrades. He had observed what the Americans considered the treacherous shrewdness of the Japanese. What did he imagine a skipper might do if he saw a single light flashing in the pitch-black of the strait? Jack's bravery and foolhardiness were so seamlessly interwoven that it was impossible to tell where one ended and the other began. He could not lie on the island waiting to see what time would bring, be it Japanese or Americans, or slow death. And so he set out.

Jack swam out to the reef where he could stand up in waist-deep water. From there, he swam for an hour far into Fergusson Passage, scanning the blackness for telltale signs of phosphorescence. Far over the horizon he saw flares flashing in the blackness and realized that this was the one night when the Americans were elsewhere. Jack started back toward his comrades, but he felt weak, and the current seemed to stiffen.

Jack was a competitor, and his foe out here this evening was death itself, which was ready to take him not in a burst of blood and fire but in sweet repose, pulling him gently down into the blackness. Jack did not pray to God, at least he did not remember doing so. He was a child of fate. He stopped swimming, stopped fighting the tide. He gave in to the night, in to his fatigue, in to the endless water, and he drifted along, like a man floating through space. There are indeed strange tides in the life of a man, and as the sun rose he saw that he was back where he had been the night before. For a moment he thought he was mad, hallucinating. Then he began to swim, and he made it back to the island, where he told Ross that it would be his turn next. After saying that, he passed out on the sand.

That night Ross swam out in the strait and had no more luck than Jack in sighting a friendly boat. Jack decided that the men could not lie there any longer. They would have to swim south to a larger island closer to Fergusson Passage. So they set out again, and for three hours Jack dragged McMahon along tethered to a strap in his mouth. This new island had some coconuts lying on the ground that they used to quench their thirst, but they appeared no nearer to rescue than they had been before. Once again Jack and Ross set out swimming to another, even larger atoll, Cross Island, directly on Fergusson Passage. The two men found a canoe and supplies and startled two native watchers working for the allies. When Jack and Ross returned the next day, they found the natives there. The natives took Jack's note carved on the husk of a coconut (NAURO ISL NATIVE KNOWS POSIT HE CAN PILOT 11 ALIVE NEED SMALL BOAT KENNEDY) and sailed to Rendova to bring back help to rescue the men.

Jack lay in his bed in his skivvies in Mobile Hospital Number Four at Tulagi suffering from fatigue and from the abrasions and multiple lacerations that covered his body, especially his feet. He looked emaciated and had a slight limp. He was a hero in the eyes of the *New York Times,* the *Boston Herald,* and the other papers that celebrated him in their news columns. Out here the word "hero" was not used as often. There were those who thought that Crash Kennedy had mucked up again, by skippering the only boat in the entire war to be sunk after being rammed by an enemy ship, and that he deserved not the Silver Star he received but a court-martial. Others believed that in those waters they could have lost their boat too, and that if Jack was no great hero, he had acted admirably after his boat was struck. "I can say in all honesty, one night out there I had a similar type of thing almost happen," recalled another veteran, Bryant Larson. "The destroyer missed my boat and he chose to shoot at us and he missed. It was so black you couldn't see twenty feet."

The navy's "Personal and Confidential Report" of the incident concluded judiciously and fairly: "There is no doubt but that the officers and men . . . are deserving of much praise for the courage, resourcefulness, and tenacity displayed . . . but such conduct is general in both large and small ships operating in enemy waters and does not appear to have been of such character as to warrant special awards."

Jack's own crew—and they were the best judges of his actions—applauded Jack's courage after the sinking, but they hardly thought of themselves and their skipper as heroes. "Our reaction to the 109 thing was that we were kind of ashamed of our performance," Ross recalled. "I had always thought it was a disaster." That was too harsh a judgment, but it was a measure of Jack and his men that they judged themselves by such a standard.

When John Iles, one of Jack's navy buddies, came to see him in the hospital, he mentioned to Jack that when the crew of PT-109 was presumed lost, he had gone to Father McCarthy and asked the priest to say a mass for Jack. It was the least he could do for his fellow Catholic officer, and he wanted Jack to know about it. "He was furious!" Iles recalled. "He read the riot act to me. He said he wasn't ready to die just yet and why the hell had I given up hope? I couldn't understand it."

Jack's Catholicism had been more of a minor inheritance than a deeply felt belief, but it was not even that now. "Jack, your family is *the* Catholic family in the country," Iles lectured him. "If you lose your religion, just think how many of us . . ."

"I'll work it out someday," Jack interjected, waving off Iles. "I'll go see Fulton Sheen and get it all straightened out when I get back home." To Jack, his faith, or the lack of it, had become little more than the costuming of his public life, a matter that could be taken care of by one of the princes of the Church before he moved on to more important things.

Jack no longer seemed to believe in the moral certitudes of his church. In a draft of a letter to Inga, he wrote, "Americans can never be fanatics, thank God," and, "The Catholic Church is the only body approaching the fanaic [*sic*], and even they are having considerable difficulty expressing its belief."

Jack had always seen life from a psychological distance. That ironic shield was gone now when he sat and wrote letters to Inga. His thoughts and feelings flowed together from his mind and his heart to the hands that typed out his truths. "I received a letter today from the wife of my engineer," he wrote Inga, "who was so badly burnt that his face and hands and arms were just flesh, and he was that way for six days. He couldn't swim, and I was able to help him, and his wife thanked me, and in her letter she said, 'I suppose to you it was just part of your job, but Mr. McMahon was part of my life and if he had died I don't think I would have wanted to go on living. . . . ' There are so many McMahons that don't come through."

Jack had gone out into the wilderness of the Pacific with a young man's bravado, boasting how he could stare death down. He knew now that he was not master of this world. He was a cog in a wheel that turned without his knowledge in a direction he could not foresee. He was no hero, not if a hero is a man who grasps onto fate as if he owns it and steps out into the breach. "A number of my illusions have been shattered, but you're one I still have although I don't believe illusion is exactly the word I mean," he wrote Inga, who was married now to Nils and living in New York. "By an illusion I would mean the idea I had when I left the States that the South Seas was a good place to swim in. Now I find that if you swim, there is a fungus that grows in your ears. So I shall return with athlete's foot and fungus growing out of my ears to a heroes [*sic*] welcome, demand a large pension which I won't get, invite you to dinner and breakfast which I'm beginning to have my doubts about you coming to, and then retire to the old sailors home in West Palm Beach with a lame back."

Jack had always fancied himself a man who looked truth in the face and stared it down. He wrote Inga that he had "had in the back of my greatly illusioned mind" an idea that he would spend "the war sitting on some cool Pacific Beach with a warm Pacific maiden stroking me gently but firmly while her sister was out hunting my daily supply of bananas." Jack was hardly so simpleminded as to think that scenario was a real possibility, but he had imagined that out here he would find an overwhelming logic to the war, a rude fairness that he had not observed in what he considered the brothel-like world of Washington politics. But it was the sheer irrationality of it all that was so confoundedly maddening.

At times Jack was drowning in bitterness. "Munda or any of these spots are just God damned hot stinking corners of small islands in a group of islands in a part of the ocean we all hope never to see again," he wrote Inga. "We are at a great disadvantage. The Russians could see their country invaded, the Chinese the same. The British were bombed. But we are fighting on some islands belonging to the Lever Company, a British concern making soap. I suppose if we were stockholders we would perhaps be doing better, but to see that by dying at Munda you are helping to insure peace in our time takes a larger imagination than most men possess."

Long before—or so it seemed, though it was but two years—Jack had told his friends that to live you had to believe that you would live. So many times as a boy he had come back from the land of the dying because he knew that he would come back. Out here he had seen how poor Andrew Kirksey had the scent of death on him since the day a bomb landed near the boat and he felt his time was up. It figured that Kirksey was one of his two shipmates to die on PT-109. "He never really got over it," Jack wrote his family. "He always seemed to have the feeling that something was going to happen to him. . . . When a fellow gets the feeling that he's in for it, the only thing to do is to let him get off the boat because strangely enough, they always seem to be the ones that do get it."

But if a man could will himself to live, then he could will himself to die. Out here a man often died first in his eyes, with that empty, glassy stare, and as often as not the body soon followed. "I used to have the feeling that no matter what happened I'd get through," he confessed to Inga. "I've lost that feeling lately, but as a matter of fact I don't feel badly about it, if anything happens to me I have this knowledge that if I had lived to be a hundred, I could only have improved the quantity of my life, not the quality. This sounds gloomy as hell. I'll cut it. You are the only person I'd say it to anyway, as a matter of fact knowing you has been the brightest in an extremely bright twenty-six years."

If he lived, he would have a new life to make back in America. Inga had seen those two roads before Jack, roads that did not simply divide but headed in opposite directions. "You said that you figured I'd go to Texas, and write my experiences," he told Inga, referring to the journey westward. "I wouldn't go near a book like that, this whole thing is so stupid, that while it has a sickening fascination for some of us, myself included, I want to leave it far behind me when I go." The road west, then, was no longer a high climb into the pristine reaches of life. It led through valleys of darkness now.

Jack could have written a book resonating with the themes of such classics as *Catch-22, The Thin Red Line*, and *The Naked and the Dead*. In it he could have grasped his life whole, without parsing every phrase for its implications, political and social. He could have traveled toward the tortured heart of his truth, but he would have taken that journey alone, without his father and his family. This would have been a journey from which he could never have totally returned. He did not want to take it, to travel into the dark caverns of his life and time. The truth as he saw it was embarrassing and untidy; to explore it would have taken an intellectual daring that he did not have, or if he did, that he turned away from.

"On the bright side of an otherwise completely black time was the way that everyone stood up to it," he wrote his parents. "Previous to that I had been somewhat cynical about the American as a fighting man. I had seen too much bellyaching and laying off. But with the chips down that all faded away. I can now believe—which I never would have before—the stories of Bataan and Wake. For an American it's got to be awfully easy or awfully tough. When it's in the middle, then there's trouble."

Jack had found a commonality here in these men, no matter their backgrounds. Jack was a man of the Spee Club and the Stork Club, of Palm Beach and London, and they were men mainly from small towns and modest circumstances, and they had come together and worked together as one. He was a man, despite his education and travels, of profoundly limited experience, having gone from one oasis of privilege to the next, hardly even observing the world that lay between. This was the first time in his life that he broke the bread of life with all kinds of people, and much of what he thought of America he extrapolated from this time.

Jack had an authenticity about him now that was forged in the hot blasts of combat. He was an officer, but he had a grunt's vision of war and could be just another kvetching voice in the mess line. "When I read that we will fight Japs for years if necessary and will sacrifice hundreds of thousands if we must—I always check from where he's talking—it's seldom from out here," he wrote his parents in his anti-heroic mood. Back in Washington, the politicians tossed out words like "sacrifice" and "honor" and "courage" like cheap baubles, but out here he had learned their true meaning. For the most part, there were no flag-waving heroics in this struggle, no awesome acts of self-sacrifice, but good men were doing what they had to do. Jack was a patriot now in a way he had not been before the war. He was a patriot no more or less than most of his comrades, and in a way that Americans would not be again, not in his lifetime.

Jack believed that those who spoke of sacrifice and courage in the distant halls of power had better see to it that the peace was worth the war, "for if it isn't, the whole thing will turn to ashes and we will face great trouble in the years to come after the war." That was the one worthy road open to him—to help to see that life in the postwar world would be worth all the dread losses of war. His would be an arduous journey in its own right, if only he could take it.

When Joe learned that Jack was missing in action, he did not tell Rose. They treated each other with the courtly reserve of monarchs, civil and gracious in public, circumspect and contained in private. Joe knew that he was terribly complicit in what appeared to be his son's death. Earlier in the year, he had written Father Sheehy that while he had "pride in . . . [his] . . . heart" that his sons had opted for the most dangerous of services, he had "grief in . . . [his] mind." His was the pride that his sons were living out his vision of what a true man must be. His grief was an ominous, oppressive sense of impending tragedy that had haunted him since his days at the Court of St. James's, a grief that never left him. For once in his life his heart had triumphed over his mind. He had not stood back when his sons rushed toward the sound and smoke of combat, but had pushed them ahead.

As for Rose, his wife had the constant solace of faith, believing that if God took their sons, He would only pluck them away to a better place. In her chatty chain letters to her sons and daughters, she treated the dangers of war as if they were no worse than a pickup football game on the lawns at Hyannis Port. "Jack, you know, is a Lieutenant, J.G. and of course he is delighted," she had written the children a year before. "His whole attitude about the war has changed and he is quite ready to die for the U.S.A. . . . He also thinks it would be good for Joe's political career if he died for the grand old flag, although I don't believe he feels that is absolutely necessary." What was a lighthearted repartee to Rose contained a kernel of truth in the lives of her sons. The field of competition had moved from the fields of play to the fields of war, but Jack and Joe Jr. had set out as if the game were much the same.

Joe could not go to Rose and share his fears. And so he held them within himself for three terrible days, pretending to Rose and to everyone he met that life was as it had always been. On the fourth day, he was driving his car back from his morning horseback ride, listening to the news on the radio. When the announcer said that Jack had been found, Joe drove his car off the side of the road.

When Joe Jr. heard the news and read the stories of his brother's heroism, he did not share his father's pure elation. "When I returned home the other day, Mother told me she had finally heard from you," his father wrote him on August 31, 1943, in a passage that in the lexicon of the Kennedy family was stinging in its rebuke. "We were considerably upset that during those few days after the news of Jack's rescue we had no word from you. I thought that you would very likely call up to see whether we had had any news as to how Jack was."

Joe Jr. had learned that Jack was missing when a friend wrote him from the Pacific. Three hours later he saw the headlines about his little brother being rescued. Despite that, he could not bring himself to call Hyannis Port to ask about a brother who had won the public praise that he sought so desperately.

All that Joe Jr. wanted in family life was to be first, a condition that he thought was his natural right. His letter to his parents cried out, *Look at me, look at your other grown-up son, your first son.* "With the great quantity of reading material coming in on the actions of the Kennedys in the various parts of the world, and the countless number of paper clippings about our young hero, the battler of the wars of Banana River, San Juan, Virginia Beach, New Orleans, San Antonio and San Diego, will now step to the microphone and give out with a few words of his own activities," Joe Jr. began his letter.

Joe Jr. was struck full in the face by the sheer unfairness of it all, and in his letter he mentioned his brother's name only once. His words reeked of bitterness that a younger brother whom he considered second in everything but name had supplanted him. Joe Jr. had entered the service while Jack was malingering. He had gone through a merciless gauntlet to win his wings, while Jack had been given yet another free pass. Joe Jr. had not lost a ship to the enemy. He may not have sunk a German U-boat, or even seen one on his endless patrols out over the Atlantic, but it was planes like his that had helped drive the U-boats back from preying on merchant shipping; in the Caribbean arena the Germans had been sinking a ship and a half a day during the first half of 1942. The risks he courted daily had won him not a line in the nation's papers. He had served his country in a series of posts all over the East Coast and Puerto Rico, and now he was ferrying planes out to San Diego to be heavily armed, an endeavor that in the end would go further to defeat the Axis than anything Jack had done.

"In their long brotherly friendly rivalry, I expect this was the first time Jack had won such an 'advantage' by such a clear margin," Rose wrote in her autobiography. "And I daresay it cheered Jack and must have rankled Joe Jr." Rose may have believed that Jack reveled in his victory as much as Joe Jr. agonized over his perceived loss, but Jack no longer saw the world as the playing fields of Hyannis Port writ large. Joe Jr., for his part, still lived partially in that boy's world. He was obsessed not only with his little brother but with the struggle for fame and noble accomplishment. That was the page in the book of life that his father had opened to him and told him was life itself. He was mired in this obsession.

Joe Jr. knew how deeply his father cared for him. When Joe Jr. won his wings in Jacksonville, his father had come up from Palm Beach to address the graduates. Joe had been in the middle of his speech when he looked at his beloved son there in that row of splendid young men, and he had come close to crying. That was a measure of how proud Joe was of his son and of how deeply he feared for his life.

Joe Jr. was a better brother to his siblings than Jack. He thought about them. He wrote them. He cared. He may have been a stickler for regulations, yet when seventeen-year-old Bobby came down to visit his big brother, Joe Jr. snuck him onto the base and took him out for a patrol, letting his kid brother handle the controls from the co-pilot's seat. Joe Jr. might have been court-martialed if the brass had found out, but he was willing to chance that to let Bobby feel what it was to fly.

Joe Jr. blocked out much of the world that did not lead him toward high honor. As a pilot, he had no time for pleasantries. He strode across the hot tarmac not even acknowledging the airmen and mechanics with a nod. He had little use for most of his flight instructors, seeing them as little more than means to an end. To his crews, he was a merciless perfectionist who treated them like imperfect machines.

Whatever his detractors thought, Joe Jr. was no brass-polishing sycophant who thought that through social guile he might succeed. He sought action, and he did what he felt he had to do to stand at the front of that line where he believed a man could prove himself. It was not his fault that he was piloting flying boats, lumbering craft ideal for long reconnaissances far from the retort of angry Nazi guns or menacing Luftwaffe fighter planes. On one of his trips to Virginia Beach, where he had rented an apartment, he had a talk with Mark Soden, a fellow officer, about why he felt he must stand one day in the full line of fire. He talked about what it had been like in Madrid at the end with the bullets so near, and how so much was expected of him in this war. He was a child of wealth and privilege, and in this war he felt he had to prove himself worthy of the rich bounty of his life.

Joe Jr. was not all self-absorbed purposefulness. As long as he stood far from the arenas of heroism, he made the most of his leaves, turning his apartment in Virginia Beach into fertile ground for assignations. "He had a special friend in Norfolk, a married woman whose officer husband was away," recalled Soden. "I think sometimes Joe felt it was safer with a married woman than with a single person, no pressure."

"Joe was what we liked to call a prime cocks man," reflected another officer, Robert Duffy. "He was down there in Virginia Beach as often as he could."

In mid-July, Joe Jr.'s commander, Jim Reedy, called his crews together in a hangar at Norfolk to tell them the dramatic news. Reedy would be heading a new command, Patrol Squadron 110 (VB-110). Their job would be to hunt down Nazi submarines as they left their bases in the southern French ports on the Bay of Biscay and headed out to maim allied shipping. Their weapon would be not the vulnerable flying boat but the B-24 Liberator. The plane, rechristened the PB4Y-1 by the navy, was a squat, four-engine, thirty-ton bomber. "No one is compelled to accept this assignment," the men were told. "Undoubtedly it will be a dangerous one." The pilots knew that they faced a double challenge—not only the dangers of combat but the immediate task of learning how to fly the confounded thing. Not a man backed away, and none of the pilots was as gung-ho as Joe Jr.

During the next six weeks, just when Jack was first tasting combat, Joe Jr. was up in the sky above Norfolk learning how to fly the unwieldy monolith. He would probably have been a better PT-boat captain than a pilot of a PB4Y, but he was a dogged trainee who kept his doubts and fears wrapped up tightly in his own psyche. He was a young man in any other world but this one. At twenty-eight, Joe Jr. was older than most of the other pilots. He was persistently upbeat, never letting on to his family the uncertainties that he felt. "They really don't give us quite enough instruction . . . before turning us loose, and this has brought about quite a few accidents," he wrote the family, coming as close to admitting his anxieties as he ever did. Only to one of his father's friends, John Daly, did he admit the darkest of his doubts. "In my talks with him during his training . . . I always felt he had a premonition that he would not be one of the fortunate ones to come out of the war," Daly reflected. "Disregarding this feeling, even though approaching his thirtieth year, he fought hard to excel as a pilot and become as proficient as any of his younger cadet associates."

Once Joe Jr. checked out in the plane, he flew several of the squadron's new PB4Ys to San Diego, and then back again with their new bow turrets installed, making five trips cross-country in eight days, a schedule that would have done in a commercial pilot. He would have no second chances flying above the Bay of Biscay, and the navy prepared him and his crew with merciless rigor. Radar. Strafing. Guns. Instrument training. Tactics. Plane recognition. Physical conditioning. Antisubmarine work. Dawn to dusk, day after day, the training continued.

When the regimen finally ended, Joe Jr. got a short leave to travel to Hyannis Port in time for his father's fifty-fifth birthday. Although the birthday celebration on September 6, 1943, was for his father, Joe Jr. had reason to think that he would be honored that evening as well. In a few days he would be flying off to England to test his mettle against the savagery of Nazi arms. Although he knew himself to be a child of fortune, this might be his last great occasion with his family. He had come through. He had done everything he was supposed to do and more. He had written his father about preparing a will. He wore his navy whites, and as the group sat down, no one could look at young Joe Jr. without thinking at least momentarily of the war.

Judge John Burns stood up to propose a toast. The judge looked down the long table and raised his glass. "To Ambassador Joe Kennedy, father of our hero, our *own* hero, Lieutenant John F. Kennedy of the United States Navy." The silence hung in the air like a condemnation. Joe Jr.'s face froze in a smile that was more like a grimace. Then he drank a sip of the bitter wine. Later that evening Police Commissioner Timilty heard Joe Jr. crying in the bed next to him. "By God, I'll show them," Joe Jr. said as he clenched and unclenched his fist. "By God, I'll show them."

Bobby and Teddy were too young to prove themselves on the fields of battle, and their lives continued much as they had before the war. For years Rose had been complaining that whatever her third son did, be it reading, sailing, or collecting stamps, he did by rote, without the enthusiasm that to Rose's way of thinking defined a young Kennedy. Bobby hadn't liked to go to the young people's dances in Palm Beach, where he would have picked up the social graces that he would need as a young man. He appeared singularly disaffected by the world where his family thought he belonged.

Bobby arrived at Milton Academy as a junior in the fall of 1942. After attending seven schools in little more than a decade, Bobby had never stayed long enough at one school to set his roots deep into the nurturing soil of friendship and place. The students at the Massachusetts prep school were for most part conservative in dress, Republican in politics, and High Church Protestant in faith.

As a Roman Catholic transfer student, Bobby would have done best to slip silently into the school, hoping slowly to win acceptance. Instead, Bobby arrived in a checked coat that looked as if it could have been played as well as worn, gray pants, white socks, and a flamboyantly loud tie. If there was a choice between being viewed as a hapless mediocrity or a self-proclaimed misfit, Bobby had declared that he preferred the latter. Nor did he try to hide the fact that he was one of perhaps half a dozen Catholics at the Massachusetts school. There were no Catholic churches in Milton, but Bobby invited friends like Sam Adams to go with him to Dorchester to worship in a cathedral among people of all ages and classes. Bobby was a youth certain of few things, but one of them was that his church was true and his faith was deep.

His older brothers were handsome men with an ease of manner that helped propel them through the world. Bobby was the least physically appealing of the Kennedy brothers. His teeth were too big, his ears extended out, his body was scrawny, his voice a girlish tenor, his wit savage. The smile that once had marked a boyish cuteness now seemed an embarrassed grimace, emblematic of his shyness.

At Milton, Bobby was as mediocre on the football field as in the classroom. But how he tried on the gridiron, in practice attacking the blocking dummies as if they were glowering opponents, waiting for his chance in a big game. "I played in the football game against St. Marks," he wrote his father, "but I was quite nervous and did not do very well."

When Joe answered his son, it was not to send him hustling back onto the field, bloodied or not, to tackle and charge with mad fury. Joe had such a profound impact on his sons because his was a love tempered with insight. To him, each son was a different kind of jewel, and if Joe Jr. and Jack shone brilliantly on their own, Bobby had to be polished until he glittered like his older brothers. He reminded Bobby that his brother Joe had only made the Harvard team his senior year, and that Jack hadn't made it at all. Bobby was hopelessly behind his big brothers in his social development. That was what Joe sought to bring out, telling his son that the whole idea of football was "the opportunity to meet a lot of nice boys." These "nice boys" would turn out to be powerful men, and Joe told his son that "the contacts you have made from boyhood on are the things that are important to you in your own life's development."

Bobby was far more self-aware than his brothers and fiercely honest. He was nervous about everything that mattered to him, football, studies, faith, and girls. Everything was complicated, endlessly analyzed, pondered over, criticized.

When Bobby arrived at Milton, there was one young man who everyone at the school would have wanted as a friend. His name was David Hackett, and he was the most celebrated football player and athlete at the school. There is no fame like youthful athletic fame, unsullied by compromise and uncomplicated by any motivation other than to play well and fairly. And of all the people who could have become his closest friend, Hackett chose Bobby, the most unlikely of all. Hackett, progeny of a poor family, thought of himself as a misfit as much as his new friend did, and that apparently drew them together.

David Hackett became to Bobby what Lem Billings was to Jack—a co-conspirator in all the trials of growing up. When he wasn't around his best friend, Bobby wrote him letters. Bobby's writing had none of the grace, wit, and literate detail of his older brother's frequent missives to Lem, or any of the sheer exuberance. Bobby was blunt, but rarely purposefully vulgar. He was emotionally honest in a way that Jack was not. Jack usually kept Lem at a self-conscious distance, treating even his closest friend as an audience.

Bobby's friendship was as much about the pain and difficulties the two adolescents experienced as any joyful kinship. Bobby was a demanding friend ("if this *friend* of mine was the Son of a Bitch he seemed to be the world would catch up with him eventually and he would have to live the rest of his life with himself"). For Bobby, everything was a struggle, from studying to making friends, to seeking some measure of control over his future. Whereas Jack led Lem into precocious, bawdy adventures, there was a shy innocence to Bobby's approach to the opposite sex. While Jack bragged in X-rated detail about his putative conquests, Bobby was delighted merely to have a date. He had a certain moral rectitude that was more than a way to hide his timidity with young women. He couldn't abide hearing the dirty jokes that the boys swapped back and forth, making it clear to everyone how offensive he found such humor. "He would turn away, with almost a snarl," Adams recalled.

Bobby sought an authentic experience to mark him as a man in the same way that the war marked his brothers. In the summer of 1943, he wanted to work on a fishing boat on Cape Cod. He asked a family friend, the political operative Clem Norton, to get him a job where "nobody would know who he was and he would have to work just as hard as any fisherman." That was a manly endeavor, but his mother would hear none of it, invoking the names of her two soldier sons. "This boy will have to go soon," she said of her seventeen-year-old son, "so I want him as long as I can have him around." Instead, his parents relegated Bobby to a pallid pursuit, working as a clerk in the East Boston bank founded by his grandfather P. J., where Joe had been president. That didn't take the whole summer, and Bobby invited Sam Adams down to the Cape for a visit. Like so many other visitors, Sam had a splendidly idyllic time at Hyannis Port. He and his friend splattered each other with paint when they worked on Bobby's boat, ran off for picnics with box lunches ordered up by Rose, and sang show tunes in the evening with Rose playing the piano.

Bobby's sister Jean felt that a deep core of sadness resided within her brother. When Bobby returned to Milton for a visit after graduating in 1944, he wrote Hackett of a reunion that was more melancholy than gay. He was hardly the returning hero whose visage conjured up myriad images of athletic success or achievement. "Of course they were overwhelmed with happiness upon seeing me," he wrote, his words etched in irony, "but I can see if I went back six times in as many weeks that they would get just a little tired of me." He walked the old haunts, and to him it "all seems so inconsequential changing to blue suits and being on time for chapel and all that sort of thing." He went on with the kind of emotional candor that Jack would never have exposed. "Things are the same as usual up here and me being my usual moody self I get very sad at times."

Little Teddy had ample reason for sadness too. He had been shuttled from one private school to the next, eleven different institutions in all. "I was paddled fifteen times at Fessenden," Teddy recalled, his one detailed memory of his tenure at the Massachusetts prep school. Through it all, this plump, impish boy remained resolutely good-natured, with an inordinate interest in all things chocolate.

One night at Fessenden, Teddy decided that he would sneak down in the darkness into the pantry and steal some chocolate. He was there reaching for several of the treats when he felt a hand grasp his neck and hold him tight. The one blind master had been lying in wait to capture the stealthy thief. As punishment, Teddy spent the night sleeping in a bathtub.

Teddy learned there were other ways to turn a chocolate trick. "Your youngest brother, Teddy, the merchant in the family, is as he says running a black market at Fessenden," Joe wrote Kathleen. "He goes downtown to his Catechism Class, buys himself some chocolate bars at five cents apiece, comes back and sells them at ten to fifteen cents apiece to the boys who can't get out. . . . There is a sneaking suspicion, I imagine, among some of the parents of boys who have done trading with Sir Edward that somewhere in the long dim past there was a little Jewish blood got in with the Irish and it is all coming out in him."

Jack had warned Joe Jr. that if he insisted on shipping out to the Pacific, after a week there he would wish he had not been so rash. The warning held equally for the Cornish coast in England's West Country, where Joe Jr. arrived in September 1943. The region had struck a devil's bargain—exquisite, verdant landscape in exchange for rain, rain, and more rain. In the nearly daily barrage, Joe Jr. had plenty of time to peck out letters on his portable typewriter. Jack had been transferred to the United States for health reasons, and Joe Jr. wanted little Jackie to know that his big brother was still ahead in the game of sexual conquest. "I succeed in dispersing my first team in such various points that it will be impossible to cover all the territory," he bragged. "If you ever get around Norfolk, you will get quite a welcome if you mention the magic name of Kennedy so I advise you to go incognito."

In his letter to Jack, Joe Jr. didn't mention that he was doubly grounded, by the miserable weather and by the equally miserable ambiance of south England. The pubs closed at nine, when Joe Jr.'s hunting season was only beginning. The only diversion was dinner at the Imperial Hotel, breaking bread with a dispirited collection of evacuees from London, the room devoid of any of the smart young things who had enlivened Joe Jr.'s time in prewar London.

Joe Jr. wasn't about to write a fan letter to little Jack, congratulating him on his heroism. Nor would he even admit that he might have a grudging respect for a brother who had held up the family name so well and so high. "I understand that anyone who was sunk got thirty days' survivor leave," he wrote, as if Jack stood to receive a vacation for his ineptness. "How about it? Pappy was rather indignant that they just didn't send you back right away." That was the nastiest cut of all. The truth was, as Joe Jr. knew from his father's letters, that Joe believed Jack had given all that a man should be asked to give and had vowed to try to get Jack out of the war theater for good.

Joe Jr. managed a flight to London so he could see his sister Kathleen. She was serving with the Red Cross at the Hans Crescent Club in an old hotel in the center of London. The young Kennedys picked up with one another as if all the time between had been desultory nonsense. They headed out to the 400 Club, where before the war Joe Jr. had attended many a glorious soiree. As they descended into the nightclub, Joe Jr. saw that the kind of young gentlemen who had frequented the place in the old days were still there. But now they wore not evening clothes but officer uniforms. The champagne was the same, if far more expensive, but whereas in 1939 pleasure had been an amiable pastime, now it was pursued with no regard for tomorrows that might never come.

The next evening William Randolph Hearst Jr., a foreign correspondent, invited Joe Jr. and Kathleen to have dinner with him at the Savoy. Hearst had also invited an exquisite brunette, Patricia Wilson, whose husband was a major serving in Libya. Joe Jr. did not let those minor impediments stand in the way as he turned on his inestimable charm. He learned that Patricia was an Australian. She had come to London to make her debut. There the seventeen-year-old debutante married the earl of Jersey, a relationship that lasted six years and brought her a child and endless embarrassment over the activities of her pure rake of a husband. Shortly after their divorce, she had married Robin Filmer Wilson, with whom she had two more children. Now she was doing her bit by working part-time in a factory.

This woman with the cascading, carefree laughter was in some respects not unlike Jack's Inga. Patricia was at that age when beauty is exquisitely refined by life. Like Inga, she was married and she was daring, at least daring in the glimpses she gave to Joe Jr., daring in the way she invited him and his friends to her cottage for the weekend in Woking, not far from London.

Joe accepted Patricia's invitation and began an affair that on each weekend sojourn, each sweet homey interlude far from the muck and cold of duty, became more passionate and intense.

Joe Jr. had cursed the mud and rain of St. Eval, but that had been a sweet oasis compared to Dunkeswell, where his unit was now stationed permanently during the wettest season in memory. It would have been miserable enough if the squadron had been billeted in town, but they were out on a barren flat. The airdrome was nothing but a bunch of big hangars and oval Nissen huts set there to serve as offices, the rain beating a steady tattoo on the metal. To protect themselves from enemy attack, the 64 officers and 106 enlisted men of VB-110 lived a good distance away. "Mudville Heights" they called it, a pathetic group of Nissen sheds set in an ever-deepening pit of mud that appeared likely one rain-soaked night to swallow up the sheds and the men forever.

Joe Jr. had been deputized the squadron secretary, keeper of the diary. He wrote about the mud with the passionate detail that he had never mustered in his articles about the Spanish Civil War. "During the winter months an intermittent drizzle, occasionally whipped into a solid wall of water by the capricious winds, made it almost impossible to stay dry," he wrote, as if he were preparing a prosecutor's brief against the weather. "Inadequate heat—miniature coke stoves sparsely scattered around the base—made it almost impossible to get either dry or warm. Plumbing was early stone age and even more widely dispersed than the living sites or aircraft. There was no toilet paper, although rolls of what seemed to be laminated woods were provided plainly stamped with 'Government Property.' Ablutions were located near the officers' mess which, unfortunately, was about a half mile, as the herd grazes, from Site one."

Mudville Heights was the squadron's main place of repose after their often debilitating, frigid, twelve-hour flights in search of German submarines. The U-boat men called the Bay of Biscay the "Valley of Death," and so it had become. As Joe Jr. flew his plane at fifteen hundred feet, scanning the ocean for the telltale sign of a periscope, he knew that this water was a graveyard not only for submarines but for allied planes as well. He traveled alone, without a fighter escort. He could easily have become bored, sweeping the empty ocean hour after hour. But at any moment a pack of Nazi fighters might appear from above, falling upon the slower-moving plane with their deadly sting.

Joe Jr. had hardly begun flying his patrols when Commander Reedy spotted six German JU-88s on the horizon and had to lumber up into cloud cover before the Germans got close enough to fire. The next day Lieutenant W. E. Grumbles broke radio silence to say that he was being attacked. He called again and again, his poignant messages heard by the other planes, and when he called no more the pilots knew the squadron had lost its first plane.

The very next day Joe Jr. was approaching Junkers Junction, the Atlantic waters off the coast of northwestern Spain, his head hunkered down over the radarscope. He spotted a blip and looked up. He knew immediately that it was a German plane a little over seven miles away and closing. This was a moment to curse the cloudless sky and wish for all the fog and drizzle of Mudville Heights. The German fighter drew within eyesight and tried to shepherd Joe Jr. coastward, further away from any help, like a wolf separating a deer from the herd. A second plane joined its comrade, near enough now that Joe Jr. recognized them as Messerschmitt 210s. One fighter moved in for the kill, now no more than six hundred yards astern.

"Commence firing," Joe Jr. shouted. The gunner whirled his bow turret toward the streaking plane and fired a devastating barrage at the oncoming plane. The ME-210 dove on, homed onto Joe Jr.'s plane, and at the last moment peeled off and retreated up into the sky. The pilot could have downed Joe Jr.'s plane, but his guns must have jammed, for he never fired. On this day the blue sky had conspired against Joe Jr., but he had reason to believe that he was still that child of fortune.

When Joe Jr. got back to the base, he had the next day off. He wrote letters, read, and in the evening headed off to the only diversion within leagues, the Royal Oak, a pub where many of his colleagues attempted to see how many pints they could down before the closing bell sounded. Joe Jr. had a politician's calculated gregariousness and an intuitive understanding of just when and with whom to turn on his blazing charm. The Royal Oak was not such a place that merited his charm, and since he drank hardly at all, he was not always the most convivial companion at one of the battered, dark oak tables.

After that day of so-called rest came a day of briefings, and then the following morning he went up into the gray skies once again. As often as not, Joe Jr.'s most implacable foe was not the Germans but the confounded climate of southern England. He would return after ten hours of fruitless searching to find the airfield socked in. Low on fuel, tired and cold, he would have to find somewhere else to land. At its best, landing was a hairy business, and it had gotten some of his colleagues killed. The worst of it had occurred a couple of weeks before Christmas, when they had been sent out even though the weather reports said that the airport would be closed by the time they got back. When he reached the base, dark clouds blanketed the airport, and he was ordered on to Beaulieu Airdrome outside Southampton. He headed east in the black, rainy night and found himself in the midst of the Southampton barrage balloons, a weapon against his plane as much as they were against the Germans. He descended beneath the five-hundred-foot ceiling. Later he noted in his report that "the wind at the time made it doubly hazardous and trying to watch the field and make a two-needle-width turn at 500 feet in the rain made it quite difficult. I made a short circle and came in."

Sooner or later the Dunkeswell dampness got through not just to the bones but also to the souls of everyone on the base. Joe Jr. was a man of matchless high spirits, fueled by his incomparable Kennedy energy, but his letters home sounded wistful and melancholy. He wrote about the weather because that was the reason he was sitting in his dank quarters writing his family instead of flying. He had only been there a few months, but he was already talking of going off on another assignment. "My love life is still negligible," he complained to his parents in a letter written at the end of January. "Around here, I am taking out a Waff [*sic*] who is very cute but nothing very exciting. There seems to be quite a little talent around the town, but it's such a bother to get in and out, and the added difficulty of obtaining reservations. It's really not much fun unless you know someone."

This was not the boastful Joe Jr. of a few months before who had bragged at the wide swath he had cut through the young ladies of Norfolk. This was a Joe Jr. who, when he opened his letters from home, discovered inevitably that Jack was the big news—Jack's sickness, Jack's appearances, Jack's publicity, Jack's future. His little brother was such a phenomenon that the *New Yorker* ran an article about him and PT-109 by John Hersey; later in 1944 it was reprinted in *Reader's Digest*. Joe Jr. could hardly bring himself to mention his little brother without inserting at least a small dig. "Several people have called my attention to my dear brother's portrait in News week [*sic*], and my apologies for his appearance have been profuse. Who hates him on that paper?" It wasn't enough just to put Jack down; he had to push himself forward in the next sentence. "There have been some articles written over here on our work which I shall send you." Joe Jr.'s problem was a simple one. "I have done nothing to make myself outstanding, but manage to get back, which I suppose is the important thing."

Even in the muck and the cold, Joe Jr. still had his vision of a political ascent ahead of him in America. "What you gonna do when you get back to Delaware?" he asked Duffy one day.

"Got no idea, Joe," Duffy shrugged, looking at what he considered "a big amiable Irishman."

"Politics! That's the game, Bob," Joe Jr. enthused. "We gotta go back and run this country."

Although Joe Jr. still fancied himself as a candidate for high office, he confided to Angela Laycock on one of his trips to London that a Kennedy would be president one day, but that it would be Jack, not him. He declared that Jack was simply smarter, and now with all the glamour and honor that he had won as spoils of war, Jack stood far ahead on the road the two brothers had for so long been traveling.

Among Joe Jr.'s meager effects in the Nissen hut was no creased, much-read letter from his father telling his son that he was a hero too, and that when he returned to Hyannis Port, they would all stand and toast his honor. That was the letter that he needed, some paternal notice that he had done right and good. His father had nearly cried when Joe Jr. won his wings, and his love for his sons was the deepest passion of the man's life. He did not see, though, that Joe Jr. needed, if not a letter then at least a paragraph, or if not a paragraph then one simple line telling him that his father was as proud of him as he was of Jack.

Joe Jr. was not a man who thought that irony was a worthy lens through which to look at the world, but he was full of irony these days. Jack had not courted danger to win a hero's medal. Danger had come to him in the shape of a black destroyer slicing through a blue-black sea. Jack and his men had survived, and for that he was sung a hero's song. Unlike Jack, Joe Jr. was wooing heroism, stalking it as if it were something that a man had to pursue. He had done what he had been asked to do, and he had done it well. He had won no singular medal, but he had been one of a few hundred men who had turned the Bay of Biscay into killing grounds, dropping Nazi submarines into the deep waters so they could not go forth to disrupt shipping, prolong the war, and kill hundreds of merchant marines and sailors. It did not matter that Joe Jr. had no notch on his belt for a sunken submarine or a plane that he had sent down. He was doing his share, if only he could see it.

In May, a *New York Sun* reporter showed up to write about the squadron and said that he wanted to interview the best pilot. He was shown into the officers' hut, where Joe Jr. sat before a warming fire. That was a singular honor to be so chosen, but Joe Jr. was obsessed by what he had not done. "I'll still take carrier duty with a fighter," he said, bemoaning the fact that he had seen no submarines. "Things happen. You don't fly 1,700 hours and see nothing. You don't make twenty-nine trips, ten to twelve hours each, and see nothing. Yeah, I've made twenty-nine. The next one is my thirtieth. Know what happens after you've made your thirtieth? You go out on your thirty-first."

"That is," the reporter concluded his article, "if, like a guy named Joe you just don't have any luck."

Whatever solace Joe Jr. received was in the tender arms of Pat Wilson whenever he could get away from Dunkeswell. Her cottage was one of those rustic redoubts favored by the nature-adoring British upper class. She had chickens on her tennis court, a chauffeur whose duties included milking the cow, and an ever-evolving set of smartly attired weekend guests that often included not only Joe Jr. but Kathleen and her beau, Billy Hartington.

The war had led some to wenching and boozing and mindless games, but gave others the strength to cut through all the silly proprieties and narrow moralisms to live as they wanted to. Joe Jr. was having an affair with a married woman. She had not seen her husband for over two years, and she might not see him until war's end or perhaps never again. Hearts were the least of the things being broken in this war, and no one condemned the couple for their adulterous affair. Kathleen had fallen in love with Billy Hartington before the war. He was a man worthy of love, a combat officer in the Cold Springs Guards, a good and gentle man, bearing one of the great names of England, and he would be the next duke of Devonshire. His only flaw, as Kathleen saw it, was that he was a Protestant, from a family known over the centuries for its hatred of the Catholic Church. In a different time, Kathleen would have fled from his heretical embrace, but she hurried toward him now. Not settling for a paltry affair, she was contemplating marrying the man outside the Church.

The possible marriage risked splitting the Kennedys. Rose stood on one side, a mother who had forfeited much for her family and her faith. For her, the matter had already been decided because it was simply unthinkable. "I wonder if the next generation will feel that it is worth sacrificing a life's happiness for all the old family tradition," she wrote Kathleen in February in a sentence that could have been Rose's own epitaph. As the weeks passed, Rose's entreaties ranged from shrewd arguments to near hysteria.

Kathleen's father adored his daughter and valued her happiness more than all the proprieties of faith, but on this of all matters he dared not stand against his wife. In February, he wrote Joe Jr. that his sister was "entitled to the best and with us over here it's awfully difficult to be as helpful as we'd like to be. As far as I personally am concerned, Kick can do no wrong and whatever she did would be great with me." Joe always cut through the externalities, be they of politics or faith, and he was all for his daughter getting on with her life and happiness, but he could not be seen confronting his wife. As far as he was concerned, he wrote Kathleen, she and Billy should work it out together and "let all the rest of us go jump in the lake." He had heard that his daughter was making converts to the Church, surely a mark of her deep and honest faith. "Maybe if you make enough of them a couple of them could take your place," he wrote his daughter in March 1944. "If Mother ever saw that sentence I'd be thrown right out on the street. . . . I'm still working for you so keep up your courage."

Kathleen endlessly pondered the question of whether she dared to turn away from a Catholic religion that would cast her out of its sanctuary of certainty if she married this good man. Billy had come back to run for the House of Commons in his family's bailiwick. His defeat had brought the question of duty and happiness even more to the fore.

Joe Jr., a man of deep Catholic faith and natural conservatism, might once have sided with his mother. Now, with the tacit blessings of his own father, he stood with Kathleen. "Never did anyone have such a pillar of strength," she reflected later. Joe Jr. had always accepted the dogma of his church, but on this matter he stood up to it, and to his own mother. His was a courage that others outside the family would not see or understand. It was a courage that merited him nothing but the deep gratitude of his sister, and a private satisfaction that he had helped to contribute to what might be Kathleen's happiness.

Rose thought that Kathleen, like all women, was susceptible to the evil blandishments of men, seducers of mind and body. Kathleen astutely realized that her own brother "might be held largely responsible for my decision." Indeed, when she announced her engagement, Rose cabled Kathleen from the hospital bed where she had gone in distress: "HEARTBROKEN. FEEL YOU HAVE BEEN WRONGLY INFLUENCED." On May 6, 1944, Joe Jr. stood next to his beloved sister at the Chelsea Register Office, where he gave her away in marriage to Billy Hartington. He joked that by doing so he had ruined his political future, losing the votes of Boston's Irish. But on this day he hardly seemed to care.

As the firstborn Kennedy son, Joe Jr. had always assumed that he belonged nowhere but at the mountainous heights of life, celebrated for his courage and applauded for his awesome accomplishments. The war had convinced him, though, that human happiness was a worthy goal, not a shameful avoidance of life's high struggles. First, he must complete what he had come here to do, to win a high and noble honor.

"I have finished my missions and was due to start back in about two weeks," Joe Jr. wrote his parents two days after the wedding, "but volunteered to stay another month. I persuaded my crew to do it, which pleased me very much. We are the only crew which has done so."

That was a lie. Perhaps Joe Jr. was trying to reassure his family, but he had not asked his crew to stay on after their thirty-five missions. Joe Jr., however, wanted to be there for D-Day, when he might have his moment of heroism. And so, on June 6, he flew as part of a massive grid protecting the invading forces from German submarines that might work their way into the midst of the mightiest armada in history, wreaking havoc and death and threatening the whole operation. His orders were to fly at fifteen hundred feet near Cherbourg, a city in northwest France far from the Normandy beaches, but the flotilla was so enormous that he could spot some of the ships far out on the horizon. He saw no submarines that day or on the many sorties that followed during the invasion month. On one sortie he spotted five German torpedo boats. He could easily have swooped down and destroyed them. He was just one square in the gigantic grid, however, and his orders were to keep to his patrol and leave it to other planes. He did what he was told, as did the other pilots, and in the end not one ship out of the entire invading force was lost to the German submarines. Joe Jr.'s month was coming to a close, and on his final flights he treated danger as his mistress. On his last mission he flew his plane so near the German guns at Guernsey that he arrived back at Dunkeswell with bullet holes in the fuselage as his souvenir.

Joe Jr. usually had letters from his family to help relieve the tedium of waiting for the next flight. But back home everyone expected his imminent arrival in time for his birthday on July 25, and the family had largely stopped writing him. His father had even stopped sending him the candy that was Joe Jr.'s only addiction. On July 19, his father decided that maybe Joe Jr. wouldn't be home so soon, and he wrote another letter. The major news was that Jack was in the hospital. "He seems to be getting along better, although he has had his ups and downs," Joe wrote. "Rumor hath it that you are very much in love, but we have not had any of the details. . . . Anyway, let us hope you are on your way and we shall see you very soon."

Joe Jr. knew that he would not be returning a hero, not in the exalted way he defined the word. He had lost that sprightly step that he felt would lead him to the highest reaches of political office, and yet his father expected so much of him. "I think at this point you are about the only one of us many that has the good graces of the head of the family but I give you about 4 days after your arrival in this country before you'll join the rest of us," Bobby wrote his brother, warning Joe Jr. of what would face him in America.

Joe Jr. was already packing up his gear when he was called into the squadron office and told about an extraordinary secret mission code-named "Anvil." The allies faced a terrible danger, the Nazis' newest weapon, the V-1 rocket. Since the week after D-Day the V-1s had been slamming into London by the hundreds, wreaking death and destruction and, worse yet, generating a new kind of fear. In France the Nazis bunkered their rocket bases in defenses that so far had proved impregnable to allied bombs. Out in a hangar sat a specially fitted PB4Y bomber that the navy hoped would end all that. This plane that had just been flown over from Philadelphia would be filled with explosives. The pilot would fly the craft up to two thousand feet and then switch the piloting over to one of the two mother ships trailing behind. Then he and his co-pilot would parachute out, while the deadly cargo flew on, to be guided directly toward one of the Nazi V-1 bases.

Joe Jr. volunteered immediately for the assignment, without a moment's reflection and with boyish enthusiasm. In raising his hand high, he was making the kind of choice envisioned by the men of Harvard half a century before. Life was a journey of character through time. A man was true and strong and brave, and one day, if life was as it was supposed to be, he would have a chance to test those virtues on a great and bloody field of combat, and what was important was not whether he lived or died, but that he proved true.

Joe Jr. had finally found the mission that he had sought since entering combat. This was his last great chance. He was going to be thirty years old. He had his first tufts of gray hair. As he saw the world, he was not yet a true man until he had fulfilled his one transcendent heroic act. He figured that he had a fifty-fifty chance of surviving, and those odds were good enough.

Some would argue later, from the safe distance of time and place, that Joe Jr.'s act was not heroic but mindless bravado, a twisted working out of the Kennedy family drama. What must be said, though, of the other officers who volunteered? Would these young men have been held to the same merciless scrutiny? What monsters lay in their childhoods? What demons dogged them? How could they come forward so boldly when they shared neither Joe Jr.'s blood nor his heritage?

Great and noble acts of heroism are born of complicated, ambiguous motives. It is best to use restraint in analyzing them, lest one make cowards of heroes and heroes of cowards. There was perhaps no one better suited to ponder the nature of Joe Jr.'s heroism than his own brother. Years later, when scribbling notes for a letter to Ernest Hemingway, Jack had deep, telling insights into the nature of such self-conscious acts of heroism as his brother's. Jack had lived too long and seen too much to imagine that heroism was a simple, idealistic act. A hero's courage came from "pride—his sense of individuality—his desire to maintain his reputation for manliness which may be more important to him than office—the desire to maintain his reputation among his colleagues as a man of courage, his conscience, his personal standards of ethics—morality—his need to maintain his own respect for himself which may be more important than his regard of others—his desire to win or maintain the opinion of friends or constituents." Even after all the motivations were duly noted, it remained a mystery why in World War II many men lay huddled, shivering in their foxholes, not even firing their weapons, while a few brave men stood up and rushed forward through a ring of fire.

That was a mystery that lay at Joe Jr.'s core, a mystery to those who sought to understand him, and a mystery perhaps even to Joe Jr. himself.

Joe Jr. transferred immediately to the base at Fersfield, where the U.S. Navy was preparing Anvil. The immense dangers of the project were not just theoretical. The Army Air Corps had already lost one man, and two others were seriously wounded in developing a similar program. The navy, which thought itself better, smarter, and quicker than its sister service, was forging ahead, hoping to be the first to strike a fatal blow at the Nazi bases.

During the day Joe Jr. flew the new plane, empty of its deadly cargo, on testing missions. In the evenings he cycled off the base to a phone booth and talked to Pat for twenty minutes or more. Before he went to bed, he got down on his knees in his room full of other men and said his prayers. If another officer had done that, somebody would probably have made a dismissive aside, but no one said anything. And one night, when the other men were still playing cards at two in the morning, and Joe Jr. got up and told them enough was enough, no one told him to buzz off. Some of them outranked him, but they went somewhere else to play their game.

Two of the other officers confided to each other that Joe Jr. was their model of what a man should be. He did not wear his life on his sleeve, spewing out all the details like a tour guide to his own autobiography. At times he would venture a word or two, saying that his father wanted him to enter politics after the war, but he wasn't quite sure what he wanted to do. One night, as he lay half asleep in his bunk, he mumbled a few words about getting married and heading off to Scotland on his honeymoon. But these were just snippets of dreams, memories floating through the night.

Another man would have written letters to his family and friends to be delivered only if he died. Joe Jr. was not a man to take ominous precautions. He wrote a letter to Jack that he knew would not be delivered until after the mission, a letter full of untruths.

The brothers communicated in only one idiom, a jocular bonhomie more appropriate for boys at Choate than for brothers in a war-torn, tragic world. This was the only emotional language among men that Joe Jr. knew. He told Jack that he thought the *New Yorker* article was "excellent" and that "the whole squadron got to read it, and were much impressed by your intestinal fortitude." But he gave that niggardly spoonful of praise only to set Jack up for another merciless put-down. "What I really want to know, is where the hell were you when the destroyer hove into sight, and exactly what were your moves, and where the hell was your radar."

Often Joe Jr. found that the only way to deal with fears and emotional truths was to say the opposite of what he meant. "Tell the family not to get excited about my staying over here," he wrote. "I am not repeat not contemplating marriage nor intending to risk my fine neck (covered in the back with a few fine silk black hairs) in any crazy venture."

Joe Jr. was at the start of his adult life and career, but he had an overwhelming sense of his own mortality and age. All the letters from America were full of stories of Jack and his health problems. Joe Jr. might have contrasted his own vibrant health with the ill health of his brother. But he saw Jack as youth and vitality and sexual vigor. He figured he'd be back in the States around the first of September, and then maybe, if Jack had the time, he "will be able to fix your old brother up with something good." Jack would have to get on his stick, for his big brother had "graying hair," and he would have to come up with "something that really wants a tired old aviator."

Jack was the winner in all the competitions that mattered, from the honors of war to the spoils of women. "My congrats on the Medal," Joe Jr. concluded his letter. "It looks like I shall return home with the European campaign medal if I'm lucky." Joe Jr. knew that was not true. If he succeeded, he would fly home to America not only with the medal given to all who had served in the campaign but with the navy's highest honor, the Navy Cross, and he would once again be ahead of Jack on that road on which he believed they both traveled.

If Joe Jr. could only have seen Jack, he would have realized that his younger brother was no longer trudging manfully up the road their father had paved for his sons. Jack sat at the side, dazed, looking out on a world that hurried on without him. The previous winter he had suffered relapses of malaria, his body shaking uncontrollably, his face peaked and yellow, his condition so perilous that his father's friend Joe Timilty thought that he would die. The malaria, however, was nothing more than a sweet souvenir of the South Pacific compared to Jack's back problems.

Chronic back pain is not only a physical condition but a philosophical assault that turns even Pollyannas into doomsayers. In June 1944, Jack's back had been operated upon at the New England Baptist Hospital, where the surgeon from the Lahey Clinic achieved a "thorough removal of the degenerative portion of the cartilage." But that seemed only to worsen matters. "He . . . is obviously incapacitated by pain in low back and down L. (left) leg," Jack's August 1944 medical report stated. "He cannot bend forward supporting weight of trunk. Lat and post. bending limited. . . . This is a high-strung individual . . . who has been through much combat strain. He may have recurrent disc or incomplete removal but better bet is that there is some other cause for his neuritis." This is the only place in Jack's long medical history where there is a suggestion that there may have been a relationship between his psychological condition and his health. Another possible contributing factor was impending Addison's disease, a potentially deadly malady that would not be diagnosed for several years.

As Jack lay in his bed, a group of his old PT-boat mates burst into his room at Boston's Baptist Hospital with a basketload of good cheer and endless stories of life in the Pacific. Jack was heartened at the sight of Joe and Lennie and Johnny and Al and by their enthusiasms. But as consummately as Jack could portray himself as far healthier than he was, now he was so weak that he could do little except lie there. Lenny had brought his bride, Kate, along. She was a nurse, and she found it ominous that Jack's bed had not been cranked up and that he lay there perfectly flat. The visitors had hardly been in Jack's room a few minutes before a nurse walked in and told them that they would have to leave. Jack was so tired, so terribly spent, that he did not protest, but bid them adieu.

For Joe Jr., these last two weeks in August 1944 had been full of baleful signs. Training flights had gone astray. The project was raked with silly bureaucratic ineptitude. Serious warnings about the faulty electrical system had been smothered by the command. The whole project reeked of the military's institutional stupidity and sterile arrogance that had made Jack so despairing of the war efforts. On the evening before the mission, one of the men, Earl Olsen, tried to warn Joe Jr. that the arming panel and so-called safety pin might blow up the plane. Joe Jr. cut him dead. He did not want to listen to the man. When the mission was delayed a day, Olsen had another chance to talk to Joe Jr. and tell him explicitly about the problem. Joe Jr. might have asked some more questions, but he turned from the man and his menacing truths. Joe Jr. admitted to another officer that he was sorry he had volunteered, but he believed that it was too late to do anything but go on.

Joe Jr. could have gone to his superior and asked that the mission be postponed until the plane was properly checked out. That would have required a different kind of courage: if he had done so, some men might have wondered whether Joe Jr. was a coward, and that was an appellation that he would allow no man to connect with him. A brave man is often as fearful as a coward. It pains him to be called a coward, for he knows what lies within himself and fears what he might have been or might well be.

August 12, 1944, dawned a cloudless day, and Joe Jr. knew that his mission would be delayed no longer. Late that afternoon he and his co-pilot, Wilford Willy, boarded the shiny new PB4Y drone loaded with 23,562 pounds of Torpex, an explosive nearly twice as powerful as TNT. This was a mission to be chronicled to the last detail and to be celebrated as one of the triumphs of the war. It was so important that Eliot Roosevelt, the president's son, was on hand taking photos of the men before they took off, then flying off himself in a Mosquito plane to memorialize the flight even further.

They were a mini-armada, with Joe Jr. at the center of it all, piloting his silver PB4Y, the two mother ships, Roosevelt's plane, sixteen Mustang fighters as fighter protection, plus a B-17 ready to fly to another airport to pick up the parachuting navy men. Joe Jr. took the plane up to two thousand feet and leveled off. The first step was to let the mother ship take control of the drone and run through a few maneuvers. Once that was over, Joe Jr. and Willy would bail out. It was as simple as that.

The mother ship gently guided the robot plane into a left turn. At that moment, Joe Jr.'s plane exploded in an immense yellow circle of flame, like the sun blazing in the evening sky. Then the light was gone, leaving the sky filled with black smoke and small fires in the woods below.

Book Two

12

A New Generation Offers a Leader

On the second Sunday in August 1944, Joe lay upstairs in his bedroom in Hyannis Port taking his afternoon nap. For a man whose sons were the unmitigated joy of his life, this weekend was a blessed time, almost the way it had been before the war. Jack had come home on leave from Chelsea Naval Hospital in Boston. Bobby had made it to the Cape for a couple of days. He was wearing navy blue, but as Joe saw it, Bobby was still blessedly thousands of miles from combat. Little Teddy was his irascible self, running around the grounds, he and Joey Gargan filling the house with a joyous youthful tenor. Best of all, Joe Jr. would be home in a few days too, charging into the house with hugs and hollers, belittling his father's morbid fear that he would lose at least one son in the war. Eunice, Pat, and Jean were there too, and they adored their father and showed their love in demonstrative ways that their brothers could not.

Rose woke up her husband and said that there were priests downstairs who insisted on talking only to him. Joe went downstairs and led the priests to an anteroom. Joe queried them again and again until he knew that Joe Jr. was gone, gone forever. Then he came out of the closed room and with his arm around Rose told his three surviving sons and his daughters that Joe Jr. was dead.

"Children, your brother Joe has been lost," he said, looking at Jack, Bobby, Teddy, and their sisters, his eyes gleaming with tears. "He died flying a volunteer mission. I want you all to be particularly good to your mother."

Rose turned to her church and found the solace available only to a woman of profound faith. As for Joe, he had no such faith and could offer his children nothing but clichés. The most profound tenets of faith and philosophy were to him just mindless platitudes, propping up the soul. "We've got to carry on," he told his children. "We must take care of the living. There is a lot of work to be done."

There is no device calibrated to judge the magnitude of a father's mourning over his lost son, but those who saw Joe said they had never seen a man suffer more and feel more deeply. When Joe called his sister, Mary Loretta, his mournful sobs were so deep that she feared he would never stop.

"Joe's death has shocked me beyond belief," Joe wrote James Forrestal, the secretary of the navy, responding to his letter of condolence. "All of my children are equally dear to me, but there is something about the first born that sets him a little apart—he is for always a bit of a miracle and never quite cut off from his mother's heart. He represents our youth, its joys and problems."

"He was a real man," Joe wrote Forrestal. Joe had brought his sons up to be true men, to pursue lives of courage, and his eldest son had lived as his father had wanted him to live. Joe could have manipulated his sons' military careers so that they would have been away from the cannons' roar. He had not done so, and he allowed them to fight in a war in which he did not believe. And what did he have for his misbegotten nobility but a dead son, and a second son who was half dead, a pallid invalid?

"And so the story of young Joe, some may think is at an end," Joe wrote the official. "But is it? I cannot but believe, as you so nicely said in your letter, that 'his character and extraordinary personality' will be a legacy to others." Joe would found a foundation in Joe Jr.'s name. He would have a ship named after his firstborn son. But the memorial he cared about was flesh and blood, Joe Jr.'s life of true manhood living on, memorialized in the lives of others.

Jack wandered the beach at Hyannis Port that Sunday afternoon and then returned to his hospital bed in Boston. Jack was the sickly brother. As he saw it, his own death would have been far easier to bear. "It came at a time when I was really awfully, you know, I was weighed at 122 or 123 and sick as hell, gray and green and yellow," he recalled years later. "If something happens to you or somebody in your family who is miserable anyway, whose health is bad, or who has a chronic disease or something, but anybody who is really living at the top of the peak, then to get cut off, it is always more of a shock."

Jack had time now to contemplate his brother's loss and to puzzle out his own uncertain fate. Gone was all his ambivalence toward his big brother. He remembered him first of all as a man who "enjoyed great health." It was a characterization that would occur only to a brother who was himself sick. Jack saw that Joe Jr.'s health was the mother of his other traits, his "great physical courage and stamina, [and] a complete confidence in himself which never faltered." Jack believed, rightly, that Joe Jr.'s virtues "were in the end his undoing." The physical courage, stamina, confidence, verve, and gusto had led him to stay on in England and to fly that last fateful mission.

As Jack lay in his hospital bed, his father did not stand outside the door with Joe Jr.'s fallen banner, waiting to thrust it into Jack's unwilling hands as soon as he was able to walk unaided. Joe surely wanted his oldest surviving son to continue on the pathway set out on by his brother, but there were other pressures on Jack as intense as his father's expectations. One of them was the profound question that every surviving combat soldier asks: Why me? Why was I saved?

Jack received many deeply felt letters of condolence, futile attempts to ponder the imponderable. "You are the Kennedy fame," Mike Grace wrote him, "but it becomes all too plain that Joe was the heart behind your name." The friend meant to honor Joe Jr., not to demean Jack, but there was the devastating reality that he faced.

Jack thought that "the best ones seem to go first," and that there was "a completeness to Joe's life, and that is the completeness of perfection." It was part of the solace of war to believe that the bravest died young, their virtue confirmed by their deaths. "There must be a reason why the good are called upon while the bad are left to rot," his old flame Harriet Price wrote him.

There in its starkest form was the psychological reality that Jack faced. He had not volunteered for a hero's role. He was a man of intellectual honesty and saw that there was a profound distinction between what he had done to live and his brother's actions that had led to his death. As sick as Jack might be, he was alive. No matter how he had suffered, how bravely he had acted, he dared not think of himself as being as good as his brother who was gone. And now he had the additional burden of picking up a banner that he felt he could hardly lift. "You will have some of your brother's unfinished business to do during the long years that face you, so get thoroughly well before you start," Barbara Ellen Spencer wrote him.

The night turned even darker a month later in September 1944 when a friend, Kathleen's new husband, Billy Hartington, died leading his company in a fight against the retreating Germans in Belgium. That October, as Joe sat in Hyannis Port thinking of Joe Jr.'s death and Billy's death and of poor Jack lying in the hospital in Boston, he wrote his friend Lord Beaverbrook that he was writing with his "natural cynicism."

That was as close to a moment of psychological insight as he would allow himself. He would never say that he was "depressed"; the emotional lexicon of the Kennedys did not contain this word. "To have boys like ours killed for a futile effort would be the greatest reflection on us all," he wrote Beaverbrook.

"Yet, if you would ask me what I am doing to help, I would tell you nothing. However, I assure you it is not by choice but rather by circumstances."

"Why does no one come to see me?" Joe asked his first cousin Joe Kane, a shrewd political operative close to the family.

"Who'd you ever go to see?" replied Kane. A man's goodness and generosity are his capital, and Joe had spread precious little of it beyond the precincts of his home. All he seemed to have left was his cynicism, and that he had in bountiful supply. He was an old sailor adrift on a lifeless sea.

Joe had let it be known that he was thinking of "making a speech for [Governor Thomas] Dewey," Roosevelt's putative Republican opponent in the 1944 election. That was surely high on the list of reasons that the president called his former ambassador at the Waldorf on October 24 and invited him down to Washington for a visit two days later. When Joe walked into the Oval Office, he was shocked at the aged man who put his hand out and waved him to a seat. The president had always had a splendid memory, but now events seemed to be lost in a fog of memory, names half-remembered, numbers skewed.

As Joe sat there realizing that the president was ill, he had a dreadful premonition that "the Hopkins, Rosenmans, and Frankfurters could run the country now without much of an objection from him." Joe was still obsessed with the Jews and what he took to be the conniving, unscrupulous way they rose to power and wealth. He told Arthur Krock afterward that he warned the president he could not abide "the crowd around you—Niles, Hopkins, Rosenman, etc. They will write you down in history, if you don't get rid of them, as incompetent, and they will open the way for the Communist line. They have surrounded you with Jews and Communists and alienated the Catholics." Joe warned the president, as he wrote in his diary, that the old-line Democratic voters "felt that Roosevelt was Jew-controlled."

Those who defended Joe's attitudes pointed at Krock and Baruch and argued that a man with Jewish friends couldn't be anti-Semitic. He had his Jewish friends and associates—journalists, lawyers, doctors, and political advisers—but only because he considered them smart and useful. He played golf at the Palm Beach Country Club, a Jewish club, not because he chose to make some kind of statement, but because it was close to his home.

His children listened to their father long and well, and Kathleen shared her father's keen ability to spot a Jew. Six million European Jews had died in the Holocaust, but after the war, Kathleen found them to be ubiquitous in Paris. "The people one sees at the collection aren't a bit chic and the shops complain the people who buy now are black market profiteers," Kathleen wrote the family on September 15, 1946. "The Jews are in evidence in all the shops and restaurants."

For Jack, the pain and illness never seemed to end. He had been back in the States for over a year now, shuttled between hospital beds, cut open and shot up with drugs, and he appeared no closer to good health than when he had arrived. The doctors tried procaine, and while that made his back and leg pain tolerable, he was still in pain. In addition, in November 1944, they diagnosed him as having "Colitis chronic." The doctors had tried everything they knew to try, and in the end they let him go, telling him that his convalescence might last another year.

Writing in a hospital bed from which he had good reason to believe he would never rise to full health and well-being, he sent a letter to his friend from the Pacific, Paul "Red" Fay Jr. He did not tell Red of the despair that he surely must have felt but covered his emotions with joking bravado.

"Sometime in the next month I am going to be paying full price at the local Loews," he wrote. "I will no longer [be] getting the forty-per-cent off for servicemen—for the simple reason that I'm going to be in mufti. This I learned yesterday—as they have given up on fixing me up O.K. From here I'm going to go home for Christmas, and then go to Arizona for about a year, and try to get back in shape again."

Jack headed out to the mountains of Arizona to see whether the western air could do what scalpels and medicines had not done. In his old navy fatigue pants and shoes, he was hardly the East Coast dandy. To a new friend, J. Patrick Lannan, Jack looked "yellow as saffron and thin as a rail," and he gently bemoaned the fact that he could not digest much food. He was not much of a horseman, but he galloped through the high desert as if he thought he could outrace his illness and his own uncertain future.

His beloved sister Kathleen had written him: "It just seems that the pattern of life for me has been destroyed. At the moment I don't fit into a design." He could have replied to her in much the same way.

Jack worked on a memorial book for Joe Jr., patterned in part on his favorite book, John Buchan's *Pilgrim's Way*. Since Jack's work on the book primarily involved editing the reminiscences of others, he had plenty of time for other pursuits. He talked to his cottage mate, Pat Lannan, about world affairs. Jack told his new friend that he was thinking of running for Congress from Massachusetts. He probably knew that his father was attempting to provide financial incentives to Congressman James Michael Curley sweet enough that the indicted politician would retire, opening up his Boston seat to Jack. "I am returning to Law School at Harvard in the Fall, and then if something good turns up while I am there I will run for it," he wrote Lem rather cryptically. "I have my eye on something pretty good now if it comes through."

Jack told Lannan that he hoped to enter public service. "Oh, you mean politics," Lannan said, but Jack treated the word as if it were an epithet. "Politics" still had that immigrant stigma to it. It was the pathway upward. At Choate and Harvard he had heard the understated disdain for the motley business of politics, and he was not yet ready to shout out the name of his intended profession. Men of the upper class might go into the State Department or OSS, but they did not for the most part run for office. They did not choose to have their careers depend on the will and whim of people they neither knew nor, for the most part, wanted to know.

Every afternoon at five, Jack made sure that he was in the cottage waiting for a call from his father. Joe sent him books that he thought his son should read, as well as steaks and chops to bolster his strength. In their daily conversations, Joe tried to decipher the truth filtered through Jack's cheery dialogue.

"Frankly, don't think it [his health] is any too good," Joe admitted in a letter in March 1945 to Red Fay, Jack's friend. "On the phone he is still his gay self so it is very difficult to give you any more definite information from this."

Jack didn't have to depend solely on his father to learn what was happening in Hyannis Port. Bobby wrote his big brother, providing Jack with intelligence about life back home. Bobby did not lace his letters with the joshing put-downs that had run through Joe Jr.'s correspondence but wrote with simple sincerity. Bobby knew that he and his brother shared the mixed blessings of a mother and father who hovered over them, egging them on more like coaches than traditional parents. "Everyone evidently thinks you're doing a singularly fine job out there, except mother who was a little upset that you still mixed 'who' and 'whom' up. Get on to yourself. . . . I suggest you stay out there for as long as possible. That staying away will also be good for keeping you a fair headed boy around the house, for I know it takes you almost as short a time as it does me to finish yourself off when you're home."

Even though Jack's back troubled him so much that he thought he might have to return to the Lahey Clinic in Boston, when he drove into Phoenix to stay at the Arizona Biltmore, the sight of attractive women, including one of Hollywood's leading stars, made him forget his pain. "Anyways their [*sic*] was some pumping which interested me," he wrote Lem, "and I did take Veronica Lake [for] a ride in my car. . . . I don't mean by all this that I pumped her or that if you should ever see her you should get a big hello. . . . I am heading out of here to Palm Springs where I expect to tangle tonsils with Inga Binga among others." He was not without his conquests: he bragged to Lem that he had slept with forty-three-year-old Lili Damita, a former film star and Errol Flynn's recent wife. "I took a piece out of Lili

Damita just for the sake of Auld Errol Flynn," he wrote Lem, "but did not come back for a second helping."

Jack ended up in Los Angeles. Since he was a little boy, and his father had returned from Hollywood with Tom Mix cowboy suits and wondrous tales, Jack had been fascinated with the movie world. On the Hollywood screen the sick were made whole, the barnacles of age removed, and cares exorcised. Some of these illusions were so powerful that nothing erased them, no merciless houselights, no rude truth-sayers, nothing.

His friend Chuck Spalding was working for Gary Cooper. The lanky, taciturn Montanan conveyed a sense of manly heroism and dignity that nothing he did or said could obstruct or diminish. Jack was what they called a war hero, but what was he compared to Cooper, who portrayed *Sergeant York*, the great hero of World War I, inhabiting his very life, and played the role of Robert Jordan in *For Whom the Bell Tolls.* During those weeks in Hollywood, Jack studied Cooper as if he were learning an exotic language, attempting to see whether he could speak it without an accent.

Jack saw Cooper as a consummate artist of charisma and endlessly pondered how the man did it. For hours, Jack discussed with Spalding the nature of stardom. How was it created? Why was it that crowds swarmed around Cooper? What was it about Spencer Tracy, a man of undistinguished looks, that made him such a star? Did he, Jack Kennedy, have what it took? Did he have this quality of stardom? Or could he create it?

The star-making machinery took obscure young actors and actresses and set them up on the pedestal of fame to be worshiped and emulated. The process may have seemed patently transparent, churning out nothing but one fraudulent creation after the next, an endless parade of interchangeable performers. The reality, however, was that the public had to connect to that person on the screen, and if it did so in a profound and visceral way, then a star was born. That process could be manipulated only so far. That was what obsessed Jack—whether, in Spalding's words, "he had it or he didn't have it."

As candid as he was with Chuck, Jack always held back a part of himself, the self-critical, manipulative self that sent the other Jack Kennedy on his public pathway. Of all his male friends, Spalding probably saw more of the spectrum of Jack's personality than did anyone else and had deeper insights into his friend. Chuck saw that Jack was a natural seducer and that women were merely the temporary objects of his game.

Spalding observed that Jack's obsession with bedding every attractive woman who even momentarily passed across his horizon was not primarily about sex. The act itself was usually nothing more than a quick release. It was about testing his power and charm and will. It was not sexual impotence he feared but the possibility that he, the Jack Kennedy he had created, wouldn't work anymore.

What better test of his powers, though, than to seduce Olivia de Havilland, who had played the angelic Melanie Wilkes in *Gone With the Wind.* Jack was usually proudly disheveled, but this afternoon, as he was about to meet Olivia for drinks at her house, he prepared his toilet at the Beverly Hills Hotel with detail and refinement.

Jack fixed Olivia with deep, glowering eyes. He flattered her with subtle phrases. He paraded his wit before her. Nothing could ruin his little recital more than speaking his lines too long or too loud. He rose to say good-bye to the star, finally turning and opening what he remembered as the front door. As he turned the knob and moved to head out into the world, tennis rackets and luggage came tumbling down out of the closet he had mistakenly opened.

The closest most veterans got to Olivia de Havilland was a front-row seat at the Bijou, but Jack fancied he could have whatever he wanted, from an assignation with a movie star to a privileged seat at the founding of the United Nations in San Francisco in the spring of 1945. His peers would have been delighted simply to have been there, but Jack wrangled an assignment as a reporter for Hearst that would have been the dream of any journalist.

Jack wore his seriousness as lightly as it could be worn. One evening Arthur Krock found his erstwhile new colleague propped up on his hotel bed in impeccable evening clothes trying to reach the managing editor to tell him, "Kennedy will not be filing tonight."

When he arrived in San Francisco in late April, Jack had neither the giddy optimism of the more vocal internationalists nor the paranoia of right-wingers who believed that this so-called United Nations would end American sovereignty. In his first articles he had the cautiously hopeful attitude of most of his fellow veterans. He compared what was going on to "an international football game with [Soviet foreign minister Vyacheslav] Molotov carrying the ball while [Secretary of State Edward] Stettinius, [British foreign minister Anthony] Eden and the delegates tried to tackle him all over the field." He continued to employ sports metaphors, accusing Molotov of "throwing curves" while the United States "juggled the ball."

The Soviets, the Americans, and the British, however, were playing different games with different rules, and Jack struggled to understand. During these three weeks he was chronicling not the inspiring beginning of a world organization but some of the first mistrustful bickering of the cold war. He tried at first to comprehend the world from the Russian position. "Americans can now see that we have a long way to go before Russia will entrust her safety to any organization other than the Red Army," Jack wrote. "The Russians may have forgiven, but they haven't forgotten. . . . This being true, it means any organization drawn up here will be merely a skeleton."

Jack understood that it was inevitable that veterans would be disillusioned. They were men of dual simplicities, those of youth and those of war. Now was the time for the old men to make the peace, building a world organization that was "the product of the same passions and selfishness that produced the Treaty of Versailles." Jack discovered "one ray of shining bright light. That is the realization, felt by all the delegates, that humanity cannot afford another war."

As the days went by, even that ray of hope dimmed. He saw the spectacle of the Russians dickering to win admittance for White Russia and Ukraine as separate states, while the United States backed the admittance of pro-Fascist Argentina. He saw Britain struggling to maintain its empire. And he saw a Russia whose armies sat ensconced in Eastern Europe, imposing their own new tyranny on subject peoples.

Jack had arrived in San Francisco sharing most of the affinities and aspirations of his generation, but as he left he sounded more like the old men who had made the peace than the young men who fought the war.

"Our preoccupation with the war and our desire to remain on good terms with our Red allies has prevented us from taking a strong stand against Russian infiltration through Europe," he wrote shortly before leaving. "That time is over and it is becoming evident that the Big Three relationship is at the crossroads."

Jack's fling as a journalist was not over. From California he traveled to London, where he covered the British election for Hearst. Churchill was one of Jack's authentic heroes, the personification of his nation's noble struggle against Hitler. It was almost unthinkable, then, that Churchill's victorious compatriots would turn him out of office. It pained Jack to hear the great man booed by a surly, ungrateful populace. Yet when Labor won, he saw that the British people were tired of nothing but "toil and sweat." He saw the profound class nature of British society and how it had created, in Disraeli's words, two nations.

Jack spent his time with the nation of wealth and privilege and observed a society in which at times the nobility and purpose of the battlefield seemed just another casualty of war. One weekend he was at one of the great houses of England. Some of the other guests had driven over to the races at Ascot, where the upper class was resurrecting its prewar social rituals, but Jack preferred to sit at the country estate talking about war and peace. Up on the third floor, beneath a harsh light, he observed a group of young men playing poker. A year before, the lights had been out in London, and many of them had been risking their lives. Now the lights had come on, and these gentlemen turned their risks to the gaming table. They played with abandon, and that afternoon one of them lost a fortune, somewhere between 100,000 and 200,000 pounds.

When Jack was not musing with his male friends, he was squiring beautiful women around London's nightspots. His thin face had a wan, poetic cast, suggesting sensitivity to the will and desires of women that his conquests soon learned did not exist. One of the women he saw in London, the tennis champion Kay Stammers, found him British in his casual disregard. "He really didn't give a damn," she recalled. "He liked to have them around, and he liked to enjoy himself, but he was quite unreliable. He did as he pleased."

At the end of July, Jack traveled with Secretary of the Navy James Forrestal to war-ravaged Germany. As he flew above the once-great cities, he wrote in his diary: "All the centers of the big cities are of the same ash gray color from the air—the color of churned up and powdered stone and brick." He scribbled down no unique insights about the ravaged land, but he had a journalist's realization that the gritty facts came first, no matter how unpleasant.

Jack recorded that the Russian soldiers, on entering Berlin, had spent their first seventy-two-hour passes largely "raping and looting" and were now stripping the land of everything of value, from factories to manpower. He also recorded that in the allied-occupied port of Bremen, the American and British troops "have been very guilty of looting," and that navy personnel said that a congressional fact-finding committee had been interested only in "lugers and cameras." It was as if much of the selfishness, greed, lust, and pettiness that had been dammed up during the war had broken loose, washing over the desolated landscape of Europe.

On the last stop of his German tour, Jack traveled to Hilter's legendary mountain redoubt, high above Berchtesgaden. Allied bombs had gutted Hitler's chalet, and the allies had stripped his "eagle's nest" of rugs and paintings, but the aura of the Führer remained. Jack stood there in the immense round living room from which he looked down on all sides on endless expanses of forest. He saw no pockmarked landscape, no smoking ruins, no haunted faces, only pristine forest. Jack had just been in Potsdam, where he had seen President Harry S Truman, a matter so inconsequential to him that he did not even mention it in his diary. He noted that General Dwight D. Eisenhower "easily won the hearts of those with whom he worked," but no one impressed him as did Hitler.

"You can easily understand how that within a few years Hitler will emerge from the hatred that surrounds him now as one of the most significant figures who ever lived," he noted in the final page of his diary. "He had boundless ambition for his country which rendered him a menace to the peace of the world, but he had a mystery about him in the way that he lived and in the manner of his death that will live and grow after him. He had in him the stuff of which legends are made."

Jack was arguing that if a leader is judged simply by how much he changed the world, then Hitler was the great man of his age, a political artist who transformed the world through his evil. Although Jack's comment displayed an appalling insensitivity to the ravages of Nazism, he was no apologist for Hitler. He was studying politics and politicians, picking up bits and pieces of people and ideas that he might find useful in his own career, even if they came from Hitler.

Since researching *Why England Slept*, Jack had been dissecting the nature of leaders in a democratic society. Could a democratic leader transform the world for good as much as Hitler transformed it for evil? He observed Englishmen voting their bellies, not their souls. He watched German girls selling themselves for a lipstick. A month before the atomic bombing of Hiroshima and Nagasaki, he mused about "the eventual discovery of a weapon so horrible that it will truthfully mean the abolishment of all the nations employing it." In such a world, was greatness possible?

When, in April 1946, twenty-eight-year-old Jack declared that he would seek the Democratic nomination for Congress from his grandfather's old eleventh congressional district, he was doing more than putting on clothes his father had laid out for him. Inga Arvad probably understood Jack more deeply than anyone. Even before the war she had prophesied that Jack would enter politics. During the war, he had joked to Red Fay that he could feel "Pappy's eyes on the back of my neck," and told his friend that "when the war is over I'll be back here with Dad trying to parlay a lost PT-boat and a bad back into a political advantage." That was the cynical posturing that was one part of Jack. Red rarely heard the more serious Jack who for hours discussed running for office with Lannan and Spalding. Jack scrutinized the prospect from many angles. He observed other politicians. He courted publicity in Massachusetts for months. He waffled back and forth in front of friends. Lem thought his closest friend would be entering law school. But Jack was no more interested in enduring the tedium of case law than he was in leading the outsider's life of a journalist or entering a business world that he largely disdained.

Jack had grown up in Bronxville and prep schools and knew nothing about his prospective constituents and their ways. The district included middle- and upper-class people around Harvard and parts of Beacon Hill, but most of the residents were working class and poor. They were so solidly Democratic that the winner of the primary would inevitably win in the general election. There were some new arrivals and many second- and third-generation Americans who had not been able to move away from their immigrant ghettos.

Whether they were Italian, Irish, Polish, Portuguese, or Chinese, these new Bostonians dreamed of the day when they would live somewhere else. But for now they lived among the shipping yards, freight depots, oil tanks, and factories, and close by the state prison. Jack's deepest political interest was not these inner-city residents and their mundane problems, but the world far beyond. Jack's biggest job, as *Time* phrased it, "was to convince the 37 different nationalities in some of Boston's grimiest slums that he was not just the wealthy son of Joe Kennedy . . . but rather an attractive individual in his own right."

In the triple-deckers that lined the long gray streets, the kitchen was the center of the home, especially in these cold winter days, for the flats did not have central heating. A knock on the front door was a stranger's knock. Someone apprehensively hurried through the frigid, shut-off rooms to greet a visitor who as likely as not brought bad news, not good.

These days the door often opened on a pallid, rail-thin, tousle-haired young man standing there with a nervous smile, a faltering greeting, and a manner and dress that instantly signaled that he was not one of them. One of the flats Jack visited was that of Dave Powers, a navy veteran whom Jack hoped to enlist in his cause. He stayed in Powers's cold living room for a half hour chatting away, asking the Charlestown man to accompany him to a speech for Gold Star mothers at the American Legion the following week.

Powers had a sense that Jack was "aggressively shy." Any man of even modest sensitivity would be shy knocking on the doors of people he did not know and asking them to support him for reasons he was not even fully sure of himself. Jack's was not the ham-handed, eyes-averting shyness of insecurity, but the shyness of a man thrust into a world in which he would be dependent upon the kindness of strangers, a world he did not know, and one in which he did not feel comfortable.

Jack half mumbled his prepared speech to the honored mothers in a virtual monotone, looking up only occasionally as if to be sure the audience was still there. "I think I know how all you mothers feel, because my mother is a Gold Star mother, too," he droned. The audience knew that the speech had ended only because the young candidate stopped speaking. By any measure, the speech had been a disappointment, but the mothers rushed forward to greet Jack as if he had been the most stirring of speakers. To them, Jack was no longer a privileged outsider. He was a young man who needed them, and they responded not simply with their votes but by buttonholing their friends and relatives and telling them about this fine young Kennedy.

Powers sensed what was happening. Despite his pledge to work for a fellow Charlestown man, John F. Cotter, Powers signed on with Jack. Powers led the candidate up the back staircases of his neighborhood where Jack was greeted not as a stranger but as a friend.

Jack may have impressed the Gold Star mothers, but he was far from a stellar candidate. His own campaign manager, Mark Dalton, would probably have made a better candidate. The thirty-one-year-old Harvard Law School graduate was a true son of Boston. One of his brothers was a priest; the other was the political editor of the *Boston Traveler*. Dalton had fought a war that few veterans had fought. He had been at D-Day in the sixth wave and landed on Okinawa in the last major land battle of the war.

Dalton had no money to run. Beyond that, he had a kind of moral rectitude that struck the hard-edged politicians around Jack as naivete. Dalton was a man who took the rhetoric of the campaign as literal truth. He was a passionate speaker when he believed in something, and the first time he introduced Jack, he delivered such a forceful speech that poor Jack's speech sounded, by contrast, as thin and weak as his appearance.

After that address, Jack and Dalton drove over to the Ritz-Carlton Hotel, where Joe sat in the living room of his suite with Joe Timilty, the Boston police commissioner. The two older men had listened to the radio address, and by the time the candidate and his campaign manager arrived, they had had ample time to evaluate Jack's pallid performance. They told Jack and Dalton that never again would they appear together on the same platform; the contrast was too great.

Joe had let his sons take their chances standing in the front lines with other young Americans. That was the last time he would play what he considered a fool's game, abiding by the simple rules that guided others. Mark had the title "campaign manager," but he was simply a handsome face on the platform and a name on a letterhead.

Joe ran the campaign, but he did it in such a surreptitious way that no one knew just which strings he had pulled and how hard he had pulled them. Even the fact that the congressional seat had suddenly opened up probably was the result of Joe's manipulation. The incumbent congressman, James Michael Curley, was facing an indictment for mail fraud; Joe Kane, who was intimately involved with the campaign, later asserted that Joe had paid Curley twelve thousand dollars to retire and had promised more money when Curley decided to run for mayor of Boston.

Joe called Dalton every few days, spending hours seemingly going over every detail of the campaign. As the weeks went by, however, Dalton realized there was another secretive campaign about which he knew nothing. Many of the old politicians who had first opposed Jack now had smiles on their faces, a lilt to their walk, and the name "Kennedy" on their lips.

Joe spent an estimated three hundred thousand dollars on the campaign. That was enough to blanket the area with billboards, distribute one hundred thousand reprints of John Hersey's PT-109 article, and place scores of radio and newspaper ads and placards on the trolleys. Joe believed that every man had his price, and he would pay that amount, but not a dollar more. He believed, moreover, that if you set your money in plain view people would steal it, your friends as easily as your enemies. Most campaigns had all sorts of unaccountable cash expenses. Joe had Eddie Moore set up an elaborate accounting system, detailing even the smallest expenditures on triplicate forms.

Those who hoped to work with the Kennedys learned that parsimoniousness was next to godliness. Dave Powers didn't rent chairs for the campaign office; he borrowed them from a funeral parlor. When he took Jack around to the bars, he offered a round of drinks to the assembled. The patrons, who had been drinking beer all night, immediately developed a taste for liquor from the top shelf, ordering a gentleman's drink, scotch or bourbon whiskey. When Powers offered them drinks again, he pointedly offered them a round of beers. That was the Kennedy way.

Joe was merciless in wresting out any possibility of failure. Joseph Russo was a minor candidate, hardly of concern to Jack. But Timilty and an associate visited another Joseph Russo, a janitor, and paid him to put his name on the ballot and split the Russo vote. "They gave me favors," Russo recalled years later. "Whatever I wanted. I could have gone in the housing project if I wanted. If I wanted an apartment, I could have got the favor. You know?"

Jack had been brought up to think of such gambits as part of the colorful panoply of urban politics, fancied by such irascible players as his own grandfather, Honey Fitz. When the chuckles ended, though, the fact remained that his father had stolen candidate Russo's votes as surely as if he had stood in the polling places tearing up ballots. If Jack did not know about it beforehand, he surely did when he learned that there would be two Russos on the primary ballot.

Jack was no innocent either in the way the campaign was exploiting his war career. He had never been comfortable when the newspapers called him a hero, and now his own people exaggerated even those exaggerations. "Naval hero of the South Pacific" he was called in one campaign news release, as if he had single-handedly defeated the Japanese.

Not only did Jack have his father masterminding the campaign, but the whole family, except for little Teddy, was out working the hustings. Rose was a Gold Star mother who could speak mother to mother about her beloved Jack. Eunice was a woman of fierce intelligence and energy who irritated Jack at times by standing on the platform soundlessly mouthing the very words he was speaking. Nonetheless, Eunice, Pat, and Jean were a formidable trio, setting up teas and meetings, working as hard as any of the volunteers, heading out each morning from their suite at the Ritz-Carleton. Jack's beloved sister Kathleen was missing only because she was living in London.

In the last weeks of the campaign, Bobby showed up too, still dressed in navy blue. Jack deputized Red Fay to take his brother to a movie and show. Red was a talker, and he found the taciturn, morose Bobby a formidable burden, even for a few hours. Red had a risqué wit and a devilish interest in good times. His charge was a self-righteous Puritan, who wrinkled up his nose at an off-color joke as if he smelled something foul.

Jack had seen that part of Bobby shortly after Joe Jr. died. Over Labor Day, Bobby had come upon Jack and his old PT-boat buddies and their wives sitting drinking forbidden booze in the kitchen in Hyannis Port. His father rationed family and visitors to one drink before dinner. When Bobby, a scrawny little Savanarola, lectured them, Kathleen laced into the would-be snitch, telling him to get lost and tossing him out of the room like a mongrel pup.

Bobby was attempting to don the clothes worn by his father and big brothers. Upon graduating from Milton, Bobby had entered the navy's V-12 officer training program. He headed off to Harvard while most young men his age were drafted. Bobby was hardly a shirker. Even after Joe Jr.'s death, he fancied himself a navy aviator; he would honor his big brother by following in his oversized shoes. But he didn't seem to know what he was, or what he should be, what was authentic and what was not.

"I am not sure, between you and me, just how much I go for flying but I guess that's the best thing to do," he mused to Dave Hackett. "There are so many complications and decisions to make and I am so mixed up." He had a politician's self-consciousness about what the world might think of him, and he wanted it on his résumé that he had been a navy pilot. "I know that there will be a great deal more risk in this, but I think that it will be a lot more exciting, stimulating, and will do more good when I get out," he wrote his father.

His father pulled no special strings to help Bobby in that quest, and when he failed the flying aptitude test, that particular dream came to an end. The fact that his two older brothers had made their own heroic contributions to the war effort did not diminish Bobby's desire to stand within sight of the flash of combat. Like Joe Jr. and Jack, he had been instilled with the ideal that a true man rushed forward to the sound of strife, neither retreating nor pointing to his brothers' wartime service as evidence that his family had done enough. His father, though, would risk no more calls from priests waking him from a fitful sleep. Instead, as the war ended, Joe saw to it that Bobby was assigned to one of the nation's newest destroyers, the *Joseph P. Kennedy Jr.*, on which he would win no combat ribbons.

For the rest of his life, Bobby would sail with his brothers' legacy. On the destroyer, the sailors did not know that the scrawny young seaman was the younger brother of the hero for whom their ship was named.

It was fitful, unsatisfying duty, sailing down the Atlantic coast to Guantánamo Bay in Cuba. The best part of it, as far as Bobby was concerned, was that he was meeting the kind of fellows he hadn't met at Harvard. They were mainly uneducated southerners who had a strength that his fancy friends rarely had. That didn't make his duty any more palatable. After six weeks he was transferred, rising, as he wrote, "from the lowest grade of chippers, painters & scrubbers, the 2nd Division, up into one of the highest grades of chippers, painters, & scrubbers, the 1st Division."

When Bobby arrived in Boston late in Jack's campaign, he was sailing on another ship with his brother's name on it. Jack was eight years older than Bobby, a man who had experienced a world that Bobby had scarcely visited. Nevertheless, Bobby was not going to swab the deck, ordered around by the likes of Lem Billings and Red Fay.

Bobby insisted on taking over three of the toughest wards in East Cambridge, the lair of Jack's strongest opponent. He worked with the residents in a way that Jack and his highborn friends never could. He ate spaghetti with the adults and played football with the kids, and maybe Jack didn't win the majority of the East Cambridge votes, but he did far better there than he would have without Bobby's efforts. Bobby's own little entourage included his sister Jean and her friend and fellow Manhattanville student, Ethel Skakel.

Joe Kane coined Jack's campaign slogan—"A New Generation Offers a Leader"—and it was with the new generation of veterans that Jack was at his best. He may not have spoken in the firm, resonant cadences of a great speaker, but when he talked to his fellow veterans in the primary and later in the general election campaign, he spoke as truthfully and authentically as he ever did.

Jack could have offered himself up to the returning soldiers as "their" man, set to head down to Washington to unlock benefits that they deserved for their rich sacrifices. He did not do that. He saw these eighteen million Americans, 43 percent of the adult male population, as representing "mentally the most able, physically the best . . . in truth members of a citizens' army." These veterans were his natural constituents in a sense that no other group could ever be. He had shared with them what was for him, as for most of them, the most profound and the most formative experience of his life.

After World War I the authentic patriotism of the returning veterans had been transmuted in the American Legion into reactionary, jingoistic pseudo-patriotism, hostile to new ideas and immigrants. This time Jack thought it could be different if only the veterans looked beyond their own narrow interests to those of all Americans. "I have noticed among many friends I knew who have come home . . . one common reaction since their return—a letdown," he said.

> I have seen it in many of their faces, I have heard many of them mention it—the realization that home is not what it was cracked up to be. . . . In civilian life, many of them feel alone. In the last analysis they feel they have only themselves to depend on. What they do not always understand and what all of us in this country sometimes forget is that the interdependence is with us in civilian life just as it was in the war, although perhaps it is not as obvious. . . . In a larger sense, each one of us is dependent on all the people of this country, on their obedience to our laws, for their rejection of the siren calls of ambitious demagogues. In fact, if we only recognized it, we are in time of peace as interdependent as soldiers were in time of war.

Would Americans divide themselves in interest groups by age, class, and race, or was there truly an interest common to all Americans? Jack believed that the veterans should ask what they could do for their country, because they would benefit that way more than from narrow special-interest bills. America was their special interest.

Jack thought the veterans should lead the nation, with concern for policies that would maintain the peace and build a strong and prosperous nation. In his most passionate speeches, he was saying little more than he had in his hurriedly written letters from the Pacific when he vowed that the men who had died would have given their lives for something more than maintaining the lives of easy compromise and moral squalor that had seemed to him so prevalent in political Washington.

Jack was pleading for America's veterans not to retreat into private life, leaving the public arena to the predators, the self-interested, and the narrow parochial interests who shouted only their own names and their own causes. "If we turn our veterans organizations into mere weapons for obtaining special benefits for ourselves at the expense of society, we shall be sending ourselves down the rocky road to ruin."

Jack was fighting not only to end up first in the primary but also to create the illusion of health. He admitted to no one how much he suffered. When he had to walk up three flights of stairs for the Massachusetts Catholic Order of Foresters Communion Breakfast, he was limping. "You don't feel good?" asked Thomas Broderick, a friend, solicitously. "I feel great," Jack said.

Whatever the pain, he always said he felt fine, but at night in his little suite at the Bellevue Hotel he sat soaking in the tub, hoping the hot water would ease his back pain. One afternoon, another of the old politicians, Clem Norton, found the exhausted candidate in his room at the Bellevue crying, bemoaning the fact that he had agreed to this race.

Jack may have felt half dead at times, but he was nonetheless able to project a magical aura of stardom. It wasn't just the billboards and the pamphlets that did it, but that ineffable quality that had somehow attached itself to his wasted frame. When he talked to the students at East Boston High School, the girls rushed up to him afterward shouting "Sinatra! Sinatra!" comparing him to another emaciated-looking sex symbol.

A few days before the primary, fifteen hundred women showed up at the Hotel Commander in Cambridge for a tea party in Jack's honor. They swooned over him, flashed their eyes, and smiled, and presumably thought of things other than the housing shortage and the unemployment rate.

Joe had brought in sophisticated outside pollsters, and the candidate knew that he was well ahead. For his birthday at the end of May, the family got together in Hyannis Port in a celebratory mood. They all were taking part in the campaign, a mini-armada of Kennedys. Only fourteen-year-old Teddy had nothing to do with the campaign and seemed a spectator to the compelling drama.

Joe looked down the long table and asked each of his children to proclaim a toast to the future congressman. In the family, this was the time for juvenile put-downs. Each Kennedy attempted to top the last in the outrageousness of their toasts and the rudeness of their words. Then, finally, it was Teddy's turn.

"I would like to drink a toast to the brother who isn't here," Teddy said solemnly. They all stood then and toasted Joe Jr., and if they did not cry, it was only because they did not believe in shedding tears, not any longer. They remembered afterward that it was little Teddy who had made the toast, little Teddy who had his own sense of family.

By mid-June, Jack had to wear a back brace and arrange his schedule so that he could fit in half a dozen scalding hot baths a day and back rubs by "Cooky" McFarland, a boxing trainer. By rights he should not have even considered marching in the annual Bunker Day parade through Charlestown the day before the June 1946 primary. But the other candidates would be there, and he could hardly advertise his health problems that until now he had hidden so convincingly.

It was a hot Boston day, and by the time Jack reached the reviewing stand he was staggering ahead, nearly collapsing. State Senator Robert Lee happened to live right there, and Jack was carried into the politician's home. Lee called Joe, who told him to wait until a doctor arrived and his son could be moved. Lee stood and watched the twenty-nine-year-old candidate turn yellow and blue. "He appeared to me as a man who probably had a heart attack," Lee remembered. "Later on I found out it was a condition, which he picked up, probably malaria or yellow fever. We took off his underwear, and we sponged him over, and he had some pills in his pocket that he took. That was one of the questions his father asked, did he have his pills with him."

No news of Jack's condition got out, and the whole Kennedy family was there the following evening in the headquarters on Tremont Street to hear the happy results. Jack had scored a formidable success in the primary, defeating the other nine candidates with 22,183 votes, 40.5 percent. His closest challenger, Michael J. Neville, the mayor of Cambridge, stood far behind with 11,341 votes. The authentic Joseph Russo received 5,661 votes, while another 799 votes went to the faux Russo.

Of all the people who were there that evening at headquarters, only Joe seemed strangely out of sorts. "I got the impression that night that Joe was disdainful of us all," Dalton recalled. "I just couldn't understand it. He wasn't going around saying, 'Thank you, thank you, thank you for what you've done for my son.' He wasn't doing that at all."

Joe may well have been haunted by his dream of what might have been. Joe Jr. would have walked along the route of the Bunker Day parade, firm of stride, strong of manner, shaking hands, and slapping backs. He would have needed no microphone this evening, no urgent hands to push him up to the platform, no sitz baths, no doctors to monitor his steps, no sad reticence.

13

A Kind of Peace

The Harvard College that Bobby returned to in the fall of 1946 was bursting its walls with veterans wanting an education. The university took over the Brunswick Hotel in downtown Boston to house 115 lucky married couples and mandated that any student who lived within a forty-five-minute commute would have to live at home.

Now with the largest student body in its history, many of the totems of civilized life at Harvard seemed like silly rituals, a waste of money and time. Men found themselves sleeping in beds stacked on top of each other. In the dining halls where waiters had always served Harvard men, students stood in line grasping metal trays. Some of the frosh hardly knew what to do, but to Bobby and the other veterans these double-decker bunks and newfangled cafeterias were standard issue.

There were 659 Harvard men who died in the war. Those who returned safely were several years older in age, and decades older in experience. They might listen to scholarly pretenders tell them how to think, but not what to think nor how to live or how to drink.

In classrooms, a man who had parachuted into Burma or flown a bomber over Germany listened with both a hunger for knowledge and a hardy reserve of skepticism. When one veteran who had lost a couple of fingers in the war got drunk, no proctors dared to point theirs at him in rebuke. Another student had had his ear burned off. When he came rolling in drunk at dawn, no one said anything.

Bobby was a veteran but he was far different from most of the former GIs matriculating at Harvard. He bore a name that was almost as honored at Harvard as it was elsewhere now, and he was voted into the Spee Club with an ease that Jack had not experienced.

Bobby could easily have passed his days in the world of clubs and class and privilege within its gilded enclave. But he was not comfortable sitting among heirs who thought that what their families had done in the past was more important than what they might do in the future. He was bored with the endless social palaver, and he spent little time at Spee. One of his friends, a Catholic, had been rejected and Bobby took that as an assault on his very faith.

Bobby went out for football and on the playing field met many of his closest lifelong friends. He was no more a natural athlete than he had been at Milton, but he had awesome quantities of determination and mindless physical courage. As a 165-pound end, he shared the honor with two backs of being the lightest players on the team. In scrimmage, he kept blocking Vince "Vinnie" Moravec, the 200-pound first-string fullback, a task that was akin to running into a slab of granite.

"For Christ's sake, would you tell that little bastard to stop hitting me so hard!" Vinnie yelled at Wally Flynn, an end. "He's gonna get kicked!"

Wally looked up the field at Bobby, hustling back to the line. "Vinnie, he's gotta prove himself to his family, to those kids at Milton, to just about everybody. You gotta tell him. I'm not tellin' him."

As hard as Bobby hit in practice, and as much as he threw the ball around in the evenings with his buddy Kenneth "Kenny" O'Donnell, he was far down in the rankings of Harvard ends. In the first game of the 1947 season against Western Maryland, Bobby finally got his chance when the two starting ends were so sick that they couldn't even make it to the bench. The game was a 52–0 rout, and Kenny, the quarterback, threw a touchdown pass to Bobby.

If Bobby had played forcefully before in practice, the next day out he was even more of a human missile, blocking with abandon, tackling with fierce resolve. The first-string ends were scheduled to come back, but with one word, Coach Harlow, sitting high above the play in his elevated chair, could change all that. Like his father, Bobby had a perfect memory when it came to slights, and he remembered that Harlow had refused to let his brother Joe win a letter by playing in the Yale game.

Harlow was now an aging, unhealthy man who looked out on the squad as if they were a bunch of malleable preppies. He shuttled players in and out for little reason except personal whim and indulged in pep talks that motivated no one but himself. Harlow was still the man, however, who would decide whether or not Bobby played. Bobby ran down the side of the field and crashed against an equipment car, crushing his leg. The accident would have been enough to end any other man's practice, but Bobby got up and with a slight limp came back onto the field.

Three days later Wally stood across the line from Bobby in scrimmage. As Wally waited for the play to start, he saw an impossible sight. Bobby seemed to be crying. Wally stopped the game and hurried over to his friend. "Hey, wait a minute, Bobby?" Wally implored. "What the hell's the matter with you?"

"I think my leg's broken," Bobby replied, embarrassed to be holding up the game.

Wally didn't know a thing about medicine, but when he took a look at Bobby's leg, he shook his head authoritatively: "Yeah, Bobby, it's broken."

The rest of the year Bobby had a cast on his leg, but despite that he managed to play in the Harvard-Yale game and win his letter.

The math was simple. Harvard tuition was four hundred dollars a year, and the GI bill paid a stipend of seventy-five dollars a month. A man could earn a thousand dollars in the summer and during the school year work a few hours for his room and board. And so the veterans came, and they changed Harvard forever.

Bobby spent his free time with his friends at the Varsity Club, a rambling old two-story brick house with a pool table, a television set, and knockabout furniture. They rarely talked of the war, but when they did they told heroes' tales. "Oh, those guys were tough cookies," Flynn recalled. "Vinnie Moravec got torpedoed in the Atlantic and spent all night saving the lives of about ten kids, getting 'em on board. Oh, Jesus! The navy wanted to give him a medal, but he wouldn't take it. Said, 'I didn't do anything.' Leo Flynn jumped out of a burning B-17. So did Kenny O'Donnell. And I could go on and on. It was the finest group I ever met in my life."

Despite all the clichés about jocks as mindless semi-illiterates who read nothing but the sports page, these men had a fierce concern for the world in which they lived. Many evenings they debated the political future of their country. Despite its image in certain quarters, Harvard had a predominantly conservative, Republican student body. The Varsity Club was different. Sam Adams, Bobby's friend from Milton, was the one Republican bobbing up and down in a sea of Democrats. The others took special delight in pillorying poor Sam as the relic of a dying class.

"We're going to take the country away from you, Sam," opined Kenny, the son of a football coach. "You guys have had it long enough."

Bobby had not jumped out of a burning plane over Belgium or lost his fingers in the Pacific, and he knew far less of life than most of his teammates. He might have been a subject of mild derision among his mates, little Bobby painting his sailboat while others fought the war. It said much about Bobby and the caliber of these men that he was no more a figure of amusement at the Varsity Club than anyone else. Half a century later, when the men who had been there those many evenings were asked what had been Bobby's special role, they all said the same thing. "He was one of us." There were no leaders. There were no followers. "He was one of us."

Even among this group, however, the man was a controversialist. Bobby enjoyed nothing more than slithering into a genteel conversation, increasing the intensity of the argument to a level only slightly below physical combat, and then slithering away again as his friends went at it, red-faced and blustering.

"Nick, I think he does that on purpose," Wally told Nick Rodis one day. "He's a little son of a bitch."

Bobby was more conservative than most of his friends, often mouthing the self-serving homilies that he had learned at his father's knee. There are few signposts on the road between righteousness and self-righteousness, and Bobby marched briskly ahead, heedless of where he might be headed. What made Bobby something more than a rich boy who took his good fortune as proof that all was right with the world was his sense of justice. He didn't like unfairness. He didn't like bullies. And he wasn't afraid.

Bob had many Catholics among his friends. Unlike the prep school Harvard football team of his father's day, public school men, most of them Catholics, dominated the postwar team. When the team played Holy Cross, there were more Catholics on the field for Harvard than among the Catholic college's eleven.

Father Leonard Feeney, a charismatic Jesuit priest, headed the St. Benedict Center, just down the street from Harvard Yard. Feeney was a man of fierce faith who converted scores of Harvard students, turning philosophy majors into Catholics, and newly born Catholics into priests and nuns. He walked with the banner of faith into the very citadel of secular rationalism. He was the kind of man Bobby might have been expected to follow. But Bobby was not a man to accept received wisdom, whether it came from the dons of Harvard or a popular priest.

Bobby listened to Feeney and he did not like what he heard. The priest took literally Boniface VIII's *Unam Sanctam* that "it is wholly necessary for the salvation of every human creature to be subject to the Roman Pontiff." Feeney was a Catholic Billy Sunday, excoriating the infidels and prophesying spiritual death for all those who stood outside the Catholic faith.

Bobby's football teammate Chuck Glynn recalls how Bobby confronted Father Feeney and then went straight to Archbishop Richard Cushing to complain about the priest. Bobby was only one of many voices that in the end condemned Father Feeney. Shortly before he was excommunicated,

Father Feeney and his followers stood in Boston Common on Sunday afternoons. There he condemned "people like the Jews [who] are loudmouthed and stupid like Archbishop Cushing," and yelled out at those who spoke out: "I hate you—a dirty rotten face like that! You dogs. Go home! I will ask the Blessed Virgin to punish you."

These men around Bobby were not like Jack and Joe Jr., who thought that it was a man's natural calling to score with as many women as he could. They had been out in the world, however, and among them Bobby seemed haplessly naive. On one occasion, Bobby somehow managed to set up a date with Shirley Flower, who bore the exalted title of Miss Lynn, Massachusetts. Bobby was terrified by the prospect.

"What am I gonna do with this girl?" he asked a group that included Nick, Kenny, and Chuck. "Well, she lives in Lynn," Chuck told her. "Take her out there, give her a couple of drinks, buy her a sandwich, and see what happens. And report to us."

"Well, what'd you do, Bobby?" Chuck asked the next day at the Varsity Club. "We went to a movie and had a soda." "Then what happened?" "Well, geez, I kissed her and she opened her mouth." "Oh, heavens, Bobby, that's terrible."

Bobby drove his friends down to Hyannis Port, racing his old Chrysler with its bare tires down the narrow roads, frightening men who had jumped out of burning airplanes. His parents were used to having their sons bring college friends to the Cape, but for the most part, Joe Jr. and Jack had brought young gentlemen.

But now hordes of roughnecks descended on the pristine precincts, men who in Eunice's eyes were "tough and rough . . . all big and bulky and very unsophisticated," wearing army fatigues and wrinkled shirts, so unlike her exquisitely dressed father. Rose and Joe learned the background of men such as Nick Rodis, whose father was in the produce business and never made more than thirty-five dollars a week in his life. Then there was Paul Lazzaro, whose dad worked in a factory.

Joe hardly spoke to Bobby's friends, not even nodding hello to Wally when he sat next to him in the private theater watching Al Jolson in *The Jazz Singer*. When Joe did speak, as often as not it was to push the pedagogical imperative of the family. One weekend Kenny O'Donnell and Bobby sat over dinner joking about how they had come in last in a sailing race at Harvard. Joe blustered and fumed, getting more and more infuriated at the young men.

"What kind of guys are you to think that's funny!" he exclaimed as he got up from the table, unwilling to sit any longer and listen to such blasphemy.

As for Rose, her face was a mask that rarely displayed its displeasure, but she could not contain herself. She and Joe had not sent their son to Harvard to have him socialize with working-class ruffians. If they belonged at Hyannis Port, it was in the kitchen with the cook and the chauffeur. Bobby's friends were so bemused by his mother that they gave her a nickname, "Billie Burke," after a Hollywood actress known in part for her high-strung, scatterbrained roles.

One weekend, Rose lectured her son on his woeful excuse for friends. "Bobby and his mother were having this conversation," Flynn recalled. "She wanted to know why he didn't have some other friends from other places. And one of our guys was standing there where she couldn't see him. He could hear her. And it was too bad."

Rose and Joe may have blamed Bobby's friends in part for their son's abysmal academic record, which landed him on probation in 1947. In his postwar tenure at Harvard, Bobby received thirteen Cs, two Ds, and not a single B or A. When he graduated in March 1948, nothing in his record suggested that he was anything but a shadow of Joe Jr. and Jack.

Jack wasn't taking his role as the newly elected congressman from the eleventh district with ponderous seriousness. On the morning of his swearing-in, Jack sat at a drugstore counter and ordered two soft-boiled eggs and tea. As he waited for his order, Billy Sutton, his driver and general factotum, paced nervously. "We've got to get up there! Mr. McCormack is anxious that you get there." The most powerful Democrat in Massachusetts was sitting there waiting for Jack to finish his eggs.

"How long would you say Mr. McCormack has been here?" Jack asked as he toyed with his food.

"Twenty-six years," the detail-minded Sutton answered after an instant's calculation. "Well," Jack replied, still eating his eggs, "don't you think Mr. McCormack wouldn't mind waiting another ten minutes?"

Most twenty-nine-year-old congressmen would have attempted to appear older by dressing with funereal seriousness and speaking ponderously. Jack was so nonchalantly boyish that he was sometimes stopped when he attempted to walk out on the House floor and harshly warned that a page had no place there. This happened enough times to Jack and two of his colleagues that the House instituted a new rule that pages had to dress uniformly in blue blazers and white shirts.

Jack may not have affected the gravitas of his political elders, but he knew that he was entering a Congress in which the problems were as large as they were intractable. Cold winds blew across Europe, which was experiencing its worst winter in memory, a spate of blizzards and frigid days that cursed the efforts to clear away the rubble of the war and build new lives.

In Britain alone, unemployment stood at six million. Across the expanses of Europe, the last echo of the bells of victory had died out, and uncertainty was the most positive characterization of the prevailing sentiment. Winston Churchill had given a prophetic speech in Missouri the previous March that helped define the new era. "From Stettin in the Baltic to Trieste in the Adriatic, an iron curtain has descended across the continent," he thundered ominously.

In March, as Jack sat in the House chambers, President Truman addressed a nervous, querulous Congress and asked for a $400 million aid package for Greece and Turkey. His historic speech set forth the themes that would be developed later in the Marshall Plan, an unprecedented package of aid to a beleaguered Europe.

The Marshall Plan stands now as one of the strongest achievements of American diplomacy. At the time, however, politicians and publications of both the Right and the Left challenged the president's initial proposal. What was most ominous to those who thought that they had warehoused their weapons for good was Truman's proposal "to authorize the detail of American civilian and military personnel" as supervisors and advisers.

Time asked what would happen to the glorious ideals of international cooperation if the United States bypassed the nascent United Nations. "Was it politically wise to support the government of Greece, which was hardly a model of democracy?" the weekly asked. "Wouldn't this program lead to the same kind of imperialism which Great Britain had followed so long and which Americans had so sternly criticized in the past?"

The Kennedy who spoke out loudest on these issues was not the young congressman from Massachusetts but his father. Joe remained as isolationist as he had ever been, and he considered the pathetic state of the once-great Britain ample evidence of his prescience. He was a man who took pleasure in bad news as long as it justified his own prophetic fears. Now an impoverished Britain had to dismantle its empire and pull back from its involvements in Europe, just as he had said it would in what the island nation had dared to call victory.

As Joe saw it, America had stepped in before to rescue Europe from itself, and he did not want his nation to step in again. He opposed the Marshall Plan as a massive giveaway of American wealth. Joe was all for letting the tired peoples of Europe have their desperate fling with communism if they chose. America would stand back from it all, its gold still in its vaults, growing rich and happy away from the dark dreams of the rest of the world.

At the same time as his father was beating his drum of isolationism, Jack was making a different sound. Congressman Kennedy was a lowly freshman, and he did not have the platform his father still could occasionally command. But that spring, in a speech at the University of North Carolina, Jack stated that he was in favor of aid to Europe, specifically supporting the controversial proposal to Greece and Turkey.

Truman talked of "free peoples who are resisting attempted subjugation." Language was one of the first victims of the cold war, for these were not pristine democracies full of "free peoples" but authoritarian regimes whose primary virtue was that they were not Communist. In this darkening world, an enemy of the Soviet Union was a friend of the United States. Jack was resolved to support the authoritarian rule of the Greek military, just as he was resolved that his government should take dark and secret measures to combat Communist subversion. Jack supported whatever measures were necessary to contain a Soviet Union that he believed pushed relentlessly outward, driven by traditional Russian expansionism and fortified by Communist ideology.

Jack accepted the reality that the Soviet Union would soon have its own nuclear arsenal. And he envisioned, decades from now, a possible nuclear Armageddon. "The greatest danger is a war which would be waged by the conscious decision of the leaders of Russia some twenty-five or thirty-five years from now," he told his youthful audience, whose age was defined by the atomic bomb. "She will have the atomic bomb, the planes, the ports, and the ships to wage aggressive war outside her borders. Such a conflict would truly mean the end of the world, and all our diplomacy and prayers must be exerted to avoid it."

The young congressman had touched on the darkest irony of the age: the massive power of nuclear weapons made their possessors unwilling to use them, but the nuclear arsenals would build up over the years to create a confrontation that could end life on the planet. Two or three decades hence, as Jack saw it, an American president might have to make decisions of awesome finality and moment.

In his early months in Congress, Jack attempted to cut his own private path through the endless bramble of politics. He had set out publicly in a direction far from his father's on international issues, but on the economy he advocated a balanced budget in a hectoring voice that could have been Joe's.

For a pro-labor Democrat, the major issue of the day was the attempt by the Republican Congress to push through the Taft-Hartley Act, which would radically limit the role of unions in American life. Much of the American public was tired of the endless strife and rancor of the union movement, from the belligerent arrogance of the United Mine Workers' John L. Lewis to the obstructiveness of leftist militants such as Harry Bridges of the West Coast International Longshoremen's and Warehousemen's Union.

Jack saw that while the excesses needed to be ended, the proposed bill was a piece of conservative deception that sought not only to reform what was wrong but also to cripple what was strong. Jack had Mark Dalton take the train down from Boston to work with him preparing alternative proposals and writing his first important speech in the House. Jack was an intellectual tinkerer. He liked to turn problems upside down and shake them every which way until he had figured out the solution that he believed was always there.

Freshman members were rarely heard on such matters, but Jack got up and made an impassioned and daring speech. He proposed stopping secondary boycotts and excessive union dues, but he also opposed ending the closed union shop and industrywide bargaining.

"There is no need to—there is great need not to—smash the American labor movement to rid ourselves of 'featherbedding,' racketeering, and similar evils," he said on the floor of the House on April 16, 1947. "This bill does not assure the worker freedom, and the men who wrote this bill must have known that it does not. . . . This bill in its present form plays into the hands of the radicals in our unions, who preach the doctrine of the class struggle."

In international affairs the whole political lexicon had changed from the era just two years before. Then Joseph Stalin was "Uncle Joe," and many Americans believed that Russia and America would walk arm in arm and army to army into the postwar world. The public mood was shifting radically, and Jack was entering a Congress in which politicians of both parties had begun to flay away at Communists and communism.

Jack, of all the newly elected members of Congress, was the first to berate publicly a putative Communist, months before Senator Joseph McCarthy or even Congressman Richard Nixon had begun their own attacks. On the day after Truman's historic speech, Jack interrogated Russell Nixon, the legislative representative of the United Electrical Workers, before the House Labor Committee. The former Harvard instructor had taught Jack back in his college days, but Jack showed him no deference.

"Could you tell me whether Julius Emspak is a Communist?" Jack asked, referring to the union's secretary-treasurer. This was the kind of question that Senator McCarthy would soon make notorious, asking witnesses to cough up the names of others.

"You do not need to rely on me to tell you that," Russell Nixon replied. "He testified before the Senate Labor Committee . . . and said he was not."

"For your information, he was a Communist and proof can be provided,"

Jack said. This was the kind of exchange that led a Catholic magazine, *The Sign*, to call Jack "an effective anti-Communist liberal . . . more hated by Commies than if he were a reactionary."

Later that month Harold Christoffel appeared before the House Labor Committee. Christoffel was the honorary president of United Auto Workers local 248 at Allis-Chalmers Manufacturing Company outside Milwaukee. The union was in the midst of a brutal, nearly yearlong strike in which almost half the workers had given up their membership and returned to work. That in itself was a rebuke to the leaders, and left alone, the rank and file might soon have condemned them. Nonetheless, they were paraded before the House Labor Committee.

Christoffel may have considered fidelity to a foreign ideology more important than loyalty to his own government; in 1941, at the time of the Soviet-Nazi alliance, the union leader had led a seventy-six-day strike at a war production plant. That was an unseemly business, but to those with deep concern for civil liberties there was something equally unseemly about Jack berating Christoffel and asking him whether he was a Communist or had been a Communist. And there was something unsettling about Jack's recommending perjury charges; in fact, the labor union leader was eventually sentenced to sixteen months to four years in prison.

The young congressman was so furiously anti-Communist that when China fell to Mao's army, he blamed the Democratic administration, giving a speech that any Republican would have gladly given. "The responsibility for the failure of our foreign policy in the Far East rests squarely with the White House and the Department of State," he said on the floor of the House.

Jack might attack the State Department and take a few swipes at alleged Communists, but he was not the kind of politician who reveled in controversy. When James Michael Curley was convicted and sentenced to prison for mail fraud, the Massachusetts Democratic pols lined up to sign a petition to get Curley pardoned. Jack owed Curley for having retired from Congress, leaving his seat open. Moreover, Curley was wildly popular among Massachusetts Democrats. To ignore Curley now might make Jack an outcast in his own party. Yet if he joined with the others, he became just another political hack, a self-righteous hypocrite no better than the man he had succeeded.

He turned to Mark Dalton for advice, though, like most men, Jack often sought not advice but confirmation of what he already intended to do. "Jack was fearless," Dalton recalled admiringly. "He would listen to you, and if he decided you were right, he would go with you. Everybody wanted him to sign the Curley pardon. He came to me and asked me what I should do. I said, 'Listen, you haven't got your seat warm down there. And these bastards are putting you over the barrel. Tell them to go to hell.'"

On the floor of the House, Jack went up to John McCormack, the head of the Massachusetts delegation, and asked about the pardon. "Has anybody talked with the president or anything?" he asked.

"No," McCormack replied, knowing full well what lay behind the question. "If you don't want to sign it don't sign it."

Jack did not sign the petition, though he knew it was a bad thing to make enemies in politics, for enemies remember when friends forget. McCormack didn't stand up publicly against Jack. He just held on to every scrap of the political patronage, and worst of all, he remembered.

As a freshman congressman, Jack carried with him the bitter memories of a wartime capital full of what he considered the self-serving and the self-seeking. He was hoping to find something else in Washington this time. In September, he traveled back to Choate for the school's fiftieth anniversary and gave one of the most heartfelt speeches of his young life.

Until recently he had been at best ambivalent about the word "politician," disdaining the title, particularly in a gathering like this one. This evening, though, he proudly wore that appellation so as to inspire some of these youths as well as other distinguished alumni not to retreat into private enclaves of privilege but to come forward and stand for election, as he had.

He pointed out that despite all the contributions of Choate men, there was "one field in which Choate and the other private schools of the country have not made a great contribution and that is the field of politics. . . . In America, politics are regarded with great contempt; and politicians themselves are looked down upon because of their free and easy compromises."

The speech was reprinted in the alumni bulletin as "Jack Kennedy's Challenge," but it was a challenge that Jack himself met only partway. He could stand before the Choate youths and fill the room with inspiring rhetoric, but when he returned to Washington, he was like Gulliver in the land of Brobdingnag, so close to the realities that he smelled only the stench and saw only the endless compromise. "Everything you said is true—only more so," Jack wrote Clare Boothe Luce, a woman of studied cynicism and disdain who had long since forgiven Joe for betraying her in the 1940 presidential campaign.

As the months went by, Jack became less and less a forceful advocate and more of an ironic, disinterested bystander. He brought the same emotional distance that he kept from everyone in his life to the world of politics. He was living in a small, disordered house in Georgetown with his sister Eunice. Thanks to their father's intervention, twenty-six-year-old Eunice had an important position, the first executive secretary in juvenile delinquency at the Justice Department. Eunice did not have that common liberal malady of loving humankind more deeply in the abstract than the specific. She not only talked endlessly about the problems of troubled youth but also often brought groups of troubled girls home for Sunday dinner.

Although Jack tolerated his sister's eccentricities, he much preferred spending his evenings in the company of an endless parade of women. In some of the newspaper photos of the time, his eyes appeared dark and hollow, his features seemingly twice their size in his thin face. Yet he had a sexual aura that made him nearly irresistible. The Associated Press named Jack one of the fifteen most eligible bachelors in America. Even if Jack ran well behind Cary Grant and Clark Gable, he was the only politician on the list alongside the biggest stars in Hollywood.

Jack enjoyed the company of another Democratic freshman, Congressman George Smathers of Florida. Smathers had already met Jack's father in Florida, and he had more than an inkling of what the Kennedy men were all about. Joe had purchased a part ownership in Hialeah Racetrack and he generously invited Smathers, a young assistant U.S. attorney, to his box.

"Joe was using me to sit with these pretty girls," Smathers recalled decades later, full of admiration for the sheer duplicity of the man. "Joe was married, and it made him look all right. He didn't look like a real old guy with all these pretty young Florida girls following him around. So that's what he used me for. There were always these good-looking girls. And they would snap pictures, and I was always there."

Smathers's office was just down the hall from Jack's in the House Office Building. The Florida legislator was a handsome man of elegant bearing who as often as not preferred pulchritude to politics. Smathers adhered to the southern tradition that a successful politician better dress like one, and he was a veritable dandy in fancy suits.

Jack, for his part, didn't mind a wrinkled suit, scuffed shoes, or white socks. The two young politicians did share, however, a common interest in women. Like his friend, Jack did not believe that quality and quantity were mutually exclusive and he ran through any number of young women. On one occasion, Jack's old friend Rip Horton went out with Jack and his date, a blonde from Florida. The woman was stunning enough to have kept most men occupied for at least several dates, but Jack got rid of her before the evening was over.

Rip was staying at Jack's house, and soon afterward another woman showed up. Rip was amazed at the audacity of his friend, but even more so the next morning when Jack emerged from his bedroom with yet another woman, who apparently had arrived during the night.

Jack's friends might have been impressed, but in America of the late 1940s, the boyish Catholic politician could not afford to be seen as overly happy with bachelorhood. "Guess I just haven't found the right one yet," he told one interviewer, in words that Jimmy Stewart could have spoken in a Frank Capra film.

"I really prefer the homebody type of girl. One who is quiet and would make a fellow a nice, understanding wife and mother for his children. The color of her hair or her height wouldn't make much difference. Just as long as she's a homebody is all that counts. When I find her, even politics will take a back seat then." The interviewer never asked the wistful bachelor why he did not date women who came at all close to his supposed ideal.

Washington was full of tedious, somber men who prattled on with what passed as seriousness. Jack's friend Smathers could play that game too, but when he came off the platform, and the reporters had put away their pencils, by all appearances the man didn't give a damn. It was all a splendid joke, and at times he and Jack stood on the sidelines ridiculing much of the spectacle that took place before them. Politically Jack was virtually schizophrenic: at times he was seriously and passionately concerned about his country, and at other times he dismissed the whole business as a silly circus.

"He was a guy whose father had a lot of money and he wasn't quite sure how Joe got it. It didn't interest him too much where he got it," Smathers recalled. "He just knew he had it, it was easy to come by, and he didn't think about it. And people who worked around him were the ones who had to talk to him about poor people. But Jack was very sympathetic. He was a very sweet guy. Just a real sweet man."

Early on the Florida congressman noticed that his compadre was not at all what most of the women he bedded took him to be. As freshman legislators, the two bachelors had been relegated to the distant reaches of the House Office Building. Smathers walked with such long easy strides that it was nothing more than an invigorating jaunt to the Capitol. For Jack, getting to the floor of the House was often a different matter.

"Jack was crippled, and he couldn't walk well," Smathers remembered. "So the bells would ring for a quorum call or a vote. And I would go over and say, 'Come on, Jack.' And he would lean on me, and we would trudge our way to the elevator, which was about seventy-five yards, and catch that and go down three floors to the basement, then have to walk again and get on a tramway car that would take us across underground from the old House Office Building to the Capitol. Then we would get off there and have to go up three floors, and then again, you had to walk about half a block to get to the floor to vote.

"So that was a long walk for a guy who had a bad hip and a bad back and a bad everything else then. So he would lean on me, and we would get over there. And we'd vote. So that's really how we became very close friends, just through the labor of getting from our office over to vote. He was a pretty brave guy. He didn't complain about hurting. But you could see the hurt."

Smathers recalls Jack being sickly from the first day he met him but there came a point when his condition dramatically worsened. In August 1948, Jack traveled to Ireland to spend some time with his sister Kathleen before heading off to the Continent on a congressional junket. Jack was an Anglophile who had no Irish mud on his English boots. He stayed with his sister at Lismore, the Irish castle owned by her deceased British husband's family. The group spent most days playing golf, as close to the common sod of Ireland as most of Kathleen's aristocratic friends ever traveled.

Jack had a bad back and wasn't about to test it by swinging a golf club. He decided to go off to see whether he could find the old Kennedy homestead. It was a peculiar journey for a man who had once found it expedient to say that his father was born in the more fashionable Winthrop, not the largely immigrant community of East Boston, and who had managed to get into Harvard's Spee Club only by latching onto friends whose forebears had not been born in humble Irish cottages.

One of the other houseguests, Pamela Churchill, the shrewd, socially astute wife of Churchill's son, Randolph, did not play golf either, and she agreed to accompany him. Pamela was a connoisseur of upper-class men, and she found Jack rather disappointing in his immature boyish ways, a woefully less sophisticated man than his brother Joe. Pamela had a courtesan's adeptness at casual conversation, and she carried on in full form on the hundred-mile drive through the Irish countryside. Whether or not Pamela expected to see a Kennedy castle rising out of the heather, she surely had not anticipated a thatched cottage outside of which stood a menagerie of pigs, goats, and chickens, and a sturdy outhouse, while inside resided an unaccountable brood of children and a modest couple who called themselves Mr. and Mrs. Kennedy.

"I spent about an hour there surrounded by chickens, pigs, etc., and left in a flow of nostalgia and sentiment," Jack recalled a decade later. "That was not punctured by the English lady turning to me as we drove off and saying 'That was just like Tobacco Road.' "

Pamela might have wanted to hold a perfumed handkerchief over her nose, but for the first time in his life Jack had looked straight on at his heritage and admired it. It was disappointing to him that his sister did not feel the same way.

"When we got home, we were very late for supper, but Jack was very excited," Pamela recalled. "He said, 'We found the original Kennedys.' I remember Kick saying, 'Well, did they have a bathroom?' And he said, 'No, they did not have a bathroom.' And I think that she was interested but not in the same way that Jack was."

Jack traveled to London before heading off on his congressional junket. He had hardly arrived when he collapsed in his hotel. A British doctor gave Jack the potentially devastating news that he had Addison's disease, a condition marked by an insufficiency of the adrenal glands. These glands, the size and dimensions of a small strawberry, sit on top of the two kidneys. They play a crucial role in physical and psychological health, pumping steroids into the system. These hormones regulate metabolism, sexual characteristics, and the ability to handle stress and injury.

The disease has no distinctive signs itself and masquerades in many guises. Its victims often just seem vaguely tired. They sometimes have stomach troubles, diarrhea, or vomiting, symptoms they blame on bad food or nervousness. Jack had been ill so often, and had overcome so many maladies, that he tended to ignore complaints that would have sent anyone else to a doctor. He may well have already been suffering from Addison's for a year, or perhaps considerably longer. The weakened victims of Addison's often do not die of the disease itself but usually of something else, such as getting a tooth pulled or the flu. By the summer of 1947, the introduction of a new wonder drug, cortisone, had changed the prognosis of the disease from almost certain death to a manageable condition.

The press was told that young Jack had suffered a temporary relapse of the malaria that he picked up at the end of his service in the Pacific. It was a war hero suffering from his war injuries who was carried from a London hospital in pajamas to an ambulance and through the streets of London to the hospital on the *Queen Elizabeth*, where a private nurse ministered to him. It was a war hero who was carried by ambulance from the New York docks to a private plane to fly him to another ambulance that took him to a private room at the New England Baptist Hospital.

Although the family attempted to minimize the seriousness of the matter, saying that Jack's condition was only temporary and that he would spend only a few days in Boston for "observation," the photos of Jack entering the hospital belied all this. He was dressed in a suit and tie, as if these clothes would mask the seriousness of his condition. But the man lying under a blanket on the stretcher wore a death mask, his face haunted and bony, his chin slumped against his chest.

Addison's disease turns its victims into medical Dorian Grays. Jack often had gloriously tanned skin that seemed to exude a sailor's health. He had thick brown hair that probably would never turn gray. This was all due to

Addison's disease. The malady is so insidious that at first it seems not like a disease but like a slow draining of the spirit. Its victims lose weight, muscular strength, and their appetite, slowly declining toward death.

At the Lahey Clinic, Jack had a series of appointments with the prominent endocrinologist Dr. Elmer C. Bartels, who would treat him for the next thirteen years. By then Jack was already injecting himself with cortisone. "He had to take medicine," Bartels said. "He'd forget to take it, or not take it with him on trips."

Jack needed these injections to live. Patients were advised to increase the dose during periods of high stress, not as a medical crutch but as a physiological necessity, since in crises and danger healthy adrenal glands produce more of the essential steroids. Jack thus acquired the capability of manipulating his own health, or at least manipulating his psyche. If he took too high a cortisone dosage, however, he risked worsening such potential side effects of long-time use as muscular weakness, high blood pressure, and "agitation, euphoria, insomnia and, rarely, psychosis." As it was, many of the illnesses and physical problems Jack had had over the years, from his back pain to his stomach troubles, may have been either caused by or exacerbated by his Addison's disease.

Jack had been brought up to believe that a man was a vibrant physical being who had the strength to flick off adversity. A man's whole sense of himself, and everything he did, felt, and thought, was based on good health. Women were not attracted to whiners or weakness. Neither were men.

Jack's great creation, then, was not some piece of legislation but himself, a man of apparently endless vigor and health. He let no one stand close enough to his pain to betray his illusion. He let no one know what he felt when he jabbed the needle into his leg for the umpteenth time. On one occasion, though, Red Fay was standing there when Jack was injecting himself. Red was adept at playing the amiable buffoon, and Jack called him into his presence for moments of diversion.

"Jack, the way you take that jab, it looks like it doesn't even hurt," Red opined as he stood over his friend. Red's insensitivity was usually one of his charms, but on this occasion Jack plunged the needle into Red's leg. He yelled in pain. "It feels the same way to me," Jack told his friend.

Jack easily could have affected a demeanor of such seriousness that it would have seemed natural for him to give up such silliness as touch football and adolescent roughhousing. But that was a part of him he was not willing to lose.

When Bobby came down for a visit, they got a gang together and went out in the park for a game of touch. On a sunny weekend afternoon the mall and parks in Georgetown were full of ad hoc games, the players sketching out their imaginary diamond or gridiron in the grassy expanses. The Kennedys had their own peculiar game of touch that gave them what they wanted, endless action, but took up a great deal of space. It was Kennedy football, passed on from brother to brother, and eventually from father to son. "They actually didn't play the game as you normally play it," reflected Dick Clasby, a Harvard football star who married Jack and Bobby's cousin, Mary Jo Gargan. "They could throw the ball all over the field. You could throw it behind the line of scrimmage . . . and if you caught it you could throw it again." The offensive team could keep passing the ball forward beyond the line of scrimmage until the player holding the ball was finally touched. Then, on the next play, as likely as not, one of the Kennedys would change the goal line.

Just down from Jack's touch football game, a group of younger men played baseball. They thought that they had as much right to the grassy field as the football game, and on occasion a batter hit the ball into the midst of the football game. "You better stop that," the young men were told.

The next time a baseball came flying over, one of the football players tossed the ball away. The young men came running over. Bobby squared off against one of the baseball players and they battled each other with bare fists. Other men joined in. Jack stood nearby, afterward rationalizing, "I was too dignified to fight." Jack might pretend that he was a sturdy quarterback for an afternoon, but it would not have done for the congressman from Massachusetts to be arrested in a melee in the park.

Even a few years later, when Bobby was married and had children and a responsible position on a congressional committee, he still treated a football game as ritual war. On Sunday mornings, he often joined in the pickup touch football game at the Volta Street playgrounds near his Georgetown home. On one play, he came barreling in at the opposing quarterback from his blind side, decking the man before he had time to release the football.

"Excuse me. Do you know what game we're playing?" asked the two-hundred-pound quarterback.

"What are you talking about?" Bobby sneered. "Can't you take it?"

"Yeah, I can take it," replied Bruce Sundlun, the hefty quarterback. "I just want to make sure you understood what game we were playing."

A few plays later, Bobby came tearing in and again knocked Sundlun to the hard turf. This time Sundlun rushed toward Bobby and had to be pulled back. "Now wait a minute!" exclaimed Jim Rowe, a burly Washington lawyer. "Let's stop this. This is touch football, and let's just play touch football."

"Yeah, and if you do this once more, sonny, you're gonna get hurt," Sundlun said, shaking himself off and knowing by the look in Bobby's eyes that he would be coming again.

On the next play, Sundlun cocked his arm to pass and waited until his tormentor got within three feet of him, and then threw the ball with all his strength directly into Bobby's face. Bobby's head snapped back so forcefully that Sundlun thought he had broken his neck. Bobby lay there for a while, his nose and mouth bloody, and then jumped up to fight the quarterback. "Now wait a minute," Rowe said, interjecting himself once again between the two men. "He told you if you did that once more, you were gonna get hurt. You did and you got hurt. Now stop this. If you can't play according to the rules, then get the hell out of this game and go home."

Bobby stood there thinking for a moment while wiping the blood off his nose. "Yeah, he did tell me, didn't he? Okay, let's go."

"Years later when Bobby got to be attorney general, he used to use that story," recalled Sundlun, who was governor of Rhode Island from 1991 to 1995. "If I was at a function and he was there and it suited his purpose, he'd send somebody over to say, 'The attorney general wants to see you.' So I'd go over and he'd throw his arm around me like I was his new best friend. 'See this guy? He's the only guy in Washington who's got guts enough to knock me on my ass. The rest of you are a bunch of wimps.' Or he'd insult them by some expression. But if it didn't suit his purpose, hell, he'd walk by like he'd never seen me before in his life. But that was what Bobby was like. I'm sure he was very loyal to his brother. And to his family. But the rest of the world didn't make much difference."

When the family got together at Hyannis Port, Teddy played with his big brothers and their friends. He was eager enough but so slow of foot that Bobby or even Jack could dance around him and run down the expanse of lawn for a touchdown. At Catholic Cranwell, a Jesuit school where he spent eighth grade, husky Teddy challenged a priest in robes to a foot race. The father left Teddy eating his dust. As sluggish as he might appear, Teddy was so strong and vigorous that he seemed the very definition of good health. Jack looked at his youngest brother with awe at the precious gift that Teddy had been given, a gift whose value the kid could not possibly understand as Jack did.

In the first two summers after the war, Teddy and his friend and overseer Joey Gargan worked on the Kennedy farmland on the Cape, clearing the bridle paths and cutting hay for Joe's horses. The two boys earned thirty-five dollars a week for their endeavors, and as with everything else, Teddy's father attempted to turn their sweaty labors into a series of life lessons. Teddy had a subtle memory that tossed out much of what passed as a childhood, but these summers he remembered.

Teddy had a rapport with his maternal grandfather unlike any of his brothers, and it was on those long summer days that he first became close to Honey Fitz. The old man did not get along with his son-in-law, who had long ago tired of the ebullient Fitzgerald and his oft-told Irish tales. Honey Fitz found in little Teddy a worthy repository for his endless anecdotes. He was a vibrantly healthy octogenarian who fancied that he ingested immense quantities of iron and bromides by lying on the beach covered with seaweed.

Much of the time, though, Honey Fitz sat on the sunporch telling stories. "He was a marvelous storyteller," Teddy recalled. "I heard my first off-color story from Grandpa. He was laughing so hard that I don't think he ever did get to the punch line."

In the fall of 1946, Teddy transferred to Milton Academy, where he spent all four of his high school years. He went out for football and made the team as the regular end for his junior and senior years. He did not have the speed to lope along the sidelines and stretch out his hands for a pass soaring thirty yards down the field. But he was big and tough, the choice for short five-yard passes, grasping the ball with certain hands and steeling himself for the tacklers who tried immediately to knock him to the ground.

Teddy was not only a poor student but a sloppy one, with little regard for such educational basics as grammar and spelling. He was a great talker, though, and made his mark on the Milton debate team. When he spoke, his arguments were perfectly organized, the way they were not in his term papers. No one checked his spelling, and he had such rich verbal gifts that his twisted syntax slid by on a burst of eloquence.

Teddy already had the makings of a politician, with an inordinate interest in getting his smiling picture in the Milton yearbook as many times as possible. Both Jack and Bobby had had a special close friend, but Teddy was like many politicians in that he had many acquaintances and no close friends.

At Milton, Teddy borrowed one of Commissioner Timilty's cars so that he could drive into Boston for his dental appointment. Teddy was in a hurry. He was always in a hurry. And the confounded car kept stalling on him. "Finally it stalled for the last time at Mattapan," Teddy wrote his father. "And so I left it there. Commish still thinks it's my driving most likely, but it really was the car."

Teddy seemed always to be getting in one sort of minor trouble or another, and it was always someone else's fault. "This morning I served mass with a boy that said he knew how," Teddy wrote his parents. "The boy kicked the bell over and stood and knelt at the wrong times. After mass we were practically chased out of the church, but even after I told the priest I didn't even know the fellow, he mumbled . . . 'that he would rather have no one serve than me.' This was a definite blow to my pride."

Another quality of Teddy's was revealed in his letters to his parents. He didn't try to impress but laid out his life as it was. Unlike many teenagers, including Jack, he didn't have a need for a private life apart from his parents. He had been shuttled back and forth between so many schools that he could have been a complete emotional outcast, but that had not happened.

Teddy might not have Jack's intelligence or Bobby's intensity. Unlike Joe Jr., he did not boast to his peers of his intention to be president. Ambition was not his secret mistress. He left that to his brothers. But with all the pressures in the Kennedy family, it was no mean accomplishment to have the same dreams as other young men. He found joy in a dance at the celebrated Totem Pole, a large dance hall outside Boston. He took pleasure in buying a boat and paying for it with money he had earned, and delight in sitting around with his friends or family. Of all the siblings, he had the greatest prospect of a life full of what most people call happiness.

While Teddy struggled through Milton with few thoughts of the burden of the name that he bore, his father was attempting in a subtle, even brilliant way to institutionalize the Kennedy place in American life. During the war, Joe had gotten involved in real estate, becoming a major player in New York City and elsewhere. In 1945, he made what was probably the most successful business deal of his life. Without even one visit, he purchased the gigantic Merchandise Mart in Chicago, then second only to the Pentagon as the biggest building in the world, for $13 million. The building had been built in 1930 for $35 million, an indication in itself that he had struck a good deal. Beyond that, he knew that the government tenants of the four-million-square-foot building would be leaving after the war. Then he would be able to lease space at top commercial rates, generating several million dollars of profit a year.

Joe had no intention of having *his* money siphoned off to Washington in the unholy name of taxes. He considered laws, whether they dealt with the stock market, taxes, or political contributions, the instruments on which he played the tune that he wanted to play. And thus in 1946, he turned the Merchandise Mart into a device to help perpetuate his family and their legacy.

Joe gave one-quarter of the Merchandise Mart to a new family foundation named after his fallen son, the Joseph P. Kennedy Jr. Foundation. This money would grow tax-free, and Joe would be able to spread his bounty around to whatever charity, or semi-charity, he saw fit, stamping every gift with the Kennedy imprimatur. He turned the other 75 percent into a partnership that paid lower taxes than a corporation, keeping one-quarter of the company for himself and Rose and putting the rest in trust for his children. In 1946, Joe divested himself of Somerset Importers, his liquor holding.

Joe set out to follow in the exalted footsteps of the old Boston elite, turning himself into a philanthropist. The original mandate for the Joseph P. Kennedy Jr. Foundation was full of a nineteenth-century paternalism with its purpose of "relief, shelter, support, education, protection and maintenance of the indigent, sick or infirm; to prevent pauperism and to promote by all lawful means social and sanitary reforms, habits of thrift, as well as savings and self-dependence among the poorer classes."

Joe did not use the foundation merely as a cynical device to foster goodwill for his family. If he had wanted to do only that, he would surely have begun by handing out grants during Jack's primary campaign. Instead, he waited to announce the first major grant until two months later, in August 1946. Jack was there to hand out a $600,000 check to the Franciscan Sisters of Mary to construct a convalescent home for children of the poor named after Joe Jr.

Joe heard applause for his largess, but wherever he looked he saw outstretched palms. Some were desperate, heartfelt pleas, others the most cynical of gambits. There was a letter about an impoverished blind and deaf brother and sister. An Alabama woman was desperate for a place to live. A bishop in India wanted money to build a leper asylum. The priest at the Most Holy Redeemer Church in East Boston, the old family parish, asked for an ad in the commemorative program book for the church renovation, no more than forty dollars. Joe was discovering that in his attempt to memorialize his beloved son, he was saying no a dozen times or more for every yes, and to those he turned down, he hardly appeared the great man of endless largess.

Joe used this engine of beneficence to lift the Kennedys high above Catholic society to a place where saints and angels trod. The Kennedys were the most celebrated Catholic family in America, and Joe and his minions made sure that message was shouted to all within hearing. Whenever the Kennedys gave a gift, there was sure to be a picture of Jack, Bobby, or Ted, or perhaps all three, handing a check to a beaming priest. And every time Joe gave a major gift he insisted that the children's home or the hospital wing be named after his martyred son. Life is made of mixed motives, and although Joe demanded his full quota of publicity, he gave his money with care and precision, attempting to create not only goodwill but also good.

In 1956, when the Kennedys' gifts to Catholic charities in the archdiocese had reached $2,609,000, Archbishop Richard Cushing devoted an entire section of the weekly *Pilot* to celebrating the family's exemplary Christian charity in a spread approved by Joe. On one page was a portrait of Joe Jr. in his naval officer's uniform, with his hands outstretched above a list of all the gifts, the picture similar to one often seen of Christ. "They [the gifts] are made possible by an extraordinary man of God, a great American and his charming wife and children," the bishop wrote in an article reprinted in the *Congressional Record.* Cushing did not wish his less generous parishioners to miss the point that Joe was buying his way into God's good graces. "When each charity, from faithful hearts, carries with it the seal of Christ's Church, who can begin to estimate its heavenly reward?"

In most aspects of his life Joe was proudly, passionately cynical, yet he understood that his wealth, at least part of it, could be an engine of social goodness. "I have always thought that people who for some reason or another are not willing to risk giving away money while they are alive lose one of the greatest joys of their whole life," Joe wrote Cushing in February 1956. "If one has been fortunate enough, with God's help, to amass a fortune, one comes to a sense of realization that God must have meant him to give so that he could make it possible, in a measure, for his noble workers, like yourself, to carry on His charity."

In the spring of 1948, Joe traveled to Europe for only his second trip since leaving the Court of St. James's. In Paris, he stayed at the Hotel George V, where he talked to Bill Cunningham, a reporter for the *Boston Herald.* The man talking this day was not the Joseph P. Kennedy that history remembers, a man great only in his cynicism, a malevolent, grasping patriarch who pushed his sons mercilessly onto the stage of history. This was a man who had at least one noble idea, an idea that in some measure had cost one of his sons' lives and cast his second son, with his broken body, into the fray. It was an idea that Joe continued to profess.

Joe believed, as he told the reporter, that in America the sons of wealth had a special obligation to serve their country. Even as he said it he knew his words might sound sentimental, or worse yet, that he might seem to be promoting his own brood.

"But it's not just my children," he insisted.

> I think it should apply to all children of parents who can afford it. What we need now is selfless, informed, sincere representation and service at home and abroad. . . . Please don't try to make any heroics out of this, but you asked me an honest question and I've given you an honest answer. That has been our family plan for our children from the first. If it doesn't work out with them, it could work out with some others. I'm naturally tremendously proud of John. I think Joe, if the Lord had seen fit to spare him, would have been a fine man, and would have taken his place somewhere. Robert has yet to prove

> himself, but he's bright, conscientious, and he seems to be tremendously interested.

Joe mentioned Eunice's work in Washington, but he didn't mention Kathleen. If she had been his son, he would doubtless have considered her life self-indulgent, a trivial pursuit of pleasure. Kathleen had made herself a part of the upper-class British world that shunned him. Kathleen had an endless supply of wit and a wondrous self-possession that never left her. At the age of twenty-seven, she had fallen in love again, this time with thirty-seven-year-old Earl Fitzwilliam, who suffered the dual disabilities of being both Protestant and married.

Fitzwilliam was a perfect exemplar of the parasitic life of wealth and privilege that Joe abhorred for his own sons. Kathleen and her beau were scheduled to meet Joe in Paris to discuss their future. If Kathleen could convince her father that Fitzwilliam offered her happiness, he would have to decide whether he would stand with her against all the onslaughts of Rose and the Church.

Joe did not sit idly in Paris waiting for his daughter and her lover to arrive. He had taught his children that time was the rarest commodity in life. Some men squeezed a half-dozen lives into their given time. Others diddled through their days in what was scarcely half a life. "Time is man's dominant foe," he said. "All man has on earth is the present moment. . . . To make proper use of your time is life—to waste it is merely to exist."

Was it any wonder, then, that Kathleen lived her life like a Fourth of July sparkler, flashing brilliantly in the night? Kathleen and Fitzwilliam had decided to fly down in a private plane from London to southern France for one day. Then they planned to fly back to Paris for a Saturday luncheon at the Ritz with Joe.

On their way to the Riviera, the couple stopped in Paris for *dejeuner*. Their meal ran late, and when they returned to the airport, the weather had turned so threatening that all commercial aviation was grounded. Despite the late hour and the menacing reports, Fitzwilliam insisted that the pilot take off. Kathleen agreed with her lover and they flew toward the dark storm, considering it nothing but a momentary diversion, holding them back for a few nervous moments from the sun and warmth of the Côte d'Azur.

Early the next morning, Joe received a call in his suite from a *Boston Globe* reporter who told him that the plane had crashed. Kathleen and the other passengers were dead. There are as many ways to grieve as there are to die, and Joe turned immediately to what he considered the task at hand: to see that the truth was buried even before his daughter. No one was to dare to suggest that Kathleen had died a merry widow blithely flying off for a weekend with her adulterous lover. Joe told the reporters that his beloved Kathleen, who had stayed in England to be near her husband's grave, had hitched a ride with Lord Fitzwilliam, a mere acquaintance.

Jack was usually out somewhere for dinner, but this evening he was sitting at home listening to a recording of the Broadway musical *Finian's Rainbow*, with its haunting, nostalgic "How Are Things in Glocca Morra." Eunice took the phone call and told her brother that Kathleen had died. Jack was a man who did not cry, but this evening he cried and ran off to be alone in his bedroom. The next morning, when a family friend arrived at the house in Georgetown, Jack and his siblings had reverted to the stoic family mood. "As far as I remember, Eunice, Pat, and Jack were there," Mrs. Christopher Bridge recalled, "and there was a grim, tragic restlessness about the atmosphere, with the gramophone playing, and a closing-in of the ranks of family and friends, but no emotional collapse."

The next day Jack and Eunice traveled up to Hyannis Port to be with the rest of the family. Teddy was already there. He had left Milton Academy when he heard the terrible news and had taken the next train to the Cape. Jack was outraged that the masters at Milton had let his distraught little brother leave by himself. But there was a profound homing instinct in Teddy that in moments of grief, uncertainty, and doubt brought him to the restless seas of the Cape and the solace of a family and a house that resonated with memories of good times.

Kathleen was buried in the small cemetery in the Cavendish burial grounds at Edensor. She had been a well-loved woman and two hundred of her friends came to bury her. Of the Kennedys, only Joe was there that day. Rose had not felt right about the way her daughter was living her life, and Jack seems to have been so stunned that he could not bring himself to fly to London.

Joe stood at a distance from all those who had known and loved Kathleen, greeting no one and saying nothing. "He stood alone, unloved and despised," recalled Alastair Forbes. And he left that day as silently as he arrived, not even stopping to pay the priest.

Joe was haunted by the image of his gay, ebullient daughter whom he would never see again. He knew that some people thought that the death of one child did not make him terribly depressed, for he had so many children. How little they knew. He thought to himself that no one had any idea that he was depressed, not even Rose, who had faith as her solace.

When Kathleen died, Bobby had been at the Grand Hotel in Rome in the midst of a six-month-long European and Middle Eastern tour. He was suffering from a bout of jaundice, and when Jack called to tell him of the accident, his companion, George Terrien, recalled that he "broke down like a little kid." Despite his illness, he could have traveled to the crash site to accompany his sister's body back to England, or he could at least have flown to the funeral. He did not, however, and that said much about how his father saw the world, and the obligations he had put on his sons to be forestalled not even by the death of a sister.

Bobby's journey replicated the youthful journeys that Jack and Joe Jr. had made. Like his older brothers, Bobby was accompanied by a friend, George Terrien. He stayed in the best hotels, but that was the least of his privileges. Joe had seen to it that his twenty-two-year-old son would meet with ranking diplomats and leaders. Despite Bobby's total lack of either experience or apparent interest in journalism, he carried reporter's credentials from the *Boston Post.*

What saved Bobby from being simply a spoiled son of wealth were two qualities he could claim as his own: courage and physical daring. His brothers were brave too, but they did not extend courage as the defining virtue not only of men but also of nations. Jack perceived the postwar political world as a massive conflict of ideological systems in which too much of what was once called courage might destroy the world in a nuclear holocaust. Jack, moreover, astutely perceived why democratic man in the mass was often less than brave.

Bobby, however, saw the world through a prism of courage. He had at times a simpleminded belief in the absolute virtue of courage, not recognizing that physical and moral courage were not the same, and that a man could be physically brave and a moral coward.

When Bobby arrived in Palestine, he saw the Arabs and the Jews as admirable peoples of courage and the British as a corrupt, cowardly regime attempting to keep them from killing each other. The streets were full of dangers. Twice authorities picked Bobby up. Once he was blindfolded and when finally released warned to stay inside. That would have kept all but the most intrepid reporters from venturing no farther than the hotel bar. But Bobby headed out again to meet with prominent Jews and Arabs.

Bobby did not struggle over the transcendent question of ends and means that bedeviled the majority of Zionists, who, as much as they wanted a Jewish state, disapproved of the terrorism of some of their brethren. "Met officers in the Irgun who were responsible for blowing up the train and killing 50 British soldiers as well as the blasting of King David Hotel," he wrote in his diary, disclosing not an iota of disapproval. "They'll fight any soldiers no matter what uniform they are wearing if they attempt to administer their homeland."

In the series that Bobby wrote for the *Boston Post*, he portrayed the conflict as a tragic struggle between two peoples fighting for what they believed was right. They were voting with their own blood. He had a far lower opinion of American Jewish leaders who spoke from a sanctuary of safety.

"Many of the leading Jewish spokesmen for the Zionist cause in the U.S. are doing immeasurable harm for that cause because they have not spent any or sufficient time with their people to absorb their spirit," he wrote in a passage cut from the published article. Bobby might admire the intrepid spirit of Israeli Jews, but he still shared some of his father's stereotypical views toward American Jews.

Bobby concluded his series with a call to arms for the rest of the world. "I do not think the freedom-loving nations of the world can stand by and see 'the sweet water of the River Jordan stained red with the blood of Jews and Arabs.' The United States through the United Nations must take the lead in bringing about peace in the Holy Land."

Bobby was not deeply versed in the terrible ironies of history. He looked at the world not to understand it as much as to change it, and as a young man at times he did not see that understanding is the beginning of change. He took dangerous and simple pleasure in the profession of rhetoric.

In the struggle against communism, he considered dispassion a pallid and dishonorable excuse, a pathetic sheathing of swords. He called for the protection of Cardinal Mindszenty, the imprisoned Hungarian priest, for "all eyes will now be turned our way to see if he will be betrayed, or if resistance to evil can expect support. . . . For if we fail, the fault as with Julius Caesar's Romans will be 'not in our stars but in ourselves.' LET US NOT FAIL!"

Bobby's rhetoric may have sounded strangely overwrought, but as a deeply religious, conservative Catholic, his sentiments were honest. These were largely Catholic countries, and American Catholics felt deeply about their religious brethren being unable to worship God as they chose. At mass on Sunday, priests conveyed an image of Eastern Europe as a gigantic Communist prison in which millions of people lived on their knees but could not pray on their knees.

For Bobby, as for most Catholics, Mindszenty was a symbol and martyr, a fearless, saintly priest who would not kneel down before the alien Marxist faith. That he was so tortured and debased that, emaciated and broken, he finally confessed to crimes against the Hungarian state did not diminish his martyrdom in the slightest, but only enhanced it. For Bobby, as for millions of his co-religionists, issues of faith and politics came together in their unyielding anticommunism, to them the transcendent issue of their time.

Those who stood stalwartly against evil deserved at least the wreaths of memory and Bobby traveled to Heppen, Belgium, where Kathleen's husband, Billy Hartington, had fallen. Like Bobby, Billy had been a man who felt that he had to prove himself worthy of his heritage, yet however much he did was not quite enough. On the day he died he had led his troops and tanks wearing a white mackintosh and bright pants and he fell with a bullet through the heart.

Bobby went to Billy's grave and what struck him was how routine the death was to the villagers. "They went through all the motions of how Billy was killed and how he fell, etc., which was a little much for me," he noted in his diary. "The farmhouse he was killed attacking and into which he threw his grenade now houses a couple of little children and parents with the man having difficulty keeping his pants up. The people all tell their stories of the war as if they just came out of a wild western movie."

When Bobby reached London, he stayed in Kathleen's townhouse. One evening he went to see *The Chiltern Hundreds*, a play by William Douglas-Home. The playwright, who had been in love with Kathleen before the war, had re-created her on stage as a central character of the play. Bobby's beloved sister was resurrected as an exquisite, ethereal character walking across the stage of the Vaudeville Theater. The actress playing Kathleen had soft, vaguely aristocratic features and was beautiful in the way that Kathleen was only in the memory of her friends.

Bobby met Joan Winmill, the twenty-one-year-old actress playing his sister, and fell in love. His father and brothers had all had flings with actresses; to them, the very word "actress" had a vaguely erotic connotation and was the preferred category for casual dalliances. Joan belied all that. She took the moral mandates of the Church of England as seriously as Bobby took those of the Church of Rome. Her mother had died in childbirth, and she had been brought up largely by relatives. She had a trusting innocence and was a woman of the strongest character. She was as taken with this idealistic young American as he was with her.

For the next weeks, Bobby met Joan every evening after the play for dinner, finishing the evening in his late sister's house. Joan did not see Bobby as a shy, socially inept young man, but as a dashing American with a "freckled face and white, toothy grin." For Bobby, this romance was as great an adventure as any of his escapades in Palestine or Berlin, and as healthy a tonic to his sense of well-being. No woman had ever had such an exalted conception of him. No woman had ever listened to his dreams, not as unlikely reveries, but as the road map of his tomorrows. "He talked about wanting to do good things for his country," Joan Winmill recalled. "He wasn't more specific about that, but that's what he wanted to do."

As soon as Joe heard about the romance, he sent up every distress signal. He had lost one daughter to England; he was not about to lose a son. When Bobby's father heard the word "actress," he thought of Gloria Swanson and the romantic misery of that misbegotten affair. Beyond that, Bobby was unlike Jack, who moved from woman to woman so rapidly that Joe rarely had time to worry about whether one momentary alliance was appropriate or not. Bobby was serious about this woman and Bobby was a young man who used language as a tool of his truth. Bobby dismissed his father's concern with a chuckle to Joan, but he could hardly remove Joe's grasp on his life. "He was so controlled by his father," Winmill reflected. "He was clearly in the shadows of his brothers, and his father dominated much of his life."

Bobby's other problem was more immediate, and that was his relationship with Ethel Skakel. His sister Jean had introduced him to Ethel during the winter of 1945/46 on a ski trip to Canada, but he had been less impressed with the skittery, intense Ethel than with Pat Skakel, her earnest, serious, older sister.

Both young women were students at exclusive Manhattanville College in upper Manhattan. There they were taught by the nuns of the Sacred Heart that educated Catholic women should not only be good wives and good mothers but also do good work in the world. Both women were deeply religious, Ethel so much so that she at one time contemplated becoming a nun. Pat became president of the student body and after graduation worked for the Christophers, a Catholic activist organization whose slogan, "It is better to light one little candle than to curse the darkness," could have been penned by Bobby's father. Pat could sit at the Kennedys' dinner table and discuss current affairs with as much confidence and vigor as anyone in the family.

While her older sister was downtown helping to integrate Schrafft's Restaurant, Ethel stayed uptown, a prankster full of mischief to enliven the tedious regimen of learning at Manhattanville. Ethel had an endless number of friends and was blessed by a bounty of energy and impish delight in the world. Ethel might have been plain, but her sheer joy in life gave her a radiance not seen in many far prettier women.

Ethel flung open each of life's doors that stood before her and rushed forward, fearing nothing and no one. She had the Skakel sense of humor—rude, raucous, and daring. When a handsome young rider on the Irish national team spurned her affection, she sneaked into the stable and painted his horse green. One evening she sat in her dorm room momentarily remorseful at the fact that she had so many marks in the nuns' demerit book that she would not be able to leave campus. Suddenly she came upon the solution: steal the confounded book and throw it down the incinerator.

Ethel's endless pranks all had an edge to them that left their victims outside the circle of gaiety and at times rendered almost as much hurt as laughter. At Manhattanville, she did not come forward and own up to her not-so-practical jokes but preferred to stand with the other students and have them punished along with her. She treated the school library like her own bookstore, not bothering with the tedious business of returning books.

Ethel might have seemed like a frivolous young woman, but she was deadly serious in her pursuit of Bobby, who was two and a half years her senior. She was a fierce competitor on the tennis court or in a horse show, and she was a doubly fierce competitor when the opponent was her own sister. That said, she still might have lost the game had she not had Jean, her closest friend and Bobby's sister, as her co-conspirator. Ethel had no interest in politics, but she went to Boston to work alongside Bobby in Jack's primary campaign. So had Pat, and soon afterward it was Ethel's sister who went with Bobby to the Manhattanville senior prom.

That Christmas Bobby invited Pat to come down to Palm Beach and spend the holidays. Bobby did not invite a young woman to the Kennedy family home without the highest and most honorable of intentions. Jean and Ethel understood this full well, and they showed up together at the Palm Beach estate.

Whatever the sport, Ethel played with fierce competitiveness and a casual disregard for the more onerous rules. By the time the family left Florida, Ethel and Bobby were the couple, not Pat and Bobby. The fact that Pat did not marry Bobby probably had less to do with Ethel's relentless maneuvering than with her own desires. Soon afterward, she married an Irishman and moved to a modest house in Dublin, a world away from the Skakel estate in Greenwich, Connecticut.

When he wasn't playing football at Harvard, Bobby spent as many weekends as he could with Ethel, often staying at the family estate. Coming up the circular driveway on Lake Avenue the first time, Bobby was instantly confronted with the immensity of Ethel's home. The Kennedys lived well, but nothing like this. The entrance hall had twenty-five-foot ceilings, and there were eleven bedrooms in the main house alone. The floors were of polished teak. There were enormous black marble fireplaces. The decor announced its refinement authoritatively enough to tone down even the earthiest and most vulgar of visitors, but it quieted the Skakels not at all.

George Skakel, Ethel's father, was a self-made multimillionaire who made his fortune with what became the Great Lakes Carbon Corporation. Ann Brannack Skakel, his Irish-Catholic wife, had given him seven children, four daughters and three sons. "Big Ann," as she was known, was oversized in everything, from her enormous frame, the intensity of her faith, and the decibels of her voice to the rudeness that she inflicted on underlings and shopgirls. The Skakels did not live country club lives of sedate games and gatherings but partied and played with wild, frenetic energy.

Bobby knew all about a self-styled, self-made father who was often not home, a deeply religious mother surrounded by nuns and priests, lives of endless privilege and energetic good times. But that was the end of the similarities. Bobby's father was a man of immense social ambition. Insecurity is often the father of achievement, and Joe had propelled his sons to make their proper places in the world.

George Skakel, by contrast, didn't give a tinker's damn that in Greenwich his family was considered boorish, uncouth, Catholic interlopers. Unlike Joe, he didn't have a fancy Harvard education. He taught his children that wealth was their preserve, and they could do whatever they wanted.

Joe, the son of a liquor dealer, had made a fortune in the liquor industry. He was willing to sell it as long as he didn't have to drink much of it. He saw liquor not so much as the curse of the Irish as their pathetic cliché. Joe would no more sit around drinking whiskey than he would go out on the street carrying a shillelagh.

Joe felt so strongly about the pernicious impact of drinking that he had promised his sons one thousand dollars if they did not drink until they were twenty-one. Bobby had earned the money without regret or temptation. He might take a drink occasionally, but at social events he did not head nervously to the bar. Although Jack didn't quite make Joe's deadline, he did not drink much either.

The Skakels, however, considered liquor the fuel of good times, and they liked nothing more than good times. They had none of what they considered the Kennedys' prissy, puritanical fear of liquor.

Big Ann didn't drink sherry from a thimble but whiskey from a crystal tumbler. She did not live the corseted life of Rose Kennedy, speaking in clipped civil terms and keeping her house in such impeccable condition that at any moment she could happily have invited a prince of her church into her living room. Fancy furniture did not intimidate Big Ann. She liked dogs and gave them the run of the house. So be it if they stained the old rugs and chewed up the furniture.

For a person who preferred more than one weak drink before dinner and considered politics an unholy bore, an evening at the Skakels was far more amusing than dinner at the Kennedys. The Kennedy siblings thought that putdowns of each other were funny, but the Skakels turned their humor outward onto the world. God and laughter were the two great solaces of their lives.

They might cry when times were bad, but they attacked sadness, laughing, singing, and drinking until the blues gave up and joined the revelry. Ethel's three brothers, excessive in everything they did, were the engines that drove these wild good times. They drove their cars at a hundred miles an hour over country roads, and they handled everything else the same way. They liked nothing more than riding around Greenwich with one of them standing on the roof of the car while the driver tried to find a branch low enough to knock his brother off his fancy perch. They drove cars into the pond on the estate and ran from the silly cornpone cops who had the audacity to think they could catch a Skakel.

On one occasion George Jr., his father's namesake, lost control of his car running from the police and smashed into an abutment, breaking his jaw. Ethel was their sister in name and deed. Running late to a horse show at Madison Square Garden, she drove her car up onto the sidewalk in Central Park, passing the other cars moving slowly down the parkway.

Ethel, like Bobby, was lost in the midst of her family. The sixth of seven children, she struggled to compete with her big sisters or to get the kind of attention that came so easily to her pretty baby sister, Ann. Bobby and Ethel had much in common, but to those who watched them, they seemed more buddies than lovers. "It was very nice and cordial, but I never saw him except once put an arm around her or give her a kiss in public or something like that," recalled Bobby's football teammate Nick Rodis. "We were driving Ethel somewhere. Bobby and Ethel were sitting in the back seat. I turned around one time, and I talked to him and he had his arm around her and everything. And I just kept my eyes on the road from there on in."

Ethel was one of the guys, and immensely popular with Bobby's friends. "I think Ethel was great for him," recalled Gerald Tremblay, a law school friend. "She brought him out more. He became a little more of an extrovert. He was a pretty introverted person. He never discussed his feelings with people. To my knowledge, he never had anybody, like a friend, that you might discuss a pretty intimate problem with, maybe about your love life or your relationship with your father or your mother."

Ethel had made Bobby her life project. Bobby did not appear mindlessly, breathlessly in love with Ethel, electrified by her touch, morose when she was away. He had found in Ethel a rich commonality, but he found something different and dangerous with Joan. The two women symbolized the two directions that Bobby's life might take. His big brother Jack had stood on an open road too, with one road pointing westward toward freedom and uncertainty, the other road heading east toward more restricted venues of power. For Bobby, the journey toward Joan led to freedom and uncharted passages. Whatever else he felt about Ethel, the journey toward her meant that he was ready to pick up all the heavy burdens of his family name and position and walk a pathway set out by his father.

Even across the Atlantic, Ethel could smell the perfumes of romance and see the smoke of betrayal. She suddenly arrived in London with Jean, ostensibly to attend the Olympic games. Bobby prided himself on his honor and honesty, but now both he and Ethel indulged in the petty duplicities that often save a romance but destroy a love. Bobby took Ethel and Jean to see Joan performing in *The Chiltern Hundreds* and backstage introduced them simply as his sister and a friend. Joan was the only innocent.

When Bobby went backstage each evening, he was entering an exotic world unlike anything he had known in America. In her way, Kathleen had entered a different world too, and her death symbolized just how dangerous it might be to travel beyond the parameters of the known.

Joe disdained Bobby's foreign fling. Joe and his father had both married up. That was the Kennedy way. Joe did not believe in squandering his family's social capital on a romance that was as likely to be a tingle in the groin as a tug of the heart. Love was an unexpected bonus, but it was not the primary reason to marry. If a man married right, he married well. The Kennedys and the Skakels were two of the wealthiest Catholic families in America. Marrying Ethel would be a mating of dynasties, a further solidifying of the family's position in the world.

"Don't cry," Joan recalls Bobby saying as he left late that summer. "I'll be back next summer. I can't stay away from you."

Bobby sailed back to the States and entered the University of Virginia Law School in September 1948. He would have much preferred to go to Harvard, but with his abysmal college record, he was fortunate to get into the southern school. Ethel came down to visit him often, and if his love for Joan had been less than profound, he would have forgotten the British actress. His feelings were such, however, that he kept his promise to return the next summer to England. It was perhaps even harder to say good-bye this time, and when Joan stood on the ship bidding him adieu, she recalled how he talked "about my coming over to be with him."

Bobby was not the sort of man to invite Joan to the States for some hidden assignation. If she sailed westward, it would be only if Bobby had decided to stand up to the father who so dominated him, to take a first dramatic step away from the life that had been cut out for him so long before, and to marry Joan.

Joan did not stand on the beach looking seaward like the betrothed of a sea captain, but she waited just as loyally and just as long. Bobby wrote her love letters passionate in his devotion, and she knew that if she did not sail westward, he would at least be back the next summer, as he had the summer before. In May 1950, she received a letter that looked like all the others but announced that he was marrying Ethel Skakel, a woman whom Joan remembered only as Bobby's sister's quiet friend.

The force of circumstance and custom is at times stronger than that of love and honor. Bobby was a loyal soldier marching to a tune his father had composed. But then, how could he turn away from Ethel? When she descended on Charlottesville for the weekends, she was a force that nothing could deny. Bobby was a competitor, but no one played tennis the way Ethel did, as if defeat and death were roughly synonymous. And that was nothing compared to how she played the larger game of romance.

Ethel was a joyful woman who lifted Bobby's spirits and seemed to hold them up with the sheer force of her will. That she loved Bobby profoundly nobody could doubt. There was a tantalizing quality there, of which Bobby was only too aware. "My financee [*sic*] followed me down here," he wrote his sister Pat, "and wouldn't let me alone for a minute—kept running her toes through my hair & things like that."

14

The Grease of Politics

When Bobby married Ethel in June 1950, the Skakel brothers and their friends arrived at his bachelor party like a wandering minstrel troupe of such spirited demeanor that they made everyone dance to their song. Bobby's old college football teammates were ready for the party too, and by the time the thirty or so guests left the Harvard Club they had consumed twelve and a half bottles of champagne, five bottles of Haig & Haig Pinch Scotch, half a bottle of rye, most of a bottle of gin, and a third of a bottle of bourbon. On their way out, one of Bobby's friends picked up a fire extinguisher and doused the room, causing over a thousand dollars of damage, before staggering out into the New York night.

The Skakels conceded nothing to the Kennedys, considering themselves their equal in wealth and position, vastly superior in their enjoyment of life and sense of humor, and lesser only in their lack of pretense. The Skakels did not have the Kennedys' public name, but they were an immensely powerful psychological force that changed the Kennedys far more than the Kennedys changed them.

On the morning of the wedding, several bridesmaids walked out of the mansion to talk to the young Skakels and Kennedys. The men tossed the screaming women into the pool. Despite this gambit, which made the poor hairdressers' morning even more hectic, the wedding party made it to St. Mary's Church in Greenwich on time. There, close to fifteen hundred guests sat waiting in a sanctuary transformed into a garden of white Easter lilies, peonies, and white gladiolas. Bobby was the first Kennedy man of his generation to wed, and this was no mere marriage but an event that the families used to celebrate themselves politically and socially; in those pews sat many of the postwar American Catholic elite, along with other powerful Americans.

Jack stood quietly in the front of the church. The best man wore the same morning suit as the other groomsmen, but unlike most of Bobby's beefy athlete friends, Jack looked like a model out of *Gentleman's Quarterly*. As the wedding party prepared to proceed down the aisle, Lem Billings and the other ushers hurried late arrivals to their seats. Lem, a man of endless solicitousness, bent down to pick up some change that had fallen onto the floor. Ethel's big brother, George Jr., introduced himself formally to Lem by kicking him full force in the behind, sending him, gray morning suit and all, falling to the floor.

Big Ann and George Sr. didn't begrudge spending a fortune on the wedding, but it irked Ethel's family that the Kennedys were so confoundedly cheap. When the two families went out together for dinner, the Kennedys were healthy enough when it was time to order the best food and finest wines, but by the end of the evening their arms became so weak they were unable to reach out and pick up the check. This happened so many times and in so many ways that the Skakels concluded that they were not being appreciated for their largess but played for chumps.

One of Ethel's brothers-in-law, John Dowdle, found what he thought was the perfect device to exact some exquisite revenge. Dowdle had offered to set up the newlyweds' six-week Hawaiian honeymoon. Bobby would be paying for this, and nobody would be there to pick up the check this time but his father.

Without Ethel's knowledge, but with the blessing of the rest of the family, Dowdle searched through resort brochures and talked to travel agents with one goal in mind—to make the trip as absurdly expensive as possible, an endeavor with which Dowdle's travel agent was happy to comply. A first-class hotel? Of course. A suite? Why not. The bridal suite? Definitely. The presidential suite? All the better. Flowers delivered every day? Show me where to sign. Lilies or orchids? Whichever is the most expensive.

This was a classic Skakel joke, endlessly amusing to its authors but potentially immensely hurtful to its victims. In their first weeks of marriage, Bobby and Ethel could have been embroiled in the kind of petty bickering over money that destroys many marriages. Ethel, however, didn't get the joke, and neither did Bobby. To the bride, money poured out endlessly from a golden spigot; her honeymoon was not mindlessly extravagant but simply *comme il faut*.

As for Bobby, he was not much more versed in the mundane prices of life than his bride, and anyway, he was in love. He found no disparity in the fact that while his bride traveled with thirteen suitcases, he had only one. Twenty-four-year-old Bobby was still so imbued with a sense of himself as a

Kennedy that after his honeymoon he planned to spend three weeks at Hyannis Port before returning to law school at the University of Virginia.

Bobby and Ethel went to live that fall of 1950 in Charlottesville, where they rented a large, comfortable old house. One evening Wally Flynn showed up at the front door for a steak dinner with the newlyweds. "Wally, come in," Bobby said, a hangdog look on his face. "Come in. We're all fouled up," Bobby told his old Harvard football mate. Wally looked around and saw that the place was in a shambles. A pregnant Ethel rested upstairs in bed. Hearing the voice of Bobby's Harvard teammate, she shuffled downstairs. "I've got to go back to bed," she said and shuffled back up the stairs.

"Look, I'll straighten this out right now." Wally said looking at the stacks of dishes in the kitchen. "You go do what you have to do. I won't get in your way tonight or anything else." Bobby didn't argue but left and shut himself away from the worst of the squalor.

Ethel and Bobby had nobody to take care of the tedious business of domestic life. Undone chores were simply piling up around them. Wally could see why nobody had washed the dishes. The sink was stopped up. He found some tools, got down under the sink, and cleaned out the whole system before putting it back together. Then he washed the dishes, took out bag after bag of garbage, spiffed up the place a little, broiled some steaks, and set the table.

"What's going on?" Bobby asked, as he returned to the dining room.

"Nothing unusual. You wanted a steak dinner and I fixed the kitchen sink."

"Where'd you learn to do that?"

"My mother taught me."

Bobby shook his head in awe. That wasn't exactly one of the things that he had learned from Rose.

Bobby was a dogged, pugnacious young man attempting to follow the same arduous pathway that his older brothers had set out on. But what Jack performed with fluid ease and grace, Bobby managed with awkward difficulty. He had no qualms in treating much of the rest of the world as helpmates in his ascent.

One day Bobby came charging into Jack's office holding a stack of papers in his outstretched hand. "You're Mary," he said to Mary Davis, Jack's secretary. "Yes, I am," she replied, in no doubt about her name. "You've got to type this up for me right away," he said urgently. "It's one of my papers for school."

"I can't do that," Davis insisted.

"You have to," he insisted. "I'm Bob Kennedy." That was the ultimate argument and it showed Mary's gaucheness that he should even have to mention his name when it was so plainly obvious. The secretary still refused to type the paper and Bobby kept repeating his arguments. The woman still wouldn't give in.

Bobby was a man trying to open a door with a key that had always worked before. But as much as he turned it in the lock, he was left standing in the cold. In the end, Mary called in Jack as the arbiter, who told his brother forcefully that his secretary had other matters to attend to besides typing his term paper.

Most of the papers that Bobby wrote during law school gave no scope to his mind and emotions, but in one major essay on the Yalta conference he wrote with the moral certainty of a man whose palette contained only two colors, black and white. "What is the rationalization of this most amoral of acts whose potential disaster has long since become for us present day catastrophe," he asked rhetorically. "The God Mars smiled and rubbed his hands." Bobby believed that staying "friendly toward Russia" was "a philosophy that spelled disaster and death for the world." Even if Soviet armed might had allowed the Russians to march into Central Europe, "there would have been a great difference between Soviet stooge regimes set up by the Red Army and those strengthened by the acquisence [*sic*] and endorsement of the western powers. The former would have enjoyed no shred of moral authority."

Bobby cared about politics, not law, and he took the Student Legal Forum at Virginia and turned it into a lecture series that brought in a number of important speakers, including his own father. Joe could have been a memorable teacher. He was so provocative, so perverse in his thinking, that he would have forced his students to reflect and to defend themselves.

Speaking in December 1950 during the middle of the Korean War, when a narrow patriotism had quelled most voices of discontent, he daringly said that the United States should pack up and leave Korea and all of Asia. He asked bluntly what business we had supporting "Mr. Syngman Rhee's concept of democracy in Korea." He seethed at the way the United States supported the French colonial regime in Indochina. And he didn't care if all Europe became Communist. "The more peoples that are under its yoke, the greater are the possibilities of revolt."

Bobby loved this father who spoke such unparsed words and struck down conventional wisdom with a flick of his rhetoric. Bobby mimicked his father's bluntness and copied his verbal flourishes, but the two men did not see the world the same way. Joe sought to pull America back from all the sordid complexities of the rest of the world to live in a sanctuary of peace and civility. Bobby wanted to move aggressively forward. Unlike his big brothers, Bobby had not seen war. Despite Joe Jr.'s death, Bobby did not fully understand the wages of heroism. He saw politics in part as a venue for courage, where men stood up and proved the worth of themselves and their nations.

Hypocrisy is the grease of politics, but Bobby used this lubricant only sparingly. When he invited the distinguished black diplomat Ralph Bunche to speak, Bunche replied that he would not speak before a segregated audience, a stipulation that Bobby surely must have expected. Bobby knew, then, that he would be confronting a Virginia law that prohibited blacks and whites from sitting together in public meetings. Bobby called together representatives of student government and asked them not simply to put forth a resolution calling for an integrated audience but to sign the document.

The students were all for integrating the speech, but they blanched at putting their names on a document that might be widely publicized, bringing rebuke down on their families. These young men had walked most of a long hard mile, but they had pulled up short of the finish line; they could have been saluted for how far they had come, not condemned for the few feet farther they were unwilling to walk.

Bobby ranted at them, barely comprehensible, his words even less understandable as a Boston accent in a sea of southern drawls. In the end the students voted down the resolution that they would have had to sign, but the Student Legal Forum adopted it.

This was not the first time that the university, one of the most liberal institutions in the state, had been confronted with this problem. Up until then the school had liberally applied the grease of hypocrisy by prominently posting a notice stating that a hall was segregated and then allowing blacks and whites to sit wherever they chose.

Bobby would have none of that. The matter was of such seriousness, the dispute so rancorous, that it came before Colgate Darden, the president of the university. Darden declared that the lecture was not a public meeting at all but an educational meeting, and could go on unsegregated.

For the first time in his life Bobby had confronted the most terrible American conundrum of his age, the question of race. He was not a fledgling politician who saw himself as an arbiter between different interests and peoples, seeking a consensus that would push society ahead inch by inch. When he saw what he called truth, he went for it, and woe betide those who stood in his way waving what he considered a white flag of compromise and expediency.

The youngest Kennedy man entered Harvard in the fall of 1950. Teddy had none of the social ambitions of his father or, to a lesser extent, his brother

Jack. Nor had he the disdain for the narrow social elites of Cambridge that marked Bobby's college tenure.

As a boy, Teddy had been not the youngest in the family but often among the newest boys in many of the schools he attended. To get along he developed a genial, conciliatory manner. He was interested in good times more than in great ideas, and he surrounded himself with young men of similar instincts. Most of his friends were football players and other athletes, the amiable sort who would make a natural transition from the gridiron to the manly world of business.

Many of Teddy's friends had been shuttled off to prep school during their parents' unseemly divorces. They were largely trophy children paraded home on holidays. Some of them spoke disdainfully of their parents or dismissed them irreverently. Teddy's friend Claude Hooton Jr. was startled to hear one of their companions calling his mother by her first name. That was unheard of back in the Texas that he called home.

Teddy took literally the biblical injunction to honor one's parents. He always called Rose "Mother" and Joe "Dad." Whatever Teddy's friends thought of their own parents, when he took them down to Hyannis Port for the weekend, they sat a mite taller at the dinner table and watched their words more carefully than they ever would have in their own homes.

Teddy's father had taught him that he had a special responsibility as a Kennedy man. But what was that admonition to an eighteen-year-old finally free of all the constraints of prep school life and of his father's overwhelming presence? He didn't like rules, be they silly speed limits or other regulations that sought to hold him back from the life he intended to live.

Bobby's football teammate Wally Flynn recalled that Teddy asked Wally and Nancy, his wife, to chaperone a party at a Harvard club to which they no longer belonged.

"Teddy, am I going to get in trouble?" she asked, knowing full well that Teddy was up to something that was not quite right.

Teddy's friends at Harvard had their own special moral code, and it was a code that played into the part of Teddy that was weak and intellectually slovenly. These athletes considered academic course work a tedious, largely unnecessary regimen that kept them from playing sports and having a good time. They helped one another with their studies, choosing the easiest courses, passing on notes, cramming together for exams.

In the fall semester, Teddy took a course in natural science, a subject in which he and his friends had not an iota of interest. One of the fellows had taken a lot of physics in prep school. That led to the obvious solution. During the final exam the amenable friend sat up front in the amphitheater of the

Allston Burr Science Building, writing in big letters in his blue book while Teddy and his buddies sat behind copying the answers.

For Teddy, it was a morning of little moment, but it set him apart from his brothers' lives. Joe Jr. might have had a tutor priming him beforehand, or even handing him the previous year's exam, but he would not have done what Teddy was doing. Nor would Jack. Bobby would perhaps have struggled mightily with the dilemma, and if he had gone along, it would have been only to get a good grade. But Bobby's friends at Harvard were too proud and too morally straight to attempt such a thing. And so, probably, was Bobby.

In the spring Teddy took Spanish I. He had no natural instinct for languages, and he was appalled at the idea of having to study what he considered a useless subject for yet another semester. Somehow if he could get an A, he would be relieved of his language requirement.

Teddy and Warren O'Donnell, Kenny's younger brother, went for a walk the night before the final exam.

"How are you doing?" Warren asked.

"This is a tough one," Teddy recalled saying as the two men walked through Harvard Yard. "I've got to get that C minus or I can't play football in the fall."

Teddy and Warren decided to see another friend who was a crackerjack Spanish student. The young man was a scholarship student and he was open to suggestion. "Fine, hell, I'll be glad to take that thing," he said, agreeing to pose as Teddy the next morning and ace the Spanish exam.

For the rest of his life, Teddy would be surrounded by overly solicitous people who called themselves his friends and were ready to do what they had to do to get him what they thought he wanted. In this instance, Teddy stood by saying little while his friends pushed this young man, even waking him up the morning of the exam, prodding him to get dressed and fill in for Teddy.

There appeared to be a calculated passivity in Teddy, as if he thought himself less morally culpable if he had given no command. Others might have considered Teddy's conduct doubly dishonorable: if he was going to cheat, then he should at least have had the gumption to do it himself without bringing in a gullible innocent. That was a subtlety lost on the Harvard dean, who, when the cheating was discovered, treated each young man equally and expelled them both for at least a year.

For Teddy, as for his brothers, the overwhelming fear was not what he did but what their father would think of what he did. Joe was a man of limitless ambitions for his family, yet he did not rage at his youngest son for betraying the Kennedys while Jack was thinking of running for the Senate or governor. As tough and merciless as Joe could be, he cared now more about his son's life than his family's future.

"Initially, my father just thought about what the impact was going to be on my life, etc., so he was initially very very calm," Teddy recalled. Joe did his own inventory of what it would mean, learning that after a year his son could be reinstated. "And then after he got a feel for that sort of thing, he went through the roof (that was about twenty-four hours later) for about five hours and then he was all fine and never brought it up again."

For Joe, the mystery was not that Teddy had cheated, but that he had cheated for so little. "The father was terribly disappointed in Ted's doing something as foolish as that when there was so little at stake," recalled the other young man, who after being thrown out of Harvard got to know Teddy's father. Joe made a grand symbolic gesture to suggest that he believed young men should have a second chance. When members of Army's football team were thrown out of West Point in a cheating scandal, he paid their way to study at Notre Dame.

For Joe, the matter may have been behind him, but for Teddy, yanked unceremoniously out of his happy Harvard life, it was not. Teddy's father did not believe in penance, but Harvard did, viewing a term in the armed services as suitable punishment. "If I had a good record in the Army, this would resolve and satisfy them," Teddy said. One day later in the spring of 1951, he went down to the U.S. Army recruiting office and signed up. When Teddy returned to the house, Joe discovered that his son had signed up for four years, not two. Teddy was no student, but he could certainly tell two years from four, and it was probably a mark of his anxiety that he had not even noticed.

"When I signed up for four years, it was just a matter of paper shuffling," Teddy recalled. "They had three forms at the recruiting office and I had no idea. I went down with the idea to sign up for two years . . . it was just an administrative type of thing." That may have been true, but it took the considerable efforts of his father to rectify the error.

After basic training at Fort Dix, New Jersey, Teddy transferred to Fort Holabird in Maryland, where he intended to enter Army Intelligence. He had scarcely started the program when he was abruptly terminated and sent to Camp Gordon in Georgia to be trained as an MP. From there he sailed to France on the *Langfit*.

Teddy had a self-deprecating sense of humor and was perfectly willing to make himself the butt of his own jokes. On the long crossing he did not write of his female conquests back in Georgia, as Joe Jr. might have done, or of being here while his friends whooped it up in the exalted confines of Harvard and other young men died in Korea. He wrote instead of the minor misfor-

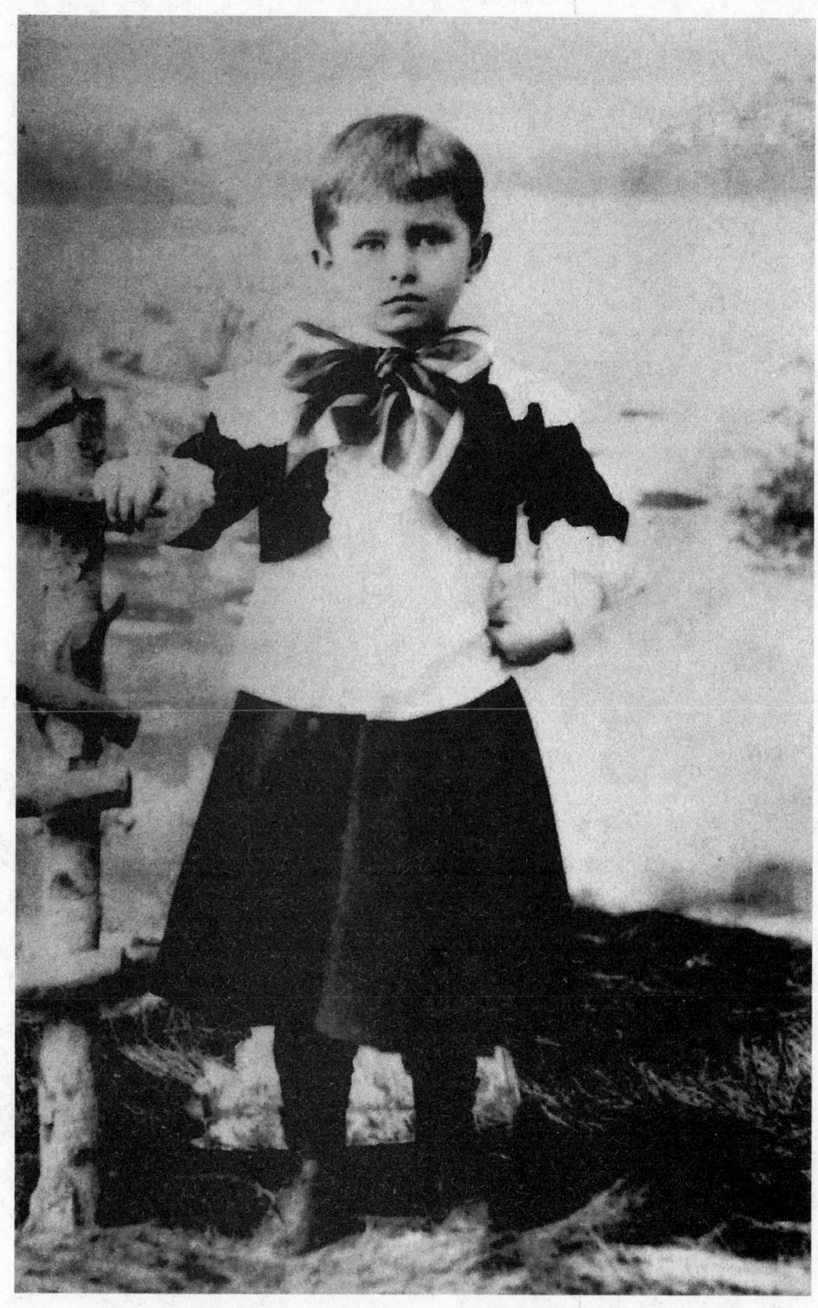

When a young Joseph P. Kennedy was formally photographed, he clenched his left fist. *(Loretta Kennedy Connelly Collection)*

A 1899 tintype of Mary Augusta Kennedy (second row, left) with her sisters, nieces and nephews, and her ten-year-old son Joseph (bottom row right) and six-year-old daughter Loretta (bottom row, center). *(Loretta Kennedy Connelly Collection)*

Twelve-year-old Joseph P. Kennedy with his sisters Margaret, 2, and Loretta, 8. *(Loretta Kennedy Connelly Collection)*

Senator Patrick Joseph "P. J." Kennedy at the beach in Nantasket, Massachusetts. *(Loretta Kennedy Connelly Collection)*

Joseph P. Kennedy playing baseball with his father, Patrick Joseph Kennedy, 1901. *(Loretta Kennedy Connelly Collection)*

Sixteen-year-old Joseph P. Kennedy (front, right) with his sister Loretta (front, left) and his parents, P. J (second row, left) and Loretta (second row, second from left). *(Loretta Kennedy Connelly Collection)*

A patriotic celebration in Winthrop, Massachusetts, with Joseph P. Kennedy (first row, second from right) and family friends. *(Loretta Kennedy Connelly Collection)*

A handsome Joseph P. Kennedy graduating from Boston Latin School, 1908. *(Loretta Kennedy Connelly Collection)*

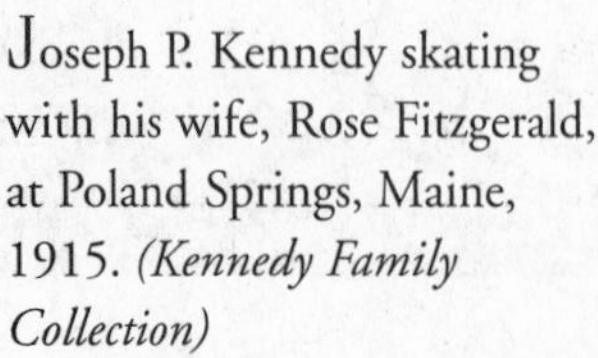

Joseph P. Kennedy skating with his wife, Rose Fitzgerald, at Poland Springs, Maine, 1915. *(Kennedy Family Collection)*

Joseph P. Kennedy holding his two sons,
Joe Jr. (left) and John (right), c. 1917.
(Kennedy Family Collection)

John F. Kennedy with his closest friend, LeMoyne Billings, in Palm Beach, c. 1936. *(Kennedy Family Collection)*

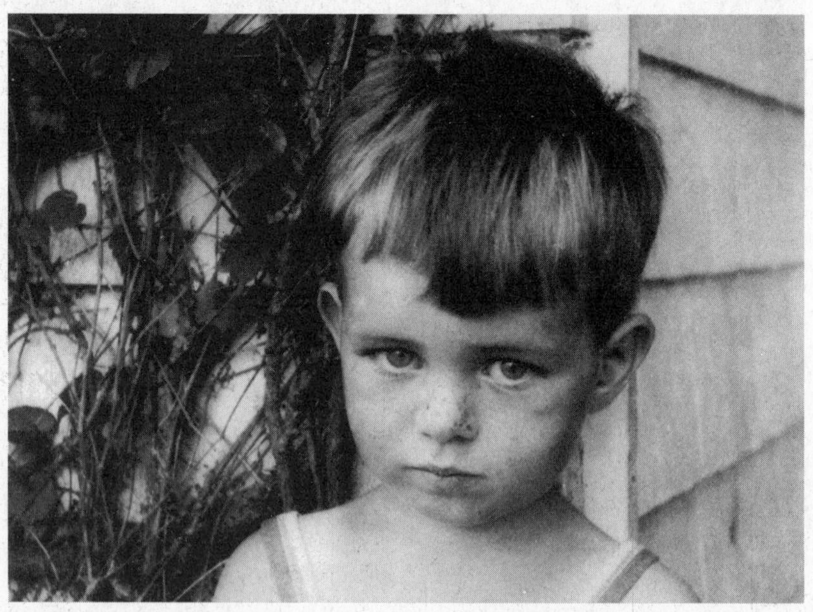

Unlike his big brothers, there was a shy, tender quality to the young Robert F. Kennedy. *(Kennedy Family Collection)*

John F. Kennedy juggling on the streets of a German city, probably Cologne, August 1937. *(John F. Kennedy Presidential Library)*

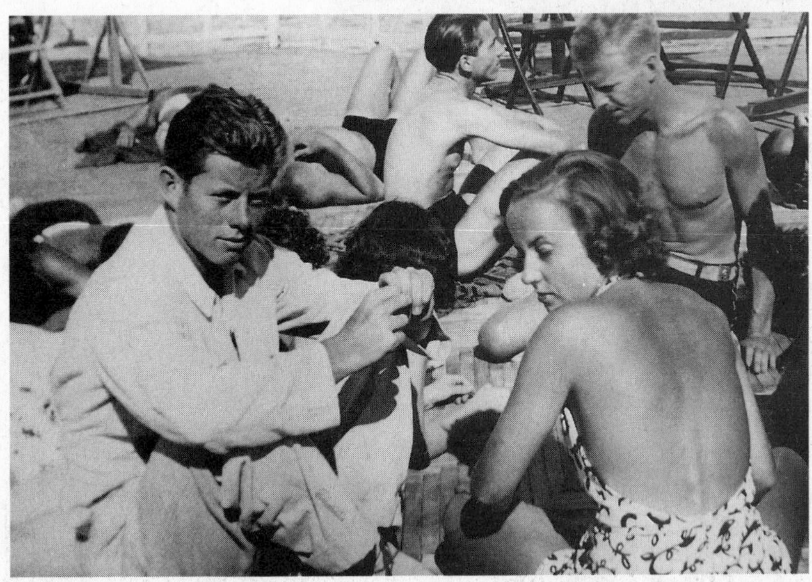

John F. Kennedy at the beach in Venice, August 1937. *(John F. Kennedy Presidential Library)*

Joseph P. Kennedy, ambassador to the Court of St. James's, and two of his sons, Joe (left) and Jack. *(John F. Kennedy Presidential Library)*

Joseph P. Kennedy and his precious sons, Bob, Jack, Teddy, and Joe Jr., on vacation in Antibes, France, 1939. *(Kennedy Family Collection)*

Three young Kennedys—Joe, Kathleen and Jack—off to the House of Commons to hear Great Britain declare war on Germany, September 3, 1939. *(Kennedy Family Collection)*

A handsome Joseph P. Kennedy as a Naval pilot in World War II. *(John F. Kennedy Presidential Library)*

Kennedy brothers Joe Jr. and Jack in their naval uniforms. *(John F. Kennedy Presidential Library)*

Lieutenant John F. Kennedy with a cigar in the Pacific. *(John F. Kennedy Presidential Library)*

Jack Kennedy (right) and his men on PT 109.
(John F. Kennedy Presidential Library)

Joe Jr. and Kathleen Kennedy in London, 1943.
(Kennedy Family Collection, Swebe, London)

Joseph P. Kennedy greeting his son Robert, serving on the destroyer *Joseph P. Kennedy, Jr.* *(John F. Kennedy Presidential Library)*

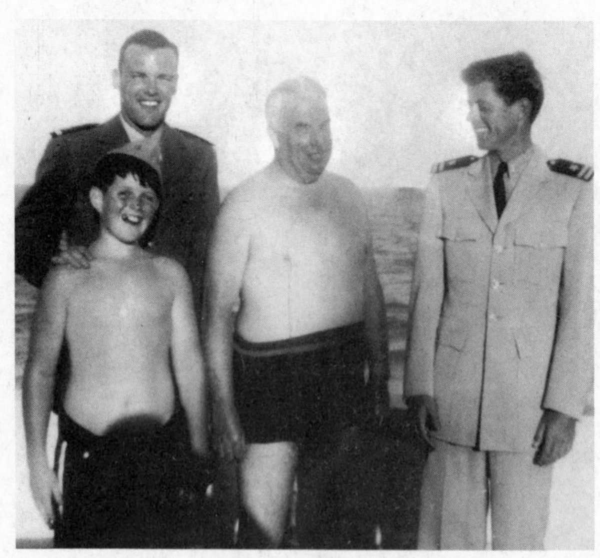

In the summer of 1944, Lieutenant John F. Kennedy (right), a returning hero, stands on the beach in Hyannis Port with his friend Lem Billings (left), his little brother Ted, and family friend Joe Timilty. *(John F. Kennedy Presidential Library)*

Teddy Kennedy jumping high after a football at Hyannis Port. *(John F. Kennedy Presidential Library)*

John F. Kennedy campaigning for Congress, 1946. *(Robert White Collection)*

John F. Kennedy on crutches campaigning with his proud mother behind him. *(John F. Kennedy Presidential Library)*

Senator John F. Kennedy feared that if married he would lose the appeal that made him irresistible to so many women voters. *(Robert White Collection)*

John F. Kennedy and George Smathers,
the two "playboy" senators, in pensive moods.
(John F. Kennedy Presidential Library)

John F. Kennedy fell in love with a beautiful young Swedish woman, Gunilla Von Post. *(Gunilla Von Post)*

The wedding of the year—Senator John F. Kennedy marries Jacqueline Bouvier, 1953. *(John F. Kennedy Presidential Library)*

tune of having been chosen for KP, a duty that had him supposedly contemplating going AWOL in Norfolk. "However, upon considering my welcome back in New York by my family . . . I concluded that to make the crossing now was the only thing left to do," he wrote, his humor still intact. His major accomplishment, as he saw it, was that he lost fifteen pounds.

Teddy did not look toward Europe as his big brothers had as a place to test their manhood and their minds. The Continent was not the dangerously inviting place it had been before the war. But Europe was still full of young Americans whose wanderlust had brought them to the cafés of Paris and the steps of Rome in search of adventure and culture they felt they could not find at home. Teddy was not the kind of young man, however, who was deeply attracted to foreign accents. He served out his time as a member of the NATO honor guard outside Paris, a lackluster, ceremonial duty, and though he went bobsledding in Switzerland, the adventures he sought were back at Harvard.

Now that Teddy's scandal was behind him and the matter had effectively been kept quiet, Joe had one major duty to perform before the family could go on as he wanted it to go on. He had been responsible for Rosemary's lobotomy. So too was he responsible for her care. Joe had sent Rosemary to several psychiatric hospitals before he had settled upon Craig House in upstate New York. The private psychiatric hospital catered to the wealthy and famous, including F. Scott Fitzgerald's wife, Zelda, who spent nine weeks there in 1934. Rosemary was in all likelihood the only patient to spend years there, however, shut away behind the barred windows high above the Hudson River.

Joe had turned his eldest daughter into the unmentionable Kennedy, as exorcised from the family dialogue as if she had been condemned to a biblical shunning. Was it possible that this man who at times still cried when his eldest son's name was mentioned cared so little for whatever was left of his eldest daughter? Did he dismiss her because she was a mere woman, bearing none of the noble manly traits and prospects of his sons? Could he simply walk away from his daughter's life, never looking back? Or did he know too well what he had done and feel too much, finding it unbearable to mention her name? Could he not stand to see what damage the scalpel had done to a sweetly compliant young woman who had once made her debut before the king and queen of England? Was Rosemary his secret torment? That answer lay only within Joe, and there it would always reside. But it was a terrible truth that lay there, whatever it was. Joe was either a monster in his unconcern or a man keeping matters within his own heart that no one should have to keep hidden.

Joe had never expressed any interest in the problems of mental retardation, but his friend Archbishop Richard Cushing had told him about a well-regarded Catholic institution, St. Coletta School for Exceptional Children in Jefferson, Wisconsin. In 1949 Joe traveled there and arranged for the building of a small home alongside Alverno House, in which resided lifelong adult residents. In August, when the construction was finished, two nuns traveled to Craig House and brought Rosemary back to the modest house where she would spend the rest of her life watched over by several nuns. But Joe did not go to visit her or see her. Nor for years did anyone else, not her mother or any of her brothers and sisters. The first visit to Rosemary probably took place in 1958 when Jack made a secret trip to St. Coletta's during a campaign swing.

Jack did not like confrontation. He found the endless battles between Democrats and Republicans not a real war but a wearisome and tedious routine. He had one of the worst attendance records in the House of Representatives, missing over one-quarter of the roll-call votes during three terms. Only some of those absences could be attributed to his many illnesses.

He simply could not tolerate sitting through committee hearings and House debates, pretending to listen to the pious posturing and platitudes of so many of his colleagues. That said, he was no more interested in wandering through the streets of East Boston or Somerville listening to what most politicians called "the real people." That to him was equally tedious.

Jack treated his district the way many British members of Parliament treated their borough, as little more than a convenient device that won them election and a place to return to primarily when they sought reelection. Reading Jack's travel itinerary for the first half of 1949, the uninitiated might imagine that he was the congressman from the right side of Manhattan, not from the wrong side of Boston.

During those six months, Jack spent at least a dozen weekends in New York at the Waldorf or St. Regis, as often as not with one woman or another. There was good theater to be seen, good restaurants to frequent, good times to be had, and it was a damn sight better than hanging out at the American Legion in Brighton.

A year later, Jack suddenly started taking a deep interest in attending such homey events as the kickoff dinner for the Home of Aged Italians in East Boston, the Lexington Minute Men banquet, the Boston Jubilee baked bean supper, the South End Post Number 105 American Legion dinner, and the Army Day Dance, 2nd Brigade, 101st Infantry, at the Cambridge Armory. He had begun wooing Massachusetts like a forgotten lover—or more accurately, like a politician who had decided to run for either governor or senator in 1952.

When George Smathers learned that Jack was going to announce for Senator Henry Cabot Lodge Jr.'s seat, the newly elected Florida senator walked across the Capitol to the House side to try to talk his friend out of the race. Smathers could have made a formidable argument that the race against Lodge would be political suicide.

The Republican senator, bearer of one of the most celebrated political names in America, was the perfect model of the old Brahmin ideal of stewardship. He was a man of prickly independence, with a quick, nervous irritability that his detractors chalked up to snobbishness. He had the kind of integrity that occasionally comes from generations of wealth. At six feet three inches tall and with lean, patriarchal good looks and public manners as impeccable as his dress, the fifty-year-old senator had a stunning public presence. Lodge had a handsome Italian-born wife, Francesca, and was most decidedly a family man.

Lodge and Jack had many parallels in their lives. As a senator in his early forties, Lodge had every reason not to serve in combat in World War II. He had done so, however, with valor and distinction, then returned to reclaim his Senate seat in the same election that sent Jack to Washington.

Lodge was a man of moderation and thoughtfulness when those virtues were not always common. He was an internationalist who on many of the issues of the day sat on the same side of the aisle as Jack. In what boded to be a Republican year, Lodge appeared impregnable.

Smathers was appalled at the way his friend was throwing away his political career. As a former House member, Smathers had floor privileges and he wandered around until he found Jack lying on a couch in the cloakroom. Jack was in such a state that he could not even stand up. Smathers reached down and pulled his legs down onto the floor. The two men walked onto the House floor and stood leaning on the bar at the back of the room. "My God, man, I don't see how you can possibly think about running," Smathers implored in a loud whisper, "when you can't even get up and down."

Smathers was not making a trivial point. A tough campaign is about as vigorous and sustained a period of physical activity as can be imagined, and running against a popular incumbent is the very definition of a tough campaign. There was a new nastiness to American politics as well. Smathers had won in 1950 by turning Senator Claude Pepper, a somewhat naive but decent reformist liberal, into a virtual traitor. ("Florida will not allow herself to become entangled in the spiraling spider web of the Red network. The people of our state will no longer tolerate advocates of treason.") Jack's friend and colleague Richard Nixon had done virtually the same thing to Helen Gahagan Douglas to win a Senate seat from California.

"I'm running," Jack replied firmly.

"Why do you say that? You can't even move. How can you run from a hospital bed? I don't understand. I don't think you ought to try. I think Lodge is too strong at this point."

"I've made up my mind," Jack replied definitively. "I'm going to run."

From everything he had experienced, Jack knew that he couldn't expect to live to a sweetly blanketed old age. He saw his plague of maladies not as an omen telling him that he must prepare for death rather than life, but as a goad pushing him out into a world that he probably would not long inhabit. Even Smathers did not know how sick Jack truly was. The columnist Joseph Alsop recalled that in the late 1940s, Jack "turned a strong shade of green: this odd skin color combined with his hair—still decidedly reddish—to make the congressman look rather like a bad portrait by Van Gogh."

When Alsop asked Jack why he exuded this strange color, he replied that he had "some kind of slow-motion leukemia. The disease, he explained, was a kind of blood cancer for which the doctors kept prescribing chemicals to cure. The latest chemical, he felt, had turned him green. He added in a flat tone, 'They tell me the damn disease will get me in the end. But they also tell me I'll last until forty-five, and that's a long way away.'"

Years later, in reflecting on what Jack had told him, Alsop concluded that Jack was talking about Addison's disease, not leukemia. The chemicals Jack was taking that had turned him "green," however, would probably not have been for Addison's disease but for some other illness. There was, moreover, another strong witness to his purported cancer. "He had leukemia at one point," Rose told Robert Coughlin in an unpublished, tape-recorded interview for her autobiography. "I remember because there was one doctor who could cure it, or who had specialized in it. . . . .They don't get over that [leukemia] very often."

Jack's mother would not have invented an illness for a son so beleaguered by illnesses. His leukemia, or suspected leukemia, was yet another secret that had to be carefully contained. No record of this adult illness exists in any medical data that have yet been made public. However, if true, it adds even deeper poignancy to the life of a man plagued by disease and weakness, a man parading in the public arena as if he were the very totem of health and youth.

Jack had another medical problem as he traversed the state weekend after weekend for close to two years, traveling with Frank Morrissey, his father's friend. He had once again begun to suffer from "intermittent slight burning on urination" from "a mild, chronic, non-specific prostatitis." He slept in the back of the car wrapped in a blanket while the chauffeur sped five or six hundred miles a weekend, getting him up for events where Jack shook half a million hands in the two years of his extended candidacy. Morrissey's task was not only to estimate handshakes but also to report back to Joe on everything his son did and did not do.

If Jack was going to run for the Senate against Lodge, he needed an imprimatur as an expert on foreign affairs. In January 1951, he set out on a five-week trip to Europe. He kept a daily diary of a journey that took him to England, France, Italy, West Germany, Yugoslavia, and Spain. Jack proceeded much like a diplomatic correspondent, interviewing American and foreign diplomats, world leaders, and American foreign correspondents. He was interested in Europeans in the aggregate, not in the star-crossed lives of individuals. He did not talk to workers, housewives, bureaucrats, businessmen, or students, noting their comments. On only one occasion in the 158-page diary did he write down a physical description of what he was seeing. That one instance suggests that he could indeed look on the world with a journalist's vivid descriptive eye:

> Yugoslavia—Belgrade—Stones cold and damp—no heating—windows bleach clothes of poor quality—the streets full of crowds—partly due to the fact that there are such few stores. The crowds seem young and energetic many soldiers among them. Tito guard with . . . machine guns over their shoulders—all with red stars in their vests. Though they look strong—they are not healthy—The disease rate particularly tuberculosis is the highest rate of any country in Europe.

Jack had come to Europe to learn, not to preach, and his diary is almost totally devoid of his own opinions. As he journeyed across this continent whose history had so defined him, he saw a world resonating with many of the themes he had observed in London before World War II. The great threat now was not Hitler but Stalin, and the European democracies faced Communist Russia with some of the same lassitude and uncertainty with which they had once faced Nazi Germany.

Jack's anticommunism was tempered by the terrible realities of war in a nuclear age, as well as his own subtle, ever-growing awareness of the complexities of the modern world. The Italians should have been doing their part, but he learned that "the Italian economy is so precarious—so poor—with the necessity of paying for food 6% of which they must export, that they hate to give up economic recovery for rearmament." He was told that "many Germans do not want their country to become Korea [and] are sick of war—feel that strength cannot be built up to stop R. [Russia] on the land." As for the French, he learned that "because of over powering strength of R. [Russia]—many French feel everything is hopeless . . . lack confidence in themselves—doubt if French who are expected to provide the mass of land troops for the defense of Europe can do so."

Joe made sure that Jack received major publicity during his trip. Upon his return he testified before the Senate Foreign Relations Committee and gave a talk over the Mutual Broadcasting Network on "Issues in the Defense of the West." Most politicians fly off on junkets to stamp authority on their firmly held opinions. Jack had gone to learn. Nothing mattered more to him intellectually than these crucial issues.

For the moment Jack threw away the little drum of anticommunism that he and his colleagues had beaten on so loudly that they had drowned out most other sounds. He picked up a different instrument now that played subtle, complicated notes. He and his colleagues had helped create an image of a Soviet monolith ready to strike, but now he had grasped the truths of nuclear détente.

"Why should they [the Russians] take the risk of starting a war, when the best that they could get would be a stalemate during which they would be subjected to atomic bombing?" he asked the Senate Foreign Relations Committee in February 1951. "Why should they throw everything into the game, why should they take risks they don't have to—especially when things are going well in the Far East? In addition, Stalin is an old man, and old men are traditionally cautious." He was all for helping Europe by adding four new American divisions to the two already there, but he was equally in favor of the Europeans stepping up and contributing substantially more to defense. These were impressive, well-considered arguments, and it was not without reason that *Boston's Political Times* headlined its article on Jack: "Kennedy Acquiring Title, 'America's Younger Statesman.'"

In October, Jack made a second, even more important journey, a twenty-five-thousand-mile, seven-week trip to Asia, traveling with Bobby and Pat. As Jack was about to set out, he mused to Lem about whether Bobby would prove "a pain in the ass." The two brothers had never spent such an extended period together, and these weeks defined their relationship for the rest of their lives. Bobby admired his brother beyond all men. He admired Jack's intelligence and grace and wit, but above all he admired his brother's courage. He admired it because, in Bobby's own words, "courage is the virtue" that Jack himself "most admired." In Washington, Jack's back had been letting him down so badly that he had been on crutches for seven straight weeks; he was finally walking freely only in September.

With his various maladies, Jack might have flown into capitals from Cairo to Tokyo largely to have his passport stamped, returning to Washington to read foreign policy speeches written by Harvard scholars, but he wanted to touch the world with his own mind. As he had done earlier in Europe, he kept a detailed 180-page diary, primarily writing down what others told him.

Jack was a relatively unknown, thirty-four-year-old, third-term congressman, but he traveled at the highest levels of political society. He did not walk into a president or prime minister's office for a handshake, a photo, and a few perfunctory words, but in many instances sat down for serious dialogues in which he held his own. While many politicians retreated into the easy truisms of Right or Left, Jack was attempting to understand the complex, dark, uncertain world of 1951. This was not easy in an America that adored simplicity.

A dangerous new world was opening up before Americans. Julius and Ethel Rosenberg had been sentenced to death in April for conspiracy to commit espionage by giving atomic secrets to the Russians. The Rosenberg trial suggested to many that a massive Communist conspiracy was alive in the land. On television Americans were mesmerized by the mobsters appearing before Senator Estes Kefauver's Crime Investigating Committee, testifying about another dark world that linked racketeers, businessmen, and public officials. A treacherous cloud rose above the atoll of Eniwetok in the Pacific on May 12, when an H-bomb was first detonated. In Korea, GIs were fighting a brutal war against the North Koreans and the Chinese.

In April, President Truman fired General of the Army Douglas MacArthur after he showed his disdain for presidential leadership by calling for a total war against China. As MacArthur made his dramatic, elongated farewell, the United Nations troops continued slogging their way back up the peninsula to roughly the Thirty-eighth Parallel, while UN and North Korean officials began negotiating a truce that none would dare call victory.

In Paris at Supreme Headquarters Allied Powers Europe (SHAPE) on the first leg of his trip, on October 3, 1951, Jack met General Dwight D. Eisenhower. Jack was impressed enough by Eisenhower to keep a detailed account of his meeting. Eisenhower appeared willing to grapple with the terrible postwar complexities and to forgo simpleminded solutions.

> Eisenhower looking very fit . . . Attacked those who criticized those who attacked settlements made during war. Said he was merely fighting a war. Had very little to do with them. States that he asked Truman at Potsdam not to beg Russians to come into war. . . . He mentioned that only one conversation he had had of importance at Potsdam and Truman mentioned there about supporting him for Pres in 1945 and had done so several times since. . . . Said $64 question was whether Kremlin leaders were fanatics—doctrinaires—or just ruthless men—determined to hold on to power—If first—chances of peace are much less than 2nd. . . . He talked well—with a lot of god damns—completely different type than MacArthur, seems

somewhat verbose as does Mac. Does not believe Russ can be frightened into aggressive war by the limited forces we are building up.

In Israel both Jack and Bobby kept extensive diaries, and there is scarcely any overlap in their accounts. Jack stood back from the accusations and hatreds and emotions and sought to understand. One of the burning questions of his life was whether a man who stood at such a psychological distance from the world could help to change it. These pages suggest that in this world detachment was a burning and necessary gift. "You can feel sense of dedication—especially in young people—willingness to endure hardship—essential," he wrote, celebrating the strengths of the Israelis. That did not mean he was any less understanding of the plight of the Arab refugees whom the Israelis refused to take back, saying that "during war [they] went on own accord." The Arabs refused resettlement elsewhere, however, because "Arabs don't want to say ok for internal reasons." As always with Jack, the omnipresent threat in the world lay in the Soviet empire. "We must convince Arab and Jews threat not each other but from the north," he wrote.

One evening in Jerusalem, Jack and Bobby went to the modest home of Prime Minister David Ben-Gurion for dinner. Jack celebrated leaders who shaped history, and in helping to create the Israeli state out of the Palestinian desert, Ben-Gurion had surely done that. Jack was an observer of men, and by the evidence of his diary, he did not impose himself on this evening as much as take his measure of those around him, including the American ambassador, Monnett Davis, several other Israeli ministers, and New York Congressman Franklin Roosevelt Jr. The late president's son was a large, handsome, and verbose political gentleman. Like his mother, he was considered a friend of Israel. FDR's namesake was the center of this evening, not Jack, whose father's reputation always went before him. "It was almost as if we weren't there," Bobby recalled of their time in Israel with the former president's son.

Roosevelt asked the inevitable question: Could there be a real peace between Arab and Jew? "It depended on the recognition of the liberal elements, responsive to the peoples wishes," the prime minister said. "Present gov[ernment] not concerned with peace but protecting own action." In this spirit of candor, Ambassador Davis boldly told Ben-Gurion that the Arab states were afraid of Israel. "How could Egypt with its large population be frightened" Ben-Gurion replied, with rhetorical flourish. "We wouldn't want to go back to Egypt again. We had enough the first time."

As the evening grew late, his hosts led FDR Jr. up onto the roof, where he and others looked out on the ancient city. The Jewish sections were all lit up, while the Arab sections were as dark as the night, a distinction that said more about the differences between these two peoples' lives and their conditions and the chances of peace than anything that had been said that evening.

This was Jack's last night in Israel. After writing in his diary about the dinner, he ended with a few lines of poetry. He kept his love of poetry private. Indeed, a few years later, when he would read poetry in his Capitol office with Jackie's cousin Wilson Gathings, the young man had the impression that Jack feared that others in the family might find his interest unmanly.

This evening, though, Jack wrote down four lines of a poem written by Percy Bysshe Shelley in 1819. That year British troops fired upon a gathering of unarmed, peaceful Manchester radicals and about fifty thousand supporters seeking the reform of Parliament. When the troops finished, eleven people lay dead and one hundred more were injured. Shelley blamed Lord Castlereagh, then the government spokesman for civil matters in the House of Commons. That may not have been a judicious rendering of the culpability, but in "The Mask of Anarchy," art has triumphed over the interminable debates of history. Castlereagh stands remembered and condemned in a work of great and savage splendor.

Was there a Castlereagh among the leaders Jack had met this evening, an arrogant, myopic, reactionary politician leading his nation into unnecessary death? Was that why Jack chose these words? Or did he mean to suggest that a bloody hand reaches out to grasp a public man who seeks to change society radically? Was it that death waited up the road, an assassin holding a cool grip on his trigger, with history herself in his gun sights? Whatever Jack meant, a half-century later one reads these words with dread foreboding.

I met murder [on] the way,
He had a mask like Castlereagh
Very smooth he looked yet grim
Seven bloodhounds followed him.

Unlike Jack, Bobby personalized politics; he always hung ideas on a human face. On his earlier trip to Israel, he had seen Jews acting nothing like the devious, avaricious race he had learned about sitting at his father's feet. Nonetheless, he was still more offended by those speaking with what he considered a pro-Jewish bias than by those who vociferously expressed a pro-Arab position. "Drove to Haifa with Franklin D. Roosevelt Jr. and had all my questions on Israel answered but of course with a very pro Jewish slant," Bobby wrote. "I think he has gotten so he believes it though. . . . U.S. consul a Jew. Talks about Arabs as if a teacher talking about child. FDR Jr. talks about Arabs 'These people must learn if they don't get on the ball we'll cut them off without a penny.' His love for common man stops at Jews and Negroes."

In this part of the world, history was written with blood and vengeance, and a man who entered politics knew that he might die if he lost, or die even if he won. Four days after Jack and Bobby met with Prime Minister Liaquat Ali Khan of Pakistan, he was assassinated. In his diary Jack noted that "assassinations have taken heavy toll of leaders in Middle and Far East," and then made a list of some of the murders. Jack tallied seventeen assassinations in the past four years alone. From Mahatma Gandhi in India to Count Folke Bernadotte in Palestine, murders often changed history the way few laws or mandates ever could. Jack was a student of power, and the lesson of this lengthy list was that in Asia the assassin and his dagger always lurked in the shadows of the throne.

Jack was far subtler in his judgment of others than his brother, and more understanding of the limitations of what some would call courage. When he sat down for lunch with Jawaharlal Nehru, he did not find the neutralist Indian prime minister to be a coward betraying the West. "Nehru—handsome . . . intelligent, good sense of humor," he jotted in his diary. "Bored by westerners believes he is right—interested in . . . bigger questions." Jack considered Nehru a wise leader who had made a shrewd assessment of the precarious position of his nation set between East and West.

Nehru repeated what was the central theme of many of the leaders and observers whom Jack was meeting: "Asia is at present the scene of great nationalist waves directed against the colonial policies of the West and seeking better economic conditions." Jack was not blind to the upraised faces of Asia, but there was another issue that took precedence. Jack explained to Nehru that his government ended up supporting colonial regimes "because of our obligation to Europe and its defense and because of our concern for Communists in Ho [Chi Minh] terms. . . . We found it extremely difficult situation." Nehru parried by saying that "arrangements could have been worked out with Ho," the Vietnamese Communist leader fighting a guerrilla war against the French in Indochina. "The poor French, they know that regardless of what happens now . . . they are going to eventually lose their positions so that they are really fighting for nothing."

Jack scratched all sorts of ominous predictions and reflections into his notebook those days in India. It was all enough to jar one's faith in rational economic and social progress. His own nation had hardly entered the Asian arena, and he was already being told that "when [Indian] independence was granted U.S. most popular country now unpopular." The United States had given generously to India, but if America gave a loaf of bread and Russia or

China only a dried crust, it was the crust that was celebrated. He was told that "USA had given so much to other countries that what it gives is no longer appropriate."

Jack saw how the rituals of faith and custom held the populous nation back from what he called progress. Here in India lived the true Brahmins, not the ersatz Boston variety, the highest rung of a rigid caste system that stood against everything Jack believed about the lives of free men in free societies. In America, Catholics, Protestants, and Jews had their quarrels, but it was nothing like the feeling between the Hindus and Muslims that had ended in bloody war and partition. One day, Jack talked to the man fixing his air conditioner, who "would not eat at home with member of different castes. I asked him why he does not like Muslims. 'Because they eat cow meat and cows Mother of Hindus.' "

As seriously as Jack studied the life and politics of India, it was lush, fertile Indochina that stood at the epicenter of Asian cold war politics and his interests. Indochina was the diadem in the crown of the French colonial empire. As Jack flew into Saigon, the heavy stench of war hung over the lush paddies in the North. There the dedicated, determined forces of Ho Chi Minh were fighting the legions of France in a merciless struggle. Jack noted in his diary that an attaché told him that while 184 members of Congress had visited Rome in the past two years, he was the first one to make it to Saigon. That neglect was a tragic dereliction, for American money was helping to finance the French effort. He was told that the United States had already spent over half a billion dollars. Where the dollar ventured, the flag might soon follow.

Jack was savvy enough to know that he had to reach below the elites of diplomacy and politics to get some semblance of truth. As he arrived at Saigon Airport, he was important enough to merit a greeting by top French officials and American diplomats. He quickly brushed by them and walked across the tarmac to a group of American journalists standing there and started chatting with them.

"I would like to have a talk with you," he said to Seymour Topping, the AP's man in Saigon.

"All right," the young journalist told him. "I'll come to see you."

"No," Jack said, "I'll come to see you."

That afternoon Jack showed up at Topping's apartment in the center of the city. He spent hours talking with the reporter, who, as Jack noted in his diary, told him bluntly: "Two years ago Americans were popular, now people feel that they are already allied with French, should have maintained that French promise independence before we bought French. Now U.S. is more disliked than the Russians in South East Asia." Here again was that persistent theme of American unpopularity. Jack listened well, and as he left he told Topping: "I'm going to talk about this when I get home. But it will give me trouble with some of my constituents."

Jack also went to see Edmund Guillion, a State Department official deeply critical of the French. Guillion was an idealistic, liberal internationalist whose thinking resonated with Jack's own ideas. "One great reason why countries are far more successful than ourselves in spreading propaganda is that they have militant organizations within countries which we do not," Jack wrote in his diary, an idea that probably came from Guillion.

That logic led to an aggressive American internationalism in which Guillion argued that "we should throw our weight between French and Vietnam in order to encourage support for war." Another observer told Jack, "Asians feel that the West has always meant imperialism under the strongest power and in the present case it is U.S. U.S. must try to achieve our aims but not expect to be well liked doing so."

Time and again Jack was told that the French were defending a colonial regime. ("Last year best year in business in history. Made almost as much as war cost them economically.") Time and again he was reminded of the determination of Ho's forces, and of the corruption of the French regime. Time and again too he heard the melancholy logic of the domino theory: if the French lost northern Vietnam, the "country probably would be lost—Burma would go—Malaya would be in a bad way and our entire position in South East Asia would collapse."

Jack and Bobby flew over the battlefields in the North where the French Foreign Legion was battling against the forces of Ho Chi Minh. In the besieged city of Hanoi, a visit from an American congressman merited a parade. Along the road the children of the city stood waving flags.

Jack admired General de Lattre de Tassigny, the haughty French general, for he admired courage and de Lattre was not lacking in that quality. Jack and Bobby had been brought up to believe that each man stood on a road with a direction sign pointing to courage in one direction and cowardice in another. The road that the two brothers were traveling in Vietnam, however, was not so clearly marked. The signposts were obscure, written at times in strange languages. Moreover, courage was a dangerous virtue when it was not tethered to the mind and the spirit.

De Lattre was a brave man, but he was fighting what appeared to be a doomed conflict, and any admiration of the general had to be tempered by that reality. The French tragedy, as Jack could not help but see, could easily become an American tragedy. One minister shared with Jack his observation "that French realize that now [there's] no hope of retaining old hold on country, now [they are] only fighting to prevent Tunisia and Morocco from feeling they can break away . . . so there is danger that French may drop it into our laps."

Bobby did not so readily understand that compromise and cowardice were not synonyms. Bobby sought to emulate his brother, but often when Jack painted in subtle strokes, Bobby splattered the canvas with a few bold strokes. Jack's diary reads like a foreign correspondent's notebook, listening and learning before he makes his own judgments. Bobby's diary is full of quick judgments and caustic asides.

In Japan toward the end of the trip, Jack fell critically ill, his temperature rising to a dramatic 106 degrees. Bobby arranged to fly his brother to a military hospital on Okinawa. "And everybody there just expected that he'd die," Bobby recalled. Bobby was there for his brother's deathwatch, and he knew that death was not some distant unmet stranger but a companion his brother might meet at any turn. No one could know how sick Jack had been.

When Jack returned to the United States, he spent time secreted in Virginia. His office staff was used to Jack's mysterious periodical disappearances and knew that his absence meant that once again he was sick. "He'd be in the hospital quite a bit, and he didn't want the reporters to know he was up here, because they'd put it in the paper," recalled Grace Burke, his Boston secretary.

When Jack recuperated, he gave a nationwide address on the Mutual Broadcasting Network. Most politicians learn not to stray too far beyond the safe confines of sloganeering and cliché, especially if they are contemplating running for the Senate in a nearly impossible race. Jack, however, had a refinement of mind and perception that he did not discard simply because he wanted to be elected.

"Foreign policy today, irrespective of what we might wish, in its impact on our daily lives overshadows everything else," he told an audience that presumably included a number of insular, skeptical Americans who wanted nothing of the duplicitous world beyond the borders. "Expenditures, taxation, domestic prosperity, the extent of social service—all hinge on the basic issues of peace or war. And it is a democratic America and not a bureaucratic government that must shape America's destiny. Just as Clemenceau once said, 'War is much too important to be left to the generals,' I would remark that 'Foreign policy is too important to leave it to the experts and the diplomats.'"

Here again resounded his greatest political insight: that foreign policy was profoundly limited in a democracy. Before World War II, he had seen how British politicians pandered to their constituents' fears, tragically slowing the nation's rearmament. Now in America, he saw the political establishment pandering to the American public's fear of communism, proudly propounding clichés and simplicities instead of elucidating complex truths.

In Washington, policymakers envisioned a blood-red tide of communism rising across South Asia. Jack pointed out that communism was a different thing in different countries. In Indochina, what Americans called communism was equally a nationalistic movement. In Malaysia the Communist guerrillas were mainly Chinese, considered alien by the ethnic Malays. The Indonesians, for their part, thought of the struggle in Korea not as one against communism but one of whites against Asians.

His was not always a welcome message to Americans taught that this was their century and their world. America was a land of quick fixes and problem solvers. Jack, however, saw a world of intractable problems whose best solution was often the lesser of two bad options. Massive poverty was the soil in which communism grew. Yet America had neither the resources nor the ability to turn millions of people away from whatever course they might choose.

"Our resources are not limitless," he warned one audience in a statement that his father could have made. "The vision of a bottle of milk for every Hottentot is a nice one, but it is not only beyond our grasp, it is far beyond our reach."

Not that Jack wanted the United States to turn away from the world. In a perception that a few years later would be echoed in the popular press, Jack saw America's diplomats not as ambassadors of diversity, but as a narrow, inbred social set "unconscious of the fact that their role is not tennis and cocktails but the interpretation to a foreign country of the meaning of American life and the interpretation to us of that country's aspirations and aims." As it was now, these gentlemen were primarily emissaries from one elite to another.

Jack had a profound gift of political empathy. From the Suez Canal to Tokyo, he ticked off one complex situation after another. In Indochina, Jack's comments were prophetic. When he talked to the Boston Chamber of Commerce, he did not pander to these businessmen but forcefully told them his rude truths:

> It is France trying desperately to hang on to a rich portion of its former empire against a communist-dominated nationalistic uprising. The so-called loyal native government is such only in name. It is a puppet government, manned frequently by puppeteers once sub-

> servient to the Japanese, now subservient to the French. A free election there, in the opinion of all the neutral observers I talked with, would go in favor of Ho and his Communists as against the French. But Indo-China is a rich country. What France takes out of it today pays most of the costs of that bigger war. . . . We have now allied ourselves with the French in this struggle, allied ourselves against the Communists but also against the rising tide of nationalism. We have become the West, the proponents of empire—carriers of what we had traditionally disdained—the white man's burden.

Jack spoke as a new kind of aggressive internationalist, offering not weapons and trade but ideas and aid. He favored a foreign policy not so allied with the imperial empires of Britain, France, and Holland. He envisioned Foreign Service officers whose lives ranged far beyond the diplomatic compounds and who spoke with a new voice of America in the languages of the peoples themselves, offering foreign aid larger in scope and directed toward the masses.

To another group in Massachusetts in December 1951, Jack talked of a new world in which "young college graduates would find a full life in bringing technical advice and assistance to the underprivileged and backward Middle East." "In that calling," he went on, "these men would follow the constructive work done by the religious missionaries in these countries over the past 100 years." This was one of the earliest suggestions by a politician of the idea that became the Peace Corps. In his own mind, Jack was taking the rituals of true manhood that he had learned from his father and turning them away from the battlefield as the ultimate testing ground to a different field of challenge.

Jack's internationalism may have borne the seed for a Peace Corps that would send thousands of young Americans to many of these same countries that he had visited and for foreign aid programs that were supposed to help the poor, not maintain the rich. His aggressive internationalism, however, carried another seed as well, one no larger than the first but of a far darker hue.

When Jack decided that it was time to announce his bid for higher office in Massachusetts, he called Mark Dalton and asked his 1946 campaign manager to return to run the new campaign. Since Jack had entered Congress, Dalton had been, in Dave Powers's words, "the closest political adviser . . . a brilliant man." Time and again in those years, Dalton had traveled down to Washington on the train, helping out with speeches and ideas, never asking a cent in expenses or even thinking of using his entrée to advance himself economically or socially. He didn't hang out with Jack, and he didn't care to. If Dalton had an obvious fault, it was that he had something of the vanity of the idealist, believing at times that his words had more moment than they did. Surely Jack valued his counsel and drew on it often, but he never made a decision without asking a myriad of people. That was not a mark of Jack's insecurity, but his technique of decision making.

As campaign manager, Dalton had access to Joe's money in almost limitless quantities. There wouldn't simply be paltry campaign circulars but tens of thousands of comic books celebrating Jack's war efforts and reprints of John Hersey's famous PT-109 article. There would be no smattering of billboard ads but Jack's face and message blanketing the state, as well as on subway trains and stations of the MTA.

In one of the crucial campaign moves, Joe signed on Batton, Barton, Durstine and Osborn, the third largest advertising agency in America. BBD&O had such a strong Republican identification that the firm was handling the 1952 presidential campaign of General Dwight Eisenhower. Joe had signed on BBD&O in good part because of their expertise in the dramatic new medium of television. Governor Thomas E. Dewey had won reelection in New York in 1950 with the prophetic use of television advertising.

John Crosby, the TV critic, had written in the *New York Herald Tribune* that "no one will ever know how much TV helped Thomas E. Dewey win reelection as governor of New York State. But no one can dispute that Dewey is the first political candidate to understand how to use TV properly."

BBD&O told Joe that if Dewey had been the first, Jack would be the second. There would be no static, poised campaign speeches on television but programs "marked by informality and action." Television audiences get bored easily. Best to keep it short, no more than fifteen minutes. The BBD&O executives decided not to advertise these political programs or even to list them as such in the schedules in the newspapers. They wanted viewers to come upon Jack unannounced. Viewers might even mistake what they saw as a legitimate news program, or at least not realize that the questions seemingly asked so spontaneously by the "man in the street" were as staged as Jack's answers.

Dalton would have been appalled if the Kennedys had proposed handing out ten-dollar bills at the polls, but he had no complaints about the higher duplicities of the emerging media age. Dalton had a wife and family now, and a legal practice to support them. Running a campaign was more than a full-time job, and for the first time in his dealings with Jack, he became a paid employee. Dalton was under the considerable illusion that being paid changed nothing, but he was now expected to couch his words in an inferior's deferential phrasing.

At a meeting with a group of Democratic politicians at the swank Fall River Social Club, Dalton made what he thought was a good solid point when he realized that Jack was glaring at him, apparently angry that he was being upstaged. As they hurried out of the club, a group of men at the bar spotted Jack. "Oh, there's Kennedy!" one of them shouted. The men jumped off their bar stools and went to embrace the candidate they thought of as a hero.

Dalton fancied himself the campaign manager, not a bodyguard or a coat-handling lackey, and he left Jack to extricate himself alone from his cloying followers. When Jack finally got back to the car, he turned toward Dalton, sticking his finger at him. "Don't you ever let that happen to me again! Do you hear me?"

Dalton was already sick of being pushed around by Jack's father, second-guessed on everything he did, receiving never a thank-you, never a note of grace, nothing but push, push, push. And now Jack was treating his campaign manager as if he owned him.

"Ball game over!" Dalton recalled. "Ball game over! The shock of recognition was complete. I knew what I was dealing with, and I was dealing with a really bad man, an absolute ingrate. And also I was insulted because he thought I was a patsy. If I had misestimated him, then he—with all the adulation and sycophants around him—had misestimated me. I was stunned. I had almost an emotional physical reaction."

A few days later Dalton sat at a meeting at the Bellevue Hotel with Joe and the eight or so top campaign aides. "The father wanted me out right from the start," Dalton said. "It was as simple as that. And so he decided to get me out, and he started pushing me around. So between John's ingratitude and the old man's actions there, the thing came to an end pretty quickly."

Dalton recalls that at that meeting "Joe Kennedy blasted the living daylights out of me, absolutely blasted me in front of these people." One of the other participants, John Galvin, recalls the day somewhat differently: "Mark didn't like the old man's style; he resented the old man, who was a son of a bitch. . . . He wasn't laying into Mark. He was just saying, he was being kind of, oh, unreasonable, saying some unreasonable things about people and whatever."

Dalton's pain amplified every slight and magnified every curt query into an assault on his very being. Everyone in the room that day saw how distressed the campaign manager had become. Dalton did not confront Joe or run raging from the room, but slunk away, leaving the Kennedys and their ambitions behind forever.

Years later, Bobby spun his own mean version of this sad tale: "Mark Dalton was going to be the campaign manager, and then he had what amounted to, I guess, a nervous breakdown about it. . . . He wouldn't come out of his room. I guess it was the pressure about it and everything. I was working in Brooklyn, so I came up."

There was at times a high selfishness to the Kennedys, a cold, impenetrable core that displayed itself to anyone who was expendable, and in the long run that was almost anyone outside the family. Those who got close to that core often found themselves pushed out a door that locked behind them. Now Dalton was gone. Billy Sutton had been shuttled aside too, his usefulness finished, a bit too much of the bottle perhaps, the same jokes told too many times.

Those on the campaign referred to the days of Mark Dalton as "before the revolution" and the days of Bobby as "after the revolution." And every revolution demands its blood. Jack was no good at firing, disciplining, or demoting, all the mundane nasty chores involved in running a political organization. Like most successful politicians, he had learned that when bad news is to be handed out, he should be elsewhere.

Any political campaign, fueled as it is by the work of poorly paid and unpaid staffers, is full of jealousy and rude positioning. Kennedy campaigns were worse; the family seemed to set subordinates against each other in a race that had no clear rules and no obvious finish line. When Jack made his rare appearances in the campaign office, he was sure to be approached by staff members complaining about their peers. Jack learned to stay away from the office, or to come rushing through, as if on a campaign stop, shaking hands and exchanging pleasantries, but leaving before anyone had a chance to take him aside.

Bobby stayed, however, and Jack learned in the 1952 campaign that his younger brother was willing to take on these most onerous of chores, castigating those who didn't measure up, pushing others with crude force. It was not Bobby's style to preface his criticisms with a few nuggets of praise or to nestle his condemnation among platitudes and pleasantries. He got right to it with brutal efficiency and what to observers looked like pleasure. When Bobby asked Sam Adams to help on the campaign, and Sam responded that he was too busy, Sam felt "like he burned my bridges." Bobby was doing a lot of bridge burning.

Bobby had his father's nearly maniacal sense of precision. Politics was a matter of inches, not feet; of ounces, not pounds. After one of the campaign tea parties, Joe asked O'Donnell, "How many people were at that tea in Springfield, Kenny?" O'Donnell said, "Oh, about five thousand." And Joe said, "I know *about* how many people were there. I asked you *how* many people were there. Didn't you have a checker there?" A chastened O'Donnell replied, "Yes, we did." Joe wasn't about to leave his little lesson at that. "When I ask you how many people the next time, I want to know *exactly* how many people."

15

The Golden Fleece

With Joe and Bobby at the helm, Jack ran a brilliant and prophetic campaign. "He was all things to all men," recalled Massachusetts Congressman John McCormack, a powerful figure in national politics. The candidate was a putative liberal to the good professors of Harvard. He was a closet conservative to the more amenable Republicans. At dawn, as he shook hands outside factory gates, he was a friend of labor. In the evening, as he shared cigars and cognac with their bosses, he was the businessman's Washington friend. He was an Irish-American, son of the sod of old Eire. He was an upper-class Catholic who sat comfortably in the great houses of Back Bay. Jack was a dream lover to the young girls who waved their handkerchiefs and God's glory of a son to their mothers.

As the campaign started, Jack read the extraordinary statistic that for the first time women voters outnumbered men. In Massachusetts, their numbers were the highest of any state, 52.6 percent. Jack was the prisoner of an upbringing that had taught him that women were for the most part giggling creatures uninterested in the manly business of politics. He did not attempt to solicit female voters by developing campaign issues that might appeal to their intellect. Instead, he developed a strategy that shrewdly exploited their social ambition and sexuality.

"His theme was to hit the woman vote," reflected Edward C. Berube, a Fall River bus driver who worked intimately with Jack. "He indicated this to me . . . that he was going to come out for the women, that he figured the woman was the one that was going to put him in. And he wanted coffee hours and tea hours and arranging coffee hours in homes."

Lodge had started the whole business of tea parties back in 1936 when he first won election to the Senate. The Kennedys democratized the stuffy, mannered tea party and turned it into a mass gathering more Barnum and Bailey than Brahmin. The Kennedys held these teas all over Massachusetts, thirty-five of them in all. Jack shook hands with as many as seventy thousand women.

Jack moved around the teas, a handsome young gentleman, both a son of Irish immigrant culture and an aristocrat, a man of the people, a man above the people. Jack's opponent was a debonair, elegantly dressed aristocrat who set the hearts aflutter among the good Republican matrons of the suburbs, but the difference between the two men was like the difference between a Broadway actor and a Hollywood star.

Lodge might appeal to the elite ladies, but Jack had the masses, the bobby-soxers, the blue-haired immigrant ladies, the aspiring suburbanites, as well as the conservative matrons. As the lean, mildly disheveled, thirty-five-year-old candidate worked the room, he was playing brilliantly on the social aspirations of these women, bringing them into an ersatz version of the Kennedy social life. Though it was easy enough to satirize these women, their desire to be with a better class of people was no different from the desire that for three generations had dominated the Kennedys themselves.

What none of the women knew, however, as they chatted amiably with Jack was that his smile was at times a grimace. In August he was sick enough with nonspecific prostatitis that he was urinating pus and had to be secretly hospitalized. Then, on a visit to a Springfield firehouse, Jack couldn't resist a dare to slide down a fire pole. When he hit the cement from his third-floor perch, he grimaced, feeling once again the terrible back pain that plagued him. From then on, he was relegated to hobbling around on crutches. When he got to an auditorium or a hall, he would leave the crutches outside and stride into the room as if he were health and youth incarnate.

The campaign was not all tea parties and handshaking. Indeed, when Lodge looked back dispassionately on the race, he realized that his largest problem was not what he called the "damn tea parties" but his endorsement of Eisenhower for the Republican presidential nomination; the offended Taft Republicans stayed home or voted for Jack in protest. The two candidates shared one common problem, Senator Joseph R. McCarthy of Wisconsin. When McCarthy got up before the Republican Women's Club of Wheeling, West Virginia, in February 1950 and said that he had a list of fifty-seven card-carrying Communists or fellow travelers in the State Department, he set out on the most dangerous demagogic campaign in American political history. It was the rare specificity of McCarthy's charges that made them so powerful. It would take several years before it became clear to many Americans that

McCarthy was playing on the fears of his compatriots, destroying not Communists but legitimate anticommunism, wreaking havoc not on American enemies but on the American political system.

McCarthy was so dangerous because he was not an aberration but rather the logical extreme of what Michael S. Sherry has called "a highly politicized form of postwar militarization." The witch-hunt McCarthy initiated was precisely what Jack's father had feared would happen. To defeat the Axis the United States had created a state apparatus of such magnitude and coercion that it risked destroying what once had been called liberty. It was not McCarthy, after all, but one of his enemies, President Harry S Truman, who in February 1947 instituted a loyalty program to fire "disloyal" employees. It was not McCarthy alone but a whole generation of postwar politicians—even including, in a small way, Jack himself—who helped create a climate of such fear that the owners of the Cincinnati Reds renamed the baseball team the "Redlegs" rather than risk being called Commies.

These politicians had turned the Communist into a mythic anti-Christ, unseen but all-seeing, ready to betray the unwary and seduce the innocent. This figure resided in the bowels of the unions and in the highest counsels of government, in the pristine groves of academe and in the newsrooms of the nation's newspapers. There were indeed Soviet spies ensconced in crucial positions in Washington, and Communist cadres in labor unions and various liberal political movements. These were for the most part Americans who in the 1930s had given their higher loyalty to what they considered a noble cause, not a mere nation. To destroy them, McCarthy and his kind called out political artillery of such magnitude and so carelessly aimed that the friendly fire ended up shooting hundreds of innocents for every true enemy of America it struck.

Jack first gained notice in the House by attacking several union officials, including Dr. Russell Nixon, a former professor of his at Harvard. Jack was roundly praised, as if he had bested an evil giant in unequal combat. The jingoistic anticommunism had been a willing servant of these politicians, helping to elect them to office over candidates who did not shout so shrilly.

During the campaign, Jack was not simply the brilliant observer of international affairs, as he had been when he returned from Asia speaking with a subtlety of mind and nuance rare in American political life. He also gave another kind of speech in which he set aside his complex and tragic worldview for florid, apocalyptic, anti-Communist rhetoric.

In the spring of 1952, in an address to the graduating class of the Newton College of the Sacred Heart, Jack presented a vision of the world full of threat and terror, a vision and indeed a language that he used on several occasions that year. He saw the dark encroachment of the state, enveloping liberty: "The theme of today—the scarlet thread that runs throughout the thoughts and actions of people all over the world—is one of resignation of major problems into the all-absorbing hands of the great Leviathan—the state."

Despite what Jack said at Newton College and elsewhere, when he talked to elderly citizens whose Social Security checks allowed them to live in dignity, he did not bemoan the encroaching leviathan but celebrated one of the New Deal's great achievements. And when he looked at the economy, he was a Keynesian who thought that government must at times intervene in the nation's economic life.

The truth has few friends, and Jack was not about to risk his political future by trying too often or too openly to elucidate the complexities of the political world. There were no easy votes to be garnered that way, as he could by shaking the tambourine of anticommunism. He took the easy way, though what appears the easy way is sometimes in the end the most difficult path of all.

In November 1950, Jack had attended a seminar at Harvard given by Arthur Holcombe, one of his old professors. In an article published in *The New Republic* in the midst of the 1952 campaign, one of the participants reported on Jack's candid statements. He had told the group that more had to be done to get rid of Communists in government and that he had a certain respect for McCarthy. He thought he "knew Joe pretty well, and he may have something." He said, moreover, that "while he opposed what he supposed to be Communist influence at home, he refused to become emotionally aroused even on this issue."

Jack was not the sort of politician who took pleasure in attacking his enemies, and even if he had wanted to confront McCarthy, which he did not, he had his own special relationship with the Wisconsin senator. McCarthy was a close family friend who dated Pat, Jack's own sister. Joe talked to him on the phone frequently and contributed liberally to his campaign. He was a guest in the Kennedy homes. On one trip to Hyannis Port, McCarthy was dragged behind a sailboat on a rope. The Kennedys considered that fun and games, but the landlubber from Wisconsin almost drowned. The two men were both Irish-Catholic politicians, and that too was a special, if unspoken, bond. Jack might seek to distance himself from that ethnic appellation, but the 750,000 Irish Catholics in Massachusetts were his bedrock support at election time; they were for the most part fiercely conservative Democrats and proud, if narrow, partisans who considered Joe McCarthy a great patriot, if not a secular saint.

Jack's one fervent moment over McCarthy during the campaign year came on February 9, 1952, at the one-hundredth-anniversary dinner of the founding of the Spee Club. An after-dinner speaker said that he was delighted that Alger Hiss was not a real Harvard man, even if he had gone to Harvard Law School. Hiss had been convicted of perjury for lying before a congressional committee and was serving forty-four months in prison. Up until his conviction, Hiss had seemed the perfect exemplar of the kind of man Jack admired above all others, a gentleman of wealth and privilege who had opted to become a public servant.

Over the years, evidence would mount to prove almost indisputably that Hiss had been spying for the Soviets during the 1930s. At this point, however, equal evidence suggested that he was nothing more than the most publicized victim of the Red Scare. It was all maddeningly unclear and obscure, and Hiss had become what David Remnick has called "the *Rashomon* drama of the Cold War," and a litmus test that had more to do with sentiment than fact.

The speaker went on to say that as delighted as he was that Hiss had not gone to Harvard College, he was infinitely happy that his beloved alma mater had not turned out a Joe McCarthy.

At that point Jack jumped up and shouted, "How dare you couple the name of a great American patriot with that of a traitor!" He was so angry that he left before hearing the rest of the speeches.

The extremes of American politics had reached such a point that some of those present may have wondered just whom Jack was condemning. To those on the right, Hiss was the traitor and McCarthy the patriot. To those on the left, the opposite was true. On a visceral level, Jack knew where he stood, and it was not with the liberal intelligentsia of Harvard.

Bobby had even stronger feelings about McCarthy. When he took time off to attend the Harvard-Yale football game in New Haven, he and his old football teammates spent Saturday evening in New York. As always, the subject turned to politics, and in the early 1950s, politics meant McCarthy pro or con. Except for Sam Adams, Bobby's friends were all the same solid Democrats they had been when they had left Harvard. They despised McCarthy and what they thought he was doing to the America that had been so good to them. Bobby was his lone defender.

"Oh, Bob, come on now," O'Donnell fumed in exasperation. "McCarthy could prove your mother was a Communist by his way of reasoning." Bobby could simply not understand that in the name of anticommunism McCarthy had created fear where there had been hope, and suspicion where there had been trust. "By using his methods of proof, the

Pope could be a Communist," Kenny shouted, his words not changing Bobby at all.

For a centrist consensual politician like Jack, McCarthy presented a dilemma. If Jack stood too strongly against the Wisconsin senator, he would lose his hard-core Catholic constituency, the very bedrock of his power. If he supported McCarthy, he would lose the liberals, the intellectuals, most Jewish voters, many labor leaders, numerous teachers, and activists, people who voted with their effort and resources and deeply voiced concerns.

Jack was a man of profound emotional disengagement, and he was no more comfortable with McCarthy's rude outbursts than with the shrill replies of the liberals. Thanks probably to his father, Jack had the questionable honor of being one of the few Democrats who did not have to worry about McCarthy coming into their state to berate him. With no such problem, he set out to convince McCarthy's detractors that he was worthy of their vote.

No group worried longer or more over this than did the Massachusetts Jewish community. Not only had Jews suffered disproportionately from McCarthy's assaults, but they were acutely aware of an anti-Semitic strain among Irish-American Catholics, especially in Jack's own father. The Jewish groups implored Jack to come forward and condemn McCarthy.

"I told you before, I am opposed to McCarthy," Jack privately told Phil David Fine, his foremost liaison with the Jewish community. "I don't like the way he does business, but I'm running for office here, and while I may be able to get x number of votes because I say I'm opposed to him, I am going to lose . . . two times x by saying that I am opposed. I am telling you, and you have to have faith in me, that at the proper time I'll do the proper thing."

If what one man had done weighed heavier than what the other man promised, then Lodge deserved the bulk of the Jewish votes. Lodge had a stellar record on the issues, such as Israel and civil rights, that preoccupied the Massachusetts Jewish community. Beyond that, Jack carried the heavy burden of his father's history.

Jack's campaign did not confront these issues but tried a more circular approach. One day in the kosher butcher shops, delicatessens, and grocery stores in heavily Jewish Dorchester there appeared blocks of free tickets for two movies that evening in the largest movie theater on Blue Hill Avenue. By the appointed hour there was not an empty seat in the entire theater, and people stood all around the back and the sides.

By the time the first movie ended—an American film about the heroic birth of Israel—the audience was full of deep emotion and a passionate sense of their relationship with Israel. As the lights went up, there stood Congressman Franklin D. Roosevelt Jr., an apparition who moved the audience only slightly less than if Ben-Gurion himself had walked up the aisle. Then Congressman John McCormack, a man nicknamed "Rabbi John," appeared, to another thunderous round of applause.

And finally, when the two speakers had made their remarks, Congressman Kennedy strode onto the stage. When he finished his talk and walked off to more applause, a travelogue on Israel played on the screen. It was a glorious evening, especially if you were working for Jack. The campaign replicated the same approach in other heavily Jewish areas.

While they attempted to manipulate the various ethnic groups and constituencies, the Kennedys maintained a well-earned cynicism toward much of the press. Jack was forever flattering journalism and journalists, but flattery is to respect what copper is to gold, the cheapest kind of currency. Jack applied it to those journalists inordinately attracted to its coinage.

The Kennedys' scorn for some members of the press was honestly won, and it presented a number of moral conundrums. When did generosity become a bribe? And who was guiltier, the supplicant with the outstretched hand or the patron who greased his palm with a few coins?

Jack had observed for years the family's relationship with Arthur Krock. His father would not think of issuing significant campaign statements without running them past Krock. The reporter was not a conniving hack seeking to supplement his miserable wages with handouts from Joe. He was the premier political columnist for the *New York Times*, the most important newspaper in America, and he was available at all times and all hours, for advice, help with speeches, or whatever sundry duty the family demanded. In the 1930s, Joe had offered to pay the journalist five thousand dollars for his work on Joe's book supporting Roosevelt, and that may have been the least of it. It was whispered around the Kennedy camp that he was on the old man's payroll. Better if he was, for if he was being paid only in access, deference, and the illusion of importance, he was a man who was bought cheaply indeed.

If a man of Krock's stature was so amenable, then certain lesser journalists and newspapers were even more so. In October, Lodge met with John Fox, the new owner of the *Boston Post*. The *Post* was a Democratic standard-bearer, at least it had been before Fox bought the troubled daily. Fox planned to endorse Lodge, however, support that Lodge thought easily worth forty thousand votes, enough to ensure his reelection.

Lodge told the good news to one of his top aides, who mentioned the extraordinary endorsement to Joseph Timilty, Joe's closest political associate. Joe went to see Fox. Joe knew that the paper was in financial trouble, and after the apparent application of a half-million-dollar loan, the next day the *Post* endorsed Jack.

"I don't know whether he arranged for him to get a loan or got him a loan or what," Bobby recalled. "I don't remember the details, but the *Boston Post* supported John Kennedy—and there was a connection between the two events. I don't know . . . specifically what was involved, but I know he was an unsavory figure." Jack was more blunt in speaking to the journalist Fletcher Knebel: "You know, we had to buy that fucking paper or I'd have been licked."

Boston's thriving ethnic press was for the most part just as mercenary as the publisher of the *Boston Post*. These papers viewed the election not as a subject for vigorous reporting but as an enviable opportunity for political advertising. When one of Jack's aides, Ralph Coghlan, went around Boston meeting the various editors, he reported that every paper, from the Armenian *Hairenik* and the Italian *Gazzetta Del Massachusetts* to the black *Chronicle*, expected ads in return for its support. In the inner world of the campaign it was all tit-for-tat, my favor for yours, a series of exchanges that had little to do with principles or ideas.

For the Kennedy men, this was not only a campaign for the Senate but a testing ground, and they were perfecting techniques and strategy that they intended one day to employ to elevate Jack to the White House.

After one particularly tough day on the campaign trail, an exhausted Jack sat in his father's apartment on Beacon Street talking with his father and Morrissey about the campaign. As difficult as this Senate campaign was proving, Joe said that Jack must think further ahead. If he won in November, he would win the presidential nomination and election to the White House.

"I will work out the plans to elect you president," he told his son, in a voice brimming with assurance. "It will not be more difficult for you to be elected president than it will be to win the Lodge fight."

When Jack won the election by seventy thousand votes, or 51.5 percent, the candidate was not the only noble victor that evening; his father, brother, and mother had triumphed as well. Joe and Rose remembered so vividly how Honey Fitz had run against Lodge's grandfather for the Senate in 1916, and how painful it had been when he lost. "At last the Fitzgeralds have evened the score with the Lodges," Rose said.

What more exquisite revenge for a century of slights than to best the senator bearing the greatest old Boston political name of them all. As he had promised he would do if he won, Jack sang "Sweet Adeline" that evening. This was not his own tune, however, but his grandfather's political theme song, and he would no more look back at his immigrant past than would the shrewd young politicos who surrounded him.

Joe had done what had to be done, and if this meant buying the *Boston Post*'s endorsement as one would buy billboard space, so be it. He had made at least one other crucial move. The man who ran for Jack's seat that year, Thomas P. "Tip" O'Neill Jr., insisted years later that Joe had pushed Governor Paul Dever to run for reelection when the man wanted to retire. Dever was a party politician who had built a machine across the Commonwealth dedicated to the advancement of the Democratic Party and its candidates and agenda. A man with high blood pressure and a heart condition, the governor had been advised by his doctors not to run.

On this election evening there was only one dour face in Jack's headquarters. That was his own father. Joe spent much of the time on the phone talking to the governor, hoping to hear that Dever had finally pulled ahead. The governor's organization had helped Jack more than Dever. At four in the morning Joe gave up. "Paul is not going to make it. I guess I'll go to bed." Joe got up from his chair and turned back once more before he left. "I wonder if Jack Kennedy will ever realize what Paul Dever did for him in this election."

Dever had lost not only the election and possibly his health but the political organization that was much of his life's work. During the campaign, Anthony Gallucio and others had traversed the state, bringing a myriad of new people into politics, the natural constituents for a Kennedy Democratic machine, similar to the Dever Democratic machine. Gallucio pleaded with Jack to keep the organization intact that they had so laboriously put together to further the Democratic Party. Jack replied curtly: "I'm going to run my own boat." Jack did not see the Democratic Party as a sea that raised all boats or none. To him, politics was more like a series of locks that his ship would work its way through while other boats waited far behind. It was of little concern to him whether or not the other boats continued up the canal.

For Bobby, the election was a victory in many ways. His sister Jean observed that in those months, he had proved himself to his father "very quickly and definitely." He was only twenty-six years old, but he had no problem leading people twice his age, often bossing them with dismissive arrogance. He moved people around as if they were furniture, shoving them into this space or that. For the first time in his life he was a man of authority, and he used it willfully. He was his brother's man. That was Bobby's proud identity, a moniker he would carry the rest of Jack's life.

Bobby went down to the Cape a few weeks after the election for a weekend of football and sailing and good times with old friends. It was time to savor the victory, like football players reliving each play of a close victory. Joe would have none of that. Life always lay ahead. "What are you going to do now?" Bobby's father asked. "Are you going to sit on your tail end and do nothing now for the rest of your life? You'd better go out and get a *job*."

In December, Bobby told a reporter from the *Cape Cod Times* that he was "aiming for the post of Massachusetts attorney general" in a few years, but that he would first work in Washington to gain some experience. He could have gained that expertise working for any of a number of Democratic senators or congressmen. Instead, his father decided that he would call upon his Republican friend, Senator Joe McCarthy, to place Bobby in what boded to be the most publicized, most controversial staff position in the Eighty-third Congress: chief counsel to the Senate Permanent Subcommittee on Investigations chaired by the Wisconsin senator.

Joe showed up in his limousine at McCarthy's townhouse on Capitol Hill one winter evening. McCarthy was out in back grilling steaks, but he hurried inside, wearing his apron and holding a cooking fork. "How do you like your steak, Mr. Ambassador?" asked George Mason, one of the senator's friends. "I have no time," Joe said peremptorily and turned toward McCarthy. "Bobby will give me no peace," Joe said. "He wants a job. He wants to come to Washington. You've got to give him a job. You've got to do something about Bobby."

Joe asked for few favors, and when he did, they were usually wrapped in velvet, but there was an urgent, imploring quality to this request. Joe was McCarthy's supporter, fellow Catholic, and friend. Moreover, McCarthy was enough of a politician to realize when he could not say no. "I'll talk to [Senator John] McClellan [the ranking Democrat] in the morning, and see what we can arrange." With that, Joe turned and walked out the door, never having even taken off his homburg.

McCarthy told Joe that he had already hired twenty-five-year-old Roy Cohn, another ambitious young lawyer, as chief counsel. Instead, the senator offered Bobby a slot as assistant counsel.

Jack found McCarthy's rhetoric vulgar and overweening but he was not about to attack him for such faults. Nevertheless, he was upset that Bobby was going to work for McCarthy, even if he thought it was for "political, not ideological," reasons.

Bobby had scarcely arrived in his new position before he made his presence felt in a variety of ways. One day Maurice Rosenblatt received word that Bobby wanted to see him. Rosenblatt was a leading anti-McCarthy activist and the anonymous author of a series of articles about Joe's anti-Semitism and dubious business dealings in the *City Reporter*, a small liberal publication. He was not used to getting calls from the Kennedys. Rosenblatt walked over to the assistant counsel's office in the Old Senate Office Building. "I walk in, and he gets up, walks around his desk, and puts out his hand, and I put out my hand," Rosenblatt recalled with the most vivid and immediate of memories. "He pulls my hand and twists it, and I'm off balance, and [he] throws me onto a leather coach. No words or anything else. I blink and say, 'What is this about?' He says, 'We have our eye on you.' "

Bobby had a dogged, tenacious quality that he applied in full measure to his investigation of American allies trading with the Chinese. He discovered that three out of every four ships carrying goods into Chinese ports flew a Western flag. Many of these shippers also had contracts to carry allied defense goods to Western Europe. And all of this was happening when American boys were dying in Korea.

This devastating information seemed to verify American feelings that the world outside its borders was a duplicitous, dishonorable place. Bobby's initial report was judicious and serious, the very model of the way the staff of a congressional committee should do its work. Bobby did not point to bureaucratic culprits in Washington to be grabbed by their disreputable necks and hauled before the justice of the McCarthy committee.

The report should have led to lengthy, spirited hearings in which a wide variety of viewpoints would be heard. It was, after all, a world of fearsome complexities. The Japanese appeared to be one of the worst violators. They were shipping only seaweed to China in exchange for iron ore, however, and that seaweed was hardly the stuff of which the Chinese could make bullets. As for the British, they had colonies in Asia and had been a trading nation for centuries; it was far more onerous for them to stop shipping to China than for the United States.

These points were made, but they were drowned out by the sheer force and fury of McCarthy's rhetoric. What could a man say—a politician, that is—when McCarthy shouted on the floor of the Senate: "We should perhaps keep in mind the American boys and the few British boys, too, who had their hands wired behind their backs and their faces shot off with machine guns—Communist machine guns . . . supplied by those flag vessels of our allies. . . . Let us sink every accursed ship carrying materials to the enemy regardless of what flag those ships may fly."

If McCarthy had been able to marry his rhetoric to Bobby's research, he might have staved off his ignominious political end for a while longer. He was, however, a man rising to his worst instincts, and no one played to those instincts better than did Roy Cohn and his new associate, G. David Schine. While Bobby was working on his shipping report, these two dapper, diminutive inquisitors traveled around Europe pulling suspicious books off Voice of America library shelves, happily exporting fear and suspicion to American officials abroad.

Bobby had an immense dislike for Cohn, an emotion that Cohn fully reciprocated. When the two ambitious young men looked at each other, it was as if they were looking into a mirror that exaggerated their blemishes and faults. Cohn did Bobby one of the most valuable favors of his life. If Bobby had not abhorred Cohn so profoundly, he would probably have stayed with McCarthy, a man he personally admired, and would have borne the heavy burden of that livery for the rest of his political life.

The history of the Kennedys might have been different if Bobby had remained with McCarthy, or had taken the position of chief counsel. The family would have been so closely identified with McCarthy that Jack would have found it difficult to get the support of enough liberal and centrist Democrats to win the presidential nomination.

None of the Kennedy men grasped the terrible danger of McCarthy. Across the nation men and women spent sleepless nights pondering whether they would be condemned for an acquaintanceship they once had, a petition they once signed, a belief they once held, a cause they once supported. This fear reached into the higher reaches of academe, into the unions, and into the bureaucracies of Washington. It even entered into the Kennedys' own family. The fact that the matter was kept so quiet shows that in those years, fear was no stranger even among the Kennedy men.

During the summer of 1954, the FBI learned that Jack Anderson, then a reporter for columnist Drew Pearson, had information that after completing army basic training in 1951, "Teddy had not been permitted to go to a school at Camp Holabird, Maryland, because of an adverse FBI report which linked him to a group of 'pinkos.'" Here, then, was just the kind of silent, unsubstantiated allegation that destroyed people. It had apparently been responsible for Teddy's abrupt departure from Camp Holabird, destroying his Army Intelligence career, and now it might destroy his public honor.

FBI agent L. B. Nichols wrote Clyde Tolson, the FBI deputy director, that Joe "stated that he sent word to Drew Pearson that if he so much as printed a word about this that he would sue him for libel in a manner such as Drew Pearson had never been sued before." Nichols reported that he had told Joe that there had been no such FBI report and that it may have been the case of "somebody confusing the FBI with some other investigative agency." There may have been no formal FBI investigation, but Nichols told Tolson that "apparently some of the information which Anderson had on his son's Army activities was accurate and Kennedy stated the army was somewhat incensed over how the information got out."

Teddy, then, was probably a victim, if a minor one, of the Red Scare. If not for Joe, Teddy might have found himself permanently tainted. He was left unscathed, but neither his father nor his brothers appear to have grasped that if the finger of accusation could point at Teddy, then it could point at anyone. Men who exalted courage above all virtues surely should have known that when the name of your own son or brother is called out, then it is time to stand up and condemn those fingers pointing so wildly, often destroying lives with the flick of an allegation.

Joe did not quite see it that way. He shared many of McCarthy's beliefs and reveled in his association with J. Edgar Hoover, who fancied himself the greatest of all Communist hunters. Joe's friendship with Hoover may have saved Teddy's reputation, and Joe took every occasion to flatter the FBI director.

The year before the threat against Teddy, J. J. Kelly, the special agent in charge of the Boston office, made one of his periodic visits to Joe in Hyannis Port. Joe told the agent that "if it were not for the FBI the country would go to Hell." Then he referred to a series of newspaper columns regarding civil rights investigations. Although the name of the columnist has been blacked out in the FBI Freedom of Information documents, Joe was apparently referring to Drew Pearson. Joe told the agent that he believed that the columnist "was angling his columns at the Jews, Negroes and the Communist element behind the Civil Liberties outfit, as well as the NAACP."

Joe's endless devotions to the FBI director were the mark not of an unctuous poseur but of a shrewd man who understood his subject only too well. As much as Hoover loved power, he loved praise even more. And in the midst of the McCarthy era, the director received accolades and acclaim so extravagant that only a man of boundless egoism could have believed it. Even among this army of courtiers and sycophants, Joe's fawning voice stood out as, in the words of the FBI special agent in Boston, "the most vocal and forceful admirer [of Hoover] that I have met."

The Senate was as close to a natural aristocracy as could be found in American politics, and Jack fit into the clubby collegial atmosphere as he had not in the rowdier, more populist House. A patina of authority descended on Jack, as it did on all members of the Senate, even one as youthful and naturally irreverent as the junior senator from Massachusetts. "Knowing him from then on was not knowing him at all, because once you become a member of the club, everything about you changes," reflected Dave Powers.

Even Jack's old friend Charles Bartlett noticed that a change had come over him. Up until then, nothing gave Jack greater pleasure than employing his wicked wit on the buffoons, mediocrities, and pretenders with whom he felt he served in Congress. Charley was a man of courtly civility who would no more have passed on Jack's indiscretions than he would have written about them as a journalist. But now Jack was forgoing his usual playful put-downs. "Dad says don't knock anybody," Jack explained, although in closing down much of his wit he was shutting off part of himself.

Jack's ambition came into focus. He was a man concerned only with what you would do for him tomorrow. Loyal Anthony Gallucio had traveled the state by bus, eating at cheap restaurants and treating his employer as if he were an impoverished candidate, not the son of one of the wealthiest men in America. He had given his all to the campaign, his formidable organizing skill, energy, wit, and integrity, and he assumed that he would be going to Washington in Jack's enlarged office, a minimal reward for his two years of relentless effort. Jack called finally to give him the news. "I've got no money," Gallucio recalls Jack telling him.

For six years, Mary Davis had not simply served as an excellent secretary to Jack but used her astute political sense to promote the congressman in a myriad of ways. For several years she had been working in the office six days a week, then finishing up her work Sunday at home. She lined up a number of new secretaries and clerical workers for his expanded staff, agreeing on salaries that could reasonably be paid out of Jack's allotment.

"Well, I can't pay any more than sixty dollars a week," Jack replied.

"Sixty dollars a week!" Davis exclaimed. "You've got to be joking. Nobody I've lined up would be willing to accept a job at that salary. And I wouldn't ask them."

"Well, that's the way it's going to have to be."

"Where are you going to get somebody competent for sixty dollars a week? You cannot do that."

"Mary, you can get candy dippers in Charlestown for fifty dollars a week."

"Yes, and you'd have candy dippers on your senatorial staff."

Mary was being paid only ninety dollars a week. Salaries on the Senate side were higher, and she asked to be raised to the one hundred fifteen dollars a week being offered her by a freshman congressman. "Mary, you wouldn't do this to me," Jack replied incredulously, unwilling to go beyond a 10 percent raise.

It would have been nothing for this multimillionaire heir to pay this loyal woman an extra eight hundred dollars a year, less than he spent during his weekends in New York. Jack, like the rest of his family, considered it part of the livery of service to be poorly paid. Those who sought market value for their services were expressing their disloyalty and they deserved to be gone, and gone Mary Davis was.

The most notable person Jack hired that January—and the most important aide he ever hired—was twenty-four-year-old Theodore C. Sorensen. That the lanky, soft-spoken Sorensen would join the staff as Jack's chief legislative aide was a mark of two great ambitions, Jack's and Sorensen's.

The position would have been a natural for a gregarious, witty Irish-American who had worked his way through Harvard Law School and could be counted on to work with loyal devotion and political savvy. When Sorensen was thinking about working for Jack, he was warned that he would have to pass Joe's scrutiny. But in half a century, Jack's father had hired only one non-Catholic.

Sorensen stood doubly disadvantaged. His mother was of Russian Jewish heritage, while his Protestant father was a member of those extraordinary progressive Republicans from Nebraska who formed around Senator George Norris. Sorensen arrived in Washington, however, with impeccable academic credentials: a Phi Beta Kappa in college, he had been first in his law school class at the University of Nebraska. He was also a talented writer who had published articles in liberal publications such as *The New Republic* and *The Progressive*.

Sorensen could have stayed in Lincoln, started a law practice, and run for political office himself, but his ambition was different from Jack's. There are those with public egos—politicians, talking-head journalists, preachers—whose pleasure is in the appearance, the speech, the sermon, the byline, the applause. And there are those with private egos—aides, editors, directors—who prefer to stand behind watching others reading lines and performing actions of which they consider themselves largely the creator.

That latter kind of ego is so disguised that it is mistaken for humility when it is often the opposite. Sorensen fancied himself a liberal idealist, but that liberal idealism ended when the young Nebraskan chose his employer, a politician to his right on most important issues, but a politician with an eye on the big prize of political life.

Sorensen would temper his ideas and his words so that he would sound perfectly like Jack. The man was so adept at mimicking Jack that he occasionally pretended to be the senator on the phone. Sorensen did this so well that the danger was that he would think that he played Jack better than Jack played the role himself. Bobby spotted this quality in Sorensen, calling him in these early years "far more interested in himself" than in the Jack Kennedy he was supposedly serving.

Sorensen was often called brilliant, but he was more the brilliant mimic, be it of ideas or styles. If he had been an artist, only an expert would have been able to tell that his work came not from the master himself but from someone painting in the same school, copying the master's brush strokes.

Sorensen and the rest of Jack's staff wrote the speeches and articles that left the office stamped with Jack's name, even if on occasion he hardly had time to glance over them. That process began in Jack's first days in office when Sorensen flew up to Boston to meet with a group of scholars and economists put together by James Landis, who had left the deanship of the Harvard Law School and was now working full-time for Joe.

Jack thought that problems were solved by calling in the premier experts in the field. You heard them out, by word or memo, and then using their wisdom you decided what was best to do politically. In this process, Sorensen was not the originator but the transporter of ideas who translated those ideas into the politically plausible, in language full of sound logic and occasional eloquence.

After this first meeting, Landis addressed a memo not to the senator but to his father. Joe had put together a formidable team of attorneys and accountants who worked out of a family office on Park Avenue in New York City, largely hidden from public view. Their sole purpose was to advance the fortunes of the Kennedy family, the most important Jack and his vision of becoming president of the United States.

In May 1953, Jack presented a series of three speeches in the Senate titled "The Economic Problems of New England: A Program for Congressional Action." He sketched a portrait of a region proud of its past and its seminal role in so much of American life. But he also described a region whose fishing grounds and forests were becoming depleted while its traditional industries, such as textiles, were moving south to a haven of cheap, nonunion labor and abundant resources. Worse yet, it was a region where "government management and labor have resisted new ideas and local initiative." Kennedy called for the creation of a Regional Industrial Development Corporation, job retraining, a higher minimum wage, increased business incentives, and the serious investigation of freight rate discrimination.

As Jack stood speaking before a nearly empty Senate, he was reading words that Sorensen had written and promoting ideas that were largely not his own. He was not a plagiarist, however, but a politician, and he deserved the accolades he received for looking not simply at his state but at his region and trying in a serious, analytical way to be a national senator.

When Jack left Capitol Hill early for a party in Georgetown, or flew up to New York for an engagement, he knew that Sorensen would most likely still be there, writing articles and op-ed pieces, speeches, and letters, all with Jack's name on them, for publications including the *New York Times Magazine, American Magazine, The New Republic*, and the *Atlantic*.

Jack liked men who were quick studies and there was no quicker study than Ted Sorensen. Within months, he had Jack down perfectly. "The Atlantic Monthly article was approved without substantial change by the senator; and both he and his father liked it very much," Sorensen wrote Landis, with whom he had apparently co-authored a piece for Jack's byline titled "New England and the South." "I am looking forward to more collaboration in the future."

Jack's political life was in competent hands, but he still had a major problem if he ever hoped to run for president. His good friend Jim Reed observed that Jack thought of women as "chattel . . . in a casual, amiable way." They were a pleasurable sideshow to the business of life, in which men were the only players.

Jack would have gone on a bachelor indefinitely if he had not been so politically ambitious. In Eisenhower's America, a perpetual bachelor was considered most likely not an asexual mama's boy or a high-living libertine, but a closet homosexual. "We used to kid Jack all the time about getting married," recalled Ben Smith, one of his Harvard roommates. "I remember in the 1952 campaign he said that if he won he would get married."

"You know, they're going to start calling you queer," Morrissey told Jack after the election. Jack decided that he would put on the velvet shackles of marriage, but he would do so only because he knew how to pick the lock.

Jack had met Jacqueline "Jackie" Bouvier at a dinner party at the Bartletts' house in Georgetown in May 1951. Twenty-one-year-old Jackie had a wispy, gaminelike voice more suitable to a geisha than a sophisticated young woman who had studied at Vassar, the Sorbonne, and George Washington University. Despite the twelve-year difference in their ages, or perhaps in part because of it, Jack was intrigued enough to want to go out with her afterward for a drink. When they got out on the tree-lined street, there sat one of Jackie's beaus asleep in her car, waiting for her, and Jack made a discreet retreat. Jack was so busy with his campaign for the Senate that he rarely saw Jackie, but he invited her to Eisenhower's inauguration in January 1953, and then started seeing her regularly.

Jackie was working as an inquiring photographer for the *Washington Times-Herald*. It was a superficial job, running around the capital, taking pictures of prominent Washingtonians, and asking them benign, obvious questions. She had incredibly wide-spaced eyes that missed nothing, and nothing of what she truly saw found its way into her column. Although her manners were impeccable, she had a devastatingly wry humor that suggested the caustic way she viewed lesser mortals. Once, while driving with her stepbrother Hugh D. Auchincloss III from Washington to Newport, police stopped their car on the Merritt Parkway. While the Connecticut trooper stood there preparing to write a ticket, Jackie innocently and oh so generously offered:

"Excuse me, officer, but your fly is undone." The policeman murmured a thank-you and hurried off without writing a ticket.

Jackie did not talk much about her own childhood. Her parents had divorced when she was a young girl, and she had sought solace in horses and poetry and hours of dreamy introspection. She had an adventurous soul and as a girl had enjoyed many books that boys generally read, from Kipling's *Jungle Book* to Sir Walter Scott's *Ivanhoe*. With her wistful imagination, she once decked out her stepbrother in a deerstalker hat, an Inverness cape, and a royal Stewart kilt so that when they traveled around Scotland he looked like the young Sherlock Holmes.

Her father, John "Black Jack" Bouvier, had lost everything in the Depression except his charm and his eye for a well-turned ankle. Black Jack worked the magic of his charm on no one more than his own daughter. When she went to see *Gone With the Wind*, she thought the irresistible Rhett Butler the image of her father, while the beautiful, manipulative Scarlett O'Hara resembled her mother and gentle Ashley Wilkes reminded her of her new stepfather.

After the debacle of her first marriage, Janet Bouvier, Jackie's mother, decided not to marry for romance the second time. She proved her case by managing to marry Hugh D. Auchincloss, a gentleman whose most important assets were his name, his wealth, and his constancy. She expected her two daughters to follow her own lead in choosing a man to marry.

Jackie was a subtle, impeccably mannered, immensely literate young woman fascinated by the rebel artistic spirits of her age. In her essay that won *Vogue*'s fifteenth Prix de Paris Contest, she wrote that the three men she would most like to have known were Charles Baudelaire, Oscar Wilde, and Sergey Diaghilev.

Jackie admired Baudelaire and Wilde as "poets and idealists who could paint their sinfulness with honesty and still believe in something higher." That was a heretical thought. Jack's mother and sisters would have found Jackie's creative heroes little more than pied pipers of decadence, hardly the models for a proper young woman. These daring artists lived on the dangerous edge of their time, and Jackie was drawn to them and their art and lives.

Jackie wrote: "If I could be a sort of Over-all Art Director of the Twentieth Century, watching everything from a chair landing in space, it is their theories of art that I would apply to my period, their poems that I would have music and paintings and ballet composed to."

Jackie knew all about Jack's sexual proclivities, and she condemned him no more than she did Baudelaire and Wilde, or her father for that matter. She was swept away by love for this handsome young politician. Jack, however, appeared more distant.

Jackie flew off to London to cover the coronation of Queen Elizabeth on June 2, 1953. On the return flight, Jack surprised her by greeting her plane when it stopped in Boston on its way to New York City. Jackie's mother was a superb judge of the male ego, and she had set up Jackie's trip and paid for it in part just so something like this might happen.

"If you're so much in love with Jack Kennedy that you don't want to leave him," she told her daughter, "I should think he would be much more likely to find out how he felt about you if you were seeing exciting people and doing exciting things instead of sitting here waiting for the phone to ring."

Even though Jack had decided to marry Jackie, he was not swooning with courtly love. When he called Jim Reed to tell him about Jacqueline Bouvier, he was so uncertain about the whole business that he told Jim that he "might" marry Jackie while at the same time asking his friend to be one of the ushers. Jack wrote Red Fay asking him to be best man at his wedding. The prospective groom made one small oversight that suggested his inquietude about the approaching nuptials: he did not mention the name of the woman he was marrying.

The women Jack had gone with over the years had understood the sophisticated game he was playing. Dinner at the Stork Club, '21,' or other elite watering holes. Lighthearted repartee. Quick, efficient sex. A few smiles. A laugh or two. No painful revelations. No cloying commitments. No midnight phone calls. No grasping emotions. No jealousies. And good-bye.

Jack had been especially attracted to wealthy divorcées who played the game as well as he did. One of the women he had dated for years was Florence Pritchett. Flo was a gorgeous model who had a laugh that rang out like struck crystal. Jack had met her back during the war when she had divorced her rich husband. Flo could always make Jack laugh and that was among the greatest gifts you could give the man. He saw her off and on over the years. He was not a man to send flowers or gifts, but he understood that with Flo jewels were in order. On her twenty-seventh birthday in June 1947, he wrote in his appointment book: "Flo Pritchett's birthday! SEND DIAMONDS." Diamonds were not quite enough and within a few months Flo had married her second wealthy husband, Earl Smith. "Florence Pritchett was a serious girl," Fay recalled. "She turned him down. She didn't think he was big-time enough. She wanted to make sure she was going to get there. It was very unsettling to him."

"I was very stuck on her [Pritchett]," Jack told James MacGregor Burns in 1959. "It was rough, but . . . I'm not . . . the tragic lover. . . . At least that was the girl I liked, and I've had some other girls that I've liked but . . . it's never been sort of a depressing experience."

Jack told Fay that he was both "too young and too old" to marry. He was not far wrong. He was too young in that he still had a bachelor's eye. He was too old in that he was not only set in his ways but embedded in them, and he was self-aware enough to know that no mere ceremony was going to change that. Beyond that, as he wrote Red, he knew the extent to which his political career depended on his appeal to women. "This means the end of a promising political career as it has been based up to now almost entirely on the old sex appeal," he told his friend.

When his son left for a trip to France in the weeks before the wedding, Jack's father worried that he might get "restless" about the marriage, for Joe knew where Jack's restlessness usually led him. "I am hoping that he will . . . be especially mindful of whom he sees," Joe wrote Torby Macdonald, who was planning to accompany Jack. "Certainly one can't take anything for granted since he became a United States Senator. That is a price he should be willing to pay and gladly."

A long time before Inga had talked of those two roads that faced Jack, the one toward freedom and love and wondrous uncertainty, the other a well-defined, arduous track leading toward power and a place in history. For all her beauty and charm, Jackie represented another long, hard step up that narrow road to power. And though that was the direction Jack had chosen, he still looked fitfully over his shoulder at the road not taken.

That August, Jack flew over to southern France for his last few days as a single man. One evening he was in Cap d'Antibes when a blue sedan stopped. "Jack! What are you doing here?" exclaimed Gavin Welby, a British acquaintance. "We're going to have dinner at Le Château at Haut-de-Cagnes. Why don't you join us?" As Welby spoke, he nodded toward the two stunning young Swedish women he had picked up in Cannes where they had been hitchhiking.

Jack hadn't said whether he could make it for dinner, but he managed to arrive at the romantic restaurant high above the valley before his host and his two other guests. A few minutes later the two young women walked into the famous restaurant wearing simple dresses that set off their fresh features, which glowed with youthful health. Gunilla Von Post and Anne Marie Linder were staying in a nearby villa. Gunilla came from a distinguished old Swedish family and would never have hitchhiked if the stipend from home had not been so late in arriving.

For the two friends, summer had a meaning that an American or an Englishman could never understand. It was short-lived and intense; a sensuous, passionate time when the sun seemed to burn all the moroseness out of the dark Scandinavian soul. For the women, it was adventure lying in the hot sand, feeling the Mediterranean sun beating on them.

Gunilla was small and delicate with refined features and deep, melancholy eyes that suggested a Garbo-like mystery. Jack liked wellborn ladies, and there was an exquisite juxtaposition between this seemingly carefree, hitchhiking Swede and her aristocratic background. Jack had traveled enough within the upper-class European world to be able to spot a poseur immediately. Gunilla was not one of those. She knew many of the same people Jack knew. In Great Britain she had even stayed with the Earl and Lady Home, whose son William Douglas, the playwright, had once been in love with Kathleen.

That evening, as Jack sat next to Gunilla in the banquette, once gently touching her hair, he never once mentioned Inga. But he could hardly have failed to think, at least momentarily, of his Danish lover and of how close he had come to walking with her up that unknown pathway to a free and open life. And now, days before his wedding, he sat with another beautiful young Scandinavian woman, a woman without a past, a woman who with her laughter and smiles beckoned him up that dangerous pathway again.

Jack was never one to talk much about his family and his past, but this evening he went on and on about his father and mother and brothers and sisters. Europeans are often appalled at the way some Americans tell the most intimate details of their life to strangers and mistake these revelations for friendship. Jack's admissions, however, were a true sign of intimacy, a mark not only of how much he was thinking about his past as his wedding day closed in on him but of how affected he was by this young woman and by these days in southern France.

After dinner, Jack led Gunilla to Jimmy's Bar, a popular nightclub, where the couple danced and talked some more. Jack usually considered sentimentality a weak man's emotion. Yet this evening he suggested that they drive to Hotel du Cap Eden Roc at Cap d'Antibes, where he had spent so much time as a boy and young man. The couple sat there looking out on the Mediterranean near the very spot where Joe Jr. and Jack had dared Teddy to jump off the cliff into the water. And there he kissed Gunilla and, as she remembers it, told her, "I fell in love with you tonight."

This was not some tired romantic verbiage that Jack used to impress a gullible young woman. Jack was not a man to say such a thing. He was, however, in a sweetly melancholic mood that was as rare for him as were these words. It was almost a decade and a half since he had trod on this grass and had swum in the ocean below with his brothers and sisters, two of whom were now gone and one of whom was locked away.

Jack was a U.S. Senator able to lead the discourse on the most serious problems of the age, but in his personal life he rankled at taking on all the tedious responsibilities of adulthood. He had chosen the road he would travel, but on this sweetly scented evening he stopped and looked back in the other direction and for a moment wished he could have chosen the other path.

"I'm going back to the United States next week to get married," Jack said suddenly. He did not have to admit that. He could have played the evening out, taking his chances at bedding Gunilla before heading back to the States, but he felt more than that. "If I met you one week before," Jack said starkly, "I would have canceled the whole thing."

Jack may have believed what he was saying. But he would never have thrown over his life and commitments so cavalierly. He was speaking, however, to something more than simply the beautiful young woman sitting next to him. Gunilla represented freedom and sensuality and an exotic European world within whose pleasures a man could disappear.

In the early morning hours, Jack drove Gunilla back to her house. "May I come in for a nightcap?" Jack asked, as Gunilla recalls. "One for the road." Gunilla knew what he was asking and part of her wanted to invite him in. She knew instinctively that if he spent the night with her, she would never see him again.

"You take your own road," she said. "And good luck, my dear."

And so Jack turned around and drove back up the road he had come.

Jackie had set the wedding for September 12, 1953, in Newport, Rhode Island, where the Auchinclosses lived in the genteel world of the Protestant upper class. Joe was openly disdainful of this aging, pretentious enclave where the residents called their mansions "cottages" and looked down their lorgnettes at déclassé tourists.

"Their wealth is from an era gone by," Joe told Red Fay, in a voice brimming with irritation at what he considered an insipid, declining social set. "Most of them are just keeping up a front and owe everybody. If you pulled the carpets up most likely you'd find all the dirt for the summer brushed under there, because they don't have enough in help to keep those big places running right."

Jackie's mother expected that the wedding would be a sedate, exclusive ceremony far from the vulgar flash of cameras. Joe flew down to Newport to disabuse Mrs. Auchincloss of that illusion and to inform her that over one thousand guests would be invited, including most of the U.S. Senate. Gliding down the steps, his face was alight with benevolent charm, his hand in his pocket. As Jackie saw her future father-in-law there, she thought, "Oh Mummy, you don't have a chance."

Jackie appreciated Joe's stylish nature. She admired his Sulka lounging clothes and the light blue gabardine suit he wore when he rode off to Hialeah in his chauffeured Rolls-Royce. She saw too that his manners were as elegant as his clothes, a subtle rendering of social nuance. His charm was not a dandy's plaything, but a device he used to extract what he wanted. "When he turned on his charm to gain what he wanted, it was great to watch," she recalled.

Thirty-six-year-old Jack was full of fitful anxiety over his approaching wedding day, concerned most notably over the political cost of his marriage. At a stag party at the Parker House in Boston for many of his cronies, he worried over the price he might pay for no longer being the golden bachelor.

"I was seated next to him on his right, and he was kind of shy," recalled John Droney, a veteran who had worked in his campaigns since 1946. "He asked if I thought he was doing the right thing and what will the women think. I said, 'Oh', you are doing the right thing, because I have a little girl and you'll get a lot of pleasure in this thing.' "

By the time the wedding weekend arrived, Jack no longer was making such public musings. "Well, the first thing you have to do, Jack," Red whispered, leaning toward his friend at the prewedding dinner at the Newport Clambake Club, "is you've got to make a toast to the bride, and you've got to throw that glass in the fireplace."

Jack looked at the superb crystal glass as if divining the future there. His future mother-in-law was a woman of shameless social ambition, narrow snobbishness, and silly garrulousness. Hughie, Jackie's stepfather, was a man who mistook cheapness for frugality, humbug for humility. Jack sensed that once he scratched their gold veneer he would find little but chintz. Here was an exquisite opportunity to stick it to the Auchinclosses while staying protected behind the shield of civility and a countenance aglow with innocence.

"To my future bride, Jacqueline Bouvier," Jack said as the guests joined him in the toast. "Everybody throw your glasses in the fireplace."

As the precious crystal shattered against the stone fireplace, Mrs. Auchincloss's countenance took on an ashen gray pallor, but the lady was nothing if not game. She motioned to the waiters to bring new glasses; after setting the crystal down in front of the guests, they filled them anew.

Jack rose again. "I realize this is not the custom, but the love that I have for Jacqueline Bouvier overcomes me," he said as he proposed a second toast to his beautiful young bride. "And now, everybody throw your glasses in the fireplace."

The guests had become adept at this custom by now, and they hurled the crystal with abandon toward a fireplace alive with shards of glass. Mrs. Auchincloss ordered more glasses, but this time they were cheap water glasses. The evening moved on, since Jack was not about to toast his elegant Jacqueline with anything less than crystal.

This was an evening more for jocular comments than sentimental musings, and when it came time for twenty-four-year-old Jackie to talk, she held up a postcard from Bermuda with a picture of a red hibiscus that Jack had sent her after his election. She read the words on the back—"Wish you were here. Jack"—and told the audience that during their courtship, this was the only correspondence that Jack had ever sent.

The audience roared with laughter, but the reality was that Jack was less than deeply solicitous of his high-strung young bride. For all of his natural charm, he was full of the high selfishness of an ambitious politician, eager to use every public moment to advance himself. That included even his wedding.

Jackie had wanted an intimate ceremony with guests who knew and cared for them, not the massive spectacle that the Kennedys had made of what after all was *her* wedding. There were 750 guests, most of whom she did not know. Worst of all, she abhorred the hordes of journalists, the photographers with the snouts of their cameras pointing at her, the reporters pressing forward in sweaty earnestness. At the church, she suffered a further humiliation when her father was too drunk to give her away.

After the couple had said their wedding vows at St. Mary's Catholic Church and stepped outside on the steps, a scene took place that was a harbinger of what much of their public life would be like. First stood rows of photographers, like a media Praetorian Guard, all pointing their cameras up the steps. And across the street, behind a police barricade, stood over three thousand onlookers, clapping, whistling, and shouting, pushing forward so fervently that they knocked over the barricades and surged forward, a human tide. Jack was amused at the circus, and he surely must have realized that by marrying, he had not lost his appeal that so transcended politics but perhaps had enhanced it.

Jack was not a man for a lengthy honeymoon filled with little but handholding and vows of devotion. The newlyweds went to Acapulco, where Jack caught a swordfish. From Mexico, the couple spent some quiet days in Los Angeles and then traveled on to Pebble Beach to play golf with Red Fay before driving two hours north to San Francisco with Red and his wife, Anita.

Jack seemed not to care that the Fays were not necessarily Jackie's kind of people, and definitely not participants in her kind of honeymoon. As much as Fay enjoyed his best friend, even he saw that this was no longer "the kind of honeymoon any young bride anticipates."

Jack, however, had apparently had enough of romantic solitude and wanted his own life back. He even may have suggested that Jackie go home early, an idea that his bride declined. Nonetheless, Jackie was so smitten with her husband that she gladly accepted whatever else Jack wanted, even if it meant on the last day of their honeymoon going off with Anita while Jack and Red attended a San Francisco Forty-niners football game.

Jackie read literature and poetry not as a pallid diversion but as life's vision written big and clear, and she saw that men were great in their failings too. She saw her Jack as she might a hero in an epic poem, as a grand romantic figure living a transcendent life. Like Inga before her, she sensed that there were two directions that Jack could travel, toward personal fulfillment or up that difficult path toward a place in history. She saw where Jack was heading when in California he admitted to her that he wanted to be president. Afterward she wrote a poem to Jack containing the lines:

He would find love
He would never find peace
For he must go seeking
The Golden Fleece.

16

Aristocratic Instincts

Jack's bride was not simply the youngest and most beautiful of the Senate wives, but also one of the most dutiful. She turned her sloppily attired husband into a fashion plate and brought him lunch that he could digest on his nervous stomach. When the couple had a dinner party in their rented house in Georgetown, Jackie led Jack and his guests into the dining room for a meal full of dishes exotic to the American palate.

Jack's taste in food went to meat as long as it was steak not gussied up with silly sauces. Jackie, however, believed that style was not something that one wore only on festive occasions. For Jack, it was an exceedingly expensive lesson, and the food bills were the least of it.

Jackie returned one day with a spectacular find in eighteenth-century French chairs for the living room. Jack could hardly contain his displeasure before his old friend David Ormsby-Gore, later Lord Harlech. "I don't know why!" he fumed. "What's the point of spending all this money? I mean a chair is a chair and it's perfectly good the chair I'm sitting in—what's the point of all this fancy stuff?" The point of all this fancy stuff, as Jack took a number of years to realize, was that it impressed the hoi polloi enormously and brought him a cachet for high style that until then he neither possessed nor valued.

During the first winter of their marriage, Jackie worked on a private little tome for her half-sister, Janet Auchincloss, called *A Book for Janet: In Case You Are Ever Thinking of Getting Married This Is a Story to Tell You What It's Like*. The book was a wistful romantic account full of gentle caricatures of the couple—for instance, Jackie looking to see whether a flag flew over the Senate chambers, signaling that her Jack was off doing the nation's business. And there was a drawing of a slightly risqué Jack in bare hairy legs saying: "I demand my marital rights." It was in some ways like a children's book yet in its way was a sophisticated fantasy.

Charley Bartlett observed a marriage so different from the one portrayed in Jackie's book that he at times regretted that he had ever introduced the couple. Bartlett felt that this exquisite young woman who had talked so fervently of art and culture had become a dispirited wife who sought in things what she could not find in marriage. He noticed "a sad look in her eyes." One day the same woman who was writing a fantasy about her marriage got into her car and drove over to the Walter Reed Antiques Shop on Georgia Avenue to sell many of her wedding gifts.

Jackie grew gloomy and withdrawn, and around Jack that was simply unacceptable. "Jack went crazy when someone sulked," said Lem Billings. "He couldn't stand the tension, and he'd go absolutely crazy trying to contrive ways to restore a friendly atmosphere. Jackie saw this almost immediately and used her sulks masterfully."

Jack, who considered faithfulness a fool's virtue, was continuing with his affairs. Like his father, Jack had learned to cloak his marital deceits in an elaborate garb of euphemism. He had an unlikely admirer of his adulterous trysts in his father-in-law, Jack Bouvier, who continued to cut his own wide swath through the female population. "Miss New Zealand isn't too bad, and it might be fun to run into her again some time," he wrote Jack, like a gourmet discussing meals he has eaten. "I would like to see that English nurse of yester-year, she of my twenty minute romance, which you and your gang so rudely but effectively interrupted. All this providing . . . I still have the 'wherewithal.'"

In the summer of 1954, while his bride of less than a year was in Europe, Jack traveled up to Northeast Harbor, Maine, for a house party. Jack's host was an old friend, Langdon P. Marvin Jr., who *Life* magazine had dubbed "Harvard's outstanding [1941] graduate," endowed with "name, wealth and brains." The godson of FDR, Marvin had played an important role in the war managing the air shipment of strategic imports. After the war, Marvin became an important public advocate of air transportation, but he enjoyed his pleasures as much as Jack. Marvin presumably knew that his dissembling this weekend was not just for Jackie but also for a public that would not look kindly on this most blessed of married men having a romp among the women whom Marvin had so graciously assembled.

Jack was on crutches, and he should have been thinking of anything but embellishing his sexual reputation. This weekend was little more than a consolation prize. Since March he had been writing Gunilla seeking to set up a rendezvous with her in the summer. His letter that month was doubly circumspect. He subtly reminded Gunilla of their meeting as if she might have forgotten ("Do you remember our dinner and evening together this summer at Antibes and Cagnes?"). As his return address, he gave his aide Ted Reardon's Georgetown home.

When Gunilla responded positively to his entreaties, Jack aggressively raised the stakes. He told her that he was willing to come to Sweden to meet her in August, but that was not his preference. "I thought I might get a boat and sail around the Mediterranean for two weeks—with you as crew," Jack wrote her. Gunilla tentatively agreed to see him in Paris, but he kept pushing her to go off on a private cruise. He was so sick, however, that in September 1954 he had to cable her: LEG INJURED AND HOSPITALIZED TRIP POSTPONED WILL WRITE MANY REGRETS JOHN.

Jack was so weak that he could not even attend the sailboat races but instead had to watch them on land sitting in a chair, observing the finish line with binoculars. That did not prevent him from more vigorous nocturnal activities that attracted the prurient attention of Mrs. Kelly, who oversaw activities at the Old Kimball House, where the group was staying. The maids started spending more time examining Jack's sheets than changing them. Marvin thought that hotel personnel were listening in on phone conversations.

As good fortune would have it, Mrs. Kelly was Catholic. Marvin and his friends decided that Jack would have to take the good lady to mass. "We plopped him into a bathtub of cold water, got him down to the lobby," Marvin recalled. "There was Mrs. Kelly with her car in the driveway. After that, no espionage, no gossip, full security, full cooperation."

Jack had what in many ways was an aristocratic conception of life. He identified with Cecil's Melbourne and the young lord's "Renaissance ideal of the whole man, whose aspiration it is to make the most of every advantage, intellectual and sensual, that life has to offer." He could be acerbic about his déclassé fellow citizens, but such views were anathema in his egalitarian nation, doubly so in a politician seeking the votes of many people he thought well beneath him.

Down in Palm Beach over the Christmas holiday, Jack read Cecil's new biography of the adult Melbourne, *Lord M. or The Later Life of Lord Melbourne*, a companion volume to the biography of young Melbourne that he had admired so much as a young man. Then he had read those eloquent pages portraying a dispassionate, elegant, hedonistic, young aristocrat who was the model of what he himself aspired to be. Now he read about a Melbourne who came as close to a portrait of what he had become as anything else he was likely to have read.

Even as prime minister, Melbourne had a temperament that was "all salt and sunshine. The world might be a futile place, but how odd it was, how fascinating, how endlessly full of interest! By now he had acquired the skill of a life-long hedonist in extracting every drop of pleasure from life that it had to offer. . . . A cynic who loved mankind, a skeptic who found life thoroughly worth living, he contrived to face the worthlessness of things, cheerfully enough."

This Melbourne was as wearily aware of the sheer futility of most human endeavor as he had been as a young man. Yet as prime minister he became one of the great leaders of his time, and a gracious mentor to the young Queen Victoria. He loved the ladies still and was named a correspondent in a famous divorce case, a charge that he beat, as he did any challenge to his honor.

Jack loved to read the story of Melbourne and other richly ornate tales of European history. Like Melbourne, he had an aristocratic conception of marriage. His was not a middle-class union in which adultery was a crime against the human trust that held the couple together during the ceaseless competition and uncertainties of life. Nor did he hold the bourgeois illusion that a mere marriage saved one from the essential isolation of life. He did not give all of himself to Jackie in part because all of himself could not be given. He had his own world-weary sense of men and women and their perpetual games. That he lay in bed reading *Lord M.* and thinking occasionally of sweet Gunilla in Sweden did not mean that he was unhappy with his wife or thought his marriage a failure.

David Cecil's book may well have been what set him to musing, scribbling notes that he surely meant only for himself. He was an American senator, but he wrote that he preferred reading European history. It was "more interesting because of [the] leisure class." American history was the "struggle to survive" and "except on Western frontier [was] not glamorous." European women were interesting because they were "women of leisure." As for American women, they were "either prostitutes or housewives . . . [and did] not play much of a role in culture or intellectual life of country." The "Civil War [was] glamorous . . . but only in Virginia because of remarkable personality of leaders."

The Civil War may have been momentarily glamorous to a gallant young officer riding a thoroughbred steed alongside General Robert E. Lee, but hardly so for a tattered, barefoot Irish peasant walking behind in the dust with his rusty musket on his shoulder and hardtack in his kit. And that is where Jack's ancestor would surely have been if he had fought for the Confederacy.

"Jack did have aristocratic instincts," reflected his old friend Charley

Bartlett. Jack, however, was a self-conscious aristocrat, and a self-conscious aristocrat is no aristocrat at all. He knew—and this rankled enormously—that he still would not be welcome in such haunts of the Brahmin elite as Boston's Somerset Club, as either a member or possibly even as a guest.

In these notes Jack pondered an essential contradiction in his own life and nature. If not for his father and the lessons of the war, he might have lived a life extreme only its pursuit of pleasure, as a snobbish dilettante, dipping into the arts, endlessly amusing himself. Even if Jack had wanted to live that sort of life, he realized that there was no stylish leisured society in America like the society he found so attractive in London.

Most of Jack's contemporaries were fascinated by the lives of the most adventurous elements of the middle class or the working class. Jack, however, found that American history "tend[s] to lack romance and drama, except the romance and drama that can be found in the story of a people and a country expanding from a beachhead to the most powerful nation on earth." For him, American political history was in part a dreary business, unlike European history, with its grand, gaudy aristocratic lives that from a distance appeared "bright with color and romance" against a gray background of the "squalor and quiet desperation" in which the masses existed.

Worldly upper-class women, especially Europeans, intrigued Jack. Even his own wife, as sophisticated as Jackie was, had a girlish, unformed quality and could not compare with the great ladies of the past. He could applaud the virtue and achievements of such exemplary American women as Jane Addams or Susan B. Anthony, but he would have preferred to have the Marquise de Pompadour or Catherine the Great as his dinner companion.

In private, Jack placed a higher value on wit than virtue, cleverness than sincerity. At times he could hardly tolerate the relentless industriousness of American life, a philosophy encapsulated for him in Longfellow's words: "Life is real! Life is earnest!/And the grave is not its goal." He much preferred the European view of Lord Byron, who, while willing to die for Greek freedom, would have wanted to do it in a properly cut coat. Jack could quote approvingly Lord Byron's axiom: "Let us have wine and women, mirth and laughter,/Sermons and soda water the day after."

Scribbling his notes in Palm Beach, Jack wrote that as bad as American women were, politicians were their equals, "rather pompous," their "humor unsophisticated." What was impressive about them was "the vituperation that surrounds them on the floor of Senate, in newspapers etc."

Jack occasionally revealed his disdain for many of his colleagues among trusted friends, blasting all the cant and stupidity that swirled around him in a cathartic purging. On one of his trips a few years later, he was flying with Ben Bradlee of *Newsweek* and Chalmers Roberts of the *Washington Post* when he set off on one of his minor rants, calling Senator Stuart Symington "Stubum" and deriding the elegant Missouri politician as little more than a well-dressed fool. If a word of what Roberts considered Jack's "gratuitous insult" had gotten out, Jack would have unnecessarily made a new enemy, even if the Washington cognoscenti chuckled at the accuracy of his invective. But it did not get out because Jack, like Bobby, had a sixth sense about which journalists could be trusted.

There were moments, though, when another Jack Kennedy rose from his seat, a man imbued with the highest aspirations of his office. In his first term in the Senate, he did few things as memorable as his speech on Vietnam in April 1954. Everything he said that day could have been extrapolated from what he had seen and felt three years before on his trip to Asia. But he said it now on the floor of the Senate, his words unparsed by expediency, his logic true, and his words audacious in their implications. He saw that in Vietnam there was no possibility of preventing a Communist takeover unless the French granted a subject people their independence. He was in favor of the $400 million aid program only if the French worked toward ending their colonial regime.

"I am frankly of the belief that no amount of American military assistance in Indochina can conquer an enemy which is everywhere and at the same time nowhere, 'an enemy of the people' which has the sympathy and covert support of the people," he told the Senate. "Moreover, without political independence for the Associated States [Vietnam], the other Asiatic nations have made it clear that they regard this as a war of colonialism; and the 'united action' which is said to be so desperately needed for victory in that area is likely to end up as unilateral action by our own country."

After the fall of Dien Bien Phu in May 1954, the French settled for an ignominious peace that split Vietnam in two, with Ho Chi Minh in the North believing that he had won only half a victory, and the South left a dissident, troubled land riven with all the scars of colonization.

In speech after speech, politicians had taught Americans to think of communism as a massive Red tide, the onrushing currents crushing everything in their wake. Jack thought, or at least part of him thought, that communism was more like a malignancy, a fungus that grew in darkness and want and could be cured or arrested by men of will and foresight. It was that image of Marxism-Leninism as "a kind of disease which can befall a transitionary society" that was promoted by MIT Professor Walt Rostow, whose thinking would influence Jack. This disease did not threaten healthy societies, or threatened them only fitfully.

Jack began to hold a very different, contradictory vision of the situation in Indochina. In some of his speeches he now saw the non-Communist

South Vietnamese as a people worthy of American help, no matter what the cost. He took a step away from what he had seen and felt and knew to embrace ideas that pandered to American political clichés and paranoia.

"Vietnam represents the cornerstone of the Free World in Southeast Asia, the keystone in the arch, the finger in the dike," Jack said in 1956. If South Vietnam was indeed the finger in the dike saving the West from drowning in a sea of communism, then its people had to be given whatever they needed, at whatever cost.

A scarlet thread ran through Jack's beliefs. This was his concern over the omnipresent threat of the Soviet Union. He saw Russian communism as a singular monolith, the world's greatest colonial power, an aggressor tempered by neither time nor opposing might. To him, this cold peace of the modern age was the continuation of war by other means.

In his private reflections, Jack could be as dark as his father. In preparation for one 1957 speech, he jotted down: "Fighting thousands of miles from home in a jungle war in the most difficult terrain in the world—man to man—with the majority of the population hostile and sullen—or fighting guerilla warfare. The more troops we send the more will pass across the frontier of the battle. It will be another Korea without the limited terrain." That was a terrifying vision, in the middle of the American century, a crippled giant slowly bleeding to death on ground it neither knew nor wanted. "The U.S. is willing to make any sacrifice on behalf of freedom," he noted, but he wondered whether "American servicemen [can] be the fighters for the whole free world, fighting every battle, in every part of the world." There, as Jack saw it, was the tortured dilemma.

Bobby was drawn to the sounds of controversy wherever he heard them, and early in 1954, they were heard nowhere in Washington louder or more stridently than on the Senate Permanent Subcommittee on Investigations, which he had only recently left. He returned this time as minority counsel on the Democratic side.

It was in some ways a curious appointment. Bobby still liked Joe McCarthy, whom most of the Democrats on the committee considered their nemesis. Bobby shared with the Wisconsin senator a hard-nosed, fundamentalist, militant anticommunism. McCarthy was a proud Catholic who stopped priests on Capitol Hill when he saw them to pay his respects. Bobby would have done the same whereas Jack would have rushed by seemingly embarrassed to be seen with them. McCarthy was tough-talking, unpretentious, fun to have up at Hyannis Port, the kind of man with whom Bobby felt comfortable.

Like McCarthy, Bobby was a hater. He hated in the way that some men loved: consumed with his hatred, he brought to it all his mental and emotional strength. He usually chose the targets of his vituperation with exquisite judgment. Seated across the committee table was one of the persistent hatreds of his life, Roy Cohn.

"Bobby did come back," Cohn recalled in his autobiography. "But . . . he didn't come back to fight McCarthy, he came back to fight me." Soon after Bobby joined the committee, Cohn writes, the new minority counsel sought out McCarthy's secretary and told her: "I want to give you a message. In these hearings, I'm going to do nothing to hurt [McCarthy]. In fact, I'm going to protect him every way I can, and I still feel exactly the same way as I always have about him. But I'm really out to get that little son of a bitch Cohn."

In what became known as the Army-McCarthy hearings, the senators were presented with compelling evidence of a conspiracy to thwart the legitimate workings of American government. The culprit, however, was not a Communist or a fellow traveler, but Roy Cohn and his boss, Joe McCarthy. Cohn had used his power to see that their colleague G. David Schine, now a private in the U.S. Army, received special treatment and was relieved of such tasks as peeling potatoes or cleaning his rifle. The more the facts were presented, the more outrageous McCarthy became in his attempts to attack those who criticized him. And the more he scowled and vilified his enemies, the more millions of Americans watching on television saw a McCarthy they had not seen before.

While this compelling drama played out, Bobby and Cohn glowered at each other across the table. Cohn recalled that "whenever I said anything or tried to do anything, he would always have this smirk on his face, which I suppose was designed to get under my skin and did get under my skin." Unlike his nemesis, Bobby had a brilliantly focused hatred that made him an immensely dangerous enemy.

Bobby understood that the sword that would reach Cohn's heart was tipped with a poisonous mixture of humor and ridicule. On June 2, 1954, Bobby wrote a memo for Democratic Senator Henry M. Jackson to help prepare him for the next day's hearings. Cohn had testified that Schine was investigating the Communist infiltration of the making of the atomic bomb. "As you were on the atomic energy committee over in the House, you might wish to pursue this matter and ask him what peculiar background and experience Mr. Schine had had to equip him to delve into this important question," Bobby wrote. "I do not think certainly there is anything against youth, but the point is that I don't believe that Mr. Schine had any experience in atomic or hydrogen bomb affairs. . . . It seems to me you could make the whole business rather ridiculous if you approach the questioning of Cohn on this matter in rather an incredulous way, if you know what I mean."

Jackson knew what Bobby meant. A week later Jackson questioned Cohn about Schine purportedly setting up a worldwide psychological warfare program. Jackson mixed his words with a fatal dose of sarcasm, his disdain for Cohn unmistakable. For weeks, Cohn had been watching Bobby sliding questions over to the Democratic senators or whispering in their ears, and he had little doubt who was the architect of this mockery. Just after the day's hearings had ended and the television cameras were shut down, Cohn walked over to Bobby and berated him.

"I want you to tell Jackson that we are going to get to him on Monday," Cohn said, as Bobby remembered. It was Bobby, though, who was the major target of Cohn's ire. "You hate me!" Cohn exclaimed.

"If I hate or dislike anyone, it's justified," Bobby replied.

"Do you want to fight?" Cohn asked, his voice loud enough that reporters turned and listened.

"You can't get away with it, Cohn!" Bobby exclaimed, standing toe to toe with the diminutive attorney. "You tried it with McCarthy, and you tried it with the Army. You can't do it."

The next day Bobby was the lead story in the largest newspaper in America. "Cohn, Kennedy Near Blows 'Hate' Clash" read the New York *Daily News* headline. Although the paper attempted to tell the story with requisite balance, the reality was that Bobby had defeated Cohn much worse than if he had fought him physically and left him lying sprawled out and bloodied.

Bobby had discovered an irresistible weapon. He had taken the rude put-downs that were the essence of humor at the Kennedy dinner table and sharpened them into a brutally disdainful sarcasm. The shaft of this weapon, though, was barbed on both ends, at times hurting the one who wielded it as much as its victims. It turned opponents on one issue into enemies who never forgot.

When Bobby returned in the evening to his rented house on S Street in nothwest Washington, he was not greeted by the kind of refined setting that Jack met when he arrived home a few blocks away. Bobby and Ethel had three children by now, and seemingly twice as many dogs. Both the children and the dogs had the run of the house, jumping up and down on furniture and tearing up and down the narrow stairs. There was no drinking or smoking allowed and some of the dinner guests would have sold their birthright for a glass of sherry.

Bobby and Ethel were living in the middle of tree-lined, cobblestoned Georgetown, the preferred bastion of the old Washingtonians known as

"Cave Dwellers." Some of these genteel folk took inordinate pleasure in spotting such outrages as a congressman's wife eating her soufflé with a soup spoon. From all appearances, Bobby and Ethel didn't give a damn. It was a mark of immense audacity for them to live as they did, though perhaps less so since they appeared totally unaware that theirs was an unusual household. There was much talk about these eccentric Kennedys, but their behavior was more Skakel than Kennedy.

Bobby had an awesome toughness of mind and body. When Teddy called him in the fall of 1953 and invited him up to watch a football game at Yale, Bobby knew that his brother would not be playing before tens of thousands in Saturday's Harvard-Yale game. His brother had just returned to Harvard. Probation meant that for a year he could not play varsity football. Instead, Teddy had joined the team at Winthrop House and traveled down to New Haven for the annual game against Davenport, one of Yale's residential colleges.

Teddy had gone through a couple of tough years, and it was a measure of Bobby's love for his brother that he drove up from New York, where he was working on the Hoover Commission's plan to reorganize government. Bobby was the least likely of spectators, however, and as soon as he got to New Haven he talked Teddy into finding another uniform so that he could join his brother on the field.

The two brothers stood stalwartly together on the line, Bobby at end and Teddy at tackle, playing with a ferocity rare in such a modest competition. Bobby was on the verge of his twenty-eighth birthday and out of shape, but his opponents across the line would not have known. "Bobby . . . could play in that league like a tiger," Teddy recalled. "I mean, he was very good, and it was great fun." Was it any wonder, then, that to young Teddy, brotherly love was the highest love, and that he saw in his two big brothers the very models of what a man should be?

Life was decidedly better now for Teddy. For the most part, the academic life of Harvard was just a dreary routine that he passed through on his way to good times. He was a drudge who managed to earn more Bs than Cs, doing far better in his grades than his father had done. He took one course with Professor Arthur Holcombe on the Constitutional Convention that excited him. Not only was Holcombe a brilliant professor teaching his last year, but he had also taught Teddy's father and brothers. The legendary professor may have moved Teddy intellectually, but he was unimpressed by his student. "I think academic activities came out third [after athletic and social activities]," Holcombe reflected from retirement. "He did just what was necessary to remain in good standing."

Teddy was not much of a student, but he had willing helpers, including Bobby, who shipped some of his old term papers up to Harvard. Bobby admitted that one of them, an essay on the Ninth and Tenth Amendments, "seems a little technical, but perhaps you can water it down a little bit and still be able to use it." Teddy had been thrown out of college for cheating, but neither brother seemed to understand that copying term papers was as serious an offense as cheating on an exam. They lived in their own moral universe and had a code singular unto themselves.

At Harvard, for the first time in his life, Teddy had good friends, most of them his fellow football players. Several of these college buddies became lifetime friends whose identification with Teddy and his life was almost total. He brought them down to Hyannis Port, where they played spirited games of touch football and ate immense quantities of good food prepared by the Kennedys' cook. Teddy didn't brag about his famous family. During freshman year, it wasn't until one of his new friends, Claude Hooton Jr., noticed a caricature in the *Boston Globe* that looked surprisingly like his classmate that he realized that Teddy was not an heir to the Kennedy Department Stores on the Cape but heir to something a bit larger. Teddy was by most measures a good and thoughtful friend whose graciousness sometimes even embarrassed Hooton. After a rugby match Hooton hurriedly showered and put on his tux for the big dance that evening. When Hooton picked up his friend, Teddy was carrying two corsages, one for his date and one for Hooton's.

Teddy could drink more beer than any of his buddies and still be up at dawn for a sail or a tennis game while his friends lay in bed, pillows over their heads, trying to quell their throbbing hangovers. He wasn't the sort who went looking for a fight, but if a fight came looking for him, he didn't duck down the alley. One summer sailing with David Hackett in Maine, he was rowing a dinghy to their boat when a smart aleck in a yacht made the mistake of shouting to Teddy that he should row faster, and then challenging him. Teddy and Hackett scampered up onto the yacht and, as the occupants hurried on deck, threw them, one after another, into the Atlantic.

Anyone who knew Teddy would have laughed at the idea that the man was a pallid inheritor, the last and least of the Kennedys, feeding off the scraps of heritage. He was living an intrepid life, one summer going out west to work as a forest ranger, another year heading out to be a crewman on a race from California to Hawaii. He and Hooton went off one summer to teach water skiing in California. Driving back east they were nearly arrested when they attempted to sell their auto in New Orleans without title papers. Teddy called Jack's office, where Bobby said he had never heard of his brother. Bobby called back later and they were freed.

Teddy did not have Jack's sophisticated charm or his subtle gamesmanship. He developed a rating system from A to F that he applied to every woman he met, including his own sisters. Even if a woman was an A, he was soon ready to move on, to go back with his buddies, before heading out another evening to score again.

It bothered Bobby that Teddy had done all the penance he was going to do and back at Harvard was majoring in football and good times. Teddy's self-indulgent conduct so rankled Bobby that on occasion he saw fit to lecture his kid brother. "I talked to Dad last night," Bobby wrote Teddy in January 1955, "and he agreed with me that you really made a fool of yourself New Year's Eve." That may have been, but it was hardly brotherly to go running off to their father, all clucking commiseration, the contrast between upstanding Bobby and rascally Teddy all too stark.

There was always an edge to Bobby's humor, a sting to any balm he applied. Jack's attitude toward his kid brother was far different. Jack was a man who, by Bobby's estimate, was in physical pain half his life. To Jack, physical well-being took on a spiritual quality of which those who possessed it were rarely aware. Even in this transitory world of youth, where health was as common as the very air, Teddy seemed like the benchmark of health. As their mother observed her two sons, she thought that "Teddy had the strength and the vitality and Jack rather envied him his health and capacity to take part in all these sports."

Not just envy brought Jack up to Boston on those autumn weekends to watch Teddy playing end on the varsity. Earning a letter was something that he and his brother Joe had dreamed of. His father sat next to him celebrating whenever Teddy took the field. During his junior year, Teddy's father and brothers spent some disappointing Saturdays watching him sit on the bench. When it came time for the final game against Yale in the 1954 season, he had played end for only fifty-six minutes, four minutes short of the hour necessary to win the coveted football letter that meant more to Teddy than any other Harvard honor. It was still the first half when Teddy got in the game for the first time, and it was almost certain that he would have his Crimson letter. As soon as Teddy took the field, however, a Yale back ran sixty-two yards around his end. It was not Teddy's fault, but he was taken out and never got back in the game.

"So I ended up at the end of the season with fifty-nine minutes, fifty-eight seconds," Teddy recalled years later with a memory he has for few other things in his past. "And at the last minute Harvard scored to win the game. So the place was euphoric, and I had to be euphoric. And my father's in the stands. I know I've disappointed him, disappointed my brothers who were up there. They're trying to cheer me up, and I'm trying to be happy because the team wins, and I'm part of the team. These enormous emotions going through a person at that time."

The following year Teddy had played fifty-six minutes when it all came down to the final game against Yale in New Haven. Out there on that field, Teddy was living the life that both Jack and Joe Jr. had so much wanted to live. To the other Kennedy men, it hardly seemed to matter that, heading into its final game, this miserable Harvard team had won only twice, while losing six games.

Teddy was inspired by the fact that his brothers and father were often up in the stands. "I think . . . [Jack's] most profound influence was not so much when I was a teenager or a pre-teenager but was the later years when I was in high school and through college and the post-college years," Teddy reflected. "I suppose the relationship with him was as much as a friend as a brother. There was a fourteen years' difference—and he was remarkably interested in all the different kinds of things I was involved in. He came to the sports events when I was in school. He was interested in the subjects I took in college and the teachers I had. I think he enjoyed vicariously a lot of the things I was doing—football in college. . . . He could look at the light side of situations which made things not look as grim."

The way the men of Cambridge saw things, a victory here in New Haven in November would mean a successful 1955 season. Jack and Bobby sat in the stands at the Yale Bowl. So did Joe, who had brought along two railroad cars full of friends, associates, his New York staff, even the family cook. More than fifty-five thousand spectators sat in the open stands on this blustery day as the young men fought their way up and down the icy turf, sliding into the piles of snow on the sidelines. Several times the referees had to stop fistfights between opposing players.

As Joe watched Teddy running up and down the frozen field, the world below did not look that different from what it was four and a half decades earlier when Joe had first entered the gates of Harvard. Teddy had to prove his courage and his competitive zeal so that one day he would demonstrate those qualities on a larger field. Teddy may have shamed his family name by being caught cheating, but down on the field on this glorious afternoon he was the man Joe knew he must be.

The stadium trembled with cheering, and for three hours nothing mattered more in life than that the Harvard men should prove not wanting. It all may have been diminished from what it was in Joe's day, but none there that afternoon could have imagined that within a few years the crowds would dwindle, the bands would play mocking airs, and those students who still bothered to come would for the most part watch from an ironical distance.

Joe had wanted few things in life as much as to see one of his sons on that field, not shuttled in for a play or two, but out there quarter after quarter. Even on the most splendidly exuberant of occasions, Joe's thoughts were often haunted by memories of Joe Jr. On just such a day, Joe had watched as his oldest son sat on the bench, never getting into the Yale game for even a play to win the letter that he had struggled so to win.

Blasts of cold wind coursed across the field. In the third quarter, as Yale led 14–0, Harvard pushed up the field to within striking distance of the end zone. Teddy stood on the end of the line, waiting for the ball to be hiked. He was notoriously slow of foot, hardly the material for a great end. But at six feet two and 210 pounds, he had a beautifully proportioned athlete's body and was fearless in the way he leaped after a ball. He also was so wondrously resilient, and of such good cheer, that even when he was viciously blocked, he jumped up and patted his opponent on the behind before trotting back to the huddle.

Teddy ran a few yards beyond the line of scrimmage, turned, caught a short pass, and ran into the end zone for Harvard's first touchdown. The stadium erupted in applause, no one cheering louder than Teddy's father and brothers. For all that the Kennedy men had tried for close to half a century, this was the first, last, and only great athletic moment in their history of playing football at Harvard. The Crimson scored no more, but when Joe, Bobby, and Jack entered the locker room after the game, they were not there to commiserate with a defeated team but to celebrate Teddy's triumph. And when they shouted too loud and boasted too much, he shushed them as best he could, lest they bother his dispirited teammates. "My brothers and I were euphoric, and Harvard was depressed," Teddy recalled. "It's a great part of life, too . . . teaching you about the ups and downs and sort of coming back, and the irony of defeat and victory."

Joe didn't have to hide his high spirits on the train ride back to New York City. He replayed the touchdown over and over again with his friends, but none of them could understand what a moment of vindication and sweet victory this was. So many years before, when little Jack had nearly died, Joe had made his bet with God that if his son lived he would contribute half of his worldly wealth to charity. And today, whether or not he had made another bet with God, he decided to give another tip of the hat to the Supreme Being. Joe turned to Janet Des Rosiers, his young secretary, who was seated beside him. "Janet," he said, "remind me in the morning to tell Frank [Morrissey] that I'm going to give a million and a half dollars to that home for the elderly outside Boston."

Joe had met Des Rosiers back in the fall of 1948 in his suite at the Ritz-Carleton Hotel in Boston, where she interviewed to be his new secretary. By the time Des Rosiers walked into the living room, Joe had already talked to at least forty candidates. Twenty-four-year-old Des Rosiers had a sweet sensuality, a gentle demeanor, modest downcast eyes, brown hair, and a soft voice that one had to strain to hear. Joe told her later that as soon as he saw her he decided to forget the other candidates.

For Des Rosiers, it was no small commitment to take on this job at a salary far from munificent. She was a single woman. Her father had died when she was only twelve, taking away most of her childhood with him. From then on, she had learned to live as an adult, helping to take care of her four younger brothers and sisters. She had worked at various secretarial jobs, but she had never had much time to date or hang out with a gang of young friends. Now she was expected to move down to a small apartment in Hyannis Port that Joe provided for her. In the winter she would have a cottage in West Palm Beach. It would be even harder to develop a life apart from her work.

Each morning at precisely nine o'clock, Des Rosiers had to be at the Kennedy house in Hyannis Port when Joe had finished his horseback ride and breakfast. He walked into the living room wearing a stylish lounge suit, sat down in a large chintz chair, and began to answer letters, dictate, and talk for lengthy periods on the phone. Since Kathleen's death, he had backed off from many of his business activities and was semiretired.

Joe called his new secretary in the evenings and began to make gentle overtures. Des Rosiers did not think that there was anything crudely predatory in a sixty-year-old man propositioning an employee less than half his age. Joe was elegantly appointed, with manners as impeccable as his dress. He was strong and well built and youthful. He was not suggestive or vulgar. He was an astute student of human beings and all their vulnerabilities and he sang a gentle song of wooing.

"He'd say, 'Well, you know, we've got our house in West Palm Beach,'" Des Rosiers recalled. "Because I had the house that every other secretary rented. It was just a little cute cottage in West Palm Beach. And he started calling it 'our house.' Well, you'd have to be pretty stupid not to understand what he has in mind. Little innuendos like that went on the first two months. I didn't see that much of him. And then I went to Florida in December, and that's when it started. He was marvelous, you know. It would be very difficult for any woman to not succumb to his charm. He had a lot of charisma. But it was a shock to me because I was young and I didn't take that job with this in mind. My God, you can imagine. It's hard to fight off a man of his authority and experience and his ability, so I was gone right from the beginning."

Des Rosiers's account of her first evening with Joe in West Palm Beach three months after meeting him has an eerie similarity to Gloria Swanson's account of her initial sexual encounter with Joe. Gloria recalled that Joe had come rushing into her suite, practically tore her clothes off, and "kept insisting in a drawn-out moan, 'No longer, no longer, now.' He was like a roped horse, rough, arduous, racing to be free." Des Rosiers recalled that Joe had come rushing into her little house one evening at eight o'clock and began undressing her. In Des Rosiers's case, Joe discovered a bounty that Gloria did not possess. His new mistress was a virgin.

Joe was possessive with his young mistress in a way he had never been about his wife. He wanted to know where she was and what she was doing and whom she was seeing. "One night I wasn't home when he called," Des Rosiers remembered. "And he said, 'Where were you?' And I said, 'Well, I stopped off at the driving range to hit a few golf balls.' He was mad. He didn't like that at all. I think he was afraid maybe that I'd get attracted to somebody else or something, I don't know. But he wanted me to work and go home, be there when he called me at night, and we'd speak for about an hour after being together all day."

Joe was a man of terrible vulnerabilities that he would never admit to another man. But Des Rosiers would never seek to barter his words for advantage. "He confided to me about how he wanted to end his life," Des Rosiers said. "He told me practically in tears that he never wanted to end up dependent on people to have to take care of him. He just wanted to die very quickly and quietly and get it over with. He didn't want to have a stroke or something like that. And he said, 'I don't want to linger and be dependent.'

"Yet he wouldn't give me any freedom. Once in Hyannis Port, Rose was not there, and I went to a party. And when I got back, I think it was maybe nine, ten o'clock, he was in bed and he was crying, and he said, 'Why did you do that to me? Why did you do that to me?' And he was so upset because I went to this party. He was dependent on me because he needed somebody around him. And in all my innocence and immaturity, I happened to fill the bill for him at the time. I was a very nice girl. And I wish I knew then what I know now in life."

Des Rosiers asked for nothing from Joe, not jewels or furs, not secret accounts or promises, not a raise or a bonus, nothing, and Joe gave her precisely what she asked for. She identified with Joe and maneuvered comfortably around Rose, whose gift of seeing only what she wanted to see had reached such rarefied heights that she appeared oblivious to the fact that Joe's mistress was in her own house. "Have a long, happy holiday, dear Janet, and be assured always of our affection and our deep gratitude," Rose wrote Des Rosiers in June 1956 as the secretary headed off on her vacation. Rose kept to herself, going off each morning to mass, keeping little notices pinned on to her blouse to remind her of what she had to do.

In relentlessly social Palm Beach, Joe and Rose rarely went out and almost never entertained. Indeed, in an eight-year period, Des Rosiers remembers only one dinner party at the house, and that was for the duke and duchess of Windsor. In large part because Joe never went out in Palm Beach, a legend grew up that he and Rose had been socially ostracized, banned from the exclusive clubs. This was simply not the case. The Kennedys had been members of the restricted Everglades Club, though Joe rarely went there.

One summer Des Rosiers was riding with Joe in the chauffeured Rolls-Royce along the French Riviera for an afternoon of gambling. Joe leaned toward Des Rosiers and whispered, "You know, I would like to divorce Rose and marry you." Des Rosiers reached over to hold Joe's hand, but he pulled it back so the chauffeur would not see this mark of intimacy. Des Rosiers took this as a sign of Joe's discretion, but if he could not even accept this gesture, he could hardly leave his wife of four decades. He spoke one other time of marriage, but it is unlikely that he was serious. He was a man who could place a value on anything, and he surely understood that there are few cheaper gifts than a promise you did not keep. He had not forgotten that years before he had made another promise to Rose, an arrangement that had allowed them together to build one of the great American families. A divorce would not only bring shame to his deeply religious wife but also might harm his sons' brilliant futures. That he would not risk, not for a mere woman.

The relationship between Joe and Rose was as proper as that between a king and queen. "She was super-concerned with her looks, her body, and her clothes were really an obsession with her," Des Rosiers reflected. "But there was no depth to her, no womanly sexuality. But she presented a very nice picture. She believed that every minute of your life had to be a learning experience. She brought up her children and grandchildren that way. She tried to bring me up that way because she would often say, 'Well, you should do this, you should do that.' And I felt like saying, 'Well, I'm doing the thing I shouldn't do.'"

Joe was a vibrant vital man in his sixties with a mistress less than half his age. It was his son whom he had groomed for the highest of offices who had an old man's gait, hobbling around like a cripple. Jack was not a man to broadcast his discomfort, but his back pain had become so overwhelming that he had become increasingly irritable. He could hardly make it over to the floor of the Senate.

To Jack, life on crutches was no life at all, and he set out to find a remedy so that he could live as he felt he must live. In the summer of 1954, he traveled to Boston to consult with Dr. Elmer C. Bartels and other specialists at the Lahey Clinic about an operation. Dr. Bartels had been treating Jack for seven years. The doctor had a realistic, if disheartening, appraisal of his patient's prospects. Bartels believed that Jack had been born with an unstable back and that his Addison's disease made an operation even less feasible. Bartels was eminently aware of the limitations of his profession and believed that Jack would simply have to exercise carefully and live a sedentary life.

"I don't know if the words 'back pain' follow," Dr. Bartels reflected, emphasizing that Jack could have lived with his condition. "I think discomfort. He never took care of his back. He'd come up to Boston and go scrimmage with the Boston College football team. It just wasn't his temperament to take care of himself. He played touch football, and you can certainly injure your back playing that."

Jack went shopping in New York for a doctor willing to attempt the dangerous, radical surgery. The physicians who agreed to perform the surgery were proposing to break the bone and then reset it, hoping that it would grow back properly. The doctors realized, as they wrote later, that it was "deemed dangerous to proceed with these operations."

Jack was thirty-seven years old. He had a dazzling political career, a beautiful young wife, and a fortune. The operation would probably improve his back only marginally; it was also possible that the operation would not improve it at all, or would kill him. Weighing all these factors, another man would have turned away and limped back to Washington. That Jack did not do so suggests the magnitude of his ambition, the strength of his identity as a vibrant sexual being, and the magnitude of his pain. For all those reasons, he was willing to roll the dice marked "life" and "death."

Jack wanted the whole business put behind him. He insisted that the two fusions be done the same day, rather than taking the more conservative approach of having two separate operations. The operation, performed on October 21, 1954, by Dr. Philip D. Wilson at New York's Hospital for Special Surgery, with Ephraim Shorr there in an advisory capacity, was deemed a success. Three days later, however, Jack was stricken with a urinary tract infection and slipped into a coma.

The priests arrived and administered the last rites. The deathwatch began. Jack's father was devastated. Joe's sons were the last half-drained reservoir of his faith, and that day Krock saw Joe cry as he had not cried since the death of Joe Jr. He had hardly wiped away his tears when Jack began slowly to revive, but with horrifying setbacks. Jackie was sitting beside the bed when he received a blood transfusion to which he reacted adversely, his whole face puffing up.

Jack suffered weeks of what only a resolute optimist would have called convalescence. He lay there with a draining, open, eight-inch wound on his back. "So I was just sitting really in bed with a lot of acute discomfort," he recalled. "I didn't read much because it was God-damned uncomfortable, and then I was being woken up every half hour for the first two weeks to do this test on my blood."

It wasn't just his back that was troubling Jack, but a recurrence of the nonspecific prostatitis that probably was a venereal infection he had picked up at Harvard and that had periodically plagued him ever since. He had been blessedly free of the problem during his honeymoon with Jackie, but upon returning to Washington he had begun to suffer again. On occasion he had a large number of "pus cells" in his secretion. When he entered the hospital, he had what his physician specialist, Dr. Thomas A. Morrissey, called an apparent "acute attack of mild prostatitis . . . the prostate was tender, somewhat swollen and there was a moderate urinary urgency." The sitz baths and Pyridium, a urinary tract analgesic prescribed by the physician's brother and partner, Dr. John H. Morrissey, seemed to have worked, but in the aftermath of the operation, the problem and all its pain were back again.

When Jack had spent weeks in bed as a teenager at the Mayo Clinic, he had been able to weave a sexual fantasy for his friend Billings that projected him away from his wounded body. As he lay in a hospital bed this time, he attempted to distance himself as best he could from his broken body and constant pain. He had an upside-down picture of Marilyn Monroe pinned to the door as if to remind himself of what lay out there in the world beyond his hospital bed. He had stacks of books on the floor to remind him of other aspects of the world outside. He was often on the phone, even talking when the nurses changed his dressing or yet another doctor poked at his dormant form. He had many guests too, especially young women. One of them, Priscilla Johnson, was a student of Russian affairs at Columbia University who had worked for a few weeks in Jack's Senate office. He had not managed to bed the twenty-three-year-old student, but he called her periodically and had met her several times in New York. "Tell them you're my sister," Jack told her on the phone. "You got to be a relative to get in here." Johnson did as she was told. "I've never seen a man with so many sisters," the receptionist said as she pointed the attractive young woman down the hall.

Jack thought at times of Gunilla, especially when a dark-haired Swedish nurse entered his room. He was terribly sick, but he fantasized about driving up to that house high above the Riviera and meeting Gunilla there. "We stay in session in Washington until the end of July and then I return to the mountains of Cagnes," he wrote her in December.

Most of the time there was no respite from his pain. "You could see that he [was in pain] by the look in his eyes," said Janet Auchincloss, his mother-in-law. "He still would always talk about the world you were in and not tell you about his operation, which is unusual."

While he lay in bed, the conflict over McCarthy rose to a dramatic call for a censure vote in the Senate. Several years afterward Jack scoffed at those who had attempted to turn the conflict over Joe McCarthy into an important moral issue. "I think your attitude toward this [McCarthy] thing right through was you seemed to want to divest it of any great moral, ethical [dimension] to a literal specific thing," Jack's authorized biographer, James MacGregor Burns, told him.

"That's right," Jack replied. "Well, I think that's right. . . . Hell, if you get into the question of just disapproving of senators, you're going to be in some difficulty. . . . I don't think, Jim, you could probably tell me very well what McCarthy ultimately was censured for."

Jack only had to look among his New England colleagues to find senators who would have told the junior senator from Massachusetts why McCarthy was being censured. Senator Margaret Chase Smith, a Republican from Maine, was spending most of her term in a lone crusade against McCarthy. Ralph Flanders, a gritty, taciturn Vermonter, was another unlikely hero of the anti-McCarthy crusade. Jack had only to look in Boston at his onetime campaign manager and former friend, Mark Dalton, to see a citizen rising up against McCarthy. Dalton had been sitting in the front row when McCarthy had brought his investigative hearings to Boston. He had been so appalled at what he saw that he ended up running for the Senate himself in 1954. He got nowhere, however, on his anti-McCarthy platform and believed that the Kennedy family was behind much of the negative press he received.

Earlier that year the television networks broadcast the Army-McCarthy hearings with Bobby sitting there as minority counsel. Americans watched mesmerized as attorney Joseph Welch told McCarthy in a voice nearly breaking with emotion: "Until this moment, Senator, I never really gauged your cruelty or your recklessness." Neither had most Americans. Although at the beginning of the year a majority had supported McCarthy, by the end of the hearings most Americans opposed him.

In Congress, the toughest votes are often delayed by tacit agreement long enough so that what once would have been a heroic stance becomes merely expedient. That was decidedly the case for most of Jack's Democratic colleagues when, on December 2, 1954, the Senate voted to censure McCarthy, 67–22. Jack shared the distinction of not voting with only one other senator, Alexander Wiley of Wisconsin. The moderate Republican wangled an invitation to an economic conference in Brazil because "it would be a nice way to get out of the McCarthy business."

For all his intellectual detachment, Jack was not a man to retreat into narrow, self-serving legalisms to defend his actions. But this time retreat he did, running away from the great moral issue of the decade into a thicket of justifications. "So I was rather in ill grace personally to be around hollering about what McCarthy had done in 1952 or 1951 when my brother had been on the staff in 1953," he rationalized to Burns. "That is really the guts of the matter." If Jack could not make a distinction between Bobby's well-documented investigation of shipping to China and McCarthy's egregious excesses, then he suffered from a political myopia of disastrous proportions. Beyond that, his brother had managed to change sides in the midst of battle and by his actions had helped push the Senate toward this historic vote.

If Jack had gone along with all the other Democratic senators, his vote would not have been long remembered, and he would surely not have been saluted for courage the way his critics condemned him now for cowardice. He bristled at the charge, though only in an unpublished interview with Burns could he display the full volcanic range of his displeasure:

> Am I going to vote in the hospital when I was God damn sick on how he treated the censure committee and how he treated Arthur Watkins on the floor of the Senate? They say McCarthy was an obvious son of a bitch, and we all should have been up making speeches against him . . . but who did it. . . . [Hubert] Humphrey never made a single speech, nor did [Stuart] Symington until the gun was put at his head . . . when it actually got to a personal dispute with he and Joe, but it wasn't on moral grounds. Nor did any other senators, except [Mike] Monroney and [Herbert] Lehman and [Thomas] Hennings and [J. William] Fulbright. . . . Now I think that anybody who is deathly against McCarthy in the beginning has a right to say that my sensitivity in regard to the abuse of the civil rights and liberties of people were not sufficiently attuned to recognize the real menace of McCarthy. That's a reasonable indictment, and I don't mind accepting it . . . though I would put myself with 90 percent of the senators, which is no excuse, but which at least puts it in proportion. But they have moved beyond that in their criticism. They almost associate me with McCarthy.

Just before Christmas, Jack was wrapped up in a plaid blanket, placed on a stretcher, and taken by ambulance to La Guardia Airport to be flown down to Palm Beach. A nurse flew with him, along with Jackie and Bobby, who hoped that he would be able to recuperate in the Florida sun. The gaping, open wound in Jack's back was not closing. Not only was he in pain, but his body remained vulnerable. He limped along the beach beside his old friend Chuck Spalding. "How is it now?" he asked. "Is any stuff running out of it?"

Spalding had an astute sense of Jack's emotional condition, and he tried to change the subject, playing the buffoon, or the wit, whatever worked to get Jack thinking about something else. Jack just wasn't getting better, and in February he was flown up once again to New York. He was so sick that the last rites were administered once again over his dormant form. And while surgeons operated, removing the metal plate in his back and doing a bone graft, his family huddled outside united in prayer.

This time the operation seemed successful and Jack was able to walk out of the hospital and return to Palm Beach for more months of recovery. By mid-February, he was able to go through his mail and dictate a lengthy memorandum to his Washington staff. He was already focused on his reelection campaign three years hence. He told his staff that for "anybody we did a favor for . . . take their name for our political file so that we will have . . . it in 1958 when we need it for the campaign . . . and go back through all the letters since we started in Congress and see the people from the correspondence . . . that would help us in future campaigns."

Jack was not a man of natural solicitousness, but he recognized that politics was a matter of relationships. The cheapest and best way to win votes was often through letters or cards or the baby books that he sent out to new parents in Massachusetts. He told his staff how to answer each letter and whether to address the person by his or her first name. He told his aides how to handle journalists: "I don't know Raymond Lajoie . . . and I don't know really what sort of story he wants to write—definitely can't use any pictures from here—no quotes— . . . I don't think there is anything in it so that Ted can turn him off tactfully."

Jackie was there with Jack in Florida. Marriages are tested in the bad times, not the good. And the bad times always come, though usually not as early as they did in Jack and Jackie's marriage. Jackie had hardly said her wedding vows before her new and much older husband became seriously ill. In that dreadful fall of 1954, she not only saw her husband almost die but miscarried what would have been their first child. In the months since then, she had spent most of her time succoring him. She changed the dressing on his open, draining wound. She put his slippers and socks on his feet and sat with him for hours. She took dictation and helped him research a book he was writing. She flew to Washington and in Virginia found a white Georgian mansion known as Hickory Hill where she and Jack would move once he recuperated. Jackie proved herself a devoted wife without a hint of the sulkiness that previously had so perturbed her husband.

As Jackie watched over her husband, Jack kept up a correspondence with

Gunilla, trying to arrange an assignation with her in the summer when he hoped to be recovered. The Swedish woman knew that Jack was married, but she believed that "he needed someone to love. It was all about love, unconditional, passionate love. I knew in the depths of my being that this was exactly the kind of love I could give him, and that he would give back to me." Jack had already found a trusting, caring love in Jackie, as her conduct these last months should have told him. The problem was not that he had a wife incapable of deep love, but that he was apparently incapable of returning that emotion for more than a few days in a distant clime.

"I am anxious to see you," Jack wrote Gunilla that spring. "Is it not strange after all these months? Perhaps at first it shall be a little difficult as we shall be strangers—but not strangers—and I am sure it will all work out and I still think that though it is a long way to Gunilla—it is worth it."

In April, Jack was ready to fly back to Washington when his crutch collapsed, and he reinjured himself. Five weeks later, on May 24, 1955, Jack walked back into the Senate chamber and received a hearty round of applause. He appeared to be fully recuperated, yet his back and neck troubled him so much that he could not turn to one side without moving his entire body. He looked tanned and fit, although he was severely anemic. His cholesterol level was about 350. His left leg was supposedly slightly shorter than the right, another potential source of back pain. He had set aside his crutches for these public moments, but they were never far from him. He had used them so long that he had developed calluses under his armpits.

For two decades now, Jack had been prodded and probed by an endless series of doctors, some of whom had made faulty diagnoses while others had prescribed treatments that only exacerbated his problems. Jack was like many chronic pain sufferers, wandering from one specialist to the next, one promise to the next, his pain traveling with him. Like the rest of his family, Jack was a believer in credentials, always asking who was best and going to them for counsel.

In this instance, Dr. Ephraim Shorr, a leading expert on adrenal insufficiency, felt that he could contribute no more to Jack's treatment. So the doctor introduced Jack to Dr. Janet Travell, a well-known expert in pain management. Only three days after his return to the Senate, he flew up to New York to meet with her.

Dr. Travell thought his condition was serious enough that she had him admitted immediately to New York Hospital, where his weeklong stay was publicly described as "routine therapy." One of his first directives to Travell was to keep his condition from his young bride. "I don't want her to think she married either an old man or a cripple," Jack told the doctor. Jack had found some momentary respite from his pain sitting in an old-fashioned cane-bottomed rocking chair in the doctor's office. Travell wrestled the chair into her big sedan and delivered it to her new patient in the hospital.

Jack had less than sanguine opinions of women who entered what he considered manly professions, but the New York physician inordinately impressed him. She combined a woman's gentle touch with an authoritative voice backed by ample credentials, and Jack felt that finally he had found a doctor who could help him. Travell was no miracle worker. In July, Jack was back in the hospital twice, for a week at New England Baptist Hospital, and later in the month for five days in New York at the Hospital for Special Surgery.

Jack became so dependent on Dr. Travell that he sometimes flew up weekly to see her. The specialist measured Jack, and when she diagnosed that one leg was shorter than the other, she prescribed lifts for his left shoes. She firmed up the seating on the upholstered chairs in Palm Beach and did what she could to make the rest of the furniture tolerable for the Massachusetts senator. Most important, though, Dr. Travell was a militant proponent of the wide efficacy of procaine injections, known more commonly as novocaine or Novocain (its trade name), a drug Jack had first used in 1944.

Novocaine is used primarily as an anesthetic in minor surgeries and dental procedures. For a person of Jack's medical history, especially with his periodic asthma attacks, the drug could cause drowsiness or tremors. Dr. Travell injected novocaine in what she called "trigger areas." It was a technique that Jack could learn to do by himself, shooting the painkiller into areas that troubled him.

"She began to really fix me up by this business of novocaine," Jack said in 1959, when he was still her patient. "I think it's so outrageous these doctors—if I had met her fifteen years ago, I probably wouldn't have any trouble. . . . She's been this pioneer in this business of muscular spasms, and of putting novocaine in, which relaxes the spasm, which eases and permits blood to flow and, therefore, she does that enough and then the muscle relaxes. Otherwise, muscles stay in spasms for years, and they gradually get so you get a stiffening."

One of the problems with the drug was that it quickly wore off, and the temptation was to inject it all over again. Although Dr. Travell was widely and positively known, she had her skeptics among the most esteemed leaders of her profession. "The use of novocaine in this way is of long standing and very widespread knowledge," wrote Dr. Alexander Preston to Jack's authorized biographer, James MacGregor Burns, in 1959.

> For some reason or other Travell has developed a special reputation. [Dr.] Bayard Williams and [Dr.] Jim Leland whom I talked to about it, feel that there is a little more than goes through the needle or meets the eye in the results she obtains . . . in many medical procedures, there is an overlay of patient-doctor identification, hope, [and] confidence . . . which brings a success to many procedures that would fall short, of themselves. Apparently . . . she has this power whether it be psychological or mystic, which brings more than average success to an ordinary procedure. . . . Of course, anybody that can take an unsuccessful orthopedic operative casualty and reconstitute it by any means whatever deserves great credit. Apparently this is what she does to many and what she has done to Kennedy. Anyone after two unsuccessful major operations on the spine has to have something to hook onto and she apparently provided the fix.

When the Senate adjourned early in August, Jack was on crutches, a condition that merited his spending the next couple of months at Hyannis Port recuperating. Instead, he sailed on the USS *United States* to Le Havre. From there he traveled to meet Gunilla at Skånegården, a hotel in Båstad, a charming Swedish resort town. His old friend Torby Macdonald, now a freshman congressman from Massachusetts, accompanied Jack. Torby was not only a boon companion but, as Joe had taught his sons, a defense against being compromised or blackmailed.

Gunilla rushed into Jack's hotel room and soon fell into his arms. "Tender? Oh, he was tender," Von Post recalled. "Very tender. He was really wonderful to me. He never talked about his brother who died. We didn't have time. It was only love. It really was. He didn't talk about Jackie. Only Torby talked about Jackie, how she had not been in the hospital as much as she should have. And his mother he never talked about either. It was his father he talked about. I think he looked up to his father very very much. And he wanted to please his father. He was pleasing him. He wanted to please everybody. And also, he was very witty and funny and amusing, and we laughed with him. Oh yes. He was on crutches and with all his suffering, he was always smiling. We drove around the south of Sweden, and he was singing 'I Love Paris in the Springtime,' and we sang together. Like a dream. And I think it was one of the few weeks in his life that he was free as a bird."

Torby met a Swedish woman who became his lover, and the two couples drove around the country. Jack had once again traveled up that open road to freedom, but when the week was over, he flew to Nice and from there to Hotel du Cap. The last time he had been there he had sat in the moonlight kissing Gunilla and telling her of his love. This time he waited in the daylight for Jackie to arrive. "I just got word today that my wife and sister are coming here," he wrote Gunilla, addressing her as "Dearest," a salutation he had not used before. "It will all be complicated by the way I feel now—my Swedish flicka [girl]. All I have done is sit in the sun and look at the ocean and think of Gunilla . . . All love, Jack." He telephoned and wrote her trying to arrange another meeting in Capri, where he was staying with Jackie, or perhaps a few weeks later in Denmark. He sent her a picture postcard from Capri, being careful enough not to sign his name ("I wish *you* could have been here").

Jack was like an actor playing theater in the round, turning toward one part of the audience to play one role, turning to another group to inhabit a whole different character, and then turning yet again to assume a different persona. To Gunilla, he played the dutiful young lover to whom their romance was the only universe. To Jackie, he played the deferential husband. To Pope Pius XII, who gave him an audience at his summer villa, Jack played the good Catholic concerned with the Church's flock behind the Iron Curtain. To the Poles, whose land he visited for the first time since the war, he played the serious political observer.

When Jack called Gunilla from Poland, he was no longer playing the ardent, impassioned young lover but the heart-sickened man struck down by all the demands of the world. He told Gunilla, as she remembered it, that he had talked to his father about divorcing Jackie and marrying her. Joe had yelled, "You're out of your mind." Jack's father, like Torby and Lem and almost everyone else, was a convenient foil. His father, if he did talk to him, was only telling Jack what he surely must have known already.

There often comes a point in a romance when the participants evaluate each other as objectively as if they are weighing semiprecious stones. Gunilla would have made a fine wife but a better mistress, and Jack suggested that she move to New York where, if she lost some weight, he would make her a "top model." Jack's friend Billings, who had never met the Swedish woman, wrote her that he would certify that she would not become a public charge, since "Jack has called me to do this and I shall be glad to."

Gunilla's parents were willing to have their daughter involved in an adulterous affair as long as marriage was at least possibly in the offing. They would not, however, have Gunilla the lesser half of a mere arrangement. They talked to Jack on the telephone, ending the romance that had so consumed their daughter and had been such a perilous adventure to Jack.

Jack was not so willing to give up, even after he learned that Gunilla was engaged. He continued to write her a few more times, but he no longer penned his letters in bold, expansive strokes but in a smaller, cramped and nervous style. "I had a wonderful time last summer with you," he wrote her. "It is a bright memory of my life—you are *wonderful* and I miss you."

Gunilla married a few months after Jack left, but he still hoped to meet her in Sweden the following summer, replicating the exquisite time he had had with her the previous August. Jack was not a man who sought to repeat sweet moments in his life, and he rarely looked back at past pleasures. But there was a certain melancholy in him now. "There must be a beach in Sweden," he wrote her, knowing full well the sensuous warmth of the Scandinavian summer. He dreamed of returning. He was always careful not to pen words of overt romance that might come back to haunt him. But there was a poignant quality in Jack's letter this time. He mentioned a friend of Gunilla whom he was hoping to meet. He ended his letter: "I am looking forward to asking her if she knows a beautiful Swedish girl with a quiet smile who lives on top of a mountain in the Cote d'Azur in August 1953."

When Jack was recuperating in Palm Beach, he spent much of his time working on a book project. Since his college days, Jack was used to having the sort of help on his literary efforts that few of his peers received. Now, as a convalescing senator, he engaged a team to foster his efforts. The book, eight studies of courageous senators bracketed by two thematic essays, lent itself to a largely collaborative effort. Jack solicited advice and suggestions from a group of academicians. Sorensen, a government employee, worked almost full-time drafting four of the profiles. Landis wrote the draft of another chapter and gave other help, including the overall theme. Jules David, a professor at Georgetown, also helped with the conception while writing drafts of four profiles and the closing essay. Jack and his associates did make dramatic changes; the final version of the essay on John Quincy Adams is a dramatic, novelistic rendering of Adams's career, unlike the more scholarly tone of the first draft.

"When I'd finished my other work each day I'd dictate into the machine," Jack told the *Boston Globe*, hardly the typical regimen for the serious writing of history. What Jack was doing essentially was taking these drafts prepared by others and working them over. He had no qualms about taking the writing of others, applying a few flourishes, and passing the work off as fully his own. Sorensen, with his gift for euphemism, says that the "opening and closing chapters, which are more personal and more reflections of his philosophy, probably were more heavily influenced by his literary style than those that were simply historical accounts." Arthur Krock recalled Jack "lying on a board in his bed, absolutely flat, with one of those blocks of yellow paper, and there he sat writing the introduction . . . and some of the biographical material."

It is in the introduction and final chapter of *Profiles in Courage* that some of the intellectual themes of Jack's life emerged, and they may be viewed largely as his own work. "This is a book about the most admirable of human virtues—courage," he begins his book. Courage. Not faith. Not honor. Not honesty. Courage. It is how he judged other men, how he judged himself, and how ultimately he would choose to be judged. His father had taught him that courage was the king of all virtues in a true man's life, and he showed that courage in his struggle against illness. He showed it when his PT-109 was sliced in two. The courage he spoke of now was political and intellectual. It was subtle, complicated, and contradictory.

Jack saw courage among his colleagues, even if others did not recognize it. And just as he primarily blamed the self-interested British populace, not their leaders, for the country's belated rearmament before World War II, now he found fault with the American people more than with his fellow politicians. He bemoaned what he considered the misguided perception that his fellow senators were midget imposters dancing around controversial issues where great men had once bravely stood. As Jack saw it, the fault lay not with his fellow senators but more with a decline in "the public's appreciation of the art of politics, of the nature and necessity for compromise and balance, and of the nature of the Senate as a legislative chamber." This man who had once been so disdainful of elected officials that he talked of becoming a "public servant" now proudly called himself by the term he once found so distasteful: politician.

Acts of political courage had become more difficult now. Jack saw that "our everyday life is becoming so saturated with the tremendous power of mass communication that any unpopular or unorthodox course arouses a storm of protests," a reaction that his political predecessors could not have envisioned. "Our political life is becoming so expensive, so mechanized and so dominated by professional politicians and public relations men that the idealist who dreams of independent statesmanship is rudely awakened by the necessities of election and accomplishment."

Jack had prophesied that one day an American president would be confronted with the immediate prospect of nuclear war with the Soviet Union. Now he saw a danger of equal but subtler form—a time when the very concept of political courage would be endangered. "And only the very courageous will be able to keep alive the spirit of individualism and dissent which gave birth to this nation," he wrote.

Jack was writing when his own political courage was as suspect as his health. He had this book idea long before his failure to vote on the McCarthy censure, but as he wrote, he may have reflected on where that lack of action lay on the spectrum from cowardice to courage.

Courage was the highest virtue, and he aspired to a moment when he would be tested and found not wanting. The last sentences of the book are an exhortation to the American people, but also surely to Jack himself: "The stories of past courage can define that ingredient—they can teach, they can offer hope, they can provide inspiration. But they cannot supply courage itself. For this each man must look into his soul."

The themes in *Profiles in Courage* resonated with the American psyche, and the book became a major best-seller. Jack used the royalties to buy more advertising, getting a double whammy for his money, a higher spot on the best-seller list, and his name linked time and again with the word "courage."

This was not enough for Joe, who wanted his son to win the Pulitzer Prize, the most prestigious literary award in America. Anyone else would have considered the game up when the modest inspiring book of portraits was not even nominated for the prize. For Joe, that was not the end but the beginning.

"One of his slogans which Joe often quoted was 'Things don't happen, they are made to happen,'" Rose reflected. "As for instance when Jack got the Pulitzer Prize for his book or when he or Bob were chosen as outstanding young man of the year. All of this was a result of their own ability plus careful spadework on their father's part as to who was on the committee and how to reach such and such a person through such and such a friend. However, Joe was lucky because his sons were good material to work with. They behaved well, they were intelligent, and best of all they always had confidence in their father's judgment, because it had been vindicated so many times."

Joe asked his friend Krock to be of service in getting the award for Jack. The *New York Times* columnist had for years been on the Pulitzer board. He considered himself "sort of the Mark Hanna of the board, a very ruthless politician." Krock was delighted to see whether he could help make the book of profiles the winner of the biography prize. If Jack's book did not deserve the prize for its literary quality, it surely might deserve it for the best politicking. After all, in past years the award had not always gone to the most deserving book but to the best job of what Krock called "logrolling." He called the board members who would be voting, members who for the most part sat there because of Krock's efforts.

In a world in which honor was more highly valued than power, the board members would have told Krock that they had not nominated the book for good reasons, and that his "logrolling" was not only inappropriate but futile.

That did not happen, though the members may well have been moved by the inspirational tone of the modest book more than by any lobbying.

A book that was not even on the screening committee's list of nominees received the 1957 Pulitzer Prize for biography. In a world in which truth was valued more than appearance, Jack would have gracefully shared the award with his collaborators. But that did not happen either.

17

The Pursuit of Power

On the first evening of the 1956 Democratic Convention in Chicago, Jack narrated a film on the history of the Democratic Party titled *The Pursuit of Happiness*. His great audience lay not in the eleven thousand delegates in the sweltering confines of the arena but in the more than one hundred million Americans who watched at least part of the convention on the new medium of television.

"I am Senator John F. Kennedy of Massachusetts," Jack said as he appeared live introducing the film. Most Americans were seeing and hearing the thirty-nine-year-old politician for the first time. He was cool but conveyed a hint of passion. His Boston accent sounded slightly exotic and aristocratic. He was a youthful politician standing shoulder to shoulder with Jefferson and Roosevelt and the other great men of the Democratic past whose names he evoked. The film ended with a close-up of Roosevelt and Jack's resonant tones. "For the proud past of the Democratic Party is but a prelude to its future," he said, as if he were willing to step forward into these giant shoes "to the leadership it offers the nation, to the faith by which we all abide."

When the applause died down, Governor Frank Clement of Tennessee came to the podium to give the keynote address. The thirty-six-year-old politician had a fleshy, handsome countenance and a reputation as a spellbinding orator. This evening Clement sounded overwrought, his emotive manner more appropriate to a nineteenth-century evangelist's tent than a mid-twentieth-century political arena, his ponderous words heavily bejeweled with metaphors.

When Clement ended his lengthy speech, demonstrators rushed forward carrying the banners boldly emblazoned "Kennedy for President." In politics there is nothing more calculated than the illusion of spontaneity, and these Massachusetts delegates had been primed for their moment in the television spotlight. Jack was a candidate for the vice presidency, not the presidency, but in this one evening he had become a major presence in national politics. The star of the evening was the man whose cool voice exemplified the future of politics as much as the florid Clement represented the past. "Kennedy came before the convention tonight as a movie star," wrote the *New York Times*.

Jack's fascination with Hollywood was not a dilettantish indulgence, but a formidable weapon in his rise to national power. The millions of Americans watching their television screens bestowed celebrity on those they chose to anoint. Celebrity was becoming the most desired new currency, readily cashed in for money or power, and by the time the convention was over no one in the vast hall had received more freshly minted celebrity than John F. Kennedy.

Jack had another unique element in his panoply of power, and that was his beautiful wife. Seven months pregnant, Jackie was suffering in all the heat and turmoil of the convention. Only once, however, did she confess her discomfort. As she looked out the window of their tenth-floor suite, a reporter asked how she liked the convention. "Not much," she sighed. But for the most part, she sat dutifully in a box, and during the week the television audience saw a tableau of marital happiness, a star married to a star.

In his three and a half years in the Senate, Jack had been sick and absent so much that he had had no major impact. By all appearances, he had neither the record, the energy, nor the ambition to attempt to win such a prize as the vice presidential nomination. The reality was that the calendar of Jack's life might have only a few pages that had not yet been turned over. If he aspired to national political office, he could not wait to be noticed. For the first months of 1956, he had coyly refused to proclaim his interest in the nomination while allowing Sorensen and others to work on his behalf.

Jack spent his time promoting Adlai Stevenson, the putative Democratic presidential nominee. Stevenson's strongest advocate within Jack's family was his own wife, who for the first time took a strong interest in politics. Jackie was one of a myriad of educated women devoted to the former Illinois governor, applauding what she considered "his intelligence, farsightedness, and reasonableness." She did not limit her words to whispered asides to her husband but used them in drafting in her own hand the very words spoken by the senator from Massachusetts in announcing his support for Stevenson's candidacy.

Ambition is often a wondrous tonic, and in advancing his candidacy Jack showed an energy and sheer pleasure in the high deviousness of politics that he had rarely shown before. For instance, in seeking to remove an anti-Stevenson man, William H. Burke Jr., from the chairmanship of the Massachusetts Democratic Party, Jack spread gossip accusing Burke of "spend[ing] the whole day with Carmine DeSapio," the New York boss and Stevenson foe.

Most politicians would have sought to replace Burke with a new chairman who would do his bidding. Jack realized that doing that would exchange new enemies for the old one. In a letter marked "Personal and Confidential," Jack told one supporter, attorney Walter T. Burke (no relationship to William H. Burke Jr.) that "if someone who is as close to me as you is named, the feeling will generally be that this whole fight has been one for personal gain rather than a clear-cut attempt to establish an entirely new, impartial, and clean organization." Instead, Jack proposed former Mayor Pat Lynch of Somerville. "While he is not overly close to me, the fact that he is acceptable to most of those who are on our side . . . makes him, I think, an ideal new face for the job."

When Stan Karson, a Stevenson campaign executive, talked to Jack soon after William Burke Jr. had been deposed, Karson left feeling "that for the first time since I have known him over the past few years, he appears to be in control of a political situation, knows it, and likes it." Karson presumably believed that Jack's interest was in masterminding the Stevenson campaign in Massachusetts, not positioning himself for the vice presidency by ingratiating himself with the former Illinois governor.

Jack's daunting problem was his Catholicism. Ever since Al Smith's disastrous run for the presidency in 1928, it had become part of political catechism that a Catholic could not be elected to national office. Jack was not about to confront this issue directly. The shrewdest partisanship masks itself in a bland cloak of neutrality. Sorensen prepared a seemingly impartial report on voting records suggesting that a Catholic on the ticket would bring votes to Stevenson, not lose them, and then leaked it to the media. The Kennedy people also did another pseudo-neutral report winnowing down the possible vice presidential choices by criteria that included marital status ("Should be married, and with no previous divorces"). Jack's picture-book marriage stood in stark conquest to the domestic lives of five other divorced candidates. When the sorting out was finished, Jack turned out to be the perfect candidate.

Jack wanted Bobby beside him at the Chicago convention in July. With Sorensen and the others there was always the niggling doubt that they would put their own interests ahead of his, that they might prefer flattery to truth. With Bobby there was none of that. Bobby saw no greater honor, no higher goal, than to advance Jack. The younger Kennedy cut through all the cant and self-promotion and grasped the nubs of truth, no matter how unpleasant they might be.

Jack called Congressman Tip O'Neill who had taken his old congressional seat, to try to get a delegate seat for Bobby. "Tip," he said, "I'd like to have you name my brother Bob as a delegate to the convention. My brother Bob is the smartest politician I have ever met in my life. Tip, he is absolutely brilliant. You know you never can tell, lightning may strike at this convention out there, I could wind up as vice president. And I'd like my brother with credentials so he could be on the floor to really work for me."

O'Neill was a politician with a consummate understanding of politics as a game of endless exchanges. "As long as you feel that way about it, Senator, okay," O'Neill said, taking his own name off as a delegate and letting Bobby go in his stead.

Delegate credentials were not like ducats to a prizefight that could be sold, exchanged, or traded. One citizen, Robert P. Donovan of East Boston, was outraged that Bobby gave his home address as 122 Bowdoin Street, where he supposedly resided with his pregnant wife and four children along with Jack and his wife in a two-room apartment. Donovan protested to the state ballot law commission, whose commissioners were hardly going to cast out the favorite brother of their favorite son.

At the convention, Jack had Bobby on the floor and Jackie in her seat, but he had one irreconcilable problem—his relationship with Eleanor Roosevelt. The late president's widow was the moral center of American liberalism. Mrs. Roosevelt found Jack's silence on McCarthy a coward's lament.

"She had seen me on a program a week or so before, and had not felt that I had been vigorous enough on McCarthy," Jack recalled three years later. "So I went up [to see her], and my explanation was under the most adverse conditions, because she was in her room, she was hurrying to go downstairs. . . . It was like eighteen people in a telephone booth, and she was giving it just half attention—not listening really."

One of the reasons Jack did not like liberals was that many of them had an overweening sense of moral superiority and were under the misapprehension that a lecture was the same thing as a dialogue. By the summer of 1956, attacking McCarthy was no longer mortally dangerous to a political career. That Jack did not get in a few blows at the fallen McCarthy was one of the more inexplicable moves in his career. Instead of agreeing with Mrs. Roosevelt, Jack reached into his little bag of legalisms and spoke as if McCarthy had already died. "My point was that . . . because I had never really been particularly vigorous about McCarthy during his life, that it would really make me out to be a complete political whore, for me to be really chomping and jumping . . . vigorous[ly] in my denunciation of McCarthy when he was gone."

Mrs. Roosevelt could not understand why the junior senator from Massachusetts would not denounce the evils that even he could now see. Jack did nothing, and this moral albatross remained tightly wrapped around his neck. McCarthy's status as a family friend may have had something to do with his reticence. Beyond that, what Jack's detractors never considered was the possibility that this man, so obsessed with courage, may have realized that he had hidden when he should have stood up, and that he said nothing now was a mark not of dishonor but of its opposite.

Instead of choosing his own running mate, Stevenson took the dramatic step of throwing the choice open to the Democratic delegates. The Kennedy people had been prepared for all eventualities but this one, and they hustled the great floor seeking votes from delegates wavering primarily between Senator Estes Kefauver of Tennessee and Jack.

Neither Joe nor Teddy was among the Kennedy people frantically working undecided delegates. Joe was on the Riviera, an ocean away from what he considered a futile effort destructive to Jack's presidential chances four years later. Teddy was on a trip to Africa, preferring adventure to politics. To give his journey the veneer of seriousness, he had set himself up with press credentials as a stringer covering the French forces in Algeria, but unlike his big brothers, he never managed to file a series of published stories. His father had warned him about spending "too much time" among the more pleasurable fleshpots. It was an admonition that had some muscle; true to the family pattern, Teddy was not traveling alone but had Fred Holborn, a Harvard instructor, for a companion.

Jack had proved to be prescient in wanting Bobby to have delegate credentials so that he could be on the floor. Bobby employed the whole anthology of emotions in promoting Jack. He pled, argued, threatened, and cajoled as he moved from one delegation to the next. He saw that among some of these people even his most salient political arguments did no good. "So many people had come up to me and said they would like to vote for Jack but they were going to vote for Estes Kefauver because he had sent them a card or visited in their home," he reflected afterward. "I said right there that as well as paying attention to the issues we should send Christmas cards next time."

Jack came a tantalizing thirty-eight votes short of winning on the second ballot before Kefauver surged ahead. During the voting, Jackie had been surrounded by well-wishers, but as the voting began to go against her husband, the adoring sycophants began sidling away until she was left alone, looking to one observer "a forlorn figure."

Jack may have lost that evening, but he was the largest victor of the week. He left the 1956 convention brilliantly positioned for a run at the presidential nomination in four years. The Boston papers celebrated their native son's near-win, but Jack's name was heard now in areas far distant from Massachusetts. The *Clearwater Sun* in Florida called him "one of the ablest and most promising young men on the American political scene today . . . substantial feeling exists among Democrats and other observers that he may have advanced himself a very long stride toward a presidential nomination in the years ahead."

Jack flew out of Chicago exhausted after not having slept for three nights. Jackie was exhausted too, and when the plane landed in New York, she took an air taxi to Newport to rest at her mother's home. The admirable John F. Kennedy the newspapers were celebrating would have been on that plane with his wife to spend the last days of summer awaiting the birth of their first child. Even the most cynical of politicians would have joined his wife; no matter how boring he may have found married life, if he wanted to be president, he would do best to play the loving husband and father-to-be.

Instead, announcing that he was going off on a two-week visit to Nice and the Middle East, Jack got on another plane to Paris. In fact, Jack was heading off on a Mediterranean yachting trip. Since March, when Jack already knew that his wife was pregnant, he had been actively negotiating for a yacht with the Mercury Travel Agency in Cannes. "I do not wish for personal reasons to have the direct responsibility or be in any way connected with the hiring of the boat," he wrote H. W. Richardson at the travel agency in April. "I am sure you can appreciate my reasons."

Jack arranged for a Washington railroad executive and lobbyist, William Thompson, to sign the contract and pay $1,750 for the eighty-five-foot *Vileshi* and its four-person crew for two weeks beginning August 21, 1956. Those who disliked the big, brawny Thompson judged him little better than a procurer, but Jack considered him a friend with whom he shared an insatiable appetite for nubile young women. Jack and Thompson were the horsemen of the night, riding out together in search of sexual adventure, be it in Florida, Cuba, or elsewhere. Beyond their mutual pleasure, Thompson doubtless understood how useful it was to have such a well-placed friend, and arranging cruises brought better access than buttonholing the senator outside the cloakroom.

For the senator from Massachusetts and potential presidential candidate, this excursion was unimaginably foolhardy. Jack was heading off on a yacht paid for by a lobbyist/executive, accompanied by several young women, and with a letter on file at the agency showing that he wanted his role kept secret.

When Jack arrived in southern France, he spent a few hours with his parents at their villa. There is no evidence that Rose, who exalted family above all else, suggested to her son that it was inappropriate for him to be vacationing away from his twenty-seven-year-old pregnant wife who had already lost one baby. The talk was of the convention and the hack politicians who Jack thought had betrayed him.

Jack had hoped that Gunilla might be with him on the yacht, but she took her marital vows literally. Jack was accompanied instead by a group that included Teddy and a number of young women. Teddy, a student of his older brother's life, had yet another opportunity to take studious notes on the man he took as his ideal.

For years, Jack had taken risks in the name of sexual amusement, but now there was a dangerous crescendo to his activities. Jack had nearly died after his operation in New York. There are those who rise out of a nearly fatal illness and its telling lesson of the transitory nature of human existence displaying a generosity of spirit that they had not exhibited before. Others leave their sickbeds believing that they must chase every bright flash of life, heedless of its costs to others or even themselves. Jack's father had taught his sons that life must be sucked of its essence. Jack's illnesses were more lessons on the same page. He sought everything now with an intensity that he had not showed before, not only political power but also pleasure, and pleasure to Jack meant sexual diversion.

Jack had just come through the most physically exhausting political week of his life. He wanted to relax, and he would have found no relaxation listening to the mindless social chitchat of his in-laws, and no solace walking the Newport beach hand in hand with his pregnant wife. He took what was best from everyone around him. When he wanted to be amused, Lem or Red appeared, all but wearing clown suits. When he wanted to discuss public policy, Sorensen sidled into the room, all solemnity. When it was his political future at stake, Bobby and his father appeared. And when he sought relaxation, he wanted sex. "Unique among them," recalled one of the Mediterranean revelers, "was a stunning but not particularly intelligent blonde who didn't seem to have a name but referred to herself in the third person as 'Pooh.' She fascinated Jack, who was wound very tight when he arrived in Europe and almost completely unwound a few days later."

While Jack was basking in the sun, Jackie began hemorrhaging. The doctors performed a cesarean section, but the baby was stillborn. When Jackie woke up, it was not Jack who was sitting there by her bedside but Bobby. It was Bobby who was always there, but as much as he watched out for his sister-in-law, he was watching out even more for his absent brother.

A hospital official stated that the baby died owing to the mother's "exhaustion and nervous tensions following the Democratic National Convention." When Jack learned that the baby had died, he sailed on. It took a call from

George Smathers to convince Jack that he should go home to his grieving wife. "I told him, 'You ought to come back,' which he did," Smathers recalled.

For Jackie, the fact that her husband had not returned was like a light so brilliant that even a blind woman would have seen its flashes. She had written a poem to a noble, star-crossed hero, but that was not the husband who returned to her.

Bobby was doing far more than advancing Jack's political career and tidying up after his romantic sojourns. In his own right, Bobby was a man of extraordinary ambition, energy, and cunning. Bobby did almost nothing without explicitly mixed motives, and made no move without figuring out all that might go wrong.

Bobby attempted to ingratiate himself with J. Edgar Hoover, the FBI chief, seeing him twice early in 1954 on what were considered personal matters. He visited the FBI chief again the next year when he was planning to take a trip to the Soviet Union. Hoover suspected Bobby not as a cryptic liberal but as someone from an equally treacherous category, a publicity lover. "Kennedy was completely uncooperative until he had squeezed all the publicity out of the matter that he could," Hoover noted about a subject before the subcommittee, Bobby having had the bad taste not to realize that Hoover was the one to do the squeezing.

In the summer of 1955, Joe arranged for Bobby to go on a six-week trip to the Central Asian republics of the Soviet Union with Supreme Court Justice William O. Douglas. Joe was forever pushing his sons to confront their truths in spirited debate with those of radically different positions. Douglas favored a less rigid, more accommodationist policy toward the Soviet Union, and the liberal jurist was the worthiest of choices to debate Bobby as they journeyed around regions of the Soviet Union where, since the revolution, few Westerners had gone.

Douglas was a proud humanist whose faith looked no higher than the peoples of the earth. He found it notable that, as they flew above routes that Marco Polo had once traversed, Bobby sat in the plane not boning up on the agriculture of Uzbekistan or literacy in Kazakhstan but reading his Bible. And wherever they went, Bobby grasped the Bible in his left hand to ward off the evil virus of communism and lectured to the Marxist true believers, trying to convert them.

In Omsk, near the end of the trip, Bobby fell sick and ran such a high fever that Douglas feared he might die. Bobby refused to see a doctor, saying, as Douglas recalled, "Russian doctors are Communists and he hated Communists." The justice insisted, and a young, white-clad woman doctor walked in to minister to a delirious Bobby. She stayed in the room for thirty-six hours straight until Bobby's delirium ended, the fever broke, and he began to get well. He arrived in Moscow in decent shape but about twenty pounds lighter than when he had begun the trip.

"As a result of this rather extensive and arduous Russian journey, I began to see a transformation in Bobby," Douglas recalled. "And in spite of his violent religious drive against Communism, he began to see, I think, the basic, important forces in Russia—the people, their daily aspirations, their humanistic strains, and their desire to live at peace with the world."

Bobby may have seen all that, but he also had heard enough to reinforce his most rigid views of Marxist totalitarianism, from the loudspeakers spewing propaganda in the fields and factories ("It just drives you out of your mind") to the absence of any voices of dissent, spirited or otherwise. He saw Russian colonialism at work in the segregated school system that separated the Asian and Russian populations. As he sat with his hosts eating such exotic fare as lamb's brain, he saw too that there were few Communists in this Communist land. He saw a Soviet Union attacking the colonial powers of the West in Africa and Asia while running its own empire in Central Asia. He showed a perceptive awareness of one of the great ideological struggles of his time, not communism versus capitalism, but Russian imperialism versus the nationalistic instincts of subject peoples within the Soviet empire.

When Bobby returned home, he parlayed the journey into a bounty of publicity and goodwill. Douglas may have believed that Bobby's experiences in Russia had begun to open him up to a world more divergent and complicated than any he had imagined. Yet Bobby successfully hid that awareness in the many interviews and speeches that he gave and the articles that he wrote.

Bobby had had an opportunity that no American official had ever had before—to meet with the largely forgotten peoples of Central Asia, whose nationalistic aspirations would explode into life three decades later. Yet Bobby remained as much an ideologue as the Communists he despised, and no more interested in the story of the individual human life than they were. Bobby had gone to the Soviet Union on a political mission, and he was not ready to display reportorial snapshots in lieu of ideas.

It is out of the specificity of human experience, however, that empathy grows. Nowhere did Bobby tell the tale of the woman doctor who saved his life, presumably because it would have given his audience the idea that a Communist could be a human being. Nowhere did he make Douglas's essential point, that these were people, not ideological stick figures. Nowhere did he draw the obvious conclusion that as offensive as the ethnically segregated schools might be, Americans could hardly condemn them too loudly until they changed their own segregated system.

Any other twenty-nine-year-old political figure in the nation's capital would have been ecstatic to have a fourteen-page Q&A in *U.S. News & World Report* and would have assumed that the then-conservative publication would treat him fairly. Bobby not only edited the transcript and suggested an added question and answer, but got the magazine to state that he had been off the government payroll during the six weeks he was in Russia. He invited a reporter from the *Boston Globe* down to the Cape for another lengthy feature article.

Bobby was a terrible speaker. His high-pitched voice was sometimes almost inaudible, and his sheer discomfort at being up on the podium communicated itself to his audience better than his words. Yet he agreed to give the prestigious Gaston Lecture at Georgetown University in October 1955, to talk about his trip. He was so dreary an orator that even his own secretary, Angie Novello, told him: "The pictures were great, but I think you could have spoken better than that." He was still mining the trip nine months after his return, writing a lengthy article for the *New York Times Magazine* on "The Soviet Brand of Colonialism."

Bobby continued to ingratiate himself with Hoover, going to see the FBI director to tell him about his trip and the following year writing him the kind of fawning letter that was de rigueur for impressing Hoover ("I hope the United States enjoys your leadership for a long time"). The next year Hoover sent Bobby a copy of his book, *The FBI Story*. Bobby had the exquisite good sense to ask the director to autograph the best-seller, saying that he would "cherish it even more."

For all Bobby's calculation and shrewd ambition, he had time for friendship and family that Jack simply didn't allow himself. He touched his children and kissed them, enveloping them in his warmth as his parents rarely had done when he was a little boy. He did not discard his old Harvard football buddies because they were no longer useful to him but kept up a correspondence in which they humbled each other with their rude put-downs. When Nick Rodis or Kenny O'Donnell came to Washington, they had to come out to Hickory Hill, the new house in McLean, Virginia, that he and Ethel had bought from Jack, or they would feel the blunt blows of his wrath.

In a matter of weeks, Ethel and Bobby had transformed a Hickory Hill that under Jackie's tutelage had been a sedate, elegantly appointed estate into an eclectic amalgam of summer camp, zoo, boot camp, college fraternity, political headquarters, and religious retreat. Less than intrepid visitors were well advised to turn around and drive back down Chain Bridge Road to Washington.

Ethel had the high skittishness of the Skakel clan. Handing off her newest baby to the latest nanny or maid like a football, she ran to be present whenever Bobby made a speech or quizzed a witness in a Senate hearing. The couple played tennis matches with all the intensity of the finals at Wimbledon and had discussions with their friends over dinners that were equally competitive.

Years later Rose mused that when a family has everything, sooner or later it must experience loss. Ethel's losses began on October 3, 1955, when her parents, George and Ann Skakel, died in a private plane crash in Oklahoma. Ann had spent much of her life with a drink in her hand, and her family mourned her with as much liquor as tears. Bobby kept his wife away from the Skakels' Greenwich home for their nonstop Irish wake, an absence that only exacerbated the growing tensions between the two families. As much as Bobby tried to succor his wife, her parents' death left Ethel with a white-knuckled fear of flying so extreme that sometimes he had to cancel trips.

The Skakels and Kennedys had two different visions of life. The Skakels were full of matchless exuberance, an unbounded generosity of spirit, and a blithe unconcern for the world beyond their gated precincts. Bobby was a Catholic Puritan who puckered up his lips at the taste of liquor. He was profoundly concerned with making his mark on the world, disdainful of mere inheritors, and impatient with those beyond his intimates whose words would not advance him. Both families kept a tight hold on their truths. But for the Skakels, there was no excuse for Ethel not joining her six brothers and sisters to mourn their parents as they should be mourned and limiting her presence to an appearance at the funeral. Bobby had a worthy excuse. As a Skakel, his wife had two possible solaces: religion and liquor, her family's secondary faith. If alcoholism could be passed on, either by inheritance or proximity, then his wife was vulnerable.

During the 1956 presidential campaign, Bobby traveled with Stevenson. He was supposed to be a campaign aide, but he spent most of his time taking notes and making observations. "Bobby accompanied Stevenson on his presidential campaign at his father's request," according to Rose, "because of course Joe thought Jack would eventually run [for president] and anything that Bobby could learn from Stevenson's campaign would be useful." For months Jack had been ingratiating himself with the Democratic candidate, but that did not mean that he and his brother still admired him. Stevenson was the archetypal liberal that the Kennedy brothers abhorred. He had what they considered a prissy, overrefined quality. "He's got no balls," Bobby told O'Donnell. "I think he's a faggot."

The liberal ladies might swoon over their precious champion, but Bobby privately sneered. The Democratic presidential candidate stood on the campaign podium saying that he was going to take on big business. A true man who had the gumption to attack Wall Street, however, would have shouted words that came from his heart. Stevenson stood there *reading* the speech. His words were eloquent, his voice was cultured, but he exuded nothing more, as Bobby saw it, than "an appearance of insincerity." Stevenson tinkered for hours over his eloquent speeches and squandered days in meaningless discussions with his campaign staff. Bobby's attitude toward Stevenson was not helped when one day Joe called the candidate and Stevenson moaned: "Oh, my God, this will be an hour and a half."

While Bobby stayed close to Stevenson, Jack traveled from state to state giving speeches that advanced his own celebrity more than they did the Democratic presidential candidate. The good liberal matrons of the suburbs might be "madly for Adlai," but Jack aroused passions in younger women. At Ursuline College in Louisville, Kentucky, the female students blocked his car shouting, "We love you on TV!" and "You're better than Elvis!" They had seen him on television presumably at the convention and had connected with his persona. Like Elvis, Jack had his own unruly mop of hair that he brushed back casually with his hand. Later that month in Queens, New York, when hordes of older women nearly fainted away during his speech, he was the one who made the Elvis connection. "The Republicans remind me of two Elvis Presley records," he said. "For three months before election they serenade the people with 'I Love You, I Want You, I Need You.' But the rest of the year they change their tune to 'You're Just an Ole Hound Dog.'"

As the weeks went by, Jack became primarily concerned with salvaging every last ounce of gain he could from the impending electoral debacle. Stevenson went down to ignominious defeat, winning only seven southern states. In politics much that passes as politeness is little more than organized hypocrisy. Bobby wrote the defeated candidate a warmly disingenuous letter. "I hate to impinge on your well-deserved rest but I wanted to write and thank you for your many kindnesses to me during the campaign," he wrote in a handwritten note on Jack's Senate stationery. "I regret that I was not able to make more of a contribution but your allowing me to travel with you was a wonderful opportunity and experience for me." He did not find it expedient to mention to the former Illinois governor that he had been so appalled at what he saw that he hadn't been able to bring himself to vote for Stevenson but had pulled the lever for Eisenhower.

Even before Bobby wrote his note to Stevenson, he had flown off to California to begin an investigation of the International Brotherhood of Teamsters as chief counsel for the Senate Permanent Subcommittee on Investigations. For months he had been badgered by Clark R. Mollenhoff, an investigative reporter, to investigate the union's corruption. Mollenhoff was a glowering hulk of a man who spoke in a hectoring, belligerent tone that offended many he sought to convince. "He had a tendency to use acid sarcasm, and he was inclined to be irritable or rudely blunt when he was crossed," Mollenhoff said of Bobby, though he equally could have been describing himself. "Occasionally, I taunted him by questioning his courage to take on such an investigation," Mollenhoff recalled. "At other times I prophesied such an investigation could do for him what the Kefauver crime investigation did for [Chief Counsel] Rudolph Halley."

Union corruption was a canvas on which Bobby could write his name in bold letters. He knew, however, that every action he took now could impinge on his brother's campaign for the presidency. It was a daring move going after corruption in the largest, most powerful union in America, though as he began his investigations he thought he would be clipping away a few precancerous growths on a healthy body. Even so, he risked alienating millions of union men and women whose votes Jack would need.

The Teamsters president, Dave Beck, was a plump, aphorism-spouting burgher more comfortable sitting around with Republican businessmen than his fellow labor leaders. As the Republican White House saw it, he was the perfect exemplar of a new conciliatory age when labor statesmen and wise businessmen could walk together. Although Bobby had no vendetta against Beck, the labor leader and his associates were bound to portray Bobby as a union-busting rich boy out to make his name by trying to destroy a powerful organization that looked out for the good workingmen and -women of America.

Bobby's father was adamantly opposed to Bobby involving himself in such a dubious enterprise. The primary result, as Joe saw it, would be Jack's loss of labor support. There may have been another reason Joe became so upset, a reason that he could not tell his son. In his hand, Joe held a map of power in America far more extensive, detailed, and truthful than anything Bobby or Jack grasped. Joe saw the dark tunnels that burrowed under America. He saw where they led and how they entered the soft underbelly of American life. He knew what a dangerous enterprise it would be even to chart these tunnels, to say nothing of attempting to close them off. Joe may also have feared that Bobby's quest would lead to his own father's activities. He argued with Bobby as he had never done before, but Bobby gave him no ground. Joe went to Justice Douglas and asked him to plead with his son to give up this crazed mission, but nothing dissuaded Bobby. Others might have thought that Bobby was nothing but his father's puppet, but Joe was not pulling the strings that Bobby danced to now.

Bobby put together a stellar team that included Carmine Bellino, the former head of the accounting unit at the FBI. Bellino could read a page of figures as if they were fingerprints, and he put in seven-day weeks. "Unless you are prepared to go all the way, don't start it," Bellino cautioned Bobby as they began. "We're going all the way," Bobby answered.

On his first trip to the West Coast, Bobby discovered what to him was another America. He was a conservative Democrat who had been brought up with a certain ambivalence toward left-leaning unions that pushed militantly against the owning class. In a few weeks, though, he was confronted with union leaders who struck out militantly not at management but against their own union brothers and sisters. These local leaders forged sweetheart deals and pocketed bribes and concessions. Their minions bludgeoned honest workers who stood up to them and intimidated the rank and file into silence. As Bobby interviewed source after source, the twisted trail of corruption led higher and higher until he was confronted with the stark suspicion that Beck himself was corrupt. In one instance alone, Beck may have pilfered $163,000 from the union treasurer to build the kind of palatial house in which once only the bosses had lived.

Bobby and his staff pored over documents and interviewed scores of people. It was often a tedious, frustrating business, running down one lead that turned back on itself, then following another that did the same, and always traveling, from Los Angeles to Portland and Seattle to Detroit and Chicago.

On a frigid December day five days before Christmas, Bobby and Bellino hurried back to their room at Chicago's Palmer House to pore over subpoenaed documents that detailed the relationship between Beck and Nathan W. Sheferman, a so-called labor relations consultant. As the two men meticulously examined the documents, they realized that they now had the evidence to prove indisputably that Beck was not the "labor statesman" he purported to be, but a criminal who had betrayed the trust of the men and women who drove trucks across America and delivered bread and beer to stores. These documents had the power to put a man in prison and to begin to shut down a golden spigot of corruption that had benefited everyone from union officials to extortionists, from politicians to thugs, and from sweetly scented, refined business gentlemen to common murderers. And this was not the end of the story, but only the preface.

In January 1957, the new Senate formed a special eight-member Committee on Improper Activities in the Labor or Management Field. The chairman, Senator John L. McClellan, was a cautious man of subtle political skills. In a move that was as shrewd as it was compassionate, the senator insisted that the largely disgraced Senator McCarthy sit on what became known as the Rackets Committee. Jack was invited to join as well, and he reluctantly agreed to serve. The select committee was likely to receive immense publicity, but serving on it might make him appear a foe of unionism.

During the next three years, Bobby served as chief counsel in an unprecedented investigation of corrupt labor-management practices. This work was part of a history of Senate investigations, including the Nye Committee's exploration of international arms trading before World War I, the Kefauver Committee's examination of organized crime, and the McCarthy Committee's inquiry into American communism.

Bobby was not the first chief counsel to play a crucial role in an investigation. He was the first, however, to so overshadow the senator in whose service he was presumably acting that few remember these historic hearings as the work of the so-called McClellan Committee. While publicly deferring to McClellan, Bobby largely orchestrated the emergence of his own celebrity. He did so surely knowing, as the *Louisville Courier-Journal* wrote in March 1957, that if he came "out of his investigation with as much credit as seems likely, the nation can count on having the Clan Kennedy on the political scene for two, or three, or four decades."

Bobby helped make sure that he was exalted in the general media. His father had taught him that nothing just happened. A few extraordinary humans made what was called history. Bobby had listened well. Calculated self-promotion did not embarrass him the slightest bit. Bobby had been so upset by Edward R. Murrow's historic *See It Now* program opposing Joe McCarthy in March 1954 that nine months later he walked out of a speech that the journalist gave in Louisville. That didn't prevent Bobby, however, from going on Murrow's *Person to Person* show in September 1957; the *New York Times* reported that Bobby and Ethel "were charming and poised . . . and their five children were delightfully disheveled."

In February 1958, Francis X. Morrissey, Jack's executive secretary in his Boston office, wrote Bobby congratulating him on the honorary degree from Tufts University in Boston that he was about to receive. "We have been working on this for the last three months. . . . It is essential that you let no one know you have received it until they announce it from the University itself." A month later Morrissey wrote Bobby again to alert him that when the Boston papers hadn't initially run a picture of the counsel for the Senate Rackets Investigating Committee with his wife and new baby, Michael Le Moyne, he had taken steps to ensure that the *Boston Traveler* ran it.

Even the most personal and seemingly spontaneous moment could be as scripted as an inaugural address. Joe wrote Bobby: "I think some reporter should ask Bobby [Jr.] 'where his Daddy is and what he is doing' and Bobby's answer should be 'he's chasing bad men like a cowboy.'"

Bobby knew many reporters who would have been delighted to ask that question. Bobby was astutely cultivating journalists, the crucial element in creating his legend. In 1957 his Christmas gift list contained thirty-seven journalists, ten photographers, and sixteen radio and TV technicians. The next summer he invited them out to Hickory Hill for a swim and touch football.

Comradeship and calculation blended impeccably. Reporters gave Bobby their stories early and told him what had been edited out. Murray Kempton, a *New York Post* columnist proud of his fierce independence, identified so much with Bobby that he even suggested a letter that Bobby should write to two attorneys who opposed him.

With Bobby, there was always a bottom line, and the bottom line was that the journalists wrote to advance the Kennedys' agenda. If they did not, if they fell short or gave aid to the family's enemies, they soon learned about it. Jack Anderson wrote a largely adoring story on the Kennedy brothers for *Parade* magazine. Yet he received a two-page letter from Bobby detailing the errors that he said "ought to be cleared [up] . . . inasmuch as you may be writing or commenting on our family again at some future time." "My objective has always been to report fearlessly, but accurately," the chastised Anderson replied, presumably promising to write more carefully next time.

Bobby was livid over a Mike Wallace television interview in which, Bobby felt, Wallace urged a guest to name as a "degenerate" a police officer with whom the committee had worked. He contacted the chairman of the Federal Communications Commission and asked "if there are some steps that the . . . Commission can take in matters such as this."

Certain journalists saw themselves as cohorts in Bobby's direct assault on the forces of evil in the labor movement and elsewhere. What he had was theirs, or so they thought, and what they had was his. These reporters knew they had crossed a forbidden threshold, but they saw a higher purpose. "Spoken bits of praise can be dropped into casual conversation without serious thought, and only because it seems a nice thing to say," wrote Mollenhoff in December 1957, downplaying any verbal bouquets he might offer Bobby. "I want to put it on the record so it can be held against me."

In the afternoons during the hearings, Bobby called in a group of sympathetic reporters and alerted them what to expect the next day. The reporters could write a first draft of their stories and add details the following day, thus keeping themselves several steps ahead of other journalists not so allied to the Kennedys.

However much Bobby sought to push his own name and accomplishments higher up in the coverage of the hearings, they in fact deserved to be on the front pages of America's newspapers, played on live television during the day, and featured on the network news programs each evening. Many

Americans had considered it tantamount to guilt when accused Communists and alleged racketeers took the Fifth Amendment. What did it mean when a parade of union leaders, consultants, and enforcers hid behind their constitutional rights, ducking questions that any honest labor official would presumably have answered forthrightly? Beck himself, when he finally appeared, was the master prevaricator, rambling on in a lengthy monologue before finally and seemingly reluctantly taking the Fifth Amendment. Bobby served not as dispassionate researcher gathering information to help Congress write laws but as a merciless prosecutor. His voice was taunting and disdainful. His questions moved from one specific detail to the next, pulling the noose so tightly around Beck's neck that in the end he had no breath left for his endless prevaricating soliloquies. By the time Beck left, he was a broken man who would soon be convicted of federal tax evasion and state embezzlement charges and sentenced to prison.

That spring of 1957, Joe was quoted in the *Boston Globe*: "[Bobby's] 'a great kid. He hates the same way I do.'" Bobby was too much of a moralist to admit that he considered hatred a worthwhile emotion that could motivate a man as much as love. But when James R. "Jimmy" Hoffa, head of the Teamsters Central State Conference, Beck's heir apparent, and the next hearing witness, appeared across the committee table, Bobby fancied that he could see "the look of a man obsessed by his enmity, and it came particularly from his eyes. There were times when his face seemed completely transfixed with this stare of absolute evilness." What Bobby did not admit, or perhaps did not understand, was that he hated Hoffa and was as obsessed by him as much as Hoffa hated and was obsessed by Bobby.

Observers noticed immediately that there were certain similarities between the two men that only exacerbated their enmity. They both had a small man's tenaciousness and constantly measured opponents who appeared larger than they were. They both cultivated the realities of power, not its mere appearance. They were men whose enemies had reason to shiver in their tracks.

Hoffa, however, had never needed the faux combat of the Harvard-Yale game to test his manhood. The son of a coal miner who had died when he was seven, the diminutive Hoffa made his way as a teenager to the docks of Kroger, where he found work unloading fruits and vegetables. He was hardly more than a kid, but he tried to organize the workers. Hoffa learned that a man spoke loudest with his fists, and louder still if he carried a club. He became a protégé of Farrell Dobbs, a Trotskyite labor leader who taught that behind the sleek gentlemen of the boardroom lay all the coercive violence of capitalism. Hoffa became a reformer who organized the unorganized, a fearless man in a time of fear, and he rose steadily up the union ranks. "Nobody can describe the sit-down strikes, the riots, the fights that took place in the streets of Michigan, particularly here in Detroit, unless they were part of it," Hoffa said.

Hoffa soon lost whatever faith he had in the Left. What remained was a rude philosophy of power that saw life as little but an endless series of brutal exchanges. Us against them. Your gang against my gang. Hoffa came to value power for himself more than power for the men and women of his union. He bragged about coercing the companies into higher wages and better benefits, but in reality he was cutting sweetheart deals with owners, throttling those who stood up to protest, and selling out his brethren with impunity. What made him so dangerous was that most of these union men and women identified with good old Jimmy and his vision of the world and stood with him, electing him their new president. He took that as a mandate to expand his union and its power.

If Hoffa had abided by a gentleman's code of conduct in the brutal brawls of the 1930s, he would have ended up a bloody pulp in the gutter. In his fight against Bobby, there was seemingly no blow that Hoffa was unwilling to strike. The committee staff had hardly begun its investigation when, in February 1957, Bobby met with John Cye Cheasty, a former Secret Service officer and investigator, who told an extraordinary tale of just how far Hoffa was willing to go to sabotage the investigation. Cheasty said that Hoffa had paid him $1,000 as part of an $18,000 payment for getting a job on the committee and serving as his spy. Bobby went ahead and hired Cheasty, then fed him information to pass on to Hoffa, sometimes on a Washington street corner while undercover FBI agents filmed the meeting. It was a mark of Bobby's obsession with Hoffa that he wanted to be present when the FBI arrested the union leader. The FBI realized how inappropriate that would be, however, and Bobby was not there when agents arrested Hoffa at the DuPont Plaza Hotel with some of the committee documents that Cheasty had given him in his possession. Hoffa seemed certain to be convicted and sent to prison.

Hoffa hired attorney Edward Bennett Williams, a brilliant, mercurial gladiator who had a lyrical love for the law and a street fighter's instincts. Williams defended Hoffa with every weapon at his disposal, from the high-soaring eloquence of his defense to the low subterfuge of suggesting that Cheasty was anti-black and having Joe Louis, the black former heavyweight boxing champion, flown in from Detroit to embrace Hoffa in the courtroom in front of the twelve-person jury, eight of whom were black.

Bobby was in the midst of conducting hearings when Angie Novello, his secretary, passed a note to him. Bobby glanced at the message and turned back to the witness as if nothing of moment had happened. No one in the room could have imagined that the message contained the news that Hoffa had been found not guilty. Later, when Bobby walked into the committee's office, he sensed the despairing mood of his colleagues. "Come on now," he said. "Let's get to work. We have a lot of work to do. No sitting around."

Bobby's father had brought up his sons to step over defeat and move on. Unlike his associates, thirty-one-year-old Bobby did not bemoan the verdict as a disgrace to the American system of justice. He did not underestimate his opponents, and his conclusion was judicious, fair, and bitter. He credited "the work of Williams, his effective defense attorney, plus Hoffa's own strong testimony, together with the unpreparedness and ineffectiveness of the government attorneys who prosecuted the case."

When Hoffa returned to testify before the committee again, he forcefully defended his stewardship of the union. At times, Bobby's questions got close to their mark. Then Hoffa flicked the queries away, feinted for a while, and moved forward to rake the chief counsel with a few tough blows of his own. Hoffa took exquisite pleasure in taunting Bobby. "Oh, I used to bug the little bastard," he recalled fondly. "Whenever Bobby would get tangled up in one of his involved questions, I would wink at him. That invariably got him."

The invective that Hoffa slung in Bobby's direction was merciless. At one point he described his adversary to a reporter as "a young, dim-witted, curly-headed smart aleck and a ruthless little monster." Bobby had never come up against anyone like this. In his attempts to bloody Hoffa and his associates, Bobby at times sounded like McCarthy at his inquisitional worst. "Is there any question in your mind that Mr. Goldblatt [secretary-treasurer of the International Longshoremen's and Warehousemen's Union] is a Communist, Mr. Hoffa?" he asked rhetorically. "You haven't got the guts to [answer], have you, Mr. Glimco?" he accused another union official who took the Fifth Amendment.

What Bobby and his associates exposed was a dark hidden kingdom in which corrupt labor unions, willing local politicians, organized crime, and amoral business executives all worked in harmony while Washington leaders and most of the press looked the other way. Hoffa treated a criminal conviction as a worthy reference for work in the Teamsters.

As the parade of witnesses passed through the Senate hearing room, there was a seamless meld between corrupt unions and organized crime. One of the many witnesses was Sam Giancana, the reputed head of the Chicago mob, who had muscled into the Brotherhood of Electrical Workers to take over the lucrative jukebox business in Chicago. If physiognomy was fate, Giancana might have run a small restaurant or dry cleaners. Instead, the nondescript, balding "businessman" with a taste for show business and pretty women had a reputation as a pathological killer. Bobby asked him whether he disposed of his enemies "by having them stuffed in a trunk." When he took the Fifth Amendment in a laughing way, Bobby taunted him: "I thought only little girls giggled, Mr. Giancana."

The exchange played well on national television, but it was silly gamesmanship. Bobby was indulging in what Yale law professor Alexander M. Bickel would later call "relentless, vindictive battering" of reluctant witnesses.

Beyond that, Giancana was a vicious man who could call up a hired killer with the nod of his head; Bobby was foolish to take such seeming pleasure in taunting such a man.

When Bobby and Jack stared across the committee table at mobsters such as Giancana, they were looking evil straight in the face. In all the hours of testimony the Kennedy brothers heard endless accounts of just how far such men went to protect their illegal turf. Intimidation. Coercion. Bribery. Assault. Murder.

Bobby took testimony and heard tales that would have made any public figure cautious about ever allowing himself to be vulnerable to blackmail. As a committee member, Jack heard much of the same testimony and surely should have learned the same lessons.

Mollenhoff sent Bobby a memo about Giancana's Chicago that would have made most politicians draw back from letting friends and associates set them up with women. Mollenhoff told of Dan Carmell, an attorney for the Illinois Federation of Labor and a former Illinois assistant attorney general, who had recently jumped out of a Chicago hotel to his death just before he was to go on trial for white slavery. "The sworn statements of the girls involved . . . show the way that these girls operated from an apartment in North Chicago, and the way various political figures including at least one senator were brought into the web of Carmell's labor influence," Mollenhoff wrote. "Indications are that labor money was used for the transportation of these girls, and that this pattern of sex parties was used to trap political figures into situations where they were under the power of Carmell and the labor organizations he represented."

This was not the kind of matter Bobby liked discussing. When he was taking testimony from a Portland madam, Bobby became so embarrassed that Chairman McClellan asked whether he might take over. Jack was not embarrassed at all by such talk. These hearings nonetheless were the most extraordinary warning that if he hoped to be president of the United States, he would have to conduct his personal life with a caution that did not come naturally to him.

Bobby and Jack were performing not only before a national television audience but also before their own family. Every day Ethel sat in the hearing room in a choice seat held for her by one of the guards. Jackie sat there many days too, as did Teddy whenever he could get away from his studies. Even Joe, as opposed as he was to this interminable spectacle, showed up at least once and sat watching his sons. It was a tableau of loyalty that helped to further the image of the Kennedys as a peerless family, steadfast to the core.

The audience watching on television and journalists writing feature articles on the Kennedys noted the youthful vigor of the two brothers seated side by side. Jack was thirty-nine, though he looked ten years younger, and thirty-one-year-old Bobby looked as if he could have just graduated from college. Of the one thousand letters the committee received one week, four hundred were what the New York *Daily News* called "mash notes," most of them asking for autographed pictures.

The Eisenhower years were not a time when youth was considered to have special merit, but youthful Bobby identified with the young and tentatively began reaching out to them as his natural constituency. "It's ridiculous to wait until a man is 40 to give him a responsible job," Bobby said. "By that age he may have lost most of his zeal."

Bobby was fearless intellectually and physically, and not just when the cameras were focused on him. A woman reporter from Joliet, Illinois, had disappeared, presumably murdered by the labor mobsters she had exposed in a series of articles. Stories like that only emboldened Bobby. He and his associate Jim McShane traveled to Joliet Prison to interview an inmate who said that he knew where the reporter was buried. The two men went with the prisoner to a farm field where Bobby took his turn shoveling in an unsuccessful attempt to find the body.

Bobby learned that all across the country lived brave people like this reporter who were willing to stand up to evil at the risk of their livelihoods and sometimes their very lives. It rankled Bobby that some of the Washington know-it-alls whispered that John Cheasty had told of Hoffa's perfidy only because he wanted a job on the committee. Bobby knew that the sickly Cheasty had lost far more than he gained, and he was only one of scores of examples.

Sometimes even Bobby was dumbstruck by the quiet, unaccredited heroism of some of the witnesses. George Maxwell, a Cleveland labor relations consultant, agreed to testify before the committee that he had negotiated sweetheart contracts for major carriers with Hoffa. "I don't understand it,"

Bobby beseeched him, realizing that even he himself was becoming cynical. "You will ruin yourself, your business, if you testify like that." Maxwell replied, "I tell the truth, Mr. Kennedy."

The great experiential moment of Bobby's life lay in these hearings, which largely defined his view of the world and of the American people the way World War II had defined Jack's worldview. Bobby saw now that there was corruption of a magnitude he had hardly imagined, an evil that threaded its way through American life. It was *The Enemy Within*, as he called it in his 1960 book about the committee's work, though it had penetrated further than even he realized. He had also seen good on a scale that he had hardly imagined, and he knew that to betray these lonely acts of courage was to betray life itself. If one day he would romanticize certain groups of Americans, his faith grew out of a specificity of experience that he kept close to his heart.

Bobby's name would always be linked with his obsessed campaign to convict Jimmy Hoffa, but his team of investigators went far beyond the Teamsters Union. He even investigated the *New York Times*, whose editorial support his brother would soon be seeking in his presidential race. He took startling testimony that the greatest paper in America had made payments to reach favorable agreements with the Teamsters. In all, the committee and its 100-member staff, interviewing 1,525, witnesses and taking testimony that filled 59 volumes of official hearings, presented strong evidence of corruption in at least 15 unions and more than 50 companies.

Bobby's critics on the left accused him of destroying unions, but he celebrated union reformers who stood up to corruption and whose victories would ensure a stronger movement. His critics on the right said that he avoided investigating Walter Reuther and the United Auto Workers, but in fact the committee looked at the UAW and found little seriously wrong. Sadly, this congressional investigation, the most monumental in half a century, achieved a result hardly commensurate with the enormous effort and resources put into it. Congress did enact a labor reform bill, and a year after the committee disbanded, twenty figures in labor and business had gone to jail. But as Bobby himself admitted, there was "appalling public apathy"; by the measure of congressional mail, practically no one in America cared deeply about labor corruption. Beyond that—and this was not something that Bobby conceded—he had helped make Hoffa a folk hero to the Teamster rank and file, a working-class Houdini who had wiggled out of all the handcuffs and jails in which his enemies had sought to confine him. The union president sat behind his nine-foot-long mahogany desk in a Washington office far more spacious and impressive than the Oval Office. He ate gourmet meals prepared by chef Jean Grihangne, and when he had eaten too richly, he worked out in the tiled gymnasium that he had built for his use. He had hopes of expanding the Teamsters into other areas of American life.

As for Bobby, he too had become a hero to millions of Americans. "At thirty-three years old, this quiet little fellow is the most exciting guest we've had in years," Jack Paar told his late-night audience on national television during Bobby's appearance in June 1959. "I think that Robert Kennedy is the bravest, finest young man I know."

Bobby was a man who never forgot, and when he left the committee after three years in 1959, he was as consumed with the Teamster leader as when he had first met him. He believed that Hoffa stood at the center of a contagion of evil that had to be stopped or the whole nature of America would change. He had other work to do now in helping to elect Jack president of the United States, but it was clear that one day these two implacable enemies would meet again on another field of conflict.

Jack did not have the same visceral feeling about corruption and evil that Bobby had. His most passionate political concern remained foreign affairs. In July 1957, Jack gave a speech in the Senate on the tragic situation in Algeria, where the French were brutally repressing a guerrilla movement seeking independence. The speech was a revelation to many. One of those listening that day was Howard E. Shuman, administrative assistant to Senator Paul Douglas. Shuman, like many of his peers, considered Jack "an extraordinary minor figure" in the Senate. Yet as the aide listened, he thought, "My God, this is really great stuff."

Jack called for the United States to take the lead "in shaping a course for political independence for Algeria." In that North African country, half a million French troops were fighting a people attempting to throw off the heavy yoke of colonialism. It was an ugly, vicious war, an endless cycle of torture, bombings, and reprisals. The bloody conflict was tearing apart not only Algeria but France itself. America had fought its own war for independence, and if not for the exigencies of the cold war, Washington policymakers might well have stood foursquare on the side of the Algerians. France, however, was a prominent member of NATO, and the Eisenhower administration stood with its ally and looked away from the tumult in North Africa.

Jack was a determined student of history, and it appalled him how little his country had learned from its misguided support of French colonialism in Vietnam. He told his colleagues: "Did that tragic episode not teach us, whether France likes it or not, admits it or not, or has our support or not, that their overseas territories are sooner or later, one by one, inevitably going to break free and look with suspicion on the Western nations who impeded their steps to independence?"

There was the essential reality. As Jack saw it, nationalism, not communism, was the unstoppable tide sweeping across Asia and Africa, and his nation, once a colony itself, had to ride with that tide, not against it. Within a few years, Jack's words would sound self-evident. But in the summer of 1957, this was a daring, controversial statement that won him more condemnation then praise. Of the 138 editorials that Jack's office clipped and saved, 90 of them criticized him.

Among prominent Democrats, Stevenson was particularly incensed at what he considered Jack's ill-timed, inopportune call for Algerian independence. Nonetheless, for the first time Jack's voice had resonated with progressive intellectuals across America and the world. He had struck a deep resonant chord among the Algerian guerrillas listening to his words in French on the Voice of America and among other young Africans and Asians, who heard the voice of an America that had long been silent.

Jack's father was appalled at his son's speech. He had no use for liberal moral posturing, hand-wringing, and loud moaning over the brutal realities of power in the world. Worse yet, if Jack was to become president, he could not be climbing out on limbs that might prove rotten or could be sawed off by his opponents safely ensconced on the ground. Despite his disapproval, Joe was as supportive as ever. "You lucky mush," Joe told Jack over the phone. "You don't know it and neither does anyone else, but within a few months everyone is going to know just how right you were on Algeria."

On a Saturday evening a month after giving his speech in the Senate, Jack appeared before the Americans for Democratic Action at the Astor Hotel in Manhattan. He had arranged to meet there with Arnold Beichman, a reporter who had just returned from an unprecedented visit with the guerrillas in Algeria that had been the basis of a *Newsweek* cover story. Jack had a genuine liking for brave reporters, and he asked Beichman to meet with him after the event. After Jack said his good-byes, the two men walked out of the Astor Hotel and along Broadway. Beichman told Jack that the guerrillas had heard his speech while sitting in their mountain hideaway. They had asked all kinds of questions as they sat eating lamb stew with their guest. Who was this man Kennedy? How come this Kennedy was so influential? Why couldn't he get independence for Algeria?

As they hurried along the late-night streets of Manhattan, Jack peppered Beichman with questions. Jack mentioned an Algerian lobbyist at the UN. It was a name that few people other than foreign policy experts on North Africa would even know, and Beichman was impressed. "What do you think of him?" Jack asked. "Is he to be trusted?"

Jack kept talking as they walked into the Commodore Hotel on Lexington Avenue. Beichman was surprised that Jack was staying in such a second-rate hotel, but he followed him over to a bank of elevators where Jack pushed a button. "I've got a very important date," Jack said, suddenly grinning. "Sorry I can't invite you up."

18

The Rites of Ambition

Jack could have stepped forward now and stood as the great champion of colonial peoples, whether they resided in Algeria or Indonesia or Poland. At far less political risk, he could have stood his ground as an articulate spokesman for eventual Algerian independence. Instead, as was his pattern, he stepped gingerly back, telling his staff that he was "wary of being known as the senator from Algeria."

Just as nationalism and the struggle for independence was the great international moral issue of his time, so civil rights was the great domestic moral issue. On this issue, Jack was even more reluctant to take a leading role. He was instinctively a moderate, tempering his progressive instincts on foreign policy and social issues with a conservative wariness of the dangers of wrenching change. He cared to some degree intellectually about the plight of blacks in the segregated South, but he did not have the liberal passion of his colleague Senator Hubert Humphrey of Minnesota, who cried out in a loud, fervent voice that enough was enough, wrongs had to be righted, and righted now.

Jack stood with some of his more reactionary colleagues on several technical matters as the 1957 civil rights bill worked its way through Congress. Unlike the vociferous Humphrey, Jack was wooing southern Democrats to stand behind his presidential banner, but his refusal to stand forthrightly with his liberal colleagues was perhaps something more than narrow pragmatism. As he had with some of the procedural votes on McCarthy, he ended up looking like a man of expediency. The charge rankled him. "It's awful . . . you know what they say about you, but they say this . . . [was] an attempt to appease the South," he told Burns. "Politically it was a mistake."

The reality was that if Jack's chances for a presidential nomination had rested on his legislative record, he would scarcely have been considered a plausible candidate. He had no stellar record of bills stamped with his bold mark, as did Senate Majority Leader Lyndon Johnson of Texas, another potential candidate. Nor did he stand at the forefront of a political issue, as Humphrey did on civil rights. Neither of these men, however, had the captivating persona that Jack was exhibiting on speaking trips that took him from Arkansas to New York City, Baltimore to Mississippi. He was deemed worthy of a *Time* cover in 1957, and major articles appeared in magazines ranging from women's periodicals such as *McCall*'s and *Redbook* to mass general-interest magazines such as the *Saturday Evening Post* and *American Weekly*, to more obscure, policy-oriented publications such as *Foreign Policy Bulletin* and the *National Education Association Journal*.

As he flew from state to state and interspersed his days in Washington with one interview after another, Jack continued to be cursed by bad health. His bad back and Addison's disease were burden enough. Then the dog that Jackie had given him for Christmas in 1956, when they were staying at his father's New York apartment, set off such a severe allergic reaction that he had an asthma attack and the animal had to be given away. For a long time afterward, he became sick again every time the couple stayed in the apartment at 277 Park Avenue.

On another occasion while he was in New York Hospital, Jack checked out to go to a friend's house for dinner. The friend had a dog, and by the time Jack checked back into the hospital, he was breathing so badly from his asthma that the hospital staff became frightened.

In the middle of September 1957, Jack developed an abscess on his scarred back that induced a high fever and such wrenching back pain that he entered New York Hospital. The doctors drained the abscess and put him on a heavy dosage of penicillin and streptomycin.

Jack wanted to minimize the whole business, but a vibrant, youthful politician did not spend over two weeks in the hospital for his annual checkup. Dr. Janet Travell, his personal physician, wanted to tell the press that her patient had "a small abscess on his back." Jack would have none of that. "You know, that's a very ugly word," he told her. "I don't want to have an abscess." Since there was an epidemic of Asian flu making its rounds, the word went out that Jack had "a virus infection."

Jack left the hospital on October 1 and flew to Hyannis Port. During his weeks in the hospital, Bobby, George Smathers, and Governor Foster Furcolo of Massachusetts had given speeches in his stead, but now Jack was about to fly to Canada to give a talk. He was so weak that he canceled a dinner that Lord Beaverbrook was planning for him on the trip so that he could rest.

Dr. Travell had the impression that Jack was "discouraged." Joe and Rose had brought Jack up *never* to be depressed, or if he was feeling low, *never* to admit it, not to himself, and certainly not to the world. It was a measure of how down he felt that he could admit that he was feeling downcast.

"You know, what you need is a real good hot-tub bath," Dr. Travell said. It was hardly a suggestion that seemed likely to bring on a dramatic change.

"You know, I haven't been in the bathtub since I entered New York Hospital because of the wound in my back," Jack said, looking at the doctor. "I can't go on with another great big gaping hole."

The words had a slight tinge of self-pity, an emotion unknown to Jack. "You don't have a great big gaping hole in your back, and there's no reason why you couldn't get right into a hot tub and soak," the doctor said as Jack stared at her in disbelief. "You haven't seen what is there. It's been covered by the dressing. You've got a dressing on it."

Jack took his bath, and Dr. Travell believed that it was "a cake of soap that saved the day and a hot tub bath." But when Jack left to fly to Canada, he took the same scarred and aching body with him, and his health was hardly demonstrably improved.

Jackie had an impeccable sense of how to memorialize special occasions. On their fourth wedding anniversary in 1957, she prepared an illustrated book titled "How the Kennedys Spoil Wedding Anniversaries." The sketches were wondrously whimsical, but as always with Jackie, there was a subtle edge to her humor. The first drawing in the exquisite book portrays Jack lying in bed with the diligent Jackie at his side. In the second drawing the couple has changed places, and Jackie is in bed, while Jack watches over her. The pictures made light of an unpalatable truth. If they disclosed a hidden irony in Jack's young wife, the gift also suggested that this was a woman who cared enough for her husband to sketch these gentle scenes of their marriage.

There were persistent rumors that Joe had headed off a divorce by promising Jackie a million dollars if she would stay with her husband. There is no evidence that any such offer was made, and it hardly would have been enough of a payoff to keep a despondent Jackie in a dreary marriage. If her fidelity to her marriage was indeed purchased, she was the consummate actor, not only keeping the Kennedy name but also displaying interest in her husband's career.

"I was alone almost every weekend while Jack traveled the country making speeches," she recalled. "It was all wrong." Jackie had an intense inner life that even Jack did not fully know. She was a woman not simply of mood swings but of dramatic changes in her perceptions of the world around her.

One day she would flirt with Jack with those gaminelike eyes, as her mother recalled, "writing him little jingles and poems and bringing him little presents with appropriate rhymes accompanying them." Then the next time she saw him, she would be so coldly uncaring that Chuck Spalding believed that her feelings toward her husband had gone from love to hate.

Only Jackie truly knew what she felt toward her distant, philandering husband, and this deeply private woman was not about to unburden her soul in the authorized American psychological fashion. "Look, it's a trade-off," she reflected later. "There are positives and negatives to every situation in life. You endure the bad things but you enjoy the good. . . . One could never have such a life if one wasn't married to someone like that. If the trade-off is too painful, then you just have to remove yourself, or you have to get out of it. But if you truly love someone, well. . . ."

Jack would not have been a Kennedy man if he had not wanted to carry on his line, and it had been painful that it was proving so difficult for him and Jackie to have a child. It was both a relief and a blessing when, on November 27, 1957, Jackie gave birth by cesarean section to a squalling baby girl who her father declared was "as robust as a sumo wrestler."

"She's easily the prettiest baby in the room, don't you think," the proud father asked the nurses. Jack took his friend Billings to the nursery at New York's Lying-In Hospital and stood looking at the newborn through a glass window. "Now, Lem, which one of the babies is the prettiest?" he asked his friend, seeing no need to point out the obvious. Jack's mother-in-law, Janet Auchincloss recalls that, when Lem made the mistake of pointing out another baby, Jack "didn't speak to him for three days."

Soon after the baptism of Caroline Bouvier Kennedy three weeks later at St. Patrick's Cathedral, Jack and his favorite carousing buddy, George Smathers, headed off to Havana for the first of two trips there. They were accompanied on one of their Cuban trips by Bill Thompson, who was often around Jack when pleasure was the aim.

The Cuban capital was a corrupt, lascivious place, the product in part of an unholy alliance among President Fulgencio Batista, American business interests, and American mobsters. Jack had seen that the deadly hand of colonialism and neocolonialism was losing its grip all over the world and that America should stand with the rising forces of nationalism. By that measure, Jack should have supported the students at the University of Havana who had been bludgeoned when they protested the increasingly totalitarian regime. Jack, however, joined the millions of Americans visiting Havana to gamble at the casinos, drink Cuba libres, and, by their presence, help to sustain the dictatorship.

Soon after his arrival, Jack met with Ambassador Earl Smith, a Palm Beach neighbor and the husband of one of his former lovers, Florence Pritchett. Smith was an apologist for Batista, full of tales of how the dictator was America's stalwart friend and an implacable enemy of the leftist guerrillas in the hills. That conversation and a talk to the embassy staff were the sum total of Jack's serious work in Cuba.

Jack was not much of a gambler, but Smathers recalled that his friend took great interest in the floor show at the Tropicana Nightclub with its parade of gorgeous showgirls and a statuesque French cabaret singer and actress, Denise Darcel, whom he managed to meet. At the Hotel Nacional's casino, Jack had his picture taken with the manager, Thomas McGinty, who had once been his father's bootlegging partner. A picture is evidence of nothing more than that Jack had stood next to a gangster running mob-controlled gambling, but Jack was appearing with distressing frequency among those whose hands he should not even have shaken. In February 1958, for instance, an FBI surveillance team noted that on a speaking engagement in Tucson, Jack had been accompanied to church by a man the FBI identified as a close friend of Joseph Bonnano, a top organized crime figure.

In Havana, Jack and Smathers went to visit Batista. Smathers had an amiable relationship with the Batista regime, but as a potential presidential candidate, Jack was foolish even to be there. Another of his father's bootlegging partners, Owen Madden, had known Batista early in the dictator's career in Cuba. Now, by his presence, Jack was giving added credibility to a dictator who kept his tenuous hold on power by a regime of increasing violence, threats, and brutal reprisals.

"Batista had a big uniform on," Smathers recalled. "And when we went in to see him, he had these two guns that he pulled out of the drawer, and laid them out on the desk so you could see them. And one of them was looking at Jack and one of them was looking at me. I was thinking, 'You better say the right thing if you want to get out of here in one piece.' And then he made us go with him out to a great big tent where it was the custom that any mother who had had a baby, they would bring them there and they'd line up, and they had a lot of the priests and bishops and Catholic hierarchy there. And Batista made Jack and me stand up there with him and they'd give us a baby and we'd pass the baby down the line. And everybody would kiss the baby. And about thirty-five babies, forty babies go by. And Jack would look at me like, 'How many more do you see out there?' So we sat there one whole afternoon kissing babies. That was fun. If you look back on it. Anyway, we had a lot of fun in Cuba."

Jack was forever off on his pursuits of pleasure, but he always returned to live within the bounds of family. Joe and Rose had created in their children a family that was like a temple into which no outsider could ever enter and those standing outside looking in could not begin to understand the rituals that took place within its precincts. No wonder, then, that Jack's sisters were no more interested in marrying young than he was. In the end, they did not so much marry into other families as bring their husbands in as members of the Kennedy clan. The first to marry was thirty-one-year-old Eunice. Her husband, thirty-eight-year-old R. Sargent "Sarge" Shriver, was already at work for Joe at the Merchandise Mart. Sarge came from a distinguished, though now threadbare, Maryland Catholic family. He had gone to Yale University, where he was a baseball star, and served as a navy officer in World War II. He was a man of deep religious and philosophical concerns who tried to live a good dutiful life as a steward of God's earth. His depth was not always apparent, for he could be a man of tiresome garrulousness with a salesman's upbeat pitch, who exhausted listeners with his sheer enthusiasm. Sarge liked fine things, silk suits, antique furniture, and first-class restaurants. He was devoted to Eunice, fortunately enough, for over their nearly seven-year courtship she at times treated him more like a hapless retainer than a worthy suitor.

Sarge and Eunice were profoundly religious, and their wedding took place in May 1953 at St. Patrick's Cathedral with Cardinal Francis J. Spellman officiating, along with three bishops, four monsignors, and nine priests. At the elegant reception Eunice told the guests, "I found a man who is as much like my father as possible." This was not true, even if Eunice believed it, but in her mind it was the highest honor she could give a man. She adored her father and was perpetually bewildered when others held him in lesser esteem than she did. Whatever doubts Eunice may have had about her new husband, her marriage to Sarge would prove to be the most successful and deepest of any of the Kennedy marriages.

Pat was not so fortunate. She had always been fascinated by Hollywood. In California she met Peter Lawford, a British-born movie star to whom she became engaged after an acquaintance of scarcely two months. Joe investigated the men whom his daughters dated even casually, and it is one of the unanswerable questions why he did not condemn this marriage before it ever took place. Joe was close enough to Hoover that the FBI director surely could have told Peter's potential father-in-law of his dossier, which included a 1946 investigation involving "White Slave activities in Los Angeles" and four years later, a call girl's statement that Peter was "a frequent trick." Peter was not Catholic, and so the wedding took place in April 1954 at the Church of St. Thomas More in New York City, before about three hundred guests.

Jean married a man more like her father than any of his sons. In Stephen "Steve" Smith's heritage flowed a rich mixture of politics and business. Steve's grandfather, William Cleary, had worked alongside his Irish compatriots building the Erie Canal. Cleary saved enough to buy a tugboat and start Cleary Brothers, the family company that employed Steve. Cleary went on to serve three terms in Congress. Twenty-eight-year-old Steve was brash, tough, and rudely charming, an elegant dresser with an eye for women, and he was accepted by Joe and his sons as one of them in a way that the seemingly prissy, morally sensitive Sarge would never be. The couple was married in May 1956 at the same St. Patrick's Cathedral where Eunice and Sarge had wed, but twenty-eight-year-old Jean settled for a more modest wedding and a far larger wedding gift, an enormous diamond pin.

They were truly a clan now, and Bobby's, Jack's, Sarge's, and Steve's families built their own homes nearby in Hyannis Port. There they lived together in the summer beneath the overwhelming shadow of Joe, who dominated his sons-in-law almost as much as he did his sons. The Kennedys were grandiose in their ambitions, and there were jealousies, pettinesses, and spats large and small, but Joe had taught them all that the world outside must know nothing of the internal workings of the family. Thanks to Joe, they were all wealthy, with scarcely a worry about the mundane details of living that preoccupied most Americans. *Fortune* estimated the family wealth at $350 million in 1957. Even if the fortune was only $100 million, as *Forbes* considered a possibility, it was still "one of the great American fortunes." The Kennedys spent exuberant, exquisite days together in the summer on the Cape and in the winter in Palm Beach, each holiday together only reinforcing their feelings for each other and their sense of how different they were from the world beyond.

In the spring of 1956, the graduating Catholic students at Harvard held a celebratory get-together. Archbishop Cushing stood greeting the students beside Harvard President Nathan M. Pusey. Cushing knew none of these students and was delighted when Teddy appeared in line. "What are you going to do next year?" the bishop asked.

"I am going to try to go to Harvard Law," Teddy answered.

"I think they will take you all right," Pusey said. If the Harvard president was saying that in public, then Teddy was an apparent shoo-in, and Cushing sent Joe a letter of congratulations. It did not matter that Teddy had a mediocre grade point average and a cheating scandal against him. He was a Kennedy, and that outweighed everything else.

Teddy was not, however, accepted at Harvard Law School. Joe did not let matters stand there but asked Cushing to intercede with Pusey. "I didn't get anywhere in my intervention," Cushing wrote Joe. "He [Pusey], himself, was just as much surprised as I was that Teddy didn't get over the 'hurdle' of the aptitude test. If there is anything else I can do for the boy let me know. Like yourself, I think he has an abundance of latent abilities and a fascinating personality."

If the Kennedys had been a poor family and Teddy had been asked to wear his brothers' clothes, he would have been unable to pour his outsized frame into Jack's or Bobby's pants or shirts. It was his brothers' lives that he was being asked to don, and he was having a difficult time even walking in the costuming of their lives.

Teddy headed down to the University of Virginia in the fall of 1956 to pursue a law degree. Wherever he went, he was an echo of his brothers who had gone before him. This time it was Bobby who had left footprints that his younger brother was trying to fill. Teddy simply did not have the quickness of his big brothers, and he had to make up for this with sheer diligence. He pored over the legal texts so many times that he risked wearing the words off the page. When he finished his long hours of study, he was not interested in changing his little piece of the world but in having a rousing good time.

Teddy had enough money in his trust fund to live more like a country gentleman than a law student. He and his friend Varick John Tunney, the son of the former heavyweight boxing champion, rented a gracious, three-bedroom, red-brick house that looked out on a splendid view of the Blue Ridge Mountains. "Cadillac Eddie" they called Teddy, and Cadillac Eddie he was, a souped-up, ebullient, twenty-four-year-old driving his battered old Oldsmobile convertible, its broken, plastic back window snapping in the air and his loyal German shepherd at his side, speeding ninety miles an hour along the country roads.

Cadillac Eddie roared along, once outrunning a Virginia police lieutenant, Thomas Whitten. The next Saturday night the officer lay in wait, and when Teddy pulled over, Whitten thought he looked "weak as a cat." Even after this incident, Teddy ran a red light and was caught again.

His big brother Joe Jr. had died a hero's death, and Jack had risked his life saving his men, but there was nothing heroic about this business. Teddy was showing willful disregard, not only for his own life, but for those he endangered by his recklessness. His father had pleaded with his sons to drive within society's laws. Teddy listened to everything else his father told him, and he followed the precepts of his family, but about this, he paid Joe no heed. Teddy was playing his little joke on the tyranny of time, and nothing was going to change him, not his father's admonitions, not speeding tickets and warnings, not pleading friends. Nothing.

Teddy was oversized in everything from his biceps and his appetites to the splendidly resonant quality of his baritone voice. He was the natural speaker in his family, not Jack or Bobby, and when in October 1957 it came time to dedicate a gymnasium in his sister Kathleen's honor at the new campus of Manhattanville College in Purchase, New York, Teddy was chosen to give the address. His mother was there, as well as his sister Jean, and his sister-in-law Ethel, but it was unthinkable that one of them would stand up and talk when a Kennedy man was available.

The auditorium was full of privileged Catholic women whose education in becoming exemplary wives and mothers would be largely worthless unless they married a spouse like the splendid specimen of Catholic manhood who addressed them. To most of them, including Teddy's mother, sister, and sister-in-law, the choice of a husband overwhelmed every other decision in a woman's life. It was understandable why Jean walked up to her brother during the reception, bringing with her Joan Bennett, a blonde, twenty-one-year-old Manhattanville student. Jean had met Joan at a party and knew that she came from what was considered a fine Catholic family. Her father, Harry, was a prominent advertising executive, and she and her sister Candy had been brought up to exemplify all the virtues of traditional Catholic womanhood.

As Teddy talked to Joan that afternoon, he could hardly have been unaware of the two characteristics that largely defined her. At five feet, seven inches, Joan was a head taller than most of her classmates, but it was her stunning beauty, not her height, that set her apart. She had worked as an actress-model on national television, and if sheer loveliness was what mattered, she could have been a movie star. She was a woman who bore herself with more modesty than many women of such beauty would have considered appropriate. Since her childhood, Joan had learned that her beauty drew people to her. Beauty was the reason she was standing here next to what she considered a "darn good looking fellow," and it was a secret code that opened all the doors of life.

Another quality of Joan's was apparent even on first meeting. That was a startling innocence and lack of guile. She was full of a breathless incredulity that made some people think that she was rather stupid. She was in fact an intelligent young woman, but with an apparent lack of insight into the darkness of much of the world around her. She looked away from whatever was dark or distressing and moved always toward the light.

Teddy was delighted to go out with women who were willing to do things the nuns told them they must not do. Joan, however, was a potential wife, and he treated her with respect and subtle deference. They went skiing and dancing. Wherever they went, they were chaperoned. He invited her up to Hyannis Port the following summer to meet his mother, who put Joan through a test that included playing some Brahms on the piano and professing her faith. The Kennedys checked Joan out further with Mother Elizabeth O'Byrne, the president of Manhattanville, who raved about the young woman. No one marked it against Joan that both her parents were heavy drinkers.

Teddy's romantic life was the one arena of adventure left to him, and he took immense pleasure in all the sexual games of bachelorhood. During the summer of 1958, he was still only twenty-six years old, and he was more like Jack than Bobby in his enjoyment of life as a single man. His mother and sisters talked about Joan, and it was clear that if he was going to obey the mandates of his family, he should follow his brothers and sisters into matrimony.

Teddy and Joan were walking along the beach on Labor Day weekend. They had rarely been alone, and they were lovers neither physically nor emotionally. They prowled around each other with wary uncertainty, both knowing the stakes of the game they were playing. "What do you think about getting married?" Teddy asked as nonchalantly as if asking for a dinner date.

"Well, I guess it's not such a bad idea," Joan answered, replying with the same shrug-of-the-shoulders dispassion as Teddy.

"What do we do next?" he asked, as if he had just made a business deal.

For Joan, this was in many respects the most crucial moment of her life, and over the years she reflected time and again on it. At first she thought that Teddy "proposed in a very off-hand manner" because he was "kind of guarding himself in case I said no." Teddy's reticence, she realized later, was perhaps not the mark of a shy lover overwhelmed by the depth of his emotion and the fear that he might be turned down. His seeming inarticulateness was an articulate expression of his feelings.

The next morning Teddy told Joan that she had to meet his father, who had just arrived home from France. The sixty-nine-year-old patriarch sat in his great wing chair in the far end of the living room, looking at Joan like a monarch holding court. Joan walked tentatively into the room and sat at Joe's feet on an ottoman.

"Do you love my son?" Joe asked. It was the crucial question, but it was rarely asked so boldly. This was no social chitchat but an intense interview. Joe had been home only a few hours, but he seemed to know everything about Joan and her family. "When the interview was over—and it was an interview—he said if we wanted to get married, we had his blessing." Joan remembered. "At the end, I felt terribly relieved. He may have been tough, but he did make you feel at ease. When I returned home on Monday, I told my parents, and they were very happy."

Joan may have been happy, but Teddy was sending out distress signals to anyone who chose to read them. "I was young and naive then, but looking back, there were warning signals," Joan recalled. "We didn't see each other from the time of his proposal until the engagement party." That evening at the Bennett home in Bronxville, he arrived when the event was half over. "So he wouldn't embarrass my mother, he chose to come in the back way, through the maid's quarters," Joan said. Teddy ran up the stairs, bringing with him an engagement ring that his father had picked out and that he had not even seen until Joan opened it.

Joan saw that Teddy was acting purposefully disdainful toward all the rituals of his engagement. She would never have behaved so rudely as her fiancé, but she too had her doubts about the approaching wedding. Joan went to her father and told him of her overwhelming fear and reservations.

Bennett thought that his daughter's happiness mattered more than family appearances. He went to Joe and told him that Joan wanted either to postpone the wedding or to end the engagement.

Joe was a man who took care of problems, and Teddy was a problem. Joe and Rose considered it their own achievement to be marrying off a son whose endless escapades worried both his parents. Marriage would settle Teddy down and minimize the prospect of any embarrassing scandals. Joe was not under any illusions that a lack of love and commitment was any reason to stop a marriage. He railed at Bennett and his sentimental nonsense and insisted that the wedding go forward. Few men could stand up to the full titanic force of Joe's temper and will, and Bennett was not one of them. The wedding would go on as scheduled.

Teddy was in charge of Jack's 1958 reelection campaign, by far the most important political duty he had ever had. All that stood between Jack and a race for the 1960 Democratic presidential nomination was a landslide reelection to the Senate. That would signal to the rest of America that John F. Kennedy was indeed Massachusetts' favorite son. The 1958 race had all the earmarks of the earlier Kennedy senatorial campaign, though the stakes were much higher and his opponent was not the formidable Henry Cabot Lodge Jr. but the largely unknown Vincent J. Celeste.

There was a self-indulgent lassitude to Jack's baby brother. Teddy could be guilelessly charming one moment and the next turn into the spoiled scion, insisting on the prerogatives of his name. He did not run the campaign the way Bobby had six years before, berating incompetents, humorlessly pushing the staff to their limits. He did not run it at all but allowed others on Jack's staff to direct matters.

Teddy was used to being handled. There was always someone there writing a speech for him, making an introduction, driving his car, fetching a Coke. He didn't require any help, however, with the female sex. He was obvious in his attentions and heedless of the Kennedy pattern of always having a beard at your side. When he went out putting stickers on automobiles, he brought with him an entourage of pretty young women. Even the usually adoring *Boston Globe* could not avoid noticing, writing that wherever he went the very engaged Mr. Kennedy was "customarily surrounded by a flock of young beauties." His mother had tried to teach her son to cultivate good careful habits, like always having another man at your side. Teddy, however, did what he pleased.

Teddy may not have had Bobby's skills at managing a campaign, but he was an exuberant, outgoing campaigner like his grandfather Honey Fitz, who would click his heels and sing "Sweet Adeline" at the hint of a request. When Teddy stood at factory gates at five in the morning, he reached out and grabbed the gnarled hands, slapped backs, and shouted his brother's name, trumpeting Jack's virtues to the very heavens. And when his car stopped in traffic once, he jumped out and slapped as many bumper stickers on other waiting cars as he could before setting off again on the endless campaign road.

Teddy was a carefree, youthful presence and seemed to carry none of the heavy burden of ambition and power borne by Jack and his father. He wooed voters not with logic or passion but with endless zeal. He went door to door in Mount Washington, a Republican town that Jack could hardly expect to carry, but on election eve, when the votes started coming in, Jack saw immediately that the town was his. Then he remembered all the hours Teddy had put in going "house to house."

The Republican candidate was an Italian-American from East Boston who was making a minor profession out of being defeated by Jack Kennedy. Celeste had gotten up after being walloped by Jack in the 1950 congressional race and was ready to be knocked to the canvas again eight years later. He had almost no money, and his tedious television spots ran five minutes while Jack's professionally produced epics went on for a half hour. Celeste thought that he at least would get some free publicity when Jack finally agreed to a debate put on by the League of Women Voters in Winthrop. That evening Ted Sorensen showed up instead, and the Republican candidate refused to debate a surrogate.

The men around Jack were a tough, intelligent, ambitious lot. They spoke in the shorthand of power, a language that Teddy was only fitfully learning. And they worked as if they were seeking not Jack's reelection in Massachusetts but the first vote in his campaign for the presidency. Those who could not keep up fell back, like Ted Reardon, the candidate's longtime aide and friend.

For all his ambition and brutally realistic assessment of his associates, Jack found it almost impossible to fire anyone. Francis X. "Frank" Morrissey would have been a superb candidate. He ran the Boston office with a slap-on-the-back, what-can-I-do-for-you congeniality. He seemed like a modestly romantic Irish-American character, until one realized that blarney is just another word for lies. He was a spy for Jack's father and considered by one insider little more than "a professional tattletale."

Jack had received two devastating letters about Morrissey. One accused Morrissey of referring constituent problems "to a law firm, and in other cases shakes them down." A second anonymous letter stated that Morrissey and his associate "have been running in and out with women, and intoxicated, and having the life of a Riley . . . it is in my opinion that they both are a couple of pimps."

The allegations in the letters may not have been true, but Jack considered the matter important enough to keep them among his confidential office papers.

Jack's newest aide, Myer "Mike" Feldman, was a tall, lean attorney with a self-effacing manner that only partially hid his high ambition. Feldman had grown up in an orphanage in Philadelphia. He had graduated at the top of his class at the University of Pennsylvania Law School and had that quickness of mind and tongue that Jack liked. If Feldman had a weakness, it was that he fit too comfortably into his defined role as Sorensen's subordinate, a quality that was probably one of the reasons Sorensen recommended him in the first place.

Feldman sent out a press release under Jack's name saying that he was calling for a higher minimum wage. Jack's two aides were used to issuing all manner of words in Jack's name, from speeches to magazine articles, that sometimes he had not even read. That was how closely Sorensen had learned to mimic Jack's idea and style. This press release was like scores of others, but when Joe saw it, he became livid. He was working with the business community raising money and seeking their support. He was conservative himself and fed up with such liberal kowtowing to the working class as guaranteeing them a basic, ever-rising minimum hourly wage when it was sheer gumption that pulled a man out of poverty. He considered the press release a debacle, tying Jack to labor and its limited constituency. He railed against the release and called for the firing of Feldman, the architect of the disaster. Joe probably felt that he had a special say in this matter. Out of his own pocket, Joe was paying Feldman $15,000 a year, doubling his government salary.

Joe marched Feldman in to see Jack and told his son that his new aide had betrayed him. Then Feldman, in his studious, low-key manner, told his side of the story, asserting that what he had done was right and good. "I think it's the right position," Feldman said as Jack listened intently. "We need the labor vote, and this is how we get them excited. Without them, we're not going to win big the way we have to win."

After hearing his aide out, Jack told his father that Feldman would not be leaving the staff. It was a rare instance when Jack refused to go along with his father's advice. Even after this experience, Feldman's assessment of Joe was generous. "The father had a lot of clout," he said. "The father was active in the campaign. Only behind the scenes, he wasn't out in front, even though he was enlisting the aid of all the businessmen. But everybody knew that he was there; everybody knew this was his function. And he performed it very well. He raised a lot of money. And he deserves a lot of credit."

While the others worked fervently in Massachusetts, Jack flew across the country giving speeches for other Democrats. He was running for the presidency of the United States, even if he hadn't announced yet. That did not mean, however, that he was going to give in easily to the repetitious nature of most conventional campaigning. As he traveled from city to city, he added to his speeches a touch of humor here, some unexpected witty dialogue there, and then on to the next performance. To stave off boredom, Jack created his own private moments of levity. When he went to Hawaii to help local Democrats, he took Red Fay with him. At each stop he introduced his old navy buddy, first as the distinguished "Congressman Fay" from the mainland, then as the former cook on PT-109, and on the last stop as "Dr. Fay," the renowned surgeon.

Jack's friend Smathers, the new chairman of the Democratic Senatorial Campaign Committee, pushed him to fly into Wisconsin for a day to help the underdog William Proxmire win his race in a state that had not sent a Democratic senator to Washington since the New Deal. "But find him some feminine companionship," Smathers told Joe Miller, the campaign director of the Democratic Senatorial Campaign Committee. "He likes that when he's on the road." Miller demurred, but he did line up the board that Jack had requested for his bad back.

"I really hate this," Jack muttered to Miller during the interminable day. "It's meaningless." But that day he was brilliant at hands-on politicking, saying one thing to the Polish-Americans, something else to a group of blacks, and still other words over a Catholic radio station. It was a compelling day, and it may have been the crucial factor in Proxmire's narrow victory.

On a campaign trip to California in October, Jack dictated a lengthy confidential letter about his own Senate race to Steve Smith, admonishing his brother-in-law, "Be sure to keep this letter under lock and key or destroy it after you have taken notes from it." Jack was running his own campaign, and he was running it with the most extraordinarily detailed concern. He saw his state as a complex matrix of peoples and groups that had to be wooed individually, each with a subtle, tailored approach.

The labor unions were crucial, and Jack had a different approach to each union leader. "I think it will be well for dad to call Dan Donovan of the Longshoremen's," he wrote. He thought Steve should consider having the Longshoremen's leader "write a letter which we could pay for to all of his union members saying what a great fellow I am." He wanted Reardon to "call Vic Turpin of New Bedford, head of the fishermen's union." Steve was to write a letter to Max Dobro, president of the Boston Taxi Drivers Association. Jack was fed up with Governor Furcolo's refusal to work with him. "You might suggest to Kenny [O'Donnell] that he make it clear to the appropriate Furcolo people that there are quite a few people disturbed by the apparent 'knifing,' and who wonder whether the Irish are going to continue to support Italian candidates when the Italians won't vote for Irish candidates."

Jack was in control of his own political destiny, even if he felt it wise to keep secret just how much he dominated every gesture and every step. He was largely the mastermind of a campaign that O'Donnell and Powers boasted was "probably as nearly perfect in planning . . . as an election campaign in an off-year could be." Teddy, the official campaign manager, was so irrelevant that in Jack's three-page memo he did not mention his kid brother except to ask, "Should we get Eunice and Jean up in order to expand our coverage, especially as we are losing Teddy?" Jack mentioned his father only once, asking him to make a phone call. Joe contributed an estimated $1.5 million to the campaign coffers, made crucial phone calls, and talked his son up, but he was no éminence grise controlling the campaign.

Joe still believed that his conservative, isolationist political judgment was impeccable. After rereading an interview he gave to *Life* in 1945 in which he outlined his worldview, he marveled at the perfect precocity of his views. "I wouldn't change a single word," he told the *Boston Sunday Herald* in April 1957. "If, as chances, facts and later developments have made me look like something of a prophet, if only a Cassandra, don't forget that time has been on my side," he said with unbecoming immodesty. "But I have no pride in any of that," he said, although his every word belied that statement. The obvious conclusion was that by disagreeing with his father and supporting domestic social programs and an internationalist foreign policy, Jack was foisting wrongness on the world. Yet Joe had decided to withdraw from public life and leave it to his sons, vowing never to speak on matters of national import.

Joe set out to use every personal relationship that he could to advance Jack's cause. Jack and his minions complained mightily about anti-Catholic prejudice, but the Kennedys had already learned to have it both ways, publicly condemning those who stood against Jack because of his faith, while privately wooing those Catholics ready to vote for Jack largely because he shared their faith.

Those Protestant ministers who talked menacingly of a popish plot to elect Jack would have given their Sunday offering to learn the extent to which Archbishop Cushing, soon to be a cardinal, was working with Joe to promote Jack's candidacy. "I have told Jack of our talk about getting out the vote and Jack agrees that that and that alone is the only problem," Joe wrote Cushing in May 1958.

If Jack was going to win the largest victory in Massachusetts history, propelling him toward the White House, he needed Catholic voters to get out there in unprecedented numbers. Later in the campaign, Cushing wrote Joe telling him that he had authorized a statement to be read at every mass in the archdiocese of Boston on September 21, 1958, making it practically a religious mandate to vote: "Among our civic obligations, the highest priority should be given to the duty to vote. . . . Those who have not registered are reminded that they must register now to be eligible to vote in the November elections. . . . No good citizen will neglect this important duty."

It was a dangerous business for the princes of the Church to promote Jack's candidacy while seeming not to promote it, and it required a political subtlety worthy of a Medici. Some Church leaders lacked all such sensitivity. In May 1958, the bishop of Covington, Kentucky, wrote Cushing: "I do hope that sometime in the not too distant future Mr. Joseph Kennedy will see his way clear to do something for us. One hesitates to admit this or to include it in a letter, but a project of this kind aided by him, south of the Mason-Dixon Line, would be a great political aid to him in a Democratic state like Kentucky." For speaking so bluntly, the bishop deserved to wear a dunce cap, not a cardinal's peaked red hat. No matter the merits of his case, the foolish bishop would not be given a farthing. Cushing replied that he doubted "if you can get any help from the Kennedy Foundation" because the foundation would now be directed toward research. Then he sent both letters on to Joe.

In the decade that Joe had overseen the Joseph P. Kennedy Jr. Foundation, he had seen how useful philanthropy could be to his family, and especially to Jack. After his son's election to the Senate, Joe had asked Lawrence

O'Brien, then the president of the Hotel and Restaurant Employees' Health and Benefit Fund, to do an analysis of the foundation's role. O'Brien concluded that during Jack's senatorial campaign "a program was carried out in a manner which created good will for the Kennedys as a family. This was bound to inure to the benefit of any member of the family who became a public figure without any attempt at actual exploitation . . . their political enemies are at a disadvantage in bringing the existence of . . . the Foundation into political discussion." The Kennedys did not have to talk about the foundation to reap its benefits, while their opponents could not even admit its existence without harming themselves.

The Kennedy Foundation was an exquisite machine for the creation of goodwill and a perfect device to help Jack in his race for the presidency. And yet, with the guidance of Sarge and Eunice Shriver, Joe now turned the foundation primarily to financing research into mental retardation. The mentally retarded did not vote, and the papers would not celebrate these grants the way they did many of the earlier ones. But so little research had been done in this area that the Kennedy Foundation would have an enormous impact. Jack or Bobby could have criticized their father for such single-minded largess, giving away a chunk of the family's political capital, but they and their siblings celebrated this effort and took part in furthering it.

Although Joe never mentioned Rosemary's name, it is inconceivable that he would have taken such a step without thinking of her life. Rosemary, then, would make her own tremendous contribution to the world, for out of the loathsome lobotomy, the savagery that had called itself science, would come research to change the lives of these often-forgotten Americans and of the unborn. Did Joe feel guilty? Was his support of research into mental retardation an attempt to balance the scale before he met God's judgment? Or was *guilt* a strong enough word to describe Joe's emotions?

Joe had hardly become a man of pure beneficence. At the age of seventy, he had begun to have an old man's jealousies and was suspicious of those who might try to pry away the power that was so much the essence of his life. He had always been a vital, energetic man. He had never accepted life's cards but had drawn from the bottom of the deck or slapped down an ace from up his sleeve. He had no cards to trump this hand. He had always looked ahead: for several years he had been planning a massive mausoleum where he and Rose would be buried in Holyhood Cemetery in Brookline, Massachusetts. Cushing had arranged for the Church to donate a prominent plot of the kind normally reserved for bishops, on which would sit a great marble structure.

Joe was beginning a decline about which he could only rage in sputtering disbelief. "I am really disgusted with the setback that I had last week," he wrote Cushing in June. "I thought that I was in bang-up shape and I had just finished a physical checkup. I sometimes get as leery of doctors as I do of politicians." Joe was suffering in his right arm from a painful neuritis, an inflammation of nerves. Six weeks later he wrote Beaverbrook: "I haven't been fit company for man or beast for six months."

Jack's aides had learned to walk warily around the candidate's father. Joe's relationship with Des Rosiers had ended, and he assumed it was one of his manly prerogatives to hit on one of Jack's campaign secretaries, a beautiful twenty-year-old woman who found his attentions unseemly and frightening.

This was one activity that the two Kennedys, father and son, had perfectly in common. Back in Boston during a campaign fund-raiser. Jack noticed a pretty young brunette looking up at him as he spoke. Afterward, he asked one of his staff members to get her name and phone number. Jack was a forty-one-year-old U.S. senator with aspirations to the presidency. The young woman was a twenty-year-old Radcliffe College student, but that difference did not prevent him from calling her. Nor did it prevent the young woman from replying and agreeing to meet the senator.

Jack had all the charismatic glamour of a movie star. He was aging the way Cary Grant did, with the years heightening his handsomeness, deepening his tanned, manly features. He was irresistible to an adventurous, sophisticated young woman bored by the narrow social rituals of her class and time. Jack was not one for elaborate rituals of seduction; from him there would be no roses, no flowery sentiment, no midnight phone calls, no impassioned vows. He asked the young woman her views of the issues of the day, and that proved seduction enough:

> At the beginning, he would ask me my opinion about politics, about the speech he'd just made, about something he'd read. What can I say? I was twenty years old and it certainly worked on me. I happened to be a kind of bluestocking, and it was important to feed my vanity in that department as opposed to just saying, "Well, gee, you've got pretty blue eyes." I wanted to be more than that. Cliffies were supposed to be smart. And in the courtship phase, it seemed to me that I was special. The fact that he wanted to know what I thought made me think he liked me and knew me as a person. Was that accurate? I doubt it. He liked me because I was pretty and because I came from the right class. My family tree had a statue on Beacon Hill and he liked that a lot.
>
> And me. I was living a novel. Years of daydreaming about

> romance had prepared me excellently for his seduction. Part of me recognized that there was a lack of connection to this man, a lack of intimacy, but it looked so wonderful. In addition, I thought, "Oh, this is amazing. He's handsome. He's glamorous. He's a senator. He's president." It seemed quite wonderful except that it didn't feel good. I was naive. I was pseudo-sophisticated. Above all, I was emotionally isolated, a truly lethal combination. It was important that the phone would ring, that I would be picked up, that we would have dinner together after some event or party and hash it over as if we were really lovers, really companions. But the reality of the connection? Not memorable at all.

Jack saw no contradiction between giving a compelling, idealistic speech and an hour later bedding a coed. He could have it all, and he did. He cynically and compulsively exploited his power in his sexual conquests. He picked up his young mistress at her Radcliffe dorm in his car. When McGeorge Bundy, the Harvard dean of the faculty of arts and sciences, and a man in whom the blood of his Puritan ancestors still flowed true, learned of Jack's audacity, he purportedly was outraged. He did not think it appropriate that Jack, a Harvard overseer, should be seducing young women under the dean's sheltering tutelage.

There was the dangerous, reckless part of Jack, but to see him simply as a hapless roué would be to photograph his loins and call it a portrait. There was still the other Jack Kennedy whose words resonated out of a deep place in his spirit and his mind.

At the beginning of his senatorial campaign, Jack gave Memorial Day talks in Brookline and Dorchester. He did not turn to Sorensen for his words but scribbled some notes on a few sheets of his U.S. Senate letterhead stationery. The words were as structured as if he had gone through half a dozen drafts.

This was a sacred day to Jack, as it was to many in his audience. Memorial Day was not a holiday for shopping or socializing, not a day to be wrenched out of its proper setting and set down next to Saturday and Sunday to make a three-day weekend. It was an occasion to stop and to remember. Jack was a son of war, a philosopher of that experience, and he reiterated the themes of his spiritual life. He disdained politicians who trafficked in patriotism and sentiment, wearing the flag as their preferred costume. That was not what he was doing this morning. His was a noble speech—short, deeply felt, and reflecting the becoming modesty of a man who believed he was a lesser being than those he honored.

"These men . . . died for honorable and eternal things—for home and family—for comradeship—and for the indomitable questions of youth," he told his audience. War was for Jack what it was for brothers—a natural training ground for youth, a field of heroes and heroisms, the plain of honor for every true man. He went on to repeat a theme that had entered his life with the war and Joe Jr.'s death. "Because their sacrifice is a constant stimulus to us on the long road forward, they didn't die in vain. . . . But the world is poor without them for they were the flower of our race, the bravest and the truest."

Jack ended his talk with a passage from John Bunyan's *Pilgrim's Progress.* The classic book of spiritual journey might have seemed far removed from Jack's life, but he apparently knew at least one passage by heart; he wrote the words verbatim and with almost perfect accuracy in his handwritten notes.

> Then said he, I am going to my Father's; and though with great difficulty I am got hither, yet now I do not repent me of all the trouble I have been at to arrive where I am. My sword I give to him that shall succeed me in my pilgrimage, and my courage and skill to him that can get it. My marks and scars I carry with me, to be a witness for me, that I have fought His battles but now will be my rewarder. . . . So he passed over, and all the trumpets sounded for him on the other side.

It was the Jack Kennedy who spoke words like these who inspired the sons and daughters of Massachusetts. This was the man for whom the people of Massachusetts voted in unprecedented numbers in November 1958, giving him 874,608 votes, or 73.6 percent of the total. They knew nothing of the Jack Kennedy who holed up in a hotel with a young woman. Nor did they know of the pain he so often suffered and the silvery needle that he injected into his body, deadening his pain. There were several Jack Kennedys, or at least several disparate parts of Jack Kennedy, and as he set out in earnest to win the presidential nomination, both the best and the worst were alive within him.

Teddy and Joan's wedding was scheduled to take place three weeks after Jack's landslide reelection. No one expected Teddy's wedding to be as magnificent as those of his big brothers. Teddy was only a law student, not a senator, and though the Bennetts were well off, they were not immensely wealthy like Bobby's in-laws or Jack's mother-in-law. But it would be a major wedding, with 475 guests filling St. Joseph's Roman Catholic Church in Bronxville.

Teddy was like a spectator at his own wedding, taking a walk-on part in an important family ritual. Teddy wanted Father John Cavanaugh, the president of Notre Dame University, to officiate at the wedding. The wry, urbane churchman was a dear friend of his father and the family. Shortly before the ceremony Teddy came to the priest and said that he had changed his mind, though it was clearly his father's decision. Teddy's was a dynastic wedding, and he had to be married by Cardinal Spellman, the most celebrated Catholic leader in America.

Teddy's bride was a virgin, and that aspect of the marriage had its natural appeal, but even their honeymoon was tied up in family ambitions. "We almost had to go to Lord Beaverbrook's house in Nassau," Joan recalled. "Joe said to Ted and me that this good friend of mine has this lovely house in Nassau. And you should go down." As the youngest child, Teddy was a natural courtier, a genial supplicant to his siblings and his elders, a young man with a perfectly honed sense of who mattered and who did not. Teddy wrote the press magnate a letter that could have been a model for Emily Post. "I remember at the closing days of Jack's campaign that you were kind enough to extend to Joan and myself an invitation to visit you on our honeymoon from November 29th to December 3rd," he began in his handwritten letter. It was a perfect beginning, reminding Beaverbrook of the invitation, and listing the intended dates. "I couldn't help but think that my brother Bob did trap you into extending this invitation, but I am much to [*sic*] ungentlemanly and entirely too excited about this prospect even to let this opportunity go by."

As the wedding day approached, Teddy's feelings of dread mounted. This was not the nervous stomach expected of a bridegroom, but a deep sense of disquiet. Joan was full of immense unease as well. Neither of them, however, felt comfortable enough with the other to voice any of their doubts.

Teddy sensed that what he was going to do was not right, not right for him, and not right for Joan. He was getting married because he was supposed to get married. He was being led up to the altar by his parents' firm hands to say vows he was not ready to say. He waited for the day like someone about to have an accident who knows that in the final instant there is nothing he can do, that he can neither veer away nor even brace himself, but must simply wait and see what damage is wrought.

Jack had been filled with his own sense of dread before his wedding, though nothing as overwhelming as what Teddy was feeling. Jack should have understood what Teddy was experiencing. Jack knew that Teddy was not the vibrant, carefree young man he appeared to be. "Teddy, who has all the physical apparatus of being rather easygoing, a rancher or something, you know really taking life easy, he got an ulcer for a while there last year because he didn't know if he was going to be able to stay in law school, because he's not terribly quick," Jack told Burns in 1959. For Jack, the ranch was his metaphor for freedom, for that road he had not taken. Jack did not seem to grasp that it was not just his studies that troubled his little brother but his impending wedding. Or more likely Jack chose not to tell his biographer that unpalatable truth.

Teddy was always doing whatever his father wanted him to do, and in marrying Joan he would be doing it again. "He stayed in and has done pretty well," Jack continued, "but that demonstrates that's really induced by outside pressures, because he's most physically well-balanced and healthy but I mean it's gotten to him now so he's doomed to this treadmill too." Teddy would be getting on the same arduous road that Jack was traveling now, and as his big brother saw it, there was no room to turn around and head back. Teddy's anguish, then, was not a trivial matter. This marriage would put an end to most of whatever secret dreams Teddy had for a life in a wild, untamed world far beyond the family compound.

During the last week before the wedding Teddy went out drinking with his friends in New York nightclubs. He had a trust in the world beyond his family that his brothers no longer had. He talked to his friends about his doubts, and they shook their heads and said that there was nothing he could do. On the evening before the ceremony, when all should have been joy and anticipation, he took a close old friend aside and said that he feared he was making a dreadful mistake.

On that late November 1958 night, a great storm fell on Bronxville. Frigid winds blew at near-cyclone force, felling trees that had stood for decades, shattering windows, knocking down power lines, leaving much of the town in darkness. Even such a merciless act of nature as this was not going to postpone the wedding. In the morning the guests arrived, driving warily through streets littered with downed trees, flooded gutters, and broken glass. Bracing themselves against the frigid winds, they ran into the church.

As the last arrivals hurried into the vestibule, they were startled by a brilliant burst of light. One of the wedding gifts was a professional film of the wedding, and the church was lit up with floodlights like a Hollywood set. All of the participants wore hidden microphones. Jack was Teddy's best man. As the two brothers stood behind the altar waiting for the ceremony to begin, Jack could see the anxiety on his baby brother's face. Teddy was about to take vows of fidelity. Jack told him that he would still be able to do what he pleased when he got married. Jack was the best example of that. Teddy could have whatever women he wanted and live much as he had as a bachelor. These were unseemly words of advice for a wedding day, but they were an attempt to bolster Teddy and to prepare him for all the obligations that he would face as a Kennedy man. For Jack, sex was a residue of freedom outside his world of obligations and ambitions and he was telling Teddy that he could have that freedom too.

After the ceremony Teddy and Joan flew to Nassau. On the third night of their four-day honeymoon, Beaverbrook decided that his houseguests needed a romantic interlude by themselves. A boat took the couple to an isolated island where they could spend twenty-four hours by themselves. "We were dumped there for an overnight," Joan recalled. "It was the worst experience of our life. It was a little cottage, practically a shack, on this tiny island, just sand. We slept on these mats. There were bugs, and it was a nightmare."

When the newlyweds returned from their honeymoon, Joan sat down to view the wedding film. She watched in appalled fascination at Jack giving his unconventional marital advice to his little brother. Joan had so wanted to believe that all her doubts were silly anxieties, but watching the film renewed her premonitions of what would face her as a Kennedy woman.

"When Ted told Jack about the 'bug,' Jack was really embarrassed," Joan said. "Jack blushed scarlet when he found out." Jack had been doubly careless, careless in the words he had spoken, and careless in the way he had left them lying around as he did his clothes, leaving them there for somebody else to pick up.

After their short honeymoon, the newlyweds moved to the house in Charlottesville where Teddy would be finishing up his last year in law school. Teddy was a wealthy young man, but he figured that being married might save him some money. "I do remember that when I moved into the house. Ted dismissed the maid!" Joan recalled. "I had to clean, cook, do the laundry, and I really learned a lot. It was fun—for a while!!"

Teddy's parents were correct in thinking that marriage would bring a new discipline to their youngest son. "Cadillac Eddie" could not go roaring through life any longer. He ended up building his own house on Squaw Island near the family compound in Hyannis Port so that he and Joan could join the rest of the family there. In Virginia, Teddy settled down to a more sedate life as a married student, and he and Joan developed a loving relationship. He poured himself into preparing for the school's prestigious moot court contest in which he and Tunney argued a mock case in front of a distinguished panel of jurists that included Stanley Reed, a Supreme Court justice, and Lord Kilmuir, lord chancellor of England. There were forty-nine other teams of ambitious attorneys-to-be, but as most of the Kennedy family sat in the hall, Teddy's voice soared above the words and logic of all the others, and he and Tunney won the competition.

Upon graduation in June 1959, Joan thought that her husband would now have time for his wife. But Teddy secreted himself away in Jack's old Boston apartment on Bowdoin Street to cram for the law boards.

When that hurdle was passed, Teddy was off again to work in the most important endeavor of his life—helping Jack win the Democratic presidential nomination.

19

"A Sin Against God"

On the first day of April 1959, Jack sat outside on the veranda of the Kennedy home in Palm Beach outlining a plan to win the Democratic nomination for president of the United States. Around him were most of the closest associates who would carry out his strategy. First of all, there was thirty-three-year-old Bobby, who, wherever he alighted or however much he paced back and forth, sat always at his brother's right hand. Jack's father was there as well, and though seventy-year-old Joe appeared at times to have reached an old man's slippered years, he could still rise up and speak with a force and perception that guided everyone, including his eldest surviving son.

Jack was forty-one-years old, on track to become the youngest elected president in American history, and all of his associates resonated with youth as well. Thirty-one-year-old Steve Smith was the other family member there that afternoon. Steve may only have been a brother-in-law, but unlike Peter Lawford and Sarge Shriver, he was so much accepted within the inner sanctum that he had nearly become another Kennedy brother. Steve had a charm and wit that rarely left him, and these qualities were valued by Jack and Bobby as much as his political savvy.

Thirty-year-old Sorensen had been anticipating this moment since the day he arrived in Jack's senatorial office. He was a crucial presence too, affecting a courtier's subtle airs, dropping his words into the dialogue with perfect acumen. Thirty-five-year-old Kenny O'Donnell was also present, as brash and caustic as ever, with his flitting, penetrating Irish-American eyes. There was yet another pair of shrewd Irish-American political eyes there that day; they belonged to Larry O'Brien, who had moved down from Massachusetts to join the operation full-time. Louis Harris, the pollster, had been invited as well; he was decidedly a Kennedy pollster, for he always seemed to have results more favorable to Jack than his colleagues. The other person present was Robert Wallace, Jack's legislative aide.

Although the campaign was not yet organized, Sorensen noted later that the meeting had a tone of "quiet confidence." As he saw it, these men had "a job to be done," and they were the people to do it. That confidence colored the whole gathering and emanated from the candidate himself.

Jack's potential competitors for the Democratic presidential nomination all had more distinguished legislative or political careers than he did, but he had taken each man's measure and found that the nearer he got to them the taller he stood. As Jack saw it, the distinguished-looking Senator Stuart Symington of Missouri had the gravity of a helium balloon. Senator Lyndon Johnson was made of weightier stuff, but he was a southerner, and that was an albatross that even the adroit Lyndon could not wrest from around his neck. Senator Hubert Humphrey was simply too liberal; his political medicine kit was full of purgatives and medicines too strong for most Americans. As for Adlai Stevenson, twice his party's nominee, the man was a proven loser.

To win, Jack knew that he could not step gingerly through the primaries but would have to run boldly forward among the people, many of whom now barely knew his name. Jack was a brilliant geographer of America's political landscape. He grasped the nuances of political America the way his grandfather Honey Fitz had understood the world of Boston's North End. Non-Catholic staff members would have to be the ones to go out to certain states. Bobby, for his part, would do a series of speeches in the South, where he was being lauded for his attack on corrupt unions. Jack's staff had already amassed a file of fifty thousand five-by-seven cards listing the names of crucial people in all fifty states. Jack had met most of these people. The cards listed the name by which he called them, where he had met them, and why they were important. In the age before widespread use of computers, this card collection was a treasure unique to Jack's candidacy.

Jack went on ticking off each state, naming its Democratic leaders and setting out a unique approach to take there. He was conceding nothing, not even Texas to its native son, Lyndon Johnson. As Jack talked on, his father interjected, "I want you to read Judge Landis's column." Joe's comment was an intrusion that no one else at the meeting would have made. Jack took the paper, tucked it under his arm, and went on with his state-by-state litany. Landis, his father's friend, had helped Jack on many occasions with ideas and articles, but there was plenty of time to read the material. "Give that back to me!" Joe exclaimed. "You'll just lose it!"

Joe was not the presence at this meeting that he had been during earlier campaigns, and it rankled him that Jack was not listening closely enough to his counsel. Joe was, if anything, even more determined that he would have a major role in Jack's victory. Later, in the study, Joe affirmed the one role that was truly his. "I'm going to tell you, we're going to win this thing," he said authoritatively. "And I don't care if it takes every dime we've got. We're going to win this thing!"

Bobby turned to his father. "Now wait a minute, Dad," Bobby laughed. "There are others in the family."

The one new face in the Monday and Friday strategy meetings in Jack's Senate offices was Joe Miller, a political operative from the Northwest. Miller had worked on a string of Senate Democratic victories from Oregon to Wisconsin. He had a cocky, joshing air that did not always sit well with Bobby. These meetings had the levity of a scene in the counting room at a Las Vegas casino. Bobby was obsessed with the campaign, and those like Joe Miller who kidded around were squandering time and attention and deserved to be shunted aside.

Jack, however, liked the man he called "Smiling Joe," a brusque former football player with the brush-cut hair of an army private. Miller traveled with Jack on a campaign jaunt to Hawaii in July. On Saturday evening at Honolulu's Princess Kaiulani Hotel, Jack was having a meeting with a group of local leaders to discuss the future of Hawaii, but the ever-present Sorensen and the other aides were missing. He had still not shown up the next morning.

"Where the hell are they?" Jack fumed. Miller explained that some of the aides had met young women.

"I brought them here to hunt delegates, not to hump hula girls!" Jack exclaimed. Jack's aides identified so profoundly with their candidate that they, for the most part, adopted his sexual habits as well, a practice that on this particular Sunday morning Jack found less than a compliment.

Miller had a tight working relationship with labor union leaders and Democratic Party principals in the West. In September he headed off on a month-long trip, starting with the AFL-CIO convention in San Francisco, and then visiting ten western states, all paid for by the Kennedy patriarch.

When Miller returned to Washington, he prepared several memos. The detailed, seven-page document addressed to Bobby was a frank rendering of the situation with the labor movement. Miller had been startled at the intensity of the hostility toward Jack among many union people. He detailed the reasons labor leaders mistrusted Jack, one of which was that they had misread his role on the labor bill. Jack's work on the McClellan Committee "created perhaps understandable resentments in the labor movement." As a result, working people and their leaders were angry and felt that Jack was "too rich, too slick, and dominated by his family . . . not one of the boys in the Truman manner."

Miller dictated a state-by-state report on his trip, seeking ways to change the perception of Jack among labor people. Then he typed another memo himself, giving the only copy to Jack's prim secretary, Evelyn Lincoln, a woman who prided herself on her absolute discretion when it came to the senator from Massachusetts. Miller told Lincoln that this sensitive document was for the senator's eyes only. The memo dealt with the single most potentially damaging problem he had encountered.

"A remarkable revelation emerged from my 100-plus dialogues," Miller wrote in an unpublished memoir. "Virtually everyone I talked to mentioned Kennedy's sex life as a barrier to his nomination. I was taken aback. Not that I hadn't heard a story or two myself. Nevertheless, in all my travels, political and social contacts with him, I had seen nothing to indicate that he was a philanderer. He was all business, the business of winning the Democratic presidential nomination."

Miller concluded that this was a serious enough matter that it could blow Jack's candidacy away. In his memo he said that the only effective way to end the rumors was for Jack to keep Jackie at his side on all his trips.

Within a day or two an angry Sorensen took Miller to lunch at the Methodist Building cafeteria. "You have been participating in some ugly talk about the senator's private life," Sorensen raged. "We will not dignify scurrilous gossip by acknowledging it. I am speaking not only for myself, but also for the senator and Bob Kennedy. Such talk will not be tolerated."

Miller was stunned. He was not saying that the stories were true. He did know that Jack's campaign should know about the widespread rumors. He railed back at Sorensen as vigorously as the aide condemned him, and when they left that afternoon neither man had backed off. Miller, though, found that he was not invited to an important strategy session at Hyannis Port, and from then on his role in the campaign diminished.

Jack had received a series of the most explicit possible warnings that he simply had to rein in his sexual conduct. For the past two and a half years he had sat on a committee that in its investigation of corruption in the labor movement showed how mobsters seduced vulnerable politicians and businesspeople with money, favors, or women, slowly enveloping them in a web of deceit. His own behavior in Havana and elsewhere had advertised his predilections to those best able to exploit them.

Already, in March 1959, Jack feared there might be a wiretap on one of his telephones. By then he was having to deal with an obsessed Georgetown matron, Florence Kater, who had rented an apartment to his secretary, Pamela Turnure. Turnure was a sensuous version of Jackie, a provocative, sumptuous presence among the dowdy professional women who largely peopled Jack's office. Kater had taken a picture of a man she said was Jack, with his hand over his face, exiting Turnure's Georgetown residence at one o'clock in the morning. She also claimed to have a tape recording of their activities in the apartment. Kater charged that in July 1958 Jack had confronted her and her husband, threatening that if the couple did not stop bothering him, Leonard Kater would lose his government job. In the months since then, Florence Kater wrote that James McInerney, Jack's attorney, had visited her seven times. She had in her possession what appeared to be a signed note from McInerney dated January 24, 1959, when she handed the attorney copies of the photo and tape.

It was a different time in American journalism, and no newspaper or magazine printed a word of the woman's charges or what Kater purported to be a photo of Jack "racing like a scared turkey bird from his girlfriend's house in his own self-incriminating pose as he tries to run out of camera range." She was, however, an obsessed woman who seemed likely to go far to expose Jack's alleged philandering.

Jack needed no more flashbulbs going off in the Washington night to alert him to the imminent danger of exposure. And now Miller had given him a memo about the ubiquitous rumors. The political operative was a forceful, highly opinionated man who could have been faulted for setting forth his truths unshorn of nuance. But Miller was no liar, and that Sorensen so quickly silenced him sent a signal that others should tiptoe lightly outside Jack's bedroom door and keep their mouths tightly clamped.

Jack believed that he could get away with his conduct without costs or consequences. He felt that others could do the same. His friend Chuck Spalding's marriage was full of the kind of volatility that never surfaced in Jack's relationship with Jackie. "Here comes the agony and the ecstasy," Jack whispered to Chuck as his wife approached. "Why don't you do what I do? Why get a divorce?"

What Jack accepted as a civilized solution was emotionally impossible to the Spaldings, who saw it as institutionalizing the rankest hypocrisy. As Spalding looked back on his long friendship with the Kennedys, he saw that Joe's sexual conduct was a malady that he inflicted on his sons, causing them damage even if they could not see the ravages of their conduct. "It just tears at the human fundamentals," Spalding reflected. "When I first saw Jack—coming from a Catholic family—it was good to see some of that animal freedom. Some of it was like a soldier home from the war who has run into normal life. How many people think they can take Gloria Swanson on their vacation and make it work? But that doesn't mean others can make it work. It left the Kennedy men with vulnerability in that area. It was like a contagious disease."

Jack was no out-of-control Lothario ready to sacrifice his political future on the sweaty altar of sexuality. He was after the greatest prize in American political life, and there were days, even weeks, when he had no time or interest in yet another momentary dalliance. But he had one favored way to relax, and nothing was going to change that.

As for his wife, Jack may not have been sexually loyal to Jackie, but he deeply appreciated her wry, mocking quality. If not for the exigencies of politics in a democracy, he probably would have enjoyed standing aloof with his wife and looking with her in disdainful amusement at what passed as humanity. He dictated a letter to her after a visit to Newport, probably in the summer of 1959, that exhibited those qualities in full measure, especially in his description of a dinner party. "I was taken into the kitchen and introduced to all the help who were just over from Ireland," he said into the Dictaphone. "I find them more attractive than the guests." He shared with Jackie an overwhelming concern for the sheer physical attractiveness of humans, wincing at the sight of ugliness. "Jenny Ryan was there with her rather squinty-eyed children for a five-week period," he said. "Mrs. Shaw [the nanny] is the loveliest figure actually on the beach and has a beautiful red-brown bathing suit that goes with her hair. She has let herself go however slightly around the middle."

Shortly after Jack formally announced his candidacy on January 2, 1960, Jack and Jackie flew to the Half Moon Hotel and Cottage Colony in Jamaica for their last vacation before the onslaught of the campaign. Jackie had at times been desperately unhappy with her marriage. She had recoiled from the heat and fire of publicity, but what she had faced until now was no more than a match compared to the bonfire of attention that would greet her in the next months. She had a strange premonition that Jack and she might die on this Caribbean island. She wrote a last will and testament and mailed it to Evelyn Lincoln. "If we don't arrive back from Jamaica will you please send whats [*sic*] below to Jack's lawyer—Jim McInerney—my will! Otherwise just tear it up!!"

> I Jacqueline Bouvier Kennedy wish to make provision for my daughter Caroline—that in the event of her parents deaths she should go to live with her fathers youngest brother Edward J.

> Kennedy and his wife Joan—to be raised as one of their own children. . . . Everything I have should be left to her—money, furniture, jewelry, etc—
>
> Signed
> Jacqueline Bouvier Kennedy
> Jan 11, 1960

Even after announcing his candidacy, Jack continued unabated in his pursuit of women. He seemed heedless of the risk, taking chances in circumstances in which he should have been more circumspect, and with the kinds of women whom he once had considered beneath him even for an evening.

Hoover's FBI began receiving unsubstantiated reports about Jack's conduct. A guard in the Old Senate Office Building told an informant that in July 1959

> he was checking offices in the Senate Building one night and noted on the top of Kennedy's desk a photograph openly displayed. This photo included Senator Kennedy and other men, as well as several girls in the nude. It was taken aboard a yacht or some type of pleasure cruiser . . . the thing that disturbed him most was that the senator would show such poor judgment in leaving this photo openly displayed and said that other members of the guard and cleaning forces were aware of the photograph and that Kennedy's "extracurricular activities" were a standard joke around the Senate Office Building.

Far more seriously, on March 23, 1960, Hoover received a memo containing allegations from an informant friendly with various hoodlums, including Meyer Lanksy, whom Jack had probably met in Havana in 1957. The man said that he had been told that when Jack was in Miami, an airline stewardess had been sent to his room. He also said that

> in Miami he had occasion to overhear a conversation which indicated that Senator Kennedy had been compromised with a woman in Las Vegas, Nevada. He stated that he knows that Senator Kennedy was staying at the Sands Hotel in Las Vegas about 6 or 8 weeks ago during the filming of a movie entitled "Ocean 11," starring Dean Martin. He stated that he observed Senator Kennedy in the nightclub of the Sands Hotel, during this period, but has no idea as to the identity of any possible female companion.

Jack had been in Las Vegas precisely when the informant said he had. There he met Judith Immoor Campbell, a stunning, twenty-six-year-old woman known now as Judith Exner. Fate takes strange shapes: a Japanese destroyer cutting through the Blackett Strait, thrusting a PT-boat captain into a hero's role; a brother blown up in the skies of England, leaving his sibling a wealth of obligations; an occasional lover scrawling her name in bold letters into the history of our time. Jack's life had no larger ironies than this—that he should be first toppled from his pedestal of public virtue not by some terrible act of misfeasance, but in large part by a woman with whom he had one of his myriad liaisons.

Jack had flown in from New Mexico on February 7, 1960, to attend Frank Sinatra's show at the Sands Hotel that evening. Sinatra, one of the biggest names in show business, was a liberal Democrat and one of Jack's most fervent supporters. The singer also had an ever-changing entourage of available women, and they were there that evening in ample supply at Jack's table. One of the other guests, the journalist Blair Clark, recalled the women sitting there as a collection of "some bimbos and some show girls," hardly the type of woman whom he and Jack had generally associated with when they had been together at Harvard, and none of them notable enough to stay in Blair's memory.

Sinatra had invited Exner to fly in from Los Angeles for the weekend, or so she claimed. The singer had ample reason to believe that Exner might well provide a midnight treat for Jack. Three months before, Sinatra had met Exner at Puccino's, one of his favorite restaurants. The next day he called and invited her to join him that day on a Hawaiian vacation. Exner flew out of Los Angeles that night and by the following evening was in the entertainer's bed. The affair was short-lived, largely, according to Exner, because Sinatra tried to involve her in a sexual threesome, an invitation that she declined. She was not distressed enough, however, to turn down Sinatra's invitation to come to Las Vegas to see his show at the Sands.

Exner recalled that it was Teddy, not Jack, who made a pass at her that evening, inviting her to fly to Denver with him. Jack, for his part, supposedly called and invited Exner to lunch the next day in Sinatra's suite. Exner asserted later that no sexual encounter took place that long afternoon. Instead, for three hours "the main topic of conversation, once I had given him my family history, was religion."

Jack was not given to lengthy dialogues about religion and may never have had a three-hour discussion with a woman in his life. Exner, moreover, was largely uneducated. She had the mascara-thin layer of culture acquired by many Hollywood actresses and would-be actresses; it largely consisted of the judicious application of a few multisyllable words and an accent suggesting that its speaker had at one time passed through London.

Jack had no way of knowing that much of what Exner said about her family history was simply not true. She fancied that she, like Jack, came from a wealthy, privileged family. She described her childhood home as an elegant, twenty-four-room mansion in Pacific Palisades so enormous that she found it "kind of spooky." There is no evidence that Exner's father, Frederick Immoor, a project architect, ever owned the house or that the family lived in it for an extensive time. The family may have stayed there for a period while Immoor was renovating the mansion, or in partial payment for his services. In any event, a more realistic version of her childhood would be painted in far more modest hues. The family traveled from rented house to rented house, in Pacific Palisades, Chicago, New Jersey, Phoenix, North Hollywood, L.A., sometimes living well, other times only a few steps from insolvency.

Exner never finished high school and says that she was privately tutored to get her degree. As a teenager, she became one of those young women who hang around the studios with vague dreams of becoming a star, a singer, a notable, somebody. She was stunningly beautiful, her face set off by thick eyebrows that another woman would have plucked but that emphasized her exquisite, dramatic features. She met Bill Campbell, an actor in his midtwenties, who married her when she was only eighteen. The marriage was a disaster from the beginning, and by 1958 Exner was a single woman again with alimony of $433.33 a month.

Exner said she was "financially independent," with "family money [that] kept her in furs and steak Diane." That was simply another illusion. In her 1958 divorce proceedings, she presented to the court a signed statement that she was "without sufficient funds or income to maintain or support herself either permanently or during the pendency of this action." Shortly after separating from Campbell, she moved in with another man, Travis Kleefeld, but left him in the fall of 1959. By then she owed $2,784 on a $3,145.50 time-plan loan at the Bank of America and was behind in her car payments. She called herself an artist and an interior decorator, though she never made a cent from either profession. In her entire life, her only daily employment was a two-month-long "public relations" position, working for the comedian Jerry Lewis at about $100 a week.

When Jack met Exner, she was living off and on with her parents. Her father, who was earning $866 a month, was in such financial straits that he took out a number of loans that year, the last one in December for $2,032 to consolidate his debts and pay for Christmas gifts.

Exner called herself an artist, but her primary creation was the illusion of affluence. She was shrewd, not smart, with an affinity for older gentlemen whose major virtues were their money and their largess. Just two weeks before meeting Jack, she had been in Las Vegas staying at the Sands as the guest of Richard Ellwood, a middle-aged businessman who later became publicly known as a "boyfriend."

Another of her friends was John Rosselli, a dapper fifty-four-year-old gangster with close connections to the film industry who frequented the same nightclubs that Exner did. Rosselli was one of the top Mafia figures on the West Coast, and his biographers have speculated that "almost certainly it was Rosselli who produced his friend to meet Kennedy." Another proponent of that position is Fred Otash, a private detective with a propensity for wiretapping. It was from Otash, who had friends in low and high places, that the FBI may first have learned that Jack and Exner were having an affair when he suggested to the agency that she was "shacking up with John Kennedy in the East."

"It was common for Sinatra and his various friends to 'pass a girl around,' that is, members of the clique were expected to introduce comrades to sexually satisfying young women of their acquaintance," Otash told the House Select Committee on Assassinations in 1978. "He stated his belief that this was how Rosselli 'planted' Judith Exner on John F. Kennedy."

Sinatra had his mob connections too, and it is possible that he invited Exner to Las Vegas to compromise Jack. But the singer was devoted to Jack, considering it a high honor to be in his company. Moreover, if Sinatra had done so, he probably would have given Exner a room at the Sands Hotel, where he was performing. Instead, she stayed at the Tropicana Hotel for three nights.

From this weekend on, Exner traveled for the next three years at the highest level of luxury, crisscrossing the country again and again. She invariably flew first-class, wearing high fashion and fur and apparently paying cash for tickets and accommodations. She had so much luggage that she paid for as much as one hundred pounds of overweight luggage. When she arrived at her destination, limousines took her to the finest hotels.

Exner simply did not have the resources herself to travel this way. As generous as her gentlemen friends in Los Angeles may have been, they did not provide her with enough to support so extravagant a lifestyle during much of her time away from California. Clearly her journeys were bankrolled by individuals who had more in mind than simply bedding the beautiful young woman.

When Jack invited Exner to meet him at the Plaza Hotel on March 7, 1960, she said that he offered to pay for her ticket. She claimed that she refused, yet flew first-class to New York City to stay at the exclusive hotel where Jack was also staying. Exner had just received a $6,000 final settlement from her former husband in lieu of further alimony. At this moment, then, she knew that she would no longer have that regular monthly stipend. The $6,000 was undoubtedly the most money she had ever had at one time, a golden nest egg that she could use to advance her fortunes in the world.

Exner arrived in Jack's room that evening looking like a privileged young woman of wealth and bearing. "It was a wonderful night of lovemaking," Exner recalled. "Jack couldn't have been more loving, more concerned about my feelings, more considerate, more gentle. . . . The next morning he sent me a dozen red roses with a card that said, 'Thinking of you . . . J.'"

A week later Exner flew to Miami, where she said Sinatra had invited her to attend his performance at the Fontainebleau Hotel. Although her host had supposedly viciously insulted her, that did not prevent her from attending the farewell party for the last evening of Sinatra's show. It was there that she says Sinatra introduced her to "a good friend of mine, Sam Flood," one of the many aliases of Sam Giancana, the leader of the Chicago mob. The next evening Exner said that she had dinner with a group that included Giancana, still, by her admission, knowing neither Flood's real name nor his profession.

Exner may have known far more about Giancana than she admitted. Jeanne Humphreys, the wife of Murray "the Camel" Humphreys, a leading mobster associated with Giancana, tells a different story in her unpublished memoirs. "Johnnie Rosselli [*sic*] who had been taken back into the fold was becoming a frequent visitor to Chicago bringing gossip about Mooney [Giancana] and his reveling with the 'ratpack.' He [Rosselli] said he had fixed [Giancana] up with a party girl that he'd taken to Florida when [his mistress] Phyllis [McGuire] wasn't looking. I said the last girl I'd seen him [Giancana] with was named Judy and she was from Chicago. He said he meant Judy from California."

Giancana was attracted to Exner in part because he had a constant need for variety. The mobster could be as generous as a pasha to his amours, but there was always a price, even if it was not obvious at first. One of his lovers at the time was Marilyn Miller, a showgirl at Chicago's Chez Paree. She bragged to her girlfriends that Giancana had spent over one hundred thousand dollars on her. A friend who had offended the mobster asked Miller to intercede with the syndicate chieftain. She tried to talk to Giancana, but he told her to be quiet. Soon after the nude body of the man was found in Chicago. Another woman Giancana was seeing at the time was Patricia Clarke, a Florida woman with four children. Although Giancana was generous to the single mother, Clarke made the mistake later that year of telling him that she was also seeing Angelo Fasel, a bank robber, and that she intended to marry him. Within a few weeks Fasel disappeared, and Clarke was told not to ask about him anymore.

Exner flew from Miami to New York, where she apparently stayed with friends before traveling to Washington on April 6. That evening she took a taxi from the Park Sheraton Hotel to Jack's house at 3307 N Street, N.W., in Georgetown. In Exner's 1977 autobiography, she described the evening as another gloriously romantic encounter. A decade later, in a 1988 *People* magazine cover story for which she was paid $50,000, and subsequently in other books, magazine articles, and television interviews for which she was almost inevitably paid substantial fees, Exner told a dubious tale about Jack and the mob.

Exner claims that Jack asked her to "quietly arrange a meeting with Sam for me" during their hours together because he thought he might "need his help in the campaign." Jack allegedly asked Exner to set up a number of meetings with Giancana and gave her a satchel containing "as much as $250,000 in hundred-dollar bills" to deliver to the Mafia kingpin. Exner believed that Jack had chosen her because "I was the one person around him who didn't need anything from him or want anything. He trusted me. I had money from my grandmother."

Exner's story is an anomaly among the many unverified tales of mob influence on the 1960 presidential election; the others involve the gangsters using their own money to influence the outcome, not the Kennedys apparently attempting to pay off the mob. Exner's whole life was a tissue of fabrication, from the story of her wealthy birth to the way she made her living. Even if she had enjoyed a reputation for veracity, her story would have been impossible to believe. Carnally was the only way that Exner knew Jack. By her own admission, Exner was seeing Jack that evening for only the third time in her life. As for Giancana, again, by her own admission, she had spent one evening with him in a large group and knew nothing about his role as a major figure in organized crime. So if Jack had made such a request, Exner would have found it a non sequitur.

As tough and cynical as Jack could be, nothing in his previous political life suggests that he was so devoid of both scruples and common sense as personally to enlist the Mafia as his partner. Even if one wanted to imagine a Jack cynically subverting American democracy, it was unthinkable that either he or Giancana would use Exner as their go-between. Neither man had achieved his power through such monumental stupidity as picking up a woman in a nightclub and deputizing her a few weeks later to foster an unprecedented act of political corruption that, if discovered, would destroy both men. Giancana came from a mob culture that kept women in the bedroom or the kitchen; he neither confided in them nor for the most part involved them in criminal acts. Jack had an equally limited opinion of the role of women in the games of power.

Giancana may have been willing to participate in corrupting the American presidential election, but he had not committed so many crimes with impunity by leaving his fingerprints on the bloody knives and the bags of loot. And even those ready to imagine a John F. Kennedy who would have attempted to subvert American democracy surely must realize that he would have kept his own clear distance from such an act.

Exner took the overnight train to Chicago, where, she asserts, Giancana was waiting at the station for her in the morning. Giancana's biographer tells a different tale of how she later made her connection with the mobster. "She came to Chicago for a week and wandered around the Ambassador West Hotel, making it plain that she was trying to get closer to Giancana and his Oak Park Home,"writes William Brashler. "Finally taking a room in the Oak Park Arms Hotel, she was seen coming and going from the hotel to the Giancana house for a few days after, then she left town."

That was essentially the version that Giancana told Joe Shimon, a Washington, D.C., police officer who became associated with the accused murderer: "She found out about Sam and tried to find him in Oak Park," Shimon recalled. "Everyone knew where he lived. She was hinting that she had great connections, that she could do a lot of good. Sam said, 'Do you remember that strumpet? She was trying to get on the payroll?' "

Jack sat in front of the Dictaphone in his Senate office musing about his life. It is unclear whether he meant these remarks for the autobiography that he would one day write or for some other purpose. Whatever he intended, he would have little time for such recollections once the campaign began. He was in his early forties, but these were an old man's words, reflecting back on the sweep of his days, seeking out the themes of his life. He was an immense puzzle of a man, his character so complex that no matter how one turned the pieces around and tried to put them together they never quite seemed to fit. There was the man Sorensen and Feldman observed, a brilliant tactician with a stunning quickness of mind and decision. There was another Jack whom Ben Bradlee, his journalist friend, and a few others saw, an American gentleman endlessly amused by the foibles of the human race. There was the man whom Smathers and some of his colleagues experienced, a Jack speaking the most vulgar argot of politicians, the language itself not nearly as low as the sentiments. There was a discolored piece of him, the sexual epicurean who let neither his marital vows nor ambition stand in the way of feasting at this sweet buffet. And there was the hidden sickly man, taking his regimen of pills and pretending he was something that he was not.

As Jack spoke into the Dictaphone, he was displaying yet another piece, and one that did not fit easily among the others. As a political philosopher he stood apart from the practical man of power. This John F. Kennedy was a man of deep contemplation. There was a distance that Jack maintained from the rest of humanity, in part because of the natural isolation of power and its pursuit, but in equal part because of his very nature. And it was out of this distance, standing back from the shrill shouts, the pettiness, the rude exchanges, the duplicities, the preening ambitions of much of Washington life, that he observed what he thought to be the natural greatness of politics and a political life.

Jack was a public man who could speak extemporaneously in perfectly ordered prose. On this occasion he welded his sentiments into an essay, stopping periodically to find the precise words, speaking with such subtlety and nuance, even pausing to signify commas, periods, and underlining, that he could have been reading from a script. Jack began by pondering the fact that politicians were held in such low regard.

> Politics has become one of our most abused and neglected professions. It ranks low on the occupational list of a large share of the American [public?]. . . . Yet it is . . . these politicians who make the great decisions of war and peace, prosperity and recession, the decision whether we look to the future or the past. In a large sense everything now depends on what the government decides. Therefore, if you are interested, if you want to participate, if you feel strongly about any public question . . . it seems to me that governmental service is the way to translate this interest into action. . . . The natural place for the concerned citizen is to contribute part of his life to the national interest.

When politicians are perceived as no more than a motley crew of corrupt careerists and cynical panderers, then the whole covenant of democracy is broken. Jack saw his colleagues at their most conniving and self-interested but he perceived an underlying nobility in the politician's life. He saw himself as part of a tradition, as a scion of a race of Irish-Americans who had made politics the chosen avenue of their advance; he was the grandson of two politicians and the son of a man who had talked to him almost daily of politics and a mother who had put her little son on her knee and told him tales of American history. He had come late and hard to regarding the word "politician" as an honorable term, but he shouted the word loudly now and was proud to be called by that ancient name.

> I moved into the Bellevue Hotel with my grandfather [in 1946], and I began to run. I've been running ever since. Fascination began to grip me, and I realized how satisfactory a profession a political career could be. I saw how ideally politics filled the Greek definition of happiness. "A full use of your powers, along lines of excellence in a life affording scope." . . . How can you compare in interest . . . [a] job with a life in Congress, where you are able to participate to some degree in determining which direction the nation will go.

This was a Jack who celebrated a political life and saw it as a public man's noblest pursuit. He may have fallen short in many ways, but that did not diminish the ideals he professed.

> In looking back, I would say that I've never regretted my choice of professions, even though I cannot know what the future will bring. . . . Particularly in these days where the watch fires of the enemy camp burn bright. I think all of us must be willing to give ourselves to this, some of ourselves to this most exacting, to the most exacting discipline of self-government. The magic of politics is not the panoply of office. The magic of politics is participation on all levels of national life in an affirmative way, of determining, of playing a small role in determining whether . . . in Mr. Faulkner's words, freedom will not only endure, but also prevail.

The first great contest of 1960 was the Wisconsin primary. This was a special challenge to Jack since it was adjacent to Humphrey's home state of Minnesota. Jack's opponent may have been the kind of overwrought, weepy liberal whom Jack usually privately despised, but there was nary a mean bone in Humphrey's ample frame, and Jack bore the man no personal animus.

A presidential campaign usually begins with as much indifference as enthusiasm. Each moment of elation is paid for in the hard cash of tedium, exhaustion, bad meals, cold coffee, predawn alarms, and late-night flights. A man of Jack's sensitivity, as much aesthetic as political, was especially vulnerable psychologically to the phlegmatic unconcern that he often encountered.

The coldness of the Wisconsin winter was at times matched by the icy reserve of the Midwestern farmers and townspeople. In one of the nameless small towns Jack had given his speech to yet another undemonstrative crowd and tried afterward to grasp a few hands in the local restaurant. He strode to the back where a group of men sat around a table playing dominoes. No one stood or greeted the senator from Massachusetts, even as Jack paraded around the table shaking hands. "We planned to come to hear you speak," one of the men said, looking up from the dominoes, "but we didn't finish our game." Jack smiled, shook hands, and left the room as if nothing gave him more pleasure than soliciting such taciturn men.

If Jack had sat down with those farmers, losing a quick game of dominoes while talking about next year's crop, he would probably have walked away with three or four certain votes. But that wasn't Jack. He wasn't a politician like Humphrey or Johnson, a toucher who thought that in the physical act of grasping a shoulder or pumping a hand he was making an indelible impression. He was as uncomfortable being touched emotionally as physically, and he still barely tolerated the gaudy sideshows of politics. If he had had his choice, he would have campaigned standing before audiences of intelligent citizens discussing with subtlety and nuance the issues of the day.

On the worst of these days Jack enjoyed no stronger tonic than a call to his father. Joe had endowed his sons with a restless, unquenchable optimism. It was a spirit that they imbibed from their father whenever they talked to him. The more down they felt, the more they faced defeat, the more Joe bolstered them as if he could lift them up with his very hands.

In the winter of the primaries, Joe spent much time in Florida going to the races at Hialeah and talking to influential people and reporters. Again and again he heard worrisome negative stories about Jack's dubious prospects, and when he talked to Jack, as Rose recalled, he bolstered him with "something positive and enthusiastic, but not necessarily in line with his own experience."

Jack shared much of his father's coldhearted political realism. He might stand on a platform mouthing idealistic paeans written by Sorensen, but he knew that the real business of power and politics often took place in private antechambers where no echo of the campaign rhetoric could be heard. On one of his early campaign jaunts, he chatted with an old-time political hand who told him of his father's role in Roosevelt's 1932 nomination. Despite Joe's "very low estimate of Roosevelt's ability," he had backed FDR, trying to manipulate him into taking less liberal positions. In a crucial moment, Joe had helped to talk William Randolph Hearst into backing Roosevelt simply because the press magnate hated another contender. Joe's conduct had been the most cynical of ripostes, and another son might have flared out at a man who would dare slander his father with such a tale. Jack was impressed enough by the story that he sent a copy of the letter to his father with a note stating, "It definitely shows the success you had in securing Hearst for Roosevelt."

As Jack campaigned on those endless cold days, there was an increasingly liberal glaze to his words. His primary speechwriters, Sorensen and Feldman, were far more to the left than the candidate they served, but it was hardly a matter of their promoting ideas that Jack would not have chosen on his own. Jack knew that to win he had to appear liberal, so as to gather in a reluctant labor movement, urban intellectuals, and social activists, despite his disdain for the priests of that particular faith. "He had real contempt . . . for the members of that group in the Senate," reflected Joe Alsop, the conservative columnist. "What he disliked . . . was the sort of posturing, attitude-striking, never-getting-anything-done liberalism."

Jack did not like being around liberals such as Adlai Stevenson, whose very manhood he doubted. As for professional diplomats, they were little more than eunuchs. "I know how they are at the State Department," he said. "They're not queer, but, well, they're sort of like Adlai." Jack hated being grouped with liberals like Adlai. "I'd be very happy to tell them I'm not a liberal," he had declared in the *Saturday Evening Post* several years earlier in an unfortunate lapse into candor. "I never joined the Americans for Democratic Action (ADA) or the American Veterans Committee (AVC). I'm not comfortable with those people."

Jack felt that there was a high prissiness to the Stevensonian liberalism of his day. As he saw it, these men preferred to profess pure virtue among their peers than to have their ideas sullied in the grimy arena of political life. Jack truly liked the sweaty, profane, cynical, street-savvy politicians whom men like Stevenson would walk across the street to avoid. When they had to, these ADA liberals would put out their right hands to shake hands with such politicians while keeping their left hands clamped over their noses. Among those progressives who accepted at face value Jack's conversion to their political faith, there was the hope that he was more principled than he appeared but as tough as his legend, and that he might make the term "practical liberal" no longer seem an oxymoron.

Jack made a brilliant move in the way he cultivated the academic community in the Harvard-MIT nexus. The engine of contemporary American liberalism was housed in the elite universities; the candidate thought of their professors not only as a source of ideas but also as a powerful group to be coopted. Among these intellectuals were many men so outsized in their professorial vanity that they could be had for the cheapest of coinages—a request for their advice. He used their ideas much less than they imagined.

The first time Jack met his Academic Advisory Group at the Harvard Club in January, he told them: "I don't want any of you to worry about the politics of the situation. You don't have that skill. Forget it. I'll do that. You just worry about the substance." The nineteen academicians left the meeting that afternoon largely converted to Jack's candidacy. They took his stipulation as a compliment. It could be seen equally as a mark of Jack's belief that "ideas" were only another category that had to be processed through political realities that he felt the academicians did not truly understand.

Back in Wisconsin, Jack thrust his hand out to factory workers in Milwaukee and stood at the bar in American Legion halls in innumerable towns and cities, and he discussed the farm problem and Social Security. The journalist Stewart Alsop, often a far more perceptive observer than his more celebrated brother Joe, found that in Wisconsin, Jack was "an unexpectedly self-conscious and diffident man." On one occasion, the journalist watched as a group of bubbly cheerleaders surrounded the candidate to give a cheer that they had spent much time preparing. "Right sock, left sock, rubber-soled shoes; we've got the candidate who can't lose," they chanted while Jack "wore a rather bemused air, as though he had unexpectedly found himself knee-deep in midgets." Alsop found Jack at his best as a political pedagogue speaking to youthful audiences. Alsop noted that in one short speech at Wisconsin State College at Whitewater, Jack quoted "Aristotle, George Bernard Shaw, Walter Lippmann, Professor Sidney Hook, President Eisenhower, Thomas Jefferson (twice), John Quincy Adams, Daniel Webster, and Abraham Lincoln."

These were the best moments, and though there may have been only a few of them, they were plentiful compared to what Bobby faced on his campaign trail. He went to the places Jack would not go, and did the things his brother did not want to do. In Milwaukee, Bobby managed the volunteers and the professional staff, prodding them and pushing them out into the homes and byways of the state.

During the winter campaign, Bobby left Eau Claire, Wisconsin, on a train with Chuck Spalding, traveling to some little burg to give a talk to a group so small and so obscure that his brother couldn't be bothered. The snow blew up in blizzard force across the frozen countryside. Down and down it fell with such ferocity that the engineer stopped the train in the countryside, still a full nine miles from their destination. While the other passengers sat and waited for rescue, Bobby led Spalding off the train. The two men walked the nine miles through the storm to a meeting hall where probably no one would even show up.

"I don't think that I could say enough to emphasize that aspect of Bobby," Spalding reflected. "Like somebody with a coat turned up, bare-headed . . . just driving from place to place. . . . You can't paint that picture too vividly. I haven't seen it matched by anybody in anything, in any other field. . . . He was searching this thing out for his brother, and he literally couldn't rest."

Young Teddy did not have the tight-jawed, humorless intensity of Bobby, but he too strode manfully into that cold Wisconsin winter in service of Jack's ambition. One afternoon while Jack was soaking his battered back in a motel bathtub in yet another obscure Wisconsin town, twenty-seven-year-old Teddy was pacing in the living room like a football player trying to get the coach's attention so that he could enter the fray. Jack, who could hardly look at Teddy without reflecting on the glories of unbridled youth, sent his little brother out to distribute handbills.

This was the most trivial of tasks, but Teddy approached it as if the election might rest on his afternoon endeavors. In this brutal winter, he was not content to place the handbills under frozen windshield wipers but instead opened car doors to set the political tracts on front seats. As he reached into one sedan, a ferocious bulldog jumped up off the floor and clamped his teeth into a meaty section of Teddy's forearm. He yanked his arm back and looked at the bloody imprint on his arm, a wound that should have sent him scurrying to the emergency room. Instead, Teddy soldiered on, putting the handbills into more cars in the parking lot.

Teddy's greatest opportunity to show his fidelity to his brother's cause came on a ski slope outside Madison. He found that the crowd was far more interested in watching the ski jumpers soaring one hundred feet into the air than in listening to Ted's little speech about his brother. Teddy was a man of endless good humor, and it was probably in that joshing vein that someone shouted out that they would listen to him well and good if he would make one jump himself. What had begun in jest became a deeply serious challenge. Teddy trudged up the slope wearing boots that were not his own, carrying skis he had never used, to attempt a sport that he had never tried or even seen. "I went to the top of the 180-foot jump," he recalled, "and watched the first three jumps. Then I heard the announcer say, 'And now at the top of the jump is Ted Kennedy, brother of Senator John F. Kennedy. Maybe if we give him a round of applause, he will make his first jump.'"

Two decades before, Teddy had stood in a similar place, a frightened seven-year-old on the cliff at Cap d'Antibes, knowing that he had to dive into the water or betray his two brothers who stood below shouting at him to jump. No brother stood below now, but the way Teddy felt, Jack might as well have been there. "I wanted to get off the jump, take off my skis, or even go down the side," he admitted, "but if I did, I was afraid my brother would hear of it. And if he heard of it, I knew I would be back in Washington licking stamps and addressing envelopes for the rest of the campaign."

Of course, that was not true, and Teddy surely knew it, but this was another of the endless challenges of a Kennedy man. He later told his friends that he was as terrified as he had ever been, but he knew that he had to jump, watched as he was by a crowd of more than eight thousand. He pushed off and flew into the air, and crashed to the earth seventy-five feet down the runway, as the spectators scattered. Teddy got up, brushed himself off, and gave his speech for his brother.

Teddy was a rousing celebrant of a campaign worker, and as much as he appreciated the women he met along the trail, he had come to love his family life with Joan. "I have never seen Ted so excited as he was on February 27th [1960] when his daughter, Kara, was born," Joan wrote Lord Beaverbrook. "Ted is away all week traveling around Wisconsin, and now West Virginia, making speeches for Jack. He phones home every night and asks, 'How is my daughter?' I love the way he enjoys using those two new words—my daughter!"

The Kennedys fanned out across the state in multitudes. Three of Jack's sisters, Eunice, Pat, and Jean, dusted off their tea sets and arranged a series of coffee klatches in which the good ladies of Wisconsin met the handsome young candidate and other members of his glamorous family. Wherever Humphrey looked, he saw Kennedys. It was like competing against the mythological nine-headed hydra: every time he thought he had sliced off his opponent's head, there was yet another smiling, garrulous Kennedy greeting voters.

A few days before the April primary, part of the state was inundated by ugly anti-Catholic literature. The pamphlets were sent primarily to Catholics, many of whom had seemed likely to vote for the Minnesota senator. The recipients were outraged by this slander of their faith. It did not occur to many of them to ask why Humphrey, that most genial and unprejudiced of men, would countenance such a foul assault. The pamphlets were enough to convince all but the illiterate that they had best vote for Jack rather than his bigoted opponent. Humphrey's people cried foul, and the conservative *National Review* charged rightly that Bobby's associate, Paul Corbin, had been behind the campaign. And behind Corbin stood Bobby, who, in the words of Edwin O. Guthman, his press secretary, "was willing to do anything to get Jack elected."

Jack won the primary by a resounding 56 percent of the vote, but he had done so poorly in the heavily Protestant regions of the state that his victory was not overwhelming enough to scratch Humphrey from the race. The Minnesota senator vowed to continue on to the West Virginia primary, even though he had only a few coins in his coffers. There was no better champion of the working man and the poor than Hubert Humphrey, and the mountainous state was full of working men, unemployed, and forgotten poor, many of them tucked up in Appalachian hollows. West Virginia was

95 percent Protestant, and if the electorate voted their religion as much as they had in Wisconsin, Humphrey had a good chance of winning. For Jack, these same demographics created a very different logic. If he won, the Democratic presidential nomination most likely would be his. If he lost, he would prove what his detractors had argued all along, that a Catholic was simply unelectable.

Jack had an added disadvantage in this vital race. While Humphrey was the most voluble of politicians, Jack was losing his voice. It was so bad that as he flew across the state in the *Caroline,* the campaign plane that his father had contributed to the campaign at nominal cost, he communicated by writing notes on pieces of paper. "My poll . . . in W.V. showed me *41%*[,] HH . . . 43–44%—rest undecided—but the rest are Protestants," he wrote on a pad of paper. "The reason I think we should do well if we get *over* 40% is *UMW* [United Mine Workers] will get word out *Byrd* is getting *meaner*. The fundamentalists are getting active [preachers]. To drag out 50% under these conditions seems optimistic."

Jack had a serious personal matter on his hands during the crucial West Virginia primary. His personal secretary, Evelyn Lincoln, put in her private files a handwritten, two-page letter dated April 8, 1960. Three days later his press secretary, Pierre Salinger, who has no memory of this incident, sealed the letter while being witnessed by his assistant. The letter stated:

> I talked today with bobby baker [a top Lyndon Johnson aide]. He informed me that three weeks ago an attorney he knew named Mickey Wiener from Newark (?) Hudson Co. called him. Wiener stated that if Sen. Johnson would give $150,000 to the wife of "a well known movie actor" (baker did not know her name or who the actor was) she would file an affidavit that she had had an affair with me. Baker said he thought it was blackmail, and did not inform Johnson of the matter. He did tell Joe Alsop that he was concerned about an attempt at blackmail of me and did not go with the details. . . .
>
> John Kennedy
> April 8th

In his sexual adventures, Jack had begun a descent into provinces he once would never have visited. He had indeed seen a woman who was married to "a well known movie actor." She was Alicia Darr, the former wife of the French actor Edmund Purdom. Darr had apparently first met Jack in 1951, when, according to FBI reports, she was running a "house of prostitution" in Boston. Darr moved to New York City, where, the FBI said, she rendered the same services but with the highest class of client. Darr made the transition from whore to mistress to wife of a movie star. Her marriage failed, however, and in the spring of 1960, Darr was in such financial trouble that she had been jailed for cashing bad checks, according to European papers.

Clark Clifford, a powerful Washington lawyer and lobbyist, recalled that in the spring of 1960 he had been asked by Jack to deal with a matter so serious that "public knowledge could have blown the Kennedy nomination out of the water." If this was indeed the matter, whatever Clifford may have done to end this threat, there is no evidence that Darr blackmailed Jack.

The Kennedys were in what they considered the most difficult, most crucial campaign of their lives, and they threw every weapon they had into the fray. Politicians trade in whatever commodity is cheapest to them, be it access, votes, promises, flattery, or money. The Kennedys were wealthiest in money, and they spread their lucre around West Virginia in large amounts, mainly in cash and doubly in silence.

The sheriffs controlled both the law and politics, and they were the ones whose palms were most generously greased. One of Joe's rich friends, Eddie Ford, got up from his table at Chicago's Statler Hotel, where he held court daily, and drove down to the Mountaineer State in a big Cadillac with Illinois license plates, carrying with him a suitcase full of money. "He'd pick up a sheriff who was powerful," recalled Former Speaker of the House Tip O'Neill, "and he'd say, 'I'm a businessman from Chicago, and I'm on my way to Miami. I think this young Kennedy would be great for the country, and I'd like to give you three thousand dollars to see if you can help him. I'll be coming back this way, and I'll be happy to give you a bonus if you're able to carry the town.'"

Jack met with Raymond Chafin, the political boss of Logan County, and tried to convince the man that he cared about West Virginia's problems; if elected, he would do more than Humphrey for the state. After Jack left, his minions worked on the man some more. Chafin had immense power in the poor county. He controlled all the Democratic election officials—amiable folks always ready to help instruct voters in how to mark their ballots. He got along with the UMW leaders and the bosses at the Island Creek Coal Company. And he was always ready to help get out the vote, whether by putting in a kind word to get someone on the welfare roles, offering a little help paying the electric bill, or supplying a pint of whiskey or a couple of dollars.

The Kennedy people asked Chafin how much would be needed to put

Kennedy's name on the slate cards that he gave voters to take into the voting booth to determine their vote. "Thirty-five," Chafin said, meaning $3,500. A few days before the election Chafin was asked to come out to Taplan Airport outside Logan and to bring a bodyguard with him. There he received two briefcases. Looking in amazement at the bundles of cash sitting there, he realized that Kennedy's people had thought he meant $35,000, ten times what he had proposed.

Humphrey's people had already paid Chafin $2,000 to have the Minnesotan's name on the slate card, and now Jack had royally trumped him. Humphrey spent $25,000 on his entire West Virginia campaign, $10,000 less than the amount in the two briefcases. Both candidates were playing the only game of politics played in West Virginia, but Humphrey was playing with a few copper pennies and Jack with bars of gold.

Chafin said that he used the money "mostly hiring people, drivers and poll workers, babysitters, people like that." The Kennedys' largess was so extravagant, so heedless of true election costs, it was likely that in many instances the money was pocketed.

The 1960 West Virginia primary was the harbinger of the modern political era, not simply in the massive amounts of money the Kennedy campaign spent per voter but in the organization, the use of television, the shrewd meshing of celebrity and politics, and the essentially negative nature of much of both candidates' campaigns. James McCahey Jr., a Chicago businessman with West Virginia roots, organized teachers and other volunteers to create a grass-roots movement for Jack.

The campaign bought television time to put out documentary-like programming that concluded that Jack was the better candidate. Other Kennedys came to West Virginia, not only to sit in the capital of Charleston for photo ops, but also to go up the rutted country roads and knock on doors and to shake hands in crossroads stores. Jackie did not go out on the hustings, but her mere presence was a revelation. Humphrey's wife, Muriel, could have been one of these local ladies, but the West Virginians sidled past her to gaze awestruck at the aristocratic beauty who had graced their modest environs. "They had a wondrous look in their eyes when they saw her," said Charles Peters, then a campaign leader in the state.

If Jackie brought grace and beauty to the campaign, Bobby brought the mailed fist. For him, life was simple. All that mattered was that Jack win, and anyone and anything that did not lead to that goal was rudely shoved aside. When the two candidates staged their television debate, the Minnesota senator was bested in the one field in which he thought he should have been the hands-down winner. Humphrey was not much of a drinker, but at the Charleston Press Club, in his dismayed disbelief, he lifted more than a few. "Bobby, I made your brother look good tonight." Humphrey said, coming up to the Kennedy group. "I'll be the first to admit he won that debate tonight. And who knows? Maybe I made him president of the United States tonight. But I've still got to campaign against you in Wheeling tomorrow morning, and I've spent so much time, I've missed the plane to Wheeling. How about letting me have the *Caroline* to whistle me over to Wheeling?"

Bobby gave Jack's opponent an answer that was as profane as it was immediate. The other men were all Kennedy partisans too, but they respected Humphrey and were embarrassed by Bobby's crude invective. "Well, Senator, I flew out here with Bruce Sundlun, and we're flying back with him, and he's sitting over there," said Kenny O'Donnell, who wasn't afraid of a Bobby who had sat on the bench when Kenny had quarterbacked the Harvard team. "Why don't you go ask him if he'll take you over? And if he can, I'm sure we don't mind . . . dropping you off."

In the last days of the primary election the Kennedy campaign added a new element, Franklin Roosevelt Jr., who for months had been discussing the campaign with the Kennedys. "When Frank came down to talk to Mr. Kennedy and Jack on the patio, and I was there for dinner, and he had that wonderful smile, that wonderful voice and vocabulary," Rose recalled. "And he was talking about West Virginia, whether Jack should campaign there and whether he would go with him. And I felt then the marvelous personality he had and the facility for public speaking and his smile and being a Roosevelt, I am certainly glad that he is not a candidate against Jack in the primaries. That thought crossed my mind then . . . you would think he would have been the one that campaigned and not my son. And that is life."

Two decades before, Joe had sat with President Roosevelt in the White House. That day he had given up what he thought was his own chance for political immortality to endow his sons with their chances for power. Joe had never traded in his marker, and though for years it had seemed valueless, he would make it pay out now.

Joe was a man who never forgot and rarely forgave. He surely was aware of the exquisite irony of this moment. The son and namesake of the president who had shoved him off the pages of history was helping Joe's own son to reach the White House. FDR was a great man and a poor father. He had given his son an immortal name, an awesome persona, and natural grace, but little of the strength, will, and ambition that was his own essence. FDR Jr. was a self-indulgent, heavy-drinking namesake. The former congressman had been a lobbyist for the Trujillo dictatorship and a distributor of Italian cars in Washington, work that Joe's own sons would have never considered.

Franklin Roosevelt Jr. bore what in West Virginia was the most glorious of names. He was the son of the man whose New Deal had, as many in the state saw it, given shoes to people who had walked barefoot, electric light to those who had sat in darkness, and bread to those who were hungry. By rights, FDR Jr. should have been having this discussion, not with the centrist Jack but with the liberal Humphrey, who was a proud, happy defender of the New Deal legacy. That's where his mother Eleanor would have placed him, or with her beloved Adlai Stevenson, not with who she considered the opportunistic Jack Kennedy.

FDR Jr. would have served Jack mightily by coming to West Virginia to speak for his candidacy and stand beside him, as if Jack too were an equal stalwart of the New Deal and its legacy. For the Kennedys, that was not enough, and they pushed Roosevelt to speak words that proved to be both untrue and unspeakable. Jack's staff had come up with documentation that Humphrey had not served in World War II. "I remember discussing it with Ken O'Donnell and with Bobby Kennedy," Feldman recalled. It was then, in Feldman's words, "made available to Franklin."

This was the kind of material that was dropped into the laps of friendly journalists, not shouted from the campaign platform by a man bearing one of the most revered names in American politics. "Bobby had been bringing pressure on me to mention it," Roosevelt recalled. "He kept calling—five or six calls a day." Bobby cared nothing about Roosevelt's reputation, or he would have backed off.

"Is FDR Jr. there tonight?" Jack wrote on his notepad. "The best thing would be some veterans group there. I have to be extremely careful however—as so many people want to stick it to me."

FDR Jr. began his assault on Humphrey with shrewd insinuations, lauding Jack as "the only wounded veteran" in the race. That was a mite too subtle and on April 27, he told an audience: "There's another candidate in your primary. He's a good Democrat, but I don't know where he was in World War II." That was so unseemly that the *Washington Star* characterized it as "a new low in dirty politics."

A good politician learns to keep his distance from the mean and the ugly, to let others dismiss those he would fire and to have surrogates speak the ugly words he wants stated. It may have been Bobby, and behind him Joe, who pushed FDR Jr. to speak as he had, but Jack knew all about it. Humphrey peeled the bark of civility off his attacks and took the desperate expedient of screaming the truth to all within hearing. "I don't have any daddy who can pay the bills for me," Humphrey shouted, his words streaked with self-pity. "I can't afford to run around this state with a little black bag and a checkbook."

"And the Star says we are guilty of 'dirty politics,'" Jack scribbled on a piece of paper. Jack could have asked Roosevelt to back off, but he did not. A week and a half later, Roosevelt had suddenly discovered where Humphrey had been during the war: a "draft dodger" hiding at home. The candidate tried to point out that even though he had been a married man with three children, he had tried to enlist in the navy but was turned down because of a physical disability. The charges had inflicted a heavy wound, however, in proudly patriotic West Virginia.

At that point, right before the election, Jack issued a statement condemning the discussion of Humphrey's war record. "There was a lot of criticism, and the Kennedys repudiated the statement and cut the ground out from under me," FDR Jr. recalled. "That was the beginning of the break between Bobby and me."

While the Kennedys stood back watching, Roosevelt had besmirched Humphrey's reputation. Although the Minnesota senator would quickly wipe off the dark spots, Roosevelt's role in West Virginia would stain him for the rest of his life. What angered Roosevelt most was that what he had said was not only unwise and unfair but untrue. "It was based on so-called reliable information which was made available to me," he later reflected. "It was used in the heat of the closing days of a vital and decisive primary, and . . . when I found it was unwarranted I went to Mr. Humphrey and not only ate crow but asked for his forgiveness."

Jack might have been traipsing across West Virginia wearing the laurels of a war hero, but he remained a Catholic. West Virginia was overwhelmingly Protestant, full of God-fearing, churchgoing folks who had probably never met a Catholic and surely had never voted for one. Most of them had heard tales of an Italian pope and his hold on the American faithful. They had heard whispered yarns of the strange language spoken and strange rituals enacted in the dark recesses of the Catholic churches whose portals they would never enter. Their ministers often told them that a Catholic president would have another master in the Vatican.

In 1960, this Pentecostal vision of Catholics was only an exaggerated version of the dominant Protestant culture's view on Catholics. Prejudice against Catholics was widespread in America, from the ignorant mouthing of the Ku Klux Klan to the no less pernicious musings of many political liberals, but the hard center of Jack's problems lay with Protestant ministers, who feared the gloved hand of Rome reaching into the White House and were ready to tell their congregations as much. Like millions of his co-religionists, Jack was only a nominal Catholic. He attended mass, but he acted as if the rituals of the Church were its essence. He was as much bewildered as irritated when he was constantly confronted with questions about his faith.

Part of the hierarchy of his own church was no more welcoming of Jack's candidacy than the fire-and-brimstone preachers of the Bible Belt. Some of the bishops and cardinals were so conservative, like New York's Cardinal Spellman, that they preferred the mock-Quaker Nixon to Kennedy. Others made the calculating and indeed accurate assessment that Kennedy would not be "their" president; he would have to distance himself so far from the Church that he would take positions on aid to parochial education and other matters that were harmful to the Church. In March 1960, Archbishop Egidio Vagnozzi, apostolic Vatican delegate in Washington, said off the record to a *New York Times* reporter that though most bishops in America favored Jack " 'simply because he is a Catholic' . . . a sophisticated current among Roman Catholics in the U.S., and in the Vatican, feels that a Roman Catholic in the White House at this moment might do more harm than good to the Church."

Most presidents invoked the name of God to justify the most secular of policies. To assuage the fears and prejudices of Protestants and Jews, Jack took an unprecedented position on the role of religion in public life. He told *Look* magazine in March 1959, that "whatever one's religion in private life may be, for the officer holder, nothing takes precedence over his oath to uphold the Constitution in all its parts—including the First Amendment and the strict separation of church and state."

In their zeal to protect the presidency from the machinations of Rome, the ministers had essentially induced Jack to promise to drive God out of the White House. This grievously offended the Catholic press, while the most prejudiced of his critics considered it a further example of the duplicities of Rome. Among the many letters of protest Jack received was one from a group of thirty-eight students at a Midwestern parochial school who lamented "the crash of an idol."

No one was more concerned with this issue than Joe and Cardinal Cushing. The Boston priest worked assiduously to promote Jack's candidacy in a manner that his Protestant counterparts would have felt proved their every fear about the heavy hand of the Church. "Wherever I go they think I am Jack's campaign manager," Cushing wrote Joe in May 1960. The previous March, when Jack's campaign was just beginning, Joe had written Cushing: "This letter really adds up to saying that if Jack stays in the fight, it will be you who has kept him in. If he wins, it will be you who has made it possible."

Two months later, Joe wrote the religious leader again, essentially giving him carte blanche as to how Jack would handle the religious issue. "I hope that we won't have the Catholic question raised again, but it might be a good idea to have some phrases worked out and handed to Jack to be added to the list of matters that he carries in his head. But we will be guided entirely by your thoughts on this."

Cushing replied two days later that "the religious issue should be taboo. Those who raise it never change their opinions no matter what answer we give them. This whole thing is very subtle. It will come to the forefront again and again but it may be a political devise [*sic*] to get us off the beam."

His father and his favorite priest had spoken from the depths of their experience that Jack had better steer as far from the religion issue as he could. But now, in the last weeks of the campaign in the heart of fundamentalist Protestant America, Jack decided to face the religion issue straight on. In doing so, he went not only against his father and Cushing but his West Virginian staff, who supposedly knew these people best, against his own pollster's studied judgment, and against the advice of most of his sophisticated Washington aides.

A great politician knows that if he stands back far enough from a problem, it may take on a manageable form and become an opportunity. He knows too that the defining of an issue is often the winning of an issue. There were many crucial decisions in Jack's quest for the White House, but few to compare to this moment. The issue would rise up again and again, but almost always in the form in which Jack had defined it as he campaigned in the hills and hollows of West Virginia. To do what he was doing took the courage that he considered the king of all virtues. But political courage rarely stands alone, and his was welded to an awesomely shrewd sense of human beings and their emotions.

"I refuse to believe that I was denied the right to be president on the day I was baptized," he told his audiences. "Nobody asked me if I was a Catholic when I joined the United States Navy. . . . Nobody asked my brother if he was a Catholic or a Protestant before he climbed into an American bomber plane to fly his last mission."

How were his listeners to respond when the matter was couched as an act of elemental fairness? Wasn't there something in their nature that didn't like bullies and thought that a fight was fair only if either man could win? How brilliantly Jack had finessed the issue so that Catholics could feel comfortable voting for him because he shared their faith, while Protestants who voted against him for the same reason became bigots.

And pity poor Humphrey rattling across the state in his sad little bus, while Jack soared above in the *Caroline.* Could the Minnesota liberal shout

out that all his life he had sung an anthem of tolerance, and that he was nothing but a foil in this whole business? If he said that, he would look like a bigot. Unable to say anything, he could only continue his bumpy ride and speak about everything but what he wanted to discuss.

On the weekend before the primary election, Jack appeared on a paid television program broadcast across West Virginia. Franklin D. Roosevelt Jr. was still of use, and he sat beside Jack, asking him questions that had been prepared by the candidate's staff. Theodore White, the famed chronicler of this campaign, recalled this half hour as "the finest TV broadcast I have ever heard any politician make."

Roosevelt gently asked the questions, and Jack ran with them, toying with them in soliloquies daring in their length. Jack was the very image of "cool," a term that was rising out of the Beat underground and the hip black jazz world. He and this new medium were one. He had all the media-anointed credibility of a television anchor, his words sanctified as truth.

There is nothing like a picture to convince another person, and there were two pictures being broadcast: the pictures on the flickering black-and-white screens in homes and bars from Bluefield to Morgantown, and the pictures created by Jack's own words.

"So when any man stands on the steps of the Capitol and takes the oath of office of president, he is swearing to support the separation of church and state," he said as his viewers fixed this image in their minds. "He puts one hand on the Bible and raises the other hand to God as he takes the oath. And if he breaks his oath, he is not only committing a crime against the Constitution, for which the Congress can impeach him—and should impeach him—but he is committing a sin against God."

Then Jack stopped for a moment. He had raised the ante a final time, placing God's own name on top of his stack of chips. It was a fierce, just, almighty God these people worshiped. Would Jack dare blaspheme against God before so many witnesses? And if he did, wasn't God's wrath worse than any judgment that mortals could mete out? Jack raised his hand from an imaginary Bible as if he had just taken that sacred oath, and then he repeated his words: "A sin against God, for he has sworn on the Bible."

By then Jack knew that the polls were looking better and better in the West Virginia primary, but victory was not yet complete. "I suppose if I win my poon days are over," Jack wrote on a notepad, lamenting the fact that his extracurricular sex life might soon be halted. "I suppose they are going to hit me with something before we are finished." He was almost certainly thinking that some sort of sex scandal would break.

On election day, when almost any other candidate would have prowled the environs of his hotel room, badgering aides for the first hint of results, Jack flew up to Washington. That evening he attended a movie with Ben Bradlee. Jack left the movie every twenty minutes or so to call Bobby at the Kanawha Hotel, each time learning that the results were not yet known. When he slumped back into his theater seat next to Bradlee, he had hardly missed any plot points; the film, a soft-core porn item called *Private Property*, consisted largely of a series of rapes and seductions.

Jack did not learn that he had won a landslide victory, 61 percent to 39 percent, until he returned to his house and received a triumphant call from Bobby at 11:30 P.M. The occasion called for a few celebratory toasts and a good night's sleep. Sleep was not even a possibility. He knew that he would have to fly back to Charleston through the night skies with Jackie to thank in person those who had helped him with the crucial victory. It did not matter that he was tired, that the hour was late, or that the air was turbulent. This was part of the natural risk of a politician's life, a backstage danger that the audiences never saw.

As Jack flew back to West Virginia for a short-lived celebration, Bobby trudged over to Humphrey's hotel and walked with the politician back to his headquarters for his public capitulation. Bobby appeared deeply touched by Humphrey's emotional concession, though his tears were like those of a pyromaniac standing back from the conflagration he has set off as his victims run from the burning building.

Jack stayed in the state capital long enough to shake Humphrey's hand, thank the voters over television, and hold a short press conference. At her husband's moment of triumph, Jackie stood alone like a shanghaied but unwanted passenger on a voyage to parts unknown. She turned and walked back to the car and sat there by herself in the darkness waiting for Jack.

As the *Caroline* flew back toward Washington in the predawn hours, the passengers were as giddy and lighthearted as a college football team returning from a victory. Only Jack was different. He sat there in the half-light looking ahead toward the Maryland primary and trying to gauge how his West Virginia victory would affect the uncommitted states. He was on the greatest journey of his life, and he was only partway there.

20

A Patriot's Song

On a Thursday evening in early July at the 1960 Democratic Convention, Wyoming cast its fifteen deciding votes for Jack, and the forty-three-year-old senator became the Democratic nominee. The candidate had secreted himself away from the convention, his whereabouts known only to his intimates. Soon after the vote, he descended on the new Los Angeles Sports Arena out of the cool night, his arrival signaled by a score of lights hurtling through the blackness.

In the cottage outside the gigantic arena stood the most powerful Democrats waiting to greet the man their party had just accorded its greatest honor. Only they had the clout to whisper a few words to Jack before he made his grand entrance to thank the delegates. The party leaders stood back as Jack got out of the sedan and greeted Bobby and Sarge Shriver, his brother-in-law. Part of the pols' reticence was the natural deference to power. As prominent as they were, and as much as some of them had done to further the younger man's ambition, there would always be a line now between them and the man who stood before them. Something else, however, kept them at a distance. As much as Jack had pretended that he was one of them, an American politician born and bred, he was different.

Jack did not have the politician's grayish pallor from a life measured out in planes, auditoriums, public meetings, and too many smoky rooms. He looked like a great star arriving to grace a Hollywood premiere where the klieg lights played across the sky and the urgent masses stretched for a glance or an autograph. Exuding movie star sexuality, he was astoundingly handsome, his perfect white teeth set off against his tanned skin. He was a vibrant, charismatic figure who seemed to radiate healthful vigor.

There was a daring, seductive quality to Jack, as if he would always be showing up out of a dark, mysterious night. "Yes, this candidate for all his record; his good, sound, conventional liberal record has a patina of that other life," the novelist Norman Mailer wrote, "the second American life, the long electric night with the fires of neon leading down the highway to the murmur of jazz."

Jack's sexuality was real and dangerous, and while all the rest of the politicians nestled down at the Biltmore and other hotels, Jack was staying in a secret hideaway on North Rossmore Avenue, off in the long electric night, a continent away from Jackie, who had stayed on the East Coast. In his apartment, Jack heard not the murmur of jazz but the sweet laughter of young women, and shook not the sweaty palms of pols, but touched the willing young flesh of the likes of the beautiful Judith Exner. The Los Angeles police guarding him did not know what to make of the young women entering the apartment. It was the kind of entourage they had previously thought the exclusive right of movie stars, not of presidential candidates.

The next morning, Jack arrived for breakfast at the ten-acre estate in the flats of Beverly Hills where his father was staying. Joe spent much of the day around the pool at the sprawling Beverly Hills mansion of Marion Davies, the former movie star and mistress to the late William Randolph Hearst. He had installed a bank of phones around the pool so that he could talk to one power broker after the next without an interruption while he basked in the California sunshine. Not only had Joe outlived most of the other powerful men of the 1930s, but he was also in the midst of the greatest triumph of his life, helping to propel his son to the presidency of the United States.

Joe would get no closer to the delegates than this. His son's enemies were whispering that Jack was nothing more than a thespian who mouthed the script his father had given him. Joe could not afford to be seen so close to Jack that he might be giving him his lines. What his detractors scarcely appreciated was the subtlety of Joe's efforts, how little he sought for himself, and how pointedly his son ignored his father's conservative thinking on most of the major issues of the day.

Joe's hands had no fingerprints, or it would have been clear that he had left his mark all over the campaign. His was the hand behind much of the money that had flowed into West Virginia and other states. "These things happened," reflected Tip O'Neill. "Jack didn't always know about them. But the old man had made his own arrangements over and above the campaign staff." Jack had tried to move beyond his father's ways. In the Maryland primary, the candidate's old friend Torbyrt Macdonald recalled, Joe wanted to pass out twelve-dollar-a-day stipends to make sure that poll workers showed up, but he and Jack vetoed the idea.

"All his way through his existence Dad had relationships and contacts none of the rest of us had," Teddy told his biographer, Burton Hersh. Joe had acquaintances, not only at the highest levels of business and politics but at the lowest levels of American life. "I remember in 1960 my brother saying to Dad, almost jokingly, 'The states you have are Illinois, New Jersey, Pennsylvania, New York.' "

With the help of New York City's Democratic bosses, Joe had helped deliver the largest city in America to his son. He worked with other bosses to whisk away northern New Jersey from under the vigilant eyes of Governor Robert Meyner. He helped to add Illinois by talking to Mayor Richard Daley, whom he had known from the time this boss of bosses was a city council member. Daley turned away from Illinois native son Adlai Stevenson in favor of a man like himself, a Catholic who looked like a winner.

Jack respected all that his father had done. He did not treat Joe as the ultimate arbiter of his political future, however, but as just another source of insight and advice that he assayed and sometimes rejected as fool's gold. As the two men sat down for breakfast, Jack had a crucial decision to make in choosing his vice presidential running mate, and father and son discussed various possibilities while Timilty and Rose listened in. "What about Lyndon?" Joe asked. That set Timilty off on a tirade against the Texas senator, mouthing words that most of Jack's supporters would have gladly amplified. The former Boston police commissioner pointed out that earlier in the week Johnson had savaged Jack in a dual presentation before the Texas delegation. The unseemliness of the politician's display was diminished only by Jack's cool riposte, which dissipated the Texan's meanness in laughter and irony.

Johnson had not come into the hot dusty political street to duel with Jack in the primaries but had sought to win the nomination by working his way through the backrooms and dark alleys of politics. As Jack had closed in on the nomination, the Johnson forces had slashed at Jack in attacks that could scar him in the general election. India Edwards, co-chairman of the Citizens for Johnson National Committee, told reporters, "Senator Kennedy, who appears so healthy that it's almost illegal, is really not a well man. . . . If it weren't for cortisone, Senator Kennedy wouldn't be alive." It was as ugly a bit of business as FDR Jr.'s attack on Humphrey's war record, but this was far worse, for the Johnson camp's allegations had the larger disadvantage of being true.

Jack knew that Johnson was a brilliantly astute legislator and the most qualified choice for vice president, but the man was a southerner, and many northern urban Democrats had their own prejudice against those born south of the Mason-Dixon Line, considering them provincial, uncouth racists, stereotypes hardly dissipated by Johnson's unsubtle, overweening persona. And yet Joe was not the first person to mention Johnson's name. Before the convention, Feldman and Sorensen, men who billed themselves as liberals, had given Jack a memo in which they listed Johnson as an "outstanding possibility." Over the weekend, Jack had seriously discussed that prospect with the *Washington Post*'s publisher, Phil Graham, one of Johnson's closest advisers.

Joe was a philosopher of power, and he looked straight on at decisions that made his sons wince. "We need Texas," Joe said, an argument that was impossible to deny. Jack listened while his father ran through the strengths that Johnson would bring to the ticket, piling more and more weight on the scale. Joe presented arguments that Jack had already heard, then added his considered judgment to the mix. In the end Jack called Bobby at the Biltmore and asked him to set up a meeting to talk with Johnson about the vice presidential nomination.

This was in many ways the most unpleasant task Bobby had yet performed for his brother, and he performed it poorly. Bobby could not accept the political charade of enemies donning the garb of friends overnight. His scorn was far more than a sneering disdain for a homespun southern vulgarian who had gone to Southwest State Teachers College in San Marcos, Texas. Bobby had his own firm reasons for loving some people and hating others, and no man ever moved from one category to the other. The Kennedys praised physical courage above all virtues, and it was one of the few qualities that Johnson did not have in excess. While Jack won his Silver Star helping to save the crew of PT-109, Johnson received the same high honor in the naval reserves, flying one combat mission as an observer. Bobby probably did not know Johnson's war record, but Bobby was a man of brilliant instincts when it came to understanding the primitive drives of his fellow humans.

Bobby sensed that Johnson was not a worthy man, as he defined the term. He knew, moreover, that Johnson bore the Kennedys no goodwill. Indeed, he learned a few months later from journalist Peter Lisagor that just before the Los Angeles convention, Johnson had berated Jack in language streaked with profanity, excoriating his brother as a scrawny, sickly mite so unable to govern that "old Joe Kennedy would run the country."

"I knew he hated Jack," Bobby admitted sadly that day, "but I didn't think he hated him that much." Even without this confirmation, though, everything Bobby knew and thought and felt told him that Johnson should not stand beside his beloved Jack as his running mate, bonded to him forever as his political brother. But Bobby was his brother's liege, and he would do what Jack asked him to do.

Those who were there that day had different tales to tell about how Johnson became the vice presidential nominee. "Well, you know, I don't think anybody will ever know," Jack told Feldman the following year. Bobby said later that Jack never intended to offer the nomination to Johnson. He was merely dangling it before the Texan's eyes, thinking the prideful politician would never accept such a cheap cut of meat, and then to his brother's dismay Johnson had simply gobbled up the offer before Jack could pull it back.

It is unlikely, however, that Jack went to Johnson's suite at the Biltmore merely to see whether the senator was interested enough to have his name firmly added to the list of candidates. More likely, Jack was surprised that Johnson was willing to accept an offer that he thought the Texan would have disdained, but only after leaving Johnson and learning how profoundly liberals and labor people opposed Johnson's candidacy did Jack wish he could somehow back off.

Jack waffled back and forth, wondering whether he had made one of the shrewdest judgments of his career or, in the name of expediency, had ripped out the very heart and soul of his party. This was largely the way Salinger and O'Donnell viewed it. "In your first move after the nomination, you go against all the people who supported you," O'Donnell raged, speaking in the intemperate language in which the newly nominated Democratic candidate was not used to being addressed.

Johnson had humbled himself to accept the nomination, and now suggestions were being made that he was unworthy and unwanted. Bobby came visiting with the unpleasant chore of asking Johnson to back out, but Johnson was not one to regurgitate meat he had swallowed whole. Not only did Bobby fail in his mission, but he did the most dangerous thing a person can do to a powerful politician. He humiliated Johnson. He made him do everything but beg on his knees.

In the end, it seemed better for both men and for the party to go ahead with the nomination, but these wounds would fester. Johnson did not blame Jack. He blamed Bobby, whom his campaign aide, Jim Rowe, told him was "a ruthless son of a bitch," an appellation that just as easily could have described the Texas senator.

After the decision was finally made, the two tired brothers repaired to the Beverly Hills mansion. Bobby's children cavorted in the pool, giving no thought to what was going on down at the Biltmore. And the elegant Joe sat there in a velvet smoking jacket and formal slippers with the embroidered initials JPK.

Bobby worked the phones, trying to pull some of the outraged liberals back into the fold. Jack, his face drawn, worried that he might have destroyed his chances at the presidency. "Jack, I don't want you to worry," Joe said, his voice still tinged with a touch of Boston Irish. "In two weeks they'll be saying it's the smartest thing you ever did."

During his acceptance speech on the last evening of the convention, Jack's face was streaked with fatigue, his deep eyes embedded in darkness. This evening he had a crucial message to convey. The somnolent Eisenhower years were ending, and the nation was heading into what Jack believed would be four of the most difficult, challenging years of its history.

Everywhere Jack looked, he saw omens pointing to the rightness of his predictions, and a range of disquieting, dangerous problems. The Soviets had downed a U-2 spy plane over Soviet territory and captured its American pilot, Francis Gary Powers, alerting the American people to a dangerous, covert action. The papers were full of stories about young black Americans "sitting in" at lunch counters and cafeterias across the South, demanding rights that were clearly theirs and had long been denied. In Cuba, Fidel Castro had overthrown the corrupt Batista and spoke a militant Marxist language, condemning *Yanqui* imperialism. Across Africa and Asia, a new generation was ready to attempt to throw off colonialism. In South Vietnam, Viet Cong revolutionaries were killing village leaders by the hundreds, while in Saigon, President Ngo Dinh Diem and his brother, Ngo Dinh Nhu, and sister-in-law, Madame Nhu, hunkered down, sequestered with their armies.

In the most important speech Jack had yet given, he needed to set forth forcefully and eloquently the themes of his campaign. The speech, originally written by Sorensen and then passed around from aide to aide, had some compelling phrases, but they were lost in a compendium of clichés. Jack stood before the delegates and in a hurried voice told those assembled: "We are not here to curse the darkness, but to light the candle that can guide us through that darkness. . . . Today our concern must be with that future. For the world is change. The old era is ending. The old ways will not do."

Despite the banality of much of the speech, one memorable phrase perfectly defined Jack's aspirations as president and became the slogan of the campaign. "But I tell you the New Frontier is here whether we seek it or not," he told the delegates and the millions watching on television. "New Frontier" brilliantly evoked the world that Jack saw ahead for America. New Frontier suggested both the romance of the American past and the dangerous future, opportunities as well as solutions, and passionate alertness, not passive acceptance. It was an irresistible slogan, with no taint of liberal perfectionism, no grand flourishes of rhetorical excess.

All the Kennedys were there that evening in the sports arena to savor Jack's victory, all except for the pregnant Jackie and Joe, who more than anyone deserved to be sitting behind his son on that great platform high above the delegates. Two nights before, Chuck Spalding had gone over to the Beverly Hills estate to congratulate Joe. He had wandered around until he found Jack's father in an upstairs bedroom. "Where are you going?" Jack's friend asked, startled to see that Joe was packing his bags.

"I have to get on a plane tonight and get back to New York and get working on this thing," Joe replied. "We've got to keep moving." Spalding knew all the stories about the old man's cynical grasp on his children, but he thought that Joe's action was "an example of incredible restraint for somebody who has always been characterized as kind of a Machiavellian figure, moving his children around."

Even his most vociferous critics would not have begrudged Joe at least this moment with his son, but he clearly wanted no new photos of father and son standing together shoulder to shoulder. And he had important matters on his agenda. In New York City, Joe called Henry Luce, the most powerful publishing magnate in America. Luce felt that Joe was angling for a dinner invitation on the evening of his son's acceptance speech. Luce proffered what was supposed to be proffered, and Joe arrived at the Manhattan home an hour or so before Jack's speech.

Clare Boothe Luce had probably been Joe's lover while he was ambassador to the Court of St. James's. That was long ago, and what the two men shared now was not a woman but similar views on the world of power. Luce knew that a man like Joe did not sit across the table from him this evening for aimless social chitchat. Luce published *Time*, probably the most powerful magazine in America. Luce was a conservative Republican, and *Time*'s take on Jack could have a crucial impact on the campaign.

Instead of waiting for his guest to raise the only subject that mattered this evening, the publisher spoke directly. "Time Inc. realizes Jack will have to be left of center to get the Democratic nomination, and will content itself with arguing domestic economic matters politely."

"How can any son of mine be a goddamned liberal?" Joe retorted, as if any fool could see that Jack was moving leftward only to win and after his election would return to his natural conservative home.

"But if Jack turns soft on communism, *Time* will cut his throat," Luce said, as he remembered later.

"Don't worry about him being a weak sister," Joe replied. It was all about toughness and manhood, and his son would not be found wanting.

Luce was one of the greatest power brokers in America, and early in the campaign, Jack met with the publisher. Luce's immensely popular picture magazine, *Life*, had done more than any other medium to create the image of the Kennedys as a gloriously romantic, handsome, energetic clan, while selling millions of copies. *Time* had not been so kind.

Jack was aware of the subtlest nuances of journalism. "I see Otto Fuerbringer got well and is back at work," Jack said. Luce was startled that by reading the publication, Jack knew that the profoundly conservative editor was putting his imprint on the week's news, inculcating opinion into the columns in such a seamless way that even the cognoscenti could not tell where fact ended and editorializing began. Of course, nothing flattered a journalist more than knowing he was being read, and read closely. Jack's message went beyond that. Jack was signaling that there would be no subliminal ministering to his opponent without Jack understanding what was being done.

Jack knew that this would be a close election, but after listening to Vice President Richard Nixon give his acceptance speech on July 28 at the Stockyards Amphitheater in Chicago, he became even more aware of the challenge he faced. Jack realized, as a more narrowly partisan politician might not have, that Nixon's speech was "a remarkable political demonstration."

Like Jack, forty-seven-year-old Nixon was a veteran of World War II, a man of the new generation ready to assume power in the America of the 1960s. Both candidates were well versed in international affairs and strong anti-Communists who believed that the major challenge of the new administration would probably lie outside the nation's borders.

"Our next president must tell the American people not what they want to hear but what they need to hear," Nixon told the Republicans. Nixon, like Kennedy, realized that the political idiom of his day was full of half-truths, and half-truths were often worse than lies, for one could never parse the truth from the lie. "Why, for example, it may be just as essential to the national interest to build a dam in India as in California," Nixon asserted, a message that many Americans did not want to hear. A politician who spoke such truths too loudly might not win election, but one who spoke them not at all did not deserve America's highest office.

"Mr. Khrushchev says our grandchildren will live under communism. Let us say his grandchildren will live in freedom." That was one of Jack's themes, and he could have spoken that line and most of Nixon's speech that day. The two politicians had nearly as many affinities as differences. Underestimated by many of their enemies, patronized by some of their friends, they both had the knowledge and the experience to lead America into a new era. It was the character of each man that had not been tested.

The wildly popular Eisenhower had overseen what most Americans considered a blessedly comfortable era of growing affluence and peace. That was the record that Nixon had the happy duty of defending. Jack had the more difficult task of not overtly criticizing the revered, grandfatherly Eisenhower while talking of a new troubled world full of what *Time* called "anxiety and discomfort." For the most part, this was not only a posture shrewdly calculated to elect him president by playing on the natural anxieties of Americans in the age of the cold war; it was Jack's own deeply felt judgment of what his nation faced in the decade ahead.

"I think that in the 1960s you're going to have a terribly difficult time, whoever is president," Jack told James McGregor Burns privately in 1959. "I think Eisenhower is probably going to get home relatively free. . . . So I would say in a sense it's almost like Calvin Coolidge and Herbert Hoover. I would say that in 1961 or '62 all these matters—changes in weapon structure, changes in NATO, and all the rest—are going to come to a head. And I would say that the job the president is going to have from '61 on is going to be the most difficult of any, certainly since Roosevelt, and I think Roosevelt had the most difficult, except for Lincoln."

In international affairs Kennedy might wave the rhetorical anti-Communist banner as high as any of his opponents, but he knew that he would not suddenly transform the world. He knew too that if freedom meant what he thought it did, then time was America's greatest ally. "I don't think there is any magic approach," he told Burns. "There's no secret source. . . . The magic power really is the desire of everyone to be independent and every nation to be independent. That's the basic force which is really, I think, the strong force on our side. That's the magic power . . . that's what's going to screw the Russians ultimately."

Jack was the elegant troubadour of this new era, celebrating the unprecedented challenge that he believed his nation faced, yet knowing that these would be years of immense uncertainty and danger. The restless spirit of America would not be contained much longer, and the new president would either ride the whirlwind or the whirlwind would ride him.

Jack believed that the next president would have to act with unprecedented vigor, projecting onto the nation some of his own energy and will. During the tough months of the campaign, he could not stumble even once. As Feldman realized, Jack could dispel the rumors only by running a campaign that was an endurance contest merciless in its intensity and brutal in its scheduling. Beyond that, his minions would have to continue doing a brilliant job of disguising his health problems and preventing his enemies from learning the truth.

Jack's campaign had already been in the unpleasant position of having to find an acknowledged authority to say something that was untrue about Jack's Addison's disease. In early June 1960, Dr. Janet Travell and Dr. Eugene Cohen sat down to write a health certificate for Kennedy. Dr. Cohen was a prominent endocrinologist who had taken over Jack's cure after the death of Dr. Shorr in January 1956. It was Dr. Cohen's analysis of Jack's adrenal condition that mattered. Dr. Travell recalled that Dr. Cohen said that he "didn't like publicity" and "didn't want to get mixed up in this," and "we fought over every word. . . . We spent 3 or 4 hours on it." The fighting was over the truth. Dr. Travell was a woman of immense political ambition who sought to advance herself through her relationship with Jack. Dr. Cohen was concerned only with the care of his patient.

In the end, after all the negotiating, Dr. Travell and Dr. Cohen signed a joint letter stating: "With respect to the old problem of adrenal insufficiency, as late as December 1958, [when] you had a general check-up with a specific test of adrenal function, the result showed that your adrenal glands do function." That was a legalistic way of getting around the hard truth that Jack had a serious condition, but the statement was hardly strong enough to make the suspicious turn toward other subjects.

Later that month, the two doctors traveled to Boston to see Dr. Cohen's colleague and another of Jack's doctors, the equally highly regarded endocrinologist Dr. Elmer C. Bartels at the Lahey Clinic. Dr. Travell described the trip as "an important tactical move." She wrote Dr. Bartels afterward to express her concern "about the security of the nurses' notes at the Lahey Clinic" and noting that he planned to write his own letter about Jack's health. Dr. Bartels may have written such a letter, but if he did, it was not considered strong enough to release, and no copy of such a document has surfaced.

India Edwards's dramatic attack on Jack's health was just what he had feared might happen. Immediately Dr. Travell wrote another letter to Bobby in Los Angeles in her own name, stating baldly: "Senator Kennedy does not have Addison's disease." Dr. Travell knew little about Addison's disease, though she presumably knew enough to realize that her statement was at best a half-truth. Even the doctor realized that she had gone too far in her disingenuous attempt to protect Jack. After the nomination was secure, she qualified her statement in another letter signed also by Dr. Cohen ("You do not have classical Addison's disease") and asked Bobby to destroy her previous letter. When Dr. Travell and Dr. Cohen's offices were broken into by someone apparently searching for Jack's medical records, she sequestered the documents and went around to the hospitals where he had been a patient, gathering up all his medical reports. "I tracked down almost everything that was available," Dr. Travell recalled. "I think that was very important."

Jack flew to Hyannis Port after the convention. Johnson joined his running mate on the Cape, where the two men sat together discussing strategy. The vice presidential candidate was an oversize, looming character who dominated almost everyone within range of his booming Texas voice. If he had spoken with the sharp cadences of a New Yorker or with a Chicagoan's flat tones, he would probably have been the presidential candidate instead of Jack. He attempted, as best he could, to shove his outsize form into the livery of a vice presidential candidate, but he was not yet comfortable wearing such a diminutive outfit.

"Now what we gotta do, Jack," Lyndon said, speaking rightfully as a brilliant political tactician, "is I work the South where I'm strong. You're big up here in the East, and we'll both do the Midwest, and you take the West too, go with our strengths."

"No, Lyndon," Jack replied, speaking with a quick urgency, as if he had to squeeze as many sentences as he could into each moment. "You're a monster up here, Lyndon. You have to come to Boston and show them what a nice fellow you are. And I've got to go South where they think I'm the pope's creature and show them that I'm my own man. That's what we're doing."

Jack had not even made his first campaign speech before the religion issue began entering the campaign like a foul tributary that, if not stopped, would flow into the political mainstream. For the candidate, it was an unpleasant reality that he knew he had to face. Bobby, however, took the anonymous pamphlets that had begun showing up across America and the whispered tales of papal conspiracy as a triple attack: on the church he revered, the free society he loved, and the brother he worshiped.

These critics considered Protestantism the natural faith of the American people, and they feared what would become of America under a Catholic president. Dr. Ramsey Pollard, president of the Southern Baptist Convention, who declared proudly, "I am not a bigot," asked that the papist church "lift its bloody hand from the throats of those that want to worship in the church of their choice." Those preaching such words the loudest were not cultural riffraff trotting down barefoot from some backwoods hollow, but many of the most powerful clergy in America. "In every Catholic-dominated country today, non-Catholics are not permitted full privileges of citizenship," declared a tract put out by the Baptist Sunday School Committee. "Illiteracy is high. Morals are low."

On August 18 in Montreux, Switzerland, at an evangelical conference, the Reverend Billy Graham, perhaps the most revered evangelist in American history, hosted a private gathering of twenty-five prominent ministers, including the Reverend Dr. Norman Vincent Peale. The ministers spent much of the afternoon planning how they could ensure Jack's defeat.

Peale believed that "American freedom grew directly out of the Protestant emphasis upon every man as a child of God." To give over the presidency to this Catholic interloper would mean the slow end of freedom and the beginning of what the Catholic weekly *America* called "Post Protestant Pluralism."

Shortly before this August meeting Peale wrote Nixon offering to help his campaign in any way that he could. "Recently I spent an hour with Billy Graham," he noted, "who feels as I do, that we must do all within our power to help you."

At the Swiss meeting one of the participants recalled Graham offering up details about the "moral character of one of the presidential candidates." It was surely not Nixon whom Graham derided. Graham and Peale knew Nixon as a friend. To them, he was a deeply moral man who believed in the sacred trilogy of Protestantism, Republicanism, and conservatism, and they saw it as their Christian duty to work for his election.

It was not the "bloody hand" of Rome that sought to place its mark on this election, then, but the nimble fingers of evangelical Protestantism. When Nixon became president, Peale and Graham would presumably greet him at the steps of their churches, signaling a return to what Peale called "the old, strong, narrow Protestantism that made America strong."

Bobby did not know about the secret meeting in Switzerland, but wherever he went in those first days of the campaign he was bombarded again and again with religious questions, some of them sincere doubts, others snide attacks masquerading as queries. Bobby not only wanted to talk about the issue, he had to talk about it. He had to stand up before his brother's critics and fend off these attacks, daring them to throw the strongest charges full in his face. In Cincinnati early in September, he decided not to wait for the inevitable questions but to charge out in front of his attackers.

"People have often said that Bob Kennedy is, you know, without emotion," reflected William A. Geoghegan, a local attorney. "Well, this time Bob Kennedy did cry and he broke down as he was giving that talk. I was sitting right beside him and got up and had to take over. It was a very emotional experience I think for everybody there present. . . . I can recall the words that he spoke . . . following which he could not go on with his talk. He said, 'I can't imagine that any country for which my brother Joe died could care about my brother Jack's religion when it came . . .' Then he stopped."

These words were similar to those that Jack had spoken so many times in West Virginia. Jack had not cried the first time he said them, or the last time, but that was Jack, and this was Bobby. Again and again in the next few days, Bobby confronted the issue. In Toledo he told a crowd of twenty thousand: "The religious problem is hurting us badly now." The next day at the University of Illinois airport in Champaign, he boldly answered a Methodist minister who asked whether Jack would owe fidelity to a Roman master. "I say you are questioning his loyalty if you question whether he would take orders from a third party. I say he's proved his loyalty in the United States Navy and in Congress."

Jack was not one to try to outshout detractors but preferred to move away to where his rational, ironic voice could be heard. He had proved in West Virginia that he could play the religious fiddle as well as any of those preachers, but to him that was not politics as it was supposed to be. He agreed reluctantly to face up to the issue in one crucial, definitive speech. He had told Johnson in Hyannis Port that he had to go down to Texas to campaign, and Jack was a political preacher who listened to his own sermons. For the occasion, he chose not only the Longhorn State, but the most difficult, hostile audience imaginable: the Greater Houston Ministerial Association. Sorensen wrote a speech that he knew could mean the presidency.

Waiting for Jack in the ballroom of Houston's Rice Hotel on September 12 were three hundred preachers and lay leaders and an equal number of guests. They would not have been Texans if they had been less than hospitable, and they greeted Jack civilly, but most of them had righteous suspicions of Catholics. They came that morning not to learn but to observe, not to change their ideas but to seek further confirmation of their beliefs.

At his best, Sorensen did not simply write speeches but channeled himself into Jack's psyche and intellect. Nine times in the short speech Jack spoke the words "I believe," witnessing to his own political faith in a manner familiar to every evangelical Protestant in the room. Jack was not an emotional speaker; what gave his words special resonance was his heavy emphasis on each syllable, as if he wanted the sheer truth of his words to prevail.

> I believe in an America where the separation of church and state is absolute—where no Catholic prelate would tell the president (should he be Catholic) how to act, and no Protestant minister would tell his parishioners for whom to vote—where no church or church school is granted any public funds or political preference— . . . and where no man is denied public office merely because his religion differs from the president who might appoint him or the people who might elect him.

Jack went on to say that he would do what his conscience told him to do, not his church, but "when my office would require me to either violate my conscience or violate the national interest, then I would resign the office; and I hope any conscientious public servant would do the same." Only a few months before, the Vatican newspaper, *L'Osservatore Romano*, had stated that the pope and his cardinals had "the right and duty to intervene" in politics and that "the Catholic may never disregard the teaching and direction of the church." Jack stood before the Houston preachers that morning in as much opposition to these certitudes, as he was to the certitudes of the Protestant clergy.

Jack knew one simple fact. There were twice as many Protestants as there were Catholics, and if most Americans voted for a man of their own faith, he had no chance of winning this election. Although Jack may not have won any votes that morning, he partially dammed off that foul tributary of religious prejudice. It was dammed off too by Americans who may not have heard his voice that day but showed that they did not want their political system polluted by attacks on a man's faith.

Jack's speech that morning was magnified a hundredfold. Bill Wilson, a television producer hired to oversee the debates and other live events, had insisted that the lengthy speech be kinescoped. He cut the speech down to a half hour to be shown again and again on television stations across America. This crucial element of the campaign was essentially a political infomercial unlike anything Nixon's campaign was doing.

Most Americans were not about to vote against Jack simply because he worshiped at a different church. The Reverend Peale learned that in early September, when he chaired the Washington meeting of a new group, Citizens for Religious Freedom, whose intent was the opposite of its name. The most pernicious of prejudices are not shouted in the street but spoken in sonorous tones, garnished with reason and plausibility. The ad hoc group issued a statement with a disingenuous disclaimer that the "religious issue" was "not the fault of the candidate." It was his church that was the problem, and as much as Jack might swear obeisance to the First Amendment, he could never be free of the Roman church's "determined efforts . . . to breach the wall of separation of church and state." Peale said that he doubted that, as president, Jack would be able to stand up to what was "both a church and also a temporal state."

Peale and his followers had every expectation that the mainstream American clergy and prominent newspaper editors would greet these comments positively. That would have been a strong signal for the ministers to make a public issue of Jack's faith, possibly ensuring his defeat. Instead, the ministers' statement roused the dormant leaders of liberal Protestantism. But it was not just Democratic-leaning clergy who were outraged. The *Methodist Outlook*, the *Presbyterian Outlook*, and the *Christian Century* all condemned Peale as well. Many newspaper publishers were appalled, and almost 10 percent of the newspapers carrying "Confident Living" canceled Peale's popular column. Even the editor of the *Saturday Evening Post* criticized him.

Peale backed off and retreated to the Marble Collegiate Church in New York City, where he preached to a white, well-fed, increasingly suburban congregation who were a fair representation of what the minister thought America to be. After the election Peale said that "Protestant America got its death blow on November 8."

Jack had more than his faith to worry about. He was so apprehensive about his father's potential negative impact on the campaign that when the British journalist Henry Brandon asked to interview Joe, the candidate told him: "Henry, if you do, you'll never speak to me again." Jack's worries were well founded. Later in the campaign a secret report on Jewish voters noted their apathy toward Jack based in part on "anxiety about Joseph P. Kennedy and his alleged America First leanings."

Joe was as sensitive as Jack to the detrimental impact he might have on the campaign. Once Joe had taken care of his business in New York City, he flew over to France and his rented villa, Bella Vista. There on the Riviera, Joe gave his only extensive interview of the campaign to *U.S. News & World Report*. "Americans are wondering why one of the most powerful men in his [Jack's] camp is sitting out the campaign 3,000 miles from U.S. shores," the weekly asked. Joe was going to be back in a few weeks and he gave a shrewd, duplicitous answer. "I just think it's time for 72-year-old men like me to step aside and let the young people take over," he said. "Since 1952, when Jack went to the Senate, I've never campaigned for him, never made any speeches. You know, I've never heard Jack make a speech except on television."

Joe had brilliantly distanced himself from Jack. His only faux pas was to have his beautiful young caddy in residence. Joe was described as "teaching [her] English during the golf rounds," lessons that presumably continued later when he invited her to visit him in the United States.

While Joe was playing golf and lying in cabana 513 at Hotel du Cap, his new secretary, Bonnie Williams, arrived at Hyannis Port shortly before Labor Day. The Kennedys lived in a village within a village, their homes scattered around their own town square. The children frolicked on a broad plot of grass, and the parents hurried from one house to the next.

Jackie, Ethel, Eunice, and other adult Kennedys flitted in and out of the residences, but when Joe arrived in early September, Williams sensed an electric charge of excitement. Her boss was truly the patriarch, greeted with subtle deference, his smallest want taken care of by the help and the family.

Joe cut a splendid figure as he rode his horse each morning, impeccably attired in riding clothes. And as she sat next to him during the day he talked to some of the most powerful people in America. Jack called too, sometimes several times a day. Williams sensed that Joe wasn't completely happy with her, and she thought that she knew why. Joe had impeccable manners, and he made no overt advance toward her, but she knew what he wanted and what she would not give.

Joe had been back scarcely two weeks when the chauffeur drove Joe and Williams to Boston for a flight to New York. All through the long ride Joe sat silently. On the plane to New York he said not a word. Finally Williams turned to him and spoke urgently: "You know I came to do one job, and that's the way I want it. And if it's going to be some other way, now is the best time to end it."

Joe shrugged and gave no answer, but from then on Williams had no more difficulties with her seventy-two-year-old boss.

When Jack headed out on the campaign trail, he had an enlarged staff to service his needs. Among the new arrivals was Archibald Cox, an austere, cerebral Harvard Law professor who came down to Washington to oversee a group of academics writing speeches and preparing policy papers. Jack knew Cox as an adviser on labor questions, and it was fitting that Jack opened his campaign on Labor Day, giving a speech that Cox had written, before a vast crowd of sixty thousand in Detroit's Cadillac Square. About halfway through the turgid lecture studded with facts and figures, the candidate pushed the pages away and ad-libbed his way through the rest of speech.

That was the last time Jack read one of Cox's speeches. Cox and his colleagues were grievously disappointed at how few of their words Jack ever uttered, though their research and ideas often worked their way into his addresses, compressed into pithy phrases and slogans. For the most part, however, it was the prodigiously industrious Sorensen who wrote almost all of Jack's speeches on the road, based in large part on Feldman's research.

Cox began taking each speech that Jack gave in cities across America and comparing them with the talk that Roosevelt had given in 1932 in the same city. "I was just aghast from an intellectual point of view at the lower level of all Kennedy's speeches," Cox reflected. The Harvard professor was so appalled with these speeches full of wind and sloganeering that he wrote a futile letter to Sorensen suggesting that Jack give at least two or three truly substantive speeches each week.

Jack would gladly have given grand, serious addresses, but such speeches no longer held an audience. Even before remote control devices became ubiquitous among television-viewing Americans, the public already had learned to change the channel on whatever was boring, complex, or tedious.

Americans would never again sit still, either in person or in front of their televisions, listening to the kinds of formal speeches that once had defined presidential campaigns. On the campaign trail Jack saw what moved the vast, restless crowds and what made them anxious and nervous. His speeches now had less substance than the talks he had given during the primaries.

Jack did not shirk ideas when they clearly benefited his candidacy. Fritz Hollings, the former governor of South Carolina, called Feldman praising a speech that General James Gavin had given in which he broached the idea of Americans volunteering in the underdeveloped parts of the world. The speech itself was hardly a seminal document. The general had no written speech, only notes that were sent up to Washington. In Michigan, Jack tried the idea out to an enthusiastic reception, and then developed it further into a full-scale speech given in San Francisco.

The Peace Corps, an idea with which Jack's name would forever be linked, was in a sense the democratization of the ideal that he had expressed at Choate in 1946, that those to whom much had been given should give back in public service. In the affluent America of the 1960s, everyone seemed privileged or at least everyone in the middle class, and it was the children of the middle class who heard this bugle call.

When Jack arrived in Chicago for the first of four unprecedented televised debates on September 26, 1960, he was slightly behind in the polls, and many political handicappers thought that this occasion would worsen his chances. Nixon's supporters could claim that for eight years he had been groomed in the White House to assume the presidency, while his opponent had been a legislative misfit in the Senate. Although Jack had developed into a fine public speaker, the vice president was a stellar debater who had famously stood up to Khrushchev in Moscow the previous summer in a spontaneous debate. Beyond that, the two candidates' acceptance speeches provided devastating evidence of Jack's apparent inferiority.

Jack created his own universe around him, and everyone in his circle served a purpose, from Dave Powers, who woke the candidate each morning and knew when to pump him up with Irish yarns or with a terrifying vision of Nixon already up, shaved and showered, out on the campaign trail; to Janet Des Rosiers, his father's former mistress, now the stewardess on the *Caroline,* who gave him everything from coffee to massages; to Pierre Salinger, his ebullient press secretary; to O'Donnell, O'Brien, and his campaign staff.

This weekend, Jack brought Sorensen and Feldman to prepare him for what he knew might be the most important moment of the campaign.

These intense, sharp men spoke the candidate's idiom perfectly, cutting to the core of any issue. Throughout Sunday evening and the Monday morning of the debate, they peppered Jack with the questions they thought he might be asked, and he answered back in his staccato rhythms, sharpening his arguments.

Jack told his two aides that since the Democrats were the majority party, he would proudly position himself as heir to that tradition. Nixon fancied himself an expert on foreign affairs, but Jack believed that this area was the vice president's weakness and his own strength. If Jack could lure his opponent onto these dark shoals, he believed that he had a chance to run Nixon's campaign aground. Unfortunately, the first debate was supposed to be on domestic issues. The second and third would be question-and-answer sessions, and only the fourth debate would focus on foreign policy.

Part of the time, Jack's back was bothering him so badly that he lay on the bed in obvious pain while the two aides continued their questioning. Feldman was amazed that he could even go on. Jack may well have had a secret weapon that afternoon that lifted him high above the deadening fatigue of the campaign. In recent months, Jack had noticed that his old friend Spalding, who was stuck in the dread routine of a bad marriage and a nine-to-five job, had a new bounce to his step, a lilt to his voice, and a lowering of his querulous complaints. His friend said that it was all due to Max Jacobson, a wondrous doctor who gave his patients magical vitamin injections that contained, among other things, the blood of young lambs.

The first time Spalding had gone to Jacobson's New York office, he had been taken aback by the unkempt quality of the place. Jacobson had yellow spots on his smock and to some might have appeared slightly mad. Yet his office was full of patients, some of them famous, all happily waiting for their shots. The moment the doctor injected Spalding, he felt his body fill with life, a pure energy of such magnitude that he stayed up for three days without sleeping. Spalding got his own supply of this liquid life and at home started injecting himself. It was a time when few Americans knew what amphetamines were or the dangers they represented. Spalding had no idea that the magical ingredient was Methedrine, an amphetamine that would in a few years become notorious as "speed."

Dr. Jacobson worked his chemical miracles as much as one hundred times a day, injecting celebrities, socialites, and politicians with a happy mixture of vitamins and amphetamine. They came sometimes once a month, others once a week, and some every day. Truman Capote might be sitting there in the small outer office, or Eddie Fisher, former Senator Claude Pepper,

Alan Jay Lerner, or Cecil B. DeMille. Jacobson filled his syringe with unique mixtures drawn from half a dozen bottles filled with various liquids and injected them almost anywhere in the body.

For Jack's first visit to Jacobson in September just before the debate, he entered an office suite cleared of other patients. "The demands of his political campaign program were so great that he felt fatigued," Jacobson wrote in his unpublished autobiography.

> His muscles felt weak. It interfered with his concentration and affected his speech. . . . If not attended to these complaints could not only become more severe but would probably lead to more serious discomforts in the future. I took a short case history, and previous diseases he had, accidents, and treatments he had been given. I asked him about his present condition and what medication he was presently taking. The treatment of stress has been one of my specialties. After his treatment he told me his muscle weakness had disappeared. He felt cool, calm, and very alert. I gave him a bottle of vitamin drops to be taken orally, after which he left.

Jack had a long, laugh-filled luncheon with Bobby and several others. Then he planned to take a nap, but he was so restless that he got up in his bathrobe and walked out on the terrace of his suite. There was a confidence in Jack, not a dumb, blustering arrogance, not a dangerous, willful pride, but a subtle understanding of his own abilities and those of his opponent.

Jack paced back and forth in his bathrobe that long autumn afternoon before the first debate, hitting his fist again and again, like the challenger in his dressing room before the championship fight. As Jack dressed, he told Powers that he felt the tense excitement and nervousness of a prizefighter getting ready to enter the ring in Madison Square Garden. "No, Senator," Powers replied, with his exquisite sense of what to say to the man he served with such perfect fidelity. "It's more like the opening-day pitcher in the World Series—because you have to win four of these."

At the studio, Jack waited with a small group that included Bobby and Bill Wilson, his television adviser. Wilson had observed the contrast between Jack's elegant demeanor and Nixon's stolid frame. Citing the historic Lincoln-Douglas debate, Wilson had convinced the Nixon people that the candidates should stand behind small lecterns that would expose much of their bodies.

As the final minutes approached, the candidate turned to Wilson. "I got to take a leak," Jack said.

"It's two minutes before air," Wilson replied, looking at his watch.

"I got to take a leak." The two men hurried to the men's room.

"Kick him in the balls," Bobby said, when Jack returned. Then Kennedy strode into the studio to face Nixon, sitting there already under the hot lights.

Jack won the right to give his opening remarks before Nixon, only the first of his many victories that evening. As Jack sat there so still, so certain, so straight, Nixon began sweating under the hot lights, first imperceptibly, then looking as if he had stumbled fully clothed into a sauna.

There was something dreadfully unfair in the visual contrast between the two candidates. But it was hardly Jack's fault that in late August Nixon had hit his kneecap on a car door in Greensboro, North Carolina, and had ended up spending twelve days at Walter Reed Hospital. And now, earlier in the day, Nixon had bumped his troubled knee on a car door on his way into CBS station WBBM and almost passed out from the pain. He then had his face doused in a pancake makeup called "Lazy Shave," which was inadequate to hide his stubble or mask his sweat.

Nixon should have been able to hold the rhetorical high ground using Eisenhower's immense popularity as a shield against Jack's attacks. It was Jack, however, who in his opening statement defined the evening, talking about the great Democratic tradition of Roosevelt and Wilson, making himself appear their logical heir. He took all the glorious truisms by which Americans lived and made them his own, locking them away from his perspiring opponent. He talked about the role of foreign affairs in American life, though that was not supposed to have been broached at all this evening. "In the election of 1960, and with the world around us, the question is whether the world will exist half-slave or half-free, whether it will move in the direction of freedom, in the direction of the road that we are taking, or whether it will move in the direction of slavery."

Nixon began with the excruciating necessity of agreeing with almost everything Jack had said. "The things that Senator Kennedy has said many of us can agree with," Nixon said, reverting to a schoolboy debating technique. "And I subscribe completely to the spirit that Senator Kennedy has expressed tonight, the spirit that the United States should move ahead."

While Jack referred to the vice president as "Mr. Nixon," his opponent called Jack "Senator Kennedy." The two candidates stood close together on many of the major issues of the day. They spent most of the hour debating nuances. Jack talked not about changing the direction of America but simply getting the country moving on or moving ahead, tacitly admitting that he agreed with the basic thrust of the past eight years. "This is a great country, but I think it could be a greater country," he said, "and this is a powerful country, but I think it could be a more powerful country."

Television and radio stations across America carried the debate, and the largest political audience in history, seventy million Americans, heard the two men discuss serious, even esoteric issues in a responsible, reflective way. When the hour was over, both sides could reasonably declare a victory. But the fact that Jack still stood in the ring after debating Nixon had elevated the Democratic contender to a new position, proving that he was a challenger who deserved to be in the same heavyweight class as the vice president.

A poll taken the next day showed that television viewers by a slight majority felt that Jack was the winner, while radio listeners overwhelmingly declared Nixon the victor. This was not because Jack was a synthetic creature of a new media age. He had matched Nixon idea for idea and complexity for complexity. On radio, however, Jack at times sounded strident and overwrought, while on television his words wedded to his cool presence took on a different meaning. Those exposed to the debate only by reading the transcript would have had a third verdict, that the debate was a dead heat. These were all truths, but the television sets that in 1960 were already in 87 percent of American homes were the dominant medium of the new age, and from then on a president who had not mastered TV would find it difficult to effectively lead the nation.

Jack was still a person in creation, constantly tinkering with his public persona. He watched his appearance on television as if he were looking at another person. "'Party,' not 'pawty,'" he said one evening, watching his image on the black-and-white screen, like a speech teacher admonishing his pupil.

The following three debates for the most part only solidified the verdict of the first. Just before the second debate, J. Leonard Reinsch, Jack's media adviser, realized that Nixon's people had turned the thermostat down to a chilling sixty-five degrees at the NBC studios in Washington, D.C., hoping that a cool room would stem Nixon's embarrassing propensity to perspire. Reinsch hurried through the studio until he found a janitor who, after ample browbeating, turned the hidden thermostat up as high as it could go.

Jack had a subtle, sophisticated understanding of America's role in the world, but he was elected by people who largely did not have his knowledge or insight and did not necessarily share his views. There had always been a gap between his truths and the realities and limitations of practical politics. In the Senate when he was speaking on Algeria or Vietnam or working on a labor bill that would be fair to both unions and management, he had attempted to bridge that gap. He knew that what he considered political courage was largely the act of making that leap knowing that one might fall into the chasm of defeat. It was a leap that he was not willing to attempt during the campaign; he preferred standing rooted in the firm and narrow grounds of seemingly practical politics.

In the third debate Jack said that the profoundly anti-Communist Nixon had "never really protested the Communists seizing Cuba, ninety miles off the coast of the United States." He continued that same assault in the fourth and final debate, blaming the Eisenhower-Nixon administration for losing Cuba to communism.

"In 1957 I was in Havana," Jack said from the ABC studios in New York while his opponent debated him from the network studio in Los Angeles. "I talked to the American ambassador there. He said that he was the second most powerful man in Cuba." That was a devastating admission of the true nature of the American role in Cuba. The seeds of Cuban communism had grown during the Eisenhower years in a soil of corruption abetted in part by American businesspeople, the American government, and American mobsters. As Jack had stated previously, the corrupt "dictatorship had killed over twenty thousand Cubans in seven years." That was in large measure why so many Cubans cried out fervently against the hated *Yanqui,* not because they were the mindless dupes of communism.

Jack criticized Nixon in the debates for traveling to Cuba in 1955 and "prais[ing] the competence and stability of the Batista dictatorship." Jack, for his part, had treated the Caribbean island as a glorious playground. He had stood beside the dictator kissing babies and frolicked in the mendacious capital with seemingly nary a thought of the American role. And now, instead of trying to explain that government policies would have to change or American foreign policy would become a recruiter for the Communist movement, he played to the most narrowly chauvinistic instincts of the American public. "I have seen Cuba go to the Communists," he said. "I have seen Communist influence and Castro influence rise in Latin America."

This was bad enough for him to say without a hint of context, but the day before Jack had told one of his new aides, Richard Goodwin, to prepare "a real blast at Nixon" on the Cuban issue. Sorensen and Feldman were brilliantly attuned to Jack's thinking. Goodwin was a brash, pugnacious young man with the overweening confidence that can come from having served as an editor of the *Harvard Law Review*. Goodwin wrote a press release in the candidate's name stating that "we must attempt to strengthen the non-Batista, democratic forces in exile and in Cuba itself who offer eventual hope of overthrowing Castro. Thus far, these fighters for freedom have had virtually no support from our government."

One of the reasons Jack was so leery of academics and liberals in politics was that when they attempted to engage in what they thought was realpolitik, they were playing with weapons that often blew up in their own faces, and those of their friends, rather than hurting their enemies. When Goodwin called Jack at the Carlyle Hotel and learned that he was already asleep, the aide went ahead and issued the press release.

The headline in the *New York Times* ("Kennedy Asks Aid for Cuban Rebels to Defeat Castro, Urges Support of Exiles and Fighters for Freedom") unsettled American liberals still uncertain about Jack's bona fides. James Reston, the *New York Times* columnist, a man of studiously judicious opinions, wrote: "Senator Kennedy made what is probably his worst blunder of the campaign."

In the fourth debate Jack marked Nixon as an impotent bystander looking on hopelessly as Castro took over Cuba. Nixon said later that for the first time he felt personal animosity toward Jack. He knew that Jack's remarks and his press release were even more unfair than they seemed to Republican loyalists. Since March, the government had been planning a large-scale covert action against Cuba run by Cuban exiles, an operation that he assumed Jack had learned about in his CIA briefings. Nixon was not only a fervent supporter but a prime mover of the action—an understandable position considering that, if the agency had kept to its original schedule, the action would have taken place a few weeks before the presidential election.

The Eisenhower-Nixon administration that Jack was condemning for its weakness in fighting communism was the first peacetime American administration to mandate assassination as official government policy. Eisenhower had done so, or so it appears, at a National Security Council meeting on August 19, 1960, dealing with the left-wing Congolese leader Patrice Lumumba. Robert H. Johnson, the official note-taker, recalled that the president turned to CIA Director Allen Dulles "in the full hearing of all those in attendance and [said] something to the effect that Lumumba should be eliminated . . . there was a stunned silence for about 15 seconds and the meeting continued." Within a few days, Dulles authorized $100,000 to kill the new president of the Congo. Dulles preferred hiding the sting in euphemism. He told Station Chief Lawrence Devlin that "we wish to give you every possible support in eliminating Lumumba from any possibility of resuming government position."

At the same time this action was going on, a CIA agent was meeting with Johnny Rosselli at the Brown Derby in Beverly Hills, asking the mobster's help in assassinating Fidel Castro. In Miami, Rosselli brought in a group of his associates that included Giancana, the Chicago syndicate leader, and Santos Trafficante, a Florida mob boss, who agreed to use their contacts in Havana to attempt to kill the Cuban leader.

As Jack stood next to Nixon in these historic debates, he appreciated one of the conundrums of democratic government in the modern world. Time and again he had pondered how self-interested democratic man could possibly win against the regimented legions of totalitarian regimes. He knew that in World War II the magnificent qualities of his fellow Americans had come through, but would it happen again in the silent, twilight war against communism?

The Eisenhower who had apparently chosen to authorize murder was a different leader from the greatest general of American's greatest war, who led 150,000 men into combat at D-Day saying: "The hopes and prayers of liberty-loving people everywhere march with you." This was a president apparently giving secret orders that allowed the CIA to send assassins to attempt to garrote, poison, shoot, or strangle Lumumba and Castro. Did the hopes and prayers of a liberty-loving people go with them as they sidled silently out into darkest night? Was this to be the fate of any mass leader around the world who rose up carrying a Marxist textbook in his hand while waving the flag of nationalism? If that was to be, was not this new statecraft easily learned? If would-be killers could stealthily make their way to the Congo and Cuba, couldn't other killers rush the throne of power in Washington?

No evidence exists that at this time Jack knew about the assassination attempts, but in this debate Nixon had good reason to condemn Jack for his duplicitous act of pretending to know nothing about the training of Cuban exiles. To do so would have blown the cover on the CIA operation, and that he would not do. Instead, Nixon decided to lie. "I think that Senator Kennedy's policies and recommendations for the handling of the Castro regime are probably the most dangerously irresponsible recommendations that he's made during the course of this campaign," Nixon said. "Now, I don't know what Senator Kennedy suggests when he says that we should help those who oppose the Castro regime, both in Cuba and without. But I do know this: that if we were to follow that recommendation, that we would lose all of our friends in Latin America, we would probably be condemned in the United Nations, and we would not accomplish our objective. I know something else. It would be an open invitation for Mr. Khrushchev to come in, to come into Latin America and to engage us in what would be a civil war, and possibly even worse than that."

In lying, Nixon had spoken the truth. This tragic misadventure whose existence he denied was everything Nixon said it was, though he was one of its major backers. As for Jack, his goading of Nixon was an unseemly business. Although Jack later asserted that he had not been briefed by the CIA about the invasion plans, he had probably learned of them from several other sources. Manuel Artime, the political leader of the brigade of Cuban exiles, met Kennedy in July 1960, when they almost certainly discussed the prospective invasion. In October, just before the final debate, John Patterson, the governor of Alabama, said that he flew to New York and at the Barclay Hotel secretly briefed the candidate about the members of the Alabama Air National Guard who were training the Cuban exiles in Guatemala for an invasion. "I want you to promise me that you are never going to breathe a word about this to anybody if you do not know about it, because there are a lot of lives at stake," he recalled telling Jack. A few weeks later, when President-elect Kennedy was formally briefed on the operation for the first time, the CIA's Richard Bissell realized that Jack was already knowledgeable about the invasion. "I think Kennedy had obviously heard of the project," Bissell reflected. "Just how, I don't know, but that wasn't surprising; a fair number of people knew about it." The CIA would admit later that "the project had lost its covert nature by November 1960," and surely the Democratic candidate for president had been one of those to know about it.

Jack won his debate points, but in doing so he helped to entomb himself in a prison of rhetoric out of which he and his administration would never fully escape.

It said much about Jack's marriage that in his acceptance speech in Los Angeles he had violated one of the most sacrosanct traditions by not even mentioning his absent wife. He had fit Jackie into a compartment in his complicated life, but now that he was running for the presidency, she had become a potential problem. "She . . . absolutely curled up at people shouting, 'Hello, Jackie,' who had never seen her before," reflected Lord Harlech. "She'd just turn away, and you could feel a strong resistance to this kind of a life."

The Democrats did not know quite what to do with Jackie, an exotic orchid set among daisies and marigolds. They feared that Jackie lived so far outside the world of middle-class America that most voters would be uneasy having her in the White House. The Republicans trundled out their "Pat for First Lady" campaign. Presumably most Americans could identify more with this cloth-coated, modest matron, a veritable Betty Crocker of a politician's wife, than the elegant, sexy, baby-talking, foreign-sounding Jacqueline Bouvier Kennedy. That contrast had not yet sunk in when the Democrats announced that the Kennedys were going to have a second child. A pregnant mother trumped even a Bible-touting, cloth-coat-wearing matron. But there were still cynical whispers in some quarters "that it [the pregnancy] was planned that way to keep that supposedly lethal glamour out of circulation."

Jackie's seeming extravagance was such a problem that she appeared for one interview in a $29.95 maternity dress, wearing it as if it were a uniform of the middle class. "I'm sure I spend less than Mrs. Nixon on clothes," Jackie said, a subject that, fortunately for the Democrats, was not pursued.

Jackie could not talk to Jack about just how difficult she found it being forced into the confining mold of a public person. She turned to Joe Alsop, whose insights into Washington social life were often deeper and truer than those found in his political columns. Alsop contrasted his cousin Eleanor Roosevelt, who in public managed "oddly humorous and even downright fantastic things," with Pat Nixon, who created "false homey touches" that reeked of an "adman's phoniness." Alsop wrote Jackie of "things that can be done for public purposes without any departure from or falsification of your private self."

Jackie was seeking to maintain her authenticity and to preserve her rich inner life. "You have been of more help than you can imagine," she wrote Alsop. "And there is one more thing you have taught me—to respect power. I never did—possibly because it came so suddenly without my having had to work for it—(power by marriage I mean). But if things turn out right—I will welcome it—and use it for the things I care about."

Jack would have preferred to spend his campaign days talking largely exclusively about foreign affairs, but there was one domestic issue that would not wait any longer, not for him or for anyone else. That was the continued disenfranchisement of black Americans. Bobby at least had played beside a black teammate on the Harvard football team, but except for his gentleman valet, Jack had practically no contact with Americans of other races. That made him no different from most men of his class, but he was neither intellectually nor emotionally attuned to the great domestic moral issue of his time. He was, moreover, the titular head of a party that included as a crucial element segregationist southerners.

American blacks understood that fact full well, and in the 1956 presidential election, 60 percent had voted Republican, considering the party of the sainted Abraham Lincoln, if not the vehicle of their deliverance, then at least the lesser of two evils. They had voted Democratic during the New Deal, but that bond was now broken, and black voters appeared woefully slow in moving toward a Democratic Party that vowed to serve their interests. The prominent Atlanta minister Martin Luther King Sr. was one of a large number of the older generation of black southern preachers who had come out for Nixon, in large part in opposition to Jack's faith.

Martin Luther King Jr. had followed his father's calling, but unlike many of that generation, he considered his faith not a substitute for justice but the very engine of its enactment. The Reverend King was a disciple of Mahatma

Gandhi, the martyred Indian leader, and like his mentor, King was a man dangerous to all who looked on the world as it was and thought that was the way the world would always be. Despite what his enemies believed, in the first year of this new decade the Reverend King was not leading the black students across the South who had suddenly risen up, sitting in at lunch counters and picketing those who denied them their rights. He was wary of their endlessly confrontational politics and in some senses was pushed forward by the sheer energy and will of these young people. For King, the whirlwind of the 1960s had already begun.

Over his political career Jack had learned that it was at times impossible to shake the hands of black and other ethnic leaders in equality when many of them offered nothing but outstretched palms. Louis Martin, the one important black aide in the campaign, said that the way to win the black vote was through his colleagues at black newspapers. "And I know those papers aren't going to do a damn thing for you unless you pay us some money."

The most admired figure among the black minority was not the controversial Reverend King but the legendary baseball player Jackie Robinson, and he had already endorsed Nixon. Adam Clayton Powell, the Harlem congressman, was the most powerful political figure in black America, and the second-most-desired endorsement. His golden tongue could be had for a golden price, in this case an offering in the Reverend Powell's collection plate of $300,000 cash to get out the black vote. Kennedy's people knew that Powell would take most of that to get out his own vote, and they countered with $50,000 for ten endorsement speeches.

King was a leader of a different kind. He sought a harder currency for speaking positively of Jack's candidacy. King was a figure with various constituencies to satisfy, and strategies beyond the grasp of almost any of them. He couldn't afford to squander his moral capital by publicly endorsing Jack. Still, he knew that for his people Kennedy was a far better choice than Nixon. He wanted the minions of youthful protest to lie low during the election campaign, or they might give Nixon the election.

King asked for a meeting with Jack in the South, but Jack turned down that modest request when he learned that the civil rights leader felt that he should offer to meet with Nixon as well. "The hell with that," Jack told Harris Wofford, a lawyer and campaign aide concerned with civil rights matters. "Nixon might be smart enough to accept. If he does, I lose votes. I'm taking a much greater risk in the South than Nixon, but King wants to treat us as equals. Tell him it's off."

King had been looking for a worthy excuse to be out of Atlanta, where the new wave of protests had already begun and his absence was so evident.

Jack's intransigence led King to do what he did not want to do, to lead where he did not want to lead, and to become the leader he might not have become. He was left with no good excuse to stand apart from his young brethren in protest. On October 19, he joined eighty student protestors asking for service in the segregated Magnolia Room at Rich's, Atlanta's premier department store.

That morning the Atlanta police arrested King, and for the first time in his life he spent the night in a jail cell. King refused bail, and the matter threatened to create an embarrassing dilemma for a candidate who was trying to draw southern blacks to his banner while holding on to southern segregationists.

When Jack sought counsel on this matter, he turned to Sarge Shriver and Harris Wofford, who led the campaign's civil rights efforts. Wofford reflected years later that Jack was not completely comfortable with his two "super idealistic" subordinates. Neither his brother-in-law Shriver, "a radical Catholic," nor Wofford, who had become an advocate of Gandhian nonviolence in India, were likely to accede easily to the inevitable compromises of politics. Wofford wrote a strong statement for Jack to release in defense of King and condemning his arrest. Jack read the passionate statement with dismay. He had been talking to Ernest Vandiver, the governor of Georgia, politician to politician, and now, a few days before the election, Jack was not about to make such a statement.

"Look, our real interest is getting Martin out, right?" Kennedy told Wofford. "Well, I've been told that if I don't issue that statement, the governor of Georgia says, 'I'll get that son of a bitch King out.'"

The Kennedys were not the only politicians worried that the political sky was falling. William Hartsfield, the mayor of Atlanta, deplored the dismal publicity his progressive city was receiving. He went ahead and arranged to have King and the other activists released and the charges against them dropped, an action he was taking, he said, only because Senator Kennedy had implored him to do so. Wofford had done all the imploring, and the Kennedy campaign staff backed away from the matter as best they could, lest they end up freeing King but losing the solid South.

That would have been the end of it, but just as King was to leave Fulton County Jail, Judge Oscar Mitchell in Georgia's De Kalb County had him arrested again on a bench warrant for violating his parole on a traffic violation and carried away in leg irons. On the following day, less than a week before the election, the judge sentenced King to six months at hard labor. The judge had hardly banged his gavel a last time before King was taken to Georgia's toughest prison at Reidsville to begin to serve out his term.

Early on the morning that King arrived at Reidsville, Jack called Governor Vandiver. "Governor, is there any way that you think you could get Martin Luther King out of jail?" Jack asked. The two men had a perfect commonality of interests. They both wanted to end the relentless publicity about King's imprisonment and focus on other matters. They both wanted to win the election. They both wanted to minimize the civil rights issue in the campaign. And they both wanted to keep their role in this matter as quiet as possible. The governor made no public statements but called a close friend, Georgia Secretary of State George D. Stewart, who in turn called his close friend Judge Mitchell, who in the end arranged to release King on bond.

While this was going on in Georgia, Jack had just finished speaking at a breakfast in Chicago and was in a suite at O'Hare Airport, where the candidate's plane would soon be taking off. Wofford had talked to Mrs. King and knew that she was terrified after her husband's midnight ride to a new prison. Wofford called Shriver. "The trouble with your beautiful, passionate Kennedys is that they never show their passion," Wofford said, imploring Shriver to get Jack to call Mrs. King. "They don't understand symbolic action."

"It's not too late," said the terminally optimistic Shriver. "Jack doesn't leave O'Hare for about forty minutes. Give me her number and get me out of jail if I'm arrested for speeding."

Shriver sped to the airport and rushed into Jack's bedroom, where he had retreated to get a few minutes rest. He made sure that he was alone, so he would not hear the other aides nay-saying his proposal.

"Why don't you telephone Mrs. King and give her your sympathy," Shriver said. "Negroes don't expect everything will change tomorrow, no matter who's elected. But they do want to know whether you care. You will reach their hearts and give support to a pregnant woman who is afraid her husband will be killed."

Jack knew how close Wofford and Shriver were, how fervent their concerns, and he was not about to tell his brother-in-law that he had already called the Georgia governor and hoped that King might be out of prison in a few hours. He trusted Shriver only so far, for he was an ideologue committed to this issue. Instead of confiding in his brother-in-law, he listened, as if he were hearing this matter for the first time this day. "What the hell," he said, as if he were acting impulsively. "That's a decent thing to do. Why not? Get her on the phone."

Jack talked to Coretta Scott King, a call that in other circumstances would have been considered little more than a minimal act of decency. When Bobby heard of the action, he was infuriated. "You bomb-throwers have lost the whole campaign," he told Wofford and Louis Martin. He warned them to issue no more statements. When Bobby was at his angriest, he had a frigid fury, his eyes cold, his voice tightly controlled, his fists clenched. "Do you know that three Southern governors told us that if Jack supported Jimmy Hoffa, Nikita Khrushchev, or Martin Luther King, they would throw their states to Nixon?" he said as the campaign plane flew to Detroit. "Do you know that this election may be razor close and you have probably lost if for us?"

Bobby decided, however, that King would have to be released. That evening from a phone booth in New York, Bobby called Judge Mitchell in Georgia because, as he told his aide John Seigenthaler, the judge was "screwing up my brother's campaign and making the country look ridiculous before the world." It was an intrusion into the judicial process that many lawyers would think violated professional ethics. Judge Mitchell recalled Bobby saying that if King stayed in prison, "we would lose the state of Massachusetts."

Even after King was released from prison, he was still unwilling formally to endorse Jack, but his influential father was not so hesitant, saying that "if Kennedy has the courage to wipe the tears from Coretta's eyes, he will vote for him whatever his religion." Kennedy knew how important that statement would be among black voters, though he was not aware of the irony. "Did you see what Martin's father said?" he asked Wofford. "He was going to vote against me because I was a Catholic, but since I called his daughter-in-law, he will vote for me. That was a hell of a bigoted statement, wasn't it? Imagine Martin Luther King having a bigot for a father. Well, we all have fathers don't we?"

While the Kennedy campaign did everything to downplay the role that Jack and Bobby played in King's release, Wofford and Shriver went ahead and prepared two million copies of a small blue pamphlet to be handed out in black churches on Sunday, two days before the election. As extraordinary as it may seem, Wofford passionately asserts that neither he nor Shriver cleared this effort with Bobby or Jack. The pamphlet talked of Jack's and Bobby's calls and quoted King's statement that he was "deeply indebted to Senator Kennedy, who served as a great force in making my release possible."

Teddy had been given the West as his territory, and he roamed freely, spreading as much mirth as politics. On a Texas campaign trip, he and his college buddy Claude Hooton Jr. decided to play an inspired prank on Bobby, who was also in the state. They put together a collection of women's undergarments worthy of Frederick's of Hollywood, doused the frilly underwear in perfume, and hid them in Bobby's suitcase, along with an equally perfumed love letter.

Bobby had the most trusting of wives, for when he arrived home and

Ethel unpacked his bag, there were no screams heard from Hickory Hill. Bobby vowed, however, that given his chance he would get revenge on his brother and his friend.

On another campaign trip to Hawaii, Teddy attempted a measure of circumspection by having his companion, a foreign beauty queen, sitting back in tourist while he flew first-class. But he was so momentarily enamored of the stunning European woman that he kept roaming back among the plebes, signaling his attentions to all but the sleeping. Later in Honolulu at a dinner party with Sinatra, he walked into Don the Beachcomber with the woman on his arm, a gesture that infuriated the singer, who worried that Teddy's public indiscretions would hurt his brother's campaign. That did not happen, but if the other campaign managers had done as poor a job as Teddy, Jack's chances for the presidency would have been minimal.

All during the fall campaign, Jack had worried about his health becoming an issue. He sought to dispel the rumors by running the most vigorous, demanding campaign imaginable. Jack's own running mate had tried to kill off his candidacy by leaking the story of his Addison's disease. His enemies said that he was a deceitful near-invalid whose health alone should have disqualified him from high office. His medical records were inches thick, but he embarked that last week on a seventeen-state campaign that would have tired a marathon campaigner. In that final week he traveled to more states than any presidential candidate had ever visited in a scant seven days.

Jack had brilliantly presented himself in television debates, sound bites, photo ops, and shrewdly calculated advertising. Yet in 1960, there remained an unspoken compact that each voter had the right to shake hands with the candidate, to touch him, to hear his words in person, to hold a banner with his name boldly written on it, to shout his name as he drove by. Tens of millions of Americans voted twice, once with their presence, and once again with their ballot. These were not neatly orchestrated days of media events in which Jack could retreat for afternoon naps and evening sitz baths. It was still the essence of politics to get out to the people and to talk to them, creating the illusion that this moment was as important to the candidate as it was to them.

Jack drove through the depressed coal mining areas of Pennsylvania, where voters were scarce, in an open car and waved to onlookers in tiny Republican towns. Even if at times he was collecting one vote at a time, it was still one vote he had not had before. In a given day, sometimes even an hour, he would encounter enormous, boisterous crowds, and then arrive at a half-empty union hall or a rally notable only for being sparsely attended. And everywhere he went there were what the journalists called "jumpers," young women who screamed their ardor as they would have for Elvis, creating orgiastic moments that had nothing to do with what was once called politics. In the final days the crowds grew larger and applauded him with greater and greater intensity. The next to the last day of the campaign ended at three in the morning before thirty thousand supporters who had waited half the night to see their candidate in the town square of Waterford, Connecticut.

There was one other Kennedy who had worked as hard and as single-mindedly as Jack since the convention, and that was Bobby. Subtlety and grace were the first casualties of Bobby's role as campaign manager. Bobby felt that many of those who opposed his brother were not merely misguided men of good intentions but scoundrels and rogues. He had the considerable misjudgment to attack Jackie Robinson for the capital crime of backing the Nixon candidacy, impugning the black leader's honor more than his judgment. "If the younger Kennedy is going to resort to lies," the legendary sports star replied, "then I can see what kind of campaign this is going to be."

Bobby wisely backed off from that particular attack, and for the most part his campaign judgments were acute and judicious. "He gave people their head," said Wofford, "He kept saying we want to get every horse running on the track." He nonetheless left the queasy feeling in some of the campaign staff that he was ready to discard them the moment they faltered or misjudged, or that he needed them to take a fall for an error that more rightfully should have had a Kennedy name attached to it.

The brothers were perfectly in tandem on this quest: Jack, the sonorous voice and subtle mind of the campaign, Bobby the fist and the muscle. As the brothers flew around the country, they rarely saw each other in person but talked often on the phone in their own private code. On one occasion, they happened to be passing through the same airport.

"Hi, Johnny," Bobby said, as nonchalantly as if they were two barnstorming salesmen drumming up their wares. "How are you?"

"Man, I'm tired," Jack said, looking into a pair of eyes as exhausted as his own.

"What the hell are you tired for?" Bobby exclaimed. "I'm doing all the work."

As Jack moved through the last days of the campaign, Bobby was on his own frenetic schedule, as if by sheer will he could push up the vote tallies. On the final Saturday of the campaign, he took a trip to make speeches in Ohio, Kentucky, and North Carolina. In the Cincinnati airport, Bobby spied a gigantic six-foot-tall stuffed dog. His children loved animals, and though the gigantic dog was a display gimmick, not up for sale, he fancied himself walking into Hickory Hill with a dog larger than he was.

"How much is that?" Bobby asked. It was a question that bored travelers had probably asked before, and the vendeuse gave him the answer that always drove them away, down the airport corridor. "A hundred dollars."

"I'll take it," Bobby said. The gigantic dog could not even fit into the private plane without having its head unceremoniously removed. The weather was foul and the time was late, but Bobby insisted that the pilot fly back to Washington so that he could spend the night with his family at Hickory Hill. And so the plane soared north through the murky black skies, carrying Bobby and an enormous, beheaded stuffed dog.

All those weeks of the campaign, Jack's father continued to work quietly for his son's election. When Arthur Krock spoke unkindly about Jack's candidacy, Joe shut the *New York Times* columnist out of the Kennedy lives forever. Krock had become a prickly near-reactionary, and Joe was not wrong in believing that the journalist changed to a softer lens when he turned to look at Nixon.

Joe had an understanding of the relationship between money and political power that was mercilessly realistic. From the time he had first worked for Roosevelt, he had understood that big contributors did not want passbook-size returns on their money, and that what they considered good for the country was often their own personal or corporate good writ large. Cash had no fingerprints on it, and in 1960 campaign laws were still loose enough that large amounts of untraceable money moved around Kennedy's campaign, as it did in Nixon's.

Joe told one of the aides to go pick up a valise full of money and to carry it to another destination in the campaign. The aide returned, confused. "You know, there was ten thousand less than you said," he told the man whom they knew as "the Ambassador."

"In politics, money does not grow in the passing," Joe said, always the philosopher.

There has been much conjecture that the Kennedy patriarch made an unholy alliance with the mob to ensure Jack's election. In these stories Sam Giancana shows up at all the crucial moments, a shadowy presence, his face barely visible. Exner has Giancana standing on the platform at Chicago's Union Station waiting for her to arrive with a valise full of Kennedy money. Tina Sinatra has Joe asking her father to meet with Giancana to solicit his help. Tina says that Frank Sinatra met the Chicago mobster on a golf course, where he appealed to an undiscovered strain of patriotism in the pathological killer. "I believe in this man and I think he's going to make us a good president," Sinatra supposedly said, one good citizen to another. "With your help, I think we can work this out."

A different story has Joe meeting the mobster the first time in the Chicago courtroom of Judge William J. Tuohy, where the two men conspired to subvert the election. Another story has Joe meeting Giancana, other Chicago Mafia figures, and the Los Angeles mobster Rosselli at Felix Young's, a New York restaurant, while their bodyguards waited outside. In yet another scenario Joe met at the restaurant that day with an even more notorious array of mobsters that included Carlos Marcello, the New Orleans don.

Joe had contacts with the darkest part of American society. In June of that election year, he stayed at the Cal-Neva Hotel at Lake Tahoe, owned by mob interests. The FBI said later that, according to informants, he met there with "many gangsters with gambling interests and a deal was made which resulted in Peter Lawford, Frank Sinatra, Dean Martin and others obtaining a lucrative gambling establishment, the Cal-Neva Hotel." This did not mean that Joe had brokered the deal, but Mafia interests owned the Cal-Neva Hotel, and it was an unlikely place for the candidate's father to have chanced upon.

Joe was an immensely shrewd man who had for the most part kept his underworld connections quiet. Joe may well have met with individuals who had mob connections and asked for their help in the campaign, just as he was meeting with party bosses, power brokers, labor people, and others. It was unlikely he would flaunt that interest by meeting mobsters in prominent restaurants.

There were, however, limitations on what he was likely to ask and what they were likely to be able to do. Joe was one of the wealthiest men in America. It was an unnecessary risk to have tainted Mafia money pouring into the campaign in vast amounts with all the chances of discovery and the possibility of blackmail. His own son Bobby was one of the vociferous enemies of Giancana and his kind. By working out an elaborate agreement with Giancana, Joe would have betrayed his own blood.

That the Mafia had its tentacles deep into American life did not mean that it had strength enough to lift a man to the White House. The mob, moreover, was ecumenical in its political concerns, purportedly making a half-million-dollar contribution to Nixon's campaign. As for Giancana, he had power in Chicago, but for every vote that the Chicago Mafia controlled, Mayor Daley controlled a hundred. Across America the mobsters had their hooks into union locals and officials, but the unions largely backed Democratic candidates anyway, and in those instances the mob could only confirm what the unions were already doing. Even if Joe utilized his mob connections, as he probably did, it is simply unthinkable that Giancana and his cohorts could have been the crucial factor in the election.

The syndicate bosses, nonetheless, had reasons to think that in a

Kennedy administration the FBI might back off from its interest in organized crime. Sinatra provided the campaign theme song, "High Hopes," and Giancana and his associates had their own high hopes in Kennedy. Sinatra had boasted to Giancana that his friend Jack would ease up on the mob, and with that in mind Giancana had pushed Kennedy's election.

The final rally is always a sentimental moment, even in the most dispirited of presidential campaigns, for whatever the polls might say, no matter what the aides might fear, it is not they who decide, but tens of millions of voters across the land. Jack's campaign was ending in Boston, where the Kennedy saga had begun over a century before. The motorcade crawled through downtown streets so slowly that it took ninety minutes to move the two miles from his hotel to Boston Garden. Jack waved to sidewalks crowded with over half a million citizens who hoped their votes would help lift this son of Massachusetts to the White House. They shouted and they hooted and they yelped the way Boston Democrats had fifty-four years before when the torchlight parade celebrated his grandfather Honey Fitz's ascendancy to the mayor's office.

The car passed the old State House and rolled over the ground where the Boston Massacre took place, and where Rose had stood lecturing Jack and his brothers and sisters on the five heroic Americans shot dead by the British redcoats, their martyrs' blood drenching the street. At times his mother had been a merciless pedagogue, but she had woven history into Jack's very sinews, and he could see himself as someone walking in the firm steps of American patriots. His eyes were red, but as he waved and smiled, he gave no sign of collapse. Jack had his Methedrine-laced shots to provide their ersatz lift, but to a politician there was no shot like this, the pure amphetamine of politics, which lifted him beyond even the exhaustion of the two-month-long campaign.

The platform behind Jack at Boston Garden was jammed with every Massachusetts Democratic pol from Pittsfield to Yarmouth. There were those who had loved Jack from the day the scrawny veteran had first walked through the streets of East Boston. There were those who had always despised him and his benighted family and thought them the bane of party politicians. And there were many who had been indifferent. But they were all Kennedy men this evening, pressing toward their candidate. No one remembers, though, like the Kennedys—when you had signed on, what you had given, how long you had worked. A smile tonight would not make up for years of shrugs. As they stood there waiting for Jack's appearance, Kenny O'Donnell surveyed the platform with its preponderance of Irish-American faces. Then he turned and whispered to Bob Healy of the *Boston Globe* and Theodore White: "You know, they all think that Kennedy has a trick and if they can learn it they can become president of the United States too."

Jack's campaign was supposed to have ended at this triumphant rally at Boston Garden. The Republicans, however, had purchased four hours of national television time on this final evening. At the last moment the Democrats bought their own half hour on ABC directly following the Nixon marathon. So Jack's quest for the presidency ended at historic Faneuil Hall in a nationally televised speech before a smaller, less raucous gathering than the one he had just addressed at Boston Garden.

Fifteen years before, a young Jack Kennedy had stood not far from here giving his first campaign speech. Now, campaigning for himself for the last time in his life, he was elegant in dress and manner, his wit exquisitely honed, his phrasing eloquent, his speech resonant, his voice firm.

"This is the campaign, and it's now come to an end," he began as he looked out on an audience full of Bostonians who had seen him first win election to the House. "I think this old hall reminds us of how far we've been as Americans and what we must do in the future." American history resonated through Boston in ways it did in no other city in America, and that history resonated through Jack as it had through few other presidential contenders.

This was Jack's last chance to send supporters bursting into the streets ready to rally their forces to the polls. But that was not the kind of speech Jack was giving this evening. This was a solemn, somber speech, as if he realized far better than anyone else the sheer magnitude of what the next president would face. He invoked Lincoln's election of 1860, though one hundred years later the nation was not on the verge of civil war. He invoked Wilson's election of 1912, though the world did not face world war. He invoked Roosevelt's election of 1932, though America was no longer a land in which one-third of the people lived in poverty.

"The campaign is now over," he said at the end of a long speech, at the end of a long day, at the end of a long campaign. "You must make your judgment between sitting and moving. This is a race not merely between two parties, the Democratic Party and the Republican Party, or between two candidates. It is a race between the comfortable and the concerned. Those who are willing to sit and lie at anchor and those who want to go forward. This country has developed as it is. We are here tonight because in other great periods of crises we have chosen to go forward."

It is one thing to call a man dispassionate when he seems unaffected by what happens to others. It is something else when he exhibits an impenetrable cool when it is his own future at stake. As Jack sat in his parents' home in Hyannis Port watching images on the television set and hearing late returns from various aides who scurried in and out of the living room, there was not a glimmer of anxiety in his voice or a moment's irritability at the twists and turns of what would prove in the popular vote to be the closest presidential election in American history: in the end fewer than 120,000 ballots, out of 69 million, separated the two candidates. Jack was no more elated at an early computer projection on television that he would win than he was by a later projection that he would lose. He was slightly ahead at close to four in the morning when Nixon appeared before the cameras to say that if the present trends continued, his opponent would win. Jack's aides poured out their wrath on the television screen, upset that Nixon did not do the honorable thing and concede. "If I were he, I would have done the same thing," Jack said, effectively silencing those throwing epithets at the screen.

Jack went to sleep while Bobby and the others kept watch. All night long, Bobby worked the phones, helping to run up a long-distance bill of about $10,000. In Illinois, Mayor Daley had proved his fidelity to Chicago's dubious politics by not tabulating the city's final votes until the heavily Republican downstate ballots had all been counted and he knew what it would take to win. California came in for Jack finally, and by dawn it was clear that he had won, but no one considered waking Jack. Although it would become part of American political myth that, with Daley's help, Jack won fraudulently, even without the questionable Illinois votes, Kennedy still had 276 electoral votes to give him a bare majority.

When Sorenson entered Jack's bedroom at 9:30 A.M., he addressed the man sitting in pajamas in his bed as "Mr. President." Early in the afternoon, at the Hyannis Armory, the president-elect went before the cameras for the first time. Jack had insisted that his father be there, and Joe had reluctantly agreed. Jack stood there now with his wife, two brothers, three sisters, and his mother. The president-elect spoke with self-confidence and control, and almost no one noticed that out of sight of the cameras and most of the reporters his hands were trembling.

Book Three

21

The Torch Has Been Passed

On the day before the inauguration of the thirty-fifth president of the United States, a fierce storm fell upon Washington, blanketing the capital with eight inches of snow, stranding ten thousand cars across the city, grounding planes, and slowing trains. Crews worked during the night cleaning the main roads, and by noon on January 20, 1961, a crowd of twenty thousand stood on the Capitol grounds in the twenty-two-degree cold, braced against the eighteen-mile-an-hour winds and looking up at the portico where the nation's leaders sat outside to witness the ritual of passage. One million other Americans began gathering along Pennsylvania Avenue to greet the new president as he traveled in a parade that would carry him to his new home in the White House.

The onlookers had bundled themselves up against the fierce cold wearing an eclectic collection of wool coats, snowsuits, fur hats, ski jackets, hiking boots, galoshes, mufflers, face masks, hoods, and scarves. Up on the podium, seventy-year-old Dwight Eisenhower, then the oldest president in American history, sat wrapped in a heavy topcoat and scarf. The other largely aging politicians and officials were equally protected against the cold, many of them in top hats or homburgs. In the row upon row of largely indistinguishable black and gray coats, hats, and pale faces, there was one tanned, radiantly healthy-looking man.

Forty-three-year-old John F. Kennedy was the youngest elected president in American history, and he stood there the personification of the nation's energy and ambition. The new president tended to speak quickly, but this noon he spoke with a careful, deliberate pace:

> Let the word go forth from this time and place, to friend and foe alike, that the torch has been passed to a new generation of Ameri-

> cans—born in this century, tempered by war, disciplined by a hard and bitter peace, proud of our ancient heritage—and unwilling to witness or permit the slow undoing of those human rights to which this nation has always been committed, and to which we are committed today at home and around the world.

The audience listened closely, the words seemingly resonating so deeply that they did not even applaud until Kennedy had spoken nearly five minutes. He appeared to be a strong young leader for a difficult new age. Kennedy had many virtues to bring to his presidency. He had a political mind as sharp as that of any of the politicians who surrounded him, but he understood nuances and subtleties on a deeper level than did most of Washington's narrow men. This was a brilliant attribute, nurtured by his father's influence, cultivated at the Court of St. James's and in his extensive travels, honed in the House of Representatives, where he was as much an observer as a participant, and seasoned in the Senate, where he had to deal with a complex, contradictory world. He had a sense of history that was not an academic's abstract vision but a vivid sense of human character moving through time.

Kennedy's father had taught him that he could make history and write it in his own name. And so he was setting out on this day as if he had a massive mandate and his youth was an added virtue. He had always been a cautious leader, and he knew that the path ahead was fraught with dangers both known and unknown, but he set forth an uncharacteristically bold and daring message:

> In the long history of the world, only a few generations have been granted the role of defending freedom in its hour of maximum danger. I do not shrink from this responsibility—I welcome it. I do not believe that any of us would exchange places with any other people or any other generation. The energy, the faith, the devotion which we bring to this endeavor will light our country and all who serve it—and the glow from that fire can truly light the world.

Many who heard the new president's words thought of him as a hero. Kennedy was a philosopher of courage who had written a book on the subject. Although those listening did not know it, during his life Kennedy had struggled against physical disabilities that would have hobbled most men. He considered politics at its highest level an arena for heroism, a colosseum where a few good men performed noble acts whose merits were often only dimly perceived by the rancorous, fickle masses.

> And so, my fellow Americans: ask not what your country can do for you—ask what you can do for your country. My fellow citizens of the world: ask not what America will do for you, but what together we can do for the freedom of man.

As much as Kennedy wanted to sit in the company of greatness, he knew that no longer could civilized men stand on fields of battle fighting each other with bullet, sword, and fire when at any moment they might be enveloped in mushroom-shaped clouds vaporizing all humanity. The young leader's great test would be in part to see whether he could define political courage in a new way for a nuclear age.

Kennedy called upon the citizenry to face the new era with intrepidness. The president wanted millions of Americans to rise out of their privatism, shake off their passivity and cynicism, and move forward in acts of sacrifice and selfless service. Since there appeared to be no great war to be won, and no immense frontier to conquer, it was unclear just where this journey would lead, or what this leader would light with the torch he raised.

Kennedy believed that as president, his overwhelming concern would lie in international affairs, and his entire speech dealt with America's relationship to the rest of the world. He said nothing of the greatest American moral dilemma of the age: the political and economic disenfranchisement of the majority of black people. Sorensen had added the only vague reference to civil rights on the eve of the inauguration when Wofford had lobbied the speechwriter to add the words "at home" to the president's commitment to human rights "*at home* and around the world."

Kennedy talked "to those people in the huts and villages of half the globe struggling to break the bonds of mass misery," but he said nothing of the four million Americans who would have starved if not for the surplus foods the government gave them or of the tragic lives of millions of migrant laborers across America whom Edward R. Murrow had poignantly portrayed in his recent documentary *Harvest of Shame.*

Kennedy's eloquent idealism was a cup overflowing, and the new president's auditors heard what they wanted to hear and needed to hear, be they black ministers in the South, the poor and hungry of his own land, the peoples of Europe, or the masses of Asia and Latin America. Even Castro had reached out to the new administration, saying that he was willing to "begin anew" in his relations with the United States, while Nikita Khrushchev, general secretary of the Communist Party of the Soviet Union and the chairman of the Council of Ministers, hoped for a "radical improvement" in the two countries' relations. As Kennedy stood there on this day in which the sun could not heat the earth, he seemed to be promising warmth where there was cold, and light where men lived in darkness.

On the ride up the broad expanses of Pennsylvania Avenue, Kennedy insisted that he and Jackie ride in an open car so that people could see their new president in the clear cold air of the winter afternoon. Kennedy was not only a leader but also a leading man, and thirty-one-year-old Jackie a stunning model of a first lady. She was a woman of certain mysteries that would not be easily unraveled. She smiled with coy grace and waved her gloved hand.

There was yet another reason why so many Americans greeted this new president and first lady with such joy and anticipation. Just two months before, on November 25, 1960, Jackie had given birth to a son, John F. Kennedy Jr., and for the first time since Theodore Roosevelt's residency, the White House would be full of children's shouts and laughter. Since the day of John Jr.'s birth, the Kennedys had been inundated with telegrams, flowers, booties, sweaters, and a zoo of stuffed animals, the start of an immense fascination with the president's namesake, as well as a delighted interest in his sister, Caroline. It added to the president's aura of youthful vitality that his parents were still alive and healthy, and with so many brothers and sisters, nieces and nephews, uncles and aunts, he seemed to belong not to a family but to a clan.

The presidential limousine finally reached the reviewing stand in front of the White House. There sat Kennedy family members, esteemed officials, and close friends. As the president drove by, Joe rose up out of his seat to salute his son. The Kennedy patriarch had been his children's great enthusiast, but no matter what honors they merited, what race they won, he had never stood to pay tribute to their achievements. But today he stood, saluting the son whom he had always called "Jack" but who now, in public or among outsiders, would be "Mr. President." Joe's simple gesture was not only a profound act of deference and respect for the office of the presidency, but equally a symbol of the passing of the generations. Kennedy looked up at the reviewing stand and saw his father standing there saluting him. The new president took off his hat and tipped it to his father.

While the president's greatest destiny was just beginning, Joe's was nearing its end. He had achieved what few men do, his transcendent dream embodied in this president bearing his name, but in doing so he had lost part of his son. "Jack doesn't belong anymore to just a family," he reflected. "He belongs to the country. That's probably the saddest thing about all this. The family can be there, but there is not much they can do for the President of the United States."

During the campaign, Joe had bragged that while he would keep quiet until election day, afterward he would have his say. "I assure you that I will do it after that, and that it will be something worthwhile," he boasted to *Newsweek*. "People may even see a flash of my old-time form." Once the election was over, however, Joe seemed not to be concerned anymore with embroiling himself in all the minutiae of politics, and he never made the statement he had so vociferously promised. When the president-elect asked his father to suggest candidates for secretary of the Treasury, Joe replied: "I can't."

Joe cared primarily about his sons' futures now, and he had just one request to make of his son: that he name Bobby as his attorney general. As much as Kennedy wanted to reward Bobby for his endless work in the campaign, he would no more have made his brother attorney general than name an intern as America's surgeon general. It was unthinkable to make the nation's premier attorney a man who had never practiced law. Kennedy's critics would argue that thirty-five-year-old Bobby was too young, too brash, too ambitious, and too rude.

Kennedy did not dare confront his father with these truths; instead, at the swimming pool at the Palm Beach mansion, he deputized Smathers to suggest gently to Joe that Bobby would make a formidable assistant secretary of Defense. Joe would not even listen to such drivel. "Goddamn it, Jack, I want to tell you once and for all. Don't be sending these emissaries to me. Bobby spilled his blood for you. He's worked for you. And goddamn it, he wants to be attorney general, and I want him to be attorney general, and that's it."

As ambitious as he was, Bobby had his own doubts about the political wisdom of becoming his brother's attorney general. Unlike the president-elect, Bobby had not cordoned off his inner emotions from the world. One of those who had become privy to Bobby's thinking and feeling was John Seigenthaler, a reporter for the *Nashville Tennessean*. Seigenthaler had first covered Bobby during the McClellan hearings. Like Charles Bartlett and Ben Bradlee and a few other reporters, Seigenthaler had incomparable access to the Kennedys and got stories many of his colleagues could never get. Yet as the months went by more and more of what he heard and saw never made its way into his journalism.

Seigenthaler was sitting with Bobby after he had spent a long, discouraging day running around Washington talking to various people about whether he should become attorney general. That was the Kennedy way. Seek out the most knowledgeable people, get their best judgments, and then make up your own mind. In this instance, everyone from Supreme Court Justice Douglas to Senator McClellan had shaken his head in dismay at this harebrained idea. Only Hoover, who at the FBI would be working most intimately with Bobby, said that he should accept the appointment.

Bobby called his brother to tell him that he had decided against it. "We'll go over in the morning," Bobby said as he set down the telephone. "This will kill my father."

Early the next day, Bobby and Seigenthaler drove over to the president-elect's home in Georgetown. Over breakfast, the three men discussed the appointment. Bobby detailed the reasons why he had to turn it down, and his brother told him that he had to say yes.

"You want some more coffee?" Kennedy asked.

"Look, there're some more points I need to make with him," Bobby told Seigenthaler as his brother walked into the kitchen.

"I think the points have all been made," Seigenthaler said.

When Kennedy returned, Bobby set off again. The president-elect would not have put up with this endless palaver from most men. He had heard everything he needed to hear, and he had heard it tenfold. It was time to get on with things and walk outside into the cold morning and tell the waiting reporters what would become the most important appointment of his administration. "That's it, general," Kennedy said, cutting his brother off and calling him by his new title. "Let's grab our balls and go."

As Kennedy walked into the White House as president for the first time, he believed that he had surrounded himself with loyal strong men richly prepared to carry out his mandate. He had kept the obvious holdovers from the campaign, including Sorensen as special counsel in charge of domestic policy and speechwriting. Sorensen used words as the vehicle of policy. He not only wrote almost every important speech the president gave but often handed the address to the president only minutes before he spoke. Sorensen's deputy, Mike Feldman, wrote most of the other speeches and dealt with Israel, regulatory policy, and whatever other matters came his way. Feldman may have been only half the writer that Sorensen was, but he was twice the attorney, and he became the de facto legal counsel.

Kenny O'Donnell, the appointments secretary, was the gatekeeper to the presidential person, along with Evelyn Lincoln, the president's personal secretary. Larry O'Brien was in charge of congressional relations, yet another critical post. O'Donnell and O'Brien were the leaders of a small group in the White House that became known as the Irish Mafia; their ranks included Ted Reardon, Dick Donohue, and Dave Powers. These Irish-Americans shared two faiths, Catholicism and politics. Sorensen may have provided the eloquent public voice of the administration, but these tough-minded men fed the belly, and there was a natural, understated tension between the two groups.

As his chief foreign policy adviser in the White House, Kennedy brought in McGeorge Bundy as special assistant for national security affairs. The Harvard dean had arrived by bicycle for his first meeting with the president-elect at Arthur Schlesinger's Cambridge house, but there was nothing casual about either Bundy's manner or his mind. Bundy had a crucial mandate. Kennedy believed that he could not make innovative foreign policy employing the rigid, militarylike structure that Eisenhower had created with the National Security Council. The president thought the only thing to do was to pull the structure largely down, and Bundy was his engine for doing so.

Kennedy could not run his own foreign policy if he had a powerful secretary of State such as Adlai Stevenson, who lobbied for the position and was shuttled to the United Nations ambassadorship. Instead, Kennedy chose Dean Rusk, the head of the Rockefeller Foundation and a former assistant secretary of State, who willingly wore the shackles of subordination.

There was one man whom all the pundits thought would have inordinate power in the White House, and that was the president's own father. "I want to help, but I don't want to be a nuisance," Joe confessed to Steve Smith, his son-in-law. "Can you tell me: do they want me or don't they want me?"

Steve told Bobby what his father had said, and Bobby thanked his brother-in-law and said nothing. A few days later, seventy-two-year-old Joe sailed for Europe, and that entire year visited the White House only once.

Kennedy's closest White House aides had a fierce, loving loyalty to the president they served and comradely joy in what they were doing. "We had this confidence about ourselves that seems lost from the world of power now," reflected Feldman. "We thought we could do anything. We wrote over a hundred messages to Congress in our first hundred days. Those days were filled with so much excitement and such a feeling of euphoria because we achieved our goal and now we were doing what we looked forward to and you have a superhuman ability when you feel that way." The working atmosphere was one of nonchalance and wit. Sorensen occasionally sent serious memos to Feldman in rhymed couplets, and Feldman, not to be bested, replied in kind.

The humor often had a serrated edge, however, that left its mark. When Kennedy decided to find a place in the White House for his young Boston mistress who had graduated from Radcliffe, he placed her in the office of her former dean. "Kennedy put the knife into Bundy by putting her on the staff," recalled Marcus Raskin, who was only twenty-six years old when he entered the White House to serve as the resident liberal gadfly on Bundy's staff. "And since I was the junior-most person on the staff, she was put to work for me, and Bundy said to me, 'Well, I have a present for you.' I knew something was going on because the president called my office a couple times not to speak to me but to speak to her. So even I figured it out at that point. And eventually she personally told me about it."

The Kennedy humor featured put-downs in which the victim proved his mettle by quickly attacking with an even ruder counterblow. In such matters Kennedy and his friends had decorous limits that Bobby and his friends did not observe. What daring, taunting irreverence was it that allowed Claude Hooton to cable the new attorney general to remind him of the time during the campaign when he and Teddy had salted Bobby's luggage with ladies' underwear (I AM SURE THAT THE ATTORNEY GENERAL HAS NO RETROACTIVE POWERS CONCERNING PERFUMED UNDERGARMENTS INSERTED IN SOMEONE ELSES BAGGAGE). And what of Bobby, who did not fancy himself too powerful or too important to reply in kind: "There is some talk that I might turn the FBI loose on you and Teddy and that would be a full time task for all of their agents."

Bobby was not about to be imprisoned in the dignity of office or to use his exalted new position to distance himself from friends he had known all his life. On the day after the inauguration, Bobby insisted on a football game, even though his old Harvard teammates had only their good clothes. After the football game came tobogganing. The men vied for the high honor of sharing a sled with Kim Novak, the movie star, who wrapped her long legs around her momentary companion. Ethel stood on the sidelines, not amused that her husband was competing for this honor. "I don't understand, Ethel," Bobby said, as he stood holding his daughter Kathleen's hand. "Why can't a father go sledding with his daughter?"

As Kennedy was staffing his New Frontier, he talked to an old family friend, Kay Halle. She was one of the few women who spoke to the president-elect on terms approaching equality. Halle suggested that he should choose more women. He abruptly changed the subject, for as Halle observed, he considered women largely "decorative butterflies and lovely to look at." Kennedy was simply not comfortable being in a room with women who sought to be equal partners in the political process. Women tended to clutter up meetings, forcing a tedious decorum on the manly, often profane lingo of political endeavors. The best way to deal with the problem was simply not to have women present at all.

The Kennedy staffers were mainly in their thirties and early forties. They were for the most part veterans of World War II who, like their leader, had served in combat. They had the stamina to work twelve-hour days, six days a week. Like soldiers in the front line, they worked all night when they had to, and through the next day. They shared a deeply rooted patriotism and a can-do attitude about endeavors large and small. Kennedy was fond of quoting the famous St. Crispin's Day speech from Shakespeare's *Henry V* ("We few, we happy few, we band of brothers;/For he today that sheds his blood with me/Shall be my brother"). Kennedy paid each member of his band of brothers the same salary, $21,000, and would gladly have given them all the same title, special assistant. He wanted no staff meetings, no thicket of bureaucracy. He wanted his men to come to him.

Shortly after the election, Kennedy's staff had sat with the president-elect trying to figure out how they could make sure that only important information got to the Oval Office. Kennedy was obsessed with the fear that he would be locked off from knowledge. "Listen, you sons of bitches, I want you to remember one thing," he exclaimed as his neighbor and friend, Larry Newman, sat listening. "You know there's a guy right behind each of you who's working for me. And there's a guy behind *him* who's working for me. So there's not a goddamn thing any one of you guys can do to keep things away from me. So if you try to pull any bullshit, the next thing you know you'll be out."

Kennedy set up a system so that there would be no crucial information that he did not hear. He was interested in the most arcane nuances of policy, in the details of initiatives, and in the most trivial gossip. Although he appeared to take no pleasure in reading the FBI reports on his aides, he told Feldman, "I never knew my staff led such interesting lives."

Bundy's deputy, Walt Rostow, observed that Kennedy "was capable because of his great energy and human capacity to maintain more reliable bilateral human relations than any man I have ever known." He rarely praised. These were his men, and it was praise enough that they served him. They may have been his band of brothers on the field of political combat, but he would no more have socialized with them than Henry V would have sat down to dinner with his soldiers.

For the first time since the New Deal, sizable numbers of people wanted to come to Washington to work in this new administration. The president-elect deputized Shriver to seek out the best whether or not they appeared ready to come to Washington. The word went out that the Kennedy administration sought not only men who were intelligent and honest but also those who had a quality that had never been one of the necessary credentials for public service. They wanted men who were tough. "By 'toughness' I meant 'tough mindedness,'" recalled Adam Yarmolinsky, one of Shriver's aides interviewing candidates, "but when the list inevitably leaked to the press, candidates for appointment appeared in the talent search offices at the Democratic National Committee, flexing their muscles, and proclaiming, 'I'm tough, I'm tough!'"

The man who probably exemplified the ideals of these Kennedy men better than anyone was Robert McNamara, the new president of Ford Motor Company. McNamara and his "whiz kids" had transformed the automobile industry with their acumen. The incoming administration had the audacious idea that McNamara could do the same with the Defense Department, though he professed ignorance about defense. "Well, you better give me a day to familiarize myself with this," McNamara said. He began shortly after dawn in a room at Washington's Shoreham Hotel reading memos, books, and briefing materials and talking to knowledgeable sources. He worked until late that night and began again the next morning. Within two days he could give the reasonable impression of a man deeply versed in the theory and practice of defense policies. By these efforts, McNamara had become a legend even before Kennedy took power.

McNamara had one other quality that Kennedy found essential in his associates. He spoke the fast-paced, urgent shorthand that was the natural language of Kennedy and his siblings. It was a cinematic way of talking, following the basic rule in scriptwriting: always enter the scene as late as you can. Everyone knew the back story, and if you didn't, if you asked for it to be repeated, then get out, get away, be quiet. "People, even if they were brilliant and even if they had things he was very interested in, if before they came to the point they had to explain the whole build-up and background to what they had to say, these people in the end bored him," reflected his old friend David Ormsby-Gore, the British ambassador in Washington. And if they bored him, if they courted him with their meandering soliloquies, they were often soon gone, exiled to some region where he would not have to endure their endless pedantries. Most of these men may have been numbing bores, but at times men are ponderous because there is much to ponder, and slow making decisions because the decisions are hard and close.

On his first day in office, Kennedy walked into the Oval Office early in the morning. Only a fool or a megalomaniac—and the new president was neither—would have entered what was now his citadel of power without a momentary sense of inadequacy, uncertainty, or self-doubt. Despite his accomplishments as a politician, he had never administered anything grander than the PT-109. The new president sat in an office stripped of photos, paintings, and memorabilia behind a desk suitable for a middling insurance executive. He stared at a floor that looked as if it had been savaged by a regiment of termites who had begun gnawing their way from behind the president's desk, continuing their meal out to the door. The holes, as Kennedy soon realized, were left by the golf shoes that Eisenhower wore when he used the putting green outside his window. By the evidence, he had practiced regularly, a fact that brought the figure of the former president down to a mortal level.

"We ought to have a list of all the promises we made during the campaign," Kennedy said as he sat there for his first hours of work. His inaugural address had been singularly devoid of specific proposals, but now was the time to begin. "Didn't we promise West Virginia that we would do something about poverty? We ought to do something about it now."

"We thought about increasing the food allotment to those getting surplus food," Feldman said.

"How do we do that?" Kennedy asked, still unsure about the mechanisms of governance.

"You write an executive order," Feldman replied.

"Well, do it," Kennedy said, and turned to other matters. Feldman left the room to write Executive Order No. 10914 dated January 21, 1961, and then gave it to Salinger to issue as a press release. The release was hardly on the wires when government bureaucrats alerted the new administration that that was not how it was done. The president had to publish his orders in the *Federal Register* for thirty days, get comments, and then perhaps hold hearings.

Later that day, Kennedy met with John Kenneth Galbraith to discuss balance-of-payments problems. The Harvard professor seemed scarcely aware that a "briefing" is called that for a reason. On and on he went, in his professorial monotone. Kennedy had one of the greatest gifts with which a human spirit can be blessed, an Odysseus-like enchantment with the world around him. Even now, in the midst of Galbraith's lecture, he could not abide sitting any longer when there was a world to explore. He suggested that the professor continue his monologue as the two men took a tour of the White House.

Kennedy's interest in music reached no higher than the Broadway musical. His knowledge of art was limited to the greatest hits of Western culture that his mother had drilled into her sons. His curiosity about antiques stopped at the price. For the most part, his cultural taste developed by osmosis, from living with Jackie. Yet he did not envision himself living in a White

House that was decorated with all the panache of a businessman's hotel. He roamed through the rooms, criticizing the lackluster furniture, the sad reproductions, the dreary decor. Despite his bad back, he got down on his hands and knees and looked underneath some of the tables. He moved from room to room, even entering storerooms where presidents rarely or never ventured. On another one of his early explorations, he discovered what appeared to be two large covered portholes upstairs in the wall of the Oval Room. The mysterious coverings opened up to disclose matching his and her television sets that the Eisenhowers enjoyed in their cozy evenings at home.

Kennedy was no more willing to live in what he considered a pedestrian decor than he was to surround himself with pedestrian human beings whose ideas were as much reproductions as the furniture. "I won't have this," he said. "We must replace these with the correct pieces."

He took the derivative, mediocre furniture as a perfect metaphor for what he considered the derivative, mediocre presidency of his predecessor. "I'd like to make this White House the living museum of the decorative arts in America," he said, a task that Jackie would brilliantly fulfill.

As Kennedy settled into office, the White House was inundated by phone calls, few of which reached Lincoln's secretarial desk. One of the few calls that did reach the president's office on his sixth day in office was from Marguerite Oswald, whose son, Lee Harvey Oswald, had defected to the Soviet Union. Mrs. Oswald had come to Washington seeking help, and though apparently the president did not talk to her, Lincoln noted the call in the official list of calls.

Kennedy had been in office for less than a month before those officials who could not speak the president's idiom were pushed to the antechambers. "Jack feels that Stewart Udall [Secretary of Interior], though very bright, talks too much and that Arthur Goldberg [Secretary of Labor], also very bright, goes on and on," Arthur Schlesinger Jr. wrote on February 22 after a small private dinner at the White House.

Schlesinger, a liberal Harvard history professor, had been brought into the administration in part to write its official history and to provide a liaison with his close friend and ideological colleague, Adlai Stevenson, the UN ambassador. "He [O'Donnell] has caught Adlai Stevenson in two lies regarding agreements that he's made with Jack [Kennedy] as to personnel at the United Nations," Schlesinger wrote after the dinner. "As Kenny [O'Donnell] said, the people that he has got around him now at the United Nations are mostly queers and I don't think that is far from the truth." Whether true or not, "queer" was the ultimate epithet in the Kennedy White House, for queers were weak sissies, the complete antithesis to the bold men of the New Frontier.

As the first director of the Peace Corps, Sarge Shriver was forming an advisory council to include the novelist Gore Vidal, who was not only a Democratic activist but also Jackie's stepbrother. "I can't remember whether it was the president or his brother," Wofford recalled, "but one of them got the full story that he was gay . . . and they canceled him from being on the advisory council."

Shortly before the inauguration, Allen Dulles, the legendary director of the CIA, had dinner with a small group of Kennedy aides, among them Sorensen and Feldman. Dulles told the men that during the Eisenhower years the president had not known everything the agency was doing. Dulles's seemingly casual remarks were often the vehicle for his most crucial, calculated utterances. Although Kennedy's men were unsettled by Dulles's comments, the CIA director was suggesting to the new administration that his agency should be left alone to work its will on a dark, troubled world.

A week after the inauguration the president and his top foreign policy advisers met with Dulles for a briefing on Cuba. Kennedy felt an emotional affinity with Dulles and other top CIA officials. The CIA leaders belonged to the old upper-class Protestant world to which the Kennedys had long aspired. These men were doubly elite: members of the American establishment, they were also from a private world that worked its will without following any of the prissy necessities of law and politics that governed other men. Their successes, be it overthrowing governments in Iran or Guatemala or manipulating elections in France or Italy, were all secretly accomplished and privately celebrated. They were men who walked as easily into a secret rendezvous in Tehran or Lima as they did into the Somerset or Metropolitan Club.

These CIA leaders were for the most part sophisticated men who were not terrified by words like "socialist" and "social democrat," so-called progressives who seemed to want the same world that the president wanted. One of the president's favorites, and his putative choice to become the next director when sixty-seven-year-old Dulles retired, was Richard Bissell, the CIA director of covert operations. An economist, Bissell was an accomplished man who had come into government during the New Deal as a protégé of the liberal Chester Bowles in the Office of Price Administration. Fifty-year-old Bissell had developed the U-2 program, which, until the Russians shot down Francis Gary Powers in 1960, had been indispensable in getting accurate information about Soviet defenses and missile sites.

Dulles described a Cuba that had become "for practical purposes a Communist-controlled state" in which there was "a rapid and continuing build-up of Castro's military power, and a great increase also in popular opposition to his regime." For months the United States had underwritten a series of covert actions, including sabotage, infiltration, and propaganda, while training an insurgent guerrilla force in Guatemala. Most of this Kennedy already knew before the election, and he had learned the rest immediately afterward when Dulles briefed him.

The tweedy, pipe-smoking CIA head explained that there was a renewed urgency to these efforts. The Soviet Union was shipping tons of munitions to the island. Cuban pilots had gone off to Czechoslovakia to train. Castro was steadily garroting his people's liberties until soon the populace might hardly have the strength to rise up, and his agents were provoking revolution throughout Latin America. As the agency realized the magnitude of the challenge, the program of covert actions kept expanding: from the original $4.4 million the year before Kennedy took office, the budget for fiscal year 1960/61 grew to more than $45 million.

Kennedy had met with Eisenhower on the day before the inauguration, when the Republican president had bequeathed his covert Cuba program and admonished his successor to push on with the plans. These were no longer guerrilla infiltrations that Dulles was proposing to the new president, however, but a major amphibious invasion seeking to establish an impregnable enclave that would set off an uprising across Cuba, or at least be a symbol of resistance that would grow until finally the whole island was rid of Castro and his Marxist regime. The Republican president had never authorized an invasion that might involve American troops. Kennedy was being asked to authorize a far more dangerous venture than the one that Eisenhower had signed on to, and far beyond anything the Republican president had authorized the CIA to attempt during his two terms in the White House. The CIA was hoping that its paramilitary force and its agents on the island would foment a "continuing civil war," setting brother against brother in the streets and fields of Cuba, a struggle in which the United States or its Latin surrogates could then intervene and play savior.

On January 4, 1961, before Kennedy was inaugurated, Colonel Jack Hawkins, the head of the CIA's paramilitary staff, prepared a crucial memo outlining what the United States would have to do for the operation to be successful. Hawkins was a poster-handsome marine officer who had fought on Iwo Jima and at the Yalu River in the Korean War and was on the fast track to become a general. Hawkins knew little about Cuba or intelligence and was depending on what the agency told him about Cuban realities. "My belief from the intelligence provided by the CIA was that the place was ripe for revolt," said Colonel Hawkins. "As it turned out, the CIA intelligence was hugely wrong, based largely on Cuban émigrés in Miami saying things that would promote their cause."

Hawkins's CIA plan called for the brigade to liberate a small area, then to dig in, waiting for "a general uprising against the Castro regime or overt military intervention by United States forces." If the Cubans did not rise up against Castro, a provisional government would be established on the small territory and "the way [would] then be paved for United States military intervention aimed at pacification of Cuba."

At the first Special Group meeting on Cuba, held Sunday morning, January 22, two days after the inauguration, Dulles told a number of Kennedy men, including the attorney general, that he thought "our presently planned Cuban force could probably hold a beachhead long enough for us to recognize a provisional government and *aid that government openly.*" A few days later, at another meeting on Cuba, General Lyman L. Lemnitzer, the chairman of the Joint Chiefs of Staff, said that he felt that the envisioned force of six hundred to eight hundred was inadequate, and he anticipated that "final planning will have to include agreed plans for providing additional support for the Cuban force—*presumably such support to be the U.S.*"

As Kennedy listened to the arguments and perused the memos that passed across his desk, trying to decide what to do, he was so new in the Oval Office that he told Ormsby-Gore, the British ambassador, "You don't even know which of your team you can really trust from the point of view of their judgment." He knew this was an important decision, but neither he nor anyone else had any idea that his actions here would be one of the defining moments of his administration. His decisions would lead him up the pathway to nuclear confrontation with the Soviet Union. His judgments would dramatically affect administration polices from Bolivia to Vietnam and help create fiercely held attitudes that would determine American policy toward Cuba until the end of the century and beyond. His decisions in early 1961 may even have inspired an assassin who would be waiting to meet the president on the saddest of November days.

What Kennedy could not have known as he attempted to make the first crucial decision in his administration was that so much that mattered was not only left unsaid but also probably unthought, unfelt, and unseen. Spanish was hardly an esoteric language, but crucial CIA officials in Miami spoke only a few words of the language and knew nothing about Latin American culture or politics. "They were a strange bunch of people with German experience, Arabic experience, and . . . most of them had no knowledge of Spanish . . . and absolutely no sense or feel about the political sensitivities of these people," recalled Robert Amory Jr., then the CIA's intelligence chief, the top official concerned with gathering and evaluating information. "I think we could have had an A team instead of being a C-minus team."

The CIA team exemplified the cultural arrogance that for decades Latins had considered the mark of the gringo. It was the same kind of parochial superiority that had rankled Kennedy when he saw it among American diplomats when he traveled through Asia in 1951. Among the officers working most directly with the brigade were men who felt superior to their Latin associates, whose character they knew only through ethnic stereotype, but there were others who bonded with their Cuban charges and respected their courage and patriotism.

The CIA operatives in Miami put together the Frente Revolucíonario Democratico (FRD) to provide the illusion of a united anti-Castro front, and then a five-man Cuban Revolutionary Council (CRC) to provide the makings of the provisional government of a new Cuba. These Cuban leaders were nonetheless considered so indiscreet by their handlers that when the time of the invasion arrived, the plan was to put them under de facto house arrest. Yet these same men were supposed to be seen by their people, not as the tainted creatures of the CIA, but as worthy independent leaders of a new Cuba.

Kennedy probably did not know that so many of the initial attempts at sabotage, infiltration, and paramilitary airdrops had failed. Out of the sixty-nine thousand pounds of arms, ammunition, and equipment dropped to operatives on the island, between December 30, 1960, and April 21, 1961, forty-six thousand pounds ended up in the hands of Castro's forces. These errors were attributed to the incompetent Cuban exile pilots, never to the strength of the pro-Castro Cubans.

The agency believed in March 1961 that the "hard core" of Castro adherents on the island of 7 million people consisted of "not more than 5,000 to 8,000 [who] would fight to the end for Castro," while there were supposedly 2,500 to 3,000 anti-Communist guerrillas and insurgents whose numbers would grow "at least ten times that size once a landing is effected." A month later the CIA's figure had grown to "nearly 7,000 insurgents" actively fighting Castro.

The figures were little better than the wishful fantasies of operatives blinded by their mission. The CIA had other more realistic analyses of Cuba, but they did not reach Kennedy's desk. CIA Director Allen Dulles kept the agency's own intelligence chief and his operation out of the loop. "I was never in on any of the consultations, either inside the agency or otherwise," Amory recalled. Although this was supposedly done for security reasons, Dulles and Bissell apparently did not want a more detached assessment on Cuba reaching the president.

Castro's government was something more than a tyranny imposed on a hapless people. The upper class had left the island, much of the middle class was leaving, and Castro was promising schooling to those who had none, cheap medicine to those who went without, and work for those who sat idle. These Cubans had somber memories of Batista's regime, during which time Americans had owned almost all of the country's mines, most of its utilities, and close to half of its sugar industry. When Castro appropriated foreign properties, those Cubans left on the island for the most part thought it little more than the righteous return of their property. Nor did the one hundred thousand tenant farmers and squatters who had been given land pine for the return of their landlords.

When the CIA's operatives set fire to three hundred thousand tons of sugar cane, forty-two tobacco warehouses, two dairies, four stores, and a sugar refinery, then bombed the Havana power station, a railroad terminal, a railroad train, and a bus station, they not only made many pro-Castro Cubans firmer in their resolve but also gave Castro further justification for his steady intrusions on Cuban liberties. Castro's opponents were growing in force too, as citizens of the Caribbean island saw the inevitable totalitarian trajectory of the regime. But if an invasion came, there were still likely to be hundreds of thousands of peasants, workers, students, and others who believed in Castro and would defend their country with fierce determination.

In a March NSC meeting, Kennedy asked that the United States push the Organization of American States to call for "prompt free elections in Cuba, with appropriate safeguards and opportunity for all patriotic Cubans." Four days later, Schlesinger wrote the president that the State Department had decided "the risk is too great that Castro might accept the challenge, stage ostensibly free elections, win by a large majority and thereafter claim popular sanction for his regime."

As loath as many in the CIA were to admit it, Castro was an authentic revolutionary who inspired millions not only in his own land but also in the barrios of Mexico City, the wretched slums of Lima, and elsewhere across Latin America. So this enterprise that the CIA was planning would be seen in much of the world not as liberation but as invasion.

Kennedy realized that "Castro has been able to develop a great and striking personality throughout Latin America and this gives him a great advantage." The president also saw that it was only a matter of time before all but the politically blind would see that Castro was a determined Marxist who employed subversion, conspiracy, and state-sanctioned murder, all in the name of people's revolution, and that the Cuban people would pay a terrible price for the illusion of equality. Kennedy had seen in Asia how communism grew best in poverty and tyranny, and the new administration would promise an Alliance for Progress for Latin America to help create a climate in which progressive democratic regimes could grow. At the same time, with the new administration's knowledge, the CIA continued to appropriate the weapons and means of its totalitarian enemies and was using them with willful determination.

Truth is the ultimate weapon in a democracy, but truth does not work as quickly as deception. On CIA-sponsored Radio Swan, truth became only an occasional visitor on the broadcasts to Cuba, elbowed aside for propaganda blown up into gigantic cartoons. In its planning for the invasion, the CIA intended "to intimidate so as to obtain local support," subverting the will of the Cuban people much as Castro did.

While in the White House, the president and his foreign policy advisers discussed the proposed Cuban operation, the CIA went ahead with its own deadly scheme to try to assassinate Castro. Bissell, the mastermind of these plots, told the historian Michael Beschloss: "Assassination was intended to reinforce the plan. There was the thought that Castro would be dead before the landing."

The best known of the many schemes to murder Castro involved employing Mafia figures with Cuban connections to orchestrate the deed. The intelligence community had a long-term relationship with American mobsters going back to World War II and continuing in some measure in the postwar years. In late August 1960, during the last months of the Eisenhower administration, Bissell had asked Sheffield Edwards, the CIA director of security, to contact people with gambling interests in Havana, who by implication might be interested in murdering Castro. Edwards then asked Robert Maheu, a former FBI agent associated with Howard Hughes, if he had any contacts. When Maheu introduced James P. O'Connell to John Rosselli in mid-September, the mobster immediately guessed that O'Connell represented the CIA. Rosselli then introduced the CIA agent to other Mafia figures.

These agents were in Miami in early 1961, meeting with their collaborators Sam Giancana and Santos Trafficante, when they supposedly came upon the mobsters' pictures in *Parade*, the Sunday newspaper supplement, on the Justice Department's list of the ten most wanted gangsters. This newfound knowledge affected the CIA not at all. As for these mobsters, they had discovered a newfound patriotism, for the death of Castro would mean that they could return to Havana and rule once again over the hotels, casinos, brothels, and drug rings. That, to these gangsters, was what freedom was all about. In Florida, Rosselli was working to give botulinum-toxin pills from the CIA to Juan Cordova Orta, a Cuban close to Castro and willing to poison the leader. Trafficante was attempting to do much the same.

The CIA-Mafia plots were so much the stuff of pop novels and cinematic fantasies that they have largely crowded out awareness that the agency made many other kinds of serious attempts to kill Castro. The non-Mafia-associated assassination attempts in the early months of 1961 included those led by Felix Rodriguez, a Cuban exile and CIA operative, who wrote in his memoirs that his teams twice tried unsuccessfully to infiltrate Cuba by ship. John Henry Stephens, another official, testified to Congress that he had led five-man teams of "Poles, Germans, and Americans" in failed attempts against Castro. On March 29, 1961, a cable arrived at CIA headquarters from an operative code-named "NOTLOX": "(Plan [for] 9 April): Fidel will talk at the Palace. Assassination attempt at said palace followed by a general shutting off of main electric in Havana." Raphael Quintero, a prominent Cuban exile leader, recalled: "I was part of that [NOTLOX] plot. There was going to be a big boxing match and we knew Castro was supposed to be present. We planned to have him hit with a bazooka." The CIA asked Quintero's group to back off, apparently giving the go-ahead to another group of exiles who were unable to pull the assassination off a few days before the planned invasion of Cuba by an exile brigade.

The men who loved Kennedy and revere his memory most deeply are convinced that the president knew nothing about the assassination plans. They recall how repulsed the president appeared whenever such talk was even broached. Kennedy, however, was a man of many faces and possessed of knowledge far beyond that of any one of those who served him, and his denials, as passionate as they may have been, are by no means definitive. Deniability was the CIA's god. There would be no paper trail, no bureaucratic witnesses, probably not even much detail, perhaps just a nodding suggestion so vague that the president could claim that he had not heard.

The deadly language of euphemism hung in the air. Sometimes words meant more than they seemed to mean. Other times they meant less. On still other occasions, they meant nothing at all. When some in the CIA talked of "disposing of Castro," they meant ending the Communist regime. When others discussed "doing something about Castro," they meant murdering him. In the end, these clandestine men spread a fog so thick that it was impossible to tell definitively who knew and who didn't know about their acts, or even just when their murderous attempts began and when they ended.

When Bissell and his subordinates talked about murder, the 1967 CIA *Inspector General's Report* on the assassination attempts noted that "the details were deliberately blurred and the specific intended result was never stated in unmistakable language." There was, in the report's telling phrase, "the pointed avoidance of 'bad words.'" These were deadly serious men, but they did not want to be caught speaking such murderous words as "poison," "shoot," or "garrote."

In February 1961, Bissell called into his office William Harvey, a top clandestine CIA officer, to discuss "executive action capability," bureaucratic language for assassination. "The White House has twice urged me to create such a capability," Harvey remembered the CIA covert director saying.

A decade and a half after these first assassination attempts had ended, Senator Charles Mathias was among those sorting through the charred remains of the policy. "Let me draw an example from history," the Maryland senator said. Then he recalled how Henry II, the twelfth-century British monarch, had grown angry with Thomas Becket, the archbishop of Canterbury. "Who will free me from this turbulent priest?" the king asked, and four of his knights traveled to the cathedral at Canterbury to strike down Becket at the altar.

Henry's enemy was dead, but there were few in Christendom who did not condemn the king. In the end he would put to death the loyal men who had done the deed, walk a gauntlet of monks who whipped their king, and see his wife and sons turn against him. The Mafia chieftains recruited by the CIA were unlikely readers of royal history, but they proceeded against Castro in such a desultory way that they might as well have read about the fate of those who served King Henry too well.

Senator Smathers of Florida recalled walking with his friend the president on the White House lawn in March, when Kennedy told him that "someone was supposed to have knocked him [Castro] off and there was supposed to be absolute pandemonium." Smathers's recollection is not definitive either, but there are several other reasons to suspect that Kennedy knew.

The CIA was not immune to the bureaucratic imperatives of Washington, and a bureaucrat's highest imperative is to protect himself and his agency. This was a major long-term operation involving a number of CIA operatives, technicians, outside advisers, and mobsters; by the spring of 1961, at least twenty people knew of the Mafia plots, and many others were aware of the other plots.

As early as October 1960, Giancana openly talked about the assassination at a dinner at LaScala Restaurant in New York City with his mistress Phyllis McGuire and her sister, Christine, another of the three famous singing McGuire Sisters. The FBI learned about the discussion from Christine's husband, John H. Teeter, a confidential source. Teeter told the FBI that Giancana said "he had met with the 'assassin' on three occasions . . . [and] that he last met with the 'assassin' on a boat docked at the Fontainbleau [*sic*] Hotel." Teeter told the FBI that "neither he nor his wife could vouch for the truth of the information." Hoover religiously read FBI reports and memos, and even before Kennedy took office, he would have already suspected the mob's involvement with the assassination plots.

Although it began during the Eisenhower administration, the assassination plotting had become part of the Kennedy administration's policy toward Cuba. "There's a question in my mind as to whether John F. Kennedy generated that and under no circumstances do I think he did," said the FBI's former deputy director Cartha DeLoach, who religiously mirrored Hoover's opinions. "I think that Bobby Kennedy, in conjunction with the CIA, generated that and told the president about it, and the president went along with it. I think that Bobby, who was almost willing to play any game to accomplish a purpose, regardless of who it involved, went along with that very willingly and involved the president himself." Bobby's defenders state adamantly that he would not have countenanced employing the very Mafia figures he was attempting to put in prison; they ignore the reality that most of the assassination attempts did not involve mobsters.

Assassinating Castro was the crown jewel of the CIA's policy toward Cuba. Would the agency have run the risk of Kennedy's finding out about the assassination plots? And if he did so, and if this was the man the president's admirers proclaimed him to be, what revenge would he have enacted on those who betrayed his fundamental beliefs? Beyond this, could the agency afford not to tell the skeptical president about a crucial element of their plans that made this operation so much more plausible?

That, however, is little more than knowledgeable speculation. The extent of Kennedy's knowledge and approval of the assassination attempts are matters of endless debate and uncertainty, with no definitive proof offered or presumably available. That is the very nature of the most sensitive covert operations, hidden finally not only from others in government, but also from history itself.

Kennedy's overwhelming concern was not the probability of the invasion's success or the moral efficacy of America staging an undeclared war, but the secrecy of the administration's involvement. Kennedy concluded the first NSC meeting on the proposed Cuban operation by saying that he "particularly desires that no hint of these discussions reach any personnel beyond those most immediately concerned within the Executive Branch."

Kennedy hardly had to remind these men to remain silent. The free press, however, was another matter. In his obsession with keeping the Cuban business confidential, Kennedy devoted much of his energy to a cause that had already been lost. Articles about the training had appeared in such diverse publications as *La Hora*, a leading Guatemalan newspaper, *The Nation*, a liberal weekly, the *New York Times* ("U.S. Helps Train an Anti-Castro Force at Secret Guatemalan Air-Ground Base"), and the *New York Herald Tribune* ("Invasion Is Planned in Spring"). Beyond the CIA employees, there were 124 members of the National Guard from Alabama, Arkansas, California, and Washington, D.C., directly involved in training the Cubans and working with military supplies. They could not be expected to be quiet forever. Beyond that, the president had memos on his desk telling him that journalists such as Howard Handleman of *U.S. News & World Report* and Joseph Newman of the *Herald Tribune* knew specific details of the American involvement. Though the reporters promised not to reveal anything now, that pledge was not open-ended.

The president could look at the headlines on a front page and read the words on a memo and seem to deny what was before his eyes. It was unthinkable that an operation of such magnitude could take place without the United States' involvement becoming known, and it was the immense failing of the former journalist who sat in the White House to believe otherwise.

Kennedy had concerns far greater than those of the men who advised him. The CIA focused on bringing down Castro. The Joint Chiefs directed their attention to ensuring that the military plans were plausible. The State Department occupied itself primarily with the ramifications of an invasion on the rest of Latin America, world opinion, and the United Nations. The president had to think not only about all those matters but also about the broadest geopolitical concerns. He was hoping that the United States and the Soviet Union might walk at least a few steps back from the nuclear brink, forestalling the confrontation that he had prophesied so long before. In Laos, however, Communist insurgents threatened the government of Prince Souvanna Phouma, and in Vietnam the Communists pressed forward too. West Berlin was, to the president's mind, the West's most vulnerable outpost, held hostage behind 110 miles of East German Communist territory. If Kennedy moved too strongly or too obviously against Cuba, Khrushchev could be expected to make his own bold move, and the prospect for détente might well be doomed.

Kennedy also had domestic political issues that hardly concerned the CIA, the Joint Chiefs, or the State Department. During the campaign he had accused Eisenhower and Nixon of not daring to stand up to Castro. If he brought Castro down, he would be lauded by most Americans. If he backed away from plans instigated by the Republican administration, he would surely hear Nixon's and other Republicans' righteous rebukes. They would probably point out Kennedy's duplicities and seeming cowardice, as would many of the president's former colleagues in Congress.

Dulles and Bissell sensed Kennedy's political vulnerabilities and his reluctance to approve their plans. Bissell tried to put the president's very manhood in play, warning that if the United States turned away from the invasion, "David will again have defeated Goliath."

At another meeting after the president left, Dulles commented about the brigade training in Central America: "Don't forget that we have a disposal problem. If we have to take these men out of Guatemala, we will have to transfer them to the U.S., and we can't have them wandering around the country telling everyone what they have been doing." Dulles played the games of power astutely, and he surely realized that his comments would get back to the president, their power resonating even louder because they were passed on to him. And so they were in a memo written by Schlesinger, who had attended the meeting.

Dulles had presented a bitter prospect: fifteen hundred disgruntled Cubans of Brigade 2506 and their colleagues and friends condemning a craven president for being afraid to let them fight the tyrant who controlled their beloved land. "I would have called them some bad words and said we are not going anywhere," asserts Erneido Oliva, deputy military commander of the brigade training for the invasion at a secret Guatemalan base. "But the problem that we would have created in Guatemala would have been so great, Cubans fighting the Guatemalan army, taking over Guatemala . . . the Americans were the advisers and they were 15, maybe 20. That would not stop us, because we were the guys with the weapons. . . . I am telling you that the disposal problem was more than a problem. It was a BIG problem."

A great leader in a democracy must be a great politician, but a great politician is not always a great leader. As Kennedy contemplated his actions in Cuba, he was thinking preeminently not as a world leader, or as a military strategist, but as a politician who had just won the closest presidential election in American history. He was obsessed by the fear that if he did not allow the exile force to invade Cuba he would be considered a cowardly appeaser, and that view would be amplified by disgruntled brigade members shouting their slogans.

22

The Road to Girón Beach

On March 11, the CIA presented Kennedy with its detailed plans for a daytime amphibious invasion with tactical air support at Trinidad, on the south coast of Cuba. The area was supposedly full of anti-Castro Cubans who might be expected to join Brigade 2506, but if they did not and the invaders were unable to hold even a few acres of Cuban territory, they could disappear into the Escambray Mountains, where guerrillas were already operating.

To the president, this operation appeared to have all the hallmarks of a World War II–type amphibious invasion and looked nothing like a guerrilla infiltration masterminded by the Cubans themselves. He called, instead, for a much less spectacular plan in which the Cuban brigade would disembark at night without air support on an area that included an airfield from which brigade planes could be launched.

Kennedy's concern was more than a spineless reluctance to give the Cuban exiles the support they needed to end Castro's reign. The CIA director may now have been soft-pedaling the possible American military effort, but it was assumed by almost everyone involved on the operational level from the CIA officers to the brigade members that the American government would not allow the fighters to die alone on the beach. Kennedy was trying to back off from the possibility of direct American intervention, not simply in his disavowals but by structuring the plan in such a way that he would not possibly be called upon to send in Americans to save beleaguered Cubans.

The CIA had spent months planning the Trinidad operation, but four days later the agency returned with revised plans for a less "noisy" operation to take place roughly one hundred miles to the west in the Zapata region at Baia de Cochinos, the Bay of Pigs.

Bundy enthusiastically reported to the president that the agency had "done a remarkable job of reframing the landing plan so as to make it unspectacular and quiet, and plausibly Cuban in its essentials." The lightly populated Bay of Pigs was one of the most remote parts of Cuba and had a landing field within reach of the beaches. The chartered ships bearing the brigade would drop them at night and be gone before the first light of day. The forces would move largely unopposed to take over the airfields, from which "Free Cuban" planes could be launched, or at least said to have been launched, on strikes against Castro's air force. The Bay of Pigs lay so distant that it would take Castro twenty-four to forty-eight hours to counterattack along roads that could easily be defended.

One of the major criteria for the revised plan was that the terrain be "suitable for guerrilla warfare in the event that an organized perimeter could not be held," a crucial matter for both military and political reasons. On that point Bissell was strangely silent; as he knew, an immense sea of swamps surrounded the Bay of Pigs. "We were standing in the hall, and I told Bissell we could capture the airfield but that it would be hard for the landing force to get out of there through the swamps that surround the area, and moreover they would not be able to reach the Escambray Mountains eighty miles away," recalled Colonel Hawkins, the paramilitary head of the operation. "He said it's the only place that satisfies the president's requirements, and that's what it has to be."

As Kennedy saw it, the CIA had come up with an alternative that answered all of his concerns. He was nonetheless still so uncertain that even while he gave his tentative go-ahead, he insisted that he "have the right to call off the plan even up to 24 hours prior to the landing." The Joint Chiefs of Staff gave their considered opinion that in the absence of major elements of Castro's army, "the invasion force can be landed successfully in the objective area and can be sustained in the area provided resupply of essential items is accomplished." The Joint Chiefs, like the CIA, did not mention that in a losing engagement the forces would die, surrender, or be hunted down in the trackless swamps. Admiral Arleigh Burke told the president that the operation had a fifty-fifty chance of success, odds that probably would not have been considered good enough if Americans had been landing on those uncharted shores.

Of the voices that Kennedy heard those days opposing the invasion, none was listened to more closely than that of Thomas C. Mann, the assistant secretary of State for Inter-American Affairs. Mann was probably the lowest-ranking State Department official who knew about the invasion plans. A career diplomat, he had served as ambassador to El Salvador and had a hard-won, gritty sense of Latin American realities. Mann was perhaps the only diplomatic exception to what Robert A. Hurwitch, then the Cuban desk officer, called "a divorce between the people who daily, or minute by minute, had access to information, to what was going on, and to people who were making plans and policy decisions."

Bundy sensed that the president sided with the diplomat's position more than with Bissell's hawkish views. "Since I think you lean to Mann's view, I have put Bissell on top," Bundy wrote in a memo to the president. Mann believed that the fragile, embryonic laws that sought to govern the actions between sovereign nations could not be ignored or a great and terrible price might have to be paid. He pointed out that the OAS charter, the United Nations Charter, and the Rio Treaty all "proscribe[d] the use of armed force with the sole exception of the right of self-defense 'if an armed attack occurs.'" He envisioned that in the case of an obvious invasion the "Castro regime could be expected to call on the other American States . . . to assist them in repelling the attack, and to request the Security Council . . . to take action to 'maintain and restore international peace and security.'" Mann told the president bluntly that most Latins would oppose the invasion, and that "at best, our moral posture throughout the hemisphere would be impaired. At worst, the effect on our position of hemispheric leadership would be catastrophic."

Kennedy's insistence on downscaling the invasion was in part an attempt to deal with such criticisms. On March 29, Kennedy had what was supposed to have been the last major meeting before the April 5 invasion. When the president asked Bissell whether the Cuban brigade would be able to fade into the bush, the CIA covert chief told the president that the soldiers would have to embark again on the ships that had brought them.

Kennedy insisted that the brigade leaders be told that U.S. forces would not be taking part and then be asked whether they still wanted go ahead. It was not until the day before they set out for Cuba, however, that the leaders were told that the Bay of Pigs was surrounded by swamps and that in a failed invasion they would either die, be captured, or would have to reembark. For "morale reasons," the CIA decided not to tell the volunteers themselves that trackless swamps surrounded their destination, and the brigade members packed their kit and shouldered their rifles with no idea that in defeat they would be unable to join their guerrilla comrades by disappearing into the foreboding wilderness.

Mann was not the only disapproving voice that Kennedy heard. When the president flew down on Air Force One to Palm Beach for Easter, he invited Senator J. William Fulbright to join him. The chairman of the Senate Foreign Relations Committee was the most articulate student of foreign policy in Congress, and a persistent critic of America intervention. On the flight down Fulbright handed the president a lengthy memo opposing the operation, arguing in part that "to give this activity even covert support is of a piece with the hypocrisy and cynicism for which the United States is constantly denouncing the Soviet Union." Kennedy said nothing, but he surely weighed not simply the words but the political weight of their author, an acerbic intellectual who was perfectly capable of standing up in the Senate to condemn him.

For Kennedy, Palm Beach had always been a respite from the work and woes of politics, but the obligations of the presidency never left him even there. He had feared that he might be shut out from a wide range of counsel, but he was hearing at times a cacophony of voices. He had lengthy conversations over the weekend with Smathers, who was in favor of any action that would unseat Castro. Former ambassador to Cuba Earl Smith was at the Kennedy house as well. He was an equally strong supporter of military action against Cuba. Joe talked to his son as well.

On the flight back to Washington on April 4, Kennedy was still so unsettled regarding his course of action that he invited Fulbright to go with him to what would prove to be the decisive meeting of his foreign policy advisers at the State Department. They were all there in the drab meeting room, including three members of his cabinet: Rusk, McNamara, and Douglas Dillon, the secretary of the Treasury; Dulles, Bissell, and Colonel Jack Hawkins, chief of the CIA's paramilitary staff; and General Lyman L. Lemnitzer, chairman of the Joint Chiefs.

Most major presidential decisions are funnel-shaped. If broad philosophical issues are debated, they are debated first. Then the discussion is limited to general policy matters. In the end, the participants focus on the narrow details of the agreed-upon plan. In this instance Fulbright was bringing up grand moral and political questions that had never been placed openly on the table before, while the military brass and CIA officers sat restlessly wanting to go over the specific details of an operation that they had believed was already largely decided. To many of the policymakers, it appeared self-indulgent and sloppy that Fulbright should not only be present but endlessly pontificating.

Kennedy went around the table asking each official to vote yea or nay, treating each person as equal in status and equal in vote. If this was a family, it was like the Kennedys, in which some considered themselves the crucial members. Rusk was particularly upset that as secretary of State he did not receive the deference he believed he deserved. Despite all the nervous rumblings and worries, Rusk was the only administration member to voice a dissent, and he did so in the disappointed tones of a spurned lover.

Kennedy said that he preferred to have the brigade infiltrate Cuba in units of 200 to 250 men, though close to 1,500 men were about to embark.

Hawkins told the president that such a truncated plan would not work: Castro's forces would pick off the men. Kennedy replied that "he still wished to make the operation appear as an internal uprising and wished to consider the matter further the next morning."

Kennedy had opened a spigot, and he was discovering, to his increasing discomfort, that it was next to impossible to close it down to a meaningless trickle. He was the commander in chief, but this policy was now riding him as much as he was riding it. He was trying to signal that he had not yet decided whether he was willing to unleash these forces that he had allowed to build up. He was discovering, however, that indecision is sometimes the greatest decision of all. He left the two-hour meeting still saying that he had not made up his mind, but it would take an immense, wrenching force to shut down the invasion now.

"I hear you don't think much of this business," Bobby told Schlesinger on April 11 at Ethel's birthday party. "You may be right or you may be wrong, but the president has made his mind up. Don't push it any further. Now is the time for everyone to help him all they can."

Despite what Bobby said, the president was still full of doubts. Kennedy was, if anything, looking for those who would confirm his doubts, but he had no one left in the White House who dared to speak to those misgivings. Schlesinger had been the most articulate and persuasive skeptic among Kennedy's aides. In the White House he was a surrogate for tough-minded, anti-Communist American liberals everywhere. But even before his talk with the attorney general, Schlesinger had proudly taken his first flight as a hawk.

The day before, Schlesinger had written a nine-page, single-spaced memo for the president. The former Harvard professor saw himself as representing a humane counterbalance to the CIA and the Defense Department, but if he had ever been that, the historian had quickly learned a different language. Despite what Schlesinger believed, there was no epic struggle going on in the White House, with the humane academic idealist standing on one side and against him the evil twins: the State Department, with what Schlesinger called its "entrenched Cold War ways," and next to it the all-powerful, duplicitous "military-intelligence complex." There was a struggle, but it was for power among aides like Schlesinger who were stumbling over each other in their rush to get close to Kennedy and, in the end, narrowing the spectrum of voices that reached the president's ear.

In his important memo, Schlesinger wrote of what he called the "cover operation." Schlesinger and other American liberals idealized Adlai Stevenson, the noble prince of their faith. Yet Schlesinger called for his beloved Stevenson to get up in the United Nations and say that though "we sympathize with these patriotic Cubans . . . there will be no American participation in any military aggression against Castro's Cuba." The historian said that if forced, Stevenson would "presumably . . . be obliged to deny any such CIA activity."

Schlesinger realized that the Cubans would have strong arguments to make: "If Castro flies a group of captured Cubans to New York to testify that they were organized and trained by CIA, we will have to be prepared to show that the alleged CIA personnel were errant idealists or soldiers of fortune working on their own." These captured soldiers would presumably have been only a sample of others still in Cuba clinging to their American patrons to save them from execution or years in prison. And Schlesinger was willing to tear away their fingers.

"When lies must be told, they should be told by subordinate officials," Schlesinger wrote. "*At no point should the President be asked to lend himself to the cover operation.*" Schlesinger was in agreement with the secretary of State who apparently had first proposed that some other official should "make the final decision and do so in his [Kennedy's] absence—someone whose head can later be placed in the block if things go terribly wrong."

Schlesinger thought it imperative that the president stand back and allow others to lie for him. Nevertheless, there remained the melancholy reality of a free press that had the distressing inability to stay on message. Schlesinger prepared a group of possible questions and answers so that in his press conference Kennedy could dissemble with willful ease.

Truth is democracy's greatest weapon. It is a painful, difficult weapon that often seems to turn on those who use it, and it is terribly tempting to jettison it in dangerous times. But it was the one unique value that America held in the struggle against communism. In his memo to the president, Schlesinger called in the end for a progressive, liberal, post-Castro Cuba, but the road map he handed the president did not lead there.

As Kennedy contemplated whether to go ahead with the invasion plans, he had two different political constituencies that he had to placate. One was the right wing, which would condemn him if he did nothing; the other was the left-liberal coalition, which might rise up against him if he proceeded with the invasion. The president thought of Schlesinger as Stevenson's great advocate in the administration, and a bearer of American liberalism. The president considered Stevenson, the UN ambassador, as a man whose weakness and moral vanity made him dangerous. He was the perfect exemplar of everything about liberalism that the president deplored. Kennedy saw that in going ahead with the invasion he would not have to worry about Schlesinger, and perhaps not about Stevenson and other liberals either. Schlesinger, moreover, was telling the president that he could use Stevenson as his agent of deception.

As the plans moved inexorably forward, the most fervent doubters were not liberals like Schlesinger, or the diplomats at the State Department, but the two officers in charge of the operation. Jake Esterline and Colonel Jack Hawkins felt that Kennedy had fatally compromised their plans. Even though Esterline was now a civilian CIA officer, he was, like Hawkins, essentially a military man. They were both part of that sacred unspoken covenant between politicians and the professional military that is one of the glories of American democracy. American leaders do not have to worry that they will be overthrown by a restless military. In exchange, the military expects that the president will call them into combat only when their nation is truly at risk and they have the means to do the job they are asked to do.

Esterline and Hawkins were risking Cuban, not American lives, but they saw no difference between the two, and they felt that what they were being asked to do was wrong. They believed that as the plan now stood they would be leading these men into disaster and death. Over the weekend, they went to Bissell's home and threatened to resign. Bissell placated his subordinates by promising that he would convince Kennedy to add more air power to protect the brigade.

"Bissell said he felt sure he could persuade the president to increase the air force participation which we said was absolutely essential," Hawkins recalled. "Instead of doing that, without letting it be known to Esterline and me, he agreed with Kennedy in his private conversations to cut the whole thing even further."

There was an added urgency to these efforts since the plans to assassinate Castro did not appear to be working out. The Americans had so demonized Castro that they believed that once he was dead, the docile Cubans would line up behind the anti-Communist brigade and its powerful American champion. The CIA gave poison pills to Rosselli, who gave them to Trafficante, who gave them to Juan Cordova Orta, who worked in Castro's office. Instead of putting the pills in Castro's drink, the Cuban returned them to his CIA contact. The agency then gave poison pills and probably between $20,000 and $50,000 to Manuel Antonio "Tony" de Varona, one of the five members of the Cuban Revolutionary Council that the American government had deputized to form a post-Castro government. Varona, as the CIA knew, was associated with American racketeers ready "to finance anti-Castro activities in hopes of securing the gambling, prostitution, and dope monopolies in Cuba in the event Castro was overthrown." Varona had no more luck than his predecessors, and as the day of the invasion approached, it appeared that Castro would still be alive.

"I know everybody is grabbing their nuts on this," Kennedy told Sorensen, referring to those weak sisters like Stevenson and Rusk who presumably quaked at the thought of combat. Kennedy used profanity like many former prep school boys, as if it stiffened his mettle and reinforced his manhood. He had decided to go ahead, and he gave Sorensen a political reason for doing so: he "felt it was impossible now to release the army which had been built up and have them spreading word of his action or inaction through the country." Even as he decided to go ahead, the president was attempting to burn any bridges that might still carry American soldiers to the bloody sands of Cuba. At his press conference he pointedly said, "There will not be, under any conditions, an intervention in Cuba by the United States armed forces. The government will do everything it possibly can, and I think it can meet its responsibilities to make sure that there are no Americans involved in any actions inside Cuba."

In private, Kennedy would not consider supplying American soldiers in a supporting role even after a successful initial invasion. "The minute I land one marine, we're in this thing up to our necks," he exclaimed at a high-level meeting on April 12, though if he had looked down he would have seen the muddy currents already lapping around his waist. "I can't get the United States into a war, and then lose it, no matter what it takes. I'm not going to risk an American Hungary. And that's what it could be, a fucking slaughter. Is that understood, gentlemen?"

Even as the chartered freighters prepared to embark with their holds full of the young men of the brigade, Kennedy still appeared uncertain; unwilling to give the final go-ahead, he was waiting until the last possible hour to give his assent. Finally, on April 14, Kennedy agreed that the planned air strikes flown by the Cuban Expeditionary Force against Castro's air force should set off from a CIA base in Nicaragua two days before D-Day.

These planes were part of an elaborate ruse. They were B-26s with Cuban air force markings, piloted supposedly by defectors from Castro's air force flying one final mission against the Communist regime before heading toward freedom.

"Well, I don't want it on that scale," Kennedy told Bissell when he learned that sixteen planes would be taking off from Nicaragua. "I want it minimal." As he had done in so many other ways, the president once again sought to diminish the risk that American involvement would become apparent. He was tinkering with the number of planes, applying an aesthetic sensibility to the gray art of propaganda, accomplishing nothing but keeping alive for a few more hours his illusion that he could keep the American role quiet.

Although Kennedy was thinking about the propaganda battle, he was making a crucial military decision: cutting in half the surprise attack on the Cuban planes on the ground. Bissell knew that the president might be putting lives in unnecessary jeopardy, breaking the tacit pledge the CIA had given the men of the brigade. He knew too that he had promised Esterline and Hawkins to insist on more air power, not to cave in to more compromises.

The CIA officer said nothing in part probably because he was not about to give Kennedy an opportunity to end the whole operation. Beyond that, the president had made it clear he was a leader who disliked men who "grabbed their nuts," who whimpered and complained. So far, Kennedy's decision making had all the vices of the informal—sloppy, improvised, ad hoc—and none of the virtues, such as a trusting congeniality in which the participants felt welcome to say whatever had to be said.

At dawn on April 15, 1961, eight planes flown by Cuban expatriates took off from the CIA base in Nicaragua and flew toward Cuba. With their Cuban air force markings, they flew unopposed to their targets. They destroyed five planes and damaged others, but left the other ten planes of Castro's tiny air force intact. Seven Cubans died on the ground and fifty-six were wounded. Castro used the funerals as a public occasion to commemorate the martyred and condemn what he considered a perfidious attack. The raid gave Castro ample reason to continue his roundup of anyone he thought might threaten him. He would put tens of thousands Cubans in jail, while preparing the Cuban people for the invasion that he knew was coming.

After the attack, a bullet-ridden B-26 landed at Miami International Airport. The excited pilot said that he and three of his colleagues had defected from Castro's air force and staged an attack. Suspicious reporters could not know that the CIA had shot up the fuselage and that the plane had not been one of the planes making the attack. They did, however, observe that the machine guns had not been fired and that the B-26's nose was metal, not plastic, as in Castro's planes.

The grand deception had begun to unravel even before all its elements had been set in place, and it now became a matter of placing lies on top of lies on top of lies. Even before the brigade landed, the Cuban ambassador to the

United Nations was denouncing an American invasion, and Stevenson was sullying his reputation by unwittingly lying in his country's defense. Stevenson had the quaint idea that without honor a public man was nothing, and he was outraged that Kennedy had let him stand up before the world and say that America was not involved.

Kennedy went out to Glen Ora, the home that he had rented in the hunt country of Virginia, to try to make Jackie happy. Steve Smith, a weekend guest, thought that the president appeared moody. When Kennedy talked to his brother-in-law, he confided that even though he had given the go-ahead, he was still worrying about whether he should proceed with the invasion. The president seemed to be still in control, but he had lost that control the moment the ships carrying the brigade left Nicaragua. What Kennedy did not know was that many of the men of the brigade had vowed that if the president called off the mission, they would take over the boats and stage their own invasion in name and fact.

Rusk called to discuss plans that called for the B-26s to fly another air strike at dawn against Cuba just as the brigade finished landing on the beaches at the Bay of Pigs. The story would be put out that the planes had flown from the airstrips that had just been liberated by the brigade forces.

Rusk had begun receiving urgent reports from Stevenson. The UN ambassador was full of justifiable rage that he had not been told about his own government's involvement in the air strike. "If Cuba now proves any of [the] planes and pilots came from outside we will face [an] increasingly hostile atmosphere," the UN ambassador cabled the secretary of State. "No one will believe that bombing attacks on Cuba from outside could have been organized without our complicity."

Stevenson told Rusk about the severe damage already done to American prestige, warning him that if the administration went ahead with this new air strike, he would no longer be able to sustain his nation's position in the UN. Rusk decided that it was time, whatever the military price, to limit the political costs.

The president listened as Rusk told of Stevenson's unbridled anger and explained that if the planes were launched, the cover story would hardly last until the craft had returned to their Nicaraguan base. Kennedy had kept the UN ambassador uninformed about the invasion. Stevenson personified the liberal political animal that both Kennedys abhorred. Even Joe Alsop, hardly a liberal but a man of good manners, was appalled at the way the president "regularly harassed and even teased the virtuous Adlai Stevenson. The president disliked Stevenson nearly to the point of contempt. . . . It did not seem to me good style or, above all, useful to take delight in making him miserable." The president's disdain for Stevenson was full of sexual invective. "He used to drive him out of his mind," Bobby reflected in 1964. "Just because he was a girl, complained like a girl, cried like a girl, moaned, groaned, whined like a girl. Every time he'd talk on the phone, he'd whine to us. He used to drive the president out of his mind."

Bobby agreed with his brother. He observed that Stevenson's admirers considered the Illinois politician the "Second Coming," but Bobby considered him a second coming "who never quite arrives there; he never quite accomplishes anything."

In many respects, Kennedy had no problem with liberalism itself but with liberals, and it made him irrational when he was considering the progressive ideas that came from their mouths. It was Stevenson, the personification of 1950s liberalism, who irked the president beyond all men in his administration. Both men had gone to Choate, where Kennedy had been a devilish goad to all the rights and rituals of the place and Stevenson had been a little gentleman. He had never run the gauntlet that Kennedy's father believed a boy had to run or else end up nothing more than a eunuch in pants. Kennedy believed that his UN ambassador was surrounded in New York by chattering, adoring ladies who pandered to his limitless vanity. He could not see beyond his belief that Stevenson was weak to appreciate the value of many of the ambassador's ideas and the strength it took to profess them among men who ran off to defame him.

Stevenson was very aware of these slights, and now the president had handed him a reason to attack the administration. Stevenson called Senator Wayne Morse and said that he was flying down to Washington and to meet him at the ambassador's Georgetown hideaway. "I'm going down to resign," Stevenson told the Oregon politician. "I've been destroyed. No one will ever believe me again in the United Nations. They all think I've lied. I did not lie." Morse talked Stevenson out of going over to the White House to submit his resignation, but the outrage boiled within him.

"I'm not signed on to this," Kennedy replied to Rusk on the telephone, as if the decision lay elsewhere. This was hardly the response of a resolute leader, confident of his decisions. He was indeed the only one "signed on to this," the only one capable of measuring the complex matrix of political and military decisions and deciding what should be done.

Kennedy told the secretary of State that the strikes should be canceled unless there were "overriding considerations." The president later suggested to brigade members and others that he had pulled back because of international considerations, fearing that the Soviets might threaten him somewhere else. That was a rarefied reason, but from what Kennedy told Feldman and others later, he felt Stevenson had bludgeoned him into the wrong decision with his nagging self-righteousness. Kennedy realized the immense danger his young administration would be in if Stevenson resigned dramatically and condemned him. Kennedy never publicized that reason, for it hardly would have made sense to the Cuban patriots waiting to go ashore at the Bay of Pigs to face danger of another magnitude.

When Rusk told Bissell and Charles P. Cabell, Dulles's deputy, about the president's decision, they instantly tallied up what they considered the devastating military cost. The two CIA leaders were so vociferous in their protests that Rusk agreed that the planes could fly later that day at the beachhead, but not attack Cuban airfields.

At 4:00 A.M., Rusk called the president yet again and put Cabell on the phone. For hours, the CIA deputy chief had listened to the enraged, beseeching screeds of CIA officers who believed that the president's actions were dooming brave men to death soon after dawn, when Castro's planes would fly unchallenged across the Bay of Pigs. Cabell told Kennedy that at this hour only American planes could arrive in time to protect the brigade. Kennedy responded by ordering the American carrier *Essex* moved even farther away from what in a few hours would be the battle scene.

From the moment the brigade landed on the beaches on Monday, April 17, 1961, the news that Kennedy heard in Washington was not good. Just as the president had been forewarned, the skies were largely clear to Castro's planes. Castro ordered two Sea Furies and a B-26 into the air at dawn and admonished the pilots to attack either enemy planes or the motley armada still at anchor offshore. During that first day, Castro's planes sank two ships, one smaller craft, and damaged three other vessels. They also downed three B-26s and damaged two other planes, while a third crashed in the Nicaraguan mountains. Castro's air force lost two planes, a B-26 and a Sea Fury.

Kennedy had been told that Castro's troops would not be able to reach the seacoast for twenty-four hours or more, but the day was not yet over when soldiers of Cuba's First and Third Battalions launched their first attack. The brigade fought with courage and tenacity, as did Castro's forces, and the insurgents largely held the line. The brigade had only limited ammunition, however, and the munitions ship lay far out to sea, beyond range of Castro's planes.

All during the day, the Kennedy men sat around trying to monitor the events, reaching desperately for bulletins and scraps of information. Kennedy knew all about the dark uncertainties of war; in the Blackett Strait a man could not tell foe from friend, island from vessel. But even here in the White House, the fog of war had seeped under the door of the Oval Office. Men who thought themselves powerful were becoming little more than impotent bystanders. The Cabinet Room had been turned into a command post, with large highlighted maps and magnetic ships on a representation of the Bay of Pigs, scores of reports, data, radio messages, and intercepts. Men scurried in and out, yet there was little they knew and little they could do.

For Kennedy, the call of blood was the deepest call of all. That was his father's ultimate lesson. Beyond the boundaries of the family lay deception, ill will, and danger. In this moment when the president had reason to be full of a sense of massive mistrust, betrayal even, he called for the one man to whom he could tell everything and know that the bounds of secrecy would never be broken.

"I don't think it's going as well as it should," the president told Bobby on the phone. Kennedy's brother was down in Williamsburg, Virginia, giving a speech, and he flew back to Washington immediately.

Bobby had an endless fascination with the covert. Even before the inauguration, he had been involved with clandestine aspects of the Cuban situation, meeting with an attorney who told him that Raúl Castro might be turning against his brother's revolution. He had gone to the first Special Group meeting on Cuba and heard Allen Dulles make his presentation, but for the most part he had stayed away from conferences dealing with the invasion. He may have been a hawk but he flew so high above that none saw his talons.

Bobby was there beside his brother at just before noon the day after the invasion when the president met with his top advisers. The tense, impatient men gathered there were making the most crucial decisions of this young administration. As Kennedy walked into the meeting, he had just received a memo from Bundy outlining the insurgents' plight, saying that if the brigade was to have any chance at all, Castro's air force had to be destroyed, if necessary by unmarked American planes. Kennedy's top national security adviser concluded, "The real question is whether to reopen the possibility of further intervention and support or to accept the high probability that our people, at best, will go into the mountains in defeat." There were no mountains, only endless swamps, and it was symptomatic of the whole operation that Bundy did not even know that simple but crucial fact.

Men were dying on the sands of Cuba, continuing to thrust themselves into the bullets' path believing that they might still win. Here in these councils of power, however, the scent of recrimination was already in the air, and men tried to sit as far away as they could from blame. Bobby cautioned all those in the room that they were to say nothing to suggest that they were not completely behind the president, nothing that indicated they questioned his judgment.

"Can anti-Castro forces go into the bush as guerrillas?" the president asked the Joint Chiefs. Kennedy and the officers had known the answer weeks before, but Admiral Burke agreed to seek an answer.

At this meeting Burke felt that "nobody knew what to do nor did the CIA who were running the operation and who were wholly responsible for the operation know what to do or what was happening and we the JCS [Joint Chiefs of Staff] . . . have been kept pretty ignorant of this and have just been told partial truths. They are in a real bad hole because they had the hell cut out of them. They were reporting, devising, and talking and I kept quiet because I didn't know the score."

Burke recalled that he had left his contributions primarily to the judicious sprinkling of such words as "balls" to show his manly displeasure. But he was the crucial service officer and had to speak. Harlan Cleveland, undersecretary of State for International Organizations, remembered the admiral "pressing for every yard of ground he could get in the direction of more American participation by American forces in at least limiting the damage."

As Kennedy turned back all Burke's entreaties, he picked up one of the little destroyers and moved it across the map. Cleveland thought that he observed Burke's face stiffen as the president touched what was by all rites and rituals only the admiral's to touch. By his passivity, the president had allowed Bissell to push the Joint Chiefs of Staff away from hands-on input into what was a military operation about which they knew infinitely more than the CIA. There were navy ships on the map, but it was not the military's plan that was falling apart on the sands of the Bay of Pigs.

After the meeting, the attorney general called the admiral. "The president is going to rely upon you to advise him on this situation," Burke recalled Bobby saying, standing now in the White House for the first time as his brother's right arm and enforcer.

"It is late!" Burke exclaimed. "He needs advice."

"The rest of the people in the room weren't helpful," Bobby said, dismissing all the other counsels of war, as Burke remembered.

In trying to learn what was going on, the White House was hampering operations in the Caribbean. In times of combat the command aircraft carrier, the *Essex,* had to take down the vertical antennae that were needed to get distant radio communications. If they were left up, planes could not be safely launched, since the electromagnetic radiation from the antennae might set off their rockets. So much high-level radio communication was coming in, not only the high command prompted by the White House but the White House itself, and possibly Bobby in particular, that the air boss was having trouble finding time to launch planes. "There's no doubt in my mind but that communication was coming from the White House," asserted retired Captain William C. Chapman, the *Essex* air boss. "I don't know firsthand, but everything I heard was that Bobby was calling too."

While the fighting continued, the White House was concerned in part with tidying up the bureaucratic debris and pushing the evidence into history's closets. Burke cabled Commander in Chief (Atlantic) Admiral Robert Dennison, asking whether the brigade could disappear into the bush: "Authorities [the president] would like to be sure CEF [the brigade] could become guerrillas whenever they desire so that point could be emphasized in our publicity, i.e., that revolutionaries crossed the beach and are now operating as guerrillas."

The brigade would have to bring out their wounded, but these broken, bloodied men hardly represented the photographic image of victory. Burke cabled: "Wounded should be kept in Essex until suitable hospital arrangements could be made on beach in some place inaccessible to news hawks."

About six hours later, Dennison replied: "Evacuation of wounded is completely out of the question without overt involvement of U.S. forces. Furthermore, I know of no haven in some place 'inaccessible to news hawks.'"

The president attended the annual congressional reception that evening resplendent in white tie and tails. He waltzed with his wife and danced the social minuet. Just before midnight, he left and arrived to meet with the same tired men he had seen so many times these past two days.

The situation was desperate. Bissell and Burke implored the president to take some action. He could send two jets from the *Essex* to down Castro's planes, or at least have them fly over the beachhead as a sign of support, or if not that, bring in a destroyer to shell the tanks that were killing the men of the brigade.

"Burke, I don't want the United States involved in this," Kennedy said, his voice charged with anger.

The president went to his office and sat there downcast, as close to crying as O'Donnell had ever seen him. As for Bobby, he was on a manic rant, pacing back and forth, consumed with emotion. "We've got to do something," he told his brother late at night. "They can't *do* this to you!"

There precisely in those words lay the difference between the two men. For Bobby, politics always had a human face, and he saw Castro as attacking not the brigade but his brother, as surely as if he were stomping on Kennedy. Kennedy, as desperate and hopeless as he may have felt, was never a man to indulge in open self-pity and perceived the potentially devastating impact on his whole presidency.

This was precisely the moment Kennedy had feared most when he had signed on to the CIA plans. He did not see himself as a cowardly, cynical politician who would let men die in the sands, but he had preoccupations beyond those of the men here this evening. The president was playing with a chessboard as big as the world, not just with one or two pieces. He knew that the Soviets could make their countermove anywhere. "We are sincerely interested in a relaxation of international tension, but if others proceed toward sharpening, we will answer them in full measure," Khrushchev had just written Kennedy.

At around four in the morning the president got up from his desk and walked out into the darkness of the south grounds and paced by himself there for forty-five minutes or more. Kennedy believed that democracy's most crucial creations were its leaders. A great man made history and was not merely its temporary steward.

Kennedy could have heeded the call of the right wing and backed the brigade to the hilt, with shiploads of marines standing by and the might of America ready to crush Castro, running the risk of a world war and brushfire conflicts elsewhere. He could have listened to those on the left who would have sent an olive branch to Castro instead of fifteen hundred fighting men and ended the covert operations in Cuba. Instead, on national security issues he was a moderate conservative worried about communism in Cuba but worried about nuclear conflagration too. He had treated the invasion plan as if it were a piece of legislation, placating the CIA here, moving toward the State Department there, seeking an accommodation that would please everyone. He had attempted to pare the operation down, and even as it was, if much of the brigade had been able to escape into the Cuban wilderness, he would not have had his whole administration held hostage to this issue.

No matter how one turned the issue over, looking for light, all one found was darkness. He felt he had to do something or he would suffer the Republicans' accusations of dishonor and worse. That was perhaps a tinsel treasure to weigh against the lives of these men, but Kennedy was convinced that it was better "to put the guerrillas on the beach in Cuba and let them fight for Cuba than bring them back to the United States and have them state that the United States would not support their activities. The end result might have been much worse had we done this than it actually was." If these patriots died in Cuba fighting for their country, he would have considered them worthy martyrs as long as they did not die too loudly or shout too long, in their death throes, for American help.

Kennedy could have held to the CIA's original plan and let the brigade stake its flag on Trinidad, or he could have decided to at least go ahead with all the planes. Then the CIA and the Joint Chiefs would have admonished him to send in American soldiers and aid, and he might well have found himself leading his country into a war he did not want. He knew that was the game plan from the day he walked into the Oval Office. He seemed now like a craven compromiser, but he was dancing along this line with fire on either side, trying to finesse a matter that in the end could not be finessed.

If the plan had worked out as it was supposed to, Castro would have been assassinated and the Communists overturned. Kennedy would have considered that the best of all outcomes, but it is doubtful that Cuba would have blossomed into a Caribbean democracy. More likely, the island would have been a sullen, subdued land, overseen by bickering politicians, all clients of the United States. And the triumphant CIA would have felt that it had proved the broad efficacy of covert activity and assassination, not only in Cuba but also across the world.

Brigade 2506 had no air cover, no ammunition resupply, and no promise of aid to its beleaguered forces. By the third day, it was not a question of whether they would win, but how long they could possibly hold out. That morning Pepe San Román, the brigade commander, told the Americans that many of his men were standing in the water on the beach "being massacred" by Cuban fire and by three enemy Sea Furies strafing the brigade positions. He looked up and saw high in the sky four American navy jets, and he called the *Essex* and asked that those planes descend and fight. He was told that the navy command was doing everything possible to get permission. "God damn it," he swore. "God damn you. God damn you. Do not wait for permission."

In the annals of American military history—and the brigade's story surely belongs there—there are few more pathetic messages than this: "Am destroying all equipment and communications. I have nothing left to fight with. Am taking to woods. I can not wait for you."

As those soldiers who had not yet surrendered were being tracked down in the endless swamps, Kennedy ordered air cover to try to save at least a few of them. In the Cabinet Room, Bobby was merciless with those officials sitting in dismay and shock around the massive table. He acted as if he and his brothers had been mere bystanders to this debacle. He was not ready to assume any burden of blame for the president, but sought to parcel out the chunks of responsibility among all the other major players. He seemed incapable of understanding where most of the responsibility lay. He called on them "to act or be judged paper tigers in Moscow" and not simply to "sit and take it."

It was not Khrushchev who had sent these men in without air cover and it was not Khrushchev who had fought them on the beach, but from the fury of Bobby's rhetoric it might have been. It was almost as if he wanted these middle-aged officials to jump out of their seats, take up arms, and run into the street in a heroic counterattack.

The president had no such fire in his voice. As Kennedy sat in a rocking chair in the Oval Office, he looked at the *Washington News* trumpeting the disaster and let the paper fall to the floor beside him. As the Cuban exile leaders entered the room, he did not display his despair. Kennedy kept doodling on a pad of paper, writing the phrase "Soviet Cuban" over and over again, and enclosing the two words in boxes.

These exile leaders were proud men who had been kept under de facto house arrest in Miami so that they would not betray the secrets of the invasion. The Cubans sat on two long couches beside the president in his rocking chair. The group included Varona, who had failed in his attempt to kill Castro. Kennedy told the men that he too had been in war. He had lost a brother, and he knew what they felt. It was not easy saying what Kennedy had to say, but he said it, and by all accounts the Cuban leaders listened and believed.

When these exile leaders had come together, they pretended that they were in the Cuban Revolutionary Council on their own and were not Washington's creatures. And now, as they attempted to justify their ersatz coalition, they exited Washington with another lie. The CRC released a statement that the events at the Bay of Pigs did not amount to an invasion but "a landing of supplies and support for our patriots who have been fighting in Cuba for months . . . [which] allowed the major portion of our landing party to reach the Escambray mountains."

That same day, Kennedy received a memo from his brother. Love is expressed in many tongues, and few who read Bobby's terse words would imagine that they were reading not only a serious political document but also an act of devotion. Kennedy was despairing, and Bobby was speaking to him in the one idiom that mattered to him now, justifying what he had done and not done. "The present situation in Cuba was precipitated by the deterioration of events inside that state," Bobby began. His brother the president must not blame himself, but understand that everything that happened had been because of Castro. "Therefore, equally important to working out a plan to extricate ourselves gracefully from the situation in Cuba is developing a policy in light of what we expect we will be facing a year or two years from now!" Bobby went on, underlining this sentence in his own hand.

Bobby wrote his brother that what had "been going on in Cuba in the last few days must also be a tremendous strain on Castro," as if the Cuban leader were suffering too. Even as the recriminations crescendoed, Bobby sought to direct the president toward the future, and a battle with Castro that he was sure would come again. For Kennedy, Cuba was an unseemly nuisance, but to Bobby it was the most important and most dangerous country in the world. "Our long-range foreign policy objectives in Cuba are tied to survival far more than what is happening in Laos or the Congo or any other place in the world," he wrote his brother.

Bobby was investing in the island enormous amounts of psychic energy, fierce anger, and intensity. He seemed willing to do anything to bring Castro down, even staging false provocations. Bobby wrote: "If it was reported that one or two of Castro's MIGs attacked Guantánamo Bay and the United States made noises like this was an act of war and that we might very well have to take armed action ourselves, would it be possible to get the countries of Central and South America through OAS to take some action to prohibit the shipment of arms or ammunition from any outside force into Cuba?" He was not for waiting either. "The time has come for a showdown for in a year or two years the situation will be vastly worse," he wrote. "If we don't want Russia to set up missile bases in Cuba, we had better decide now what we are willing to do to stop it."

Everything Bobby had seen told him that communism was an evil malignancy that had to be attacked without qualms and without waiting. Not only did he believe this, but now he shouted it with a force and confidence not shared by any of the other bruised players in the White House. He was the attorney general, and if he had not been the president's brother, the other cabinet officers would have been tempted to hush him up, dismissing his remarks as the mindless impressions of a man who knew less about foreign affairs than anyone in the room. But in this malaise of uncertainty, he stood boldly and made this Cuban issue his own.

At the NSC meeting the next day, April 20, the cabinet members and other officials got their first rich taste of the Robert F. Kennedy who had terrorized faltering subordinates during the campaign. The attorney general saved the worst of his rebukes not for those who had been most wrong, but for those who had shown a modicum of prescience. Bobby savaged the State Department, directing his greatest wrath at Chester Bowles, who had been opposed to the invasion from beginning to end and had articulately and passionately said so. The man had had the audacious bad judgment the previous day to come up to Bobby and say: "I hope everybody knows that I was always against the Bay of Pigs." That rankled the attorney general beyond measure and probably doomed Bowles's tenure in Washington.

"I understand that you advised against this operation," Bobby replied at one point, drumming his finger on Bowles's chest. "Well, let me tell you as of right now you did not. You were for it."

"When I took exception to some of the more extreme things he said by suggesting that the way to get out of our present jam was not to simply double up on everything we had done, he turned on me savagely," Bowles recalled.

The undersecretary's fear was not groundless. On that very day, the president asked the Joint Chiefs to come up with a new plan to overthrow Castro, this time not with Cuban surrogates but with the full application of American military. They replied with a plan to invade Cuba in the west with an amphibious force in an operation slated to take place in sixty to ninety days, beginning before the hurricane season but no later than July 9, 1961.

Two days later, at the next full-scale NSC meeting on Cuba, Bobby once again dominated the thirty-five policymakers. As for the president, his questions led only one place—back to the bloody shores of Cuba. Of all the men who sat there, only Bowles dared boldly to speak otherwise. The undersecretary was a pedantic gentleman whose sonorous moralizing was hardly the best way to convince the president. That said, he spoke important cautionary words, warning that as bad as things were now, percipitent action "would almost surely be ineffective and . . . would tend to create additional sympathy for Castro in his David and Goliath struggle against the United States."

"That's the most meaningless, worthless thing I've ever heard," Bobby raged at Bowles. "You people are so anxious to protect your own asses that you're afraid to do anything. All you want to do is dump the whole thing on the president. We'd be better off if you just quit and left foreign policy to someone else."

Kennedy sat listening, tapping his pencil against his pearly white teeth.

The president was running out of stamina. The evening of the surrender, he went to a dinner at the Greek embassy, accompanied by Jackie and his mother. Kennedy smiled at the ladies and made small talk. Rose had taught her son never to show public weakness, and even in the limousine back to the White House he did not drop his stoic poise for a minute. Only after Kennedy left did Jackie tell her mother-in-law about her son's sad spirits. "Jackie walked upstairs with me and said he'd been so upset all day," Rose wrote in her diary afterward. "Had practically been in tears, felt he had been misinformed by CIA and others. I felt so sorry for him. Jackie so sympathetic and said she had stayed with him until he had lain down that afternoon for a short nap. Said she had never seen him so depressed except at time of his operation."

While the president morosely contemplated his losses, pilots from the *Essex* flew mission after mission looking for survivors. As the pilots skimmed fifty feet above the endless swamps, they were fired upon by the Cuban military, at times returning to the carrier with bullet holes in their planes. The pilots, who had orders not to fire back, flew back again and again to the shores of Cuba. When the brigade soldiers saw the American planes above, they knew they were saved, and they hurried to the beach, where American teams took them to an unmarked destroyer just offshore.

On one of the last days when the bedraggled, desperate men came out of their hiding places and stumbled toward the beach, Castro's forces chased after them. The American commander of the planes that had set the brigade survivors scurrying toward the beach, Commander Stanley Montunnas, asked permission to fire warning bursts above the heads of the Cubans, just enough to keep them back. "Hold!" he was told. The *Essex* had to seek permission from "Black Walnut," the code name of the White House command post. "Negative," the reply came back a few minutes later. "One of the intelligence men on the *Essex* told me they had contacted the White House," Montunnas recalled, "and they said they didn't want to open up a bag of worms."

Montunnas flew above, watching Castro's forces capture men he had brought out of hiding. He felt no better than a Judas sheep. There were many sailors, from admirals to swabs, who felt the same way: that they had led the Cuban brigade to slaughter. "We were all pretty well disgusted," Montunnas reflected forty years later. "We could have stopped that Cuban operation cold if they'd just turned us loose. I had the feeling that the Kennedy administration was gutless. They got the brigade in that predicament, and they let them hang out to dry."

These Americans had been trained to fight their nation's wars, and they felt they had been prevented from doing so by Kennedy. They said nothing, and they wore their uniforms with ramrod pride, but they carried within them feelings that would grow in the next few years, feelings of devastating consequence to America in the next decade. Craven politicians were holding America's brave men back, preventing them from fighting against communism the way they believed they must fight. And America's soldiers were bearing the onus of defeat, not the men in Washington who were its architects.

By the end of the week Kennedy was, in Bobby's words, "more upset this time than he was any other." In public he stood and accepted blame, but privately he mused about those who had let him down, blaming everyone but himself. The president blamed Bowles and what he considered the other limp-wristed, effete diplomats at the State Department, though Rusk's worst sin was not in being wrong but in speaking his doubts only in quiet tones. Kennedy blamed the "fucking brass hats," though the Joint Chiefs had given realistic appraisals of the military prospects based on what the CIA had told them. Kennedy seemed to place less blame on the CIA, which had masterminded the operation, though more likely he had decided not to alienate the agency. When Dulles arrived, his shoulders slumped despondently, Kennedy buttressed the CIA chieftain by putting his arm around him, a gesture he did not make to Rusk or the Joint Chiefs.

The two Kennedy brothers walked back from the East Room together after the president had taken public responsibility for the debacle. "Let's go in and call Dad," Bobby said. "Let's see if he can find something good about this."

That was the role Joe had always had. No matter how grievous the problem, how deep the tragedy, how intractable the dilemma, they could turn to their father and he would find some way to bolster them. "This is the best thing you've done," he told the president. "A person that takes responsibility, the American people love you. A person that's going to be responsible and accountable, you're going to find out you did real well."

The public opinion polls showed that the president was more popular than ever, the irony of which Kennedy was perfectly aware. He had said during the invasion that he was going ahead because "he'd rather be called an aggressor than a bum," and now he was being called both. The papers were merciless in their vivisection of the debacle, roundly condemning Kennedy for duplicity while endlessly repeating the details of the military humiliation.

On Friday evening, April 21, Kennedy might have found more fruitful uses of his time than watching the *CBS Evening News*, but he had to understand how the story was playing. He and Bobby stood in the family room a few feet in front of the television set. They watched a gesticulating, triumphant Castro standing beside twisted, smoking, unrecognizable wreckage. It was too soon for CBS to have film of Castro standing over a B-26 or other armament at the Bay of Pigs, but the public wouldn't know that, and it scarcely mattered. Then Walter Cronkite said that Castro had called Kennedy a "coward."

"Fuck!" Bobby exclaimed. He turned away from the set as if he had been struck in the face and hurried from the room. A blow against his brother was a blow against him. The epithet resonated deeply within him. Bobby's life was in part an all-consuming, never-ending struggle to prove that he was not a coward. He could play football on a broken leg, confront Communists in their very lair in Central Asia, and destroy this bearded interloper who called his brother such an unspeakable epithet and dared to dance on wreckage strewn where men better than he had died. Bobby hated Castro as the personification of evil. He believed, wrongly, that Castro had personally flown across the swamps of Zapata to "pick out these fellows who were in the swamps—and just shoot them." From now on, Bobby was like an ancient knight who had taken a vow to slay his brother's enemy.

The president named a commission of inquiry headed by General Maxwell Taylor and including Dulles and Burke. Bobby, his brother's representative, was the fourth member of the Cuban Study Group, and the only one who had the motive and strength to move the inquiry beyond the narrow parameters of military policy. The president had thought about naming his brother the new director of the CIA, but he decided that he had even better uses for him.

Bobby slouched in his chair like a disgruntled teenager, his hair messy, his tie askew, but he showed subtle deference to these men. He valued physical courage above all virtues, and these men all wore badges of bravery that he did not wear. He admired Taylor, one of America's most decorated war heroes, so much that he would name his sixth son Matthew Maxwell Taylor Kennedy.

The accused rarely are their own best judges, and by naming Dulles and Burke to the study group, Kennedy was making it clear that he did not want to look too deep or too hard into the debacle. Bobby had the same mandate as the others. He was protecting his brother, and no one dared suggest that Kennedy had made all the crucial decisions, from changing the invasion site to limiting the aircraft that were to have cleared the skies of Castro's planes.

Failure is often a great teacher, but those who rush through its pages, turning away from unpleasantness, do not easily understand its lessons. It was Cuba that the study group would be focusing on, but the lessons here, whatever they were, would be used elsewhere. This was one of the most crucial moments in the Kennedy presidency. If the administration could not figure out what had happened so close to its shores, how would it deal with these quiet, purposeful men who were moving through the jungles of Laos and Vietnam carrying guns and revolution?

Bobby put in a full day at these sessions before heading over to the Justice Department to do another day's work. The witnesses came on one after another, justifying their actions as best they could, rarely being pressed, at times making modest mea culpas. Privately, Bobby felt that both the CIA and the Joint Chiefs had in some measure betrayed the president. Bissell, as the attorney general saw it, had been so inept as to base his judgment that the Bay of Pigs was guerrilla country on a survey done in 1895. He did not dare contemplate the possibility that the CIA had willfully led the president into approving its plans knowing that swamps surrounded the Bay of Pigs. As for the military chiefs, he believed that the generals had deliberated for no more than twenty minutes before signing on to the plan. These men were not quite so cavalier in their actions, but their judgment deserved the closest scrutiny. Nevertheless, Bobby asked only a few pointed questions but showed none of the prosecutorial zeal that had made witnesses fear him so on the Rackets Committee.

During one session, Dulles sat down on the other side of the table to tell his tale. He told his fellow commission members that he thought there had been a better than 50 percent "probability of being able to effect a beachhead and to hold it for a considerable period of time," but that he "never gave a great deal of weight to the idea of a large popular uprising." Neither Bobby nor anyone else asked Dulles what was to happen then, or why there had been up to thirty thousand extra weapons in the holds of the ships for Cubans who were supposed to join their brigade comrades in the fight against Castro.

The CIA director said that he believed that the operation could not be a disaster because "quite a number . . . would go through the swamp and take up guerrilla activities." Most of the brigade members who had fled into the swamps had had their clothes ripped from their bodies by the prickly, cactus-like vegetation and their flesh lacerated. They had stumbled ahead, half starving, running away from the Cubans that pursued them and on occasion shot at them like hunters.

For the most part, the attorney general, the other study group members, and the witnesses had only the most demeaning and patronizing comments to make about Castro's Cuba. Castro's tiny air force had rendered savage damage. The "aircraft were probably flown by 50 Cuban pilots that had been trained in Czechoslovakia and returned to Cuba a few days before the invasion." It could not be, as it was, that only nine pilots flew all these missions, and that none of them had any training outside of Cuba. Castro's forces had arrived a day earlier than they were supposed to and fought more fiercely than any one had imagined they could. The Cubans had fought so well that the Americans believed that the action had been "spearheaded by Czechoslovakians . . . indicated by the report that one of the tanks knocked out had three persons aboard that were not Cuban. Further, another report said that some of the command chatter was in a foreign tongue."

There had been a foreign tongue on the beaches, but it was English, not Czech. Grayston Lynch, one of two American CIA operatives who had gone ashore, led frogmen onto the beach and fired the first bullets. Lynch had blown away a jeep with a burst of fire when its militia driver turned on its lights, thinking he had seen a lost fishing boat. And on the second day, when most of the brigade pilots in Nicaragua had their stomachs full of Cuba and were refusing to fly and the battle was lost, Bissell had ordered two American CIA pilots into the air with planes full of bombs and napalm. They flew north from the CIA base along with those brigade pilots willing to fly back once again.

As the planes reached the Zapata Peninsula, they saw that the road leading to Girón Beach was full of traffic backed up for miles. Among the vehicles on the road were twenty Leyland buses filled with the militia of Battalion 123 from the town of Jaguey Grande consisting of "men of all ages and professions—except the military: masons, carpenters, schoolteachers, sales clerks, shopkeepers, dock workers, bank and office employees, telephone company workers, musicians, artists, writers, witch doctors, surveyors, doctors, architects, house painters and others." They bombed the front of the column, and then they bombed the back of the column until the Cubans could go nowhere except into the swamps. And then the planes swept back and forth, machine-gunning and napalming, and when they came to one end of the great column, they swept back again, and they did not stop until they had no more bullets and no more napalm.

When they turned back to Nicaragua, they left seven burning tanks, trucks, and buses and a column of smoke rising for seven miles. Lynch says that informants told the CIA that they counted nearly 1,800 grave markers, more dead in that one attack by far than died everywhere else on both sides in the entire three-day battle. Hawkins recalls that it was only American pilots flying that day and the intercepts from the Cubans referred to 1,800 *casualties.* This would have included not only the dead but also the burned and the maimed. That latter figure was listed in the official Taylor Report. A Cuban writer asserts that Battalion 123 "lost nearly 100 men—dead, wounded and those hurt in the napalm attack." Castro had reason to minimize the losses his forces suffered. Even if that low figure is correct, the White House *believed* that American CIA pilots burned to death or wounded 1,800 Cubans. The pilots were flying, moreover, on a punitive mission in planes carrying the markings of Castro's air force. This was not the way Americans thought they fought their wars, and the episode merited a serious investigation. But Bobby and his colleagues flew as quickly away from this matter, as did the planes leaving the scene of carnage.

The next morning other American pilots flew back after being promised that the navy would be flying cover for them. But the navy pilots were still in the ready room when the planes arrived over the Bay of Pigs. Castro's planes shot down one of the planes and later in the day shot down another plane piloted by Americans.

The White House had given approval for the use of napalm, but an inquiry might have asked whether these Americans should have flown and died in what was to have been a Cuban struggle. There were all kinds of questions that could have been asked, but Bobby did not ask them, and neither did anyone else. It would be years before Americans learned that four of their fellow citizens had died in what they had been told was a Cuban battle.

Grayston Lynch was the first witness who had been at the Bay of Pigs. Lynch was not a politician. He was a soldier trained to fight his country's clandestine wars. He went ashore at the Bay of Pigs in large part because he would not think of leaving his men alone. His was a pure soldier's voice that asked only that he and his comrades be unleashed, not harnessed to the will and whim of politicians. His was a voice without which battles are not won, and a voice that when it drowns out all others leads to the loss of wars. He believed that, if only Kennedy had not called off the planes, he and his men would have won not only the day but also the country. He was as fearless in what he told Bobby and the others as he had been fearless at the Bay of Pigs.

"I told 'em to kiss my ass practically. 'You people are the ones who failed, and this is why you failed,'" Lynch recalled. "Why, they tried to interrupt me and I wouldn't even stop talking. I laid it all out, and finally Maxwell Taylor turned and said, 'Well, it all goes back to the planes.' He did that a second time, and when he did Bobby stared across the table at him like, 'Buddy, you don't know what you're talking about.'"

Lynch had the dog soldier's version of things, but there were matters to consider that went far beyond what happened on the beach. The reports the CIA had given Kennedy before the invasion suggested that Castro was barely tolerated by most Cubans. In these secret hearings, Hawkins said that as late as the previous September, the agency "thought he still had 75 percent of the population behind him, although his popularity was then declining." That was a startling admission, and if true suggested a willful duplicity on the part of the CIA leaders to maneuver the president into an exercise in which he would have no choice but to involve American troops.

"I don't specifically remember that statement," Colonel Hawkins said. "I had been given the understanding that everybody was ready to revolt. That must have come up after the operation was over. All I know is that I came away from the Bay of Pigs feeling that I had been cheated as well. Afterwards, what they were all trying to do, including Taylor, was covering Kennedy's rear end."

At minimum, Hawkins's statement should have signaled a serious investigation. Bobby and the others, however, were not about to look under that dark carpet. These men treated the Cuban revolution as little more than a conspiracy. They did not grasp, as did others in the administration, that Castro's revolution had been born out of a compost heap of poverty, inequality, and foreign exploitation, and that the sooner social justice reigned across Latin America, the sooner the appeal of Castro and Castroism would fade away.

One hundred and fourteen members of the brigade had died. Castro had imprisoned 1,189 of those who survived. A few escaped, including Roberto San Román, whose imprisoned brother, Pepe, had been the brigade's military leader. San Román had been sick and weak when a cargo ship picked him up in an open boat that had been drifting for nineteen days. By then, 10 of the 22 men on the boat had died, and Roberto was still in a daze when the CIA flew him up from Miami Naval Hospital to Washington to meet the president and testify before the Cuban Study Group. And when he testified before Bobby and those other men of power and majesty, he started to cry, and through his sobs, he looked up at them and said: "How could you send us and leave us there?" And Bobby came down from the platform and held his hand, saying he would be there for him whenever he needed somebody, and San Román thought that was true.

On June 1, 1961, after the last of the twenty sessions of the Cuban Study group, Bobby wrote a memorandum: "He [Kennedy] was taking them at face value and when they told him that this was guerrilla country, that chances of success were good, that there would be uprisings, that the people would support this project, he accepted it . . . Their study of the Cuba matter was disgraceful." This was all true, but he didn't ask why these judgments had been false, why the CIA's intelligence officers had been silenced, and why his brother was led into the Cuban swamps.

The president's back was troubling him when Bobby and the other members presented their final report to him in mid-June. Kennedy listened carefully and accepted the thick portfolio of documents. The report, one of the most important documents of the young administration, not only supposedly detailed why the Bay of Pigs had gone so wrong but set out a blueprint of how the government should act in the future toward Cuba and much of the rest of the world.

All his adult life, the president had wrestled with the puzzle of how a democracy could win in the struggle against totalitarian regimes. In England, he had seen how the "pacifists" had worked against rearming Britain. He thought as little of them as he did of those American leftists who now saw nothing but benign goodwill coming from the Soviet Union. On the right, he had seen those who he believed were equally wrong, men like Senator Barry Goldwater of Arizona, who was calling for the air force to go in and bomb Castro away. Kennedy tended to stand somewhere in the middle because he believed that was where the best answers lay.

Bobby and his study group colleagues, however, were for getting on with it within the context of an overall cold war strategy, taking what they called "a positive course of action against Castro without delay" and destroying the malignancy that lurked so close to American shores. The group made a few perfunctory criticisms of the decision-making structure in the administration. They suggested no reorganization of the CIA; the only change in that organization immediately contemplated in the White House was to change the agency's name to a less recognizable one. They said nothing about the underlying social realities that nurtured Castro. They said not one word about Castro's surprisingly strong support but wrote only that "Castro's repressive measures following the landing made coordinating uprisings of the populace impossible."

In the report that sat on Kennedy's desk, Bobby and his three colleagues presented ideas that, if they had been accepted, would possibly have changed the entire nature of American democracy. They wrote that this struggle against Castro and the seven million Cubans was "a life and death struggle" that would require America to fight with wartime intensity and means. They called for the consideration of "such measures as the announcement of a limited national emergency" and "a re-examination of emergency powers of the president." They gave no details, but such an announcement would probably have led to the partial militarization of America and at least some limitations on civil liberties. They called for "the review of any treaties or international agreements which restrain the full use of our resources in the Cold War." They presumably meant questioning the OAS treaty and America's involvement in the United Nations.

If the president had relied only on this report, he most likely would have called for an invasion and challenged those Americans who were asking what they could do for their country to help in its militarization. Kennedy read the report, but he had other crucial sources of information from the State Department and the CIA that presented the reality of Castro's popularity, the difficulty of any attempt to overthrow him, and the fact that the Cuban revolution was much less a threat to Latin America than it had once appeared. In the end Kennedy backed away from most of the recommendations that Bobby and his associates proposed.

Although Dulles and Bissell retired, Kennedy was left with much the same CIA and military leaders he had before. He still read their memos, but he no longer fully trusted their judgment. The president was like a pilot flying a plane with a feathered engine: always looking nervously out the window, afraid that the plane would burst into flame. What he had now that he had not had before was his brother sitting beside him as co-pilot.

23

A Cold Winter

In the middle of May 1961, Kennedy flew up to Canada for his first state visit. Travel was a tonic to a president who had been living too long in the compressed world of Washington. The crowds along the boulevards were sidewalk-wide, shouting his name out as if it were a magical talisman. As much as the Canadians saluted his visit, they were even more enthusiastic about Jackie, who had only reluctantly made the trip. She had become a problem for the president: only fitfully willing to apply herself to all the banal rituals of her life as first lady, Jackie much preferred her fancy friends in New York, Charlottesville, and Palm Beach. On this visit, however, Kennedy's elegant wife was celebrated as an American icon, giving panache to his every public moment while deflecting attention away from the Bay of Pigs.

When it came time for the ritual planting of a tree on the grounds of Government House, it would not do for the vibrant young president to join his wife in turning over a few grains of dirt with a silver trowel. Instead, he hefted up shovel-load after shovel-load. And when he was finished, he felt the ominous twinges of back pain. It got so bad that he had a difficult time walking, and in the following weeks he spent much of his time in private on crutches.

All his political life Kennedy had managed to keep the question of his health problems to no more than an uncivil murmur. So it was quite unacceptable now, of all moments, to be seen hobbling along. He was about to set off for a summit conference with the rotund, aging Russian leader Nikita Khrushchev; it would be a war of symbols, and neither Kennedy nor his nation walked on crutches.

Even though the young president looked like the very definition of good health, he often walked on what amounted to invisible crutches. He had no problem allowing photographers to picture him sitting in a rocking chair because it seemed so incongruous. When the photographers left, he often stayed in the rocking chair with his feet propped on his desk, seeking some relief that way. For most presidents, throwing out the ritual first baseball of the major league season was one of the pleasant rites of spring. For Kennedy, it was a potentially troubling moment, and each year, as the day approached, he had his right shoulder checked out to make sure he could throw the ball.

Even before his trip to Ottawa, Kennedy's back had been bothering him. The problem created an unwelcome surprise for Judith Exner, his occasional mistress, when early in May she checked into the Mayflower Hotel and visited the president. Exner recalled: "This was the first time that he remained completely on his back. He was having trouble with his back, but there is something about that position, if not arrived at naturally, that makes the woman feel that she is there just to satisfy the man."

The president strained his back anew falling out of his old spring-back Senate chair. On another occasion early in his term his chair mysteriously splintered, hurting him again. On yet another occasion he had been playing with Caroline and John Jr. when he cracked his head against the corner of a table. The three-quarter-inch wound required the services of a plastic surgeon, who covered the stitches with a thick bandage.

The president's friends occasionally caught a glimpse of the inner world of Kennedy's health. He never took off his stoic mask, but they knew that he was in pain. "You could see it sometimes," recalled his friend Ben Bradlee, then the Washington bureau chief of *Newsweek* and later the executive editor of the *Washington Post*. "He would have to lie down, watching the movies lying on the bed."

In one of the most famous photos of his presidency, Kennedy stands leaning over a table lost in thought. The reality was that Kennedy was resting his painful back. His back had become so troublesome that he had not only a special mattress of nonallergenic hair set over a heavy bed board in the White House but another bed board on *Air Force One*. Because he often woke up with a puffy face, the legs of the head of the White House bed were set on three-inch-high blocks so that he could sleep with his head elevated. He was so allergic to horsehair that the one time he attended the Washington horse show he had to leave in the middle of the event and return to the White House. To help immunize him from his allergies, he took weekly or biweekly injections of a vaccine made of dust gathered around the White House family quarters.

That did not help with his allergy to milk. That allergy was particularly irritating to Kennedy, for like many of his generation he considered milk a natural wonder drug that he would have loved to drink by the quart, not to mention the pints of ice cream he would love to have eaten. The calcium supplement that he took to make up for the lack of milk was only one of the many pills he was supposed to take each day.

George Thomas, his valet, set a box of pills before him at breakfast and lunch. There were six pills in all—Cytomel, Meticorten, hydrocortisone, Florinef, calcium, and vitamin C—that he took all at once, swallowed down with a quick swig of orange juice or water. He also took 500 milligrams of ascorbic acid once or twice a day.

Cytomel is the trade name for a T3 thyroid replacement drug that Kennedy took in 25-microgram pills twice a day for thyroid insufficiency. He also took 25-milligram cortisone pills for Addison's disease each day, and for a number of years injections of 150-milligram desoxycorticosterone acetate pellets every three months.

The Kennedy administration was one of the most crisis-filled periods of American history, and the president was weighed down by an overwhelming burden of decision. He dealt with issue after issue that resolved itself as history usually does, not in monumental triumph or brutal defeat but in ambiguity and uncertainty. He was under such stress that healthy adrenal glands would have been pumping adrenaline into his system to give him the force and stamina to prevail. Instead, in difficult times he upped his hydrocortisone to improve his general functioning. Without the daily medicine, he would have died. The cortisone was that essential to his life.

Other doctors who had not even examined Kennedy sensed that he was not the healthy man he pretended to be. When the president stood so boldly against the elements giving his inaugural address, he was watched on television by Major General Howard M. Snyder, Eisenhower's White House physician. "He's all hopped up," Dr. Snyder commented, seeing beads of sweat on Kennedy's forehead. He hypothesized that the new president had taken a double shot of cortisone that morning. "I hate to think," the doctor reflected, "of what might happen to the country if Kennedy is required, at 3 A.M., to make a decision affecting the national security."

Dr. Snyder was suggesting that the new president was a cortisone junkie, shooting himself up with the drug. Kennedy had taken the drug orally for several years, and he *needed* more of his daily dosage of cortisone in times of stress. Just as a person who begins to exercise needs more calories, so in times of increased tension Kennedy needed more cortisone. He was seeking not a euphoric high but merely the feeling of being a healthy human being. If he did not take enough, he risked feeling an overwhelming sense of lassitude and exhaustion. If he took too much, he risked feeling a rush of manic invincibility that in some instances spilled over into despair and depression.

One of Kennedy's physicians, Dr. Eugene Cohen, was worried that even if the president survived a plane crash, he would go into shock without his cortisone. Thus, the endocrinologist devised an ingenious mechanism through which Kennedy could keep a safe supply of cortisone. Dr. Cohen put a syringe full of the drug in a cigar holder. He surrounded the holder with cigars and placed it in a sealed humidor to be carried on *Air Force One.*

Dr. Cohen had been largely responsible for the president choosing Dr. Janet Travell as the primary White House physician, but Kennedy also received medical advice from several others, including Dr. Cohen himself. In her first few months in Washington, Dr. Travell made herself into one of the most celebrated figures in the administration and was featured in a *New York Times* profile and celebrated in *U.S. News & World Report* and the *Washington Post.* Kennedy did not like his aides to get publicity. He found it an indulgence that served neither him nor them. Moreover, there was a high irony in the fact that the May issue of *Reader's Digest* celebrated the matronly, silver-haired, fifty-nine-year-old doctor's successful treatment of "thousands of persons with disabling back conditions, including the president," while Kennedy hobbled around the White House, unable to get relief from his pain.

Dr. Travell was a doctor who knew only one chamber in the temple of medicine. She attempted to wall the president off from other medical advice as she continued to inject him with painkillers and stick needles in his lower back, a treatment that had always given him respite from his pain. It was not working now, however, precisely when Kennedy desperately needed relief before his European trip.

Dr. George G. Burkley was the other day-to-day source of medical advice in the White House. Dr. Travell was so possessive of her access to the president that the assistant White House physician had been on staff for two months before Dr. Travell even introduced him to Kennedy. Dr. Travell had little choice but to let the rival doctor see the president eventually since the navy captain, soon to be named a rear admiral, always traveled with the presidential party.

In Ottawa, Dr. Burkley told the president that he had aggravated his back by the way he held the shovel. The navy doctor suggested to Dr. Travell that the president see Dr. Hans Kraus, who believed in the efficacy of exercise and physical therapy in treating Kennedy's pain. Dr. Burkley's proposition was a subtle criticism of the now-celebrated White House physician. Dr. Burkley and his colleagues at Bethesda Naval Hospital were becoming increasingly worried by Dr. Travell's promiscuous use of novocaine. It was not a drug to be injected two or three times daily, day after day. Eventually its effect would be deadened and the president might move on to employing narcotics to assuage his pain.

Dr. Burkley and his Naval Hospital colleagues were not the only serious critics of Dr. Travell. Dr. Dorothea E. Hellman, a highly regarded doctor at Georgetown University and the National Institutes of Health, was growing dismayed. "I had the opportunity to see a great deal of Mr. Kennedy and was aware of the fact that he was not only pigmented but often 'Cushingoid,' " she recalled. "This meant that he was alternating between periods when he received too little cortisone and periods during which he received too much cortisone. Clearly he had chosen a physician [Dr. Janet Travell] who was an expert on his back but knew little or no endocrinology and simply did what she thought was best."

Travell's ministrations to the president had also begun to worry Dr. Cohen, who was treating Kennedy's adrenal problems. "Dr. Cohen got her the White House job," reflects a former close colleague. "He liked to be a kingmaker but he didn't always recognize the frailties of those he chose and he sometimes came to regret what he had done."

Dr. Cohen reluctantly concluded, as he wrote his colleague Dr. Burkley later, that Dr. Travell was "a deceiving, incompetent, publicity-mad physician who only had one consideration in mind and that was herself." Dr. Cohen was not an argumentative man, but he was a passionate doctor whose main divertissements were his wife, Regina, his German shepherd, a white Cadillac convertible, and his dahlias. "He was totally devoted to his patients," reflected Dr. David V. Becker, his former student at the Cornell University College of Medicine and later his colleague there. "As a teacher, he made major points about always listening to the patient and responding to the patient's need."

Dr. Cohen looked on with growing concern as Dr. Travell refused to invite Dr. Kraus down to consult with the president. Instead, she agreed to call in another prominent physician, Dr. Preston Wade, who in 1957 had drained the soft-tissue abscesses in Kennedy's back. Dr. Wade prescribed treatments that did not include Travell's novocaine injections, but she continued nonetheless with her dangerous treatments. "Then, as you know," Dr. Cohen wrote Dr. Burkley in 1964, "there were just repeated series of injections without any response—injections that were not to have been given as outlined by Doctor Wade."

Dr. Travell's jealous hold over the president's care was threatened even more ominously by the arrival in the White House of Dr. Max Jacobson. The New York doctor, who had first treated Kennedy during the campaign, was a kindred spirit to Dr. Travell. He too promised absolution from physical pain, traveling always with a magical syringe that contained not novocaine but a mixture of amphetamine, vitamins, and other drugs. Kennedy had called Dr. Jacobson when Jackie was suffering from depression and headaches before his Canadian trip. His beautiful, young, outwardly healthy wife continued to be so disheartened with her life as first lady that she might not be able to accompany her husband on his European trip. Dr. Jacobson flew down to Palm Beach, where he spent four days, and gave Jackie injections that miraculously perked her up.

Kennedy fooled his constituents, and he fooled many of the women he slept with, rarely playing the invalid. Now he would have to play the healthy man on the largest stage in the world, and he had only a few days to effect this transformation. The president invited Dr. Jacobson to Washington. Dr. Jacobson's visits to the Kennedys were not going unnoticed. "Does JFK have a new personal physician?" a gossip columnist asked in the New York *Daily News* on May 12. "Dr. Max Jacobson, with offices at 155 E. 72nd St., has been making frequent phone calls to the White House."

The Manhattan physician flew down to Washington on May 23, only a week before the beginning of Kennedy's European trip. The doctor once again gave Jackie an injection that immediately cured her migraine and ended her pouting uncertainty about whether she would accompany Kennedy. Then Dr. Jacobson saw the president, whose condition, to his eyes, appeared to have worsened: Dr. Travell had sprayed freezing ethyl chloride on his back, numbing it for a few minutes, but causing longer-term problems.

Dr. Jacobson was a German Jew who had fled Berlin before the war and had a thick accent, a comforting manner, and a mystical, knowing aura. He applied to the president the same psychologically astute approach he took to all his patients. He first talked to Kennedy about scrubbing away all the detritus of his life—giving up alcohol (though that was hardly the president's vice) and taking no opiates or dangerous drugs. He then lectured Kennedy on various exercises he should do to help his back. Only then did he administer what his patients had come to him for, a treatment that after all this worthy advice seemed just some more healthy business.

Dr. Jacobson injected his own medical cocktail into the buttocks of his patients, talking about it as if it were no more than a laying on of hands. The treatment varied from patient to patient, but what Dr. Jacobson said he gave his patients was a mixture of hormones, vitamin B complex, vitamins A, C, D, and E, novocaine, enzymes, steroids, and amphetamines. Dr. Jacobson later claimed that he gave twenty milligrams of amphetamines to his patients. Most doctors prescribed only five milligrams for legitimate medical reasons, but Dr. Jacobson argued that his dosage was too small to be addictive and was made not toxic by its interaction with the other ingredients.

The reality is that amphetamines cause extreme psychological dependency and a craving for larger and larger amounts in order to obtain the short-lived burst of euphoria. "That dosage would have made Kennedy very energetic for three or four hours," says Dr. Mauro Di Pasquale, a world-renowned expert on steroids. "That begs the question of how often he was given these injections—often enough to make him an addict, say at least a few times a day, or just intermittently to give him that boost?"

For his preferred patients, Dr. Jacobson mixed up batches of individualized dosages that they could inject themselves with when he was not there. Kennedy had long been adept at giving himself injections. If he did give himself the treatment, there were no markings on the vials to suggest just what he was shooting into his system or how the ingredients may have been changing. The doctor may have been particularly generous with the leader of the Free World, for whatever it was in the magical cocktail, the president stood up and pranced around the room like a Harvard quarterback ready to be called into the game.

The next morning after treating Kennedy, Dr. Jacobson returned to the White House. He was becoming a familiar face to the Secret Service agents, who saw him walking in and out of the private quarters with his black bag in his hand. Dr. Jacobson recalled that Jackie showed him a vial of Demerol that she had found in Kennedy's bathroom. The drug is a narcotic analgesic; while it would have deadened his pain, it was nothing that the president should have been taking without medical direction. The fact that Jackie showed Dr. Jacobson a vial suggests not only that the president was injecting the drug into his own body, but that Dr. Burkley and his colleagues were justified in their fear that Dr. Travell's promiscuous use of novocaine would drive the president to try narcotics.

Dr. Jacobson wrote in his unpublished autobiography that he told the president that this was a dangerous step he was taking. The drug was addicting and might affect his performance in the White House. If Kennedy did not throw the drug away, the doctor would no longer treat him. Dr. Jacobson's ultimatum was doubly important, for not only was it the kind of admonishment that Kennedy rarely heard, but it surely suggested that in taking Dr. Jacobson's own injections, the president ran no such risk. As it was, Kennedy was being treated by several doctors—Dr. Travell, Dr. Burkley, Dr. Wade, Dr. Cohen, and Dr. Jacobson—who were not all on good terms with one another and may not have informed one another about just what drugs they were administering to the president.

While Kennedy created the illusion of health, Bobby was assuming burdens of authority he might not have had if his brother had been healthy. At nine o'clock on the evening of May 9, the attorney general walked along the dark empty space of the Washington Mall. Alongside him walked Georgi Bolshakov, a gregarious intelligence agent of the Red Army posing as a Russian journalist.

The Bay of Pigs was hardly a month-old memory, and Bobby was not only taking a lead role in the entire Cuban issue but was also involving himself in secret contact with a Soviet agent. Bobby had set up the meeting through an American journalist, Frank Holeman, who had known the Russian for a number of years and had served as a conduit between the Soviets and the Eisenhower administration. Holeman had been the one originally to suggest the meeting, and both men had agreed.

Bobby was entering into a chamber of echoes, where sounds bounced back and forth and words meant more than they seemed to mean, or less, or maybe nothing at all. He could not be sure just who it was he was talking to this spring evening, whether these words would reach Chairman Khrushchev, or if they did, whether they would be translated with all his nuance and intent. Nor could he know who might be secretly observing the two of them walking together, be it other Soviet agents or the CIA or the FBI. "We were appalled when Bobby had a [relationship with] known espionage agents who were in the Soviet embassy," reflected the FBI's former deputy director, Cartha DeLoach. "And it scared the hell out of us. . . . We knew about them because we were tailing those people. And we also had wiretaps on the Soviet embassy. It hurt us too, you know. Here we are breaking our butts working four and five hours a day overtime, and the attorney general of the United States, our proverbial boss, [is] going out and wining and dining these characters."

Bobby began by warning the Russian that his nation must not dare trifle with America. "If this underestimation of U.S. power takes hold," he warned, "the American government will have to take corrective actions, changing the course of its policies."

The president did not need his brother to warn the Russians. Bobby was pushing forward an agenda for the summit, telling Bolshakov that the new administration was seeking a "new progressive policy . . . consistent with the national interest." He said that the president had not lost hope for a test ban treaty and was willing to compromise so that the two leaders would have a document to sign at Vienna. He said that the two superpowers could sign an agreement on Laos as well, and that the new U.S. government would show a new face to the developing world, even borrowing "good ideas from Soviet aid programs." America would reach out to the Third World. "Cuba is a dead issue," he said. Bobby asked that Bolshakov tell his "friends" what he had said and let him know their reaction.

After the debacle in Cuba, the president needed to fly out of Vienna with tangible agreements tucked under his arm. But in meeting with the mysterious Bolshakov, the Kennedys were taking risks that are rarely taken before such a crucial diplomatic encounter. They were letting the Soviets into their strategic thinking, allowing Khrushchev to know where the give was in the American positions. Beyond that, they were bypassing the entire government apparatus that dealt with the Soviet Union. The Soviet experts at the State Department were not pallid bureaucrats embedded in the past, hostile to the initiatives proposed by the administration, but some of the most deeply knowledgeable diplomats in government. Many of them were operating now without knowing that the Soviets had been given this unparalleled entrée into the American positions.

In the Soviet Union, paranoia was the higher sanity, and no one in the Kremlin would have thought to take Bobby at his word. They had their own rich dossier on his 1955 visit to their country, and if ever a man was an enemy of the people, it was Robert F. Kennedy. He had, as the KGB learned in dogging his every step, "mocked all Soviets" and pointedly managed "to expose only the negative facts in the USSR," photographing "only the very bad things . . . crumbling clay factories, children who were poorly dressed, drunk Soviet officers, old buildings, lines at the market, fights, and the like." He had "attempted to discover secret information" and asked about those in labor camps and prisons. And he had shown another decadent capitalist weakness by asking his Intourist guide to send a "woman of loose morals" to his room.

When a man like this proffered an olive branch, either his arm was trembling in weakness or he held a pistol behind his back. The Soviets, however, prided themselves on not personalizing politics, and they saw Bobby as a representative of his class and time who was selling the latest American line. They instructed Bolshakov to meet with Bobby again, and rather than respond to the specific initiatives to offer him only the bland proclamations that were the tedious essence of their propaganda. He was to tell Bobby that on the crucial issue of Berlin, there would be no compromise: the Western powers would have to accept the fact that the Soviet Union was going to sign a peace treaty with East Germany, meaning in effect that West Berlin would be locked up within a sovereign state.

Bobby's worst mistake was to say that Cuba was a "dead issue." It was a dead issue to his brother around the White House; his aides had learned that the Bay of Pigs was "almost a taboo subject." That was the Kennedy pattern, to move on and away from anything unpleasant or negative. But if the Soviets might risk placing nuclear missiles on the island, as Bobby had written his brother after the Bay of Pigs, then Cuba was one of the most crucial issues of all to be discussed at the summit. It was irresponsible and dangerous to exclude the Cuban issue primarily because it was painful, immediate, and raw. Bolshakov's bosses told him that they did not "understand what Robert Kennedy had in mind when he said that Cuba was a dead issue." He was told to tell Bobby "if by that he meant that the United States will henceforth desist from aggressive actions and from interfering in the internal affairs of Cuba, then, without question, the Soviet Union welcomes this decision."

Khrushchev had studied under the toughest of masters in his apprenticeship to power. As a Ukrainian miner, he had fought in the Russian Revolution when the armed might of the Soviets won the day. In World War II, it was neither diplomacy nor grand strategy that defeated Hitler, but a generation of Russian manhood throwing itself forward against Hitler's lines. Khrushchev was righteously proud of the Soviet sacrifices, and his ideological suspicions of men like the Kennedys were only enhanced by a Russian mistrust of outsiders. Bobby's meetings with Bolshakov may have reinforced Khrushchev's conviction that this new young president was a man of weakness, not strength, a man who could be pushed and bullied and played.

It was one of those malevolent New England nights when the wind rattled the most secure shutters and the rain slashed down hard enough to keep all but the intrepid and foolish indoors. Inside the big house at Hyannis Port, Joe's mood was as foul as the weather. He had become terribly conscious of his age, and worried that he was being shunted aside.

"Goddamn it!" the old man cursed in the half-light to Frank Saunders, the new chauffeur. "He's the president of the United States! You'd think he could at least order somebody to make a telephone call and tell his family what goddamn time he'll be home—wouldn't you, Frank—goddamn it!"

Kennedy was coming home to Hyannis Port for the first time since he had taken office, to celebrate his forty-fourth birthday and relax a bit before flying to Europe and the summit. Joe had never waited for anybody and was damned if he was going to begin now at the age of seventy-two for a son who did not value his counsel the way he had before he entered the White House. He had gone to the considerable trouble of placing nude and seminude pinups on the wall of Kennedy's room, the bathroom, his pillow, and dresser, and his son wasn't even here yet to see his efforts.

"The weather be damned!" Joe fumed in a manic rage inexplicable to the chauffeur. "He's the president. I came all the way back here from France just to see him."

The only reply was the rain beating in staccato rhythms against the old house. "The hell with him," Joe said in disdain, "I'm going to bed."

Joe was long asleep when the first headlights appeared out of the haze, the procession of black vehicles making its way to the door. A light went on in the limousine, and Kennedy stepped out into the rain wearing a felt hat and overcoat. Saunders grabbed the luggage and followed. "Send in the broads," he thought he heard someone say, but did not turn back to look.

Kennedy walked into the guest bedroom on the ground floor, followed by Lem Billings, who bounced up and down on the bed, making excitable sounds. The chauffeur couldn't understand, but this was the kind of moment for which good old Lem lived. Kennedy's friends calibrated their time with him as if it were a precious elixir that could be bottled and sold. They cut their personalities to fit the role that he preferred. Lem had seen how the president-elect treated subordinates, and he was not going to accept anything so menial and time-consuming as a mere position, even as assistant secretary of Commerce. He much preferred the proud title that he had given himself of "first friend," the perpetual guest who showed up each Friday, to Jackie's dismay, to provide endless amusement. On weekends Lem happily donned a jester's robes and spoke the fool's lines rather than attempting one of life's major roles. He was amenable, genial, charming, witty—whatever cartwheels of personality amused his friend. He was not a true court jester, though, who in rhyme and song and wit utters hard truths that none of the king's men dare speak.

Lem disliked Red Fay, the new undersecretary of the navy, as much as Red disdained Lem. Red did his own routine of buffoonery, attempting to shove Lem to stage rear with a little reminiscing here, some posturing for the camera there. His wit, like Lem's, had been honed so as never to risk pricking Kennedy's skin.

"Frank!" the president exclaimed enthusiastically. "How about a glass of milk!" Allergies be damned. "And don't mind Lem; he thinks he's still in prep school."

"Yeah, well, who's asking for milk," Lem quipped.

Jack took off his clothes and walked out into the kitchen in his shorts. "It's good to be home, Frank," the president said as he stood there chugging down the milk and rubbing his bad back. Frank did not want to stare, but he was startled when he saw Kennedy's surgery scars.

The next day the president was supposed to go on a cruise on the *Marlin* and enjoy some fine New England lobsters, but the day was so blustery that he spent much of the time reading, sitting on the front lawn wrapped in a

Notre Dame blanket. His back was still bothering him, and Dr. Travell hovered nearby; her patient was hobbling around much of the time on crutches.

As the president prepared for the summit with Khrushchev, he was inundated with memos, briefing books, letters, and advice from all quarters. Unlike much of the intelligence he had been given on Cuba, this material was sophisticated and realistic, stripped of ideological cliché, flattery, and bombast. Washington's five-year "National Intelligence Estimate" of the Soviet Union did not present an image of an irrational, expansive, chance-taking Russia, but described a country acting with "opportunism, but also by what they consider to be a due measure of caution." Khrushchev seemingly could afford to wait. In January, the Soviet leader had given an address to the meeting of world Communist leaders in which he said, "To win time in the economic contest with capitalism is now the main thing." The Soviet economy was growing at an extraordinary 8.6 percent a year. The Soviets were investing one-third of their gross national product back into the economy, as opposed to only 20 percent in the United States. Although the Soviet economy was less than half that of America, it was growing twice as fast. The experts believed that the Soviet military was already roughly on a par with the West. Politically, throughout the developing world, many of the most intelligent and idealistic young leaders looked toward socialism as their model and linked the United States with the other colonial powers of the West.

"From the particular vantage point of Belgrade, it is evident that [the] noncommitted world now stands at [a] very crucial parting of the ways," wrote Ambassador George Kennan. "If some relaxation of over-all world tensions is not achieved, it seems to me very likely that there will be a serious split between that group of unaligned nations which is violently anti-Western and anti-American and that which would like to preserve decent relations with the West."

The president was coming off the disaster of the Bay of Pigs. If he had not been the one to propose this summit originally, he would probably have postponed the event. The stakes for Kennedy were even higher: unlike most summit meetings, where some agreements have been worked out beforehand, the two leaders were arriving with nothing but a vague agenda, and it was possible that Kennedy would walk away with nothing but the echoes of rhetoric. He clearly could have used something more than generalizations about what to expect, but as the State Department rightly told him: "In an exchange of this type, particularly with so outspoken a leader as Khrushchev, it is not practicable to expect that the course of the talks can be charted in advance."

Such analyses, as realistic and thoughtful as they were, were not constructed to fill the young president with immense confidence. Of course, Kennedy's advisers were not supposed to be trainers, massaging him and whispering encouragement to him. Yet many of the documents prepared for Kennedy were marked by a startling defensiveness and a fear of the future. These experts seemed not to understand that the Soviet Union had much more to fear in the future than did the Western democracies.

In February the president had met with his top advisers on the Soviet Union. He asked Llewellyn Thompson, the astute ambassador to the Soviet Union, what had to be done to win the war against Soviet communism. The ambassador had not replied with an arcane discussion of weapon systems, covert actions, and propaganda campaigns. He had talked about the human spirit. "First, and most important, we must make our own system work," Thompson said. "Second, we must maintain the unity of the West. Third, we must find ways of placing ourselves in new and effective relations to the great forces of nationalism and anti-colonialism. Fourth, we must, in these ways and others, change our image before the world so that it becomes plain that we and not the Soviet Union stand for the future."

An immense question was just what approach the young president would take with the Soviet Union. There was fear in some quarters that he might overreact to the Soviet challenge, a fear expressed most articulately and passionately by the most unexpected of spokesmen. In his farewell address President Eisenhower had said, "We must guard against the acquisition of unwarranted influence, whether sought or unsought, by the military-industrial complex." Critics would attempt to turn Eisenhower into a critic of this military-industrial alliance, but he was warning against only what he considered its potential excesses. "We recognize the imperative need for this development," he said. "Yet we must not fail to comprehend its grave implications." Eisenhower had overseen an immense military buildup, yet he saw that the cold war was an endurance contest in which economic strength was as creditable a weapon as mass deterrence. He believed Kennedy's campaign rhetoric and feared that the new president would listen to the blandishments of the defense contractors encouraging him to build military systems of unprecedented expense and complexity while attempting to cut taxes. Eisenhower had warned in a campaign speech that "when the push of a button may mean obliteration of countless humans, the president of the United States must be forever on guard against any inclination on his part to impetuosity; to arrogance; to headlong action; to expediency; to facile maneuvers; even to the popularity of an action as opposed to the rightness of an action."

"Vive Jacqui! Vive Jacqui! Vive Jacqui! Vive Jacqui!"

As the president and first lady drove in a long motorcade through Paris, a chant echoed through the elegant boulevards and the old quarters. The worldly Parisians greeted Jackie as one of their own, celebrating her the way they had no previous first lady. Jack realized that though this was his state visit, it was his wife's triumph.

Jackie's success was in part the result of calculation and planning. Many of the extraordinary series of gowns and dresses that she wore in Europe and America might have borne the label of her American designer, Oleg Cassini. In truth, many of them were primarily the result of collaboration between Jackie and Cassini's assistant, Joseph Boccehir. As first lady, Jackie could not be seen in the French haute couture that was her preference. Instead, she perused fashion magazines, cutting out pictures and sketches of French clothes that she admired, suggesting changes, a new fabric, a bow, a cummerbund, and sending her ideas to Boccehir, who brilliantly created her vision. In Paris an exquisite pink lace gown, though officially dubbed a Cassini creation, proved to be similar to a Pierre Cardin dress in his spring collection. It was "so identical," *Women's Wear Daily* noted, "that the Paris couture couldn't believe their eyes."

Kennedy had dreamed of the great ladies of history, a Catherine the Great or Marquise de Pompadour, ladies of wit, grace, and nuance who talked of art and music and politics, moving seamlessly from subject to subject. He had thought only European women were capable of that, but the thirty-one-year-old woman sitting between him and President Charles de Gaulle at the brilliant dinner at Versailles was proving herself precisely such a woman. "This evening, Madame, you are looking like a Watteau," the French leader greeted her, but it was not only her looks that proved exquisite. The two leaders dispensed with their official interpreter, and instead Jackie translated. That was no mere ceremonial honor, for Kennedy and de Gaulle discussed substantive matters during the elegant repast. "She played the game very intelligently," the French leader reflected after she left France with a president who called himself "the man who accompanied Jacqueline Kennedy to Paris." De Gaulle said that "without mixing in politics, she gave her husband the prestige of a Maecenas," referring to a Roman diplomat and counselor to the Emperor Augustus.

In most circumstances, Kennedy would have found the grand fete at the Hall of Mirrors at Versailles a wondrous diversion, but he was momentarily so disoriented that he did not recognize his sister-in-law, Princess Radziwill, and shook her hand politely as he would that of any other guest. That faux pas appeared to startle him, and as he walked on with the imperious President de Gaulle he turned away a glass of orange juice and asked for champagne instead.

During his three days in Paris Kennedy was suffering immensely from pain. Not only Dr. Travell and Dr. Burkley accompanied the president, but Dr. Jacobson as well. Although Kennedy called in Dr. Jacobson to deal with his back pain, he appears to have used Jacobson's treatment primarily when he needed to be especially alert. After the long flight the doctor attended to the president on the morning of his first long, event-filled day in Paris. Dr. Travell shot him up with novocaine two or three times a day, but that was not enough to relieve his back pain, and when he was in his suite at the Quai d'Orsay, he got into the golden bathtub to see whether hot water could dull his pain.

It was raining as *Air Force One* flew into Vienna, raining as the motorcade moved through the old streets full of cheering crowds, and raining as Kennedy arrived at the American embassy residence for the start of the two-day summit. At around noon, just before Kennedy's first meeting with Khrushchev, the president called in Dr. Jacobson. "Khrushchev is supposed to be on his way over," Jacobson recalled Kennedy saying. "The meeting may last for a long time. See to it that my back won't give me any trouble when I have to get up or move around."

When Khrushchev arrived, he dispensed quickly with the inevitable politesse of diplomatic gatherings and began to lecture Kennedy like a professor trying to force some knowledge into a stubborn student. The Soviet leader led the American president on a journey that swept from the feudal past, through the French Revolution, to the Soviet present. "Once an idea is born, it cannot be chained or burned," Khrushchev said with the conviction of his ideology. "History should be the judge in the argument between ideas."

However much Kennedy's health dragged him down, at moments like this he was able to will himself into a sharp focus. He did not challenge Khrushchev's sweep of history by suggesting that others might view communism as a journey back into history's dungeons, not the triumphant march of the future. Such a response might have been emotionally satisfying, but this was not a debate in which he would be scored on the points he made, and he had a crucial agenda that he had come to promote.

Instead of confronting Khrushchev, Kennedy suggested that this struggle of ideas had to be conducted "without affecting the vital security interests of the two countries." The Russian leader took that to mean that "the United States wanted the USSR to sit like a schoolboy with his hands on his desk. The

Soviet Union supports its ideas and holds them in high esteem. It cannot guarantee that these ideas will stop at its borders."

Kennedy was a student of history, but the past had taught him different lessons. When he finally got Khrushchev alone, he tried to impart his own vivid sense of history to the Soviet leader. Khrushchev might believe that feudalism led to capitalism, and capitalism to communism. As Kennedy saw it, history was not made up of abstract movements but of human lives moving through time. The president pointed out that history was not won without immense costs in blood and turmoil. He talked of all the "great disturbances and upheavals throughout Europe" at the time of the French Revolution, and all the "convulsions, even interventions by other countries," at the time of the Russian Revolution. The president admitted that he had "made a misjudgment with regard to the Cuban situation," and that the reason the two of them were sitting here today was "to introduce greater precision in these judgments so that our two countries could survive this period of competition without endangering their national security." Khrushchev countered that when a subject people rose up to throw off a tyrant, that was not the hand of Moscow at work but the will of a subjugated people. The Russian leader paraded some of the more obvious examples of Western hypocrisy, including the support of the fascist Franco in Spain.

Khrushchev did not seem interested in this "greater precision" if it meant signing new agreements in Vienna. He was full of endless Marxist platitudes and bromides, while Kennedy was an American technician, attempting to strap a few more safeguards onto the machine of death that the two leaders controlled—or perhaps more accurately, that controlled the two leaders.

After lunch on the second and final day, Kennedy asked to talk to Khrushchev again in private. This was his last chance to achieve at least some of that "greater precision." Berlin was arguably the most dangerous place in the world. There, where the Soviet bloc and the West touched so menacingly, was the tinder to set off World War III. Khrushchev could pretend that history was a dove that rode on the Soviet shoulder, but the people of East Germany were turning away from the Communist future by tens of thousands a year, escaping into the free city of West Berlin. The Russian leader could suggest, as he did to Kennedy, that the famous American kitchen in which he had debated Nixon was unlike any kitchen in America, but the West was a siren song and the walls around the Soviet Union and its satellites were there to keep people in more than to keep spies out. Khrushchev needed to stop the flow of so many of East Germany's most talented people into the West, and by signing a peace treaty with East Germany he would have an excuse to do so. Soon afterward, supposedly of their own volition, the East Germans would close off the entrée points along the autobahn, and this intolerable, decadent, capitalist sore would be cauterized.

Kennedy pushed Khrushchev hard, trying to get him to back off from signing a peace treaty with East Germany. He wanted the Soviet leader to promise that whatever happened, West Berlin's rights of access to the West would be maintained. And every time Kennedy pushed, Khrushchev pushed back, giving nothing. "The calamities of a war will be shared equally," Khrushchev said in a statement with which the president could scarcely disagree. "The decision to sign a peace treaty is firm and irrevocable and the Soviet Union will sign it in December if the U.S. refuses an interim agreement."

"It will be a cold winter," Kennedy said, ending his dialogue with the Russian leader.

Air Force One flew out of Vienna that afternoon, taking Kennedy for his first visit to London since becoming president. It should have been the most glorious of reunions, Kennedy returning in triumph to the city that his father had left in disgrace two decades before. The president was indeed doubly welcomed, as the leader of Great Britain's most important ally and as a lover of the nation's peoples and culture. But no matter how fine the champagne with which Kennedy was toasted, the taste of ashes stayed in his mouth.

On his one full day in London, the president attended the christening of Anna Christina Radziwill, the daughter of Jackie's sister and her husband, Prince Radziwill. For a hundred years wealthy American women had been marrying impecunious European noblemen, appropriating royal titles, and if one of the ambitious Bouvier sisters was a first lady, the other was now a princess. The splendid chamber was full of the great names of England, solemnly witnessing this occasion. It was a glorious ceremony that had all the patina of ancient rituals, commemorating blood as the most sacred of inheritances.

Among the guests was columnist Joseph Alsop, a friend of the president and a man comfortable with the elites of Europe. While the glittering palaver went on elsewhere, Kennedy pulled Alsop off into a corner and for fifteen minutes unloaded on him in a tense, urgent voice. "I had no idea when I was at Vienna how serious it was," Alsop said, nor did the American public. Alsop listened, thinking that for the first time Kennedy had to "really face up to the appalling moral burden that an American president now has to carry."

When Kennedy arrived back in Washington, he was exhausted and went to bed a sick man. Except when he had little choice but to attend a public gathering, he barely stumbled out of bed for the next week. For the first time the White House announced that the president had hurt his back on his Canadian trip and that he was being treated with novocaine shots and swimming. He flew down to Palm Beach, where he took over the estate of Charles Wrightsman and swam laps in the heated pool.

Kennedy tried to play the healthy man, walking briskly down the ramp and entering his limousine without aid. But the public knew that their young president was hobbling around on crutches. This elicited a myriad of advice: the owner of the Bodark Crutch Company was appalled at the "*cheap* undependable pair of adjustable crutches" Kennedy was using; a British astrologer offered a "Natural Cure"; a Miami physician drove up to Palm Beach hoping to physically manipulate the president's back; a Pulaski Foundation pilgrimage prayed for his recovery at the shrine of Czestochowa; chiropractors were ready to put their hands on the offending bones; the Sleeper Lounge Company offered an electrical Sleeper Lounge Adjust-A-Bed; and a molded-shoe manufacturer wanted to send a special shoe to the president. In America, genuine solicitousness meshed perfectly with opportunism, but the larger problem was that it was unthinkable that the youthful president should be seen as limping along.

At the end of June, Kennedy became seriously ill with a cough, chills, a sore throat, and a high fever. By the time Dr. Travell was called in the middle of the night, the president was running a temperature of 103 degrees, rising to 105 soon after she arrived. The doctor diagnosed his condition as "beta hemolytic streptococcus" and immediately gave him penicillin, an intravenous infusion, and a sponge bath. The next day, when his temperature was coming down, Dr. Travell announced to a press briefing that the president had a "mild virus infection."

Kennedy had spoken bold rhetoric at his inaugural address concerning how he and his administration would "pay any price, bear any burden, meet any hardship, support any friend, oppose any foe to assure the survival and the success of liberty." Since he had returned from the summit in Vienna, however, his words did not have the same resonant ring. As a young man, he had prophesied that one day there would be a nuclear confrontation between the two great powers, but he had not imagined that he would be one of the two leaders holding the power of destruction in his hands.

Kennedy was deeply concerned about the nature of Soviet power. In July he sent a memo to his secretary of Defense asking about the Soviet air show. "Were any of the exhibits surprising?" he asked McNamara. "Do we believe their planes are superior to ours?"

He knew from all the data he had received since taking office that not only was there not the missile gap that he had talked about during the campaign, but that the United States had overwhelming nuclear superiority over the Soviet Union. Kennedy knew too that in some ways that did not matter the way it always had: in a nuclear war the Soviets would be able to turn American cities into charred, unlivable ruins even as American nuclear bombs were destroying their own cities. By this terrible new logic, it was as if all his life he had been told that the world was round and suddenly he realized that it was flat, and that he stood at a precipice beyond which lay only darkness.

On July 25, 1961, Kennedy went before the American people on television—the medium for which his cool, elegant demeanor was perfectly crafted—to talk about the Berlin crisis. He repeated to the millions watching him much of what he had said to Khrushchev in Vienna. He would "not permit the Communists to drive us out of Berlin, either gradually or by force." He backed up that promise by calling for an even greater military buildup and putting half of the nuclear bomb–carrying B-52s and B-47s on ground alert. He called too for a vastly expanded civil defense program, including building and expanding bomb shelters.

There was an eerie unreality about this speech to an America in which good times and private concerns dominated most lives, and in which the nation's enemies were both far away, over distant oceans and ice caps, and far too near, only a thirty-minute missile journey from the Soviet Union to American cities. "I would like to close with a personal note," Kennedy said. "When I ran for the presidency of the United States, I knew that this country faced serious challenges, but I could not realize—nor could any man realize who does not bear the burdens of this office—how heavy and constant would be those burdens."

Kennedy may have boldly unsheathed his rhetoric, but the reality remained that Khrushchev could not go on allowing the constant drain of many of East Germany's best-educated citizens, drawn westward by the siren call of freedom and affluence. By the summer of 1961, the exodus had become a virtual stampede: thirty thousand East Germans a month were walking into West Berlin.

Early Sunday morning on August 13, 1961, while most of Berlin slept, East German troops and police began setting up barbed-wire barricades all along the route between East and West Berlin. Within two days they had started building the permanent concrete wall that would become one of the essential metaphors of the cold war.

Kennedy was a leader who spoke loudly and carried a big stick that he flailed boldly but rarely used. "With this weekend's occurrences in Berlin there will be more and more pressure for us to adopt a harder military pressure," Kennedy wrote the secretary of Defense as the East Germans built their wall. The president had not vowed to defend *East* Berlin, and for a leader who thought that World War III might start here in this beleaguered city, the wall was not without its blessings. "The fact is that the wall was the de facto solution of the Berlin crisis, and as such, it was darn welcome," Joseph Alsop reflected. "I think the president really viewed it that way too."

With Khrushchev's greatest architectural achievement spread across the German city, the Soviet dictator would not push so hard over Berlin while the festering discontent was sealed up inside. Kennedy could not admit that the solution satisfied him, since doing so would contradict his rhetoric and perhaps inspire Khrushchev to take even more dramatic measures. As it was, Kennedy used the Berlin Wall again and again as a shameful example of Marxist failures, poking a finger in the eye of the Russian bear.

Kennedy had not been indulging in a hyperbolic exercise when he told the American people how much heavier his burdens were than he had ever imagined. One evening that summer he was talking in the Oval Office with Hugh Sidey of *Time*. Kennedy was in a morbid mood that had always been foreign to his being; now he was a dark Cassandra musing on the human tragedy. "Ever since the crossbow when man had developed new weapons and stockpiled them, somebody has come along and used them," he said. "I don't know how we escape it with nuclear weapons." That was his theme that summer. Another evening Bobby told Sidey that he had been seated next to his brother in his bedroom that August when the president began to cry. Decades later, when Sidey reflected on what Bobby had told him, it seemed to him that the attorney general had meant that the president's eyes were wet with tears, not that he had been sobbing. What was so extraordinary was that Bobby had never seen his brother like that before. "It doesn't matter about you and me and adults so much, Bobby," he said. "We've lived some good years. What is so horrible is to think of the children who have never had a chance who would be killed in such a war."

Kennedy might have had teary eyes, but this haunting, newfound concern with nuclear death had been set off in large part by the president himself and his newfound obsession with civil defense. In September, *Life* had a special issue whose cover featured a model dressed in a fallout suit. The administration knew this "fallout suit" protected only against dust, not against true nuclear fallout. A record called "The Complacent Americans," purporting to give the official civil defense survival instructions, begins after the bombs have just fallen: "The H-Bomb! The H-Bomb! The H-Bomb! Flash of brightness. A tremendous roar. . . . And I, the complacent American, thinking that no one would ever dare attack an American city. And I told my friends that nuclear war would never happen . . . but it did. I always thought I was a good American—patriotic and civic minded. But I was wrong. I failed myself and my country." In churches, ministers debated whether it was Christian to shoot down a slothful citizen trying to push himself and his family into his neighbor's fallout shelter. Other ministers were appalled at the whole idea of a nation hunkering down in private shelters. The Episcopal bishop of Washington, D.C., the Right Reverend Angus Dun, said, "The every-family-for-itself approach to fallout shelter construction is immoral, unjust, and contrary to the national interest."

The administration worked on developing a pamphlet about fallout shelters. Fifty to sixty million copies would be printed, enough so that most American families would receive one. The initial draft was full of indomitable American optimism. Those Americans with the gumption, foresight, and patriotism to build their own bomb shelters would emerge from their safe havens after a nuclear attack. Although the pamphlet was nowhere so explicit, the reality it implied was that the working class and the poor who did not have bomb shelters would be largely gone, but the upper middle class and the wealthy would survive. In this moral triage, no one would have a better chance of survival than the creators of this policy in the White House. "Some of us had been issued helicopter passes," recalled Adam Yarmolinsky, special assistant to the secretary of Defense, "so that we could be spirited away to an impregnable underground fortress in the event of nuclear attack. It was not at all clear whether the passes would be used, leaving families to cope as best they could. But the passes did help to create an atmosphere of unreality."

During a glorious Thanksgiving weekend at Hyannis Port, Kennedy had to decide if he wanted to go ahead with the pamphlet. In September, Father L. C. McHugh, a retired professor of ethics at Georgetown University, had faced the moral dilemma of the nuclear age straight on, advising that those farsighted enough to have their own bomb shelters had the perfect right to shoot their foolish neighbors who sought admission. "There's no problem here," Bobby quipped. "We can just station Father McHugh with a machine gun at every shelter." Kennedy decided to have the pamphlet rewritten to emphasize public shelters. Before long the obsession with bomb shelters receded, but it rested in the American consciousness, part of the natural paranoia of the nuclear age.

During that long difficult summer, Dr. Kraus began coming to the White House to give Kennedy an exercise program, and Joe and the president started taking a swim before lunch. But there still remained dangerous tension among the various doctors involved with the president's care, the most unsettling aspect of which was the presence of Dr. Jacobson. He had the professional patina of a celebrity doctor, and he was socially accepted in the president's circles the way none of Kennedy's other physicians ever had been. Unknown to the public, he had been there during the two days of Kennedy's serious viral illness, doubtless called in by the president. In the six months from mid-May to mid-October 1961, Dr. Jacobson spent thirty-six days with the president, based on his billing for incidental expenses and travel. Dr. Jacobson attended the president in Palm Beach, Washington, New York City, and Hyannis Port, seeing him on average more than once a week. The bill did not even include the days Dr. Jacboson had been with the president on his European trip.

Dr. Jacobson had become in some respects Kennedy's most important doctor and was injecting the president with his treatment on a regular basis. For the most part, even those around the president had no idea of his apparent dependency on Dr. Jacobson's injections. Some were struck at times by the president's unusual energy. "He wasn't a real healthy guy," said Joseph Paolella, a Secret Service agent. "On the other hand, I used to be amazed because we'd go on trips, and he seemed to be almost inexhaustible. I'd wonder, 'How can this guy do this?' The Secret Service guys would be dragging. He just had unlimited energy."

Kennedy did not mind that an old friend like Senator George Smathers knew about his drug use. "I wasn't sure what it was he was taking," Smathers reflected. "I don't think Jack knew. He just knew it made him feel better. So, he'd say, 'Give me the injection.' So I'd say, 'Now, did you really . . .' He said, 'Be damn sure that you take alcohol first and clean it off. I don't want to get infected.' I said, 'Okay.' Then just like a pro I'd stick him and shoot it in there. On these physical things, he could stand pain better than anybody I ever saw. I mean, he endured, when you stopped to look at what really bothered him, he endured pain like you wouldn't believe."

The unspoken conspiracy was to pretend that Kennedy was a vibrantly healthy man. It began not with his friends or his doctors but with the man himself. Woe betide anyone who brought to him sad tidings, reminding him that a dark ship was out there somewhere.

On one occasion Caroline came bouncing into the Oval Office carrying a bird in her hand like one of her dolls. Her pet bird had just died, and she wanted to show it to her father before she gave it a proper burial. "Get it away from here!" he shouted at his daughter, as if she had arrived with a terrible omen.

Dr. Jacobson's increasing presence in the White House was a direct and immediate threat to Dr. Travell's authority as the official White House physician. Dr. Jacobson recalled that once when he was in the White House personal quarters, the first lady's maid, Providencia "Provi" Paredes, came running to him saying, "Dr. Travell has daringly entered the second floor." No one entered the private residence without permission, but Dr. Travell was so incensed at Dr. Jacobson's access that she had dared to seek him out even there.

Dr. Jacobson slipped away and at his next meeting gave Kennedy a letter saying that from now on he thought he should see the president outside of the White House. "That's out of the question," Kennedy replied, and tore up the letter.

Dr. Jacobson felt that, thanks to his ministrations, the president was a healthy man. Others close to Kennedy, such as Dr. Cohen, worried endlessly about Kennedy's condition. "He felt strongly that a man he idolized was not getting appropriate care," reflected Cohen's former student, colleague, and friend, Dr. David Becker. "And so he really wanted to correct the situation as best he could."

In November, Dr. Cohen talked to Kennedy about Dr. Travell. "I am sorry that you were burdened with initiating a housecleaning in your medical staff," he wrote Kennedy on November 12, 1961. "In spite of repeated advice against her personal publicity, this was and is rampant. But above all (and this is a serious accusation) her own interests were placed above yours."

When Kennedy returned from the funeral of Speaker of the House Sam Rayburn a week later, Dr. Cohen had another fitting opportunity to talk to him. Dr. Cohen told the president that Dr. Travell, who he believed had no business treating anything beyond muscle problems, had been treating the Texas politician for back problems and anemia when the man was dying of cancer. "I further told him that inspite [*sic*] of my attempts of getting Doctor Kraus in that through subterfuge and direct lies, she had prevented this," Dr. Cohen wrote later. "He told me that there should be no further delays. . . . I pointedly told Doctor Travell . . . how she had for her own interest obstructed the proper therapy for the president. She had given him many injections that had long since been shown to be futile. She had lied about these injections."

Dr. Travell was not going to walk quietly off to her office and sit there ostracized by the president she had come to serve. She believed in her treatments as much as Dr. Cohen and the others believed in theirs, and hers was certainly far easier than the regimen of exercise that Dr. Kraus had prescribed for the president.

When matters did not improve, Dr. Cohen sent the president another letter. Dr. Cohen considered himself as much an authority in his field as the president was in his. Though he signed his one-page letter "Humbly and respectfully," this was a man who did not defer to Kennedy. He told the president that he could not "wait and see" any longer, but that he would have to act against a doctor who "is a potential threat to your well-being." Dr. Travell was playing to Kennedy's lassitude and self-indulgence, and Dr. Cohen warned him: "The program requires constant sacrifice on your part—not only in the present but in the future."

When Cohen had finished lecturing Kennedy that he should turn over his medical treatment to a team headed by Dr. Burkley, he turned to an even larger problem:

> You cannot be permitted to receive therapy from irresponsible doctors like M. J. [Dr. Max Jacobson] who by forms of stimulating injections offer some temporary help to neurotic or mentally ill individuals. I should state that these individuals are mesmerized to the extent that they believe they cannot perform without these injections. With such injections they may perform some temporary function in an exhilarated dream state. However, this therapy, [which] conditions one's needs almost like a narcotic, is not for responsible individuals who at any split second may have to decide the fate of the universe.

This esteemed doctor accused Kennedy of risking the future of the world by indulging himself in drug use. The president was being put on the firmest notice of the risks he was taking. And he did nothing. Dr. Jacobson continued to come to the White House, and the president continued to get his treatments.

Dr. Jacobson was not a junkie's doctor, writing out his prescriptions and giving injections with no thought of his patient's well-being. Dr. Jacobson believed that what he was doing for Kennedy was good and noble. Over those many months he undoubtedly mixed a special brew of amphetamine, steroids, and other elements especially for his exalted patient, from whom he was taking no payment other than expenses. Dr. Cohen's letter apparently had no impact on the president. He had the addict's mindset that without his injections he could not go on.

Not only did Kennedy continue to see Dr. Jacobson, but his drug use apparently became enough of a problem that Bobby eventually felt obligated to become involved. In June 1962, the attorney general's assistant, Andrew Oehmann, left a vial of orange-colored liquid at the FBI for analysis. The bottle had Dr. Jacobson's name and address on it and appeared to be the sort that he gave to patients so that they could inject themselves at home. Bobby had shown such "obvious intense personal interest" that Oehmann speculated that the medicine may have been for the attorney general's father.

The FBI laboratory received not that one small vial but four vials containing probably between thirty-five and forty milligrams of liquids, far more than would have been in the original container. Moreover, the FBI technicians described the liquid as *yellow*, not the orange-colored liquid that Bobby's assistant had delivered. Amphetamines were usually dispensed in brown-colored vials, and either an orange- or yellow-colored liquid would have been quite peculiar if these samples were indeed Dr. Jacobson's injection. These may be meaningless anomalies, or they may suggest that the FBI did not analyze the sample originally given to its laboratory. In any case, the FBI analysis showed that the specimen contained a solution of vegetable oil and water and did not contain "any barbiturates or narcotics, such as Methadon, Demerol, or opium alkaloids."

It is unclear whether the FBI tested the liquid for amphetamines, or if it did, whether the test was positive. The vial was undated, and in his memoir Dr. Jacobson wrote that when tested after two weeks, "there was no trace of amphetamine in the solution."

Surely it was reckless of Kennedy to continue with Dr. Jacobson's injections, which Dr. Cohen feared could have led to his destruction and which also represented an unimaginably dangerous indulgence for the most powerful man in the world. The president's chronic "back pain" was probably in some sense a generic term for all the suffering he was going through. By now his adrenal glands had completely atrophied. He needed to take naps in the afternoons, though most men of his generation considered naps a mark of self-indulgent weakness.

Kennedy was not using Dr. Jacobson's injections as a recreational high, but to help him get through his days with demonstrable energy. To Kennedy, true manhood was everything, and a man was a vibrant, active, physical, sexual being or he was nothing. Even with his treatment, when John Jr. rushed forward to the helicopter to greet his returning father, Kennedy could not bend down to pick the boy up but had to foist him off on an aide.

Those who worked with Kennedy would consider it slander to suggest that the president they served may have been a drug addict. He may have needed a nap every day, but he did not slur his words, wander in and out of conversations, rant on in manic jags, or do anything that would have alerted them to his drug use. He may have used the injections the way some long-term alcoholics drink, never disgracing themselves, managing their lives and business seemingly as well as if they were sober. In that case, he did a brilliant job of disguising his problem, for his awesome sense of detail, insatiable appetite for facts, and intense curiosity, which took him to the nooks and crannies of government where some presidents rarely ventured, never seemed impaired.

Kennedy's back once again became so painful the week before Christmas that in Palm Beach he had to go to bed. Outside his door his doctors began their minuet in earnest again, Dr. Burkley, Dr. Travell, Dr. Kraus, and Dr. Wade all standing there offering medical advice. Dr. Travell, as always, was ready with her injections of novocaine to deaden the pain, a treatment that appalled the other physicians. When the doctors were shown into the bedroom, Dr. Kraus dramatically raised the subject with the ailing president. "I will not treat this patient if she touches him again," he asserted, staring at Dr. Travell. "Even once." Kennedy appeared to nod his acceptance.

On Christmas day, Dr. Travell went into Kennedy's bedroom to talk to him about a front-page story in that morning's paper with the headline: "Dr. Travell Quitting as Kennedy's Doctor." After listening to the doctor, the president apparently decided against formally accepting her resignation but decided to let her keep the title while taking away most of her authority. "I hate to use the word blackmail," Dr. Cohen said, "but essentially this is how she kept her tentacles stuck to the White House."

After this meeting Dr. Burkley became in effect Kennedy's White House physician, though the change was not made public until 1963. Dr. Travell kept her office in the White House, though many members of the staff no longer visited her. She had the title that she wanted above all things, but she no longer had authority over the president's medical treatment.

As for Dr. Cohen, he kept secret all that he knew about the president's health. When the physician died in 1999, he had no idea that his letters had been saved by Evelyn Lincoln. "I asked him many times to dictate his memories, particularly about Kennedy's care, and then at least there would be a straight record of it," said Dr. Becker. "He was not interested. He was afraid it would get in the wrong hands."

The medications that Kennedy was taking, along with his various medical problems, might have diminished his sexual drive or even rendered him impotent. That is why Dr. Jacobson may have included one of the newly discovered anabolic steroids such as nandrolone or testosterone in his medical cocktail. For the president, sex had always been a life force, an assertion that he would never be tied down to all the routines of the sick and the dying. Beyond that, rapacious sex was part of his father's definition of a true man. If anything, Kennedy was even more interested in the sweet touch of female flesh, in laughter that had no reason but to please, in meetings that had no purpose but pleasure. He was greedy for it, frolicking in the White House pool with two young aides known as Fiddle and Faddle, insisting that there be women available when he traveled, scheduling his women in the White House when Jackie was away.

The haunting question is, to what extent did the president's sexual practices affect his administration? A successful leader minimizes the natural vulnerabilities of the political life, in votes cast and in actions taken. Kennedy cavalierly exposed himself to a number of people who made their livings in part by trading on the vulnerabilities of the weak and susceptible, be they a Mafia figure like Sam Giancana or a corporate lobbyist like William Thompson, who was one of Kennedy's procurers. To these men, overt blackmail was only the final and usually unnecessary step along a dark road that Kennedy had only begun to travel.

One of the FBI's informers was an upscale prostitute who told of receiving a phone call from a friend of the president and being asked to go to the Waldorf Towers. She was shown to a suite where a second woman sat waiting. Kennedy entered the room, and together the two women performed their specialties on the president. That was just another sordid little tale, fit for nothing more than to be dropped into the FBI's bottomless files of undigested, unverified facts and mindless allegations. The woman, however, had another client, a Russian diplomat. She was willing, her FBI handler said, "to give him up with pictures, the whole bit, if that's what we wanted." The Soviets had developed sexual blackmail into a dark art, and a woman who was willing to give up the Russian to the FBI might have been willing to give up the president to the Russian. That did not happen, but Kennedy had made himself endlessly vulnerable.

There was an obsessiveness in Kennedy's sexuality, unlike that of other presidents whose adulterous trysts could also have been chronicled. The handsome, debonair Kennedy had an erotic quality unlike any of his predecessors, and it made of his assignations pleasurable vicarious reading, amply supplied in books, articles, and documentaries. "You must always remember that sex is something which gives every journalist, every writer, an equal start," said Professor John Kenneth Galbraith, as wryly witty at ninety as at forty-five. "If you're talking about economics, foreign policy, or war and peace, you have to have information. On sex, everybody is equal. Therefore, sex is the avenue by which the most incoherent gain attention."

In the White House, the Secret Service agents were trained not to look at the president but only outward toward those who might harm him. Kennedy's aides learned to do much the same thing. They looked away, and yet they knew that things were not right. A serial adulterer is rarely one of humankind's noblest specimens, yet the men around Kennedy, the men who knew him best, loved and revered him deeply. Paolella sensed what was going on, but that did not diminish what he and his colleagues felt. "I would say I think everybody loved him," the agent reflected, his voice etched with feeling. "I mean, there's no doubt, he had charisma, he had a kind of self-deprecating sense of humor. And he never let you think that he was above you."

Jackie abhorred what she considered the prisonlike atmosphere of the White House and was spending as much time away as in Washington. She had the feeling that those around her husband had "hit the White House with their Dictaphone[s] running." It was as if his aides and advisers were seeking to live twice by memorializing their every moment when they ended up not living at all, or only half a life. She thought, "I want to live my life, not record it." Her perceptions of the events and people in the White House were sometimes savagely penetrating, but she kept all her impressions largely to herself, pointedly never even keeping a journal.

The president found it difficult to understand Jackie's fey reticence. At the first state dinner, the president held Jackie's arm behind her back and pushed her toward a group of women reporters in the Blue Room. "Say hello to the girls," Kennedy said, to which his wife muttered a perfunctory "hello." As she turned back out of the Blue Room, the imprint of Kennedy's fingernails was still visible in Jackie's arm. Kennedy may have been the most powerful leader in the world, but he was discovering that he had become at least partially hostage to the will and whim of his wife, soliciting her time and bargaining with her over the functions that she would grace with her appearance.

The president's other women were much easier to handle. Most of them were so overawed by Kennedy's sheer presence, so caught up in the moment, or so narrow of mind-set, that they had no rich insights into the man and his psyche. His young Boston mistress was one of the few who observed with depth and sagacity. She did not see the president very often, but when the call came, as it did every few weeks, she was available.

> There was a tremendous optimism about him that was very attractive, and a sense that good fortune would smile on him. But in the end I think that all that vitality became a trap, masking or even obliterating a more nuanced way of being. His image of high-performing achievement robbed him of his connection to his interior life. His light and

> energy could be stimulating, but it could also be intimidating and competitive. There was a magic circle, but there was always the threat of being cast out of that special place. So what do you do to maintain your position? Well, in this situation, you especially did not want to be boring or insipid or wishy-washy. I was always afraid of losing color, dwindling into invisibility. So I would try to be smooth and perfect, and then I'd be resentful that I was trying so hard and would be sullen and annoying. One day I was wearing a red-and-white-striped T-shirt, and he said, "Don't you have anything better than this to wear in the White House?" I was glad he noticed that I wasn't trying to look good the way everybody else did. And I was furious because he didn't understand that I was "making a point." All those feelings and nowhere for them to go. Because he, magical he, prized smoothness. There were so few colors, and yet human beings have so many colors.

Kennedy set up his Boston mistress in the White House, parading her past such bluenose academics as Bundy and Rostow. He had a heedless disregard for the chances he was taking, though just as Dr. Cohen had warned him about his drug use, so too did he continue to have ample warnings that his sexual indulgences might become public knowledge. Florence Kater had continued trying to expose what she was convinced was a sexual relationship between Kennedy and Pamela Turnure, sending out letters to prominent journalists, even picketing one of his campaign speeches. The president could have tried to quiet Kater's campaign by shuttling Turnure off into some obscure sinecure in Washington. Instead, he made the startling decision to bring Turnure into the White House as Jackie's press secretary.

Kennedy had few limits on whom he would proposition and where he would proposition her. At one dinner party, with Jackie present, he passed a note to a guest asking for her phone number. He called her later in the evening. "I'm sending a car for you," he said. "For Christ's sake, I'm your wife's friend," the woman replied, though to Kennedy that was apparently a non sequitur.

Another Washington woman who accepted the president's invitation talked to Kennedy's old friend John White afterward. "It all happened in such a hurry that she couldn't analyze her emotions," White recalled. "But the feeling of this power was like a hurricane, wham, you're swept off and left lying on this beach. And she said, 'That's unique in my life. No man had ever done that to me before, and there was a weird, wild pleasure in it. It wasn't an ordinary matter of affection.' "

The most dangerous of Kennedy's many liaisons was with Judith Campbell Exner. She was shuttling between her occasional visits to Kennedy and her much more frequent sojourns in Chicago with her newfound friend Sam Giancana, as well as dating Johnny Rosselli and others in Los Angeles. She had not seen Kennedy for seven months from the summer of 1960 until she reconnected with him at the Ambassador East in April 1961. "A moment later we were in each other's arms and it was like we had never been apart," Exner recalled. There was hardly time for lengthy reminiscences since Kennedy had only twenty minutes before he had to leave for his next engagement.

In her later years, when her beauty was gone and she was sick with cancer, and she had no money, Exner made her living by telling and retelling the story of her life. She treated her life story like costuming that she would wear until no one was looking at her anymore; then she discarded it and tried on a new, more outrageous outfit that would bring her more attention and money. In 1988 she first realized what new versions of her life were worth, earning $50,000 doing interviews for *People* magazine in an article written by Kitty Kelley ("I lied when I said I was not a conduit between President Kennedy and the Mafia"). From then on journalists often gave her what were called "expenses," a term of limitless elasticity, so that they could say they had not paid her.

In 1997, during a legal deposition in a libel suit, Exner recalled under oath that she had received an amount that "could have been in the $20,000" range for a television series with Anthony Summers, a British journalist. She said that he had paid her more money for a newspaper serial for the *London Sunday Times*. "I, Anthony Summers, have never paid Judith Exner a bean," Summers asserted. German television purportedly paid Exner around $6,000 for an interview. She swore that *Frontline*, the public television series, paid her. Exner said that the Japanese paid her between $10,000 and $20,000. For an article by Gerri Hirshey in *Vanity Fair* in 1990, the journalist agrees with Exner that the magazine paid her $5,000 in "expenses." For another piece seven years later in the same magazine, telling a different tale, she said in her sworn testimony that the gossip columnist Liz Smith personally paid her $10,000. "I did try to help her," Smith said. "I was very concerned for her." Exner said under oath that for a program featuring another dramatic new version of her life hosted by Peter Jennings on ABC in 1997, based on Seymour Hersh's exposé, that she was scheduled to receive a total of $20,000; this is denied by both Mark Obenhaus, the producer, and Hersh. She said that Tribune Entertainment paid her $25,000 for an option on her life story, the BBC paid another $25,000, and Showtime paid $200,000 for rights.

Whatever Exner's critics thought of her ever-evolving revelations, for a decade there was overwhelming sympathy for a woman who was suffering from bone cancer that made her every move painful and who claimed to be standing at death's half-open door during that time. In her sad last years, though she did have breast cancer, she apparently knowingly lied about the extent and nature of her illness, as she had lied about so much else in her life. Kim Margolin, her doctor and a partner in this deception, admitted in a 1998 deposition that she had a number of times made false statements when she said that Exner suffered "from extensive bone metastasis, including destruction of the spine," and that she had "made an error several times trying to help this patient. I didn't realize it was going to get us into trouble in a court of law."

Unlike some once-beautiful women, Exner did not exaggerate her makeup and her clothes as she grew old, trying to maintain some vestiges of her youthful allure. It was her own life that she exaggerated. Her true life was theatrical, and her pain genuine, but she insisted on rouging up her life, each time spreading a thick coat of drama over the few rich moments of her days, treating her occasional sexual relationship with Kennedy as a treasure that could endlessly be exploited and exaggerating even the painful circumstances of her fatal disease.

Much of the truth of her life and her relationship with Kennedy probably lies somewhere among the discarded garments of her life. Giancana was unlikely to have fancied Exner's regular company if he was not having a sexual relationship with her. "Whenever she'd come to Chicago, Mooney [Giancana] would fuck her," recalled Robert J. McDonnell, an attorney who has defended mob figures and married and divorced Giancana's daughter, Antoinette. "Mooney was one horny guy. And I don't understand why she'd have money problems. Mooney always took care of her. Always."

Giancana was fully aware that Exner was journeying from Chicago to be with Kennedy, using cash that probably came from his criminal ventures. He was content to share Exner with Kennedy, knowing that his knowledge of their affair might prove to be a free pass that he could use one day, perhaps avoiding prosecution. Some called that blackmail, but to a man of Giancana's evil sentiments, it was merely trading one item of value for another.

Love was not cheap. For his occasional dalliance, the president was trafficking with a woman who journeyed between his arms and the darkest elements in American society. She was a woman with endless illusions about herself. She came and went like all the others, but she would share the pages of history with him more than she shared the moments of his life, sullying much of what he did, and much of what he hoped he would mean to future generations.

24

Bobby's Game

Bobby vowed that he would not end up like Attorney General William Rogers, so cowed by the hatred he engendered in the South that he hid on the plane when traveling with candidate Nixon below the Mason-Dixon Line. Bobby was not a man to hide from his enemies, but on the day he first walked into his stately office in the Justice Department, he was already far more a target of hatred than Rogers ever had been.

Bobby knew that there was potential danger to himself and his family, but it was impossible to sort the real threats from the malicious gossip, the honest concerns from the pernicious rumors. During his first year as attorney general the governess was returning to Hickory Hill one evening with several of the children when, on a nearby road, she surprised a man in the bushes. The intruder jumped into his car and sped off without bothering to turn on his lights. After interviewing the governess, a Fairfax County police officer concluded that "on the basis of the man's action at the time he apparently had gotten out of his car to urinate and on being surprised rapidly left the area."

That was a matter for hearty laughter, but Bobby was hardly extreme in his fears. He had no entourage of Secret Service agents, no state-of-the-art security around Hickory Hill, and with every controversial step he took and every threatening letter he received, the drama of his life increased. His mother had her own secret anxieties. "The attorney general was fighting against Hoffa," Rose recalled. "Dealing with criminals was tough. They said they would throw acid in the eyes of his children. Bobby kept all those dogs there and they were very apprehensive at times."

Bobby was a brave man, but it diminishes the nature of courage itself to imagine that he was without self-doubt, fear, or anxiety. "My father, all of us really, were always around brave men, great athletes, people who had done great things," reflected Bobby's son Christopher. "We had constantly to risk ourselves to feel that we could be with them. Risk was a way to feel God."

When it came time to appoint the assistant attorney general for civil rights, Harris Wofford was the obvious candidate, but Bobby did not trust him to subordinate his passionate convictions about civil rights to the president's or the attorney general's agenda. Wofford was an odd mixture—an attorney who graduated from both Yale University and Howard University Law Schools, spent several years at the prestigious Washington law firm of Covington and Burling, and taught at Notre Dame Law School. But his other side had drawn him to study Gandhi in India, to work closely with Martin Luther King Jr., and to advocate civil disobedience in the civil rights movement.

When Bobby first met Wofford, he was puzzled that the attorney had attended Howard as the first white male student. "I can understand going and teaching there, but why would you go to Howard Law School to learn law?" Bobby asked. "It's not a great law school. Why would you go to a Negro law school as a student?"

Wofford had been one of those responsible for helping to deliver about 70 percent of the black vote to Kennedy, a voting bloc that was as important as any one factor in his narrow victory. He deserved an important position in the administration, but Bobby considered Wofford "in some areas a slight madman." Men who care injudiciously about human liberty are often thought crazy, and Wofford was slightly mad about injustice. It was acceptable to have an undersecretary of Commerce concerned solely with business interests, or an undersecretary of Agriculture pushing only farmers' causes, but Bobby considered it imprudent to have an assistant attorney general for civil rights whose advice would be "in the interest of a Negro or a group of Negroes or a group of those who were interested in civil rights." This domestic issue was the most volatile and dangerous to his brother's presidency, and Bobby did not want an uncompromising activist as his chief civil rights advisor. Instead, Wofford went to the White House as the president's special assistant for civil rights. At the Justice Department the new attorney general hired Burke Marshall, a reserved respected corporate lawyer who would see civil rights as a question of law, not of emotions.

Bobby agreed to make his first major speech as attorney general on May 6, 1961, at the law day exercises at the University of Georgia School of Law. Even if he had not been confrontational by nature, this would have been the time and the place for him to talk about civil rights. He knew that words themselves were action, and he and his aides spent five weeks working on the address, weighing words and phrases, making sure he would say precisely what he wanted to say.

There had been racial rioting at the university in January to protest the admittance of two black students, Charlayne Hunter and Hamilton Holmes, and they had been reinstated only through a federal court order. As Bobby prepared to fly down to Atlanta, the FBI learned that the Ku Klux Klan was vowing to picket him and to drive through Athens with a loudspeaker shouting, "Yankee go home!"

Bobby's law school audience that day was full of young people who would soon help define how the South reacted to the federal government's push for the integration of the region's public institutions, whether there would be blood and burning crosses or accommodation and goodwill. A measure of the challenge was that in the audience of sixteen hundred people there was only one black, Charlayne Hunter, who was there because she had press credentials.

Today as Bobby began his speech, he had a feeling about the injustice shown to black Americans that was more than an attempt to commandeer their votes and their allegiance. The previous fall, during the campaign, Bobby had flown into Savannah, Georgia, to give a dinner speech. On the drive into the gracious city he had asked how many blacks would be in attendance. When he was told none would be there at the segregated hotel, he said, "Well, we're not going to have the dinner unless you get some blacks there, okay?" He was being what some considered his hectoring, impossible self, but he got his way that evening, and for the first time blacks sat in that hotel like other citizens.

Bobby did not inherit a deep concern for racial justice from his father. Nor did it come from his brother, the president. He had not learned it from his professors at Harvard or at the University of Virginia Law School. He seems to have somehow extrapolated this deep feeling out of his own experience. Bobby's life could scarcely compare with the lives of most blacks in America. Yet he had struggled against those who thought him second-rate, and he could empathize with blacks as most other wealthy white scions of his generation could not. Injustice rankled him, especially when it was directed against American citizens of a different color.

Bobby was a man who not only wore his emotions on his sleeve but also forced them on those he confronted. Political figures are usually brokers, trading among various interests and constituencies. There was little of this in Bobby. Nor did he love humankind only in the abstract, stepping back from the podium where he had expressed universal love to avoid the touch of a single person. He loved people, though he embraced certain groups more than others and sometimes endowed them with qualities they did not always possess, but the man was no hypocrite. He respected ability, and by bringing aides into the Justice Department who were singular in their abilities, he infused much of the organization with a splendid new esprit.

Bobby was not some northern preacher come south to condemn only southern racists. Earlier than almost anyone in American public life, he understood the sheer hypocrisy of such racial moralizing. He saw that North and South, liberal and conservative, redneck and ascot-garbed, the whole nation bore the burden of racism. "The problem between the white and colored people is a problem for all sections of the United States," he told his Georgia audience. "I believe there has been a great deal of hypocrisy in dealing with it. In fact, I found when I came to the Department of Justice that I need look no further to find evidence of this."

Bobby was not a natural orator, and he seemed uncomfortable before the University of Georgia audience. His hands shook, and he spoke in a high-pitched voice. He told his audience that at the Justice Department only 10 of the 950 lawyers were blacks. What he did not say, and he knew as well, was that at the FBI there were only three black agents, and they were the director's chauffeurs. "Financial leaders from the East who deplore discrimination in the South belong to institutions where no blacks or Jews are allowed and their children attend private schools where no Negro students are enrolled," he continued. "Union officials criticize southern leaders and yet practice discrimination with their unions. Government officials belong to private clubs in Washington where blacks, including ambassadors, are not welcomed even at meal times."

Bobby was a truth-sayer, and his words resonated deeply within his audience, which responded in the end with sustained applause. He was so consumed with the cold war, however, that he attempted to relate everything to the struggle against the Soviets, as he did when talking about the first two blacks graduating from the university. "In the worldwide struggle the graduation at this university of Charlayne Hunter and Hamilton Holmes will without question aid and assist the fight against Communist political infiltration and guerrilla warfare," he told the audience.

As Bobby spoke, a group of black and white civil rights activists were traveling south in a Greyhound bus, not thinking that they were assisting in "the fight against Communist political infiltration and guerrilla warfare." The Freedom Riders were confronting the South's segregated system by sitting together, black and white, traveling deeper and deeper into danger. As they set out, they were not accompanied by reporters ready to celebrate their acts and, by their presence, help to protect them. Nor did helicopters fly overhead or federal marshals lead them down isolated macadam strips.

For a man who appreciated courage above all virtues and sought to exemplify both political and physical bravery, Bobby should have seen these activists as heroic figures. Instead, he saw them as endangering what he hoped would be the acceptance of federal laws by the South. As Bobby saw it, politicians like Alabama governor John Patterson might shout, "Never!" but if the federal law steadily nudged the South forward, the people would grudgingly, reluctantly assent.

To the Kennedy brothers, what was so maddening about the Freedom Riders was that instead of simply making their point and moving on, these young men and women pushed and pushed and pushed, sending new groups of nonviolent activists to pick up the bloody banner from those who could carry it no further. The first group of Freedom Riders got as far south as Anniston, Alabama, before their bus was overturned and burned. Those who were not hospitalized got onto a new bus and headed south again. When these Riders were brutally assailed in Birmingham, they flew to New Orleans while a new group arrived to get on a bus and travel farther south.

"Stop them!" Kennedy ordered Wofford, his special assistant for civil rights. "Get your friends off those buses!" The president's command showed a woeful lack of understanding of the civil rights movement. Neither Wofford, King, nor anyone else had control over the Freedom Riders or could contain the moral imperatives of this movement. Kennedy believed in the advancement of civil rights in orderly ways but this was not a transcendent matter on which he wanted to rest his presidency.

During this period Bobby met with King for a luncheon at the Mayflower Hotel. The young administration was worried that the minister might attempt to lead millions of his people to the barricades, destroying Kennedy's plans for steady incremental advances and producing a white backlash that would cost the president his congressional majority.

King could be as uncompromising as any man when great moral principles were at stake, but this luncheon was not one of those times. He understood Kennedy's political realities and knew that half a loaf of bread was as ample a meal as he was likely to receive at this table. King nonetheless wanted Kennedy's support, and he was assured that he would have a private meeting with the president. "The meeting kept being delayed," Wofford recalled, "And King's patience began to run out."

The president felt that he could afford to have his brother dine with King in a private dining room. But civil rights was the most volatile domestic issue, and the politician in the White House was worried that if he had a public identification with King, fears would arise that Kennedy was conspiring with the civil rights activists who so threatened the Old South.

King had his private meeting with Kennedy a few weeks later, and was led into the White House family quarters by Wofford. Kennedy was not comfortable around priests and preachers who spoke in high biblical parlance, and he would never have a deep rapport with King. He listened to King's high discourse and then took the conversation down to a nuanced, realistic portrayal of the American social and political realities. "It lasted at least an hour and was the most effective I ever saw Kennedy on civil rights," Wofford recalled. The president explained to King in immense detail the problems the administration faced and why, though he agreed with King on this great issue, he could not move forward now with bold strides. He would put forth a civil rights bill and issue an Executive Order on housing, but not yet.

If King and the other civil rights leaders had had Kennedy's profoundly realistic vision, then there would have been no great movement, no Freedom Riders, no sit-ins, and no massive confrontations. If Kennedy had had King's moral passions and commitments on this issue, he would most likely not have been elected president. As they sat talking in the study by the Lincoln bedroom, each man had something to learn from the other. King was traversing a maelstrom of politics that no one had attempted before; he would have to be not only brave and true but also as calculating as Kennedy. As for the president, if he was to be the great leader that he aspired to be, he would have to show something of the moral passion of the man who sat across from him. King understood that as well as anyone. "In the election, my impression then was that he had the intelligence and the skill and the moral fervor to give the leadership we've been waiting for and do what no other president has ever done," he told Wofford afterward. "Now I'm convinced that he has the understanding and the political skill but so far I'm afraid that the moral passion is missing."

Kennedy was a general being forced to fight a battle on ground he did not want. He thought that his combat lay elsewhere, in confrontations with the Soviets, not in the streets of the South. But American history had a different timetable. In 1954, in *Brown vs. Board of Education*, the Supreme Court had ruled that segregated education was inherently unequal. American schools would have to be desegregated with "all deliberate speed." A few school districts in places such as Texas proceeded with admirable dispatch to follow the nation's laws, but as Kennedy took office, only about 214,000 out of more than 3 million black schoolchildren in the South and border states attended integrated schools. Southern blacks drank water from fountains labeled "COLORED ONLY," ate at restaurants relegated to their race, had their hair cut only by black barbers, and sat in the backs of buses. This "South" of institutionalized segregation spread far north of the Mason-Dixon Line in a belt of states that included southern Ohio and southern Illinois.

The structure of American liberty stood only half finished. If the administration moved too timidly, the president might find himself the overseer of the breakdown of American freedom, crumbling in the storms. Yet if he moved too harshly, too precipitously, Kennedy might have to line the streets of the South with federal soldiers, creating a second Reconstruction and destroying that house of liberty by the sheer weight of the roof that he sought to erect.

Most politicians rarely stray from the arithmetic of democracy. The Democrats had lost two seats in the Senate and twenty-one seats in the House in 1960, and the conservative Republicans and Dixie Democrats could probably filibuster any civil rights legislation that the president tried to push through a reluctant Congress. The president thought that he would have been doubly a fool for pressing civil rights legislation: not only would he have lost but he also would have unnecessarily confronted members of his own party whose votes he needed on other vital matters. Early in his term he was doing what even a man of Wofford's passion thought was the only thing to do. He was limiting himself to executive actions to push civil rights along, hiring blacks in unprecedented numbers, authorizing the attorney general to push for school desegregation, championing voting rights in the Justice Department. And while the administration was trying to uphold civil rights in this fashion, young men and women in the South who were trying to integrate lunch counters or register voters were getting not meals and ballots but jail cells and clubbings. And the Freedom Riders continued their journey.

Bobby called the Greyhound superintendent to try to get Greyhound to find a driver and a bus to carry the Freedom Riders out of Birmingham. "I think you should—had better be getting in touch with Mr. Greyhound or whoever Greyhound is and somebody better give us an answer to this question," he said. "I am—the government is—going to be very much upset if this group does not get to continue their trip."

To many southerners, this was evidence that the attorney general was siding with these troublemakers and orchestrating this assault on what they considered their very liberty. Bobby was trying to get the Freedom Riders out of Birmingham and on to Montgomery and to end the shameful photos disgracing America in the world's newspapers.

As John Lewis got off the Greyhound bus at the station in Montgomery, Alabama, with his fellow Freedom Riders, the young black activist began to address the reporters standing there. He stopped suddenly and stood transfixed. Coming toward the bus was a mob of whites carrying baseball bats, lead pipes, and bottles. To this rabble, the reporters and photographers were as much the enemy as the Freedom Riders, and the white mob battered away ecumenically, striking out against anyone who came within reach of its weapons. Lewis and several others were brutally assaulted and knocked to the ground. When Bobby's aide and emissary John Seigenthaler arrived in the midst of the melee, he attempted to use his federal authority to stop a group of white women from beating a white woman activist. As Seigenthaler insisted that they stop, a man smashed a lead pipe down on his head, rendering him unconscious. Floyd Mann, the chief of the Alabama state police and an official with no jurisdiction in the city, stopped the attacks by pulling his gun and ordering, "Stand back!"

Bobby's close associates were even more a band of brothers than the president's men. They were devoted to the attorney general and the causes he championed. They had not anticipated, however, that they were risking their very lives by signing on to come to Washington. As the former Tennessee journalist lay inert, half sprawled under a car, the stage on which this drama was taking place had grown immeasurably, for Bobby and his men as for everyone else.

Bobby was still interested in dimming the lights and hurrying everyone offstage, but this theater had become irresistible and new players arrived who sought to speak their parts. King had chosen not to join the Freedom Riders on their journey. He knew that he had to speak his lines or in the next act of this drama he might have no lines at all. He saw this as "a turning point and a testing point. . . . If we can break the back of opposition here, public facilities will be desegregated tomorrow." Much to Bobby's discomfiture, King insisted on flying into Montgomery, where he was met at the airport by roughly fifty federal agents who escorted him to the home of the Reverend Ralph Abernathy, a civil rights leader. King was a believer in moral witnessing, and he needed dramatic moments on which to make his stand. That evening fifteen hundred crowded into Abernathy's First Baptist Church, while outside a crowd of whites twice that size stood and jeered and threatened, held off by a cordon of federal marshals. As the night went by, the mob pushed nearer and nearer, heaving Molotov cocktails and trying to break down the church doors. Time and again they were pushed back by the seriously outnumbered federal marshals.

While this was going on Sunday evening, Bobby sat in his sports clothes in his office trying to barter some semblance of peace. His brother wanted his administration's history to be written on a world stage in a bold, steady hand, not scribbled in blood by a racist mob who would gladly have set fire to that church or bludgeoned King and others to death. In that event, others probably would have also picked up the gun and written their own bloody pages in the history of the time.

Governor Patterson was a segregationist and a savvy politician. The

Alabamian was trying by word and act to tell the Kennedys that if they sought to impose what he considered liberal mores on his state, they would have an uncivil war on their unclean hands. Patterson no more wanted bloody streets and marching troops than did the Kennedys, but he could not afford to be seen as caving in to the hated northern intruders. As the governor saw it, King's arrival on the scene increased the dangers tenfold, for the black troublemakers were now led by the most hated black in the white South.

In the church hundreds of parishioners for whom the church was the sheltering center of their lives sat among the Freedom Riders and other activists. While King stood in the pulpit preaching to them words of faith and moral politics, outside danced a mad mob, dark, malevolent bearers of anarchy.

Late that night Patterson agreed to send in the Alabama National Guard to protect the church. Bobby was immensely relieved, but for King, the sight of Alabama National Guardsmen instead of the hoped-for federal troops surrounding the church deepened his fears and dismay. "You shouldn't have withdrawn the marshals!" King yelled at Bobby with such force that he pulled the telephone back from his ear. "Now, Reverend," Bobby rejoined, "don't tell me that. You know just as well as I do that if it hadn't been for the United States marshals, you'd be dead as Kelsey's nuts right now."

Earlier in the evening Bobby had tried to strike up a rapport by finding a commonality: he talked with King about his grandfather Honey Fitz's stories of anti-Catholic agitators burning nunneries in Boston. That may have resonated somewhat with the minister, but the obscure reference to Boston Irish history passed right by him. "What are Kelsey's nuts," King asked.

Bobby's night of listening to screaming, ranting players in this dangerous game was not at an end. "You are destroying us politically!" Patterson yelled at him a few minutes later. That was a charge guaranteed to resonate. Patterson had been a Kennedy supporter, and if he was destroyed, what malevolent figures might rise in his place?

"John, it's more important that these people in the church survive physically than for us to survive politically," Bobby replied. This was true, but in the physical survival of the Reverend King and the Freedom Riders, the Kennedy administration's civil rights policy survived as well. Bobby had shown himself an adept, determined negotiator in achieving that result.

When the Freedom Riders had finished their journeys, the attorney general took the action that if taken earlier might have made the rides unnecessary. He directed the Justice Department to collect photographs and other evidence of the pervasive racial discrimination in the South. Then he prevailed upon the Interstate Commerce Commission to issue an order prohibiting all such discrimination. Within a year the Department of Justice reported that segregation in interstate public transportation had ended. It had been ended by a group of courageous young people deliberately confronting segregation and by the skill and determination of the attorney general acting under the president's orders. What had started as dangerous tension between the administration and the civil rights movement had resulted in a successful meeting of popular protest and public power.

As for the South, the region was far more complex in its peoples, and far more diverse in its reactions, than many northerners understood. "But when all has been said, we Montgomerians and Alabamians are left in loneliness and with a grievous problem," the *Montgomery Advertiser* editorialized. "The agitators will be tried for defiance of Alabama law. But what of the mobsters who have defied Alabama law? They were not duly quelled and that failure is ours alone. In fact, the mobsters were encouraged."

Bobby had a heavy schedule dealing with civil rights, organized crime, and other legal matters that traditionally concern the attorney general. He was so obsessed with the debacle at the Bay of Pigs, however, that he became the crucial policymaker in the attempts to end Castro's regime.

Bobby was there at the crucial meeting on November 3, 1961, setting up "Operation Mongoose," the multi-agency plan to harass Cuba and destroy the revolutionary government. When the attorney general took notes at the meeting, he named all the other participants, then wrote: "Lansdale (the Ugly American)."

Brigadier General Edward G. Lansdale, the chief of operations of Operation Mongoose, had served in the early 1950s in the Philippines, where he was involved in anti-insurgency efforts against Communist Huk guerrillas. That experience became the subject of an adulatory fictional portrait in William J. Lederer and Eugene Burdick's 1958 best-seller, *The Ugly American*. In those pages Lansdale became the idealistic Edward Hillandale, an air force colonel who is genuinely concerned about the lives of the people in a Filipino province threatened by Communist guerrillas.

Lansdale had also been immortalized as Alden Pyle in Graham Greene's classic novel *The Quiet American*. The book is a primer on the dark side of the American character—the deadly innocent who sets out to do good, oblivious to the murderous devices he employs. Lansdale had served in the CIA in Vietnam in the midfifties, when Greene's novel takes place. In Greene's novel the protagonist is a CIA agent in Vietnam, a crew-cut, friendly fellow who, with all the best of intentions, "comes blundering in, and people have to die for his mistakes."

After the Bay of Pigs, the Kennedy brothers called on the celebrated counterinsurgency expert to lead where the CIA had faltered, taking over as chief of operations of the renewed efforts against Cuba. The Kennedys had no real trust in the CIA, and Lansdale ran the operation out of the Defense Department, using CIA and other resources. Although Bobby oversaw Operation Mongoose, he did nothing important without the clear direction and knowledge of the president.

Bobby was enamored with the mythic Lansdale, though he was no more *The Ugly American* than he was *The Quiet American*. He was, however, a brilliant propagandist, among other things, and he promoted nothing better than himself. In that crucial meeting he was a sterling proponent of a passionate plan to have the Cubans fight their own battles, a plan that resonated perfectly with Bobby's own thinking.

"My idea is to stir things up on island with espionage, sabotage, general disorder, run & operated by Cubans themselves with every group but Batistaites & Communists," Bobby wrote. "Do not know if we will be successful in overthrowing Castro but we have nothing to lose in my estimate." Lansdale took Bobby's emotional imperative and turned it into a policy that seemed to embody all the can-do spirit of America.

Lansdale condemned the CIA's ineffectual "harassment" techniques with a flick of rhetoric. He wasn't for imposing some American solution on the backs of the Cubans. As he said later, he was for having "the people themselves overthrow the Castro regime rather than U.S. engineered efforts from outside Cuba." He was going to find leaders among the Cuban exiles who opposed both Batista and Castro, bold men who would lead their people to overthrow the Communist tyrant. He knew Communists and their ugly acts, and he was not about to "arouse premature actions, not to bring reprisals on the people there and abort any eventual success."

While Bobby listened to Lansdale, he also had before him a memo on Cuba from the Board of National Estimates, representing the best judgments of analysts from the CIA, the Joint Chiefs, and the State Department. It was a rational, realistic portrait of Castro's Cuba, detailing not only the oppressive measures the state had taken but also the legitimate support it still maintained. The leadership had institutionalized the revolution in such a way that Castro's death would not end the regime. The report concluded on the deeply ironic note that "a dead Castro, incapable of impulsive personal interventions in the orderly administration of affairs, might be more valuable to them as a martyr than he is now." That was a daringly prescient analysis, for when the revolutionary leader Che Guevara was executed by American-trained soldiers in Bolivia in 1967, he became in death a symbol to Cuba and a goad to revolution that he probably never would have been in life.

Lansdale understood rightly that this report represented a profound challenge to his own aggressive, daring schemes. He warned Bobby that this "special intelligence estimate seems to be the major evidence to be used to oppose your project." He criticized the report for drawing conclusions based on inadequate intelligence, but he had nothing to offer in its stead but rhetoric and unbridled passion.

Robert Amory, the CIA deputy director for intelligence, believed that the past was a guide to the present. Amory pointed out at a later CIA meeting that "no authoritarian regime has been overthrown in the 20th Century by popular uprising from within without some kind of support—war or otherwise."

When Roger Hilsman, another expert and proponent of counterinsurgency, looked at Lansdale's plans, even he had profound doubts. Hilsman wrote in February 1962, "we may be heading for a fiasco that could be worse for us than the ill-fated [Bay of Pigs] operation."

In listening to Lansdale's entreaties, Bobby, then, willfully turned away from many of the best minds and judgments of his intelligence and diplomatic services to accept a policy based largely on one man's deadly idealism. Operation Mongoose was, as Lansdale called it, Bobby's plan, but it was also Bobby's revenge, Bobby's private war, a war that he staged to vindicate his brother, the president.

Lansdale had always been terribly eclectic in the means he employed. On the one hand, he intended to train idealistic young Cuban students to infiltrate their homeland, and on the other hand, he would also employ some of the same Mafia figures who had already been involved in assassination attempts. "This effort may, on a very sensitive basis, enlist the assistance of American links to the Cuban underworld," Lansdale wrote in a top-secret memo in December 1961. "While this would be a CIA project, close cooperation of the FBI is imperative." It is simply unthinkable that Lansdale would even suggest such a thing if Bobby, his protector and champion, had not been fully conversant with the plans.

Lansdale could speak bureaucratic language at White House meetings, but what appealed to the Kennedys was that the man was a democratic shaman, full of tales of how he had defeated communism by wit, magic, and endless courage. Whatever he did, Lansdale was to the Kennedy brothers' thinking a brave, true man who had made the fight against communism his chosen field of combat, one on which he again and again proved his mettle and his manhood.

Lansdale, who had started out in advertising before World War II, was an endless spinner of tales. One of the stories that embroidered his legend took place in the Philippines, where his men captured a guerrilla in an area threatened by the Communists. Lansdale supposedly ordered the men to puncture the man's neck as if a vampire had latched onto him, to turn the corpse upside down to drain the blood, and then to set it out on the trail to be found by the Huk guerrillas. When the peasant soldiers came upon the gruesome corpse, they were so terrified that they moved out of the area.

In February 1962, Lansdale set out a precise timetable that was as reliable as an Indian train schedule. He proposed the active fostering of revolution within Cuba to take place by October "with outside help from the U.S. and elsewhere." The guerrillas would move into operation no earlier than August, leading to the revolt and overthrow of Castro in October. "A vital decision still to be made, is on the use of open U.S. force to aid the Cuban people in winning their liberty," Lansdale noted, his plan as dependent on American involvement as was the CIA's scenario for the Bay of Pigs. He showed no awareness that Castro might have been a popular leader and arrogantly assumed that most Cubans would welcome this new "revolution" imposed by the hated gringos.

The new CIA station in Miami, code-named JM/WAVE, was set up at an abandoned naval air station in Richmond, just south of the city. It was fitting that JM/WAVE should be run from an old military base, for this was indeed a war that the CIA was running out of these office buildings and warehouses set on a secluded 1,571-acre plot. Its resources included ships and planes and hangars full of munitions, and above all, the Kennedy imperative to do something.

The president and the attorney general had told Bissell "to get off your ass about Cuba." His successor, Richard M. Helms, picked up that banner, as did the new director, John A. McCone, a conservative Republican and deeply religious Catholic. Bobby wanted the CIA operatives to stop their endless excuses for inaction and to get on with it, and he bludgeoned the supposedly slothful bureaucrats with Lansdale's plan. Even four decades later, those involved at the CIA still remember Bobby's voice on the telephone, berating them for laxness, lashing them onward. He rarely seems to have called with any specific directive but merely to whip them on. As they galloped blindly ahead, avoiding the lash, they ran down the innocent, the unaware, and the unlucky.

"Bobby's project," as Lansdale called Operation Mongoose, involved in part taking young men from among the flower of the Cuban bourgeoisie, encouraging them with rhetoric and promises, and sending them on missions that at times would lead to their deaths. It is true enough that the roots of liberty are nourished by the blood of patriots, but this blood was being shed in futile, desultory, dangerous efforts whose only result was to further United States foreign policy to destabilize Cuba. As for the mobsters' continued involvement with the CIA, there was no longer the dubious saving grace of innocence. The CIA knew their names now, who these men were, and what they did. If the agency insisted on trying to kill Castro, it did not have to do business with thieves and professional murderers. There were patriots who would have given their lives for a cause greater than reestablishing their casinos, brothels, and drug operations.

Many of Lansdale's proposed initiatives treated the Cubans like Latin hayseeds whom he and the other American tricksters could bamboozle with their magic and technology. One scheme was to use chemical weapons to sicken Cuban workers so that they could not harvest sugar cane. The plan was abandoned not because it was stupid and repugnant, but because it was deemed "unfeasible." Another scuttled plan was to spread the word throughout Cuba that the second coming of Christ was imminent, and that Christ would not like Castro. A submarine was to appear off the coast and send star shells high into the sky, a light show that would be viewed by the superstitious Cubans as a sign that Christ had arrived.

The administration did everything it could to isolate Cuba from the rest of Latin America, using its power over neighbor states to exclude the island from the Organization of American States and instituting a trade blockade as well. It was not only Cuba but the rest of Latin America that Lansdale considered rich territory for his initiatives. When these other governments seemed unwilling to follow the American lead against Cuba, he was all for enacting "a major psychological and political campaign within the country among labor, student and political groups to 'force' the government to change its mind."

"The top priority in the United States government—all else is secondary—no time, money, effort, or manpower is to be spared," Bobby admonished the top CIA leaders. "There can be no misunderstanding on the involvement of the agencies concerned nor on their responsibility to carry out this job. . . . It is not only Gen. Lansdale's job to put the tasks, but yours to carry out with every resource at your command." Bobby was there dominating meetings that proposed to destroy the Cuban sugar crops, poison shipments of goods coming to Cuba, and induce crop failures.

In April, immediately after the Bay of Pigs, Kennedy gave a speech in which he condemned a "monolithic and ruthless conspiracy" that depended "primarily on covert means for expanding its sphere of influence—on infiltration instead of invasion, on subversion instead of elections, on intimidation instead of free choice, on guerrillas by night instead of armies by day." It was nothing less than tragic to those who cared about American democracy that though the president was describing communism, he might as well have been talking about Operation Mongoose.

Senator John F. Kennedy and his brother Robert, the chief counsel, question a witness at the Senate Racket Committee hearings.
(John F. Kennedy Presidential Library)

At the 1956 Democratic National Convention, "spontaneous" demonstrations touted Senator John F. Kennedy for president.
(John F. Kennedy Presidential Library)

Janet Des Rosiers, Joseph P. Kennedy's secretary, with whom he had a long-term relationship. *(Janet Des Rosiers)*

Senator John F. Kennedy's closest aide was Theodore Sorensen. *(Robert White Collection)*

Joseph P. Kennedy with his wife, Rose, and sister Loretta Kennedy Connelly at Hialeah, January 1954. *(Loretta Kennedy Connelly Collection)*

Three Kennedy brothers sing in victory as Jack Kennedy wins reelection to the Senate, November 3, 1958. *(John F. Kennedy Presidential Library)*

On a frigid January 20, 1961, the newly inaugurated President John F. Kennedy watched the parade in his honor. *(John F. Kennedy Presidential Library)*

The elegant presidential couple attending the inaugural balls.
(John F. Kennedy Presidential Library)

Robert and Ethel Kennedy and their seven children.
(John F. Kennedy Presidential Library)

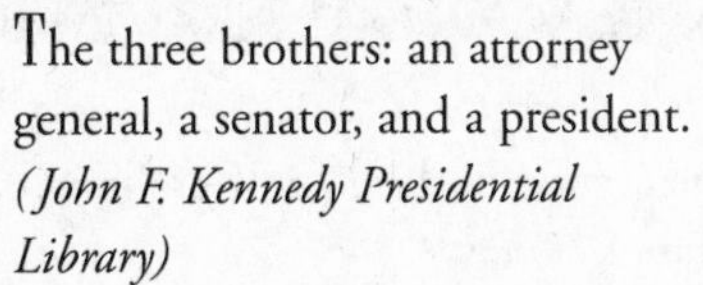

The three brothers: an attorney general, a senator, and a president.
(John F. Kennedy Presidential Library)

After Joseph P. Kennedy's stroke in December 1961, the greatest moment in his life was waiting for his son the president to arrive for a weekend visit. *(John F. Kennedy Presidential Library)*

President Kennedy enjoying a tender moment with his daughter, Caroline. *(John F. Kennedy Presidential Library)*

President Kennedy's closest aide by far was his brother Bobby. *(John F. Kennedy Presidential Library)*

Caroline and John Jr. loved nothing better than romping in their father's office. *(John F. Kennedy Presidential Library)*

Robert Kennedy with some of his children.
(John F. Kennedy Presidential Library)

President Kennedy followed by Caroline on the White House grounds.
(John F. Kennedy Presidential Library)

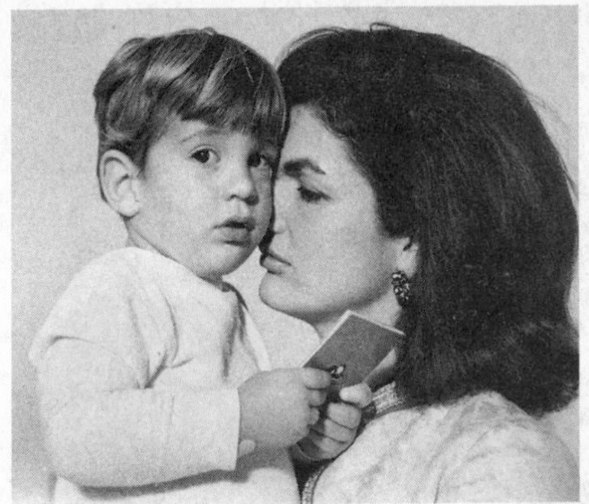

Jackie's love for her son John Jr. was evident to all who saw them. *(John F. Kennedy Presidential Library)*

On September 8, 1962, the entire Kennedy family got together at Hyannis Port to celebrate Joe Kennedy's seventy-fourth birthday. *(John F. Kennedy Presidential Library)*

The first family at rest in Hyannis Port. *(John F. Kennedy Presidential Library)*

The president and the attorney general had a difficult relationship with J. Edgar Hoover, the chief of the FBI. *(John F. Kennedy Presidential Library)*

President Kennedy conferring with deputy special counsel Myer "Mike" Feldman. *(Myer Feldman)*

When President Kennedy appeared to be in deep thought, often he was suffering from back pain. *(John F. Kennedy Presidential Library)*

A president and his son.
(John F. Kennedy Presidential Library)

Robert Kennedy was almost always in a hurry.
(Robert White Collection)

President Kennedy relaxing at Hyannis Port. *(Robert White Collection)*

President Kennedy and his brother-in-law Peter Lawford. *(Robert White Collection)*

Little John Kennedy following his father, the president. *(John F. Kennedy Presidential Library)*

Songs were in order at Joseph P. Kennedy's birthday celebration, September 1962. *(John F. Kennedy Presidential Library)*

Even a president celebrates Halloween with his children, November 1962. *(John F. Kennedy Presidential Library)*

Joseph P. Kennedy's seventy-fifth birthday celebration, September 1963. *(John F. Kennedy Presidential Library)*

The president and first lady arrive in Dallas, November 22, 1963.
(John F. Kennedy Presidential Library)

Robert and Edward Kennedy walking with Jacqueline Kennedy at the president's funeral. *(John F. Kennedy Presidential Library)*

The eternal flame. *(John F. Kennedy Presidential Library)*

"Life for him was an adventure, perilous indeed, but men are not made for safe havens," Bobby wrote in his daybook, quoting Edith Hamilton on the Greek dramatist Aeschylus. Bobby had a compelling need to be part of heroic endeavors, since only in such pursuits could he prove himself worthy. Life at Hickory Hill or in the summer at Hyannis Port was no sedate, mannered respite from the ceaseless battles of Washington, but another arena for challenge and adventure.

One weekend on the Cape, José Torres arrived for a visit. Torres was the world light-heavyweight boxing champion, and after the standard regimen of football, tennis, swimming, and sailing, Bobby decided to mix it up with the champ. "Hit him in the head!" the children screamed, exhorting the champ to down their father. "Hit him in the belly!"

"Let me knock you down," the attorney general whispered to the champ. Bobby threw a haymaker, and Torres fell to the ground, apparently knocked out. Anything was possible to Bobby's kids if your name was Kennedy, even knocking out the champion of the world. It was a life lesson that was simply wrong, but it was taught again and again—that anyone could do anything, there were no limits, and courage was the trump card besting ability, training, wind and storm, gravity and wisdom.

During one Hyannis Port summer, Bobby imported groups of Green Berets, the army's elite new counterinsurgency fighters, and had them perform their derring-do in front of the Kennedy children. The young, would-be heroes in the war against communism swung from trees and jumped over barricades. They were intrepid soldiers whose enemy was the routine, the bureaucratic, and the predictable. Bobby's brother-in-law Sarge Shriver was general of a different army, the Peace Corps; he sent young women and men out to fight with other kinds of weapons. Shriver felt uncomfortable about Bobby proudly parading soldiers on the summer fields of play. These were games in which the black card of death was often dealt. Shriver found something so disquieting about the whole business that he kept his brood away until the soldiers had left.

Ethel was her children's instructor, not by rote but by example. She taught them that for a Kennedy the only answer to "too much" was "even more." At the Washington International Horse Show, her children watched as she suddenly decided to compete in an exacting competition in borrowed clothes on a borrowed mount on which she had practiced for only five minutes. Others had practiced for months, even years, to guide their horses over the hurdles, but such tedium was not for Ethel. She did not win a ribbon, but she finished, and she had shown her sons the blood that coursed through their veins.

At Halloween most mothers tried to restrain their rambunctious sons from thinking of the annual event as a respite from the rule of law and considered it a time when a treat was infinitely preferable to a trick. Not Ethel. She drove her children to the sedate precincts of Georgetown, where bewildered neighbors watched as she led her motley brood in a "trash barrel assault" on the home of her sister-in-law Jean.

Jokes, to Ethel, always had an edge to them. She not only liked to win, be it tennis, charades, or politics, but believed that it was her right to win. Ethel was looking for an edge, no matter what it might be, and she taught her children this as well.

If Ethel had been a middle-class mother in Baltimore or Binghamton, she probably would have been considered dangerously manic. But she was celebrated in the Washington of the New Frontier. The journalists adored her, for she enlivened even the most mundane of moments. In Rome at the end of one European trip, the reporters gave their beloved hostess a Vespa. Ethel jumped on the scooter she had never driven before, roared around the block again and again, and stopped only when she went banging into a car.

At Hickory Hill there was only one song that was sung, and only one dance that was danced, and that was whatever song and dance Bobby and Ethel had chosen. Ethel was a spiritual cop, always scanning the crowds, looking for any traitors less than loyal to her Bobby and to their lives as Kennedys. Bobby's wife could smell the taint of betrayal where others sniffed only roses.

Bobby's favored reading material ran to popular biographies and political books, not the heavy tomes he thought had to be lifted to become a heavyweight intellect. He created the Hickory Hill seminars in which the ladies and gentlemen of the New Frontier heard some of the major intellectuals of their day, a regimen decidedly easier than having to read them. After listening, the Democratic gentry were supposed to question these great minds in eloquent discourse. The ladies, it turned out, were supposed to listen and learn and keep quiet; when Mrs. Nicholas Katzenbach was so bold as to ask a question, she had the feeling that Bobby cringed at her bad taste.

Even Bobby at some point seemed to realize that these seminars were exposing not great minds but vapidness and pretensions. On one occasion the host left the living room and sat outside. "Don't you want to come in?" he said to Seigenthaler, who also had walked out. The aide had been willing to go to Alabama, where he had been beaten unconscious by a mob, but he was unwilling to listen to any more of this. "What do you think?" Bobby asked. "This should be on tape," Seigenthaler said, his irony nicely played. "You should have television cameras in there."

As often as not, no matter what time of day or night, there was some great momentary drama unfolding on stage central at Hickory Hill. It was perhaps a cook tired of morn-to-midnight orders who had thrown her apron to the ground and stomped out. Or possibly one of the kids, Joe II or Bobby Jr., was lying there with a bloody arm, victim of some daring combat in the recesses of the five-acre estate. Freedom was the byword, not of course for the harassed staff, but for the children and animals, including Meegan, the two-hundred-pound Saint Bernard, and the legendary Brumus, a dog who expressed his liberty by urinating on guests. Sandy the sea lion felt so free that one day he left the swimming pool where he feasted on fish and headed down the road, presumably preferring the more sedate National Zoo. The honey bear was frightening to those who had seen bears only in picture books, but the animal was gentle enough, preferring to curl up in one of the bookcases.

Only the bravest of hearts dared to walk down into the basement. That was Bobby Jr.'s terrain, and he was not simply a lover of nature in its most popular forms, but of the exotic and even dangerous species. There the visitors had to be alert to one of Bobby Jr.'s falcons, or some great snake suddenly slithering across the cold concrete. When Ethel led two reporters down into the darkness, she was attacked by a coatimundi. The reporters, throwing their notebooks to the wind, pulled the anteater-like animal off their hostess, and led Ethel upstairs, where she was bandaged.

The guests began their evening by warily petting Brumus. Then they were expected to become exotic animals themselves, to perform some act that had not been performed before, to tell an outrageous tale, to do everything but slap their flippers and jump for fish. It was all very well for the ebullient Ethel, pregnant with her eighth child. Art Buchwald could reel off his latest humor column. The columnist Rowland Evans could pass on some political gossip. Professor Arthur Schlesinger treated the world like a classroom. But for those who were shy or just socially reticent, it could be the most excruciating of evenings. "When I go to Hickory Hill, I always feel like a chimp having to perform," Joan Kennedy lamented, her beauty not attribute enough. "Kennedy life can be rough on human beings."

Whereas most Washington hostesses set out their engraved seating cards after laboriously determining status and interests, at Hickory Hill the most disparate peoples were thrown together. That was the idea, to shake things up so that the only predictable outcome was that anything might happen short of fisticuffs and duels.

In that most famous of Hickory Hill soirées, Ethel extended the dance floor all the way to the pool and set out tables around the water, including one rather precarious perch on a wooden plank out over the pool itself. Ethel, a woman who never saw wet paint she didn't have to touch or a can of shaving cream that she did not consider a potential spray gun, was such a leader of pranks that she went into the pool in her gown first. Schlesinger and another guest were soon pushed in to join her. It was a thumb in the eye to all the sanctimonious cave dwellers of Georgetown who acted as if civility and boredom were synonymous.

Bobby was a quiet partner in his wife's pranks. He thought that he could have it all, college high jinks and serious endeavors, a life as intense and passionate in its moments of play as in the long serious hours of public service. Jack enjoyed good conversation, lovely women, and loyal friends, but he stayed a thousand leagues away from the frenetic fun and games at Hickory Hill. "I never once saw the president and Jackie out at Hickory Hill," recalled Rowland Evans. "And I know I would have been there." As close as he was to Bobby, the president compartmentalized his younger brother as much as he did everyone else, from his wife to his lovers.

On one of the few occasions when the president came out to Bobby's for dinner, the children rampaged through the house, creating a cacophony of shrieks and shouts. "I say out!" Kennedy exclaimed. "I'm the president of the United States and I say out!" He may have been president, but at Hickory Hill, Kennedy was just another tedious adult whose orders went unheeded.

25

Lives in Full Summer

Jack and Jackie were impresarios of style. In the emerging society of the last years of the twentieth century, style was substance, and the famous Kennedy style was as much a part of history as legislation and summit conferences. Norman Vincent Peale and his ministerial colleagues had condemned Jack's faith, fearing that if he were elected, the pope would stand behind him whispering orders. The ministers would have been closer to the mark if they had condemned Kennedy's style, for that was the engine that would help lead America away from the narrowly Protestant society of the midcentury.

Washington was the most provincial of the world's great capitals, a somnolent, socially suffocating city full of the prejudices and mores of the South and the wary, conservative attitudes of any company town. Any culture beyond the road shows of Broadway musicals and the local opera and ballet companies that performed in a movie theater and a college auditorium was a dangerous, foreign business. To Kennedy, the Eisenhower years had reeked of flabby self-satisfaction. As the young president saw it, in the name of conservative Americanism, the Republican administration had celebrated the mediocre and elevated the second-rate. Eisenhower, however, had lived perfectly comfortably in the White House; what Kennedy viewed as hopelessly pedestrian decor was to the Republican president a statement of stolid virtues.

"You know, we really ought to have the nicest entertaining here with the greatest distinction," the new president told Letitia Baldrige, the social secretary, on his first day in office. That comment was a subtle critique of the Rotarian-like atmosphere of the last eight years. Baldrige took the president at his word, starting with the first party for the staff. The Eisenhower administration had served only lightly spiked fruit punch, but that particular abomination hardly fit the definition of "greatest distinction," and Baldrige ordered that alcoholic drinks be served.

The president's grandfather and father had both made part of their fortunes in the liquor industry, and though he was not much of a drinking man himself, Kennedy hardly saw a scotch and soda as the devil's drink. It was only after the convivial staff event that he learned that liquor was not served at major functions in the White House and he had outraged many of the Baptists in America.

The fundamentalist teetotalers may have been sincere, but across America they were achieving not high morality but low hypocrisy. Though prohibition was long gone, the country was riven by a patchwork of bottle laws and drinking regulations that many Americans had learned to evade or avoid. As the Baptists mounted their assault, Kennedy's first reaction was irritation that he had unknowingly attacked one of the silly double standards of the age; if a politician started attacking all the petty hypocrisies of his constituents, he would not be reelected.

"You know, the fact that you put one over on us and served hard liquor that first party—and we have been serving it ever since—was the greatest thing that's ever been done for White House entertaining," he told Baldrige a few months later when the controversy was forgotten. "It's relaxed the whole thing, and you've proved it to be a success, and I just want to say thank you."

A stylish social life served his administration's image, but that alone did not explain Kennedy's involvement in the intricate details of White House events. In part, he enjoyed the respite from a desk full of woes and worries. The president did not like to be bored, and his social life was an attempt to transform the tedious rituals of Washington into amusement.

With his sense of detail, the president enjoyed nuances that many men neither appreciated nor even noticed. For a ceremony at the first state dinner for President Habib Bourguiba of Tunisia, two hundred soldiers would not do. Three hundred was not enough. Even a thousand was not enough. Kennedy wanted two or three thousand. He wanted them tall and handsome, plenty of blacks too, and nobody overweight or wearing glasses. And he wanted bagpipers too, hundreds of them as well, weaving their way across the South Lawn.

These grand ceremonial functions were not simply trifling asides to the business of government, but symbols of the nation's greatness. And never in American history had they been carried out with such grace and elegance. The most memorable gift the president gave each visiting head of state was the event itself, symbolizing in its uniqueness a detailed concern for that nation and its leader.

In July the state dinner for President Mohammad Ayub Khan of Pakistan took place at George Washington's home at Mount Vernon, the guests arriving down the Potomac River by boat. Although most commentators celebrated the event, some critics complained about Americans honoring this heathen and his entourage by drinking gin and tonics on Washington's grave. Those who criticized the exquisite party and said it was a wasteful extravagance could not have imagined that forty years later the evening would still be remembered in Pakistan and America while the issues discussed between the two presidents had long been forgotten.

These White House evenings had become the best parties in America. Jackie was the beautiful hostess of these events, blessing them with her elegance. There was good wine and quick wit. The guest lists were eclectic, including business leaders and artists, philanthropists and athletes, politicians and authors.

Kennedy clearly understood the political advantages of being seen as a benefactor of the arts, and he had the intellectual integrity to honor not just those who served his party or his ambition but also those who served the high truths of their art. In November 1961, he invited the incomparable Spanish cellist Pablo Casals to dinner at the White House. Casals had stopped playing publicly as a protest against the fascist Franco who held his beloved Spain in bondage.

Great artists and great art do not serve the state and might indeed be seen as dangerous, even subversive. "We believe that an artist, in order to be true to himself and his work, must be a free man," Kennedy said that evening. He invited Igor Stravinsky, the Russian composer, for his own evening, André Malraux, the French novelist, for a different event, and on another inspiring evening all the Nobel Prize winners in the Western Hemisphere. He resurrected the Presidential Medal of Freedom so that he and his administration could celebrate accomplished Americans for their contributions not to him and his administration but to American life and culture.

The *Washington Star* reported that "for the first time in almost 50 years, Washington has no wealthy extrovert with social position who can rightfully claim the local title of society queen. The truth of the matter is that the nation's first lady . . . actually is the town's best party giver." The president and first lady were the progenitors of a new kind of Washington social life. The very society that had spurned Kennedy's parents and that he had been groomed to enter was beginning to die. In Washington it was the beginning of an era when achievement and celebrity brought one to the head of the list, not lineage and formal manners.

The president and first lady consciously celebrated what was best in American life, in antiques and music, in art and food. They signaled that this nation of nations should celebrate its freedoms and diversity not merely in speeches but in its culture. Jackie was a true admirer of the arts, and her husband rightfully deferred to her in these matters, enjoying credit for a taste that was largely his wife's.

The president was no more the creator of this immense opening up of the American spirit than he was of the other social movements sweeping the world. But he identified with it, even though there were potential political costs in doing so. Freedom was dangerous, and a man who chose his own church and his own faith might choose none at all, or mindless nihilism, or despair. As for cultural pursuits, many Americans viewed fine art, fine foods, and fine objects apprehensively, as if their passionate pursuit suggested an effete, un-American quality.

The Kennedys promoted cultural and social events as if without them life was only partially lived. They made the White House the nation's model of taste, the celebrant of American culture. Kennedy's critics would later suggest that this was nothing more than gauze over the camera, softening the harsh lines by which his actions deserved to be viewed. It was all, they sniffed, a triumph of style over substance. The Kennedys' achievement was to turn style into substance and to celebrate the opening up of broad new cultural and social vistas that would never again be shut down. Unlike any administration since, the Kennedys turned the White House into an inspiring symbol of culture and style without equal.

Kennedy's style, substance, and savvy came together in his televised press conferences. Since the Kennedy years, presidents have learned to squeeze all the juices of spontaneity out of their media conferences, but Kennedy was the first president to give regular, live, televised press conferences. At the time it seemed a daring departure from the controlled appearances of previous occupants of the Oval Office.

Kennedy had a bemused way of looking out on the assembled press as if no one in Christendom, and certainly not his aides, could ever imagine what questions he might be asked. He played on the reporters, knowing those who wanted their moment of preening time on television, those who had serious questions, and those who had a special cause or area that they always asked about. He used his disarming wit and grace to turn back the more mean-spirited questions on the reporters who asked them. He was at his best, however, when he was asked philosophical questions, the ones he could answer by expressing the existential dilemma of a president dependent on aides and officials and yet standing alone, responsible for his own hard judgments.

"Well, I think in the first place the problems are more difficult than I had imagined they were," he said when asked after two years how his experiences had matched his expectations.

> Secondly, there is a limitation upon the ability of the United States to solve these problems . . . and I think that is probably true of anyone who becomes president, because there is such a difference between those who advise or speak or legislate and . . . the man who must select from the various alternatives proposed and say that this shall be the policy of the United States. It is much easier to make the speeches than it is to finally make the judgments, because unfortunately your advisers are frequently divided. If you take the wrong course, and on occasion I have, the president bears the burden of the responsibly quite rightly. The advisers may move on to new advice.

Kennedy looked at everything in his life through the prism of politics, not only the most monumental questions of life and death and the endless array of issues, from civil rights to education, but even the smallest social events and the media coverage of his children. The photo spreads, be they of the children or Jackie or the loving family at play, enhanced his popularity, spilling over onto his political image. He pushed Caroline and John Jr. forward to be photographed by the popular magazines *Look* and *Life*, an enterprise that was immensely profitable to their owners and may well have subtly tempered some of their political criticism. Jackie tried to shelter her children from the flashbulbs and the public's preening obsession, but the president always managed to work around her. As soon as his wife left Washington, it was not only other women who entered the White House to amuse the president but at times photographers to capture images of his children.

A political architect of the emerging media society, Kennedy created a model that his successors would try unsuccessfully to match. He did not believe, as some would have, that image was everything, but he did think of image as a kind of costuming that allowed him to walk where he wanted to walk and to do what he wanted to do. More than any previous president, his wife and family were portrayed extensively and lovingly in the social pages of the newspapers, the back-of-the-book coverage in the news weeklies, and elaborate spreads in the women's magazines and on such television programs as Jackie's spectacularly successful tour of the newly restored White House in February 1962, watched by forty-six million Americans.

With this kind of coverage, Kennedy did not have to worry about carping journalists who thought it was their professional obligation to attack him. This was not a trifling business to him, and he negotiated over the filming of his children and his wife's program as if an important treaty were at stake. The danger was that such coverage risked trivializing his presidency, turning him into a star and denying him the natural gravitas of his office.

A week after Jackie's tour of the White House, the president had another splendid triumph when the astronaut John H. Glenn Jr. spun around the world three times in America's first orbital space flight. On the morning when Glenn was being picked up from the tranquil waters of the Bahamas, however, the president was in a foul mood. He was sitting at his desk reading *Time* when he called Hugh Sidey, the magazine's White House correspondent, into his office.

"Where did you get this goddamn item about me posing in this suit for *Gentleman's Quarterly*?" Kennedy asked, throwing his copy of *Time* on the desk. Sidey fancied himself a serious reporter who had nothing to do with the trivia that sometimes found its way into the back of his magazine. He had no idea that Kennedy's press secretary had allowed the *Gentleman's Quarterly* photographer to represent his pool photo as an exclusive, and that *Time* was merely reprinting it.

"I . . . I . . . I don't know, Mr. President," Sidey stuttered. "I'll try to find out."

That was the stock answer that Kennedy heard too many times a day. He came around from the back of the desk and shook his fist in Sidey's face. "You sons of bitches are out to get me," Kennedy said, his face red and distorted. "You do this stuff, this personal stuff, as much as you can. You're out to discredit me. People are remembered in this life for only one thing. They remember Coolidge because he appeared in that Indian war bonnet. They remember Arthur Godfrey because he buzzed the tower at Teterboro Airport. They'll remember me as the man who posed for this."

Kennedy would have raged on indefinitely, but Tazewell Shepard, the naval aide, gently injected himself to tell the president that Glenn had been picked up and was on the line. "Sidey, you son of a bitch, stand there and see if you can get this right.

"Oh, Colonel Glenn, what a great day!" Kennedy said into the phone as if all morning he had been waiting for this moment.

Sidey might be invited into the Oval Office for both exclusive interviews and condemnation, but no journalist in Washington was closer to the president than Joseph Alsop. Even though Kennedy had known Alsop for years as a social friend, the columnist had the audacity to write the president-elect saying that he viewed Kennedy's election with "mixed feelings." Instead of being angry at what others would have considered an intolerable impudence, Kennedy brilliantly co-opted the columnist. It proved to be among the most useful of the president's many seductions. Alsop turned his column into a bully pulpit for the administration and used his considerable social power to advance Kennedy. Ambassador John Kenneth Galbraith, a connoisseur of power, watched on with admiration at the way the president led Alsop away from the paths of journalistic righteousness and turned him into his steward and shill. "Kennedy used Joe," reflected Galbraith. "Joe assembled the Washington establishment for Kennedy. He convened them for Kennedy."

Most of his life Kennedy had striven to be part of the upper-class, old-line Protestant world that Alsop so perfectly exemplified. The president and first lady were regular guests at dinner parties at Alsop's splendid Georgetown home, but even now the columnist needed to display his supposed social superiority. To Alsop, the stiletto of class was a shiv always ready to be drawn, in a quick feint invisible to the untrained eye. He used it eclectically and was perfectly willing to stick it into the president of the United States.

Proud vintner that he was, Alsop brought forth precious bottles of 1945 Château Lafite Rothschild. The wine, as Alsop would be glad to tell you, was clearly too tannic to be drunk and needed years more in the cellar to be worthy of its heritage. Alsop noted that the hapless man who was his president paid more attention to reading the label than to tasting the wine. Alsop would have been better off filling the bottle with *vin ordinaire* and recorking it than to waste his prized vintage on Kennedy.

This was a modern version of the fairy tale of the princess who, when asked to sleep on a mattress under which has been placed a pea, passes a restless and sleepless night, thereby proving her noble blood. Kennedy, as Alsop saw him, would have slept through the night. Although Alsop was probably correct that the president could not taste the subtle nuances, it is also possible that he thought the celebrated wine tasted like undrinkable swill but in deference to his host drank it down.

Although Kennedy had his own friends in journalism, Bobby played an instrumental role, monitoring the press, doing whatever he could to see to it that the Kennedys were portrayed in the most admirable of terms. For the most part this was not an onerous task. Many of Washington's premier journalists were eager collaborators, as much the seducers as the seduced. These reporters rationalized that they were advancing a desirable political agenda, but what they were advancing for the most part were their own careers.

The Kennedys liked to create a collegial relationship with those who wrote about them, so that the author and the subject seemed to be working together, like two artists painting the same portrait. After writing a draft of his classic book *The Making of the President 1960*, Theodore White sent the manuscript to Bobby that he said was "full of errors . . . but also full of affection and respect."

"Do keep this as a personal document for your eyes alone," the journalist wrote. White doubtless would not have sent the manuscript to Bobby if it had been less than brimming with what he called "affection and respect." Bobby, for his part, did not attempt to paint a thick coating of pastels over any harsh coloring. He was smart enough to write only narrow, fact-based criticisms, telling White that he "would be delighted to discuss" larger matters of emphasis on some occasion.

Many journalists thought of themselves as helpmates of the administration. Several of them, such as Ben Bradlee of *Newsweek* and Charles Bartlett, a syndicated columnist associated with the *Chattanooga Times*, were among the president's closest friends, and many of their competitors also nuzzled up to the administration as best they could, suggesting possible cabinet appointments, tempering unfavorable stories, offering unsolicited suggestions.

The president was adept at using his journalist friends to reward and punish his enemies. One of the few matters on which Jack and Bobby did not agree was in their attitude toward Paul Corbin, who had been such a controversial figure during the campaign. Corbin's mailing of fake anti-Catholic letters during the Wisconsin primary was only a tiny sampling of his deviousness. Bobby had a close friendship with Corbin that neither the president nor many of the attorney general's other friends understood. "Kennedy's notion about Corbin was that he belonged to Bobby," recalled Bob Healy, a *Boston Globe* reporter and a family intimate. "That was his favorite line."

Corbin looked upon the world as an interlocking series of conspiracies, his paranoia projected onto everything he saw. He was a man of intelligence and disarming candor, as well as a prodigious researcher, and he spun his tales out of a flax of truth, half-truth, conjecture, and dark fantasy. There was a dark side of Bobby too, and Corbin was in a sense the manifestation of that part of the attorney general's complex psyche.

"Corbin was abrasive and wild, but for Bobby he was a loyal lapdog," recalled Larry Newman, a White House Secret Service agent whose fiancée was one of the attorney general's secretaries. "Corbin could do things for Bobby that had deniability. But he was denied access [to] lots of things. Kenny [O'Donnell] blocked him. And he resented that, and he blamed Kenny."

Bob Healy, who had served in the air force in World War II in O'Donnell's bomber unit, was privy to his friend's thinking. "Kenny just didn't trust Corbin," Healy said. "He was always telling them [the Kennedys] to beware of Corbin."

O'Donnell fancied himself the administration's leading hard-nosed political operative, and Corbin's endless pushing represented a threat to him. Corbin's supposed closeness to Communist activists when he had been a CIO organizer in the early 1940s was enough of a problem that he had not been offered a position in the administration. Instead, he had been shuttled over to the Democratic National Committee, where he worked as a special assistant to Chairman John M. Bailey.

One afternoon in late August 1961, according to John Seigenthaler, Kennedy called his good friend Ben Bradlee. The president had just learned from O'Donnell that Corbin was not at his DNC office but hanging out at the pool at the Marriott Hotel in downtown Washington. Years later Bradlee admitted wistfully that "whereas I think Kennedy valued my friendship . . . he valued my journalism most when it carried his water." As Bradlee talked to the president, it became clear that this would be one of his water-carrying days. "I may have talked to him on the phone," Bradlee recalled. Both Seigenthaler's detailed recollections and the evidence of the *Newsweek* article largely confirm that Bradlee did indeed call Corbin at the hotel. Seigenthaler, moreover, believes that both the president and O'Donnell were probably listening in on the conversation.

Corbin had a daring disregard for all the small-scale dissembling of daily politics, and when the *Newsweek* editor asked him what he was doing, he rashly told him the truth. "Sitting by the swimming pool," Corbin said, "with a scotch in one hand and a blonde in the other."

Bradlee asked the Democratic operative what his future plans were, a not-unreasonable question, especially if the president was listening. "Stay here for sixteen years," Corbin said. "That's what I'm going to do. Eight years with Jack and eight years with Bobby. And if Jack doesn't do better, we'll run Bobby in '64." After quoting Corbin's devastating comments, the article in the September 4, 1961, issue of *Newsweek* concluded: "Jolly Paul Corbin sticks to his jokes, and his friends. But whether his friends can afford to stick with Jolly Paul is something else again."

Bobby was infuriated when he learned of Corbin's boasts. "Fire him!" he told Seigenthaler. "Get him out of there. I don't want him working over there tonight." Bobby calmed down and Corbin kept his job. Bobby had a deep visceral loyalty to his friends and aides that the president simply did not have; if there had been any doubt about that, the attorney general proved it by even now not disavowing his friend. As for Corbin, he guessed from which direction the knives had come and who had wielded them. He could not afford to make the president his enemy, but O'Donnell was a different matter. Corbin took his time, but he planned a revenge complicated and subtle enough to overcome perhaps even his formidable foe.

Bobby had another political matter on his mind, as did Jack. That was the political future of their youngest brother. After the election, Teddy had thought about moving west and starting a new life for himself there with Joan, one-year-old Kara Anne, and a second child due in September 1961. It was precisely the dream that the president had once held: heading out into the anonymity and space of the West, that richest of American metaphors for freedom. That road west was, if anything, even more appealing to Teddy than it had been to young Jack Kennedy.

Of all the Kennedy brothers, young Teddy had the greatest of possibilities for human happiness. He pursued pleasure exuberantly and laughed so deeply that cries of tragedy were obscured. Teddy had run the campaign in the West and could have been bitter about the region voting so dramatically against his brother's bid for the presidency. But he was not a man who held grudges, and he thought himself and the West a natural mix.

When Teddy went out to Wyoming for a week during the primaries, he and a local politician, Teno Roncalio, searched for Kennedy delegates from morning until evening. No matter how late they worked or how much they caroused, at six the next morning Teddy was already up, ready to head out for a vigorous hour of horseback riding. Roncalio, who preferred the bunk to the saddle, had the unhappy chore of galloping alongside his eastern visitor. On one of those rides Roncalio reflected on Jack's earlier visit. Roncalio could not imagine the senator from Massachusetts getting up to ride at dawn across landscape that he considered uninhabited for good reason. At the end of that visit, Roncalio drove the candidate to the airport in Casper. The wind was blowing a good thirty knots, but Roncalio had the top down. The candidate looked out on a landscape empty of everything but occasional sagebrush racing alongside the car. "Good Lord, why do you live here?" Kennedy exclaimed, a question he would not have asked if Wyoming voters had been within hearing distance.

Teddy knew why westerners lived where they did, and for a few intense hours he was ready to pack up and join them. The time of Conestoga wagons and homesteaders was over, but for Teddy the dream was the same. "I was there the night that it was decided to move Teddy's residence out west," recalled Evelyn Jones, the housekeeper. "And then all of a sudden in the same night the decision was changed." Teddy's father had convinced him to follow in his brother's footsteps.

For Joan it was a melancholic moment whose full import it would take her years to realize. "We wanted to move to Arizona," she recalled wistfully. "We thought we'd have fun and live our own lives, just the two of us, and

Kara and the baby on the way. Ted loved his family and his father but I think for him it was freedom from his father. Ted felt he was being pushed into public life. He could not do what he wanted to do. Nobody disobeyed grandfather."

On a day after the 1960 Thanksgiving holiday, twenty-eight-year-old Teddy went to see the president-elect to tell him of his plans to run for the Senate seat being held in trust by a family friend, Ben Smith, until the 1962 election. In doing so, Teddy would be making the most crucial move in his own life, setting off on a road that closed up behind him with each step forward. He asked his brother for a post in the new administration that would give him some stature before he returned to Massachusetts for the campaign. Teddy was thinking of something in foreign affairs. If the president had gone along, there would have been an Attorney General Kennedy on the domestic front who never practiced law, matched on the foreign front by another Kennedy brother whose overseas experience consisted largely of guarding NATO headquarters in Paris.

Teddy was not even thirty years old, the minimum age for the U.S. Senate. The president-elect did not intend to pander to his brother's ambitions, setting him up and guiding him until a real opponent toppled him. Teddy would have to get out there and get himself recognized. "Don't lose a day," he admonished. "Teddy, you ought to get out and get around. I'll understand, I'll hear whether you are really making a mark up there. I will tell you whether this is something that you ought to seriously consider."

That was not the president-elect speaking. That was the firm older brother who was not about to have his brother riding on his success. Teddy took his brother literally and within a few hours was off to Africa with two Democratic senators for five weeks on a junket they had been planning for months. As much as this impressed his brother, it was equally a lesson to Joan. As little as she had seen her husband during the campaign, now he was suddenly jetting off halfway around the world for a month, leaving her alone with their newborn daughter. That was a lesson reinforced a year and a half later, in September 1961, when Teddy slept through the birth of their second child, Edward Moore Jr.

While his brothers were running the country, Teddy returned to Boston to begin work as an assistant district attorney and give speeches all over the state, setting up his race. Teddy would forever after assert that he was the one who decided that he wanted to enter politics and run for the Senate. "Nobody forced me to run," he told his biographer Burton Hersh. "I wanted to." It was essential to his own sense of manhood that he think of the decision as his own.

Teddy had one small problem. This was a lack of credentials other than his name. This difficulty would only be exacerbated if the first important magazine profile of Teddy, a *Redbook* article, was published as it had been written. William Peters, the author of the article, had submitted his draft to Teddy for comments and vetting, and the putative candidate could not decide how strongly to attack the proposed piece.

Teddy wrote Bobby, enclosing a copy of the draft and telling his brother that the article portrayed him as a "wealthy personable lightweight." Teddy was indeed wealthy and personable, and many in Massachusetts thought that he was such a lightweight that if he were not tethered down he might float away. Teddy, in fact, not only was aware of his own failings but harbored an un-Kennedy-like insecurity that he displayed at times like a badge of honor.

Most politicians enjoy a moment or two of self-deprecating humor to establish themselves as modest fellows before they start trumpeting their supposed accomplishments in language that in any other field would be considered bragging. At a Temple Israel breakfast, Teddy jumped up to speak when he thought he heard the speaker say "the brother of the president." He sat down sooner than he intended, to laughter and applause, when he realized it was the "president of the brotherhood" who had been introduced. Teddy not only laughed that morning but added the story to his repertoire of tales. He was so likable on the public platform that if likability were the king of attributes he would have been carried to Washington on the shoulders of the adoring masses.

Teddy would never be particularly adept at the political skill of giving journalists the illusion of candor while subtly feeding them precisely the information he wanted them to have. In this instance, he had been honest with Peters, and as one of his advisers told him, "the article should be fair warning to us to handle similar interviews differently in the future." What was remarkable was how sensitive Teddy was to anything that tasted even mildly sour. Peters's article was a box of valentine chocolates, however, compared to the rotten fruit that would one day be heaved his way. As he got into politics, he picked up a shield of suspiciousness that would become second nature, and in difficult situations a studied inarticulateness.

Teddy was not secure in his own judgments, and he would begin his career in politics as he would one day probably end it, listening too much and too readily to those around him. In this instance, he was listening to two advisers who appeared to have more expertise about journalism than he did. One was his old college friend and Harvard teammate, John Culver. The

Iowan had played fullback, but off the field he was not one for plowing through the line but for finessing his way around end. He was all for shrugging it off, asking for only a couple of changes. Culver realized, as Teddy wrote Bobby, that there was "little reason for a 30-year-old Senate aspirant legitimately laying claim to the honeymoon glow that his brother as president currently enjoys."

Hal Clancy, a former Boston newspaper editor, represented another kind of analysis with which Teddy would grow familiar, an exaggerated, hysterical overreaction. Clancy played to the political paranoia that was always lying there just beneath Teddy's smooth veneer, the idea that there were people, most of them with smiling faces, who, if they got within reach, were ready to knife him between the ribs. Clancy felt that the article was "politically damaging in the extreme . . . the real danger is that at subsequent times with lazy newsmen this article with its general overtone of immaturity, intellectual weakness, and emotional sterility will be 'rehashed' repeatedly and the unfavorable image crystallized."

What was so striking about Teddy's anger was that everything he and his advisers objected to most strongly was the truth. Clancy fumed at Peters's observation that Teddy had traveled to Africa and Latin America to gather material for political speeches and so he might have "two continents to talk about." That was precisely what he had done, and it was no feat of investigative reporting to make that observation. Teddy, for his part, told his brother that he and his advisers felt that one of "the most politically undesirable references" was Peters's statement that "there are those in Massachusetts, as elsewhere, who are already grumbling about a Kennedy family dynasty." Even in these idyllic early months of the New Frontier, a reporter could walk down Boylston Street and hear a certain amount of grousing on that score, from Democrats and Republicans alike.

Teddy was worried about what Joan had said in the article, and that would become a theme of his life. Joan had what journalists came to admire as a priceless gift of candor but Teddy and his minions considered an endless predilection for political malaprops. She told the reporter that "the entire community" of Bronxville where she grew up "is so highly restricted that I actually never met a Jew as I was growing up." In her gushing love for the Kennedys, she extolled the Hyannis Port house that had everything, including "their own projection room for movies. If you want a steam bath, they have that, too."

Peters wrote that when Teddy was at the University of Virginia School of Law, the public schools and the university had been desegregated. "Yet Teddy, though he was there from 1956 to 1959, was unaware of either of these facts until told about them recently." Teddy wrote his brother: "I feel that the Virginia segration [*sic*] issues very possibly are factually erroneous and will check this out."

Teddy didn't seem to grasp that from now on journalists and others would be examining every aspect of his life. He would not be able to cordon off the parts that he wanted to remain private. Nor would he always be able to edit his words before they went public. There was much that he would not want publicized, from his cheating scandal at Harvard to the difficulties he had suffered over false charges that he had been a leftist security risk, from his compulsive philandering to all the wild episodes of his past. His Harvard friends loved to tell the story of the time they were all challenged to bring a woman to a party, and good old Teddy arrived with a prostitute in tow. It was rambunctious Teddy at his irrepressible best and worst all in one, but it was hardly the conduct of a calculating young man planning to become a U.S. senator.

Teddy had an instinct for candor, a trait that Bobby was teaching him was a mistake. Teddy had told Peters: "The vital thing is to be able to sell the people on the answers you arrive at, and I think I'm qualified to do that." To Bobby, the small matter of what his brother had told the writer mattered not at all. What mattered was what Bobby thought he should have said. Bobby insisted that his brother get the journalist to change the quote to this pretentious phrasing: "The vital thing is to enunciate clearly what you feel the issues are what you feel can be done in order to remedy the problems and the difficulties that we face."

Peters pruned his article the way the Kennedys suggested, cutting away unpalatable truths, glossing the image, creating a portrait that looked less like Teddy but more like a palatable candidate for the U.S. Senate. Bobby was teaching his brother to create a public character called Edward Moore Kennedy who had little to do with the playful, spontaneous, genuine Teddy loved by his friends for what he was, not for what he might be. Teddy was being taught to shove his oversized personality into the modest casing of a cautious, careful public man, a lesson that he studied with far more attention than he had given to many of his classes at Harvard.

"Saddle up, Joansie!" Teddy yelled over the phone to his wife. "We've got a two o'clock tea at Lowell, then another one at four. There's a banquet tonight in Boston and after that a coffee in Lawrence. We should be back at Squaw Island tonight. Did I tell you six are coming for lunch tomorrow? Could you get lobster?"

As she listened, Joan made sure she got it all down in her little notepad. "White piqué suit—black cocktail dress—8 lobsters." Then, when Teddy's urgent soliloquy ended, she enthused, "Wonderful, dear! I'll be up on the first plane." Joan appeared to be in ecstasy. "I'm going to see my *husband*," she told the cook. "I haven't seen him in six days."

Both Joan and Teddy were living lives that they had not yet quite claimed as their own. This was not England, and Teddy was not planning to sit in the House of Lords, an inherited position that gives its occupant little say in the governing of the nation. His résumé had only a few items on it other than his surname, but he was Edward M. Kennedy, a full-blooded Kennedy, and in Massachusetts he had every chance of winning a seat in the Senate.

Growing up, Teddy had been only a summertime Massachusetts resident, but in a short time he had managed to seem like a true son of the Commonwealth. Unlike Jack and Jackie, Teddy and Joan came across as an accessible couple whose happiness did not seem beyond the aspirations of middle-class life. There was a guileless vulnerability to Joan that set her apart from any other woman in the family. She sought to live a women's magazine kind of existence, her life as perfectly turned out to public view as her living room and her children.

The Senate aspirant and his wife purchased their own summer home on Squaw Island, only a five-minute car ride from the rest of the family at the Hyannis Port compound. Their house had a light, happy motif, the blue carpet like the sea itself, set against the white walls, the comfortable chairs, and the reproduction antiques. The children were perfect too, little Kara and Teddy Jr. As her youngest grew a little older, Joan liked to dress them in matching brother and sister outfits. Joan, unlike Jackie, did not wear European designer dresses but was content, like most of the women of Massachusetts, to buy her dresses off the rack. There was something reassuring about this youngest Kennedy couple, from Teddy's speeches and public demeanor to his beautiful devout wife and the modest way in which he set out to make his way in the Senate.

Teddy had a consuming love for life at Hyannis Port. He may have been only a holiday resident, but the hamlet would always be his one true home. This summer of 1961 would have been a splendid time even if his brother had not been president, for few things on earth compare to the pleasures of grandparents, siblings, and children surrounded by love and imbued with health, money, and above all the time to enjoy them. The sweetness of these weekends was incomparable, and as with all true pleasures, their sweetness was vividly felt by those who partook in them not only in memory but also within each moment.

When the helicopter set down on the grounds after the ten-minute jaunt from Otis Air Force Base, and the president stepped out and Caroline and her cousins rushed to greet him, there was perfection to the moment. The president went sailing on the *Marlin,* sitting there watching Jackie water-skiing. He even, against all advice, tried a few holes of golf. "You rich boys don't have to pick up the tees," quipped a family friend, Gerald Tremblay. The president said nothing, and it wasn't until much later that Tremblay realized that Kennedy had left the tees there because he could hardly bend over.

One summer Kennedy dove off the *Honey Fitz,* the presidential yacht, undaunted by the seven- or eight-foot distance from the water that would have discouraged even many younger men. On one windswept cold day he took Jackie and half a dozen others out on the bay for a sail in the *Victura.* The wind got so strong, gusting twenty-five to thirty knots, that he turned the tiller over to a visiting youth who knew little about sailboats but much about presidential wishes.

These summer days became the focal point of the greatest photographic story of the epoch. The Kennedys were a parade of exquisitely photogenic characters, from a maternal Jackie gently toweling off her naked young son on the cover of *Ladies' Home Journal* to handsome Jack on the cover of *Look* riding his golf cart surrounded by his laughing children, nephews, and nieces.

There were twenty Kennedy children, and most of them, at one time or another, clambered on for a ride on the president's gold electric cart, weighing it down as the vehicle lumbered ahead. No captions were needed to shout the message that this was a family graced with power, simplicity, and elegance. Kennedy was a man, as Alfred Kazin wrote in *The American Scholar,* who had it all and then some. Not only was he the president of the United States and a multimillionaire, but he was a man with "the naturalness of a newspaperman and as much savvy as a Harvard professor."

The president and his brothers all understood how much the bounty of their days depended on their family fortune. The Kennedys treated their wealth like an enduring miracle. There was no business enterprise that might easily have justified the immensity of their wealth, no morally edifying story of daring entrepreneurship, just a series of vague tales and anecdotes. Joe had brought back this wealth from an amoral, often brutal world, and it was best not to ask too many questions about it but to be richly appreciative that none of them would ever have to venture back to the world whence it came. On one occasion Kennedy asked Thomas J. Walsh, the accountant who helped manage the fortune, who was richer, he or young Teddy. Kennedy may have been president, but these were matters hidden from him, and Walsh told him he could not tell.

No one would have watched the president's endeavors with greater pleasure during the summer of 1961 than his father, but Joe was not there. He had rented a villa in the south of France and was gone that summer. "My father and mother wanted the children to have that house [in Hyannis Port]," Teddy recalled. "They didn't want to be constantly telling them, you know, 'Be quiet,' . . . I always felt, all of us felt, that it was because they wanted us to be there. They wanted the children to be there. I always interpreted that as being generous. My father never liked traveling. He liked that house. It was a great house, and he had a boat, and a cook, and his children. He was a very simple person in those areas."

When Joe returned to Hyannis Port in the fall, he assumed authority over the houses and the land. He was seventy-three years old, but there would be no gentle, sweet-tempered decline, no sitting in the sun-dappled afternoons on the porch telling oft-told tales to those who only half listened even the first time he told them. He put on his riding boots each morning and rode out as he always had, his back ramrod stiff, his clothes impeccable, and his grip firm. One day he fell off his horse, and another day he fell down in the house, but no one dared to comment, not to his face. He was not taking his heart pills the way he was supposed to either, but no one monitored Joe's conduct.

Joe continued to watch out for the other Kennedys. Luella Hennessey, the family's longtime nurse, came to him saying that she had been offered $50,000 to write a story of her life with the Kennedys. It was a fortune to the woman, and she bubbled over with excitement at the wonderful tale she would tell. "Why, look at that contract, Luella!" Joe told her. "If you don't have 375 pages for Doubleday and so many words that the printers are set up to do, you'll lose everything. I've seen so many people lose everything trying to write a book. If you need money when you're old, let me know. You're a nurse. Stay with that." Luella was happy she had such a true friend in Mr. Kennedy, and she turned down the book offer.

Joe was not much for outsiders, but that fall Frank Sinatra and Porfirio Rubirosa, the semicelebrated playboy, showed up for a party. The group of women who arrived that afternoon were rudely scrutinized by Frank Saunders, the chauffeur, who thought the guests "looked like whores." Joe insisted that his black riding boots be polished each day, and in the evening Saunders took the oiled, gleaming boots up the back stairs to the patriarch's room. There the chauffeur stumbled onto one of the afternoon arrivals pressed up against the wall being fondled by Joe. "I have your riding boots, Mr. Kennedy," Saunders said, as if that would make his presence less embarrassing. "My riding boots!" Joe exclaimed. "Just in time."

The best times were when the presidential flag flew over the compound. The big house was not an expansive place that insisted on formality, for despite its size it remained a cozy house whose rooms seemed diminutive. On one such occasion the president was sitting with his shoes off next to Jackie on the triple sofa. Bobby rested in a nearby chair, and another friend, John Hooker Jr., sat next to Ethel on the stairs. Hooker recalled that suddenly "everyone in the room sprang to their feet. The president of the United States, within a second, had his shoes on and was standing straight up to welcome his father into the room." At dinner it was Joe who sat at the head of the table, not his son the president.

Hooker was a decent athlete, and he was startled at the relentless competitiveness of the Kennedy men. He was a far better tennis player than Bobby, but the attorney general wouldn't let it go at that and kept insisting on set after set, as if he could eventually wear Hooker down.

Over the long Thanksgiving weekend in November 1961, the Kennedy family played touch football against the Secret Service. These men spent their days guarding the president, but they were decidedly off-duty this afternoon, and they blocked Bobby and Teddy and the rest of the Kennedy gang, banging into the defense as they ran out for passes. High up in the second-floor window stood Joe, his fist clenched, doing what he always did, cheering on his beloved sons.

Another day during that drizzly weekend Bobby, Teddy, Steve Smith, and Red Fay had themselves a bruising game of touch football, playing until it got so dark that they could no longer see the football spiraling toward them in the chill of a New England late November. "All right, everybody into that wonderful Atlantic!" Teddy yelled, looking at Fay as if he thought his brother's friend might turn tail. "Red boy, when that pink body hits that cold water, things are going to happen that you didn't believe could happen to a grown man," Teddy went on, anticipating the smack of the frigid Atlantic waters. "All right, everybody out into the black night bare-assed running as fast as you can." The four men tore off their clothes and ran naked into the bay. "How I survived the plunge only the good Lord knows," Fay reflected. "Literally for several days I felt the aftereffects."

That holiday the family celebrated both Bobby's and Sarge's birthdays. After dinner Joan sat down at the piano. This was not an evening for only sentimental old Irish ballads. The Twist had suddenly become the rage at the Peppermint Lounge in Manhattan. And so, while Joan played, Jackie gyrated before the group in her pink Schiaparelli slack suit. Teddy decided he would give the dance a go. His mother adored her last-born, but she noticed, "He is so big and has such a big derrière it is funny to see him throw himself around."

Jack was amused too as he sat smoking a small cigar, observing the proceeding as a spectacle staged for his special benefit. The only Kennedy who appeared out of sorts was the Kennedy who was in large part the creator of this evening. Joe sat quietly. He was not a man who complained, but he said that he had little taste in his mouth and felt "blah." Rose thought for the first time that her seventy-three-year-old husband looked old.

Joe's sons left as quickly as they had arrived, and the beach was soon cold and deserted. Joe flew down to Palm Beach, as he did each winter season. One day he was walking on the beach with Frank Waldrop, the former editor of the *Washington Times-Herald*. The two men's friendship went back three decades. "It's so incredible about Jack as president," Waldrop enthused. "You must just feel so great." The old editor had an endless curiosity about the human condition, and he was always probing people, but he was startled by Joe's answer. "I get awfully blue sometimes," Joe said, and the two men continued silently down the beach.

A conservative in style as well as politics, Joe liked things to stay as they had always been. Unlike many nouveaux riches, Joe and Rose did not feel the need to constantly redecorate. The Kennedy homes in Palm Beach and Hyannis Port had stayed much the way they were in the 1930s. Though some accused the family of being parsimonious, Joe and Rose viewed each sofa and chair as part of their lives.

Joe still played golf, teeing off at precisely eight o'clock every morning, not waiting even a minute for a late arrival. One day he went out to play golf with his daughter-in-law Ethel, taking along his oldest grandson, nine-year-old Joe II, as the caddy. He loved his grandchildren, but he wanted them to see that sports were not a respite from serious endeavors but a veritable graduate school of life. Joe still won at golf, and golf was the one sport he still played. Although as usual he beat the determinedly competitive Ethel, in the process he managed to lose two golf balls. Joe II could not find the balls, and for that the youth received one spectacular life lesson. "I had to stay outside for three hours after the game looking for the balls," Joe II recalled. "When I came home Grandpa didn't speak to me because I didn't find one of his balls."

The Kennedy patriarch treated Joe II's little brother David more gently. Joe was watching over his six-year-old grandson, reading him stories and trying to find something suitable on television for him to watch. Joe was supposed to go to a dinner with the president, who was in town only for the day, but he stayed for a long time with the boy, trying to see that he was not feeling too much alone.

All his life people had waited for Joe and had seen him off on his grand adventures, but now he was the one who waited and saw others off. In the morning he got up and set off to the airport to wave good-bye to his son who would be leaving on *Air Force One* after a one-day visit to Palm Beach. "I'm going to the airport with your father," Joe called out to four-year-old Caroline. "Would you like to come along?" Caroline hopped into the car and perched on her grandfather's knee. Many adults were apprehensive around Joe, nervous that they might set him off somehow or that he would freeze them with his cold stare. His grandchildren, though, saw a different man, one who had endless time for their games. He wasn't one for fawning baby talk but had a gentle concern for a child's separate world.

After the plane took off, Joe returned to the estate and played with the grandchildren for a while. He had been a more emotional parent than Rose, and he was a more emotional grandparent as well, full of an abiding, self-indulgent love, a blessing to children who would soon learn the burden and expectations of the Kennedy name.

After romping with his grandchildren, Joe headed out to play golf with Ann Gargan, his niece. Ann had been training to become a nun when she developed what was diagnosed as multiple sclerosis and left the order. She was better now and had become a companion to Joe, spending more time with him by far than Rose did. Ann drove him a few blocks to the Palm Beach Country Club.

"I really don't feel too well today," Joe said, "but it must be the cold I've had." He was hardly one to go inside because of a little sniffle. He was not a man for excuses, not for himself or for anyone else. Nor was he about to use one of the newfangled electric golf carts that sat there ready to be employed by the sedentary and slothful.

On the sixteenth hole Joe said he felt ill and sat down. The caddy hurried to bring over a golf cart, and Ann shepherded her uncle home as quickly as she could. "Don't call any doctors," he admonished his niece as she helped him walk into the house and upstairs to his bedroom.

Rose arrived home after mass and shopping to find her husband lying in bed. "He needs rest," she said after looking at him. Rose should have called for a doctor, but there had always been a gentle conspiracy between Rose and Joe. She and her husband had never mused morbidly about life but had insisted that their children get up, shake off their illnesses or injuries, and move on. They had always been the best example of that, and it was hardly the time to turn timid.

By midafternoon, though, Rose agreed to call a doctor. The doctor scarcely had to look at Joe before he called an ambulance to carry him to Good Samaritan Hospital. As the ambulance sped across to West Palm

Beach with sirens wailing, Rose headed out for her daily afternoon round of golf. "He'll be fine, you'll see," Rose told the chauffeur. "I just cannot let this get me down. I must keep up my schedule. I have my routine. I'm going to play golf now, Frank. Yes, I will play."

After Rose played golf, she returned to the house for her afternoon swim. She had faith in God and faith in routine, and she did not let this incident shake either belief. By the time the president and attorney general arrived that evening, they knew that Joe had had a serious stroke and the prognosis was not good.

Joe's son-in-law Steve Smith believed that Joe had "nothing more to live for; his sons now being in power and having no further need for him, so the stroke came as a kind of solution." Death might have been a solution, but surely not this, a proud, intractable man imprisoned in a broken body, unable to speak much more than gibberish, his intense blue eyes now his engine of communication, his arms striking out at those who offended him where once a word or two would have been sting enough. His right leg, right arm, and the right side of his face were paralyzed, and though his mind was lucid, try as he would, he could not speak coherently.

When Joe was brought back to the house in a wheelchair, everyone pretended that life was the same. His sons wanted the best for their father, but they could not make him walk or bring his speech back. Rose tried to be there for her husband within the strictures of her routine, but when she entered his room, he screamed "No!" and with his good left arm dismissed her with a gesture. For years they had maintained a relationship of mannered civility. Rose turned and walked away, keeping the distance from her husband in sickness that she had kept in health.

Joe wanted his grandchildren to be with him, but they had not learned to pretend, and some of them were afraid of this strange, gurgling, twisted old man. When he reached out to touch them, they fled from his embrace. He sat in the wheelchair crying.

Joe had all these women around him now, pushing his wheelchair, soliciting his tortured phrases, whispering to him, talking in low voices about him in the distant corridors. Luella Hennessey, the nurse who had gone with the Kennedys to London so many years before and had overseen the births of the grandchildren, arrived to help oversee his care. She took the daily calls from the president and held the receiver up to Joe's ear so that he could make his incoherent grunts into the phone. Another nurse, Rita Dallas, had been brought in to manage the other nurses and a major part of the burden of his care.

Ann Gargan, though, was in many respects the most important person in Joe's daily life. Ann and her brother Joe were the proverbial poor relatives, welcomed at the family table but relegated to the most distant seat, always at the beck and call of their betters. Ann sought to make herself indispensable, cutting Joe's meat up into bite-size pieces, interpreting his wishes, thwarting the harsh regime of rehabilitation that the physical therapists sought to impose on her uncle.

Joe was flown up to the Institute of Rehabilitation Medicine in New York City where he settled into Horizon House, a bungalow on the institute grounds. His children came, one after another, and whispered in his ear. One of them covered the old man's twisted hand with a scarf. Bobby and Ethel hurried around the little house, checking everything out, pronouncing, "Everything works, Dad." His children attempted to bolster him with their relentlessly upbeat cheerleading. They stood outside his door tense with anticipation, steeling themselves for this moment. Then they rushed into the room bursting with optimism, before rushing out again, their shoulders slumped with despair.

"For a group of people who don't want to face bad things, it was hard," reflected Dr. Henry Betts, who became Joe's full-time physician. "It was harder than death because with death they always went on. They'd go home, by God, and they'd find diversion and fun and work and they'd go on, because they had repressed it in a way. But how could you repress this? It was around all the time. It was harder than death. So it was just terrible on all of them."

Joe had grown to love Jackie more than any of his other daughters-in-law and had an affinity with her as deep as with any of his daughters. She did not try by encouraging words to pull him to his feet, but he sat there with her head in his lap. She kissed his twisted hand and caressed his face, and he seemed to grow tranquil.

Joe learned that the president was arriving later that day, and he motioned that he wanted to wear his finest suit and a beautiful tie. When he was elegantly dressed, an aide pushed his wheelchair outside on the patio to await his son. Joe could not stand up on his own, but when he saw his son striding toward him, he pulled himself to his feet, and with his crippled hand gave the president the same salute he had given him on inauguration day. Kennedy had struggled with his father's will and power all his life, but he loved him with an unabashed emotion unlike anyone else. He rushed forward and embraced the old man, kissing his face.

That day Kennedy could well believe that his father might learn to walk and talk again and return to being the great patriarch of the family he had always been. When he and Bobby returned for another visit, the old man once again started to rise out of his wheelchair. A doctor hurried forward to steady him, but Joe pushed him away. Bobby rushed forward, and Joe struck out at him with his cane, screaming and flailing at him. The doctor pressed Joe down in his chair, while the Secret Service men stood ready to protect the president if his father struck out against him. Bobby loved his father deeply too, and he kissed his father and talked in sweetly soothing tones, calming the old man.

Joe had always been a problem solver, and this was the most maddening of conditions. He followed the therapy, made some progress, and then relapsed. That was the way it always was in rehabilitation, but he had been in such good physical condition prior to the stroke that his doctors believed he might yet walk far up the road to recovery.

Joe had always managed to avoid domesticity's fatal embrace, but in this dollhouse of a bungalow, he and Rose were living together as they had not done during most of their marriage. Rose ordered meals from Joe's favorite restaurant, La Caravelle, and in the evening the couple sat together in front of the television, proper Rose in her stocking feet, watching the same programs that married couples all across America were watching. Ann Gargan had become a pushy, discordant presence, and the doctors had told the family that it would be best if family members other than Rose came only for short visits. Their mandate was mainly directed at Ann, who packed up and left. Her departure had added to the tranquillity as well. It was a reassertion of Joe and Rose's marriage and their life together. Joe was no longer prone to his terrifying fits, striking out against not only anyone within reach but seemingly at life itself and its terrible ironies.

"He was very content with her [Rose]," said Dr. Betts, the specialist in charge of Joe's daily care. "He liked having her around. That's my observation. His children told me that he did not like having her around—she made him nervous—before his stroke, she was always upset about something else and she wanted to change something. . . . Well, she didn't make him nervous when I saw him. I mean, he loved having her around."

On Father's Day in June 1962, Joe's children called in the morning as if by doing so they had fulfilled their familial obligations. Then, in the afternoon, they surprised their father by descending en masse on the cottage. The president, Bobby, Teddy, Jean, and Pat stood before their parents and put on a skit about growing up Kennedy. These were jokes for them alone, and Joe and Rose laughed until they were sore. And when the play was over, the players stood and applauded their parents, saluting their father and his life. There were gifts and food and endless reminiscences, and when the evening was over, Joe's children had the grace not to make dramatic good-byes but slipped out into the night and stood waiting for their siblings to join them.

The sweetness of this evening hung over the bungalow until Ann called from Detroit. When Joe set the phone down, he was full of the fearsome rage that Rose had hoped was gone for good. Rose did not know what had gone wrong and had no idea that Ann was beseeching Joe to let her back inside the Kennedy fold, crying over the phone. Again and again Ann called, day after day, imploring him, begging to be asked back.

Rose was used to being alone, and having to watch over Joe unsettled her established daily routine. When it came time for her to go to Hyannis Port for the summer, as she had for the past three decades, Rose called Dr. Betts to ask his advice. "I could tell Ann to stay in Detroit and I could stay here, or I could go to the Cape and let Ann come back," Rose said. "Now what should I do?"

The Kennedys readily solicited advice from experts, often searching until they found someone to tell them what they wanted to hear. The doctor had been around the Kennedys long enough, and he was not about to be drawn too far into this family affair. "I don't know, I don't think it's for me to tell you what to do," he said, "but I can tell you that your husband is most happy when you're there."

The doctor told Rose as clearly as he could that if she cared about Joe's happiness, she should be with her husband, but Rose was not satisfied with that answer. She called back several times, and each time he told her the same thing.

Rose left the institute for her summer at Hyannis Port, and soon afterward Ann returned. "Uncle Joe, the family has put me in charge of you," she told him as he sat in his wheelchair, "and you'll have to do as I say from now on." The old man had what his nurse thought was "a wild look in his eyes." Then his shoulders slumped. The prison of his marriage had become his only freedom, and even that was gone now.

After his holiday Dr. Betts returned to Horizon House, and he was startled to see that Ann was in charge now and Rose was gone. "I think she [Rose] made her decision, and she couldn't cope, she could not cope with a bad situation," Dr. Betts reflected. "And from that moment on Ann Gargan was in charge. Ann did not make him happy. She was devoted to him, but she was much more manipulative. He loved her and hated her. She stirred him up, and I personally think he would have been much happier to just settle down and be content with his wife for all those years."

Joe traveled to Hyannis Port that summer with Ann, and though Rose was in the same house, it was his niece who was never away from his side. He was supposed to return to the institute for further rehabilitation so that he might learn to walk and talk again, but he never did. Ann had him dressed sometimes in a suit and tie, and on occasion he traveled to New York and sat behind his desk. He even flew once to Chicago.

Time had always been Joe's friend, time and ritual and routine, his moments perfectly orchestrated. Now he sought the same control over his days, checking his watch to see that every minute of his diminished day was set as firmly as it had been when all was well. He turned away any therapist or guest who arrived even a minute late, and a family member who dared show up after the appointed hour was met with immense displeasure.

When the president came for the weekend, Joe would have his wheelchair moved out on the porch where he had a clear view of the helicopter carrying his son. He was always there too when the president departed and the helicopter lifted across the field where his sons had all once played football. He would sit and watch until the helicopter disappeared in the sky, and when they pushed him back into the house, sometimes the old man would be crying.

26

Dangerous Games

Outside of the president, there was no one in government as important to Bobby as J. Edgar Hoover, the sixty-six-year-old director of the FBI. Although the FBI was an empire of its own, Bobby was technically Hoover's boss, and the two men had to work together on a multitude of serious matters ranging from civil rights to organized crime. For years Bobby had been writing political valentines to the imperious Hoover, congratulating him in language that even Caesar might have found exaggerated. It was impossible, however, to embarrass Hoover by celebrating his greatness too effusively, and by dint of Bobby's correspondence and his father's equally fulsome tributes to the director, it might appear that the new attorney general and the FBI chief would work well in tandem.

The two men were so different in their approaches to government, however, that from the day Bobby stepped into office, much of what he did rankled Hoover. Although the FBI perpetuated an image of their leader as an intrepid G-man, Hoover was in essence the most brilliant Washington bureaucrat of the twentieth century, a career civil servant who could have been overseeing widgets as easily as agents.

There was precision in everything Hoover did, from the way he managed his Caucasian, conservative, white-shirted, clean-shaven agents to the niceties of his communication with the White House. He held a treasure trove of information on policymakers and presidents that had kept all presidents wary, helping to ensure his continued tenure. Most of what the agency did was documented in precise memos that ended up on the chief's desk, the final product of a machine whose ultimate output was not the capture of criminals but the endless perpetuation of Hoover's power.

The FBI director was consumed with fighting the American Communist party. He could not see that the party had been reduced to a few bedraggled militants surrounded by the FBI agents who had thoroughly infiltrated the organization. He refused to understand that the most serious internal danger the nation faced was organized crime. That said, the director had created a disciplined, professional organization that, for the most part, stayed within the broad parameters of the law and often rendered signal service that its opponents rarely acknowledged.

Bobby was an antibureaucrat who proudly flouted all the elaborate strictures of government, considering a day not won unless he had broken out of one tedious regimen or another or surprised some slumbering official with his unannounced appearance. Night or day, few Washington officials would have dared to arrive at their office without wearing a coat and tie. On weekends and evenings, Bobby was damned if he was going to dress like an ambitious bureaucrat, and he wandered the halls in T-shirts and sweaters. In meetings he flung his feet up on desks and brought his enormous dog, Brumus, into work, though the nation's top law enforcement officer could have been fined fifty dollars or sentenced to thirty days in jail for so flagrantly violating government rules.

Bobby had scarcely been in office for a week when he showed up one evening around eight o'clock to visit the FBI's printing unit, shaking hands with night-shift employees. That same evening he wanted to enter the FBI's gym but was told the door was locked. So began a flurry of memos back and forth within the FBI about the unpredictable attorney general and his mysterious visits. In subsequent visits to the gym, the FBI monitored Bobby's every move, even noting in a memo the number of push-ups he performed—ten to fifteen—during one half-hour session in the facility.

Hoover had created an agency in which no life or event went unmonitored. It was a trivialization of government, but the system he had created allowed Hoover to chronicle Bobby's actions and record his utterances. The attorney general would never be able to escape from any questionable actions he took, or from any of the accolades he gave the director.

As he entered office, Bobby shared Hoover's fixation with American communism, giving speeches and interviews that could as easily have been spoken by Hoover or Joe McCarthy. In March, Bobby went on the radio program of Senator Kenneth Keating, the moderate New York Republican, and said that the Communist Party remained as great a threat as during the Red Scare of the past decade because it "is controlled and to a large degree financially supported by a foreign government." As much as many American liberals ignored the painful truth, evidence in recent years from Soviet archives shows that Bobby was correct in his second assertion: the party had been largely controlled and financed by Moscow. However, by 1961 the Communist Party was decimated and probably controlled and financed more by the FBI, whose agents and informers may have made up about half of the minuscule party.

By the fall of his first year as attorney general, Bobby had come to the realization that the Communist Party was not a nefarious secret army of potential traitors with the power and will to subvert America. "They are not important numerically and present no grave menace to our security," Bobby said in November in Los Angeles. Even as he said so, he made sure that he kept stroking Hoover. After his return from his West Coast trip, he called Hoover. The FBI director noted "that in regard to the trip which he recently made, he was impressed with my people in all of the cities where he traveled."

Bobby believed that there was a vast underground army of betrayers within America more threatening than Communists. As attorney general, he set out to attack organized crime with a resolve and weaponry far beyond what had been brought to previous attempts. He had been obsessed with the mob since the days of his work on the McClellan Committee, and now he believed he had the power to end what he had begun. He quadrupled the number of attorneys in the organized crime division and sent them across the country attacking the gangsters where they nested, protected by corrupt local officials or a see-no-evil police atmosphere. The IRS targeted Mafia figures for audits, going after them with a merciless concern for detail that at times assaulted their civil liberties.

Many superbly dedicated FBI agents pursued this new enemy just as relentlessly, shadowing targeted individuals, wiretapping them, monitoring their businesses and homes, and tailing them wherever they went. Hoover, however, had been slow in turning them toward this implacable foe.

"Hoover was a miserable son of a bitch, but he was tough," said William Hundley, chief of the Justice Department's Organized Crime Section. "He was tough and shrewd and powerful. And you know, we got some movement from the FBI, but not what we would have liked. They hated us. And it wasn't right. At that time, you know, they still had about half of their personnel in the FBI working on the domestic Communist Party. There wasn't anything left except informants. And we wanted to move those guys into organized crime. And a lot of the individual agents, they wanted to do it. But it was a terrible struggle. I don't think that Hoover had ever had an attorney general that didn't kiss his ass. And Bobby didn't kiss his ass."

As Bobby read the files and listened to the various agents, he learned of America's secret history. He came across names of entertainers he had admired, politicians with whom he had worked, and celebrated business leaders. In May, Bobby made a trip to the FBI offices in Chicago. The attorney general was a vociferous reader of FBI files, and by now he knew that determined special agents in Chicago were dedicated to going after the syndicate, tailing Sam Giancana, the local leader, and bugging his hangout at the Armory Lounge.

As soon as Bobby was introduced to the assembled agents, Marlin Johnson, the special agent in charge, began his prepared statement. Johnson spoke the language not of a gritty, hands-on cop but of a ward politician, praising his visitor and then touting his own accomplishments. "Mr. Johnson," Bobby interrupted, "I didn't come here to hear a canned speech about how magnificent you are. . . . We'll listen to the agents who are out on the street, the men who are doing the work you think is so great."

Later in the day William F. Roemer Jr., one of those agents, played a tape for the attorney general from a bug that had been placed in the Democratic headquarters in Chicago's first ward. As Bobby sat there, he heard Pat Marcy, a Democratic organizer whom he had met during the campaign, talking to two cops about one of their colleagues who refused bribes. The three men decided that they had no choice but to murder the man. Bobby listened to the tape, asked Roemer to play it again, and said nothing more.

Bobby could have left that day and subtly scuttled an investigation that might one day touch on associates of the Kennedys. It was even possible that some of the accused would call out his father's name. That was not just hypothetical, for according to Cartha DeLoach, the deputy director, in 1961 the FBI bugs heard Giancana talking of the $25,000 he had contributed to the Kennedy campaign.

Many Democrats would have cried foul at the bugging of Democratic Party headquarters, believing that the FBI was eavesdropping on a citadel of democracy in its pursuit of the corrupt. Bobby did not do that either, but by word and deed he flailed the agents onward in their pursuit of the Mafia underground. Bobby had little use for the finer nuances of civil liberties. He did not always grasp that in defending the rights of those who least deserved them, one defended the rights of all. He had not flinched at hearing the results of the bugging of the Democratic offices in Chicago. It was a war out there, and Bobby wanted his brave soldiers to have the weapons that would let them win.

Bobby's defenders suggest that he did not know that the bug had been placed there by the FBI but was told the Chicago police had put it there. That was hardly much of a distinction. If Kennedy had been opposed to breaking into houses and offices to place listening devices, he presumably would have been equally against any other law enforcement organization doing it and feeding misbegotten information to the FBI. It was perhaps more dangerous for local police to be placing bugs than the FBI, which at least would exercise some controls.

"In one particular instance, in a field office, one of his people asked him when Bobby was personally listening to the conversation on the tapes brought on by microphone, 'Isn't this illegal?' " recalled DeLoach. "And he said, 'Yes, everything about it is illegal.' And then he demanded that we get better equipment. Later on he denied he knew anything about the usage of microphones. But he did and he encouraged that. It was explained to him on a number of occasions. And the material was brought over to him on a constant basis from microphones. He thoroughly understood that."

Early on in his term, the new attorney general asked that fifty Treasury officials each year take a course in wiretapping and bugging, rather than the fifteen who were taking it in the Eisenhower years. The officer in charge said that it would be impossible to more than triple the number. "You look old to me," Kennedy replied, dismissing the man. "You should think of retiring." The man soon left Treasury, and a new, younger official took over to oversee the vastly increased load.

The FBI chief provided raw files to Bobby that were compilations of wiretaps, bugs, and overheard conversations, as well as informants' tales of every stripe and degree of reliability. Hoover also sent his surrogates over to the Justice Department to inform the attorney general of the latest allegations and rumors involving the Kennedys, including a tale that the president had had a group of women with him on the twelfth floor of the LaSalle Hotel while the Secret Service surrounded the building. It was, as Bobby knew, "ridiculous on the face of it," but it was a cunning move on Hoover's part to pass on even the most questionable of tales, proving his loyalty and indispensability, while reminding the Kennedys that if he were a teller of tales, what tales he could tell.

Bobby might shrug off silly stories about the president that Hoover passed on, but he was judicious enough not to attack Hoover's tales that had the hard ring of truth. At the end of January 1961, when Hoover received a cable from Rome, he made a note to send a memo to the attorney general. The cable dealt with a story in the weekly magazine *Le Ore* in which Alicia Darr Purdom talked about her supposed affair with Kennedy in 1951. Darr Purdom said that she would have been first lady except that she was "a Polish-Jewish refugee." FBI records stated that in the early 1950s Darr Purdom had been a prostitute, "a notorious, albeit high-class, 'hustler.' " This was hardly the type of woman Kennedy would have contemplated marrying, but it was not the kind of story the administration wanted made public either.

Hoover understood the bureaucratic imperative of having his superiors sign on to any measure likely to prove controversial. He was not about to wiretap someone without getting Bobby's approval, and during his term in office the attorney general reportedly approved more than six hundred wiretaps. The FBI also placed nearly eight hundred bugs that picked up the words of the innocent as well as the suspected.

To place a bug, an agent usually had to break into the premises to plant a secret microphone. The transcripts could not be used in court, and the fruits of most of this surveillance not only rotted on the ground but also kept the FBI and other agencies from doing more legitimate police work. Hoover, however, could produce a piece of paper that Bobby signed on August 17, 1961, authorizing microphone surveillance, or "bugging."

Bobby's defenders claim that he may not have understood the distinction between the two forms of electronic surveillance, and that "perhaps . . . he did not want to know." Willful ignorance is the most pathetic of excuses, especially in a man of Bobby's abilities, and it is unlikely that it was true. Again and again he read transcripts or listened to conversations that could only have come from hidden microphones. He even proposed a wiretapping bill that would have stripped Americans of part of their civil liberties. In the White House, Mike Feldman oversaw proposed legislation usually by passing it on, but when he saw this bill, he was so upset that he went to the president, and the legislation never got out of the White House.

Bobby was so militant in his war against organized crime in part because the world that he was seeing was one in which the underground and the highest reaches of society at times appeared seamlessly integrated. John Mataasa, a former Chicago cop, often drove Giancana around and served as his bodyguard. He also boasted that he chauffeured Sinatra when he was in the Windy City, and that he had a letter of recommendation from William Randolph Hearst recommending his services to Otto Kerner, the governor.

Even before Bobby entered office, he knew that Sinatra had long ago soiled his name with his mob connections, and that it was his brother's weakness to be associated with such a man. At the end of the inaugural gala, when Kennedy praised Sinatra for putting together the extravaganza, Bobby turned to Red Fay and said: "I hope Sinatra will live up to the public position the president has given him by such recognition."

In recent years Sinatra had pushed his way to the head gangsters' table and fancied himself a kind of ersatz Hollywood don, with the power to call on his dangerous friends whenever he needed them. In April 1961, Hoover talked to Bobby about Giancana, telling him that one of Giancana's lieutenants, Joe Pignatello, was trying to get a lucrative liquor and gaming license in Las Vegas, fronting for the Chicago mob. Nothing happened in Las Vegas without connections, and Sinatra had gotten involved, speaking in favor of his friend. Bobby sent a clear message that Sinatra's name would not buy a free pass in his Justice Department. Again, Bobby could simply have made a few abstract remarks about justice, freedom, and the flag and let things fall as they were going to fall. But on that day he made it indisputably clear that he not only wanted the FBI to try to prevent the granting of the license, but wanted everyone to know that he, Robert F. Kennedy, was personally concerned. "The attorney general indicated we should be sure to indicate that we were speaking on his behalf and explain that he is quite concerned about it," Hoover noted.

As the FBI increased its surveillance of the mob in Chicago, Giancana and his henchmen expressed their outrage. Their fury would have been even greater had they known their tirades were being recorded. Early in December 1961, as the tape recorder whirled, one of Giancana's associates, Johnny Formosa, talked to his boss. Formosa had just returned from a visit to Sinatra's Palm Springs home. During his stay there, he said, Joe Kennedy had called the singer three times. Formosa had been trying to learn why the mob had been unable to cash in its chit with the Kennedys. Formosa recounted that the singer had explained that he had done everything he could. "I wrote Sam's name down and I took it to Bobby," Sinatra told Formosa. When that didn't work, Sinatra said, he had talked to Kennedy's father to see whether he could work his will on what the mob considered his double-crossing son.

"Well, one minute he says he tells me this and then he tells me that and then the last time I talked to him at the hotel down in Florida a month before he left, and he said, 'Don't worry about it, if I can't talk to the old man, I'm gonna talk to the man,' " a peeved Giancana replied. "One minute he says he's talked to Robert, and the next minute he says he hasn't talked to him. So he never did talk to him. It's a lot of shit."

Giancana was probably correct in his assessment. It seems unlikely that Sinatra would have been capable of persuading Bobby to back off investigations of Giancana, or if he would have dared to make such a request face to face with the new attorney general. He may, however, have talked to Joe, who might have been more amenable, though hardly likely to have asked Bobby specifically to avoid prosecuting Giancana. It is probable that the vain, boastful Sinatra promised something from his Kennedy connections that neither he nor anyone else could deliver. Although Giancana may simply have been boasting, it seems likely that he did make various kinds of contributions to the campaign and now raged at both the Kennedys and Sinatra.

Bobby was apparently immediately made aware of Giancana's allegations. On the bottom of the teletype from the Chicago bureau, Hoover had scrawled "& promptly" after his aide's notation: "Memo to AG Being Prepared."

"As the Bureau is aware, considerable information has been received . . . which reflects a serious rift between Giancana and Frank Sinatra," the Chicago field office wired Hoover on January 18, 1962, "which stems primarily from Sinatra's inability or lack of desire to intercede with Attorney General Robert Kennedy on behalf of Giancana."

The FBI agents spread their great nets wherever the mob hierarchy was likely to venture, periodically pulling in all sorts of unseen creatures from the dark depths of American life. In Los Angeles, the FBI targeted John Rosselli as a second-tier figure, running wiretaps and bugs, interviewing his friends and associates, and observing his daily activities. The FBI had no idea that Rosselli had become the CIA's agent in attempting to assassinate Castro, and they tried to explore every shadowy corner of his life. In their investigation they came upon the name of Judith Campbell (Exner), who led them to places that they had not expected to go.

FBI special agents noted in September 1961 that Rosselli was calling Exner when he came to Los Angeles, and he was later observed escorting her to Romanoff's restaurant in Beverly Hills. Her telephone records showed that she was telephoning Giancana in Chicago. She drove a 1961 Ford Thunderbird that had been driven from Chicago to Las Vegas by the Mafia chieftain's assistant, Joe Pignatello. Everything the FBI learned about Exner suggested that she had no income or substantial bank accounts, yet she had rented a fancy home in Palm Springs and a place in Malibu. Most surprisingly, Exner's telephone records showed a number of calls to the desk of Evelyn Lincoln just outside the Oval Office.

By the end of February 1962, Hoover had all this information sitting on his desk. The FBI chief had a brilliantly astute awareness of the wages of power. Two decades before, he did not directly confront President Roosevelt with information making criminal accusations against his son James and Joseph P. Kennedy. He had written a memo so that he would have a legal record of how he had handled this matter and sent it to Roosevelt's chief of staff by courier. This time he once again wrote a memo, addressed to the president's assistant, in this case Kenny O'Donnell, and sent it by courier to the White House, as well as a second memo to Bobby. Once again Hoover had proof that the information had been received, but the president could always deny that he had seen the memo.

Hoover knew all about Kennedy's sexual predilections and had previously passed on part of his knowledge to the attorney general. Hoover's memo was so bland, however, that reading the words a thousand times would not reveal whether irony, moralizing, or even veiled threats of exposure lay behind them. "The relationship between Campbell [Exner] and Mrs. Lincoln or the purpose of these calls is not known," Hoover wrote.

When Joe Dolan, serving as Bobby's acting deputy for an ailing Burke Marshall, walked into his office, Bobby shoved a folder of documents toward him. "What do you think of this?" Bobby asked, his even tone suggesting nothing of the potential importance of the memo. In his office Dolan carefully read documents that included the list of Exner's telephone calls and other information, then returned to the attorney general's office.

Dolan knew full well how extraordinary it was that Bobby was even showing him this material. Bobby was secretive about family matters, walling off this world from those who served him and his family. Bobby looked up, and Dolan spoke with the wry wit that was his trademark: "Mrs. Lincoln shouldn't take calls like that."

"So what do you think?" Bobby asked. In a normal world Bobby would have gone over to the White House and spoken confidentially to his brother, but Dolan realized that was not what the attorney general wanted to do. "I think I'll write Mrs. Lincoln a little memo," Dolan said, having astutely grasped what Bobby wanted. "Do it today," Bobby said.

Dolan wrote a memo outlining what Hoover had discovered. Sensing the importance of the matter, he hand-delivered the memo to Evelyn Lincoln in the White House. "Joe, I'm shocked," Lincoln said, recalling the famous line in *Casablanca* when Captain Renault claims to have learned about gambling in Rick's Café.

Sinatra had introduced Exner to the president. Bobby had all kinds of memos about Sinatra's contacts with organized crime, and it was hardly wise that his brother intended to spend a weekend at Sinatra's Palm Springs house. Kennedy, however, did not think of Sinatra as a political operative, but as a fellow sexual swordsman and bon vivant who offered him beautiful women and good times. Bobby realized that his brother would have to seek his pleasures elsewhere. The onerous task fell to him of calling Sinatra and telling him that in late March 1962 the president would be staying elsewhere.

Sinatra, like the president, was a man who thought he could have it all, a friendly greeting at the White House and a seat at the head table of the mob, accolades for his noble liberal politics and a personal life of excess. He had already built a helicopter pad for the president and lined up a weekend of assorted entertainments. He raged at Bobby, screaming at him into the phone, infuriated as much over the embarrassment as the insult itself. After hanging up, he went out to the helicopter pad and broke the concrete slab into pieces with a sledgehammer.

Just before Kennedy set off for his West Coast trip, he had lunch with Hoover. Neither man left a record of what was said that day. Although Exner claimed that Kennedy saw her later, by all documentary evidence he did not. He seems to have ended the relationship the way it began, as just another of his occasional trysts to be dismissed and forgotten.

In Palm Springs, Kennedy may not have had Sinatra to gather a bouquet of Hollywood's rosebuds for his pleasure, but he had an even sweeter treat in store. Marilyn Monroe arrived to spend the evening with the president. The blonde actress was the benchmark of American sexual fantasy. Even the president was not immune to the dream of sleeping with her, and for months he had been pestering his brother-in-law Peter Lawford to set up an assignation.

Monroe appears to have viewed Kennedy not only as a political star in a firmament far above Hollywood but as an epic hero. In talking to her analyst, she spoke about him as a man who walked in the shoes of Jefferson and Lincoln. "This man is going to change our country. No child will go hungry, no person will sleep in the street and get his meals from garbage cans." As for Kennedy, he may have suffered from the common male failure of taking almost as much pleasure in having his male cohorts know about his conquest as in the act itself. "Well, she loved him, and she was a beautiful girl," Smathers reflected. "He took her down the Potomac on the presidential yacht two times. And God she loved Jack. After he was president. She was all over him, of course, he liked her very much. But he was already married and everything. So he had to be reasonably discreet. But he'd take a lot of guys, take a lot of his old buddies out when he had someone like Marilyn, and they'd all be around, five of us, or a bunch of us, where he could blame it on any one of his friends."

On August 3, 1962, Bobby and his family flew to San Francisco for the start of a summer vacation. He was not one for sedate excursions, and he spent the weekend at the ranch of John Bates in northern California, driving back into the city Sunday evening to stay at Red Fay's home.

Early in the morning on Sunday, August 5, Sergeant Jack Clemmons was led into Monroe's bedroom in her home on Helena Drive. The actress's nude body lay under a sheet. Beside her corpse stood her psychoanalyst, Dr. Ralph Greenson, and Dr. Hyman Engelberg, her physician. An autopsy revealed that the body of the troubled actress was full of Nembutal.

Two years later Frank A. Capell, a right-wing journalist, published a book in which he alleged that Bobby was having an affair with the actress, who "believe[d] his intentions were serious." To cover up her murder, Bobby had used "the Communist Conspiracy which is expert in the scientific elimination of its enemies" by employing Dr. Engelberg, who was supposedly a Communist. In the years since, other journalists have expanded on this scenario, often more in the name of commerce than ideology, suggesting that the attorney general had secretly flown down to Los Angeles to play his own role in the nefarious act.

Bobby's host and others at the ranch that weekend assert fervently that Bobby never left the isolated premises. Beyond that, his whereabouts were so carefully chronicled by the FBI that not even he could have gotten in and out of Los Angeles without being seen.

Bobby, like his brothers, may well have had his own sexual encounters outside of marriage, and it is possible that Marilyn Monroe was one of those. Bobby had met her several times at parties given by his sister and brother-in-law, Pat and Peter Lawford, in Santa Monica. On one of these occasions, Ed Guthman recalled, the attorney general commandeered his press secretary to accompany him while he shepherded a drunken Marilyn home. On another occasion, Bobby may have unsuccessfully attempted to make a pass at the movie star. "I'm dancing with Marilyn," Joe Naar, Peter Lawford's friend, recalled, "and Bobby wants to dance with her, and makes some lewd remark, and she turns to me and makes a sign as if she's throwing up. He kept hitting on her all the time."

Bobby appeared obsessed with Monroe. "Eunice kidded Bobby one night at dinner," Charley Bartlett recalled. "She got up and said something about Marilyn Monroe. And Bobby got red in the face and said, 'If you ever say that again, I'll hit you.' "

Monroe had been on the cusp of middle age, nearing the end of her career as Hollywood's reigning sex symbol. It was not the time to add to her reputation as a petulantly difficult, irresponsible actress, but the Kennedys nonetheless pushed her to leave the set of *Something's Got to Give* to fly to New York to sing "Happy Birthday" to the president on May 19, 1962, at a gigantic celebration at Madison Square Garden. "The man in charge of the studio rejected that," recalled Milton Gould, a Twentieth Century–Fox movie executive. "He had a confrontation with Robert Kennedy, and he told him we couldn't do it. Mr. Kennedy was then steered to me because at that moment I was in charge of reorganizing the company. And he asked me if I would do it. I said no. He got very disturbed and very abusive. She left the next day for New York anyway. She was suspended, the picture was discontinued."

The Kennedys had their splendidly sensuous moment as Monroe turned singing "Happy Birthday" to the president into an erotic moment whose subtext was lost on no one. As much as that flattered the president, Monroe had become a problem to the Kennedys. She had always been a woman of wild mood swings and immense insecurities. She had had a tryst with the president, and she may have had one with Bobby as well. More likely, though, Bobby was once again cleaning up after his brother, trying to assuage this deranged, tragic woman who in a few slurred, drunken words could expose a sexual scandal that would be devastating to the Kennedy presidency. She spent much time at the Lawfords' Santa Monica home, often drunk or dazed on sedatives or barbiturates.

Monroe telephoned Bobby at his Washington office in those weeks as well, the calls generally lasting only a short time. On the evening of her death Peter Lawford sensed that something was wrong. He called his manager, Milt Ebbins, and told him that they should go over to the actress's house. "If there's anything wrong there at all, you're the last guy that should be there," Ebbins recalled telling his client. "You're the president's brother-in-law." When it came to choosing between the possible death of his friend and the Kennedy image, Lawford decided that he had best stay home, a decision that haunted him the rest of his days.

Joe DiMaggio came forward to manage his ex-wife's funeral. It was a measure of what he thought of the Kennedys and their pernicious influence on Monroe that none of them was invited to attend. Bobby flew on to Seattle to join Ethel and their four eldest children. Also in their company were Eunice and her eldest son, Lem Billings, and Supreme Court Justice William O. Douglas and his wife. The group went on a hiking and riding trip in Olympic National Park that was chronicled in the press like a major expedition. *Time* reported: "Bobby's three younger children . . . will stay on at Hyannis Port, presumably to train for a later assault on Mount Everest."

The Forest Service, at government expense, ran eleven miles of telephone line into the camp, removed overhanging branches along the trail, and hired a packer with twenty-four horses to transport most of the group to a campsite that included tents with stoves and portable lavatories. All the time a helicopter and pilot stood at the ready.

Bobby might disappear into the pristine wilderness of the Northwest, but wherever he traveled he carried with him a burdensome weight of knowledge of matters that would have shocked most Americans. Early in May 1962, Sheffield Edwards, the CIA's director of security, and Lawrence Houston, the agency's general counsel, told Bobby about a serious problem involving targeted mobsters and the Mafia plots to assassinate Castro.

When Giancana suspected that his mistress, singer Phyllis McGuire, might be having an affair with comedian Dan Rowan, he turned to Robert Maheu, his new friend with the CIA connections. The mobster asked for a favor he considered small compared to attempting to murder Castro. He wanted Maheu to bug McGuire's Las Vegas hotel room. Maheu used CIA contacts and money to attempt to place the electronic device. The technicians were so incompetent that they were immediately found out, and the Las Vegas police arrested one of them. That was October 31, 1960, before Kennedy had been elected. When Maheu was called, he used his connections to squelch the local case, but the FBI sought to turn the matter into a federal indictment. At the end of March 1962, Sheffield Edwards met with the FBI liaison to the CIA to ask the agency to back off the Las Vegas prosecution in the name of the national interest. That was the decision about which Bobby was being asked to give his approval.

"If you have seen Mr. Kennedy's eyes get steely and his jaw set and his voice get low and precise, you get a definite feeling of unhappiness," Houston recalled. Bobby's jaw may have been set, and his eyes steely, but he mouthed no words of criticism of the assassination plots. Bobby was not a man to hold in his rage or to seek less than draconian punishments of those who dared betray him. Bobby was merciless at ferreting out truth, but he was no relentless questioner this day. Edwards recalled that the attorney general neither expressed disapproval of what had gone on in the past nor asked that they refrain from such actions in the future. "He cautioned Larry Houston to the effect that he was to know about these things," Edwards recalled.

After hearing from the CIA about the bungled bugging in Las Vegas, Bobby met two days later with the FBI chief. Hoover knew about the CIA's role in the wiretapping, but until the attorney general told him, he suggested that he did not know that Giancana had been involved in an assassination plot against Castro, though he surely must have suspected.

Hoover treated truth as if it were a precious commodity, to be given out only in small quantities, and his memo of the meeting probably included only what he chose to memorialize. "I told the attorney general that this was a most unfortunate development," Hoover wrote.

> I stated as he well knew the "gutter gossip" was that the reason nothing had been done against Giancana was because of Giancana's close friendship with Frank Sinatra, who, in turn, claimed to be quite close to the Kennedy family. The attorney general stated he realized this and it was for that reason that he was quite concerned when he received this information from CIA about Giancana and Maheu. The attorney general stated that he felt notwithstanding the obstacle

now in the path of prosecution of Giancana, we should still keep after him.

Hoover wrote a memo to Bobby saying that the CIA believed that it could not afford to have this "dirty business" surface since Maheu and Giancana were involved in missions for the agency. The memo stated: "Mr. Bissell . . . in connection with their inquiries into CIA activities relating to the Cuban situation told the attorney general that some of the associated planning included the use of Giancana and the underworld against Castro."

This memo did not mention the word "assassination," but it was as explicit an admission of the plots as any official was likely to make. Anyone would have asked just what Giancana and his colleagues were doing. If the matter had been unknown to the attorney general, he most likely would have thunderously cried out against the CIA for employing the very figures that the Justice Department was trying determinedly to indict and put in prison. Instead, Bobby wrote in a margin his instruction to the FBI liaison, Courtney Evans: "Courtney, I hope this will be followed up vigorously."

Bobby was usually assumed to be speaking with the authority and the voice of the president. It remains unknown to what extent Kennedy approved many of his brother's specific actions. What was indisputable, however, was that there was hardly a crucial issue in the administration in which Bobby's hand did not appear. "It must be understood that the president wanted his brother involved in almost every important action," Feldman asserted. "Those who say that Bobby was spread thin have to realize that he was merely doing what his brother wanted him to do."

Bobby's scribbled note may have been an attempt to cover himself. He was the crucial player in the actions against Cuba, and he could scarcely have been unaware that Castro's possible death was an integral part of these plans. What was happening in the Kennedy administration was that more and more of the important actions of government were being kept purposely obscure, the connection between the word and the deed so smudged that no one could ever tell definitively who had given what order, and who had been deputized to carry it out.

In March 1962, Lansdale, Bobby, and a few others met with the president at the White House to discuss Operation Mongoose. General Lemnitzer mentioned "contingency plans" for the invasion of Cuba. The military would create "plausible pretexts . . . either attacks on U.S. aircraft or a Cuban action in Latin America for which we would retaliate." For months Bobby had been talking of creating an orchestrated event to justify an American assault on Communist Cuba. This was just what the Nazis had done before their invasion of Poland, and such an action was the antithesis of what democratic leadership was supposed to represent. Yet apparently nobody in that room condemned the idea on moral grounds. The president, for his part, bluntly told his top general that those forces might be needed in Berlin and he should not be contemplating sending them into Cuba.

Bobby spoke later. There are times when covert information is little more than wishful gossip, and Bobby told the group about reports that Castro was so upset at the way things were going in Cuba that he had begun to drink heavily. Bobby talked of Mary Hemingway, whose novelist husband had killed himself eight months before, "and the opportunities offered by the 'shrine' to Hemingway." Ernest Hemingway's beloved house in Havana was being turned into a museum, and supposedly Castro went there on occasion. Lansdale already knew about this possibility:

> I commented that this was a conversation Ed Murrow had had with Mary Hemingway, that we had similar reports from other sources, and that this was worth assessing firmly and pursuing vigorously. If there are grounds for action, CIA had some invaluable assets which might well be committed for such an effort. McCone asked if his operational people were aware of this. I told him that we had discussed this, that they agreed the subject was worth vigorous development, and that we were in agreement that the matter was so delicate and sensitive that it shouldn't be surfaced to the Special Group until we were ready to go and then not in detail. I pointed out that this all pertained to fractioning the regime.

The terms that Lansdale employed, such as "fractioning the regime," were those he apparently used when he talked of assassination, and this memo is close to confirmation that the president and the attorney general were participants in the contemplation of murder.

Six days later Bobby learned at a meeting of the Caribbean Survey Group that sabotage was decreasing in Cuba and the Communists were imposing increasing controls over the populace. After hearing this, he proposed that the tens of thousands of refugees flowing into Florida be "exploited" and "asked what are the chances of kidnapping some of the key people of the Communist regime?"

Bobby had his own secret imperatives hidden from those around him, their details probably lost forever to history. Bobby was working his own covert CIA operative, Charles Ford, a tall, gregarious officer with the bulk and swagger of a college fullback. Ford was given the moniker "Rocky Fiscalini" and sent out to meet clandestinely with various Mafia figures who had Cuban connections. "His job was to follow whatever Bobby wanted," recalled Sam

Halpern, the executive director of the Cuban Task Force. "We liked to control the meeting sites and such, and we never knew how the arrangements were made for his meetings. Charlie was going in naked. His orders were to meet these guys and come back and report to Bobby. Whether it meant assassination or not, I have no idea because Charlie never talked to me about it and I never bugged him. Bobby had some reason for all this. What the hell he was thinking about beats the hell out of me."

Bobby's main problem was that Operation Mongoose was not working. It had become by far the largest covert project in history, involving close to five hundred full-time CIA operatives, part-time agents, Defense Department, State Department, and USIA personnel, and several thousand Cubans as well. The Cubans had not sat by passively observing this assault on their sovereignty. After the Bay of Pigs, Castro did his own spring-cleaning, breaking up most of the CIA's covert operations while imprisoning those thought likely to join the U.S. efforts. There was a renewed militancy in Cuba, and a national pride at having thwarted the hated *Yanqui*, making it easier for Castro to convince his people that spying on their neighbors was a mark of high patriotism.

Castro was so successful that on the entire island the CIA had only twenty-seven or twenty-eight agents, only twelve of whom even communicated with their handlers, and then only rarely. On December 19, 1961, the agency attempted to add dramatically to that total by sending in seven more agents, but they were captured immediately, and two of them confessed on Cuban television.

As the months went by, it became increasingly clear that once an anti-Castro revolt began, only the infusion of military power would bring an end to the regime. It was the Bay of Pigs scenario all over again, and the president's attitude was much the same: backing away from the logical conclusion that a massive operation made no sense unless one day he was ready to involve American troops. In February 1962, Kennedy accepted that there would have to be contingent planning for an invasion, but he "expressed skepticism that insofar as can now be foreseen circumstances will arise that would justify and make desirable the use of American forces for overt military action."

Bobby continued lashing the CIA forward, at times screaming at officials who did not jump as high as ordered or stay in the air until he ordered them down. For all the clandestine aura of the agency, these men were career civil servants who were not used to being yelled at as if that were the only way to motivate them. Bobby had his own desk at CIA headquarters, and he often showed up at Langley unannounced. "Bobby came over almost acting like the acting director of the CIA," recalled Dino Brugioni, a senior officer at the CIA's National Photographic Interpretation Center. "He was a very arrogant guy. He looked down on government workers."

Bobby had a devastating impact on the morale of the men who attempted to implement his desperately flawed initiatives. He would have been dismayed to know that one of his many critics was General Maxwell Taylor, the new chairman of the Joint Chiefs of Staff. "I don't think it occurred to Bobby in those days that his temperament, his casual remarks that the president would not like this or that, his difficulty in establishing tolerable relations with government officials, or his delight in causing offense was doing harm to his brother's administration," reflected Taylor.

From his perch at the CIA's Cuban desk, Halpern observed a swaggering Bobby who was as dangerous in his ignorance as Lansdale. Halpern happened to be in the office of William Harvey, when the new head of the Cuba Task Force was talking to Bobby on the phone. The obese Harvey was one of the CIA's few full-blown characters. He fancied ivory-handled pistols stuck in his belt and masked his brilliance as a covert operative behind a flamboyant, foulmouthed manner. Harvey was the kind of daring character whom Bobby might have been expected to admire, but the attorney general had as low and disdainful an opinion of Harvey as Harvey had of him. On that score Harvey was the winner, for in calling Bobby a "fag" he had struck upon an epithet that Bobby probably dreaded above all others.

The CIA had just run an operation in Cuba that had merited headlines in the Cuban papers and stories in Florida newspapers. Bobby had wanted the agency to blow up bridges and burn sugarcane fields, but without any notice. He not only blamed the CIA for the press coverage but was deeply suspicious that the agency had acted as its own publicist. Harvey despised Kennedy, but he sat there muttering "yes sir" and "no sir" into the phone. "If you're going to blow something up, it's going to make a noise," Harvey said, as Halpern stood listening. "And if it makes a noise, you're going to get publicity."

Bobby had relationships with a number of the Cuban exiles. This intense, passionate man cared not simply about the details of laws and legislation but about the details of individual human lives. Yet Bobby was like a doctor who was so emotionally involved with his patients that at times he could not distance himself enough to make dispassionate judgments.

It was only natural that many of these patriotic Cuban exiles became involved with Operation Mongoose. It was only natural too that they would pick up the phone and talk to their friend Bobby. That they could do so created morale problems in Florida for those planning operations. Bobby, however, felt for these men and their struggle. In a sense, he was as much a hostage to these feelings as the Brigade 2506 members in Cuban prisons were hostage to Castro. Bobby cared about the brigade members languishing in Cuban prisons in a way the president did not—and truly could not given his immense concerns.

By the end of July 1962, Operation Mongoose had managed to place only eleven covert teams on the island, less than half the projected infiltration plan. In the previous four months, nineteen of its maritime operations had either failed or aborted. As bright a veneer as the agency attempted to paste on its failures, the CIA admitted that even if it continued fomenting resistance at a high rate, there would most likely not be a revolt until the end of 1963, and that would be successful only with an American military force backing up the rebels. That distressing reality faced Bobby and his brother in the summer of 1962 as they were confronted with new civil rights dilemmas as well.

"Which one is Meredith?" asked Governor Ross Barnett of Mississippi as he stood blocking the entrance to room 1007 at the Woolfolk State Office Building in Jackson on September 25, 1962. James Meredith stood there with Deputy Assistant Attorney General John Doar and Chief of the U.S. Marshals Jim McShane. Sixty-four-year-old Barnett had much more in common with twenty-nine-year-old Meredith than with many men he called his friends. Both the governor and Meredith had nine brothers and sisters and came from dirt-poor, law-abiding farm families. Barnett had struggled upward, earning a law degree, developing a down-home, ingratiating courtroom manner that had helped him win the governorship in 1959. To Meredith, the air force had been his vehicle of advance. After nine years of service to his country, he had decided to enroll at the University of Mississippi and earn a college degree.

Meredith was black, and the governor did not intend to allow a man of another race to attend what he considered a white man's university. Barnett read a proclamation that he did "hereby finally deny you admission to the University of Mississippi." The crowd beneath, listening on portable radios, heard the words and hooted and screamed as the three men departed. "Communist!" they shouted at Meredith. "Nigger, go home!"

Kennedy had tried to use the law as the engine to advance civil rights. Again and again in these crisis-filled years, moments such as this arose, when the law of the land came up against customs that were considered higher law codifying an immutable social logic. Kennedy sought accommodation, not confrontation, but not at the price of turning back from what he considered the forward march of justice and freedom. He was a politician, however, and he knew that these spectacles of protest and defiance risked tearing apart the Democratic coalition. He could win all these battles except for the last one, his reelection in 1964. Thus, justice had to be meted out gingerly, with full warning given to allow those who opposed him to back off without penalty or strife.

Those who believe that history provides a series of axioms to live by are as doomed to make mistakes as those who ignore the lessons completely. Bobby attempted to do in Mississippi what he had successfully managed in the Freedom Rider confrontations the previous spring in Alabama. At that time he had worked with a segregationist governor who realized that the high price of anarchy would be paid primarily by his fellow Alabamians. He had worked also with a few police officials to whom law and order was more than a slogan. In the end the mob had stepped back from the precipice of massive bloodshed, their anger boiling itself out primarily in taunts and shouts. Unfortunately, Governor Barnett of Mississippi was not Governor Patterson of Alabama, Colonel Thomas B. Birdsong of the Mississippi highway patrol was not Floyd Mann of the Alabama state police, and the rabble of Mississippi were far readier for a brutal, bloody confrontation than their counterparts in Alabama.

There would be those who would view these protesters descending on the college town of Oxford as beady-eyed cretins pouring out of the thickets and the swamps, the trailer parks and shacks. The reality was that the student protesters were many of the proudest sons and daughters of old Mississippi. And it was the good citizens of their parents' generation who loudly sang the song of anarchy, unlocking the doors to shadowy men willing to do their deeds.

"Thousand Said Ready to Fight for Mississippi," headlined the *Jackson Daily News*, as if this were 1861, not a century later. On Saturday evening, September 29, 1962, Barnett appeared in Jackson at Memorial Stadium, where Ole Miss played the Kentucky Wildcats. "I love Mississippi," he shouted over the loudspeakers from the fifty-yard line. "I love her people." And then, as the applause rose to a deafening ovation beyond even that merited by the triumph of Ole Miss on the football turf, the governor screamed: "I love our customs!" There were those in the stadium ready to march at that moment, and when they left after the game, they saw the rebel flag flying from homes and office buildings and turned on radio stations playing "Dixie," a marching song not of nostalgia but of war.

The following afternoon the president called Governor Barnett for the first time. As was often the case around the Kennedy men, there was jock bravado, a joshing, taunting air. "And now—Governor Ross Barnett," Bobby said playfully, a fight manager pushing his champion to the center of the ring. "Go get him, Johnny boy." The president picked up on enough of this atmosphere to play the moment as much to his brother and aides as to the Mississippi politician. As the phone rang, Kennedy performed his little comic routine, pretending to speak his first lines. "Governor, this is the president of the United States—not Bobby, not Teddy, not Princess Radziwill."

Kennedy spoke to Barnett much the way he had been talking to southern governors for years, as if they and he were secret comrades, faced with pesky problems that only they could solve. "Well, now, here's my problem," Kennedy said, as if he were confiding to a colleague.

"Listen, I didn't put him in the university," the president went on, not even mentioning Meredith's name. "But on the other hand, under the Constitution . . . I have to carry . . . that order out, and I don't want to do it any way that causes difficulty to you or to anyone else."

What Kennedy was trying to signal to the Mississippi governor was that he could bluster and rant against Washington all he wanted as long as he upheld law and order. He could make whatever symbolic protests he needed to make as long as he saw to it that Meredith was registered and peacefully entered the university.

Barnett had his problems too, and he wanted his fellow politician to understand just what they were. "You know what I am up against, Mr. President. I took an oath, you know, to abide by the laws of this state . . . and *our* constitution here and the Constitution of the United States. I'm on the spot here, you know."

The governor gave the president ominous warnings of calls he had received of citizens groups wanting to descend on Oxford. Barnett, however, was a glad-handing, favor-exchanging, soft-soaping southern politician, and he segued into a matter about which he felt far more comfortable. "I appreciate your interest in our poultry program and all those things."

Early that evening, federal officials escorted Meredith to an empty dormitory on campus to prepare to begin classes the next morning. For the men around Bobby, this represented something of a triumph, a sign that the worst was over. O'Donnell enthused that Bobby "should be Mandrake the Magician," though this was a sleight of hand that fooled almost no one. By then most of the makeshift team of U.S. marshals, border patrol officers, prison guards, and other federal officials had taken up their posts at the Greek Revival–style Lyceum, the very heart and soul of the university, part of the building dating back to 1848. That gesture was as much a sacrilege as camping out in one of the local churches, a mark not so much of willful insensitivity as of old-fashioned ignorance and bad planning.

If the milling students and their outside collaborators had needed a symbol to stiffen their resistance, they had one now. Within a few minutes, shouted epithets had given way to pebbles, and pebbles to stones, and stones to rocks whirling past the heads of the white-helmeted marshals. Under the cover of the gathering darkness, most of the Mississippi highway patrol officers disappeared into the night, leaving the protection of Meredith to men viewed by the crowd as foreigners. Molotov cocktails, Coke bottles full of flaming gasoline, sailed through the night skies, splattering at the feet of the marshals. A lead pipe struck an officer, dropping him to the ground. The marshals fired their first canisters of tear gas into the surging crowds, and among the first victims were those marshals who did not have gas masks. Then they fired more shells, and one of the casings hit one of the few remaining Mississippi patrol officers in the back. The officer had been caught between the students and the marshals, and he almost died there outside the Lyceum.

As the president, the attorney general, and their aides sat in the White House, they were reluctant to face the reality that this crisis risked becoming an insurrection. Bobby was wistful in his hopes that Barnett would back down and the tide of rage would suddenly recede, even once the blood began to spill. Although the president and attorney general talked often to Barnett, there was an overwhelming duplicity to the dialogue. "Mr. President, let me say this," Barnett had said earlier that day. "They're calling me and others from all over the state, wanting to bring a thousand, wanting to bring five hundred, and two hundred and all such as that you know."

What Barnett did not say was that he was unwilling to step in front of these mobs to try to drive them back by word or authority. He wiggled, finagled, backtracked, sidetracked, and finessed his way back from agreeing to admit Meredith to the university. "I say I'm going to cooperate," he told the president in another call. "I might not know when you're going to register him, you know. . . . I might not know what your plans were, you see." That was a slick bit of business, and when Barnett tried to back away from his words, Bobby gave him the melancholy news that he had a tape of his conversations, and if he backed down, the president would be ready to tell the American people of his duplicity.

While tear gas filled the air, the president was preparing to give a speech to the nation on television about the situation. Information is the most immediate power, and without it, the president risked looking like an ill-fated observer of a world he did not understand. Bobby knew just enough to have called for a postponement of the speech, or a quick-witted revision of the remarks, but he said nothing. He wore a blinder of optimism. So the president went on television at 10:00 P.M. to talk confidently about how "the orders of the court . . . are beginning to be carried out," while tear gas drifted across the bucolic campus. As the crowd of two thousand surrounded the besieged marshals, Kennedy's words sounded pandering and silly. "You have a great tradition to uphold, a tradition of honor and courage," he said as the rioters, perverted exemplars of this tradition, moved on the Lyceum.

After the speech the president joined the attorney general and his advisers in the Cabinet Room, where they attempted to monitor events in Oxford.

This was the attorney general's arena, and for the most part Kennedy allowed his brother to manage the crisis. Bobby seemed unable to grasp the sheer magnitude of what he faced. He and his aides came up with the less-than-inspired idea of having Johnny Vaught, the revered Ole Miss football coach, address the students. Gunshots had begun to sound in the night, and even the most inspiring words would have gone unheard. A shotgun blast tore into a marshal's neck. A shot from a high-powered rifle felled a patrol officer. These were not just students any longer, but a mob that held agitators willing to kill. In downtown Oxford, retired Major General Edwin A. Walker, a right-wing fanatic, incited his supporters to become the new minutemen, standing up to the tyranny of the federal state.

Kennedy realized what they faced well before his brother did. He knew that these endless mishaps and seeming mismanagement were signs not so much of Bobby's ineptness as of the nature of this crisis. "That's what happens with . . . all of these wonderful operations," he said. "War."

"I haven't had such an interesting time since the Bay of Pigs," the president mused ironically a few minutes later.

"Since the day, what?" Bobby asked.

"Bay of Pigs."

"The attorney general announced today, he's joining Allen Dulles at Princeton Univers——," Bobby said, making a mock announcement of his departure from Washington in the wake of this new disaster.

In alluding to the Bay of Pigs, the president had come up with an appropriate metaphor. As in that misadventure, the administration had grievously misjudged the strength of its opponent. Representatives of the federal government had gone in with a weak force, and like the brigade on the Cuban beach, the marshals were running out of their most crucial ammunition, tear gas. In the end, innocent men died who should have lived. "I knew that he [the president] was concerned about how we were going to possibly explain this whole thing because it looked like it was one of the big botches," Bobby told an interviewer in 1964. "I mean, if more people had been killed—I think we were lucky to get out of it."

As the night went on the news worsened. There were more tales of shootings and confrontations. At their maddening distance from it all, it was as if they were trying to make out images through the tear gas.

"They're storming where Meredith is," Bobby said as he put down the phone.

"Oh," Kennedy said softly. "The students are or the . . . ?"

"They're storming where Meredith is," the attorney general repeated.

"You don't want to have a lynching," O'Donnell said a moment later.

They were on the verge of a disaster. Meredith dead. Students dead. U.S.

marshals dead. Reelection dead. Bobby had been hoping that the federal officers would not have to fire their guns. "They better fire, I suppose," Bobby said on the phone. "They gotta protect Meredith."

Meredith remained safe for the moment, but outside anarchy reigned. Three more deputies fell from gunfire, and then came the news that the body of a reporter from France had been found behind a women's dormitory with a bullet in his back.

Bobby no longer looked as youthful as he had when his brother entered office. The intensity of these moments and the endless pressures had etched themselves into the attorney general's face. Despite what others thought, the Kennedys held no console of power in the White House from which they could merely press a button here, pull a lever there, and events moved and history was made.

On this evening, Barnett, whom Bobby considered a weak man and "an agreeable rogue," proved stronger than all the power of the presidency. The governor's objective, as the attorney general saw it, "was the avoidance of integration . . . and if he couldn't do that, then to be forced to do it by our heavy hand—and his preference was with troops." Bobby's objective was to get Meredith safely into the university without calling in federal troops. All these hours, and at a heavy cost, Bobby had sought to avoid calling in the army, to not invoke that terrible image of American soldiers marching against some of their own people. By midnight, though, he felt forced to call in federal troops.

Bobby was a man of righteous anger, but he could not express one note of the rage he felt toward Barnett when he talked to him this endless night. Nor could he vent his full anger at General Creighton Abrams and the army for their slow-footed journey toward Oxford. Even before the main force of the soldiers had moved onto the campus, Bobby was musing about how this would all play out. It was not a picture to be placed happily in the administration's scrapbook. Two men had died, the French reporter and another unfortunate bystander. And 160 of the marshals had been wounded, 28 of them by gunfire, a casualty rate like that of an invasion force. If Bobby had not refused to give the marshals the order to fire back except in defending Meredith, the casualty totals would have been far higher. That had been the most daring of gambles, for if any of the wounded marshals died, the finger of recrimination would have pointed squarely at him. "We're gonna have a helluva problem about why we didn't handle the situation better," he thought aloud. "Well, I think we're gonna have to figure out what we're gonna say."

At dawn, the main force of the troops arrived in Oxford. Well-armed, perfectly disciplined soldiers marched down a street where minutes before rioters had danced. One of the protesters threw a Molotov cocktail in front of the troops, a barricade of flames lapping across the street. The soldiers marched on in perfect order through the flames, and the crowd backed off for good. In the end, an overwhelming force of twenty-three thousand troops entered the college town so that for the first time in history a black man could attend the university.

27

"A Hell of a Burden to Carry"

In mid-August 1962, General Douglas MacArthur sat in the Oval Office pontificating about the world. This was one of the president's more onerous obligations, listening to the great old men of other eras imparting their wisdom. The eighty-two-year-old general told the president that America should invent a new mini-atomic weapon to be given to every infantry soldier with an "atomic cartridge that would clear ten to fifteen yards in front of him." The aged general droned on in the authoritarian manner of a man who had lived his life around subordinates. The further some men stood from power the wilder their schemes became, and MacArthur advised the president that he should begin guerrilla intrusions into China. "There also should be maneuvers by the South Korean army itself—"

"Wait just a minute," Kennedy interrupted. "My father is calling. He isn't able to talk much, but every now and then he tries to make a call." It did not matter who sat with him, or what issue consumed him, when his father rang, his son took the call.

"He can't speak . . . Hello there, how are you. . . . How are you getting along up there. . . . How is everything?"

Joe spoke scarcely more than grunts, but Kennedy imagined a dialogue, willing his father into verbal coherence. "I'm sitting here with General MacArthur, and he wants to be remembered to you. . . . How are you getting on up there? Well, I'll be up the weekend after next, and we'll go out in the boat . . . good. Let me talk to Ann."

Ann Gargan had become Joe's keeper, jealously protective of any who got too near, rationalizing whatever she did in the name of her charge's health. Gargan took over the phone. "He seemed a little upset," Kennedy said. "Tell him I'll call tomorrow. All right. Good."

Kennedy hung up the phone. "What's going on up there?" he asked more to himself than to MacArthur. There was a pain that the president felt over his crippled father that nothing could assuage. Joe had taught him that everything could be fixed, everything, it turned out, but his father's condition. On one of his weekend trips to Hyannis Port that fall, he had learned that his father had a seizure when his mother came waltzing into the room showing off an exquisite gown she was planning to wear to a ball. Rose had apparently only been trying to get her husband to nod his approval, but her visit had affected him terribly.

Kennedy had hurried up to Joe's room when he arrived in Hyannis Port. He was incredulous when he discovered as he walked into the upstairs bedroom that no one was there, no doctor, no nurse, no maid, no one. The president was not a man who shouted often, but he had shouted that afternoon, his screams of anger and disbelief sounding through the house.

"He's a great fighter," MacArthur said, an obligatory aside.

"Yeah, he is, but he can't speak anything, so he gets . . ." the president's words drifted into the air.

"The South Korean army should have several so-called grand maneuvers," MacArthur began again, segueing back into his monologue.

At the same time that Kennedy was listening to MacArthur, he faced the continued dilemma of Cuba. A year and a half into his administration, Castro was as ensconced in Cuba as the day the president had taken office. Despite all the efforts at sabotage and subversion and attempts to isolate the Caribbean island from the world, General Maxwell Taylor had to admit to Kennedy that the American government was unable to "assess accurately the internal conditions" and that there was "no likelihood of an overthrow of the government by internal means and without the direct use of U.S. military force."

That analysis could have led to a self-critical evaluation of Operation Mongoose and America's diplomatic strategy and a gut-wrenching acceptance of the possibility that Castro's Cuba might be there for years. Instead, the administration, following Bobby's lead and with the president's approval, went forward with phase two of Operation Mongoose. Gone was the heady supposition that the CIA's activities would lead to the overthrow of the Communist leader. Nonetheless, the scale of covert actions would be increased and the "noise level" would rise to such a level that "the participation of some U.S. citizens may become known."

The Cuban economy was hurting from both the natural weakness of socialist control and the artificial suffering caused by the American-led economic blockade. The CIA would attempt through selective sabotage to drive the Cubans further into economic despair. Beyond its own operations, the

CIA would consider helping Cuban exiles and other Latin governments in running their own operations against Castro.

In the first week of February 1962, Aleksei Adzhubei, the editor-in-chief of *Izvestiya*, arrived at the White House. To the Soviets, journalism was an adjunct to government, and Adzhubei brought with him the Soviet agent Georgi Bolshakov, who was also editor-in-chief of *USSR* magazine. Even if Adzhubei had not been Khrushchev's son-in-law, his discussion with the president would have been far more important than simply an interview with the most important Soviet journalist.

Adzhubei had just returned from spending six hours talking with Castro, and inevitably the discussion turned to Cuba. For the most part, according to the American account of the lengthy meeting, the two men discussed the situation like acquaintances talking about a troublesome neighbor. The Soviet official "wondered whether the United States realized that by its unfriendly attitude toward Castro it was pushing Cuba farther and farther away." The president, for his part, tried to make Adzhubei understand the American perspective. The United States had never had such an enemy so close to its borders, and when Castro yelled his disdain within shouting distance, Americans were bound to be upset.

Kennedy compared the situation to Hungary, where the Soviets had put down the Hungarian revolution. The president may only have been trying to make a casual analogy, but if Cuba was America's Hungary, then Castro would soon see American tanks in the streets. Kennedy later recalled that in the meeting after lunch, at which the American interpreter was not present, Adzhubei "wondered whether the U.S. would prefer Cuba to develop into a state like Yugoslavia or have it drift in the direction of China." That was a startling analogy for the Russian to make, since Tito's independent socialism was anathema to the Soviets. It suggested that the Soviets might not have wanted the burden of a quasi-satellite six thousand miles from Moscow. Adzhubei went on to complain to the president about all the money the Soviets were pouring into Cuba and all the unnecessary sugar they were buying, primarily because of their fears of an American invasion. The Russian asked the president whether he was planning to invade, and Kennedy said no, he was not.

In diplomacy, words must be used with mathematical precision, their meanings universally understood. Khrushchev, for his part, often sounded like an intemperate blusterer, but a careful look at his words reveals that he usually said precisely what he wanted to say, making the Soviet position indisputably clear. Kennedy, despite his immense rhetorical gifts, sometimes said less or more than he meant to, sending signals he did not mean to send.

Men could die for a word not said, or a message misunderstood. During a meeting in Moscow Salinger was told that the Soviets gave the English text to five interpreters. When the Russian text of the meeting was translated back into English by five other translators, they had five different interviews. Adzhubei's own personal take on the meeting was the most important of all, and it was given directly to Khrushchev, his father-in-law. Life is in the nuances, and it was Adzhubei's recollection of the tone of the meeting more than the words that mattered.

"Have you changed your opinion that the April 1961 invasion was an American mistake?" Adzhubei said that he asked Kennedy. The president did not like having his nose rubbed in that odorous mess any longer, not by Americans and certainly not by Soviets.

"At the time I called Allen Dulles into my office and dressed him down," the president said, pounding his fist on the table. "I told him: you should learn from the Russians. When they had difficulties in Hungary, they liquidated the conflict in three days. . . . But you, Dulles, have never been capable of doing that."

Adzhubei may have exaggerated Kennedy's comments, but by all accounts the president did in fact mention Hungary. That was a foolishly provocative analogy to throw out. By its covert activities and military provocations, the Kennedy administration had given Castro and Khrushchev every reason to believe that Cuba was about to be invaded. After reading his son-in-law's report, Khrushchev took a serious new look at Cuban security, beginning by having the Presidium, the Soviet leadership committee, approve a new $133 million package of military assistance.

In May 1962, the Soviet leader decided that he would further protect Castro by placing nuclear missiles on the island. In doing so, he would also buttress his nation's position against an America far superior in military strength. Although historians debate why the Russian leader would propose such a daring and dangerous move, what remains indisputable is that the massive CIA operations in south Florida and the feints of Operation Mongoose provided profound political justifications. Without those ceaseless efforts and Castro's legitimate fears that he was about to be invaded, Khrushchev would undoubtedly never have made such an offer, and even if he had, Castro very probably would not have accepted it.

Castro was not a hand puppet moving every time Khrushchev wiggled his thumb or index finger, and the Soviet leader was right in believing that he would have to convince the Cuban leader of the efficacy of dotting the landscape of Cuba with such deadly weaponry. "Tell Fidel that there is no other way out," Khrushchev admonished his delegation as they set out to Cuba. He would secretly place the weapons in Cuba. Then at the end of November, after the American elections, Khrushchev would arrive on the island, sign a new treaty with Castro, and announce to the world that Cuba was now safe from invasion. It was a provocative action, but Khrushchev and Castro had their justifications. In its way, politics follows Newton's third law of motion: every action creates an equal or opposite reaction. Whether a man is a Communist or a capitalist, if you burn his fields, sabotage his ships, destroy his goods, poison his wells, and attempt to kill his leader, sooner or later he will react, and he will be a different kind of enemy than when it all began.

Through its many sources, the CIA had been receiving information suggesting a major buildup of Soviet activity in Cuba. The United States had counted up to twenty Soviet freighters arriving with military equipment by August 22, and five more seemed to be on the way from Black Sea ports. Large numbers of Russian civilians had arrived as well. Secret military construction was going on in various places around the island. Since the Soviets were usually parsimonious with their client states in their economic commitments and military aid, the CIA concluded "these developments amount to the most extensive campaign to bolster a non-bloc country ever undertaken by the USSR."

On August 23, President Kennedy met with his top national security advisers to discuss the dramatically increased Soviet presence. CIA Director John McCone called for aggressive new action, including as the strongest alternative "the instantaneous commitment of sufficient armed forces to occupy the country, destroy the regime, free the people, and establish in Cuba a peaceful country which will be a member of the community of American states." Secretary of State Dean Rusk came up with his own aggressive suggestion: to use the American base at Guantánamo Bay as a staging ground for sabotage.

Kennedy and his associates mused about what this Soviet action might have to do with Turkey, Greece, Berlin, and other trouble spots. McCone brought up the Jupiter missiles in Turkey and Italy that so rankled Khrushchev. Secretary of Defense Robert McNamara listened to McCone and then made the point that even though these missiles had become obsolete and useless, politically it was difficult to remove them.

Every day Kennedy received an endless stream of information. He read crucial secret reports known only to half a dozen people in the world. He perused reams of pedestrian facts churned out by the bureaucratic machine. Mixed into this meld was brilliantly orchestrated misinformation conceived by America's enemies and sometimes by her friends, or even by those within the administration. As Kennedy pored through the endless pieces of paper, he remained deeply suspicious of the CIA and Joint Chiefs' memos. As difficult as it was to mistrust what he read in their memos, Kennedy had learned the valuable lesson that even on these sheets of paper lay the world in all its duplicity, misunderstanding, and uncertainty.

Since the day Kennedy entered office, conservatives had been condemning him for what they considered his cowardly acquiescence to Communist Cuba. On August 31, Senator Kenneth Keating of New York, a moderate Republican, got up in the Senate and savaged the administration for its willful laxness on the Cuban issue. Keating's speech was doubly troublesome, since he was not one of the wild men of the right, and he had what he called "ominous reports" that "missile bases" were under construction in Cuba.

Kennedy did not know where the Republican senator had gotten his information. He realized, though, that the opposition might try to use this issue to barter its way into majority control of Congress in an election only two months away. Before the Bay of Pigs, the president harbored the delusion that he could control the information that came out of the White House, and he had tried to hide an entire operation under a tarpaulin of deception. Now he was so worried about information leaking that he was unwilling to trust some of the people he had to trust if he was going to make rational decisions.

The afternoon of Keating's speech, Kennedy called Marshall Carter, the CIA's deputy director. The president was worried about the dissemination of photos showing the construction of what appeared to be surface-to-air missile sites across Cuba. The photos should have set off an alert throughout the government's various intelligence operations, but the president wanted them hidden away. "Put it back in the box and nail it tight," Kennedy told Carter.

Khrushchev needed to lull the Americans for a few weeks and Kennedy was helping him do so. Georgi Bolshakov, the Russian agent used as a conduit in Washington, told his American contacts that relationships between the two great powers might well improve if the Americans would end their "piratical" flights monitoring Soviet ships sailing to Cuba. Kennedy invited the Russian to the White House on September 4 and told him, "Tell him [Khrushchev] that I've ordered those flights stopped today." Afterward Bobby stood with Bolshakov outside the White House beseeching him to inform Khrushchev that whatever he did, he must not try any needless provocations before the midterm congressional elections. "Goddamn it!" Bobby exclaimed. "Georgi, doesn't Premier Khrushchev realize the president's position?"

In the charming, gregarious figure of Bolshakov, Bobby saw the possibility of dialogue across the Iron Curtain. The Russian had been useful to the

Kennedys before, though he had been more useful to his Soviet masters. Bobby did not quite grasp that if his own opinion of the Soviet system was correct, Bolshakov was as imprisoned by his handlers as if Bobby were talking to him in a cell.

On September 7, the president learned the deeply disturbing news that in analyzing their most recent U-2 photos of Cuba, the CIA analysts "suspect[ed] the presence of another kind of missile site—possible surface-to-surface." This was precisely the kind of information that the president's opponents might use to stir up political hysteria across America. Kennedy could no longer "put it in the box and nail it tight." But even as he told the analysts to continue their work, he froze the information's dissemination within a small circle of advisers.

While Kennedy was examining the U-2 photos and trying to put a damper on the whole uncertain business, Secretary of the Interior Stewart Udall was meeting with Khrushchev at his summer home on the Black Sea in Soviet Georgia. Khrushchev had decided that he needed a conduit to Kennedy, and Udall was the nearest available vehicle. The Soviet leader, a student of American politics, was perfectly aware of Kennedy's obsession with the forthcoming election. He promised Udall that he would not create a crisis over Berlin before the American election. "Out of respect for your president we won't do anything until November," he said.

If shrewdness were the ultimate attribute, then no one would ever have bested Khrushchev. His salty aphorisms sounded like homespun peasant wisdom, but they were superb vehicles with which to promote his Marxist ideology, the metaphors carrying meanings within meanings. As the Soviet leader saw it, everything was all so simple and logical. The United States had put nuclear weapons in Japan, but all the Soviet Union was doing was giving Castro defensive weapons. "You have surrounded us with military bases," Khrushchev said. "If you attack Cuba, then we will attack one of the countries next to us where you have placed your bases."

As Udall listened, he was in essence merely holding a microphone so that Khrushchev could reach Kennedy's ear. Udall pointed out that only a few members of Congress were spouting such craziness as to call for an invasion of Cuba. "These congressmen do not see with their eyes, but with their asses," Khrushchev replied. "All they can see is what's behind them. Yesterday's events are not today's realities. I remember Gorky recounting in his memoirs how he had a conversation with Tolstoy. Tolstoy asked him how he got along with women, and then ventured his own opinion. 'Men are poorly designed. When they're young, they can satisfy their sexual desires. But as they grow old, the ability to reap this satisfaction disappears. The desires, however, do not.' So it is with your congressmen. They do not have power, but they still have the same old desires."

As Khrushchev muttered his soothing shibboleths, he was orchestrating a deployment of Soviet military power in Cuba beyond anything even the most vociferous critic had imagined. Soviet plans were to install in Cuba twenty-four R-12 nuclear ballistic missiles with a one-thousand-and-fifty-mile range, sixteen R-14 missiles with twice that range, and eighty nuclear cruise missiles with a short range of about one hundred miles. From Cuba, the largest of these missiles could target American cities almost as far west as Seattle. A single missile could destroy one of the nation's major cities with a force over seventy times that of the bomb that devastated Hiroshima.

The Soviet leader also planned to send eleven submarines sailing out of a new Cuban submarine base, each probably carrying one nuclear torpedo and twenty-one conventional torpedos. Seven of these would carry nuclear missiles. There were also plans to send a squadron of IL-28 light bombers carrying nuclear bombs and a large number of tactical nuclear weapons called Lunas by the Russians and Frogs by NATO. These weapons had a thirty-one-mile range that could be used against anyone rash enough to attempt an invasion of fortress Cuba. Traveling to Cuba with these missiles would be 50,874 Soviet troops, a force that even without the nuclear weapons would change the nature of power in Cuba and the price to be paid in any invasion.

With one bold action, the Soviets would more than double the number of Soviet missiles targeted at American cities. Communist Cuba, so threatened by an American invasion, would suddenly become impregnable to all but an American leader willing to set off nuclear war. The Americans would see and feel what the Soviet people felt: enemy nuclear weapons near enough to cast dark shadows across their border. Khrushchev's nation was still no match for America in military might and nuclear weaponry, but this daring move would reap untold psychological and political benefits for Communist nations around the world.

During the first days of October, Kennedy did not know the extent of the awesome nuclear weaponry sailing toward Cuba, but he knew that Khrushchev had moved his queen forward on the chessboard of the cold war. Kennedy was fond of quoting Hemingway's definition of courage as "grace under pressure." For Kennedy, a true man acted not only courageously but with unflinching coolness. That was precisely how Kennedy himself had acted when PT-109 was cut in two, and it was how he acted now as he received reports on the situation in Cuba. He asked that the U.S. armed forces begin to prepare themselves for military action in Cuba—not immediately, but in the coming three months. He wanted American pilots to be fully prepared to take out the Soviet SA-2 surface-to-air missile sites that he knew were already in existence, and the air force developed mock-ups for their training.

If the president had not been Bobby's brother, the attorney general would most likely have accused him of a prissy reluctance to confront Castro. On October 4, Bobby chaired a meeting of the top officials overseeing Operation Mongoose in which he vented his rage, exhibiting a full measure of gracelessness under pressure. These were not midlevel officials he yelled at, but among them Lansdale himself, a man not used to being on the receiving end of such wrath. The general was no longer in the outsider's enviable position, able to condemn and ridicule what others had done before him. Now he was in the bureaucrat's uncomfortable chair, having to defend what had not been accomplished while giving what McCone took as the "general impression that things were all right."

No longer was the man Bobby had thought of as the personification of "the Ugly American" safe from the rebuke that he spilled so indiscriminately on those working on Operation Mongoose at the CIA. The attorney general fumed that "nothing was moving forward." Bobby wanted gung-ho action, militancy, and bold acts of sabotage, including possibly what would have been an act of war: secret mining of the harbors where the Soviet ships were arriving.

These few men in the room represented most of the spectrum of thought on Cuba in the administration. They knew that they would be judged not by how forcefully they spoke today but by how true their assessments proved. Bobby and McCone stood at one end as the most militant advocates of determined military action against Cuba. They shared the same Catholic faith, the same militant anticommunism, the same brutal assessment of Cuban and Soviet motives.

McCone asserted that the Soviets were probably establishing an offensive military posture in Cuba, including medium-range ballistic missiles. As the president's top national security adviser, McGeorge Bundy reflected the president's thinking, but he was far from being merely Kennedy's intellectual clone. Nonetheless, Bundy seemed, for the most part, to believe what the president believed, and this morning he believed that the CIA director's black assessment was probably wrong, and that the Soviets would not dare go so far. McCone, a man of genteel manners, admitted that many, even most, of his colleagues in intelligence would have agreed with Bundy, but they could not risk the future of the United States if Bundy was wrong.

The scholarly Bundy had history and reason on his side. Yet on this very day, at the Cuban port of Mariel, the Soviet ship *Indigirka* arrived carrying forty-five warheads to arm the R-12s; twelve warheads to be fitted on the Luna tactical missiles; six nuclear bombs for the IL-28 planes, and thirty-six warheads ready for the cruise missiles. The total firepower carried on that one Russian freighter, Aleksandr Fursenko and Timothy Naftali wrote, was "over twenty times the explosive power that was dropped by Allied bombers on Germany in all of the Second World War."

Bobby was more ideological than his brother, believing, like his Marxist enemies, that life was a battle over ideas. Yet he was not a man of abstraction. He wanted to know life at its most intimate, to touch not only some of the young Cubans sent off on missions to their native land from which they might never return but also his enemies or those who knew his enemies. So on the day after the Operation Mongoose meeting he agreed to meet once again with the Russian agent Bolshakov, whom he had first met before the Vienna summit.

Bolshakov had tantalized Bobby, telling him he had a message from Khrushchev. In the netherworld of intelligence, the Russians and the Americans cut their scripts up into many pages so that most players knew only their own few lines and little of the plot they were advancing. Bolshakov had a familiar message to deliver. He told Bobby, "The weapons that the USSR is sending to Cuba will only be of a defensive character." Bolshakov was like Adlai Stevenson at the United Nations during the Bay of Pigs, speaking lines he thought to be true to serve a purpose of which he had not been informed. Bolshakov believed that he was serving the high cause of peace, but in fact he was betraying Bobby in the name of his country.

The president needed concrete intelligence on what was going on in Cuba, not conjecture, speculation, ruminations, or gossip. On the morning of October 14, Major Richard Heyser of the U.S. Air Force flew a U-2 mission over western Cuba. He was over the island for only six minutes, but that was enough to take 928 photographs that showed three medium-range missile sites and eight missile transporters. Two subsequent flights the next day brought back photos showing two other sites and crates for the Soviet medium-range bombers.

The following evening, October 15, the CIA's deputy director for intelligence called Bundy at his home to tell him about the latest photos. The news was so extraordinary that it shifted the protocol of power. In any other important matter Bundy would have called the president immediately or left for the White House. This evening, though, he put the phone down and went back to his dinner party, not wanting his foreign guests to have any signal that something was wrong. Even after his guests left, he did not call the president. He knew that Kennedy was tired after a late-night return from New York the previous evening. Bundy decided, as he wrote Kennedy later, that "a quiet evening and a night of sleep were the best preparation you could have in the light of what would face you in the next days."

Kennedy sat propped up in bed reading the morning newspapers, including a front-page story in the *New York Times* with the headline "Eisenhower Calls President Weak on Foreign Policy." That would have been enough unpleasant news for one morning, but then Bundy entered to tell him the results of the U-2's photographic sortie over Cuba. The first thing the president did was to call Bobby. "We have some big trouble," the president said. "I want you over here."

A few minutes later Bobby rushed into Bundy's office saying that he wanted to see the photographs. It was typical of Bobby not to trust the CIA technicians who had so painstakingly examined the pictures. Arthur Lundahl, the head of the CIA's National Photographic Interpretation Center, led the attorney general into the room where the pictures were set up on briefing boards. "Oh shit!" Bobby exclaimed. "Shit! Shit! Those sons-of-bitches Russians." Much as he had during the Bay of Pigs, Bobby immediately personalized these events, seeing the face of the enemy before his eyes.

If there was ever a day that called for the deepest of perceptions and the wisest of judgments, that day had arrived. That reality was not evident that afternoon, as Bobby met at the Justice Department with Lansdale and those most concerned with Operation Mongoose. The attorney general had been upset at Lansdale's operation for a while. Today he played his trump card, invoking his brother's sacred name. He opened the meeting by talking of the "general dissatisfaction of the president" with Operation Mongoose. He bemoaned the fact that there had been no real acts of sabotage.

In the world that faced the president now, acts of sabotage hardly mattered against either a Cuba armed with nuclear weapons or a Cuba decimated by American bombs. Bobby could not yet grasp this essential fact but continued to rail so fiercely against Lansdale's desultory action that he insisted that from now on he would meet every morning at nine-thirty with the Operation Mongoose chieftain and his top subordinates, monitoring them like naughty schoolboys.

There were two meetings that day, Tuesday October 16, of the newly created executive committee of the National Security Council (Ex Comm), the group that in the days that followed became the crucial policymaking organization. Kennedy had sought to surround himself with fiercely intelligent men who would parse an issue up and down, tearing it apart, before they arrived at a sound conclusion. These men had largely failed to alert him to the dangers that lay at the Bay of Pigs. The scale of what faced them was far higher now than a year and a half ago, and this time there was depth and fierce articulateness to many of their contributions. It was a mark of the vigor of their minds that all of the great issues of what became known as the Cuban Missile Crisis were discussed on this first day.

Although Kennedy and his men often spoke in a kind of intellectual shorthand, these were not mere tactical meetings but discussions of political, philosophical, and moral complexity. "What is the strategic impact on the position of the United States of MRBMs [medium-range ballistic missiles] in Cuba?" McGeorge Bundy asked at the 6:30 P.M. meeting in the Cabinet Room. "How gravely does this change the strategic balance?"

"Mac, I asked the Chiefs that this afternoon, in effect," McNamara replied. "And they said, 'Substantially.' My personal view is, not at all."

"You may say it doesn't make any difference if you get blown up by an ICBM [intercontinental ballistic missile] flying from the Soviet Union or one that was ninety miles away," the president said a few minutes later. "Geography doesn't mean that much."

That perception was the darkest irony of the nuclear age. Russia and America were gigantic gladiators. America may have held a sharper sword, but the opponents were so well armed and so vicious that once they began fighting, not only were they both doomed, but in their death throes they would pull down the arena.

"Last month I said we weren't going to [accept it]" Kennedy stated, referring to Russian missiles in Cuba. "Last month I should have said that we don't care. But when we said we're not going to [accept it], and they go ahead and do it, and then we do nothing, then I would think that our risks increase. I agree. What difference does it make? They've got enough to blow us up now anyway."

This was in part a struggle over language. The president had not understood the extent to which his words were actions. Now he believed he could not back away. Beyond that, his language and the language of a thousand other politicians had created a climate in which judiciousness was considered a coward's calling, and anti-Communist jingoism a patriot's one true song. He and his administration had helped create this image of a monstrous Cuba that he was now compelled to slay or be considered less than a manly leader. The missiles may not have changed the strategic balance of power, but a failure to deal with them changed everything politically.

Some of these men expressed a moral dimension in their discourse that they had rarely sounded before. Around the table there were those for immediately raising swords, but of all people, it was one of the holders of those swords, the secretary of Defense, who first pondered the moral dimension. "I don't know now quite what kind of a world we live in after we've struck

Cuba, and we've started it," McNamara said. "Now, after we've launched fifty to a hundred sorties, what kind of world do we live in?" That was an essential question, and it was not quickly brushed off the table.

"Why does he put these [missiles] in there, though?" Kennedy asked, wondering why the Soviets had made such a dramatic move.

"Soviet-controlled nuclear warheads," Bundy replied, ever the professor.

"That's right," Kennedy said, though that was not quite what he had asked. "But what is the advantage of that? It's just as if we suddenly began to put a major number of MRBMS in Turkey. Now that'd be goddamn dangerous, I would think."

"Well, we did, Mr. President," Bundy replied. That exchange, if Khrushchev could have heard it, would have richly vindicated his decision. The proximity of nuclear weapons aimed at them was precisely what he wanted Americans not simply to know but to feel.

From this day forward Bobby participated in all the important discussions. Bobby was all for contemplating action, even staging an incident as a pretext for invasion. "Let me say, of course, one other thing is whether we should also think of whether there is some other way we can get involved in this," he said, "through Guantánamo Bay or something. Or whether there's some ship that . . . you know, sink the *Maine* or something."

"I think any military action *does* change the world," Bundy said later in the meeting. "And I think not taking action changes the world. And I think these are the two worlds that we need to look at."

By doing nothing, the whole nature of the geopolitical world would change almost as much as if they destroyed the Cuban missile bases and invaded the island.

The following evening, the president and first lady drove in the presidential limousine to a dinner party at Joseph Alsop's home in Georgetown. Kennedy had told his wife nothing about the missile crisis, and he was in an apparently carefree mood this lovely fall evening. The acerbic conservative columnist gave the best parties in Washington, other than those at the White House. This evening he had a sterling sixteen-member guest list that included the attorney general; French Ambassador Hervé Alphand; Phil and Katherine Graham of the *Washington Post;* the new American ambassador to France, Charles "Chip" Bohlen, and his wife Avis; and Bundy.

As the distinguished group stood chatting on Alsop's terrace, Kennedy and Bohlen sauntered nonchalantly off by themselves into the garden, walking back and forth beneath the spreading magnolias in animated conversation. Bohlen had been the State Department's leading Sovietologist, and it was exquisitely bad timing that he was about to fly off to a new ambassadorial post in Paris.

It was not only in the Kremlin that the tiniest of events were analyzed for hidden meaning. As the guests pretended to socialize aimlessly, many eyes were on the pair of guests pacing back and forth in the farthest reaches of the garden. The French ambassador became increasingly intrigued, his curiosity turning to nervousness as the discussion went on and on until finally the pair returned. Over dinner Kennedy was a charming raconteur, laughing and smiling, seemingly oblivious to anything but the pleasures of a social evening.

By the time the Ex Comm met the next morning, Thursday, October 18, at 11:00 A.M. in the Cabinet Room, CIA analysts had discovered IRBM (intermediate-range ballistic missile) sites for missiles that they believed were twice the size and twice the power of the MRBMs, capable of hitting most of the United States. By then opinion had hardened among the president's advisers. McNamara asked for swift action, and General Taylor, the new chairman of the Joint Chiefs of Staff, called for a full-scale invasion of Cuba. These were not mindlessly bellicose reactions, but reasonable military solutions, given how many more Americans would die if the military attacked after the Soviets had their missiles in place ready to launch.

As Kennedy saw the hard physical evidence of the photos and heard the calls for action, he thought of the political dimensions of this problem. "If we wanted to ever release these pictures to demonstrate that there were missiles there," he asked, "it might be possible to demonstrate this to the satisfaction of an untrained observer?"

"I think it would be difficult, sir," replied Lundahl. To the untrained eye, the missiles were no more than smudges. Here was Kennedy's immediate dilemma. He would have to justify any military action to the American people and the world, and it would not be easy. When a nations lies, it is no different from when a person lies: the loss of credibility is the same. Adlai Stevenson had unknowingly lied in the United Nations about America's role in the Bay of Pigs. There would be scores of diplomats, and not only those unfriendly to the United States, who would doubt America's word if they did not have indisputable proof of Soviet perfidy.

As most of his top aides pushed for immediate action, Kennedy continued to explore the political dimensions. The president mused aloud: "If we said to Khrushchev that 'We would have to take action against you. But if you begin to pull them out, we'll take ours out of Turkey.' " A few minutes later he came back to the same point. "The only offer we would make, it seems to me, that would make any sense, the point being to give him some out, would be giving him some of our Turkey missiles," Kennedy said. In the Ex Comm meetings these were the first mentions of this possible solution.

As these men discussed the situation, they followed the steps that led from an air attack on Cuba to a Soviet reaction in Berlin, and from there to nuclear war. These were not hypothetical war games any longer. If that point needed to be emphasized even more strongly, the top White House officials all had prearranged places in Washington where they were to go with their families to be helicoptered to an immense, nuclear-proof cave burrowed into the mountains of West Virginia. Some of the mordantly imaginative of them pictured the scene as a helicopter set down on streets full of bumper-to-bumper traffic to lift them to safety while panicking Washingtonians tried to climb aboard and flee certain death in the capital.

"We figured we would have fifteen minutes' warning of an approaching nuclear weapon to get out of Washington," recalled Feldman. "We didn't think we'd be able to carry out our evacuation plans. Thus, we had plans if all the government heads were killed. Each department had to have a list of who followed second, third, and fourth. That was something we developed during the crisis."

In the midst of this deadly discussion, Bobby expressed himself in a way he had rarely spoken before. "I think it's the whole question of, you know, assuming that you do survive all this . . . ," Bobby said, "what kind of country we are."

For his part, the secretary of State had up until now been silent when such moral questions came floating up in this extended discourse. "This business of carrying the mark of Cain on your brow for the rest of your life is something . . ." Rusk began in his ponderous fashion.

"We did this against Cuba," Bobby interrupted. "We've fought for fifteen years with Russia to prevent a first strike against us. Now, in the interest of time, we do that to a small country. I think it's a hell of a burden to carry."

At 5:00 P.M. that same day, Thursday, October 18, Kennedy kept a long-established appointment with Andrei Gromyko. The Soviet foreign minister was a worldly, subtle man with whom Kennedy felt he could negotiate, as he could not with the calculatedly brutish Khrushchev. Now, as he listened to Gromyko reading from a prepared script his mirthless Marxist scenario, Kennedy was tempted to open his drawer and pull out a sheaf of U-2 photos and expose the man for the intolerable liar that he appeared to be.

Gromyko was saying in his diplomatic way what leftist protesters were shouting in the streets of America: "Hands off Cuba." The Soviet foreign minister pointed out that "Cuba belonged to Cubans and not to the United

States," and he asked, "Why then are statements being made in the United States advocating invasion of Cuba?"

Inevitably, Gromyko brought up the Bay of Pigs. That was a subject that invariably irked the president. Kennedy interrupted the foreign minister's monologue by pointing out that he had gone through all this with Khrushchev in Vienna and had said that it had been a mistake. He had promised he would not invade Cuba, and though the Soviet Union had taken certain actions, he still maintained that pledge. But Soviet armaments entering Cuba had created a new and serious situation, and American policy rested on the assumption that these were only defensive weapons.

Bobby arrived soon after the Soviet diplomat left. "The president of the United States, it can be said, was displeased with the spokesman of the Soviet Union," Bobby wrote later in words of masterful understatement.

That evening at close to midnight, Kennedy went to the Oval Office. The meeting he had just attended with his top advisers normally would have been held in the Cabinet Room, but such an unusual event in the West Wing would have aroused reporters' suspicions. Not only had he held the meeting in the Oval Room, but to further hide matters Bobby and eight of the other participants had arrived at the White House crowded into one limousine.

Only now when the others had left did Kennedy return to his presidential office by himself. As Kennedy sat there, he turned on a hidden switch and began dictating into a secret tape recorder. During the summer, he had begun recording White House meetings and phone conversations, unbeknownst to everyone but the technicians who monitored the machines, several secretaries, and almost certainly Bobby. The president was probably doing this largely to verify events in his own mind and to provide accurate recollections for the book he would inevitably write of his years in the White House. Kennedy switched the mechanisms in the Oval Office and the Cabinet Room on and off at his own personal volition. Even in the worst of moments, he had the presence of mind to turn the switch on before the Ex Comm meetings, creating an unprecedented secret documentation.

The president had no recording device in the room where the meeting had taken place this evening. So he went back to his office after a long day to record his own audio memorandum of the meeting. All his life Kennedy had kept a psychological distance from the world around him. He boxed his friends into a corner of his life and brought them out when he sought what each could give. Even Bobby saw only a part of his brother's inner life. Kennedy brought his wife into his presence for family events and to add grace to public moments. Other women were a casual diversion, largely interchangeable. He shuttled other politicians in and out of his presence, rarely letting them know how much he disdained many of them. He savored his aides' virtues and measured their weaknesses, but them too he always kept at a distance.

Now at the most important moment of his presidency and his public life, Kennedy was an observer of himself. The stakes were as high as they had ever been for an American president. His failure could lead to nuclear annihilation or, if he flinched from the Soviet challenge, to disgrace. Yet even now, after this tense, interminable day, he recorded the details of the meeting as if he were a reporter taking down events that involved someone else, and other lives.

"Secretary McNamara, Deputy Secretary Gilpatric, General Taylor, Attorney General, George Ball, Alexis Johnson," he began, running down the names of the participants. Kennedy had a superb memory, one of the essential attributes of most successful politicians. His memory served him not merely in remembering thousands of constituents' names, though he could do that, but in mastering details of legislation and policy and remembering promises made or half made. That evening he recorded the events of the meeting as if he had been there as the official secretary taking copious notes, not as the crucial figure in the room.

"Ed Martin, McGeorge Bundy, Ted Sorensen," the president continued. Kennedy had an even greater memory for human character, which is essentially a recording of a person's actions over time. He knew each one of these men as well as many others whom he was listening to outside this circle. He knew their institutional prejudices and their political passions or the lack of them. In these meetings and conversations he sometimes listened far more than he spoke. He sought and shaped consensus among his subordinates, not as a pathetic need to have his actions justified, but to summon the full moral force of these men and those who stood behind them, not the endless recriminations that had come with the Bay of Pigs. As he always did, he weighed character as much as he did words, and he pondered what would be his decision alone.

The president was speaking a midnight soliloquy into an unseen microphone, his words echoing through the room. He was not muttering about the burdens of power or the loneliness of leadership, but every sentence spoke to that point. "Dean Acheson, with whom I talked this afternoon, stated that while he was uncertain about any of the courses, he favored the first strike as . . . being most likely to achieve our results and less likely to cause an extreme Soviet reaction," Kennedy said. Acheson had been secretary of State under Truman, and he spoke with the authority of a leading architect of cold war policy. The courtly Acheson was a revered figure whose advice Kennedy believed had to be carefully weighed.

"When I saw Robert Lovett later, after talking to Gromyko, he was not convinced that any action was desirable," Kennedy then said. Lovett was equally one of Washington's wise men, an architect of postwar international policy, and his opinion was the opposite of Acheson's. "Bundy continued to argue against any action on the grounds that there would be inevitably a Soviet reprisal against Berlin," Kennedy went on. The president had immense confidence in Bundy's judgment, and his NSC adviser came down largely with Lovett, but they were in a minority. "Everyone else felt that for us to fail to respond would throw into question our willingness to respond over Berlin, would divide our allies and our country," Kennedy said. "The consensus was that we should go ahead with the blockade beginning on Sunday night."

Kennedy reserved most of his boldness for his speeches, and here as usual he sought what he considered solid middle-high ground. He had plumbed the ideas of a score or more of his advisers, and then decided to do what most of them wanted him to do. But as he turned off the tape recorder and left the empty room, whatever decisions he made would not bear the names of some distinguished committee or panel, but his signature alone.

The next morning, Friday, October 19, in the Cabinet Room, Kennedy got together with his advisers again. The Joint Chiefs had just come from their own meeting, where they had decided that a blockade was not enough; they now strongly recommended an enormous air strike against Cuba without advance notice. As this meeting started, General Taylor sought to grab the initiative and lay out the military chiefs' plan. "I think the benefit this morning, Mr. President, would be for you to hear the other Chiefs' comments," Taylor said.

"Let me just say a little, first, about what the problem is, from my point of view," Kennedy replied, subtly deferring the military's presentation. In Washington those who set the agenda usually win. Until now the president had not dominated these sessions. His role had been to listen and to weigh. But this meeting had become potentially the crucial decision-making moment, and now the president defined the problem. He had a lawyerly ability to take a myriad of contradictory contributions and prune them away into a succinct, muscular presentation of the harsh choices that lay before them.

At his best, Kennedy was profoundly realistic and intellectually fearless about facing the foibles, weaknesses, and self-interests of men and nations.

"First, I think we ought to think of why the Russians did this," he said, stepping back to admire the skillful way Khrushchev had attempted to checkmate the Americans. "Well, actually, it was a rather dangerous but rather useful play of theirs. We do nothing; they have a missile base there with all the pressure that brings to bear on the United States and damage to our prestige. If we attack Cuba, missiles or Cuba, in any way, it gives them a clear line to take Berlin."

Kennedy gave to Khrushchev his own rational mind, finding in the Soviet actions a brilliantly multilayered strategic logic that may not have been there. The president and his advisers discussed almost everything but the one overwhelming reality of the entire crisis. The United States so threatened Castro that Khrushchev was not lying when he called this murderous arsenal "defensive." Whatever grand strategic role they played, these weapons were in Cuba militarily to defend the island country against an American invasion. And as everyone in the room knew, if many of these men had their way, that possibility was not Communist propaganda but a reasonable prospect. The best way to get the missiles out of Cuba would be to convince Khrushchev that the United States would not violate the territorial integrity of Cuba.

"We would be regarded as the trigger-happy Americans who lost Berlin," the president went on, spinning his tale far beyond the borders of the Caribbean. "We would have no support among our allies. . . . They don't give a damn about Cuba." There was the president's dark realism on full display. Nations, like people, watched out for themselves; if you wanted them to help you, you had better be prepared to pay in one currency or another.

As the president talked, there were no tremors in his voice, no irritability at the endless imponderables, no hint of ill temper. He was in an emotional zone all his own, holding the others in the room steady by the sheer magnitude of his dispassion. The situation was a conundrum, and it was a measure of Kennedy's leadership that he did not pretend otherwise.

Kennedy set forth all the impalpable, difficult alternatives. He could go in and take out the missiles, but that would surely set off the Russians somewhere else. "Which leaves me only one alternative, which is to fire nuclear weapons—which is a hell of an alternative—and begin a nuclear exchange, with all this happening." He could start a blockade, but then the Russians would probably blockade Berlin and the European allies would blame the Americans. "On the other hand, we've got to do something," Kennedy concluded. "We're going to have this knife stuck right in our guts in about two months [Kennedy probably meant two weeks when the midrange missiles would be operational], so we better do something."

"I'd emphasize, a little strongly perhaps, that we don't have any choice except direct military action," said General Curtis LeMay, as if only the weak-kneed would refuse to act. The air force chief had proven his courage and resolve repeatedly in World War II. After the war the general had revitalized the Strategic Air Command into a prime weapon against the Soviet Union. LeMay saw betrayal in compromise; he had his pistol cocked and his finger on the trigger, ready for the battle he was sure would come. He was unable or unwilling to grasp the complexities of decision making in the nuclear age. His narrowly focused patriotism bordered on paranoia. He was the cold war's perfect creation, fed on the rhetoric of anticommunism. LeMay had his own natural constituency out there across America.

"What do you think their reply would be?" Kennedy asked.

"I don't think they're gonna make any reply," LeMay said. "This blockade and political action, I see leading into war. . . . This is almost as bad as appeasement at Munich."

By mentioning Munich, LeMay had come close to insulting the president. "Munich" was not a word one mentioned casually around Kennedy. To LeMay, "Munich" was only a slogan. To Kennedy, it stood at the bedrock of his intellectual life. Kennedy had an awareness of Munich unlike that of any other politician of his generation. He had done his first serious intellectual work on the issue of appeasement, and he knew that the crowds that had welcomed Chamberlain with flowers and cheers as the bearer of peace in our time had soon come to see him as a carrier of an infection of moral cowardice and compromise. The word "appeasement" had been hung around the president's father's neck too, an albatross that doomed his career in public life. Kennedy knew that if he failed at this moment and the missiles stood, he would be labeled America's Chamberlain, bearer of the new Munich.

"I think that a blockade, and political talk, would be considered by a lot of our friends and neutrals as being a pretty weak response to this," LeMay said a few minutes later. "And I'm sure a lot of our own citizens would feel that way too." One of the fundamental tenets of American democracy is that the military stays out of politics, but LeMay was lecturing the president about the supposed feelings of the American people.

"In other words, you're in a pretty bad fix at the present time," LeMay concluded.

"What did you say?" Kennedy asked, perhaps not quite believing what he was hearing.

"You're in a pretty bad fix."

"You're in with me," the president said, his words punctuated by an ironic laughter. There was no one in the room who understood as deeply as Kennedy did that he was indeed in a "pretty bad fix," part of which was military leaders like LeMay with their restless fingers on the nuclear button. When the meeting ended, several of the Joint Chiefs stayed behind to talk among themselves.

"You pulled the rug right out from under him," said General David Shoup, the Marine Corps commandant. "Goddamn."

"Jesus Christ!" LeMay laughed. "What the hell do you mean?"

"I agree with that answer, agree a hundred percent, a hundred percent," Shoup exclaimed. There was anger in his voice that suggested the wrathful vitriol that might greet the president if he did not proceed militarily. "Somebody's got to keep them from doing the goddamn thing piecemeal!"

"That's right," LeMay exclaimed. As he and his colleagues saw it, their planes, missiles, and ships were being held back, hostage to what they considered the compromising palaver of a mere politician.

"You're screwed, screwed, screwed," Shoup said. "Some goddamn thing, some way, that they either do the son of a bitch and do it right and quit friggin' around. . . . You got to go in and take out the goddamn thing that's going to stop you from doing your job."

Kennedy shared something with the simplest of men and the most complex: a belief that words mattered, that they were the primary conduits of truth. He despised the morally slovenly way men like LeMay talked about a nuclear war they had not seen and could not feel and did not understand. "I don't think I have ever seen him more irritated than when he was describing how people talked rather glibly about the escalation that might take place—with apparently no deep understanding of just what it would entail," recalled Kennedy's old friend David Ormsby-Gore, who as British ambassador saw Kennedy several times that week.

While Kennedy flew off on *Air Force One* to speaking engagements in the Midwest scheduled months before, Bobby and the other Ex Comm members spent an intense day exploring alternatives before the president returned. Most of the civilian leaders believed in a blockade, while the generals universally called for preemptive air strikes.

The president had not gone off without leaving directives. "This thing is falling apart," he said to his brother and to Sorensen. "You have to pull it together." That was all he had to say. There was a private language that Kennedy spoke among his intimates, a lingo in which Bobby and Sorensen were the most fluent, the president's nod a command, the pursing of his lips a directive. Sorensen, who was the president's intellectual alter ego, sensed that Kennedy had decided that he must begin actions against the Cuban missiles with a blockade. "That's not what he said to us," Sorensen recalled. "But he didn't have to. He knew what Bobby and I thought."

Bundy was as close to a pure intellectual as anyone in Ex Comm. His mind turned back and forth between the alternatives, finding him at one moment for air strikes, the next for a blockade, and then perhaps for no action at all. Bobby had no regard for such Hamlet-like musings, and he did not appreciate the NSC adviser's "strange flip-flops." Bundy, for his part, was no more admiring of what he considered Bobby's quick, easy certitudes, which Bundy believed would have had a deeper flavor if they had been aged at least a day or two.

Bundy tossed and turned all night long, musing about all the imponderables. In the morning he went in and saw the president before he left to tell him that he was not quite comfortable with the blockade option. "Well, I'm having some of those same worries," Kennedy said, as Bundy recalled a few months later in a private memo, "and you know my first reaction was the air strike. Have another look at that and keep it alive." Kennedy was thinking of a *limited* air strike, not setting the island aflame the way the generals were proposing.

Sorensen said later that the president had been "a bit disgusted" at Bundy's academic pondering when he should have been leading Ex Comm to a consensus for the blockade. The president had suggested opposite things to the two men. Kennedy was not a man like Franklin Roosevelt who believed that nothing flattered one aide more than hearing the president slander another. But Kennedy used people, his brother as well as his closest aides, to further policies and issues and matters that only he understood.

Ex Comm was the stage on which much of the drama played out, and Kennedy was the unseen writer of many of the lines. The fact that he was secretly recording most of the sessions only made the theatrical nature of the committee's work more pronounced. He could not afford to have the military and civilian leaders at loggerheads, and much of this endless musing was probably an attempt to gently lead these men toward consensus. The fact that the president was instructing his man Bundy to speak positively about air strikes may have been partially an attempt to signal to the Joint Chiefs that their views had not been summarily discarded.

At the meeting Friday, October 19, after Kennedy had flown off on his midwestern trip, Bundy began by saying he had just "spoken with the president this morning, and he felt there was further work to be done." These were attention-getting words. He went on to say that a mere "blockade would not remove the missiles. An air strike would be quick and would take out the bases in a clean surgical operation. He favored decisive action with its advantages of surprises and confronting the world with a fait accompli."

That was an extraordinary statement; it changed the whole tenor of the meeting and allowed others to raise their swords. Acheson weighed in supporting air strikes. So did Secretary of the Treasury Douglas Dillon and McCone. General Taylor reiterated his call for bold, immediate action. Even

Undersecretary of State George Ball, who usually wore the feathers of a dove, admitted that he was of two minds, wavering between the two courses.

Bobby said that he had talked to his brother "very recently this morning," as if to say that he who touched the throne last carried the mantle of power. The attorney general was smiling, but he was doubly emphasizing the authority that he and his words carried. He was the fiercest of cold war warriors, a champion of sabotage against Castro and almost certainly of his assassination. Many of the men in the room had heard his endless tirades against Castro, but they had not heard the Robert Kennedy who spoke that morning. As always, he was for action, but not a sudden air attack against the missile bases. He talked of 175 years of American history, without the shame of a Pearl Harbor to darken the national honor. He said that a "sneak attack was not in our traditions. Thousands of Cubans would be killed without warning, and a lot of Russians too." He called for a blockade that would seem nearly inevitably to lead within a few days to full-scale military action. "In looking forward into the future, it would be better for our children and grandchildren if we decided to face the Soviet threat [in the Western Hemisphere], stand up to it, and eliminate it now," the attorney general told the group. "The circumstances for doing so at some future time [are] bound to be more unfavorable, the risks would be greater, the chances of success less good."

In the councils of power, moral arguments are usually unfurled primarily on ceremonial occasions and then put back into the closet. In this instance Bobby held aloft a flag of high principle, but he embedded it in a foundation of steel. If the blockade failed, he would hit Cuba with all the mighty arsenal of America.

Those favoring a blockade went off to write their position paper, while those proposing an air strike prepared a paper justifying their plan.

28

"The Knot of War"

Bobby called his brother in Chicago early Saturday morning, October 20, and asked him to come back early from his trip. When the president arrived in Washington later that morning, he faced one of the most difficult decisions of his presidency. His military chiefs were calling for massive air strikes consisting of eight hundred sorties that would hit all the suspected missile sites, the Russian bombers, and supposed nuclear storage facilities, a great storm of death and destruction descending on the island. Anything less and the surviving missiles might be launched against American cities. It would be done without any warning, or the Russians would hide the missiles, making it impossible to strike them.

Before the Ex Comm meeting, the president went to the White House pool. Swimming was one of the few things that helped his back, and he rarely missed his sessions in the water. This afternoon, though, while Kennedy swam back and forth, Bobby sat next to the pool talking to his brother. To the two Kennedys, Ex Comm was as much "them" as "us," a varied group that they sought to forge into a coalition of common purpose and strategy. There was no tape recorder by the pool that day, no stenographer, and the brothers planned their strategy with no one listening to their words. At 2:30 P.M., the brothers walked in together to the meeting in the Oval Room.

This was the decisive moment, and the generals wanted nothing left unused in their great arsenals, possibly including nuclear weapons. Taylor, usually the most prudent of military men, said that he did not fear that "if we used nuclear weapons in Cuba, nuclear weapons would be used against us." These military leaders had strong, forceful arguments to make, and they made them most articulately in the stern, patriarchal voice of General Taylor, a voice that to Bobby always resonated with courage and usually with wisdom.

By all bureaucratic imperatives, McNamara should have pulled his chair up next to General Taylor and loudly seconded his call for air strikes. But the secretary of Defense knew that massive air strikes were likely to lead to a smoldering, enraged Cuba, with dead Russians strewn across the island, the probability of invasion, and a bloody response from Khrushchev. Bobby and most of the civilians in the room held their ground, arguing that the president should first call for a naval blockade while trying to talk the Soviets into removing their nuclear missiles.

After listening to this intense debate, Kennedy called for a blockade of military goods being shipped to Cuba. This would be no Munich-like acquiescence but what was generally considered to be an act of war. The president struggled to understand all the possible consequences of the various actions. His mind reached further and deeper than the mind of anyone else in the room. He kept coming back to the idea of offering to remove missiles from Turkey and Italy, yet he bristled when Stevenson suggested a straight quid pro quo: the withdrawal of missiles from Turkey and the evacuation of Guantánamo. He scribbled a little note to himself when Dillon starting talking about the Jupiter missiles either being "flops" or in such oversupply that the United States had simply unloaded them on the gullible Turks and Italians.

Now that he had decided on a blockade, Kennedy did not want the world to know that his administration had contemplated the deadly action of unannounced air strikes. "We don't want to look like we were considering it," he said. "So I think we ought to just scratch that from all our conversations, and not even indicate that was a course of action open to us. . . . We might have to do it in the future."

As Kennedy sought to mobilize the nation behind him, the truth was an unruly companion that he did not want tagging along too closely. He asked about the possibility of "putting holds on the press" or at least requesting newspapers to still their reporters' hunger for stories.

Kennedy lived in a geopolitical world where unwary wolves became sheep, and sheep might metamorphose into wolves. The missiles in Cuba made America vulnerable to enemies and allies alike. Kennedy, like his predecessor, had opposed de Gaulle's grand scheme of rebuilding France's faded glory with an arsenal of nuclear weapons; now he pondered "that in the days ahead we might be able to gain the needed support of France if we stopped refusing to help them with nuclear weapons project."

While these events transpired, Kennedy tried to create an image of normality not only for the world but also for himself. That Sunday the president called the first lady and his two children back from their weekend home in the Virginia hunt country. His wife knew very little of what was going on or why he had asked her to return. Kennedy was not a man given to darkly pondering his alternatives by himself for endless hours. He preferred convivial company, even at a time like this. He loved the company of British aristocrats as much as his wife loved French haute couture, and this evening he invited the Ormsby-Gores, the duchess of Devonshire, who bore the title that his sister Kathleen would have had if her husband had not died, and Robin Douglas-Home, nephew of another old British friend, William Douglas-Home.

The president was constantly being called to the telephone. When he returned, it was not to muse morbidly about Cuba and nuclear war but to exchange some witty repartee. Kennedy peppered the others with questions about the lives of those he found interesting, seemingly unconcerned about anything but his charming dinner guests.

Now that the president had decided on a firm policy, he had to tell the American people on television of the magnitude of the crisis that faced them. As he left the Oval Office on Monday afternoon, October 22, where he had gone over Sorensen's words, Kennedy overheard his secretary, Evelyn Lincoln, talking on the phone to Ed Berube, who was asking for some autographed pictures. Berube had a feisty authenticity about him that had amused Kennedy when the bus driver worked for the young congressman in his first senatorial race in 1952. The president had invited Berube to his wedding and as president named him postmaster in his hometown of Fall River, Massachusetts.

"Who's that? Eddie?" Kennedy asked his secretary, as if this were the most normal of presidential days. "Let me talk to him. . . . How're you, pal? How are you, Mr. Postmaster?"

"Oh . . . uh . . . uh . . . Mr. Senator . . . Mr. Congr— . . . Mr. President."

"How's your office? Anything I can do for you?"

"No, you've done enough for me now, Mr. President. We're all so proud of you. You're doing a wonderful job."

"Well, you keep up the good work. I hear some good reports about you."

The president needed the support of Congress, and when he briefed eight senators and seven senior congressmen at 5:00 P.M., just before his television speech, there was in some of their shrill voices a harbinger of the jeers and shouts that would greet him if his policies failed. This afternoon the most esteemed and knowledgeable experts on foreign policy in Congress, Senators Richard Russell and J. William Fulbright, did little but argue feverishly for war.

"It's a very difficult choice that we're faced with together," Kennedy told Russell, "Now, the . . ."

"Oh, my God, I know that !" Russell interjected. "A war, our destiny, will hinge on it. But it's coming someday, Mr. President. Will it ever be under more auspicious circumstances?"

The Georgia senator was a thoughtful man, but today he sought only to push his nation off to war. Kennedy instructed his former colleagues on the vicissitudes of leadership. "The people who are the best off are the people whose advice is not taken because whatever we do is filled with hazards," Kennedy said, speaking an epigram of power. "Now, the reason we've embarked on the course we have . . . is because we don't know where we're going to end up on this matter. . . . So we start here, we don't know where he's going to take us or where we're going to take ourselves. . . . If we stop one Russian ship, it means war. If we invade Cuba, it means war. There's no telling—I know all the threats are going to be made."

"Wait, Mr. President," Russell said. "The nettle is going to sting anyway."

"That's correct. I just think at least we start here, then we go where we go. And I'll tell you that every opportunity is full."

Kennedy stopped. The time for his nationally televised address was near. "I better go and make this speech," he said.

Kennedy sat down at his desk shortly before 7:00 P.M. to give as dramatic a speech as any American president had ever given. Always before when a president made an important address to the American people, they had had some hint of what was to be said, be it the sight of the unemployed wandering the streets or news reports of ships sunk and planes smoldering at Pearl Harbor. But across the nation, people had little idea why the president had usurped airtime on this Monday evening.

Kennedy did not seek to soothe the nation but spoke with words that would create apprehension in even the stoutest of hearts. Kennedy laid out the threat: the Soviet ballistic missiles sailing toward Cuba were capable of "striking most of the major cities in the Western Hemisphere, ranging as far north as Hudson Bay, Canada, and as far south as Lima, Peru."

As Kennedy addressed the American people, it was the image of Munich that stood starkly before him, in an era before nuclear weapons. "The 1930s taught us a clear lesson," Kennedy said. "Aggressive conduct, if allowed to grow unchecked and unchallenged, ultimately leads to war."

In Sorensen's words lay some of the tensions and arguments of the Ex Comm deliberations condensed into a few passages. "Many months of sacrifice and self-discipline lie ahead—months in which both our patience and our will will be tested, months in which many threats and denunciations will keep us aware of our dangers. But the greatest danger of all would be to do nothing." In his inaugural address Kennedy had not promised ease and blissful peace but challenge, and he delivered on that pledge a hundredfold this evening. "The path we have chosen for the present is full of hazards, as all paths are; but it is the one most consistent with our character and courage as a nation and our commitments around the world."

When Kennedy finished, many of the residents of the great cities of America feared that death stalked them, and they looked up at the silent skies with foreboding.

Kennedy's restless, searching mind reached out, seeking contradictions, new imponderables, trying to will himself into Khrushchev's mind. He had set up a naval blockade around Cuba and vowed to stop further shipment of military goods to the island. But the Soviets already had a vast nuclear arsenal in Cuba, and if he were Khrushchev, he would have ships carrying more weapons turn around. It could be ships carrying baby food and humanitarian supplies that the Americans would attempt to stop on the high seas.

Kennedy described with painful vividness what could happen if the U.S. Navy stopped a ship, even one, full of nothing but baby food. "They're gonna keep going," he said. "And we're gonna try to shoot the rudder off or the boiler. And then we're going to try to board it. And they're going to fire guns, machine guns. And we're going to have one hell of a time trying to get aboard that thing and getting control of it, because they're pretty tough, and I suppose they may have soldiers or marines aboard their ships. . . . We may have to sink it rather than just take it."

When Kennedy was not worried about confrontation on the high seas, he contemplated death on a magnitude beyond anything America had ever known. He did not visualize the ultimate nuclear war between the Soviet Union and the United States, but a more modest scenario in which five, ten, or fifteen nuclear-tipped missiles hit American cities in the midst of an invasion of Cuba. For the citizens to flee almost certain death, they would need ample notice. The problem, as McCone had noted, was that "whatever was done would involve a great deal of publicity and public alarm," signaling to

Cuba and the Soviets that the invasion was imminent. There, then, was a moral conundrum that the president might soon face.

"It looks really mean, doesn't it?" the president said to his brother as they sat together in the Cabinet Room with only a few other advisers. "But on the other hand, there wasn't any other choice. If he's going to get this mean on this one, in our part of the world . . . no choice. I don't think there was a choice."

"Well, there wasn't any choice," Bobby said, reassuring the president. "I mean, you would have been . . . you would have been impeached."

"Well, I think I would have been impeached. . . ."

During the summer Kennedy had read Barbara W. Tuchman's *Guns of August*, an epic account of how interlocking treaties and misunderstandings had inexorably led in 1914 to a great and tragic world war. History was the president's favorite lesson book, and Tuchman's lessons resounded profoundly within his psyche. Kennedy, like Khrushchev, understood that the world was only a miscalculation or two away from oblivion. While the Soviet leader slept in his clothes in his office seeking a solution that would neither dishonor his political faith nor betray his Latin comrades, in Washington Kennedy sought his own way out of the impasse.

As a score or more of Soviet ships approached Cuba, Kennedy pondered endlessly what else he could do. The president did not trust the established channels of government as the only conduits between his administration and the Soviets. As the Kennedys had done before, they reached out to the Russian agent Bolshakov as a conduit to the Kremlin.

Frank Holeman, a former New York *Daily News* journalist now working for Bobby in the Justice Department, called his Soviet source and asked for a meeting. In such an infinitely delicate situation, Holeman doubtlessly would not have made the contact except under the attorney general's explicit instructions. This judgment is reinforced by the fact that Holeman told Bolshakov things that only a person conversant with the president's inner thinking would have known. "Robert Kennedy and his circle consider it possible to discuss the following trade: The U.S. would liquidate its military bases in Turkey and Italy, and the USSR would do the same in Cuba," Bolshakov wrote in his notes of the meeting. That alone represented the most sophisticated diplomatic suggestion that had surfaced in the Ex Comm meetings. Holeman went beyond that, adding a crucial caveat that at this time had probably been thought of only by the president and Bobby. "The conditions of such a trade can be discussed only in a time of quiet and not when there is the threat of war."

When Bolshakov did not reply within a few hours, Bobby asked his friend Charley Bartlett to call the Russian and berate him. "I called Bolshakov, and I said this is outrageous what the Russians are doing," recalled Bartlett, who may also have broached the possibility of a missile trade. "I said Bobby feels a betrayal." A few minutes later Bartlett received a call from the attorney general, who, after apparently listening to a wiretap of the conversation, felt that Bartlett had gone too far in his rage.

This was not a dispute that could be solved by calculated bursts of outrage. Bobby did not seem to grasp that there was a dangerous aspect to this ad hoc covert diplomacy. The attorney general did not have a diplomat's subtle skills. As much as he cared for his brother and loved his country, he risked stirring up the waters to such an extent that more dispassionate negotiators would not be able to see through to a clear solution. The secretary of State and the Joint Chiefs of Staff apparently knew nothing of these initiatives. It was bad business trading off part of their military or diplomatic assets without their knowledge.

Anatoly Dobrynin, the Soviet ambassador, recalls that during the crisis he and Bobby "had almost daily conversations," a relationship that Bobby later downplayed to several dramatic face-to-face meetings. That evening, Tuesday, October 23, both men agree that the attorney general went to see the ambassador in his office on the third floor of the Russian embassy. The attorney general was the least diplomatic of men sent on the most diplomatic of missions. Bobby might disdain the State Department as a haven for pinstriped prissy men measuring out their lives in teatime social niceties. The reality was that a diplomat's task was to put forth a precise rendering of his nation's positions while maintaining some semblance of civility, keeping a dialogue going even in the worst of crises. Bobby, however, was at his best when his emotions were wedded to facts and he could speak as a fiery truth-sayer.

Bobby, as Dobrynin recalled, was "in a state of agitation" that accentuated the inevitable tension of this moment: he "was far from being a sociable person and lacked a proper sense of humor. . . . He was impulsive and excitable." In his memo of the meeting, Bobby remembered telling the Soviet ambassador that his brother felt that "he had a very helpful personal relationship with Mr. Khrushchev . . . a mutual trust and confidence between them on which he could rely." When politicians praise each other, their feelings are often the precise opposite of what they say, and Bobby's words were as platitudinous as they were untrue. Bobby then laid out what he considered the whole litany of betrayal in the most vivid detail, accusing the Soviet leaders of being "hypocritical, misleading and false." The Russian could give only a diplomat's most pathetic response—that he knew nothing of what his nation purportedly had done. Bobby was full of righteous anger, which an envoy from the State Department would never have expressed so dramatically to the Russian ambassador. The Soviets believed that they had legitimate policy goals in Cuba that they were pursuing by legitimate means. But they had to understand that they had triggered an honest rage and sense of betrayal in a powerful enemy. The Soviet ambassador sent a message to Moscow that gave "an idea of the genuine state of agitation in the president's inner circle."

Despite the merciless tension, Kennedy had a preternatural coolness about him. That same evening, October 23, the president and first lady attended a dinner party at the White House for fourteen guests, including the maharaja of Jaipur and his wife, journalist Benno Graziani and his wife, Nicole, and two old friends, British ambassador Ormsby-Gore and Charley Bartlett. Every person in the room was aware of the immense drama that was taking place, and yet the tone of the evening was one of convivial sociality, without a mention of missiles or Cuba.

At around 11:00 P.M., Bobby arrived at the White House. He told the president and Ormsby-Gore of his difficult meeting with Dobrynin. The moment of climax was arriving quickly, too quickly, out on the open seas. To gain more time Ormsby-Gore suggested that Kennedy move the quarantine line from eight hundred to five hundred miles, a proposal that Kennedy accepted. At around midnight, when Nicole Graziani was scrambling eggs in the private quarters for a midnight repast before the group departed, Kennedy turned to the young woman. "He [Kennedy] took her hand," remembered her husband, Benno, "and he said, 'You know, maybe tomorrow we will be at war.'. . ."

On Wednesday morning, October 24, the United States ratcheted up its military readiness to DEFCON 2, only one step below war. The air force's massive B-52 bombers loaded with nuclear weapons flew twenty-four hours a day, refueled by in-flight KC-135 tankers, ready to enact their savage revenge even if Soviet missiles destroyed most of America's military capabilities. In the South Atlantic, two Russian ships, the *Gagarin* and the *Kimovsk,* drew near to the imaginary line Kennedy had drawn five hundred miles from Cuba. If the vessels did not turn back soon, the navy would try to stop them, and war would be a giant step closer to beginning.

Bobby took a seat across from the president at the Ex Comm meeting and looked into his brother's drawn face, with "his eyes pained, almost gray." The burden of this moment was like a physical pressure bearing down on him, "the danger and concern . . . like a cloud over us all and particularly over the president."

There was no button Kennedy could press, no gauge he could read to tell whether the situation was about to explode, and every moment there seemed new uncertainties, new elements. Now there were Russian submarines running deep near the Russian freighters closing toward the line of blockade.

"Here is the exact situation," McNamara said. "We have depth charges that have such a small charge that they can be dropped and they can actually hit the submarine, without damaging the submarine. Practice depth charges. We propose to use those as warning depth charges."

Kennedy was a navy man and knew that the admirals' most exquisitely conceived plans often became nothing more than inane doodles once combat began. The president had, if anything, too accurate an imagination about all the possibilities of this moment. And as he sat there, Bobby saw him put his hand up to his face, cover his mouth, and close his fist. That was not his brother as he had ever seen him, and for a moment he worried not about war but about Jack.

This room was full of powerful, intense men speaking ponderous words, but for a moment the brothers stared at each other and Bobby had the sense that "for a few fleeting seconds, it was almost as though no one else was there and he was no longer the president." As Bobby looked at his brother he recalled later that his mind flashed back to so much that had gone on in the family. Their father had taught his sons to see a pinprick of blue in the blackest sky, but he thought now only of the darkest of times. He thought of how Jack had been ill. He thought of the day at Hyannis Port when they learned of Joe Jr.'s death. He thought of the day Jackie lost a child, when he had been there and his brother had not. He thought "of personal times of strain and hurt," the memories flashing by with such intensity that he heard not a word of the discussion going on around him.

Out of this miasma of memories, Bobby heard his brother's voice. "If he doesn't surface or if he takes some action—takes some action to assist the merchant ship, are we just going to attack him anyway?" the president asked about a Russian submarine shadowing the Soviet freighters. "At what point are we going to attack him?"

Kennedy did not even wait for his military leaders to give their strong response. "I think we ought to wait on that today. We don't want to have the first thing we attack [be] a Soviet submarine. I'd much rather have a merchant ship."

John McCone came into the room a few minutes later after gathering the latest intelligence on the Russian ships approaching the imaginary barricade that the Americans had drawn in the Atlantic Ocean. "Well, what do they say they're doing with those, John?" Kennedy asked.

"Well, they either stopped them or reversed direction," McCone responded.

In that room there were no audible sighs, no backslapping, and no self-congratulations. "The meeting droned on," Bobby recalled. "But everyone looked like a different person. For a moment the world had stood still, and now it was going around again."

The threat of an immediate confrontation on the high seas was over, but the missiles of October remained in place. For Kennedy, the weight of his burden did not lessen, for U-2 photos clearly showed how quickly the missile sites were being built and the IL-28 bombers uncrated and prepared for flight. The longer the president negotiated, the greater the possibility that the missiles would be combat-ready and Khrushchev would walk with a bold new strut.

The next afternoon, Thursday, October 25, Kennedy walked in on a scene of Jackie photographing Caroline while Robin Douglas-Home carved an enormous Halloween pumpkin. His wife had just finished being filmed for an NBC special on Washington's new National Cultural Center. Although the Kennedys were often called American royalty, a coinage so antithetical to American traditions, there was one way in which they exemplified the finest aspects of noble blood. That was the way they let nothing, no personal unhappiness, no private pain, and no public problem, affect their performance of public rituals. By now Jackie knew why her husband had asked her and the children to return from their weekend home. He wanted her with him through this crisis, and he wanted the world to think that life was going on normally at the White House. She had not canceled the television interview today, despite the political drama and the fact that John Jr. was in bed with a 104-degree fever.

Douglas-Home stayed for dinner that evening. This was not a mannered social function but a casual meal. "For God's sake, don't mention Cuba to him," Jackie admonished her guest. Kennedy was involved in the greatest crisis of his presidency, and yet even now at this moment he was fascinated by gossip and trivia, all the flotsam of popular culture and modern society. He sat puffing on a cigar, the evening interrupted by any number of serious phone calls about Cuba. Listening to him through most of this evening, though, was like paging through a wondrously eclectic magazine. Here was an adroit essay on how Lord Beaverbrook ran his newspaper empire. And an ironic, insightful article on Frank Sinatra's way with women. Then an aside on the infamous photo of a model sucking her thumb while lying on a bearskin rug that ran in *Queen*, the British magazine. Nothing seemed too trivial or too bizarre for Kennedy this evening as he segued from subject to subject that had nothing in common except his curiosity about them.

Even at the worst of the crisis, Lansdale continued to push the covert role of Operation Mongoose. "Lansdale feels badly cut out of the picture and appears to be seeking to reconstitute the Mongoose Special Group operations during this period of impending crisis," the CIA's deputy director wrote his superior, McCone, on October 25. The next day at a Mongoose meeting, at which Bobby was present, McCone "stated that he understood the Mongoose goal was to encourage the Cuban people to take Cuba away from Castro" and that the CIA "would continue to support Lansdale."

The president seemed almost equally unwilling to face up to the melancholy reality that he might have to promise to live with Castro's Cuba. When Kennedy talked to Prime Minister Harold Macmillan about the possibility of guaranteeing the territorial integrity of Cuba in exchange for the removal of the missiles, the president told the British leader "that would leave Castro in power," as if the United States could demand his removal. The only way out of this was to negotiate, yet the experts on the NSC staff were advising "the primary Soviet tactic will be to draw the U.S. into negotiations, meanwhile getting a standstill."

Khrushchev had indeed been willing to dawdle until the missiles were in place and only then negotiate, but he was realizing that he might be faced with the imminent invasion of Cuba. He wrote a letter to Kennedy that arrived late Friday evening, October 26. If the letter rambled, it was no more than human emotions often ramble. It was a display that some in Washington feared meant that the Soviet leader had become unstable, overly emotional, and dangerously incoherent. Bobby understood, as some of his more phlegmatic colleagues could not, that "it was not incoherent, and the emotion was directed at the death, destruction, and anarchy that nuclear war would bring to his people and all mankind."

Khrushchev's letter pleaded with Kennedy to back off his blockade of Soviet ships. In exchange, Khrushchev would "not transport armaments of any kind to Cuba" during negotiations. For all its high emotional tenor, the letter proposed a subtle, carefully constructed solution. Everything would change, Khrushchev wrote, if Kennedy gave assurances "that the USA itself would not participate in an attack on Cuba and would restrain others from actions of this sort." Kennedy would have to call off his covert operations against Cuba and rein in the Cuban exiles and their attacks on Castro's regime. If that was done, as the Soviet leader saw it, "the question of armaments would disappear," and there would be no reasons for Russian missiles.

> I don't know whether you can understand me and believe me. But I should like to have you believe in yourself and to agree that one cannot give way to passions. . . . Mr. President, we and you ought not now to pull on the ends of the rope in which you have tied the knot of war, because the more the two of us pull, the tighter that knot will be tied. And a moment may come when that knot will be tied so tight that even he who tied it will not have the strength to untie it, and then it will be necessary to cut that knot. And what that would mean is not for me to explain to you, because you yourself understand perfectly of what terrible forces our countries dispose.

The next morning Kennedy received a second, more formal letter in which Khrushchev proposed a new element. The United States would not only have to make its all-encompassing noninvasion pledge but also remove its Jupiter missiles from Turkey. This latter idea had been discussed in Ex Comm meetings, broached with Bolshakov, and in other ways thrown on the table as a possibility. The Soviets announced over Radio Moscow their offer of a compromise, letting the world see what Kennedy himself realized would be regarded by most people "as not an unreasonable solution."

In this moment the cold war had reached its high point, not only in its imminent dangers but also in all the posturing of power. The military chiefs were the most adamant in their opposition to trading off these missiles for a result some would dare call peace. They stirred restlessly, fancying themselves in a death struggle against an implacable Communist foe. Stripped of their ideological veneer, Generals LeMay and Taylor were like the brightly plumed gentlemen leading the cavalry in the charge of the light brigade during the Crimean War. These military chiefs stood in their stirrups, swords raised, ready to charge through the valley of death in the name of honor.

"The missiles [in Turkey] were worthless in the Eisenhower administration," reflected McNamara two decades later. "They sure as hell were worthless and known to be worthless in the Kennedy administration. And yet, because the Soviets . . . said, in effect . . . 'We won't remove our missiles from Cuba unless you remove yours from Turkey,' there was almost a requirement that we go to war with the Soviets to preserve missiles in Turkey that were worthless."

Kennedy was in a predicament of excruciating difficulty. Khrushchev might coo his sweet song of peace, but as he did so his soldiers hurried to finish their Cuban missile bases. At least five bases already appeared operational. As Kennedy discussed a response with the civilian leaders of Ex Comm, the Joint Chiefs were preparing for Oplan 312, a full-scale air strike on October 29, followed seven days later by Oplan 316, the invasion of Cuba. The military leaders believed that they had to overwhelm the enemy. First, waves of air strikes would pulverize the missile sites, the airports, and the military facilities. Then the most massive American invasion since D-Day would move rapidly through the shell-shocked, demoralized defenders.

Bombing rarely decimates an enemy. Some of the missile sites may have survived, and surely many of the Cubans and their Soviet allies on the island would have defended it to the death. Moreover, the military chiefs did not know that the Soviet weapons included tactical nuclear-tipped missiles. Most of these Luna missiles would not be taken out in air raids. In an invasion the Soviet commander, General Issa Pliyev, had originally been authorized to use them, but on October 27, Moscow changed that directive, requiring formal authorization from officials in Russia. In the lexicon of the nuclear age, these were not major weapons, but anyone within half a mile of the center of the blast would die, most of them immediately, though a few would survive only to succumb within a few weeks to radiation poisoning. Once the Soviet launched their Lunas against the invading Americans, Kennedy would doubtless respond with nuclear weapons too, and World War III would likely begin.

The danger was that Kennedy and the others might become so immersed in the minutiae of the moment that they would not be able to stand back and see the full scale of what was at stake. Only Kennedy seemed able to distance himself enough to see this crisis set in the context of history and human conduct.

"If we appear to be trading the defense of Turkey for a threat to Cuba, we'll just have to face a radical decline in the effectiveness [of NATO]," Bundy told the president on October 27. Kennedy's NSC adviser probably did not know that, through Bolshakov, the administration had informally and secretly proposed such a trade.

"This trade has appeal," Kennedy replied. "Now, if we reject it out of hand, and then have to take military action against Cuba, then we'll also face a decline [in NATO]." Kennedy faced square on the natural self-interest of men, even if they wore the badge of allies. He knew that those same Europeans who would condemn him for withdrawing the Jupiter missiles would complain even louder if America went to war over Cuba.

When Kennedy spoke of these allies, he displayed a passion that he rarely displayed to his nation's enemies. "We all know how quickly everybody's courage goes when the blood starts to flow, and that's what's going to happen to NATO," he told his colleagues. "When we start these things and they grab Berlin, everybody's going to say, 'Well, that was a pretty good proposition.' . . . Today it sounds great to reject it [trading off the Turkish missiles], but it's not going to after we do something."

The other part of the deal was the promise not to invade Cuba, which to Bobby was almost as big a problem as Turkey. "Well, the only thing is, we are proposing in here the abandonment . . ." he began.

"What?" Kennedy said urgently. "What? What are we proposing?"

"The abandonment of Cuba," Bobby repeated.

"No, we're just promising not to invade," said Sorensen, always the wordsmith.

"Not to invade," McNamara repeated. "We changed that language."

No matter what pledges their government made, men such as Bobby, McCone, and LeMay would not accept a Communist sanctuary in the Caribbean. The diplomats might believe otherwise, but there was a misunderstanding here much like that between Kennedy and Khrushchev at the summit conference. Then Kennedy criticized the Soviet leader for tinkering in the affairs of other nations, but Khrushchev said that a liberation struggle like that against the Portuguese colonies in Africa was "a sacred war" and the Soviet Union would always support such struggles. For Bobby and his allies within government, Cuba was just such a sacred war.

As Kennedy attempted to find some peaceful resolution that would not be condemned as cowardice, events conspired to push him closer to raising war's banner. A U-2 plane had wandered off course in the Arctic and been chased out of Soviet air space by a menacing squad of MIGs. The Cubans had started firing at low-level reconnaissance planes, forcing them to turn back. And Kennedy learned that in Cuba a missile had been fired at a U-2 plane flying high above the island, bringing the plane down and killing the pilot, Major Rudolph Anderson.

Kennedy and the Ex Comm team were strong men at the height of their intellectual powers, but they had been working day and night for eleven days, living with an intolerable level of stress. They were tired, and even as they tried to use good judgment, some of them were ready to strike back at those who taunted them.

"We should retaliate against the SAM [surface-to-air missile] site and announce that if any other planes are fired on, we will come back and take it," General Taylor asserted.

"We can't very well send a U-2 over there, can we now, and have a guy killed again tomorrow?" Kennedy mused aloud.

"I think you're going to have great pressure internally within the United States too, to act quickly, with our planes always being shot down while we sit around here," Dillon said minutes later, a tacit criticism of Kennedy's apparent passivity.

In this decision-filled day, even the taciturn, restrained Dillon sounded strained. Of all people, the usually undiplomatic Bobby advanced a possible solution.

Stevenson was worried that the proposal to stop the blockade in exchange for Soviet removal of the missiles sounded too harsh. "I think it's just an acceptance of what he [Khrushchev] says [in his first letter]," Bobby said. "Don't you think?"

"Actually, I think Bobby's formula is a good one," Sorensen said soon afterward. "Does it sound like an ultimatum if we say: 'And we are accepting your offer in your letter last night. And therefore there's no need to talk about these other things'?" In doing so, they would not mention Khrushchev's second letter with its talk of Turkish missiles.

These Ex Comm meetings were the stage on which the president spoke his public lines, doubly recorded by the ever-whirling tape and the careful memos and recollections of the men around the large table. In this public forum Kennedy did not criticize the mindless bellicosity of LeMay or berate Stevenson for what he considered his endless timidity. Nor did Kennedy voice his uncertainties here or contemplate policies in words that could be used against him by men whose memories might last longer than their loyalty.

After this day of endless discussions, Kennedy met in the Oval Office with a select group of about eight that included Bundy, McNamara, Sorensen, Rusk, and Bobby. These men would keep Kennedy's secrets. They decided that Bobby should meet with Dobrynin. Kennedy trusted Bobby, not simply because the attorney general was his closest blood kin. He trusted him because Bobby was closest to him in his view of the crisis and his determination to seek a solution. Bobby was to tell the Soviet ambassador that in exchange for removing the Cuban missiles the United States would end the blockade and promise not to invade Cuba. Beyond that, he was to tell the diplomat that as soon as matters calmed down, the Turkish missiles would be quietly removed.

Dobrynin met Bobby in his Justice Department office on the evening of Saturday, October 27. The ambassador noticed a dramatic change in the president's emissary since he had met with him four days before. He appeared exhausted, as though he were living sleeplessly on adrenaline. Kennedy told the diplomat in the boldest terms that unless the Soviets took their nuclear missiles home, the United States would remove them with the full force of arms. Bobby's voice was tense with emotion when he told the ambassador that the American generals and others were "spoiling for a fight." Bobby was under tremendous strain. He knew that if this proposal failed, the hawks would rise in such magnitude that they would darken the skies.

In this meeting all that mattered were the final exchanges, not the unpleasantries and recriminations. "What about the missile bases in Turkey?" Dobrynin asked finally, after Bobby had spent his anger.

Early in the crisis, the prescient Kennedy asked himself that very question. From the American perspective, the immediate negotiations ended when Bobby laid out the deal to the Soviet ambassador and told him that the details had to remain secret. This was a final offer; if it was turned down, the generals would have their war.

The Robert F. Kennedy who sat there talking intensely to the Soviet ambassador was not the same Robert F. Kennedy of two weeks before. No one else in those endless Ex Comm meetings had so dramatically shifted his perspective. General LeMay did not lie down with the lambs, nor did Ambassador Stevenson saddle up as a rough rider leading the charge toward Cuba. Only Bobby spoke lines he had never spoken before. He was caught between emotions that took him from unrelieved fury at the Soviets for betraying his brother and his nation to an equally profound sense of the immense dangers the world faced, and then back again. In the end he understood the enormous stakes that were at play and how close they were to stepping off the precipice.

When Bobby talked to Dobrynin that final evening, the ambassador recalled that Bobby was almost crying. These were not calculated tears, but honest ones for his brother, his family, his nation, and the world. Bobby had not forsaken his personal agenda. Dobrynin recalled that at a follow-up meeting the attorney general said "that some day—who knows?—he might run for president, and his prospects could be damaged if this secret deal about the missiles in Turkey were to come out."

Khrushchev lived among hawks and doves too. Castro himself had admonished the Soviet leader to consider responding to an American invasion of Cuba with a nuclear first strike against the United States. Marxists see history as the story of immense social, economic, and political forces working their way across time. To Castro, the slaughter of millions was a noble sacrifice if the Marxist system survived and out of its ashes rose a Communist paradise. To those who view history that way, writing about an individual life is like chronicling the life of a toe, a hand, or some other appendage that means nothing without the body to which it is attached.

If ever history has been the chronicle of individual human character moving through time, though, it is in these thirteen days, and that is as true for the Marxist Khrushchev as for the Kennedys. The Soviet leader decided that he had to take his missiles home, and he saw the removal of the Turkish missiles as no better than a paltry consolation prize. "In order to save the world, we must retreat," Khrushchev told the Soviet Presidium. These were not words that a leader could speak often or loudly. Khrushchev had blinked, but it was probably not Kennedy's tough stance that had set his eyes aflutter as much as Russia's military weakness and the prospect of the world's being engulfed in a war of horrors beyond human imagination.

On Sunday morning, October 28, Radio Moscow broadcast Khrushchev's message around the world: the Soviet Union "has given a new order to dismantle the arms which you [Kennedy] described as offensive, and to crate and return them to the Soviet Union."

During the two-week crisis the great themes of the president's life came into focus. In 1947, as a freshman congressman, Kennedy had prophesied a time when the Soviets would have massive atomic armaments and there would be "the greatest danger of . . . a conflict [that] would truly mean the end of the world." During his years in Congress, he had seen the Soviet Union as the most serious threat to his nation, and as president he feared that he would face this ultimate confrontation. He had been in part responsible for this greatest of cold war crises, since without the relentless covert attacks on Cuba, the missiles of October never would have arrived in Cuba.

The other great theme of his life was courage, to him a man's highest virtue. He had written about it, and both in the war and in his long struggle with ill health, he had walked with a hero's bold stride. Political courage is struck of even sterner stuff, and many questioned Kennedy's mettle when it came to the hard questions of his age.

Kennedy had stood apart from his advisers in his intellectual detachment. He did not get lost in the emotional minutiae of the moment. His eyes searched ceaselessly for a safe harbor in turbulent seas.

Kennedy had been the first to mention the missiles in Turkey as a possible bargaining chip. That decision required intellectual courage to seek a solution so far away from the narrow straits of diplomatic combat. He had at the same time finessed Generals LeMay and Taylor and the other military chiefs, subtly procrastinating until peace could work its way slowly to the fore.

A brave man did not have to leave his shrewdness at home, and Kennedy had worked with immense sagacity and high political duplicity. He kept the deal over the Turkish missiles secret, since he knew that the Republicans would bludgeon him with it in the congressional election. He knew too that the Soviets would keep their pledge of silence; as much as the Soviets disliked him, they disliked the Republicans more.

As the president debated the Cuban situation, he did not stand in an arena with two doors before him marked "Courage" and "Cowardice." He stood before a multitude of doors, some of them barred, others half open, all marked in an obscure language. When he finally walked through one, some of those watching were convinced that it had been clearly labeled with the word "Courage." Others swore that it was marked "Cowardice," either for what they considered his gutless refusal to stand up to the Soviet tiger or conversely for recklessly provoking a needless crisis.

Whatever they thought, though, most of the audience left the arena believing that the drama had ended. But history is rarely resolved in epic confrontations, and much of the drama of the missile crisis began when almost everyone thought it was over. It is a drama that in many respects still haunts America today.

The Soviets moved quickly to remove their missiles, allowing American ships to come close enough to their freighters to photograph the massive closed crates on deck. They were less willing to take home their IL-28 planes since they were not explicitly part of the earlier negotiations and were considered defensive weapons. These planes became a major sticking point in ending the blockade. In the next weeks the Kennedy administration debated how to resolve this matter, not only the IL-28s but the overall relationship between the United States and Cuba.

"Once we've got these missiles out . . . don't we want to look at everything else against the background of our long-range objectives of eliminating communism in Cuba?" said Dillon at the NSC meeting on November 7. "Don't we want to press along to get sort of a halfway workable inspection system? . . . Or you might prefer to get a less good inspection system and strengthen your ability to get rid of them."

In the White House important new proposals were often not forcefully made but gingerly introduced in the guise of mere alternatives. The clear import of Dillon's suggestion was that the administration should consider backing away from promising not to invade Cuba and instead contemplate finishing off what it had begun at the Bay of Pigs.

Bundy quickly followed the secretary of the Treasury's lead, asking whether they were "trying in effect to get a bargain in which we have undertaken or to avoid any bargain."

Kennedy heard every nuance of his colleagues' statements, but he made his position emphatically clear. "With this election now over, it seems to me that we ought to just play it straight, say what is pleasing and what is not," Kennedy said in words forcefully spoken. "We wouldn't invade unless there was a major upheaval on the island or a reintroduction [of offensive weapons]. Otherwise, our commitment ought to stand. We don't plan to invade Cuba. *But* we are ready to give that in a more formal way when they meet their commitments."

There was yet another voice in this meeting that resonated with neither the narrow interests of one inner governmental constituency nor an ideological mindset. When the president asked Llewellyn Thompson to comment, Kennedy was listening to a mind versed in the nuances of history and the Soviet Union, in a man who pandered to no one. The former ambassador to Moscow knew Khrushchev and the evils of communism firsthand, and far better than anyone in Washington, but he also knew the realities of power in the nuclear age.

"There's another angle in this that we ought to keep in mind," Thompson said in his modest, studied manner. "We still have a European aspect of the whole question of our whole relationship with the Soviet. . . . Khrushchev . . . has gotten the missiles out quickly, and I think he had hoped as a result to get the quarantine lifted quickly. . . . It does look from their side that we have not only tried to widen this to the bombers and lots of other things and have indicated we're not even going to lift the quarantine and we're not going to come forward with the statement [promising not to invade]. So he's in pretty poor position to say what he got out of this thing for his quick action. . . . As important as the IL-28s are, they are old planes, and they are not as important as not having a showdown over Berlin or a chance of getting somewhere in negotiations. . . . We want to be very careful right now in the way we play it that [we] not botch ourselves off in order to get at Castro. Or we may lose the chance to see whether or not Khrushchev, now having had this confrontation, is ready to cut his losses. I think basically he does want to get in a position with us where he can put it on the basis of economic competition."

This was precisely Kennedy's thinking. If he held firm to that position, a no-invasion pledge to the Soviets had immense political benefits. He would still keep a wary eye on Castro, but he could walk away from this dangerously parochial obsession with Cuba and get on with his work on the broadest, most significant scale.

These NSC meetings were attended by articulate, impassioned men with very different ideas. Bobby spoke in a sweetly tempered tenor voice, but often his words had steel to them. If the Soviets did not remove their planes, he wanted the group to consider alternatives, including "conducting surveillance in such a fashion that they would shoot at us and then we would then have an excuse for going in and dropping bombs on the IL-28s."

"I think we've got a hell of a lot of cards in our hand," Bobby said a few minutes later. Once the IL-28s were gone, the game was supposedly over.

Bobby, though, was holding a new hand that he would not easily put down.

In this new crisis Bobby played a central, complex, and contradictory role. He usually sat with the hawks at the NSC meetings, but to the Russians he often played the dove, a role that his colleagues celebrated as good acting. "Bobby's notion is that there is only one peace lover in the government entirely surrounded by militarists," Bundy joked to Kennedy. "Bobby is feeding him [Dobrynin] that stuff, Mr. President."

On November 9, Bobby invited Bolshakov to his home in Virginia. There the attorney general once again began the private, second-track negotiations that during the past year and a half had been such a mixed blessing. The attorney general disliked all the pomp and posturing of diplomacy. He wanted to cut to the chase, in this case seeking a direct, immediate solution. His colleagues, however, did not always know what he was saying or how much he was giving away of their strategy. In this instance he gave what he called his "personal opinion" that if the Russians would remove the planes "as soon as possible," the Americans would accept that solution, if "the USSR gave an undertaking that these planes would be piloted only by Soviet aviators" and not by Cubans.

Bolshakov had hardly had time to drive back to Washington when he received a call from Bobby at the White House, telling him that the president would accept only "the rapid removal of the IL-28s from Cuba." This was the kind of snafu that would have gotten an ambassador sacked, but Bobby continued as an active player in this diplomatic game.

Bobby was his brother's chosen emissary on the most sensitive missions. Three days later the attorney general took the newest proposal to Dobrynin, promising that if the Russians would agree to remove the planes within a "definite schedule . . . let's say, in the course of thirty days," the Americans would immediately end their shipping quarantine. He said that the president would agree not to invade Cuba, but that would remain a verbal agreement, not part of a signed protocol. In the end Khrushchev accepted the American proposal. "We have the firm impression that the Americans actively hope to liquidate tensions," Khrushchev wrote Anastas Mikoyan, a member of the Soviet Politburo. "If they wanted something else, then they had opportunities to get it. Evidently, Kennedy himself is not an extremist."

In settling the immediate crisis, Kennedy could have pushed the Cuban issue to the back burner of international politics, turning it down to a simmer. He had an excellent device for doing so in his pledge not to invade Cuba. To be successful, he would have had to do precisely what he had done during the Cuban Missile Crisis: finesse the Joint Chiefs, the CIA, and other proponents of aggressive action almost as much as he attempted to finesse the Soviets. Instead, he diminished the scope and intent of the no-invasion pledge.

By authorizing extensive covert actions and making the threatening sounds of war, the Kennedy administration had some culpability in the Russian decision to place missiles in Cuba. Yet the president was willing to head back down that same dark road. He had signaled to the Joint Chiefs, the CIA, and the exile community that he would not sit by letting Castro luxuriate in peace and security. There were strong, willful men waiting for that signal, men who would set out again to try to kill Castro, men who would seek to poison and burn Cuban fields, men who would try to become such a thorn in Castro's side that he and his regime would bleed to death. Among these men, there was probably no one more determined, no one more willful in his resolve, no one more obsessed with Castro, and no one more committed to his downfall than the president's own brother.

29

The Bells of Liberty

Joe taught his sons that nothing mattered more than family loyalty. As the Kennedy brothers became men, they did not so much develop a newfound closeness as discover the ties that had always bound them. The three men shared a belief in public life as a man's greatest sphere, a transcendent concern for the high destiny of their family and its name, and a love that in its intensity and purity was unlike anything they bestowed on anyone else, even their own wives.

The decision about whether Teddy should go ahead with his Senate race in Massachusetts was profoundly complicated, involving not only the highest political calculations but also the most intimate emotional bonds of family. The president was certainly in favor of Teddy's advance in the world, but he did not like to spend his political capital unless he had some chance of receiving a return on his investment. If Teddy ran, many Americans might conclude that the Kennedys were not an exalted race of citizen politicians but an ersatz royalty believing itself entitled by right of birth to the highest offices in the land.

Teddy, a classic last son, was forgiven faults for which his brothers would have been harshly judged. Teddy's life did not have the hard-won authenticity that his brothers had achieved. In his quest for credibility as a candidate, he traveled from country to country as if he were plucking exotic fruit, taking a nibble of Ireland here, a bite of Italy there, seeking not knowledge but votes. In the fall of 1961, he sat at the head table at the annual dinner of a society of Italian-American attorneys, an honor rarely accorded an Irish-American assistant district attorney for Suffolk County. After a speech in which Teddy celebrated the incomparable greatness of Italy, the group saw a film titled "Ted Kennedy in Italy," an epic account of his "good-will mission." By the time the film ended, Teddy was already gone, on to his next public moment.

It was just as well that Teddy brought no film back from his monthlong tour of Latin America in the summer of 1961. When he left Panama, Walter Trohan of the *Chicago Tribune* asked Joseph P. Farland, the American ambassador, for his impressions of young Teddy's short visit. "I can tell you what I told Teddy the next morning," the ambassador said. " 'The harm that you have done in six hours will take me six months to undo.' " The putative candidate cut a bold swath across South America meeting with all kinds of people, his excursions making American diplomats nervous. He was eclectic in his interests. In Rio, near the end of his journey, he took an interest in a beautiful young woman, not even attempting to keep the matter secret.

Teddy was a young man of prodigious energy; as late as he stayed up and as wildly as he caroused, he was the first one up in the morning, ready to get on with his explorations. Across the continent he was greeted not as another hustling American politician but as his brother's surrogate. He had picked up the president's most idealistic phrases and talked fervently about the terrible plight of the poor and their aspirations. When his brothers had made their youthful foreign journeys, they had gone off without entourages and friends walling them off from the experience they sought. They had either published articles themselves or written extensively in their diaries, and their travels had marked them deeply. Wherever Teddy went, ambitious young men latching themselves onto his future surrounded him. It was unthinkable that he would sit down and do the grunt work of writing about his travels or penning extensive notes in a diary to be read only by him. Instead, when he returned to Massachusetts, the *Boston Globe* ran a five-part series on Teddy's journeys, with detail worthy of a president or a secretary of State and a notable lack of attention to his evening adventures; and celebrated his new insights, such as that "some 200 million human beings in Latin America are demanding membership in the 21st century."

Bobby was the least enchanted with Teddy's desire to run for the Senate. The attorney general was not only the most moralistic of the Kennedy men but the most moral, and the whole idea rankled him. He was the president's protector, not little Teddy's, and he was not about to sign on to an enterprise that might embarrass the president. Even Joe was not as determinedly behind Teddy's candidacy as he had been in pushing the president-elect to name Bobby attorney general.

Before his stroke, Joe asked Clark Clifford, a man of astute political judgment, to make his own assessment. Clifford, though initially opposed, came around to the idea that Teddy should be allowed to make his race. "Bobby was opposed still," recalled John Sharon, a Democratic Party campaign organizer involved with the matter. "Teddy wanted to run, but he obviously saw that his family had pulled in this outside adviser."

There were others criticizing the possible candidacy. Kenny O'Donnell worried about the damaging political implications of the race, as did Kennedy's dear friend Chuck Spalding. "The only argument I ever had with Jack was once when we were going to Camp David when Teddy was being considered to run for the Senate," recalled Spalding. "I thought it was possibly too much. Jack asked, 'What is bothering you?' He provoked me to say what I felt. I said, 'You're going to get all kinds of criticism. Teddy hasn't done anything.' Jack said, 'He's going to win in Massachusetts bigger than I did. And besides, Dad is interested.' "

Teddy went ahead and announced his candidacy on March 14, 1962, but one major obstacle remained if he was going to run the race he wanted to run. That was the Harvard cheating scandal. All across the Commonwealth there were whispers about a scandal at Harvard that the Kennedys had covered up. It had become a matter that, if not handled properly, could saddle Teddy with enough shame to doom his candidacy before it began. This was a matter of such importance that the president himself decided that he would be the maestro orchestrating every detail.

The Kennedys needed the proper publication, and that inevitably was the pro-Kennedy *Boston Globe*. Dick Maguire, the Democratic Party treasurer, called Bob Healy, the *Globe*'s Washington bureau chief, who happened to be in Boston. Healy was a fine reporter and a Kennedy partisan privy to many of the inner workings of the family. Maguire invited Healy to his suite at the Parker House. Over drinks, the Democratic leader broached the matter of the cheating scandal, asking the reporter just what he knew. The two men were well into the dance that politicians and journalists often perform when the phone rang. It just happened to be the president of the United States calling from Washington.

"Can you put it in a profile?" Kennedy asked, warming to the idea of sticking the story in the middle of a larger story, hoping that it would get lost in the rest of the article.

"No," said Healy, a word that the president heard only rarely.

"What do you mean?" Kennedy asked.

"Because that would give you five stories on this thing, day after day," he said. "I'm only going to do this piece if I can do it in one complete piece, the story at Harvard, everything."

The president decided that he could not handle this matter in a phone call. So he invited Healy to the White House the next afternoon. There the two men spent an hour discussing how the *Globe* would break the story.

Healy laid out everything he would need, from the Harvard academic records to an interview with Teddy.

"How'll you play the story?" asked Kennedy.

"I don't know," Healy said, though it seemed doubtful that a story of this magnitude would be anywhere but on the front page. "You know enough about newspapers that basically a story is played on the basis of the news."

"Well, that's unfortunate," Kennedy said, his irony intact. "I'll get back to you."

That evening the president called to say that he had decided to go ahead, and he wanted to meet with Healy again the next day. This time the president called in Bundy, who would arrange for the records at Harvard, as well as O'Donnell, Kennedy's trusted political adviser, who remained opposed to Teddy's electoral adventure. The four men discussed the smallest nuances and details of their agreement. When they had finally decided their strategy, O'Donnell turned to the president: "We're having more problems with this than we had with the Bay of Pigs."

"Yeah, with about the same results," Kennedy quipped.

As Healy was leaving, Kennedy stopped him a moment to add an afterthought. "Hey, you better call Teddy about this, too."

Two weeks after Teddy announced for the Senate, a front-page story appeared in the *Boston Globe* under the headline "Ted Kennedy Tells About Harvard Examination Incident." Healy insisted later that the publisher softened the article. It was a masterpiece of gentle euphemism in which words such as "cheating" and "expelled" never appeared. Teddy made his obligatory mea culpa ("What I did was wrong"), but there was a dangerous tone to the story, dangerous most of all to Teddy himself and his future. Cheating is a coward's ploy, and he was cheating again now by refusing to face up to what he had done. He had not had the strength and the will to handle the matter himself. His brother, the president, had taken care of it. Then Healy and the *Globe* had done their part to take care of him again. There were always people taking care of Teddy.

The president's prestige was on the line now, however, and it was simply unthinkable that Teddy would be allowed to lose in what was practically a Kennedy principality.

One of those who flew north to help with the campaign was Milton Gwirtzman, a speechwriter and attorney. "Teddy and his brothers considered a political campaign an athletic competition by another name," Gwirtzman reflected. "Teddy wanted to get in as many campaign stops as possible, just as he wanted to get in as many downhill ski runs, to get in that nineteenth run even though it was getting dark and sometimes dangerous. . . . Teddy got it down to an absolute minimum the time it took him to get up in the morning, showered, shaved, dressed, and ready to go out campaigning. He got it down to five minutes so he could be down on the wharf at six-thirty in the morning shaking hands with fishermen."

Teddy's Democratic opponent, Edward McCormack, bore another famous Massachusetts political name. His uncle, Congressman John McCormack, was speaker of the House. Unlike Teddy, thirty-eight-year-old Eddie McCormack had paid his dues the way they always must be paid, in small bills over time. He was a true "Southie," brought up in the Irish-American enclave of South Boston. This son of Boston had gone to Annapolis, then returned and finished first in his class at a proud local institution, Boston University School of Law. He had entered politics close to the ground, serving on the Boston city council. From there, he had gone on to spend four well-regarded years as the state's attorney general. It was a natural progression for McCormack to run for the Senate in 1962, in spite of the thunderous arrival of the youngest Kennedy.

McCormack was simply overwhelmed, not only by the Kennedy power and money but by Teddy himself. Teddy had mastered the lingo of Massachusetts liberal politics, professing his spirited opposition to poverty, racism, and inequality, while letting his slogan ("He Can Do More for Massachusetts") woo the realists in his crowd. He had one quality that could not be purchased, an immense likability, and a natural charisma that made many take pleasure in the mere sight of him.

Sometimes when he stood on a makeshift platform in the North End or in the town square in Quincy or Malden, the greatest public attributes of the Kennedy men came together in him. There was something of Honey Fitz in the way he stood there pointing his finger in condemnation of his opponent's faults. There was his grandfather's sheer exuberance, sweeping out across the crowds, embracing them in his enthusiasms. Like his brother Joe, Teddy was a rugged, handsome man who exuded health and happiness, a man whom other men liked for his masculine qualities. His baritone was the sheer perfection of the Kennedy voice, carrying out across the crowds, hardly needing a microphone. He may have gotten on the stage because of his name, but his name alone did not hold the crowds captive, nor did it push people forward to grasp his hand or to ask for an autograph.

Teddy won the party convention in Springfield by an overwhelming margin. The count was 691–360 when McCormack conceded. The vote was a devastating rebuke to McCormack, who was rejected by people who had been his friends and colleagues. Instead of conceding, McCormack vowed to take his case to the people in the primary, and he moved onward, sharpening his rhetoric against an opponent he had grown to despise. McCormack had the scent of a loser, however, and no special interests were donating to his campaign with the idea of covering their bet against a sure thing.

McCormack's last best chance was to lure Teddy into debates where, before the camera, he would expose the youngest Kennedy, leaving him stripped to what he considered his fraudulent essence. With McCormack's taunts resonating in his ears, Teddy agreed to two debates, the first one at South Boston High School. This was McCormack's spiritual home, the very lair of the old Irish Boston and its tribal ways, a place where loyalty was the highest virtue and familiarity never bred contempt.

McCormack had fueled his tank with vitriol when he walked on the stage before an audience weighted with his supporters. There was a devastating contrast between all that McCormack had done in his life and all that Teddy had been given, and the Massachusetts politician poured forth his bill of indictment, arguing that "it was wrong that that young man who really had not worked a day in his life, had never been in the trenches, should get off being a United States Senator."

Teddy was a man of immense physicality, and his instinct was to give McCormack a verbal licking. In preparing in Hyannis Port for the debate, he had vowed: "Eddie will get twice as much back as he gives out!" The president had heard Teddy's blustering threats and, as he stood there on his crutches, lectured him on how to proceed in the debate. "Now listen, Eddie," Kennedy said, calling his brother by the name he usually used, "You forget any personal attack on Eddie McCormack. You're going to need all the supporters that McCormack has right after the primary. Let McCormack attack you as much as he wants. You're running for the United States Senate. Stay on the issues and leave the personal attacks out." Teddy's father almost never spoke, but he grunted a few words comprehensible only to his sons. "You do what Jack says," the old man said, drawing the words out of some reserve of will and ambition that had seemed lost to him.

This debate was the first great moment of Teddy's public career, and as much as he was provoked, he did not respond. He had learned well from Honey Fitz, who saw politics as a theatrical enterprise in which each day the sets were struck, the politicians wiped off their greasepaint, and the next day they strutted onto a new stage speaking new lines. On this stage Teddy showed a quality both generous and calculating, an immensely valuable attribute that he would employ for the rest of his years in public life.

McCormack seemed not to understand that this evening he was facing a shrewd Gandhi of an enemy who would not strike back no matter how hard the blows. McCormack flailed away even harder. Before the debate he had talked to Trohan, and the *Chicago Tribune* reporter told him of the sexual swath that Teddy had cut across South America. McCormack ridiculed Teddy as a diplomatic "Mr. McGoo," stumbling around the world's trouble spots with his pants half down, leaving only disaster in his wake.

"In Israel he almost caused—he caused an international incident by straggling across the border," McCormack said, mocking his opponent. "In East Germany, he caused embarrassment by giving recognition to the East German government. In London, he caused a taxi strike. In Panama, the ambassador said to one of the reporters, or reportedly to him, 'It will take me six months to undo what you have done in six hours.' "

Teddy's advisers feared that his credibility had been shredded in the merciless attacks, though they were not so bold as to tell the candidate directly. After the debate Teddy and his closest aides drove over to his house on Charles River Square. Teddy called the president and talked to him while the others sat listening. Then the phone was handed to Gwirtzman to give his professional assessment of how Teddy had done. Gwirtzman had never briefed the president before, and he employed every iota of his lawyerly judiciousness. "Now, Mr. President, if they listened to the points McCormack was making, I think we might have been hurt some. But, of course, if they were sitting there . . ."

"Stop!" Kennedy interjected. "He's the candidate. He's the one who has to go out tomorrow and campaign. You tell him he did fantastic." It was precisely how the president's father had bolstered him during his own campaigns.

As the group sat around discussing the nuances of the debate, they turned on the radio to a talk show. "Now that McCormack," an indignant woman said in an Irish accent, "what he did to that nice young man. . . ." Teddy and the others stopped talking and listened to caller after caller berate McCormack for his nasty attack.

McCormack had gone too far, for in insulting Teddy he had insulted the electorate as well. These callers bristled at the idea of McCormack daring to suggest that Teddy was a man of such pathetic ineptitude that he should never dare stand for high office. They felt sympathy toward the handsome Kennedy scion who did not merit such rebuke. Teddy won the primary in a landslide of more than two to one. Then, in the general election, he defeated his Republican opponent, George Cabot Lodge, also in a near landslide.

Teddy had won, but his victory had not arrived duty-free. Even Joe McCarthy, as close to an official biographer as the family had, wrote in *Look* that "his candidacy has already done irreparable damage to the president's prestige. . . . His presence in the Senate could be an embarrassment to the president." James Reston of the *New York Times*, a temperate voice of the Washington establishment, said that the candidacy was "widely regarded here as an affront and a presumption," a perception that Teddy would not easily or quickly overcome.

The Senate that Teddy entered was the most traditional political institution in American political life, a place of subtle respect and reverence for the ideal of seniority: the longer a person stayed in office, the more power he deserved, and the more his voice should be heard. Teddy was wise not to come bursting into the Senate chambers with bold ideas, his arms full of proposed legislation. He showed a natural deference to his elders—a category that included everyone else in the Senate—and he was smart enough to act as though age and wisdom were the same.

Teddy did not arrive in the Senate to confront men whom most of his constituents abhorred, like the segregationist Senator James Eastland. Teddy came as he was, as a young politician with much to learn, and fifty-nine-year-old Eastland liked him and gave him a choice assignment on the immigration subcommittee. Eastland told him: "You want something, you come over and speak to me," as he would never have said in similar circumstances to either of Teddy's big brothers. In the Senate it was a high honor to be liked. It meant that your word could be trusted, that you played square and understood your colleagues' problems as if they were your own. And without question, Teddy was liked.

The president looked on in awe at Teddy's youthful vitality, energy, and health. At times, though, he took rich pleasure in knocking his brother off his newfound perch of principle. On one occasion at a White House party the new senator noted that the army's new rifle was putting hundreds out of work who had been working in Massachusetts factories building the earlier version. "Teddy, these are your problems now," Kennedy told his kid brother with rich pleasure. "Tough shit."

Teddy was an apt student in the school of politics, and he did not bore in obsessively on the president. He understood the efficacy of humor, bracketing his serious requests in ribaldry. When he called his brother in March 1963 to talk about the problem of wool imports, Teddy knew that this was a subject that mattered profoundly to Massachusetts clothing manufacturers and their workers. Instead of beseeching the president for help, he painted a savagely amusing picture of the patrician Republican Governor Christian Herter running around the Commonwealth in his German Mercedes Benz as he sought to keep out the nefarious imports. "Everyone turns around and takes a look," Teddy said as Jack laughed audibly, "[as] he drove up to that wool meeting . . . that really let the balloon air out of every balloon in there."

"But, of course, it's tough," the president told his kid brother, as if he were talking about nothing so banal as wool but life itself. "I tell you, boy, we went through that yesterday for two hours . . ."

"Yeah," Teddy interjected, as if he could feel what his brother had gone through. ". . . about what we would do on wool. You see, those guys don't want to give up that market."

The president's problem with wool imports was just one of an endless list of difficulties he was facing. For Kennedy, the spring of 1963 arrived without blossoms. "The major wheels of Kennedy's legislative program to get America moving again internally have yet to turn," *Look* stated in an article titled "Why There's Trouble in the New Frontier." Kennedy remained an immensely popular president, but there were danger signs out there. "In the field of party politics, where he was expected to be strong, President Kennedy is in trouble," wrote Doris Fleeson, a political columnist. Kennedy had not followed the fundamental axiom that the garden of politics must be tended from one end to the other. He had largely left it to his minions to work the politicians, and they had not done their job well. It was hard, though, to push a Congress that stood deeply divided on the social and moral issues of the day. Kennedy was having problems enough with the Democratic majority, who, as was their wont, were more concerned with their reelection than with his presidency. Some members of his own party were boldly criticizing Kennedy's refusal to push forward on crucial issues. "In rough issues . . . the Executive must lead the way with heavy tanks and artillery," said Senator Eugene McCarthy, taking unbridled pleasure in employing his wit against Kennedy. "What we sometimes get nowadays from the White House is just an encouraging word that goes like this: 'You go out and make a speech, and if you come back alive we'll be for you.' "

In foreign affairs Kennedy was having almost as many problems with those considered friends, such as President de Gaulle, as those deemed enemies. In South Vietnam, on June 11, a Buddhist priest immolated himself in protest against the repressive Diem regime. There were 15,600 American troops and advisers in Vietnam, and soon Kennedy would have to make some brutally difficult decisions about that growing crisis. As for Cuba, the Caribbean island remained a preeminent crisis in part because the Kennedy administration chose to make it one.

"They're not certain yet where the ship went from," Bobby said to the president enthusiastically over the phone in March 1963, "but they got the ammunition here in Alexandria."

"Virginia?" Kennedy asked, as he sat in the Oval Office talking to his brother about the operation that a Cuban exile group had just run against a Soviet ship.

"Yeah," Bobby replied enthusiastically. "And then they got a small boat with two outboards on the back. Went fifty-five miles an hour."

"Fifty-five miles an *hour*?" the president replied incredulously.

"Yeah," Bobby said definitively. "And then they got a little smaller boat . . . put a small outboard on it, filled it with explosives, and ran up alongside the ship and started the outboard motor and ran it into the ship."

"And where'd they jump off?" Kennedy asked, impressed with the derring-do.

"Well, either that or they ran it off, see, from their own little bigger boat. There were five of them. So they had some guts."

The president and attorney general celebrated these acts of bravery against an evil foe, even if they concluded reluctantly that they might have to prosecute the Cubans. There remained a boyish bravado in both men, and an unwillingness to realize the immense danger in creating an international vigilantism in the waters off Florida. Russian ships were for the most part transporting food and medicine, not military goods, and were sailed by civilian crews who had not envisioned themselves as combatants in the cold war. The president at least always drew back from the precipice of action, held in check by the exigencies of power. Bobby, however, maintained his obsession; he stood so close to the problem that Cuba loomed monstrous in his mind, far out of proportion to its reality.

In the immediate aftermath of the missile crisis the president's preeminent concern had been to make sure that the missiles and the Russian soldiers left the island, and once that happened, to make sure that the men of the brigade finally were freed. Bobby led the efforts to release the prisoners. The American government could not be perceived as paying a bribe for the men. Instead, Castro agreed to accept $50 million worth of drugs and medicine. The drug companies contributed not only worthwhile medicines but outdated drugs and a fortune in Ex-Lax, Listerine, Gill's Green Mountain Asthmatic Cigarettes, General Mills powdered potatoes, and a rich variety of menstrual supplies. The Americans placed pallets of the finest pharmaceutical goods on the first planeloads and reloaded the ships with the worthwhile goods salted on the top. After Bobby came up with the $3 million in cash that Castro demanded as a final payment, the 1,113 brigadistas flew out of Havana on Christmas Eve.

The president addressed the brigade in an emotional ceremony at the Orange Bowl on December 29, 1962. In a phrase that was not in the original speech, he told a stadium full of Cubans that they would indeed return to a "free Cuba." As if on signal, the crowd began chanting "Guerra, Guerra."

"The time will probably come when we will have to act again on Cuba," the president told an NSC meeting on January 22, 1963. "We should be prepared to move on Cuba if it should be in our national interest." The Joint Chiefs had already given Kennedy their invasion plans, and the CIA had outlined covert activities that included asking dissident Cubans to involve themselves in such activities as "putting glass and nails on the highways, leaving water running in public buildings, putting sand in machinery, [and] wasting electricity."

During the Cuban Missile Crisis, Kennedy and his men had looked at the larger picture and contemplated the moral dimensions of their decisions. They were practical men of power, and they had discussed ends and means because they knew that they belonged at the center of their discussion. Now, however, such discussions had been stilled. No one in these meetings asked what would be left of international law and the president's pledge not to invade if the United States attacked Cuba. No one asked about the cost to American democracy of running a secret war about which the citizens knew nothing. Nor did they ask, if Kennedy was not going to authorize an invasion, then what was the point of this endless, dangerous harassment of a neighboring state? If you reached out with a right hand proffered in peace and in your left hand cradled a knife, your neighbor was unlikely to come close enough to grasp your hand.

As Bobby looked out on the haunted faces of the freed prisoners, his belief in their cause grew even deeper. Many of these brigade veterans became his friends. He listened to their tales, and each sad recollection only reinforced his feelings. "He felt a very acute responsibility for the guys in the brigade," recalled John Nolan, then a young lawyer helping with the negotiations to free the brigade. "He was their best friend in a way that continued for all of his life. He would do anything reasonable for any of the leaders particularly, not all of which turned out well."

Bobby carried their cause to the highest counsels of government. No one in the administration was more militant, more relentlessly aggressive, than the attorney general. On April 3, 1963, Bobby called not for small covert actions but for the dispersal of a five-hundred-man raiding party into Cuba. In mid-April, at a meeting of the NSC's Cuba Standing Group, Bobby proposed three studies. He wanted to look into "contingencies such as the death of Castro or the shooting down of a U-2." He sought to look at a possible "program with the objective of overthrowing Castro in eighteen months" and a "program to cause as much trouble as we can for Communist Cuba during the next eighteen months." Bobby did not propose studying the possibility of trying to woo Castro away from Moscow or exploring the potential benefits of decreasing covert operations. In this meeting he did not talk openly about assassination, but he and his colleagues often talked about Castro's death as part of their contingency planning.

As Bobby pushed for fierce action against Castro's Cuba, he knew a great deal about the possibilities of exploring some kind of meaningful dialogue with Castro. Almost every week in the early months of 1963, under Bobby's direction, James Donovan, a former top OSS official and seasoned negotiator, along with a second negotiator, the attorney John Nolan, flew to Cuba to meet with Castro. Ostensibly, their reason for being there was to achieve the release of twenty or so other prisoners, and they had been warned not to push any diplomatic initiatives. But Donovan was almost as garrulous as the Cuban dictator, and there were few limits on their endless discussions. Castro took his guests to the Cuban World Series and to the Bay of Pigs, reliving his victory and talking endless hours with the two Americans. Donovan and Nolan were not naive, impressionable men, but they realized immediately that they were not in the presence of the narrow ideologue of American stereotype, but a subtle, complicated man conversant with the world.

Donovan and Nolan had struck up a lively rapport with Castro, and they returned each week full of anecdotes and insight. Bobby, however, had no interest in learning details about the man whom he considered his nemesis. He found Donovan's soliloquies tedious and merely wanted terse memos detailing how the prisoner negotiations were going. Neither Bobby nor the president spent extensive time with the two men.

Whenever Donovan and Nolan went to Cuba, they brought Castro a gift. On one occasion they gave the Cuban leader a Polaroid camera that he took with him on a trip to Moscow. Another time they brought a wet suit that Castro used when he went diving one day with his two American guests. It was not until many years later, during the Senate investigations into assassinations in the mid-1970s, that Nolan learned that the CIA had prepared a diving suit dusted with a fungus that would infect Castro with a serious skin disease and a breathing device laced with tubercle bacillus. Donovan and Nolan had unwittingly thwarted the CIA's plans by giving the gift before the agency was ready. There had perhaps been some rules when the CIA began to try to kill Castro in 1960, but there were none now in 1963, not when the agency would place a deadly gift in the unknowing hands of two Americans serving a diplomatic mission.

In early March, the president expressed himself as being "very interested" in Donovan's meetings, saying "We don't want to present Castro with a condition that he obviously cannot fulfill." McGeorge Bundy, Kennedy's top foreign affairs adviser, mentioned alternatives, including not only an invasion to overthrow the Communist government but "always the possibility that Castro . . . might find advantage in a gradual shift away from their present level of dependence on Moscow." In his memo of April 21, 1963, Bundy went on to say that "a Titoist Castro is not inconceivable," precisely the point that Aleksei Adzhubei, Khrushchev's son-in-law, had made to the president in February 1962.

Castro gave Donovan every indication that he wanted to initiate a serious dialogue with the United States, but the Kennedy administration backed away from any such discussion. To push such possibilities would have taken the president's strong initiative. Instead, Kennedy expressed his "desire for some noise level and for some action in the immediate future." The CIA proposed attacking "a railway bridge, some petroleum storage facilities and a molasses storage vessel," then to move on to bigger, more important targets later in the year. At the same time the administration developed a number of contingency plans that were anathema to everything for which a free, democratic society supposedly stood. The administration discussed provoking the Cubans into an attack that would allow the Americans to stage an invasion. If the Cubans were slow to anger, the United States "might initially intensify its reconnaissance with night flights, 'show-off' low-level flights flaunting our freedom of action, hoping to stir the Cuban military to action . . . [or] perhaps the U.S. could use some drone aircraft as 'bait,' flown at low speeds and favorable altitudes for tempting Cuban AAA or aircraft attacks." These were only contingency plans, but they were presented as reasonable alternatives, with no apparent awareness of the dangers of such provocation. For these were ideas meant not simply to provoke the Cubans but to deceive the American people, many of whom would probably die in a war that they thought Castro had instigated.

In a ten-hour interview on April 10, Castro told Lisa Howard of ABC that he sought a rapprochement with the United States. The United States and Castro's Cuba would never walk hand in hand, but there was something disturbingly cavalier about the Kennedy government's refusal even to employ secret diplomatic channels to seek a possible accommodation. There was consideration in the White House of attempting to block ABC's broadcast of the interview. "Public airing in the United States of this interview would strengthen the arguments of 'peace' groups, 'liberal' thinkers, Commies, fellow travelers, and opportunistic political opponents of present United States policy," an NSC analysis stated. These were the views of intellectual cowards so afraid of Castro's ideas and his articulate presentation of them that they would countenance limiting that dangerous thing known as liberty.

Castro was attempting to foster what he called a people's revolution throughout Latin America and what the United States considered a subversion of sovereign states. While the Cuban leader fostered revolution elsewhere, his own people were imprisoned for their ideas. In contemplating censorship or staging an incident to set off an invasion, the United States was flirting with the same tools of totalitarianism that Castro employed.

The president had first noted this dilemma as a young man when the European democracies faced Hitler's brutal regime. But Cuba was a pipsqueak of a nation, hardly set on a course of world domination. Castro's attempts to subvert Latin governments were no more than irritating. The best attack against Castro was a foreign policy aimed at fostering democracy among America's neighbors and aid programs that reached the desperate masses listening to Castro's message. In the end the administration's obsession with Cuba was fueled primarily by the Kennedy brothers' anger over the Bay of Pigs and Castro's taunts, the exaggerated rhetoric of American anticommunism, and the loud shouts of the Cuban exile community and the American right wing.

Operation Mongoose was dead. In its place had arisen what amounted to a private guerrilla army of Cuban exiles, of which Bobby was the architect and champion. These men were patriots, but they were equally what Castro called them, mercenaries, paid for by American coin. Manuel Artime, their leader, received $225,000 a month to run what was known as the Second Naval Guerrilla, or, by the CIA's own estimate, a total of nearly $4,933,293, an amount that the Cubans claim was twice that size. This organization purchased two major ships, eight smaller boats, three airplanes, and tons of weapons.

They did most of their training in Central America, particularly Nicaragua, where the Americans had made an arrangement with President Luis Somoza, son of the murderous dictator Luis Somoza. And they ran their own operations, without scrutiny, the kind of daring ad hoc adventures that Bobby admired.

"I had many chances to talk with Bob Kennedy," reflected Rafael Quintero, Artime's deputy in Central America. "Bob Kennedy was obsessed—obsessed with the idea that they had been beaten by Castro, that the Kennedy family had lost a big battle against a guy like Castro. He had to get even with him. He mentioned that to me often and was very clear about it. He was not going to try to eliminate Castro because he was an ideological guy who wanted to do right in Cuba. He was going to do it because the Kennedy name had been humiliated."

Bobby was the great patron of the anti-Castro Cubans, descending on them for secret visits, relishing their dangerous exploits, celebrating their courage. He showed up at their parties and drank with them, saluting a brotherhood of fearless men facing an implacable foe. It did not matter that so many of their actions risked hurting the innocent as well as their acknowledged enemies and did little more than stiffen the vigilance of Cuban Communists. They were on a sacred quest.

Bobby gave these men the illusion of equality, but there was always a moment when it became clear that he was a Kennedy and they were not. One of the men to whom Bobby was closest was Pepe San Román, the brigade's military leader. San Román was more suited for poetry than war, and he had struggled emotionally after his release. Bobby had set him up in a small house near his own home in McLean, Virginia. One Sunday morning Bobby rode by San Román's home in riding clothes, looking a grand seignior. He dismounted to chat for a while, and when it came time for him to leave, he clicked his fingers, thrust his boot forward, and signaled San Román to help him back into the saddle.

The CIA ran its own covert operations parallel to what the Cuban exiles were doing ostensibly on their own. On June 19, the president approved a plan of dramatically increased covert actions starting "in the dark-of-the-moon period in July." These plans included the "sabotage of Cuban ships outside of Cuban waters," hit-and-run attacks on the island itself, and the support of covert operations run by Cuban nationals from outside the United States that "will probably cost many many Cuban lives." The targets of these operations included sugarcane crops, the food, clothing, and housing industries, bridges, trains, and electric power plants.

Bradley Earl Ayers, a career army officer, was one of those assigned to train Cubans in Florida for covert missions and to accompany them at least to the shores of Cuba. Ayers was given a cover as a technical specialist doing classified research at the University of Miami. As Ayers led his Cubans out on their first missions, the whole nature of the missions was slowly expanding. Before the Cuban Missile Crisis, it had been something of a joke at the CIA that when they had burned sugarcane fields, the Cubans could easily harvest the burned sugar. That was no longer the case. "We fixed that," Ayers recalled. "We began to disseminate defoliants and herbicides, dumping it in the irrigation canals and wells pretty liberally in agricultural areas, doing a pretty good job of killing crops." After a few months Ayers learned that they would soon be going against refineries, mines, and other major industrial projects. Until now, such facilities had been off limits in part because American corporations wanted their properties back intact.

Another of the major players in the Special Operations Division was

Grayston Lynch, who had won a certain notoriety by going ashore at the Bay of Pigs. Lynch began by training Cubans for missions that primarily involved supplying arms to those within Cuba, but he too knew that things were changing. He recalled receiving word that they were to "set Cuba aflame."

On July 16, shortly before the covert operations expanded, Bobby attended the NSC Cuba Standing Group meeting. There were few actions he was not willing to pursue in destroying Castro's Cuba. During the discussion Secretary of the Treasury Dillon said that an official had blocked Canadian funds in the United States equal to the amount of Cuban funds deposited in Canada. It was a blatant infringement on the sovereignty of one of American's most esteemed allies. Bobby agreed with the others that the United States should stop such actions. Nonetheless, he said, he "hoped that it would be done in such a way as not to destroy the morale of the U.S. officer who had initiated the action. . . . There were too few officers in the U.S. Government who acted with comparable initiative."

Another problem facing the group that day was an article that had appeared in the *Miami Herald* on July 14. "Backstage with Bobby" described in detail the attorney general's meetings with Cubans planning to raid Cuba from Central America. Bobby said that the administration could handle the matter simply enough. "We could float other rumors so that in the welter of press reports no one would know the true facts," he told the group. In this instance he was all for bringing the CIA's disinformation campaign home, attempting to poison the free press with lies and half truths.

Although McCone agreed that they might do as Bobby suggested, he added "that in future dealings with Cuban exiles we must use cutouts and not deal with the exiles directly," and warned that "there be no direct contact with them in Washington." The CIA director was attempting to rein Bobby in, but it was that immediate, visceral contact with these men and the games of war that the attorney general sought. He could no more have walked away from that experience than from his obsession with Castro.

In early June, President Kennedy flew to San Diego to spend a night on the aircraft carrier *Kitty Hawk.* Kennedy was the commander in chief, but he was also a proud navy man who loved the rituals of his chosen branch of the service. At ten o'clock that evening the president sat in a leather-padded rocking chair on the bridge of this magnificent city of a ship watching small jets catapulting into the night. It was a scene of awesome splendor. The only sound was the admiral briefing the president on the details of the operation, talking into Kennedy's ear.

"The president said very quietly, and with infinite fatigue, 'Admiral, I'm afraid I can take no more,' " recalled the journalist Alistair Cooke, who stood beside them on the bridge that night. "He grabbed the arms of the rocker and began to force himself in a twisted, writhing motion, to his feet. It took about a minute, and then two officers led him to his quarters. It was his back again."

Later that month the president flew out of Andrews Air Force Base for what would prove the most memorable trip of his presidency, and in some ways the most emotionally powerful journey of his life. It was in Europe that young Jack Kennedy had had his intellectual awakening and taken up the themes of his public life. It was among European women that he had found two great loves in Inga Arvad and Gunilla Von Post, women whose subtlety and sophistication he found lacking in their American counterparts. It was in Berlin that he feared the monumental confrontation with the Soviet Union still might come.

Whatever difficulties Kennedy was having in Washington, among Europeans he found generous new constituents. He had transcended the common European stereotype of Americans as boorish provincials stomping across their heritage. He was returning to their continent bearing what are truly the greatest gifts in the world, hope and promise.

As his limousine moved slowly along the boulevards of Berlin, most of the residents had turned out to greet the American president, shouting his name, crying with happiness, cheering until they grew so hoarse that the name became little more than a whisper, pressing forward, seeking a touch, a glimpse, a wave, some souvenir of this moment. There had not been such an emotional outpouring in this city since Hitler had driven through the same streets and been greeted with cheers and Nazi salutes. Kennedy had been in Berlin in 1945 when the city was reduced to nothing but mounds of rubble and it was as if the earth had been salted and nothing would ever rise here again. But as he drove along, he could see behind the crowds a modern city that had at least a modicum of the élan of the old Berlin of the 1920s.

Kennedy went to the wall and contemplated this divided people. The wall had solved a problem, both for Khrushchev and for him, and it may have saved the lives of the very people who shouted that it should be torn down. It was not Kennedy, the cold war strategist, looking out toward East Berlin today, but a man of the spirit, and no man of the spirit could look at this wall without anger and dismay. When he talked to Sorenson about his speech, he told the speechwriter: "I need a phrase that will reflect my union with Berlin. What would be a good word for it? I really am a Berliner. Get me a translation. How do I say, 'I am a Berliner'?"

The speech that Kennedy gave that day in Rudolph Wilde Platz was an exhortation to liberty. "Two thousand years ago the proudest boast was *civis Romanus sum*," he told the massive crowd below. "Today, in the world of freedom, the proudest boast is *Ich bin ein Berliner*." The words resounded through the crowd with massive emotional resonance. These same people who a generation before had slouched onto the world stage as a force of evil now stood as heralds of liberty.

"There are some who say that communism is the wave of the future," he said as he looked out on the hundreds of thousands before him crowding the plaza. "Let them come to Berlin!" As he spoke, the crowd ceased to be disparate individuals but was one immense mass, thinking the same thoughts, beating with one heart, and ready to move with one great strike. Most orators would have felt an awesome sense of their own power, running their tongues around each syllable, playing with such a throng. Kennedy, however, hurried on with his speech, the words bumping into each other. This day, he was not the orator he might have been, largely because he was not comfortable with arousing such emotions in his audience. As he spoke, he feared, as he told Ormsby-Gore later, that "if he had said, 'And at this moment I call upon you all to cross into East Germany and pull down that wall,' they'd all have gone, [that] the German people as such at this moment in history were not totally to be relied upon, and that this rather sheeplike instinct of theirs could be very frightening under certain circumstances and under the wrong leader still." Kennedy feared not only the German masses but mass man anywhere, and where demagogic politicians might lead him.

His visit to Germany was an immense triumph, and he flew from there to the Ireland of his ancestors, but within whose heritage he had never felt fully comfortable. This Irish visit, however, was one of the transcendent experiences of his entire life. "Are you glad you came?" his Irish friend Dorothy Tubridy asked him when he was leaving. In private conversation he was not a man of flattering, meaningless pleasantries, and yet he told her, "These were the three happiest days I've ever spent in my life."

There are few things worse than an economy in which a man must leave home so that his family may eat, and there was hardly anyone standing cheering along the Dublin streets who did not have a brother, a father, or a grandfather who had made that journey to America, Canada, or Australia in search of what some called a future. Kennedy, the greatest of all the scions of Ireland whose ancestors had gone abroad, was returning as the leader of the most powerful country in the world. And was it not fitting that he should return now, in this year of 1963, the first time more Irish were returning to Erin's shores than were heading abroad?

One of those who accompanied Kennedy on his journey was the Irish ambassador to the United States, Thomas Kiernan. Before this visit Ambassador Kiernan had considered Kennedy "more British than Irish," a president whose "first reaction would be, if there were any even minor dispute between Britain and Ireland, to side with Britain." Kennedy had the quick wit, the verbal agility, and the protective self-deprecation of an Irishman, but when the diplomat discussed the partition of Ireland with the president, the cold, logical, British-hued mind took hold.

Kiernan had seen in America that anyone could become an ersatz Irishman, wearing the green on St. Patrick's Day, drinking toasts to a heritage they neither knew nor understood. It was not all songs and shillelaghs, however, to those who were truly Irish-Americans. A hard bitterness was mixed with that heritage, and it struck Kiernan how often Kennedy mentioned the notorious signs that the Brahmins posted in Boston: "No Irish need apply."

As the helicopter flew from Dublin to Galway, Ireland appeared to be a blessed, magical, verdant land far removed from the secret squalor and the ceaseless conflicts. As they soared above the land, Kennedy kept asking about the prices of houses and land. The ambassador sensed that the president was thinking about buying his own place here, a house where he would come occasionally and send his children to learn about their Irish heritage.

The president returned to New Ross and the Kennedy family homestead. As the helicopter headed to County Wexford, Malcolm Kilduff, the deputy press secretary, briefed the president. Kilduff told Kennedy about his experience working with Andrew Minahan, the chairman of the New Ross county council, in advancing the trip. Kilduff had involved the enormous, redheaded chairman in a myriad of details. The press aide told Minahan that the presidential limousine would pull in near the wharf. From there, Kennedy would be able to move easily up to the temporary speaker's platform. The problem was the massive heap of cow dung waiting to be placed on barges. "You'll have to move it," Kilduff told the county chairman. "Christ, no," Minahan said exasperatedly. "I'm going to pile it high and make that f'er think he's crossing the Alps."

When Kennedy's helicopter set down in freshly mown hay, the backwash sent cow dung flying across the field and speckling the children's chorus preparing to sing "The Boys of Wexford." Another president would have berated Kilduff for not planning better, but Kennedy chuckled impishly, amused at this added spectacle. Kennedy could go in a moment from wry ironic amusement to the most heartfelt emotion, and so he did today as soon as the children began to sing a song that touched him so deeply. Then he moved on to greet the various dignitaries. When he came to the chairman of the New Ross county council, he stopped for a special moment. "Mr. Minahan, I've heard a lot about you," Kennedy began, waiting for the man's chest to swell to twice its normal size. "What I'd like to know is, did you remove the dung or does this f'er have to cross the Alps?"

Kennedy drove out to the old Kennedy house in Dunganstown. In America when a man went to see his old family home, he often discovered that it was torn down long ago, or that it had acquired a new addition, aluminum siding, or at least a different paint job. He often ended up standing there wondering what was left of the past in America, his or anybody else's. But the thatched house that the president came to along this dirt road was almost exactly what it was 114 years before, when Patrick Kennedy left for America. It was a small place, only forty feet by thirteen feet, with tiny windows no larger than fifteen inches by seventeen inches or the taxes would have been higher. Next to it stood a farmhouse. There was neither indoor plumbing nor a telephone.

The president's cousin, Mary Ryan, and her two daughters, Mary and Josephine, stood there waiting to greet the president. Mrs. Ryan was a rotund woman who could have been one of the legions of Irish women like his great-grandmother Bridget who had cleaned the Boston Brahmin homes and washed their fine linens. She was the very image of the Irish peasant from whom his father had sought to distance himself. Mrs. Ryan hugged Kennedy and spoke to him with warm familiarity, as if he had grown up just down the road. At some times and in some places, Kennedy might have recoiled from such a gesture, but this time he embraced his cousin. When he talked to his relations, Mrs. Ryan's daughter Mary was amazed at how well he knew the family tree. "I think he felt back home," she recalled. Raising his cup of tea, he proposed a toast "to all the Kennedys who went and all the Kennedys who stayed." Then, almost as suddenly as he had arrived, he was off, flying back to Dublin in his helicopter high above the emerald fields.

As Kennedy left Ireland, he left part of himself and took more of himself than had been there when he arrived. "He had always been moved by its poverty and literature because it told of the tragedy and the desperate courage which he knew lay just under the surface of Irish life," said Jackie, whose pregnancy prevented her from making the European trip. What Kennedy took from his Irish heritage was neither the calculated buffoonery of the Irish-American politician nor the boozy sentimentality of the saloon, but the dark, ironic wit that was a shield against misfortune. He took also what he considered the indomitable Irish spirit, and its struggle for liberty. He believed that a desire for liberty was one of the essential drives of humankind, and he reasoned that if his forebears had struggled against all assaults, insults, and reprisals, then surely the other oppressed peoples of the world one day would do so too. As he saw it, the Irish had one great song, the song of liberty, and no one had ever been able to still Irish voices.

If heroism in the face of impossible odds is sentimental, then Kennedy at this moment was a sentimental man. He loved this land now as he had not loved it before. He took something else with him as he flew west—a part of his past. He had done here what most Americans do sooner or later: come to terms with his immigrant past, with his humble beginnings. "This is not the land of my birth," the president said as he was leaving on June 29, 1963, "but it is the land for which I hold the greatest affection, and I certainly will come back in the springtime."

When the president arrived back in America and spent the Fourth of July weekend in Hyannis Port, he invited the entire family to watch his movie of the trip to Ireland. They all came over and watched as Kennedy provided running commentary. The next evening Kennedy invited them back to watch again, and since not only was he president but he so rarely liked watching or doing anything twice, they gladly returned and sat through it all again. Then on the following night he invited them back a third time, and this time they trooped slowly and dutifully into the living room to watch film that they had practically memorized. And he sat and watched as if he were seeing these scenes for the first time.

Kennedy could move within a few moments from the kind of cynical political ripostes that some men called realism to acts of the most rarefied, deeply felt idealism. His idealism was not merely expressed in service of his cynicism but was on a different plane and served nothing but itself. Kennedy had a profound belief in public rituals. He had grown up in a church whose ornate sacraments set its believers apart from the Unitarians and the Congregationalists, to whom simplicity was the mark of true faith. His mother had led her sons and daughters along the Freedom Trail, from Plymouth to Concord to the Old North Church, inculcating in Jack the idea that the patriotic faith of Americans had to be affirmed in shrines and rituals.

As a lover of England and Europe, Kennedy was a lover of history, not as a tedious recitation of the past but in part as a vivid tapestry of rituals. As a young congressman, he had participated in all the rituals of democracy, including Veterans Day. Each year the numbers in the crowds seemed to lessen, the cheers became more muffled, and more and more of the speeches echoed in half-empty auditoriums. He was seeing the beginning of the secularization of American life; Sunday would become just another shopping day, and many national holidays were ripped out of the calendar and set next to

Saturday and Sunday so that Americans could have their three-day weekends. Unlike many of his contemporaries, Kennedy revered these moments, and in the midst of his tight schedule he penned some ideas for the Fourth of July proclamation. This was usually a boilerplate document put together by a speechwriter, to be released under the president's name and then forgotten, but not for Kennedy this year.

"Bells mark significant events in modern lives. Birth and death, war and peace are pealed and tolled," he wrote in his own hand during his European trip. "Bells summon the community to take note of things which affect the life or death of a community. . . . On the Fourth of July when bells ring again—think back on those who lived and died to make our country and their resolve, achieved with courage and determination, to make it greater in our day and generation."

To Kennedy, patriotism was a bell that resonated with all the sounds of American history, pealing forth the triumphs and tragedies, the deaths and dramas. It was a bell that had to be tolled repeatedly, and when it rang, those within hearing should stop and listen. To him, the heroes of the past, men like Joe Jr., lived on in the continuing history of the nation, their lives resonating in these sounds. "Heroes of the past are watching us," he said in the proclamation. "If we remember them when the bells ring out . . . it will help us to live like heroes too."

For Kennedy and his two brothers, the bell that pealed the loudest sounded out the name of their brother lost in war. They did not salute his name every day, but his life and, more importantly, his death rang through their own lives. To betray their brother's sacrifice was to betray their country, their faith, their father, and their blood.

In September, seventy middle-aged men and most of their wives gathered in the South Ballroom at the Willard Hotel in Washington. These were the men of Navy Patrol Squadron VB-110, together for the first time since their days at Dunkeswell Airdrome in World War II. Before them stood Jim Reedy, who had been Joe Jr.'s commander and who now was a rear admiral.

"Hello, Admiral," the president's voice sounded over an amplified telephone from Hyannis Port, where he and his brothers were all together on their father's seventy-fifth birthday. They were a family of celebratory occasions, and it was not easy wheeling Joe into the dining room, decking him out in a silly birthday hat, singing songs and telling jokes when he could not respond. It was not easy speaking always in the upbeat idiom that was the preferred language of the Kennedys. It was not easy for the president to stride boldly into the house that fall and then at times take up his crutches to help assuage his pain. It was not easy pretending that life was the way it always had been.

The president was not a man who liked to ponder his past. But the admiral was right in his belief that if the president had been in town, he probably would have been there with his brothers beside him to honor their brother's memory.

"Admiral, I want to express all of our thanks to you and to send our best wishes to all those who served in 110," the president began. "I know its record very well. And I know from the letters which my family received during the Second World War how much my brother valued his association with this distinguished squadron which had an outstanding record in the winter, spring, and summer patrolling the Bay of Biscay. I know something about the number of men who were lost in the dangerous service with the coastal patrol."

Kennedy had braved enemy waters too, and he was speaking to these men with a commonality that only they could fully understand. The condolence letters after Joe Jr.'s death had been a challenge to Kennedy. They were almost a rebuke that he was not the man his brother was. Growing up, he had harbored the most complex feelings toward Joe Jr., but all that was left now was love and reverence and a renewed belief that courage was the ultimate virtue. The only human immortality of which we can be certain is that a person lives on in human memory and that others seek to replicate his or her deeds. In that sense, Joe Jr. lived on in all three of his brothers.

"All of you, Admiral, who are meeting now have happy recollections of those who served in the squadron who did not return," Kennedy said solemnly. "And I know that my brother and all those others who served with the squadron who are not with you are with you tonight in spirit."

When the president finished, he gave the phone to his younger brother. Bobby spoke with that same Boston nasal intonation as his brother, but there was softness and a subtle tenderness to his tenor voice, unlike the president's. Bobby had known Joe Jr. only as a great figure descending on his life for a few brief moments, a brother who had died a hero's death, leaving a shadow that reached wherever he moved.

"I just know how much our brother Joe liked and respected you," Bobby said, as always more intimate, more emotionally raw, than the president. "I know how well all of you have served the country in the past and in the present time. . . . I hope you're going to meet again so one or all of us can come. I have another brother here that wanted to say hello to you. Can he do that, Admiral?"

"Indeed he can."

Bobby then passed the phone on to Teddy. "Hello, admiral, this is Ted Kennedy," Teddy said, making an introduction that his brothers had not found necessary. "Well, admiral, I was the only member of the family who didn't serve in the navy. I was in the army. And my other brothers never let me forget about it."

"I'm sure that's true," Reedy said, over the laughter.

Teddy would always be the kid brother, but he had assumed the confident manner of a public man. He was fast becoming the greatest politician in the family, loving the touch of people and their problems. He had become a professional politician too, in that he thought it was his prerogative to talk too long. In private among friends, he was the most charming of raconteurs. In public, however, when he was not speaking from a prepared speech, his syntax was often so mislaid that he did not speak sentences as much as an endless array of words. This moment may also have been overwhelming to him, thinking of what his brother had sacrificed.

"I think all of us have been tremendously impressed and are cognizant of the many members of your squadron. I think all the brothers met a number of them from the country certainly in the campaigns before, certainly in the different states they have always come up and said hello and I have always been delighted to meet with them and hear their stories," Teddy said. "I just in the last few weeks have received some notes from some of the sons wanting to come by. Even though I'm the one member of the family who did not serve in the last war we feel tremendous respect and a closeness."

Teddy was the most distant from Joe Jr. in age, intimacy, knowledge, and experience. He nonetheless saw himself as the proud bearer of a noble legacy and was willing more than his brothers to meet with anyone who had touched Joe Jr.'s life, to invite their children into his office, send them autographed photos, and listen to their reminiscences.

When Teddy finally finished, the veterans returned to their wartime tales, and the three brothers celebrated their father's birthday.

30

The Adrenaline of Action

Kennedy loved convivial gatherings whose only purpose was the amusement of those fortunate enough, witty enough, or pretty enough to be invited. He and Jackie held a number of dances at the White House. These were elegant soirées, the room festooned with lovely women. There was an ample charge of sexual electricity in the air, and the happy anticipation that something especially memorable would happen, or some unseemly gossip or perhaps a presidential witticism would be uttered, to be repeated across Georgetown the next morning.

At the first dance of the winter of 1962–63, among the ladies in their long, elaborate formal dresses was one guest who was wearing a gossamer, chiffon summer gown. It was clothing that would have seemed out of place on almost anyone else, but one looked at Mary Meyer and thought, yes, it was summer. Meyer was an abstract artist, and she brought her style to whatever she wore, whether it was a vintage dress that she had picked up for five dollars in a secondhand store or a designer gown that she had purchased in Paris. Her daring only began with her clothes. She had an impish, taunting manner unlike anyone else in the room. Once, at Hickory Hill, she had trumped even Bobby and Ethel's frenetic sense of fun by suggesting that everyone strip naked and plunge into the pool.

For a year Meyer had been seeing the president, always making her private visits when Jackie was away. "I think Jack was in love with Mary Meyer," reflected Ben Bradlee, who was there that winter evening. "She was a very interesting woman, but she was trouble. With one less chromosome, she would have been a perfect person for Jack to marry." Bradlee spoke with a certain authority, since until Mary's divorce from Cord Meyer, a top CIA official, she had been the *Newsweek* editor's sister-in-law.

Kennedy had once preferred elegant, educated, fashionable, upper-class women. Since he had begun his pursuit of presidential power, he had often settled for a parade of women who lacked all these attributes. Meyer was different, harking back to the women he had once pursued, women sweetly perfumed with all the allure of money and class. Yet Meyer represented something exotic too, the new bohemians of the sixties. She smoked marijuana and led him on a journey away from the emotional certitude in which he lived.

The president was host of the evening, seeing everything, and remembering most of it. "Blair, get Teddy to do the twist," Kennedy directed his old college friend. The junior senator made his entrance to social Washington not in some formal procession but gyrating his hips.

While the dancing proceeded to the sounds of Lester Lanin's orchestra, Meyer disappeared from the gathering, wandering off somewhere. Hoover had no elaborate dossier on Meyer, for most of her travels were in her head, but she was a dangerously unpredictable woman. That evening Meyer finally arrived back among the dancers. Her long evening gown was wet at the bottom.

"Mary, where have you been?" Blair Clark asked.

"I've been out walking around the White House," she said in her gaily lilting voice.

"In the snow?"

No one else that evening would have even contemplated leaving the party to walk out alone on the White House grounds deep with snow. Clark believed that Meyer might have wandered out there because Kennedy had just told her their affair was over. She may also have been the one contemplating just how far she wanted her relationship with the president to go. Or she may simply have felt like walking in the snow in her summer gown.

"What is the scandal?" Kennedy asked with delicious anticipation.

The president was on the telephone with Arthur Schlesinger Jr., who on March 22, 1963, had just returned from England with a precious piece of political tittle-tattle. Kennedy treated gossip like chocolate bonbons, a pleasant little addiction that he enjoyed tasting several times a day. Schlesinger had a particularly sweet item about John Profumo, the British war minister, who had gotten himself into a wicked fix.

"Well, he has written a number of letters to a girl who turns out to have been the mistress of a Soviet military attaché," Schlesinger said.

"Is that the girl that was written about in this morning's paper?" the president asked, already up to the moment on the story.

Kennedy treated the scandal as if it were happening to some exotic species of political animal. He himself had indulged, however, in precisely the same behavior that doomed Profumo's political career, carrying on with a woman who was also involved with one of his nation's enemies. The president, moreover, had done so with reckless nonchalance, learning nothing from the debacle with Exner, continuing with liaisons that were even more dangerous.

Later that spring Kennedy sailed down the Potomac on the *Honey Fitz* with a group that included his carousing comrade, Senator Smathers. The sexual gossip of the moment dealt not with the capital but with London, where the Profumo affair risked bringing down the Conservative government.

"How serious do you think it would be?" Kennedy asked, as Smathers remembered. "Do you think there is any possibility of anybody in my administration getting in a similar situation?"

"I don't believe so," Smathers replied. Then Kennedy went through the cabinet members one by one, describing the sexual predilections of each man, or the lack of them. By any measure, it was unlikely that the president had a Profumo in his cabinet. What was left unsaid was that the threat of scandal lay not in the cabinet but in Kennedy himself.

"It's different from the case you tried when you were assistant U.S. attorney in 1941," Jack said, his knowledge startling the Florida senator. That case arose from the question of whether state officials had illegally encouraged young women to cross state lines for immoral purposes. Smathers had won the verdict, and the case had gone all the way to the Supreme Court. Kennedy said that he had read about the matter when he had studied Smathers's 1950 senatorial campaign. What was startling was not that he remembered but that it was something that he was reflecting on at this moment.

Up until now, Kennedy fancied that he could laugh with impunity about poor Profumo. On June 29, however, the *New York Journal-American* reported that a "high elected American official" had been the paramour of Suzy Chang, another of the British call girls involved in the scandal. There apparently was no truth to the rumor that Chang and her colleague, Christine Keeler, had bestowed favors on Kennedy, but the president had for so long treated sex as an endless buffet that it sounded plausible. Bobby called in the offending reporters, and his stern lecture was enough to convince the paper to back away from the story. "Have we learned what Christine [Keeler] and her friend did here in the U.S. when they were here?" the attorney general asked the FBI. It was the kind of question that he would ask only because he was not sure of all the parchments on which the president had left his pen marks.

Kennedy was probably involved with another European woman, Ellen

Rometsch, a German whose husband was an air force sergeant assigned to the West German embassy in Washington. Rometsch was a beautiful, sensuous young woman who hung around men of power in the watering holes of political Washington. Some thought that she was a call girl, though she was probably on the borderline between amateur and professional: men gladly showed their largess for her favors, but she could claim she was only having a good time. She became known as one of Bobby Baker's girls. The secretary to the Senate Democrats was a man of many favors, and Rometsch was one of his favorite gifts. Baker asserts that Kennedy's friend Jim Thompson set Rometsch up with the president, a pleasure that he supposedly enjoyed several times in the spring of 1963.

Hoover was also interested in the twenty-seven-year-old woman, but his concern was of a different sort from the president's. The FBI chief's obsession with Communist spies played into the paranoia of American politics, but the reality was that the East German secret police, the Stasi, had sent East Germans west as moles. Hoover was not wrong in suspecting a woman who had been active in Communist youth groups before fleeing her native East Germany. It was curious too the way she had arrived in Washington with her husband, only to use her sexual charms to ingratiate herself with men of power, including probably the president himself.

On July 3, 1963, Hoover's deputy, Courtney Evans, told Bobby about Rometsch. It does not matter whether one believes Baker's assertion that the president told him that Rometsch gave him "the best oral sex I ever had." There was such a climate of sexual license in the White House that the story was believable. A month later the attorney general directed that Rometsch be deported to Germany. LaVern Duffy, one of Bobby's investigators on the rackets committee, accompanied her. He was a doubly good choice to shepherd Rometsch out of the country, for not only was he a good friend of Bobby's but he was among those having an affair with the woman.

There was pleasure and pain in Kennedy's affairs. The pleasure was the president's, and the pain was Bobby's, whose task was to straighten up matters after the women had left and to pretend that nothing had happened. It was a sordid duty, and the attorney general performed it impeccably. "The White House was a bawdy house, but the interesting thing about it is that, in his mind, I think Kennedy thought that he was keeping all of this stuff separate," recalled Mark Raskin, who early on in the administration had watched from his perspective as McGeorge Bundy's aide at the National Security Council. "But the reality was that one piece of this was collapsing in on the other, as it always happens in fact. The FBI is keeping tabs on him. They know what's going on. I assume that this is a problem for him personally. And so all this stuff now takes up time and energy and so forth and so on, all the pieces falling in on each other."

Kennedy had created an atmosphere in which many of his aides felt that it was the highest part of devotion to emulate their beloved leader by displaying their own sexual swordsmanship. Even some members of the Secret Service picked up on the atmosphere, enjoying their own assignations. Other agents worried that they were losing control over the president's activities. "I don't know if I would use the word 'disillusioned,' but it was bothersome that sometimes we had not been able to do a background on people visiting the White House," observed Joe Paolella, a Secret Service agent. "From the security standpoint, that made you feel that there could have been some security breaches, women, that kind of stuff. What if one was a Russian spy? Sometimes Mrs. Kennedy's secretary, Pamela Turnure, came up. You could conjecture, but we really didn't know why these women were up there."

Kennedy did not seem to realize that circumspection is the key to an adulterer's long-term success. Presumably he knew that Florence Kater was still obsessed with exposing an affair between him and Pamela Turnure. Even so, he brought Turnure into the White House as Jackie's press secretary.

The matter was of serious enough consequence that in October 1961, Bobby's press secretary, Edwin Guthman, sent to the FBI a copy of one of Kater's letters to the attorney general calling the president a "debaucher of a girl young enough to be his daughter and his cynical knowledge that the press will cover up for him is such that he has brought her into the White House itself as his wife's press secretary." He included a cover letter that asked whether the agency "had anything in files on the writer, Florence Mary Kater." Although the FBI had nothing on Kennedy's tormenter, the agency had a copy of the letter Kater had sent out in 1959 and noted that it had been distributed widely, including to the NBC journalist David Brinkley.

Kater finally managed to find someone willing to publish her tale. In November 1963, *The Thunderbolt* became the first publication to print Kater's charges in explicit detail. The racist newspaper that called itself "The White Man's Viewpoint" included a photo of Kater picketing outside the White House in which she carries a large placard condemning the philandering president, a photo of what appeared to be the then-Senator Kennedy shielding his face as he hurried out of Turnure's Georgetown apartment, and the transcript of a tape recording of Kennedy and Turnure made on July 1, 1958. The Katers, who had lived downstairs from Turnure, had secretly recorded Kennedy's visit late one night. The tape included sounds of a couple making love and having a testy contretemps. "I think that all men are cheaters," the woman said, her guest presumably giving her ample evidence of that.

There was other innuendo already appearing in publications more legitimate than *The Thunderbolt*. In July, the FBI had taken note of a column in *Photoplay*, a popular fan magazine, by Walter Winchell, who included a blind item concerning Marilyn Monroe and an anonymous powerful man who could have been either the president or the attorney general.

It was almost inevitable that either before or during the 1964 election, speculation about Kennedy's sexual activities would move from occasional innuendo in the gamier gossip columns and tabloids to full-scale articles in major newspapers and magazines. Guthman is convinced that Kennedy's sexual conduct, particularly with Exner, "would have been an issue in the campaign." Kennedy's aide Mike Feldman and Hoover's deputy Cartha DeLoach both believe that though the matter probably would have surfaced, the president would have successfully dismissed it as nothing but baseless campaign innuendo.

The public knew nothing about the president's sexual appetites and thought him an exemplary husband. Kennedy was an immensely popular president, not only admired but revered. Yet he was increasingly dogged by a multitude of problems. These were not grand crises in which he could stand stalwartly on the deck of the ship of state, most Americans hewing to his judgment no matter where he led them. These were pesky, petty squalls that risked slowly sapping away the energy and strength of his administration.

In the middle of July, Kennedy's secretary brought him a cover letter and a series of memos from his friend Charles Bartlett. "I know how you dislike bearers of bad news but I am passing these memos along under a deep conviction that they deal with an area which urgently requires your personal scrutiny," the columnist wrote. "An aura of scandal is building up. . . . At best O'Donnell, O'Brien . . . are performing legitimate political functions in a way that is breeding resentment and suspicion. At worst they are the heart of an extensive corruption which is reaching into many of the government agencies."

The memos dealt for the most part with matters taking place at the Democratic National Committee. Contributors seek to be reimbursed for their largess with positions, contracts, and influence, and it is only when the spoils system transcends the limits of law and appearances that it becomes a political problem. The greedy are never satiated, and there is a natural progression in this process toward corruption. That had happened to some degree toward the end of both previous administrations, and the question was whether it was happening again. In one of his memos, Bartlett recalled that in February he had told Kenny O'Donnell that "the committee would have been more balanced if someone like Siegenthaler [*sic*] had gone there." O'Donnell, who fancied himself a tough man in a tough world, reportedly replied that he [O'Donnell] "was responsible for the dealings with contractors and that he was determined they would ante up to the party."

O'Donnell may have been the chief political operative in the White House, but he had already violated one of the fundamental rules by breaking down the firewall between the president and his top people and the often nasty business of patronage. John Seigenthaler had in fact been scheduled to move over from the Justice Department to the DNC precisely because of the reasons that so upset Bartlett. Seigenthaler was to have been what he called "the buffer and the string cutter," saving Bobby from having to deal with big contributors and other favor-seeking supplicants. "Bobby had dealt with every major politician in America during the campaign, and every one of them thought that Bobby was obligated to them," Seigenthaler reflected. When Bobby's aide moved back to his home state to become the editor of the *Nashville Tennessean*, the administration lost a man who had both the moral fiber and the toughness to handle the largely thankless job of patronage.

Bartlett's charges were for the most part in the gray area where most patronage takes place. Everything the journalist had said could be rationalized with a shrug and a wink. What was most devastating, however, was an account of a conversation that Bartlett's source had purportedly had with Larry Newman, a Secret Service agent. Newman supposedly recounted an evening he had spent with O'Donnell at Lake Como in Italy at the end of the president's recent European trip. "O'Donnell's remark was that the president was in fact rather stupid and that if it were not for his assistance, he would fall flat on his face," Bartlett wrote in his memo. "Newman said that O'Donnell's power seemed to him to be total and that he was clearly a negative and evil force."

Bartlett admitted years later that Paul Corbin had been the source of the story about the Secret Service agent hearing O'Donnell blowing hard and strong at Lake Como. Newman had not even gone on the European trip. He did not know O'Donnell, and even if he had, he was hardly so stupid as to spout such blustering nonsense. He was bewildered when he started hearing rumors that O'Donnell did not like him and that for some reason he was on the outs with the Kennedys.

In his cover letter Bartlett admonished the president that unless he "put a personal priority on learning more about what is going on, the thing may slip suddenly beyond your control." Bartlett was one of Corbin's friends and sources, and he may have been pulling the strings here, attempting a sweet, subtle revenge. Corbin had been wrong about Newman, but he was a prodigious researcher, and his other revelations were not necessarily wrong, only perhaps tainted in their purposes.

In most ethical matters that came across his desk, the president had sought Mike Feldman's advice. Feldman never learned anything, however, of this matter. The president surely understood that by sending him these memos instead of investigating the administration and writing about it, Bartlett was letting his friendship take priority over his journalistic standards. Kennedy, however, was not so loyal to Bartlett as Bartlett was to him. "Charley, there are a lot of people over here very angry with you," he told his friend.

The president told a number of people about Bartlett's memos, including O'Donnell, making no apparent attempt to try to hide the source of these allegations. Bartlett was increasingly convinced that there was serious corruption involved and felt that the president was doing almost nothing to investigate the allegations. "The president never gave me the impression that he wanted to investigate this stuff," Bartlett recalled. "That was my concern, though it took me a while. I suppose that gave Kenny O'Donnell a great deal of license. In other words, he could go anywhere and ask people for money, and when Larry said that he spent this hundred thousand dollars from California on Get Out the Vote money, why, I suppose Kennedy would have shrugged. But the fact was that they did both end up looking pretty rich. Of course, you always end up with people like that in the White House, but it seems to me that Kenny O'Donnell does you a lot more harm than good. I didn't deal with him at all. I really sort of disliked him."

In April the White House announced that thirty-three-year-old Jackie was pregnant. In late August she would become the only first lady since Mrs. Grover Cleveland in 1893 to give birth while in the White House. For Jackie, this was a double blessing, since the pregnancy gave her an excuse to cancel all her public duties and live the secluded, private life that she preferred.

Jackie had been involved in building a new fifteen-room weekend home called Atoka on Rattlesnake Ridge in the Virginia hunt country that she adored and her husband tolerated. It was supposed to cost $71,000 but had already reached $100,000, and she had not begun to furnish the residence except for two 173-year-old porcelain eagles for which she had paid $1,500. "It's the only house Jack and I ever built together," Jackie recalled, "and I designed it all myself. I didn't want it to be exploited and photographed all over the place just because it was ours."

The president had a wife who was an American icon, celebrated for her beauty and class, a woman who was almost never the victim of envy, that most American of sins. She was spending about $8,300 a month in personal expenses, more than most teachers earned in a year, but far less than she had spent when she first entered the White House.

Kennedy might smolder privately at his wife's endless extravagances, but she was pregnant now, and woe betide anyone who did not understand the deference and care that she must be given. In July at his rented home on Squaw Island, he had come down one Saturday morning and asked his friend Jim Reed to reach Jackie's obstetrician, Dr. John Walsh. The Washington doctor had come to the Cape to be close to the first lady. Dr. Walsh had gone out for a walk, and it was about an hour later when he finally strolled into the president's summer home. Kennedy rarely lashed others with his tongue, but the cool tenor of his voice was enough to chill anyone so unfortunate as to have such words directed at them. "I just hope that if you do go off for a walk for any period of time that you always tell someone where you are," he said, "how you can be reached immediately in case I do have to get in touch with you." Obstetricians tend to be the most philosophical of doctors, realizing that for the most part nature is their master. The doctor hurried upstairs and soon reported to the president that Jackie was simply tired.

On August 7, Kennedy was in Washington when he learned that Dr. Walsh had Jackie taken to a specially prepared suite at Otis Air Force Base Hospital. By the time the president arrived, the first lady had given birth prematurely by cesarean section to a four-pound, ten-ounce son. The president had gone off carousing in Cuba soon after Caroline was born, and he had not even been in Washington when John Jr. was born. Now he had missed this birth too, but he appeared to appreciate the miracle of life in the presence of his tiny son the way he had rarely felt it before.

The newborn had some trouble breathing, but the president wheeled him into his mother's room, where he placed the infant in Jackie's arms. The baby continued to breathe fitfully, and Dr. Walsh decided to move him by ambulance to Boston. Before that was done, a priest baptized Patrick Bouvier Kennedy, named after the president's grandfather and Jackie's father.

The president flew back to Squaw Island to spend time with Caroline and John Jr. He then flew back to visit his convalescing wife at Otis, and then flew on to Boston to visit his newborn son. At this moment he was a husband succoring his wife. He was a father shepherding his children. He was a man whose newborn child was sick, and there was nothing he could do about it.

The baby was diagnosed as having idiopathic respiratory distress syndrome. The doctors could not do much more than the president was doing, waiting and praying, hoping that the membrane on his air sacs, which so troubled his breathing, would soon dissolve. Patrick had been taken from his mother's arms, and now he was moved to the Children's Medical Center at Harvard's School of Public Health, where he was put in a new experimental high-pressure chamber. The thirty-one-foot-long device was like a mini-submarine, a room within a room in which Patrick, the doctors, and the nurses were encased.

A healthy man sees only healthy men and he thinks he has seen the world. A sick man looks out on that same vista and sees the hurt and the lame, the crippled and the stricken. When it came to the imponderable pains of the world, the president had learned to turn away, looking toward all that was bright and gay. On this day, though, there was no turning his head away from the tiny inert form of his infant son, visible through the windows of the chamber.

Kennedy slept in the hospital that evening. He was awakened at two in the morning and told that Patrick was not doing well and that he should come to his son's room. As Kennedy stood waiting for the elevator, he saw a burned child in an adjoining room. The president was a man of the deepest curiosity, but he turned away from any door that led into darkness. Yet now he pulled himself out of his own misery to ask about this child. He was told that the child's mother came to see him every day. He wanted to do something for this woman. He was the president of the United States, but he had no power here to heal these wounds. He asked for a piece of paper and wrote a note to the woman, telling her "to keep her courage up." Though he knew much of men standing bold under cannon fire and politicians holding firm in the name of principle, he knew nothing of this humble courage that won no medals, gained no accolades in books of history—a mother walking alone each day into a hospital to see her burned child.

When Kennedy reached his son's room, Patrick was still alive, but two hours later he was gone. The president went off by himself into the boiler room, and when he opened the door again his eyes were red and wet. He had never thrown a football to Patrick on the turf at Hyannis Port, or led his pony down a trail at Glen Ora, but by the measure of his pain he might as well have.

When it came time to bury Patrick, the president went to the private services at Cardinal Cushing's residence with his brothers and sisters and other Kennedys. They were all there except for the convalescing Jackie, who would have heard a special prayer for a mother who had lost three children. The fifteen mourners filled the tiny chapel and heard Cushing offer the Mass of the Angels. When the service was over, the Kennedys filed out one by one until only the president and the cardinal remained. The president harbored the tiny casket in his arms, as if he sought to take it with him.

"My dear Jack, let's go, let's go," the cardinal implored. "Nothing more can be done."

The two men stood alone weeping and sharing a grief that for a moment even the grace of faith could not assuage.

Patrick's death was a brutal way for the president to grow emotionally, but after the tragic event he did care more and grieve more than he had previously appeared capable of doing. He had always been aware of the transitory nature of life, but he saw now more than ever that God or fate could in an instant turn a placid sea into turbulent, churning waters, and that in such moments man was powerless.

That summer he was sitting at mass one Sunday at St. Francis Xavier Church in Hyannis when he turned to the three White House correspondents sitting behind him. "Did you ever think if someone took a shot at me, he would probably get one of you first?" Kennedy asked. It was classic Kennedy irony, but clearly the president had been thinking about assassination.

A tragedy does not by itself bring people together, but it allows those affected to display a humanity, generosity, and vulnerability that they may not usually expose. This was as true for the president and first lady as for anyone else, and out of Patrick's death came a marriage they had never had before. "Jack was one of these men who was incapable of being loyal to one woman," reflected Feldman. "But in the last year of their marriage, he cared for her as he had not before, and they had a closeness they had not had before."

Their friends looked on as Kennedy displayed a gentle tenderness toward his grieving wife. "I think that Jack and Jackie both had their own particular problems," reflected Betty Spalding, who often talked with both of them. "She had the same emotional blocks and limitations that Jack had, but they were both growing up emotionally. They were catching up. Their relationship was getting better and better."

In September, Jackie was invited to go along with her sister, Lee Radziwill, for a cruise on Aristotle Onassis's 303-foot yacht, the *Christina.* The president felt nothing but disdain for the Greek shipping magnate who had been indicted for his business manipulations. When Kennedy returned from his European trip, there had been an exquisite model ship sitting outside the Oval Office. It was the kind of object that Kennedy immensely admired, but when he asked who had given him this marvelous gift, his secretary had not completed Onassis's full name before Kennedy ordered, "Take that out of here."

The president clearly would have preferred not to have his wife sailing around the Mediterranean with Onassis, but there was no other luxury yacht in the world like the *Christina,* and he figured it was just the tonic that Jackie might need before facing the rigors of his reelection. To keep up a pretense that the journey had some other purpose than amusement, and to watch over his wife, he asked Franklin Roosevelt Jr., his undersecretary of Commerce, and his wife, Suzanne, to go along.

Kennedy was consumed enough by the idea of his wife going off with the Greek magnate that while staying at the Carlyle Hotel on September 20, he doodled on a notepad "Jackie—Onassis." Nine days later he drafted the precise words of a press release that Pierre Salinger was to read at his press briefing the following noon. "The yacht has been secured by Prince Radziwill for this cruise from her owner, Aristotle Onassis," Kennedy wrote, a statement that was both untrue and unkind.

Jackie sailed off on October 5 from Athens, along with a crew of sixty, including two coiffeurs and a dance band. The ship had hardly left port when the previously sacrosanct Jackie became the subject of criticism. Was it "improper for the wife of the president . . . to accept [Onassis's] lavish hospitality?" asked Congressman Oliver Bolton, an Ohio Republican. With his reelection campaign less than a year away, Kennedy was attuned to even the most subdued criticism. He knew that the Republicans would attempt to create an image of the White House, in the words of the GOP national chairman, as a scene of "twisting in the historic East Ballroom . . . [and] all-night parties in foreign lands."

"Well, why did you let Jackie go with Onassis?" Kennedy was asked at a private party while the boat sailed the Aegean, bad publicity traveling in its wake.

"Jackie has my blessing to go anywhere that will make her feel better," he replied, leaving the matter at that.

"What's the helicopter coming in for, McDuff?" Kennedy asked Kilduff, his deputy press secretary, calling him by the nickname he usually employed. It was a Saturday morning in October and Jackie was still away.

"It's for the news hens," Kilduff said, using the president's preferred term for the women reporters who covered Jackie and the East Wing. "They're going to see Atoka."

"Well, you go call that thing off, McDuff," Kennedy said, not wanting the reporters to visit the first family's new weekend home. "That place is going to be just for Jackie and me."

"Me go out there and tell them it's off at this point?" Kilduff asked unbelievingly.

"Better you than me, McDuff," the president laughed, pointing his press aide toward his duty.

"To me, it was very meaningful," Kilduff reflected years later. "It sealed everything that I had observed since Patrick Bouvier died. They were closer now than any time in their marriage."

The president and first lady had an emotional bond that they had not had before, but even so, Jackie chose to be away from her husband for several weeks. While the couple was physically apart, they both continued to display the aspects of their personalities that had been most detrimental to their marriage. This was a pleasure-loving, luxuriantly indulgent Jackie dancing her way through the high spots of Europe. She was unwilling to temper her behavior, to consider the impact on her position as the nation's first lady, first wife, and first mother. As for her husband, the president used her time away as he always did, by bringing in other women. Whenever Jackie left town, he was like an innkeeper putting out a sign that the White House had vacancies.

At the same time Kennedy spent more time with Caroline and John Jr. The president had never had his father's attitude toward his own children, though not his grandchildren, that little sons and daughters were largely decorative objects who only became interesting when they were old enough to think, reflect, and talk in rounded sentences. As much as the president saw the political advantage of having children in the White House, he also adored moments such as when his namesake walked with him to the Oval Office each morning. That had always been a pleasant diversion, but since Patrick's death there was an intensity to Kennedy's moments with his children. This had become time when he blocked off the world.

"I'm having the best time of my life," Kennedy told Kay Halle, a family friend, when she came to visit. As he talked, looking at his two children, John Jr. rushed forward. "I'm a great big bear," he said. "I want something to eat." Halle pretended to feed the child, but the president intervened. "I'm a great big bear," he said his voice charged with false menace, "and I'm gonna eat you up in one bite." Little John Jr. replied in a cascade of laughter.

When Jackie was away, she wrote her husband of ten years a seven-page, handwritten letter that was an exquisite rendering of her complex feelings for him. Even her most perfunctory thank-you note always had some graceful, unique phrase. In these pages all of her exquisitely nuanced sensibilities came together. There was both an emotional intimacy and a near formality, as if she were reaching something within herself and in her husband that neither of them had touched but that she knew she must touch now. "I loved you from the first day I saw you and if I hadn't married you my life would have been tragic because the definition of tragedy is a waste," she wrote. "But ten years later I love you so much more." She wanted her husband to know how much she cherished their love. "I am just sorry for Caroline—all I will tell her to put into and expect from marriage—but if she doesn't marry someone like you what good will it do her."

Surely there are few things in life as unknowable and mysterious as the intimate truths of a marriage. She loved her husband, and Jackie's letter was like a rare orchid, a flower of beauty and subtlety, but a flower that needed warmth and light and shelter.

Kennedy loved his wife, but he preferred bouquets to a single stem. He wrote his own cryptic rendezvous letter, but it was not to Jackie. Although the salutation has been clipped off the letter, it was most likely written in October to Mary Meyer, although Kennedy's sex life was such that it could have been written to a number of other women as well.

> Why don't you leave suburbia for once—come and see me—either here or at the Cape next week or in Boston the 19th. I know it is unwise, irrational—and that you may hate it—on the other hand you may not—and I will love it. You say that it is good for me not to get what I want. After all of these years you should give me a more loving answer than that. Why don't you just say yes. J.

There was a sentimental yearning in Kennedy these days that had long been dormant within him. In the middle of October, he hosted a state dinner for Sean Lemass, the prime minister of Ireland. During the dinner Kennedy scribbled a few notes for his eloquent toast about Ireland's role as a beacon of liberty. The fiddlers and the bagpipers played the old Irish songs, but as so often with things Irish, it wasn't until the hour grew late and almost everyone was gone that the evening truly began. A small group of no more than ten guests went upstairs to the family quarters, the president bringing the Irish bagpipers with them.

Gene Kelly danced an Irish jig, and the others sang, clapped, and laughed. Teddy sang too with such enthusiasm and emotion that no one cared if he sang out of tune. The greatest of Irish songs are sad ones, and the gathering was infused with a sweet melancholic sadness.

One of the president's favorite songs, "The Boys of Wexford," echoed poignantly through the White House.

We are the boys of Wexford,
Who fought with heart and hand
To burst in twain the galling chain
And free our native land.

When it came time to go, they sang "Danny Boy," their Irish benediction. "The president had the sweetest and saddest kind of look on his face," Jim Reed remembered. "He was over standing by himself leaning against the doorway there, and just sort of transported into a world of imagination apparently."

In the third year of his administration, Kennedy still had an antipathy toward what he considered the prissy moral tone of American liberalism. After the Cuban Missile Crisis, the president had cooperated in an inside account for the *Saturday Evening Post,* written by his friend Charles Bartlett and Stewart Alsop. The two journalists found many top White House officials ready to savage UN Ambassador Stevenson, as long as they could do their mugging anonymously. One official said, "Adlai wanted a Munich. He wanted to trade the Turkish, Italian, and British missile bases for Cuban bases." When Bartlett gave the article to Kennedy to read for criticism before publication, the president asked that the reference to Sorensen as a "dove" be taken out. Sorensen had been a noncombatant as a young man, and Kennedy apparently did not like the implication that a pacifist had pushed him toward peace. When it came to the devastating comments about Stevenson, however, the president uttered not a word. Instead, he called in Schlesinger, his favorite conduit of misinformation to Stevenson. Kennedy told the former professor "he understood that it [the article] accused Stevenson of advocating a Caribbean Munich."

"Everyone will suppose that it came out of the White House because of Charley," the president said. "Will you tell Adlai that I never talked to Charley or any other reporter about the Cuban crisis, and that this piece does not represent my views."

Whatever weaknesses Stevenson may have had, a lack of honor was not one of them. He could have been as duplicitous as the president, calling in *his* reporters and defending himself with the truth. That would have ended the carping attacks that he suffered in the wake of the article, but it might have destroyed the American-Soviet deal, and it was not Stevenson's way. Undersecretary of State George Ball noticed a change in the man. He went "through the motions, making speeches, yet with a feeling in his heart that it didn't make any difference to the world if he fell over and had a heart attack."

Stevenson's presence rankled Kennedy so much that it may have affected policy decisions and his attitude toward disarmament. In 1963 the president told Bobby that this was the one area in which he wished he had done more. "It was personal again—if Stevenson brought it up, it irritated him," Bobby said. "To shock him and give him something he'd talk to his girls about, he'd say that disarmament was just a lot of public relations stuff."

Kennedy believed that peace in the nuclear age would be won by hard men speaking tough truths, not by what he considered gushy overwrought men of public virtue. In June 1963, the president finally made the issue of peace his own, and he did so at American University in one of the seminal speeches of his time. The graduating seniors heard an eloquent address that pandered neither to the American people nor to the Russians. Kennedy took a bold step forward, announcing that the United States would unilaterally not "conduct nuclear tests in the atmosphere so long as other states do not do so."

Kennedy described peace as "a process—a way of solving problems." It was a process in which he wanted to involve his nation, the Soviet Union, and the world. The president talked about the nature of Soviet communism, but beneath that he expressed an underlying belief in the commonality of humankind and the hope that the threat of nuclear weapons might in the end draw the world's people closer, not further apart.

The young people in that audience were living at a time when many Americans feared that the world's problems were overwhelming and intractable, with the shadow of nuclear war hovering over everything they did. Giving in to despair, however, was not what Kennedy had been brought up to do, and as much as his speech was about a strategy of peace leading toward nuclear disarmament, it was equally about the spiritual armament and strength that would be required in this world.

"Our problems are man-made," Kennedy told the students. "Therefore, they can be solved by man. And man can be as big as he wants. No problem of human destiny is beyond human beings. Man's reason and spirit have often solved the seemingly unsolvable—and we believe they can do it again."

No one listened to the president's words more attentively, and analyzed them more closely, than the Russians, who had their own goal of "peaceful coexistence." In late July, Khrushchev announced that he would agree to a limited ban on nuclear testing, ending all but underground tests that could not be verified by off-site testing. This was just a way station to a broader peace, but it was a way station that was reached in part because Kennedy had made such a deep, brave speech.

While President Kennedy sought peace in the world, the Reverend Martin Luther King sought freedom at home. During the spring of 1963, the civil rights leader staged a series of demonstrations in Birmingham, Alabama. He was continuing his politics of moral witness, seeking to confront evil segregation with the olive branch of nonviolence. The television cameramen and newspaper photographers were often the unwitting carriers of King's messages, and when they capped their cameras, his voice was not heard. It was not heard beyond Birmingham until May 2, when King sent one thousand children marching out of the Sixteenth Street Baptist Church to confront segregation by their presence in the restaurants and lunch counters where they were not allowed.

Sheriff Eugene "Bull" Connor sent out the bearers of his message too—dogs and fire hoses. The animals bared their teeth, and the sheer force of the water drove the protesters back, seemingly washed away into the streaming gutters. And while this one-sided confrontation went on, the television cameras rolled and the cameras clicked. Within a day the nation and much of the world knew Bull Connor's name and bore witness to his deeds. With those images blanketing the world, Birmingham, Alabama, became synonymous with oppression. Hoping to restore Birmingham's proud image, the city leaders agreed to begin desegregating the city's restaurants, movie theaters, and other public places. The agreement was punctuated not by handshakes but by dynamite exploding both at the motel where King was staying and outside his brother's home. The Reverend King preached nonviolence, but in a dozen cities across America there were confrontations that seemed to threaten the very fabric of civil society.

Bobby knew no more what to make of the rising black militancy than did most of his fellow white citizens. He asked the writer James Baldwin to set up a meeting with a group of blacks to talk about the situation. The Kennedy attitude was that when you had a problem, you found the most prominent experts, brought them together, heard their opinions, and took the best of their ideas. Then after solving that dilemma, you went on to the next problem. There seemed to be no better expert on race in America than Baldwin, who in November 1962 had written a passionately apocalyptic essay in *The New Yorker* (published the following year as *The Fire Next Time*). In the controversial work, Baldwin condemned Bobby for his "assurance that a Negro can become president in forty years" as a prime example of the white American attitude that "they are in possession of some intrinsic value that black people need, or want."

Baldwin assembled a group of what he called "fairly rowdy, independent, tough-minded men and women" and brought them to Bobby's apartment at the UN Plaza in New York City. The attorney general began by detailing what the administration was doing for blacks, and those in the room replied by saying that it should do more. This was all part of the civilized nomenclature of traditional American politics. Then a young man spoke out. "You don't have no idea what trouble is," he told Bobby. "Because I'm close to the moment when I'm going to pick up a gun."

The speaker, Jerome Smith, bore the scars of the beatings he had received when as a Freedom Rider he attempted to enter the white bus depot in McComb, Mississippi, and had spent time in a Mississippi prison. By anybody's account, he had paid more dues than these artistes and scholars who filled Bobby's living room, and he intended to be heard.

"When I pull the trigger," he stammered in rage, "kiss it good-bye."

For all Smith's profound feelings, if there were indeed a "fire next time," it would burn the black minority more than it would the whites who dominated the country and ran the machinery of oppression with a firm hand. In Baldwin's own essay he had accused white liberals of being full of "incredible, abysmal and really cowardly obtuseness." In 1963 many liberals enjoyed being humiliated for their sins and accepted whatever penance was meted out as their due, but even they had only so much skin to be flayed. Despite whatever hidden racism may have darkened their souls, they were the natural allies of a revolution of equality. And it was perhaps not wise to alienate them. Bobby was not a liberal, and he listened to Smith's nearly incomprehensible, raging sermon while feeling no compulsive need to flagellate himself.

Baldwin was the impresario of this drama. He sought in others an authenticity he perhaps did not have in himself. He had grown up gay in New York City, a morbidly sensitive young man. He was feted and dined at the well-set tables of highbrow liberalism, his presence an immunization against charges of racism. Like a prosecutor who knew how his witness would answer, Baldwin asked the young activist whether he would fight for his country. "Never! Never!" shouted Smith.

"How can you say that?" Bobby asked. This relationship between blacks and whites was a deeply troubled marriage in which the whites bore the brunt of the blame. But the largely blameless partner did not appear to realize that there were nonetheless certain things that could not be said, or threatened, without changing the relationship forever.

Smith was full of the pain of what he had seen and felt, but the other blacks in the room perhaps understood the world as he did not. They could have stepped forward and said that they felt the terrible injustice, but this was their country too and they would fight for it. Smith, however, held the moral trump card, and they sat and listened like a congregation taking in the weekly sermon. "You've got a great many very, very accomplished people in this room, Mr. Attorney General," the playwright Lorraine Hansberry lectured Bobby, "but the only man who should be listened to is that man over there."

Harry Belafonte listened to this rage with mounting disquiet. He cared as much about the civil rights of his people as anyone in the room, and in the years to come, no one there that evening would walk any further than he in searching for social justice. The singer knew that it would be a terrible thing to make an enemy of the second most powerful man in America. In his melodiously soothing voice, he began to tell the group about the hours he had spent at Hickory Hill talking to the attorney general about racial issues. Clarence Jones, the Reverend King's attorney, brought up some of the ideas that King envisioned, such as the president making a Roosevelt-like fireside chat about civil rights and issuing a twentieth-century emancipation proclamation. These were good ideas, but the room was so fogged with anger and misunderstanding that the two groups could hardly see each other. A few minutes before, Bobby might well have seriously entertained such ideas. Now he dismissed them with disdainful laughter.

Hansberry walked out of the gathering with arrogant indifference, believing that in doing so she had accomplished something. And that was the end of an evening that the distinguished educator Kenneth Clark called "*the* most dramatic experience I have ever had."

Bobby sat in the Oval Office talking with the president and other advisers about the civil rights situation. He had returned from New York City stunned by the level of vitriol. Baldwin betrayed the whole nature of the private discussion by telling the *New York Times* about the evening, saying that the attorney general had been "insensitive and unresponsive." Privately, Bobby slammed Baldwin as a representative not of blacks but only of what Bobby considered Baldwin's own depraved homosexual orientation. Yet he was not about to burn his brother's presidency as well as his own political aspirations in the civil rights fire. He told his aides that if he had grown up black, he would have felt precisely like the blacks who had come to his New York apartment. That was probably a deeper insight into his own psyche than even he realized. If his skin had been Baldwin's color, he probably would have fought against the evil that his people suffered and confronted racism in its very lair.

The president heard the footsteps in the night, and he knew that he could avoid a direct confrontation with the civil rights issue no longer. It was the

Kennedys' task to move the racial dialogue forward with speed and purpose, channeling that anger into engines of change.

"You know a black college graduate who applies for federal employment in the South can hope for nothing better than a job carrying mail," Bobby said, suddenly seeing inequality where he had not seen it before.

"Pretty good job for a Negro in the South, though, letter carrier," Kennedy said, quieting the room. Part of the problem was that the president's sentiments were more like those of most white Americans than Bobby's were. "I'd like to bet he had never met a black person in his life," his friend Ben Bradlee noted. He was wrong about that—as a young man at Harvard, Kennedy had a black valet.

The president felt a stronger affinity with some of the southern politicians like Governor Patterson of Alabama than he did with many of the black activists. "My impression of President Kennedy was that he wanted to do everything he could for the blacks, for the minorities," reflected former Governor Patterson. "He wanted to integrate them fully into our society. At the same time he recognized the political ramifications of that, and the necessity in some areas for not going too fast. . . . Now Robert just wanted to do it overnight."

Kennedy was uncomfortable with a passionate moral issue that could lead men to violence or the rhetoric of despair. He was living in a segregated southern city, the evil of which had largely been ignored by both political parties. For the most part, he knew blacks as deferential people who served drinks, made beds, and drove chauffeured cars. He was head of a political party that risked being torn in two between the conservative southerners and the more liberal northerners.

Kennedy had in fact gone far in trying to open important government positions to blacks and using the federal law to push ahead the integration of public facilities in the South. But to blacks, it was only a first step. The president was an astute politician who looked out across the expanses of America and saw that there was not a strong majority ready to push civil rights, not in the South and probably not in the North either. If he moved too slowly, however, he took the risk that frustrated black Americans would start setting fires in the night. If he moved too fast, he might have to put out brushfires of violence across America.

In June, Governor George Wallace of Alabama, a pint-sized demagogue, said that he would "stand in the schoolhouse door" to prevent blacks from attending the University of Alabama. Wallace had vowed when he lost his first race that he would "never be out-niggered again," and all those who heard him in those torpid days in Tuscaloosa would say that he had kept his pledge. Wallace, however, no more wanted federal troops and bloodshed than did the Kennedys. After his endless posturing in front of the cameras, he followed the order to step aside once federal officials and the federalized Alabama National Guardsmen arrived. Despite the ugliness of the scene, the covenant of government had not been broken, and Wallace bowed to the law of the land, even if he did so standing on a chalk line positioned for filming him from the best camera angle.

That evening the president decided that he had to give a television address to the nation boldly announcing his plans to put forth his civil rights bill. The political operatives in the White House opposed the legislation, but just as in the Cuban Missile Crisis, the two Kennedy brothers stood together. O'Brien and O'Donnell may not have been numb to the pain of black Americans, but they knew that the figures did not add up, not in Congress, not in the Democratic Party, and not in the life of an American president who would soon be seeking reelection. Great leadership is often not good politics, however, and Kennedy decided to go ahead.

Kennedy began to prepare for this historic address only an hour before the 8:00 P.M. televised speech. Sorensen hurried into the Cabinet Room where the president sat with Bobby and Burke Marshall. The president quickly sketched his ideas, and Sorensen left to try to piece together a worthy speech in a few minutes. As Sorensen worked elsewhere, Kennedy worried that he would have to give the address extemporaneously. While he was jotting down a bunch of notes on what Bobby remembered as "the back of an envelope or something," Sorensen returned with the draft of a speech. Once again Sorensen had managed to channel himself into the president's psyche, and the speech had all the resonance and depth that would have taken anyone else hours to try to match. As good as it was, Bobby felt that Kennedy should end with a few unscripted remarks—and so he did.

Kennedy was not comfortable with the moralizing of priests and preachers, but there was a difference between moralizing and morality, and he grasped that this was his nation's great moral issue. A leader did not temporize and cut his words and meaning when he had prepared an important address in just one hour. "We are confronted primarily with a moral issue," he told Americans. "It is as old as the Scriptures and is as clear as the American Constitution." He saw the urgency of this matter, and the tragic direction in which it might be heading. "Redress is sought in the streets," he said, "in demonstrations, parades and protests which create tension and threaten violence and threaten lives."

The Freedom Riders had irked him with their moral absolutism and their refusal to compromise, and the civil rights activists worried him still. He feared where their uncompromising passions were leading America. He believed that "a great change is at hand, and our task, our obligation, is to make that revolution, that change peaceful and constructive for all. . . . Those who act boldly are recognizing right as well as reality."

This president was above all a realist. Morality and the political imperative were one and the same, and he now proposed to the nation a civil rights bill that would help ensure that all Americans could go to hotels and restaurants and theaters and sit where they pleased, and that children could attend the schools that best served them, not facilities relegated to their race. He concluded in his own ad-libbed words, linked seamlessly to Sorensen's eloquent phrases, "We have a right to expect that the Negro community will be responsible, will uphold the law, but they have a right to expect that the law will be fair, that the Constitution will be color-blind."

The president had listened to Bobby and learned from him, assuming if not his brother's passion then at least his brother's insight into the black struggle. He had put his toe gingerly into this moral caldron, and he wondered aloud afterward whether he should have pulled it out immediately. "Do you think we did the right thing by sending the legislation up?" he asked Bobby, knowing what his answer would be. "Look at the trouble it's got us in." Bobby recalled that Kennedy feared that "this was going to be his political swan song."

When Kennedy learned that in August King was planning to bring hundreds of thousands of blacks to Washington for the largest civil rights event in the country's history, he was appalled. He could envision the event turning into a riot in the streets of the capital, or an unseemly gathering of the unkempt and unwashed. "They're going to come on down here and shit all over the monument," Alan Raywid, a Justice Department official, recalls the president saying.

Both Kennedys were worried not only about the dangers of a mass march on Washington but about the influence of Communists on King, particularly Stanley Levison, the minister's closest white associate and friend. Levison was a left-wing New York attorney, part of the diminishing circle of the American Left in the fifties. He unquestionably brushed shoulders with party members at rallies and meetings and raised money for left-wing causes. As a fellow traveler of the party, he had become, in the words of King's biographer Taylor Branch, "a financial pillar of the Communist Party during the height of its persecution."

There were a disproportionate number of Communists in the civil rights movement, as there were in almost all struggles for social justice. They may have attempted to push the agenda toward the current Communist line, but they were often brave men and women who spilled their blood and gave endless energy to the cause. Hoover had convinced the Kennedys that Levison was a Communist agent who was trying to turn King into a Soviet puppet and the civil rights movement into what would truly be the largest subversive movement in American history.

Levison may have been dangerously myopic when he looked at the realities of the Soviet Union. Hoover, however, made a case that Levison was a master spy when all the FBI director had against him was mindless innuendo. The FBI director had no evidence that the attorney was involved with espionage or that he had ever been a Communist Party member. Hoover and his colleagues probably sincerely believed that Levison was a Communist spy and thought they were acting out of only the most patriotic of motives when they warned the president. In the eyes of the Kennedys, Levison was a red Svengali leading the dangerously naive King down the pathway toward Marxist revolution.

After meeting with civil rights leaders on June 22, Kennedy led King out into the Rose Garden for a private talk. "I assume you know you're under very close surveillance," the president said. King was not sure whether the president was worrying about being bugged himself or meant that the civil rights leader's movements were being monitored even in the White House. King did not say, as another man might have, that this was supposed to be an America where a man walked free and talked free, without fear of secret monitors.

"You've read about Profumo in the papers?" Kennedy asked. It was a curious analogy for Kennedy to make, telling King that if he was as loyal to Levison as Prime Minister Macmillan was to his defense minister, it was not a government that would be brought down but a noble cause. The president went on to tell King that Levison and another of his aides were Communists, and in a voice scarcely a whisper said that King would have to "get rid of" them.

King and his associates sought some way to placate Kennedy while continuing to do their work as they felt they must do it. King had faith in Levison's judgment, both as a friend and an adviser; he would not walk away from him. Clarence Jones, King's attorney, decided that to avoid having their conversations overheard on a wiretap; Levison would have to communicate with King through an intermediary. That rankled the attorney general, and on July 16, 1963, he told the FBI's Courtney Evans that he wanted new wiretaps on both King and Jones.

This was not the notorious Hoover of the liberal imagination leading the innocent attorney general down the path of unrighteousness, but a petulant Bobby going ahead on his own. If Hoover had been simply an evil manipulator trying to ensnarl Bobby in some illegal acts, he would have immediately ordered his teams to begin their secret work. Instead, in that meeting Evans warned Bobby about the dangers of wiretapping King, telling him "there was considerable doubt that the productivity of the surveillance would be worth the risk because King travels most of the time and there might be serious repercussions should it ever become known that the government had instituted this coverage." Nine days later Bobby decided to follow Evans's advice and backed off from the King wiretap. The FBI went ahead immediately with the Jones wiretap, an act far beyond the pale of a legitimate investigation. Jones was an attorney who talked on his phone with clients such as King. Beyond that, there were no allegations that Jones was a Communist. His mistake was to have raised Bobby's ire by saying nothing at the infamous Baldwin meeting and then approaching the attorney general to compliment him on all the good he was doing.

In a democratic society, if wiretapping is used at all, it must be limited to the most serious of cases in which there is strong evidence, and the results must be held within the smallest possible circle of officials. Bobby read the transcript excerpts from Hoover as if they were tabloid entertainment, passing on the juicier bits to his brother. The Reverend King heard on these tapes was at times not the saintly leader revered by his followers. He was a bawdy gossip talking about sexual adventures of his own and discussing the sexual mores of his colleagues with immense relish. He worried that one of his colleagues, Bayard Rustin, might indulge his homosexuality too openly, get drunk, and "grab one little brother."

Bobby was so peeved by what he was reading that he spoke publicly about matters that could only have come from these tapes. "So you're down here for that old black fairy's anti-Kennedy demonstration?" he assured Marietta Tree, the American delegate at the UN. "He's not a serious person. If the country knew what we know about King's goings on, he'd be finished."

A very different Martin Luther King stood before a quarter-million black and white Americans in August at the March on Washington, giving one of the great speeches in American history. "I have a dream that one day on the red hills of Georgia the sons of former slaves and the sons of former slave owners will be able to sit down together at a table of brotherhood," he told the vast audience. "I have a dream that one day even the state of Mississippi, a desert state, sweltering with the heat of injustice and oppression, will be transformed into an oasis of freedom and justice." There were no foul manners among these protesters spread out across the broad expanses of the Mall, no violent rhetoric, no uncontrollable mob, only American citizens who wanted the justice they had not received and a place in America that they knew must be theirs.

Afterward Kennedy met with King and the civil rights leaders at the White House. Kennedy had feared that if there were not a riot, then only a paltry crowd would turn out at the Lincoln Memorial, helping to doom the troubled civil rights bill. Instead, there was a massive, unprecedented turnout. King had given what Kennedy, that student of language, considered a speech of immense power. The president knew that the equation of power had changed, that these men and women here before him had something they had not had previously. He honored them not with flattery and applause, but with his hard-won political wisdom, and for him that was the highest honor of all. He let them in on just how hard it was to work the civil rights bill through Congress, running down the congressional tallies senator by senator, and state by state. These people before him had a crusade, but he had a problem, and he wanted them to understand it.

Walter Reuther, the labor leader, talked about amending the civil rights bill with even more guarantees. Kennedy listened to the articulate Reuther, and then interrupted. "This doesn't have anything to do with what we've been talking about," he said. "But it seems to me, with all the influence that all you gentlemen have in the Negro community, that we could emphasize, which I think the Jewish community has done, educating their children, or making them study, making them stay in school and all the rest."

In King's glorious words, there had been not one sentence about what blacks should do for themselves. The overwhelming burden of black oppression lay elsewhere, but the civil rights leaders risked fostering a disquieting sense of entitlement. Kennedy understood the limits of government in changing the spirit of human beings, and he never would have suggested, as King did, that on some miraculous day the shackles of injustice would be broken and all men would walk free as brothers. Kennedy's comment was lost in all the talk of programs and strategy, but he had struck at something deep to which he might one day have returned.

The Reverend King had become one of the most important leaders in America; the spiritual and political guide to black Americans, he was admired by millions of white Americans as well. In October, when Hoover placed a request on Bobby's desk to wiretap the civil rights leader, the nature of the enterprise had changed entirely. Bobby had a multitude of reasons to look on the request skeptically. The FBI had already been taping Jones and Levison and had come up with nothing even hinting of subversion. It bothered the attorney general that despite everything he and his brother had told King, he still was unwilling to back away from Levison. It did not seem to have occurred to Bobby that if Levison were indeed a Communist agent, he would probably have communicated surreptitiously with King and not become known as his closest white friend.

The attorney general believed Hoover's worst accusations against Levison. In 1964 Bobby wrongfully called him "a secret member of the Communist party" and a member of the "Executive Committee." And so, primarily as a way of getting information on Levison, he signed the documents that allowed the FBI to wiretap King.

It was a shameful day for American civil liberties. Bobby's apologists have blamed his actions on everything from Hoover's manipulations to Bobby's concern for passage of the civil rights bill. That may all have been true, but it did not account for the attorney general's visceral mistrust of King, an emotion only exacerbated by the unseemly sexual revelations he had already heard on the wiretaps.

The attorney general may also have been so tired out by the endless struggles of government that he had begun to make bad judgments. Few cabinet members in American political history have taken on the range of activities and concerns that Bobby had during these years. He was a man who had to touch and feel before he could think and act.

Early on in the administration, Bobby had gone up to Harlem to visit young gang members. He didn't travel with a large entourage. He didn't have journalists with him. He sat on the curb listening to members of the Viceroys talking about their lives. Bobby was seeking to learn how the federal government could help combat juvenile delinquency, and he sought answers not in written reports, expert testimony, but in these individual troubled lives. He wanted to learn, and though he left these gang members to live just as they had before, he saw that beneath these lives that were such tragic admixtures of bravado, resignation, cynicism, and despair, there were flickers of hope.

When Bobby talked about black children, it was not enough to look at studies and slides. Instead, he might call in his old friend and aide Dave Hackett, and they would visit a D.C. school where he could take children in his arms and hear their dreams. In the Justice Department he developed the government's first major program for dealing with juvenile delinquency. He learned that there were twenty-three lawyers in the Justice Department working on reparations to Indians for land stolen from them years before, and he not only pushed that forward but met with American Indians and celebrated their lives and ways in a manner that few officials had done before. He kept up his push against organized crime, though involving men like Giancana in the Castro assassination attempts had made prosecution far more difficult.

Bobby had time for his Cuban friends and their attacks against Castro, and he was the secret goad and force behind these efforts. His brother called him on a multitude of issues, from trade to foreign policy, and he made his own firm mark on most of the major initiatives in the White House. He had his family too, Ethel and now eight children, and he was there for his children, out tossing a football, playing games.

At times Bobby displayed a mindless bravado. In May, when the first Americans climbed Mount Everest, Bobby's reaction was, if they could do it, he could do it too. Justice Douglas called him to talk about a climb. "I'll talk to Ethel about it," he said in what was a joke and not a joke. "Maybe we can work it in for a week in July after the baby comes."

Bobby was living on the natural adrenaline of action. But his face was gaunt, and he was so tired that he may not have realized how tired he was. He had always been abrupt, but he was more abrupt now. At a party celebrating Bobby's thirty-eighth birthday on November 20, one of his aides, John Douglas, thought that the attorney general seemed "quite depressed." That was an emotion that had always seemed foreign to Bobby. As fatigued as he was, the election lay ahead. And over in the White House sat his brother, about whose health and well-being Bobby knew far more than any man.

31

To Live Is to Choose

Kennedy's preeminent concern these days was his reelection. Barry Goldwater, the conservative Arizona senator and Kennedy's most likely opponent in the 1964 election, was far behind the president in the polls. Goldwater, though, was becoming better known and slowly rising in popularity, giving credence to the president's prediction that he would have a "long, hard fight to the White House" in 1964. The Republicans had already begun to step up their criticisms of his presidency as full of promises and rhetoric but few accomplishments. "The New Frontier is a little like touch football," Representative Gerald R. Ford of Michigan told one Republican gathering. "It touches everything and tackles nothing."

Kennedy stood well ahead of any Republican opponent, but he knew that he was dependent on the will and whim of a fickle electorate. Like any president in the last months of his first term, he had put aside most of his risky ventures. Almost every single initiative and piece of paper coming out of the White House was dedicated to the reelection of the president. His aides thought they were profoundly committed, but he was even more attuned to the subtlest nuances of politics, whether it be matters of life and death, international politics, or some trivial local political squabble.

When Governor Harold Hughes called wanting the president to review a clemency request for a murderer about to be executed, Kennedy sensed immediately that if he reprieved the man, the citizens of Iowa would not be happy, particular in Dubuque, where the victim had resided. "Of course, they are anxious, I suppose, to have the sentence carried out, are they?" he mused, sensing the political danger in saving the man's life, a danger he was not prepared to take. As he hung up, he asked, "Is your call to me, is that known?"

When Kennedy talked to Treasury Secretary Dillon about narrowing a tax deduction, he did not wonder about the equity of the decision but about the political costs. "I don't know how much money we are going to collect as a result of all this, and whether it is worth the heat that these people are able to put on," he said. When Red Fay called from the Pentagon to discuss closing down navy bases in San Francisco, Philadelphia, Brooklyn, and Boston, the president didn't press the undersecretary of the Navy about the economic feasibility or the military purpose of doing so. Instead, he told his old friend bluntly: "We might just as well go home ourselves, then. If you like it down here, you better not close down . . . any of those yards." And when he was contemplating sending Sam Beer, a Harvard professor, to Latin America as an ambassador, his concern was that Beer would be gone during the election year when he might be needed to placate the liberals "to keep them from going off the deep end."

As Kennedy tried to tidy up his political shop, placing the most inviting items in the window, he was bedeviled by one matter that threatened to overwhelm everything else. That was Vietnam, an issue that he had hoped would stay dormant until after his reelection in 1964. Of all presidents who dealt with this issue, Kennedy had the least right to plead ignorance of the realities of the situation in Southeast Asia. He had followed these realities since at least 1951 when during his visit as a congressman, Kennedy had seen that however much the French pretended they were fighting a noble cause in Vietnam, they were essentially defending a corrupt colonial empire. He had also seen that an American government that thought it could help its ally by paying in dollars, not in blood, might one day learn that it had to pay in that harder currency as well.

Both the previous postwar presidents had treated communism like a dreadful infectious disease that had afflicted part of the world body and had to be contained or it would destroy the world. Every time it broke out, it had to be excised, driven up the Korean peninsula, bottled up within Cuba, locked behind the Iron Curtain. For a decade and a half, at enormous cost, the regimen had worked, and throughout his entire political career, Kennedy had been a vociferous proponent of this theory. He knew that a president who contemplated withdrawal from Vietnam risked being accused of political malpractice, of allowing the virus to spread across Southeast Asia and the world.

A quarantine, though, is not the only way to treat an infectious disease. In these months Kennedy was like a doctor contemplating radically different treatments. In April, he scribbled a note on a pad in the White House: "withdrawal from Vietnam as requested." That was one prescription. The other was an escalation, of unknown size and duration, in American involvement. Three weeks later, in a meeting with Secretary of State Rusk, the president penned a few more notes, decipherable only by himself. "Military aid . . . aid etc . . . 4th corps. province . . . Cut off aid to Colonel Tuan . . . Congress resolutions . . . Congressional resolution . . . Politically conscious Americans." The problem was that whatever he prescribed would have horrendous side effects that would have to be treated as well, and while he contemplated his treatment, the patient was in continual agony.

Kennedy beseeched his subordinates to tell him the hard truths about whatever issue he faced. In Vietnam, however, the truth disappeared like the Communist Viet Cong soldiers, firing a volley or two, and then fading away into the vast jungles. Everyone from the ambassador to the military to the CIA put forth self-serving versions of what was happening. What these scenarios shared, for the most part, was a malignant optimism, based largely on not facing what had to be faced.

In January 1963, General Paul Harkins, the head of Military Assistance Command, Vietnam (MACV), proposed to fulfill McNamara's pledge to end American involvement by late 1965, not by phasing out American military operations but by dramatically increasing support, including doubling the air force's role. When Paul Kattenburg, a perceptive and daringly candid Foreign Service officer, addressed the Ex Comm meeting on August 31 about his Vietnamese experience, with most top officials present except for the president, he came away with the feeling that "they didn't know Vietnam. They didn't know the past. They had forgotten the history. They simply didn't understand the identification of nationalism and Communism, and the more this meeting went on, the more I sat there and I thought, 'God we're walking into a major disaster.' " By October, the Pentagon had begun the self-deluding process of winning on paper what they were not winning on the ground, concluding that the South Vietnamese were making "progress" against the Viet Cong, though the hard statistics suggested the opposite.

For the president, much of the discussion had a tiresome familiarity. In 1954, three years after Kennedy had visited Vietnam, Ngo Dinh Diem had become the first postcolonial leader of South Vietnam. Conservative Catholic Americans felt a special affinity for the prime minister. A man of the deepest Catholic faith, Diem had lived in the United States in his years of exile. He had spent years at a seminary in New Jersey and was a daily communicant.

One of Kennedy's seminal insights during his 1951 trip to Asia was how dangerous politics was in much of the world, and how often assassination stalked those who ascended to power. Diem's own elder brother, a provincial governor, had been murdered by the Viet Minh. Diem himself had almost been killed in 1962, and the scent of blood was never far from the palace.

Diem was no tailor of democracy who could sew the many peoples of South Vietnam into even a patchwork quilt of a country. Diem trusted few beyond his two brothers, Ngo Dinh Nhu, his political counselor, and Ngo Dinh Canh, who ran the northern regions. The brothers established the Can Lao Party, a secret society that dominated the country's economic and political life. Ambassador Elbridge Durbrow described it in 1959 as "an authoritarian organization largely modeled on Communist lines."

After nearly ten years in power, Diem had come around to where he began, a Catholic mandarin isolated not only from his own people but also from the Americans who thought him in part their creation. On July 11, a Buddhist priest immolated himself in front of the Cambodian embassy in Saigon, protesting the repression of his co-religionists and the deaths of eight Buddhist demonstrators who had sought to display the Buddhist flag. This was an act of moral witness that transcended all logic, all the nuances of policy and procedures, and created its own imperative. Diem's support in Vietnam had been eroding since almost the day he assumed office, but it took this immolation and the Buddhist civil rights protests to make the men in Washington realize that things could not go on this way with Diem, not when men were burning themselves alive and, in a Buddhist land, only Catholics could freely profess their faith and politics.

In August the president scribbled notes to himself for a speech he was to make to a White House seminar. "To govern is to choose," he wrote, quoting the French leader Pierre Mendes France. That was precisely the nature of presidential power, expressed as succinctly as possible. The implacable riddles of policy reached his desk, issues on which knowledgeable advisers disagreed, decisions fraught with negative side effects. He looked at each matter like an assayer, weighing all the possibilities, observing the subtleties. Knowing the danger that Vietnam represented politically, he would have delightedly willed it to lie dormant until after his reelection. He was trying not to be sucked into the swamp of Vietnam while not abandoning the beleaguered country in such a way that shouts of betrayal and curses of recrimination would sound through the rest of his presidency.

That summer the president apparently suggested to several people, including Larry Newman, a Hyannis Port neighbor, and Dave Powers, that he planned to withdraw from Vietnam after his 1964 reelection. As much as his auditors thought otherwise, this plan hardly presented as flattering a portrait of their hero as they imagined. If true, Kennedy was willing to let American soldiers die needlessly for over a year because the delay would help his campaign. In all likelihood, Kennedy told his neighbor and Powers what flattered their own desires, and part of his own instincts. The president understood, however, that history can't be stopped and started at will. Wide and treacherous water may lie between what one hopes to do and what one can do when time and circumstances allow.

Kennedy faced crucial choices *now*, and he faced them as he had all the major decisions in the White House, parsing the issue back and forth half a dozen ways. He took a few tentative steps one way and then back the other. On September 2, Kennedy gave a lengthy interview on the first half-hour-long CBS evening news program. The journalist Walter Cronkite quizzed the president about "the only hot war we've got running at the moment." In full public view, Kennedy walked backward and forward at the same time. "I don't think that unless a greater effort is made by the government to win popular support that the war can be won out there," he said. "In the final analysis, it is their war." This was Kennedy's philosophical underpinning. In Vietnam as elsewhere, a man who lets others fight for his own liberty often ends up being shackled anew, and the man who fights for him risks becoming not his liberator but his keeper.

Then, after stepping back from involvement, Kennedy moved forward. "I don't agree with those who say we should withdraw," he told the CBS anchor. "That would be a grave mistake. . . . Forty-seven Americans have been killed in combat with the enemy, but this is a very important struggle even though it is far away." There was his political underpinning. After all these years, the domino effect was not just a geopolitical theory but a fundamental axiom of American political life. An American leader who lost even a small, distant, corrupt, tortured land to a regime that carried the banner of communism risked losing his own political head as well. This was the President Kennedy who Bobby said in 1964 "felt that he had a strong, overwhelming reason for being in Vietnam and that we should win the war in Vietnam."

In October, Kennedy performed the same dance as he had in the CBS interview. In signing on to national security action memorandum no. 263, he agreed to "plans to withdraw 1,000 U.S. military personnel by the end of 1963." There was the step backward. But he did so only because Defense Secretary McNamara and General Taylor had assured him that "the military program in South Vietnam has made progress and is sound in principle, though improvements are being energetically sought," and that "by the end of this year, the U.S. program for training Vietnamese should have progressed to the point where 1,000 U.S. military personnel assigned to South Vietnam can be withdrawn." That was the step forward. By attempting to have it both ways, Kennedy was beginning the Vietnamization of the war long before that term became common currency.

The immediate problem was Diem and his brother Nhu, who had proved so ineffectual at leading their war-torn nation. And here too Kennedy walked in both directions. In this crisis, Kennedy did not have many advisers who made articulate, spirited presentations based on the knowledge of their agency or institution. Instead, he was overseeing a petty, preening bureaucratic warfare in which egos and personal ambitions outweighed the crucial issues that his administration faced. He had sent Henry Cabot Lodge as ambassador to Vietnam, hoping that the former senator would show the acumen of a professional politician. Instead, Nixon's vice presidential running mate became so alienated from the American military that he hardly spoke to them and secretly backstabbed his colleagues in Vietnam. Lodge had a savage realism when he talked of Diem, speaking a language rare to diplomatic discourse. "Viet-Nam is not a thoroughly strong police state (much as the 'family' would like to make it one)," he cabled Washington on October 26, "because, unlike Hitler's Germany, it is not efficient and it has in the Viet Cong a large and well-organized underground opponent strongly and ever-freshly motivated by vigorous hatred. And its numbers never diminish."

Lodge made sure that his own stories got into American papers, but he was appalled at the way Diem, and especially his sister-in-law, Madame Nhu, gnawed publicly at the hand that fed them. "The United States can get along with corrupt dictators who manage to stay out of the newspapers," Lodge wrote later. "But an inefficient Hitlerism, the leaders of which make fantastic statements to the press, is the hardest thing on earth for the U.S. Government to support."

Lodge called for a coup, as did several of his colleagues at the State Department, including Ambassador-at-Large W. Averell Harriman. CIA Director McCone would probably have supported the move as well, but in Bobby's words, "McCone hated Henry Cabot Lodge, and so he became an ally of McNamara," who opposed the coup. In essence, the government was split into two unlikely divisions: the State Department favoring the coup, and the CIA and the military leadership opposing it. Those who favored the coup, like Roger Hilsman, the assistant secretary of State for Far Eastern Affairs, tried to convince the president that the die had been cast and Kennedy had no choice but to go along. On August 28, Hilsman told the president that "Diem and Nhu were undoubtedly aware that coup plotting was going on and that the generals probably now had no alternative to going ahead except that of fleeing the country." Kennedy, for his part, said he "was not sure that we were in that deep."

Kennedy was infuriated that he was losing the control of government, and he suspected Harriman of leaking stories of this schism. "You'd better get Averell in, for Christ sake," Kennedy told Undersecretary George Ball.

"The fact of the matter is that Averell was wrong on the coup. We fucked that up. Even though it may have been desirable, so that the Pentagon can go on saying the State Department fucked it up, got us into a lot of trouble, so I think there's nobody in the position to be pointing the finger at anybody else."

On October 29, Lodge told the White House that a coup was in place, to be led by dissident generals, and that the United States should do nothing to prevent it. That day in the counsels of government, Kennedy and his associates discussed a possible coup with the hard-edged logic of those serving an imperial power. No one uttered any bromides about trying to help their Asian brothers find the true light of democracy. Nor did they trouble themselves over the question of whether there was a capable leader to replace Diem. Instead, they counted potential coup leaders and their troops, tallying up who would stand with Diem and who would stand against him. "The key units come out about an even match," William Colby, the CIA's former Saigon bureau chief, told the president and his other advisers at the Ex Comm meeting that afternoon. "There's enough, in other words, to have a good fight on both sides."

"So he [Diem] has sufficient forces to protect himself?" Bobby asked, only half a question. What these men were talking about could easily escalate into a civil war, brother against brother, while the Viet Cong stood by as happy spectators.

"The difficulty is, I'm sure that's the way it is with every coup, it always looks balanced until somebody acts," the president said a few minutes later, lighting the tedious discourse with a flash of insight. Kennedy's overwhelming concern, as he expressed it here and in a subsequent meeting, was whether the coup would succeed, not whether his nation should be promoting or acquiescing in such an action, or whether it would change the dynamics in South Vietnam for the better.

For the most part, this room was full of men acting as technocrats of power, tinkering with formulas, moving their pieces back and forth across their chessboard without seeming to realize that each piece represented a real human being. As they discussed a cable giving Lodge further instructions, Bobby's voice rose above the banal talk about tank battalions and paratroopers to say something that struck hard and true.

"I may be in a minority, but I just don't see that this makes any sense, Mr. President," Bobby said, employing the same deferential formality as the rest of the officials, but speaking with forceful candor. "What we are doing really is we're putting the whole future of the country, and Southeast Asia, in the hands of somebody [General Tran Van Don] we really don't know very well. One official of the United States government has had contact with him, and he in turn has lined up some others. It's clear that Diem is a fighter. He's not just going to get out of there. If it's a failure, Diem is going to tell us to get the hell out of the country. He's going to capture these people. They're going to say the United States is behind it. I would think we're just going down the road to disaster. I think this cablegram sounds as if we're willing to go along with the coup but we think we need a little more information."

The first of November is La fête des morts, the Day of the Dead, or All Saints' Day. It also happened to be the thirty-sixth birthday of Captain Ho Tan Quyen, a senior naval officer loyal to Diem. At around noon his deputy came by his house in Saigon and got him to leave his children to go to a seaside restaurant for a birthday luncheon. On the drive through the suburbs of Saigon, the deputy shot and killed the captain.

Soon afterward the dissident generals ordered their troops to seize police and naval headquarters, radio stations, and the post office, and to surround Gia Long Palace. The generals took one of their prisoners, Colonel Le Van Tung, head of the notorious Special Forces, and had him telephone Diem to tell him to surrender. The Vietnamese leader refused to give up, and after the phone call Tung and his brother were led away and killed.

In the middle of the night the insurgents attacked the palace. By dawn the battle was over, but Diem and Nhu had escaped to the Chinese quarters of Chalon. Diem probably could have fled farther into the countryside and sought to rally his own loyalist troops, but he decided instead to surrender. He called General Don and told him that he was prepared to surrender to his troops with "military honors." He said to those who were harboring him that he did not care whether he lived or died. But he knew these generals as his lifelong colleagues, and he surely expected that they would treat him better than if he had the terrible misfortune of falling into the hands of his Communist enemies.

General Don was not a bloodthirsty avenger. He and his colleagues asked the CIA's Lucien Conein for a plane to fly Diem out of the country, and Conein said that it would take twenty-four hours for a plane to fly from Guam that could carry the former president to exile in Europe. A man could die many times in twenty-four hours. It was not that the Americans wanted Diem to die, but they did not care if he lived.

In all those endless hours of discussion in the White House, no one had ever raised the question of Diem's fate in a successful coup. A top-secret October 25 State Department "Check-List of Possible U.S. Actions in Case of Coup" mentioned financial inducements to the coup plotters and military aid, but nothing of what might happen to Diem. This was not an esoteric moral subtlety best left to religious philosophers and college dons, but an essential part of the equation of power. These men had met the South Vietnamese president, and some had spent hours with him. Diem could have been fervently pushed to go into exile on his own and warned that if he did not, all American support would end. Of course, that strategy would have run the risk that Diem would fill the streets of Saigon with the blood of those who dared to oppose him.

Kennedy had learned that language was the bearer of power, and his precise words were the best way to contain his subordinates' actions. In this instance, he gave no clear directive. There are few things more dangerous than powerful men who think that they have covered all contingencies when they have merely justified what they want to see happen. Kennedy had repeatedly said that if the coup started, it had to succeed. Hilsman took that as his mandate. He let Saigon know that if Nhu and his family were "taken alive, [they] should be banished to France or any other country willing to receive them," but they could not be allowed to stay in Southeast Asia. That was justification for not flying Diem immediately to Bangkok or some other nearby city where he and his brother could have waited to board a flight to Europe. This did not mean that Hilsman was concerned about whether Diem and Nhu lived. "Diem should be treated as the generals wish," he wrote in his memorandum on August 30.

Diem was learning a lesson that many have learned. At times it was almost as dangerous to be a friend of the United States as to be its enemy, for the Americans often saw friendship as a one-sided contract that had to be filled by an endless array of deeds and that they could summarily cancel.

Diem was a man of ritual, both public and private, and though he said that he did not care if he lived or died, he wanted to be treated with the honor of his status. He was upset when the soldiers arrived at the church in Chalon where he had sought refuge and asked him to get into an army personnel carrier, not into a car worthy of the president of South Vietnam. He and his brother lowered their heads to enter the vehicle. Then their hands were tied behind their backs and the door was shut. When the vehicle arrived at joint general staff headquarters, the door was open. Diem and Nhu were still there, but they had both been shot to death, and Nhu had been stabbed as well.

"It's hard to believe he'd commit suicide given his strong religious career," Kennedy said half to himself soon after the generals announced that Diem had killed himself. Catholics believe that eternal damnation is God's judgment on those who end their own lives, and he knew that Diem was a man of profound faith.

"He's Catholic, but he's an *Asian* Catholic," Hilsman said.

"What?" Kennedy asked. It may have been that the president was off somewhere in his own thoughts at the Ex Comm meeting on November 2. It was also true that when someone said something especially stupid, the president often asked him to repeat it.

"He's an *Asian* Catholic, and not only that, he's a mandarin. It seems to me not at all inconsistent with Armageddon."

"There're several different reports here, Mr. President," Bundy said, having heard enough of this sophomoric digression. He then went on to read an eyewitness report that both Diem and Nhu were dead and had clearly been assassinated. Bundy also read a second report saying the two men had poisoned themselves in the Chalon Catholic church.

Whatever words these men spoke this day, the first order of business was to convince themselves that they could not be rightfully accused either of having ordered the assassination or of creating the climate that felled Diem and Nhu. They had to make themselves believe that they were innocent. Rusk was obviously the most worried about these accusations, and he sought to convince his colleagues before they could convince the world.

"I think our press problem is likely to be pinpointed upon U.S. involvement, and we need to get that straightened out," the secretary of State said. "The fact is, we were not privy to this plan in the sense that we really didn't know what they were going to do. . . . The fact that the coup was reported and rumored is not basically different in character than things that have been happening over the last several months. . . . It would be to our interest to indicate that this was Vietnamese and that we were not participants in the coup and try to keep that gap as clear as we can."

Kennedy listened to Rusk's quasi-dissembling, the words set forth in tedious monologue, devoid of affect, so boring that he did not so much win his arguments as numb his opponents into concession. When the secretary of State finally finished, Kennedy found himself asking questions that should have been asked weeks before. "I think one of the problems . . . is how we square a military revolt against a constitutionally elected government which we approve as opposed to our position on Honduras and the Dominican. How do we square that?" Kennedy had aides who could square anything, and though they had their answers, the overwhelming question still hung there.

The cabinet officers and other high officials twisted and turned, trying to assure one another that they had done everything possible to get Diem to seek reforms before seeking other ways to remove him, and that they had no complicity in his death. But they were like men standing in front of a small fire on a frigid day: no matter how they turned, they could never quite warm themselves.

"About this suicide, I've brought this to your attention only because there is some question in some of our minds how much we want to know about this, sir, suicide versus assassination," Hilsman said. The undersecretary may have been a man of bombastic public presentation, but he understood the uses of euphemism and gentle suggestion. He also understood that the ground on which he stood could cave beneath him. The press was already writing about a memo Hilsman had written that the reporters took as giving the generals permission to proceed. He had indeed, at the time of the Buddhist protests, given the embassy in Saigon a secret order approved by the president to "tell appropriate military commanders we will give them direct support in any interim period of breakdown [in the] central government."

"It's becoming more and more clear that it was assassination, at least I think it was, and people around here do," Hilsman went on. "Now, this is the cable which suggests that he [Conein] actually go to Big Minh [General Duong Van Minh] and really find out. But there's some doubt in some of our minds whether we want to or not. Maybe we ought to just let it alone."

As these men nervously discussed the bloody deed, McCone probably already knew precisely what had happened. "Big Minh offered Conein an opportunity to see the bodies, and he refused," McCone said, in his bloodless, bureaucratic way. "The suicide story is out. Conein is pretty conscious that it was assassination, and he didn't want to get involved with it. I would suggest that we not get into, into this story. Knowing it doesn't do us any good."

Regicide is the most horrible of murders, for if the king is not safe, then no one is safe. In the Cabinet Room these men usually discussed grand strategies, geopolitical considerations, theories of nuclear deterrence, counterinsurgency, and economic initiatives. They did not talk about a former friend lying in a pool of blood in the back of an army vehicle. There was an unsettling quality to their discourse, as if they had been condemned to look straight at where their ideas had led. Of all the men in the room, Kennedy seemed the most obsessed with what had happened. The president had always had a certain queasiness when it came to blood and death, and he was being led through a chamber of horrors.

"What happens if Conein asks to see the bodies and discovers there were a couple of bullets in the back of the . . . in the back, of this kind?" said Bundy. "We don't gain much by that." The former Harvard dean could hardly get a straight sentence out of his cultured mouth.

"I don't think we gain anything by it," McCone said, trying to push the men away from this deadly scene.

"It would look like a planned design to remain uninformed under the circumstances," Rusk said, not optimistic about the administration's chances of pleading ignorance.

"I think we're going to hear about it in the next twenty-four hours," Kennedy said. Although the president apparently knew less about what had happened than several of his advisers, he realized that the press was onto the story. He had his own curiosity about the details of the murders.

"If Big Minh ordered the execution, then, then, uh, I don't know," Kennedy mused. "Do we know that? Do we think he meant to?"

"Some suspect that," said Hilsman.

"Some think he did," said Bundy.

"He's stupid then," said Kennedy, half under his breath. No one would be able definitely to link the president to the assassination attempts against Castro, and there would always be uncertainty concerning exactly what he knew.

"I don't think we should be informed in advance of everyone else on this so that the story gets to be our story," Rusk said, seeking to push the deed further away from the door of the White House.

"I don't know why they did that," Kennedy said quietly, once again returning to the act.

Hilsman was a man who always had a ready answer. "Well, some of the reports show that the only time Minh got emotional was when Diem slammed the phone [down] on him." It was a curious justification for murder, and the silence in the room signaled Hilsman to move on to a different subject.

"Sir," Hilsman said, "this morning . . . this morning there was a discussion of a cable to go out tonight, getting on with the war." The Americans had a war to fight, and they could not indulge in this speculation any longer.

Two days later Kennedy sat in the Oval Office. "One two three four five," he said into a Dictaphone. "Monday, November 4, 1963. Over the weekend the coup in Saigon took place."

For a man who stood at the epicenter of power, the president was extraordinarily dispassionate, not only in the decisions he made but equally in how he viewed them afterward. In this private moment he did not attempt to polish his image. He was his own best historian, treating himself as but another player in the complex tragicomedy of life.

Kennedy went through the major players one by one, accurately outlining where each man had stood on the coup. At times great events were determined not by principles or ideologies but by nothing greater than personal pique. In Kennedy's assessment, McCone had opposed the coup "partly because of an old hostility to Lodge, which causes him to lack confidence in

Lodge's judgment, partly as a result of a new hostility because Lodge shifted his station chief."

If Kennedy ever wrote this history, he would have first filled his pen with irony. He saw his own culpability not in any strong, willful action he had taken but in nothing more dramatic than the sloppy drafting of a cable at the outset. "I feel that we must bear a good deal of responsibility for it, beginning with our cable of early August in which we suggested the coup. . . . That wire was badly drafted. It should never have been sent on a Saturday. I should not have given my consent. . . . While we did redress that balance in later wires, that first wire encouraged Lodge along a course to which he was in any case inclined."

Kennedy went on to talk about the military situation. "Politically the situation was deteriorating," he said. "Militarily it had not had its effect. There was a feeling that it would." For all the president's insights into the world of men and politics, and his ability to spot the justifications and self-promotion of those around him, he had fallen for McNamara and Taylor's tragic fantasy that the war was going well, and that soon the Americans would be able to go home, leaving their victorious partner behind.

As Kennedy went on, John Jr. entered the room, his entrance signaled by a high-pitched squeal. "Say hi," the president said.

"Hello," John Jr. said, speaking into the microphone. "Naughty naughty, Daddy." An endearing little boy to whom the White House was a great castle, John Jr. would be three years old later in the month and he already had the public presence of a child actor.

"Why do the leaves fall?" Kennedy asked, turning this moment into a learning exercise both humble and poetic.

"Because," John said.

"Why does the snow come on the ground?"

"Because."

"Why do the leaves turn green?"

"Because."

"And when do we go to the Cape?" the president asked. Hyannis Port was the scene of the most profound and joyous moments of his family life, first as a child and now as a father.

"Summer," John Jr. answered, though summer was far away.

When John Jr. left his father, he let out a whooping laugh. It was not like the laughs the president usually heard, calculated gestures modulated by what seemed to please him. This was a loving, taunting, wondrous laugh from a son who saw only the happiness of the world.

As his son left, Kennedy turned finally to the most painful matter of all, and he spoke of it without a hint of emotion:

> I was shocked by the death of Diem and Nhu. I'd met Diem with Justice Douglas many years ago. He was an extraordinary character. While he became increasingly difficult in the last months, nonetheless over a ten-year period he'd held his country together, maintained its independence under very adverse conditions. The way he was killed made it particularly abhorrent. The question now is whether the generals can stay together and build a stable government or whether . . . public opinion in Saigon, the intellectuals, students, etc., will turn on this government as repressive and undemocratic in the not-too-distant future.

While Kennedy looked eastward toward the jungles of Vietnam and fretted about his nation's future there, Bobby continued to be consumed with Cuba. He was so disdainful of the structures of government that he had gone far toward privatizing the American policy. As attorney general of the United States, he had no institutional right to claim sovereignty over America's Cuban policy. But he was the force behind the U.S.-funded "autonomous anti-Castro groups."

There were already important bases in Costa Rica and Nicaragua run by Manuel Artime, and a new operation headed by another exile leader, Manolo Ray, was beginning to establish its working base in Central America. The leaders were the attorney general's friends, comrades he invited to his home, and against them on the island stood implacable enemies. By November 1963, they were ready to escalate their attacks. "Bob Kennedy, it seems, was the person who was pushing them [the CIA] and making them do it," recalled Rafael Quintero, the deputy leader. "I mean, [he] put the Cubans in charge of their own operation—but they definitely didn't want to do it."

At the same time the CIA was preparing a series of dramatic initiatives of its own. That November a group of Cuban exiles led by an American CIA operative, Bradley Earl Ayers, trained for an operation against the major Cuban oil refinery in Matanzas Province. It was an ambitious enterprise in which teams of commandos were to sail from Florida on two fishing trawlers. The first group of commandos would make shore in Cuba in a small boat, to prepare the way for their comrades carrying rocket launchers. One of the men would climb the fence surrounding the refinery and enter a tin shed where a lone watchman sat. The commando would have a knife, a garrote, and a pistol with a silencer. He was to kill the watchman using the method of his choosing. Even though the night watchman was probably just an old man who needed a job, the plan made no mention of the possibility of merely tying him up.

Years later, when Halpern, the CIA Cuban desk officer, and Ayers, the on-scene CIA operative, were asked why this man had to die, they responded in precisely the same words: "We were at war." And so they were, and whether the attorney general knew the details of this operation, it was Bobby's war fought Bobby's way.

Bobby was so much the symbol of uncompromising opposition to Castro that when Dr. Rolando Cubela Secades, a Cuban army major with access to Castro, was contemplating killing the Cuban leader and insisted on meeting with a top American official, it was the attorney general whom he wanted to see. Instead, Desmond FitzGerald, the head of the Cuban Task Force—now renamed the Special Affairs Staff (SAS)—met secretly with Cubela in Paris on October 29. FitzGerald was traveling, in the words of the agency contact plan, "as personal representative of Robert F. Kennedy." FitzGerald insisted later that the agency had not sought Bobby's permission to speak in his name, and that at the Paris meeting he had not talked about assassination. Cubela recalled otherwise: "He [FitzGerald] offered me on behalf of the U.S. government the support . . . for being able to carry out either the plot attempt against the prime minister of Cuba or any other activity that will put in danger the stability of the regime."

Lies are most effective hidden in a bed of truth, but at the highest reaches of the CIA, lying was not dishonor but its opposite. In the end men like FitzGerald appeared to be dangerous renegades, but they were only carrying out what the Kennedys wanted them to do. The president recognized that perfectly well. "I have looked through the record very carefully, and I can find nothing to indicate that the CIA has done anything but support policy," the president said in October 1963. "I can assure you flatly that the CIA has not carried out independent activities."

While the CIA prepared again to attempt to kill Castro, the administration began a tentative, distanced approach to the Cuban leader by exploring the possibilities of normalizing U.S.-Cuban relations. At the president's authorization, William Attwood, the former editor of *Look* and an adviser to the UN mission, met with Carlos Lechuga, the Cuban UN ambassador. The president was more optimistic about the possibility of achieving some measure of peace with Castro's Cuba than was the State Department. Bobby, for his part, stated that "the U.S. must require some fundamental steps such as the end of subversion in Latin America and removing the Soviet troops in Cuba before any serious discussion can take place about a détente." As for Castro, he set no conditions and was intrigued enough by the prospect that early in November he expressed a willingness to sit down for discussions with an American official.

In the next months the two sides would have to travel a treacherous pathway. Kennedy was facing reelection, and he could not melt down the swords of war as long as Castro shouted the shrill slogans of world revolution. The Cuban leader, for his part, was the leader of a young revolution, and he could hardly turn away easily from his ideals or his patrons in Moscow. Both leaders were far more realistic than their rhetoric and saw some measure of hope in these discussions. Peace is not won in a day, however, and death comes in an instant. The CIA's FitzGerald was convinced that Castro would be dead by the end of 1963.

The president spent the weekend of October 20 in Hyannis Port. Kennedy was as restless of body and mind, and with his bad back, he rarely sat down longer than was necessary. He sat quietly with his father watching the football games on television. Kennedy was supposed to leave Sunday morning, but Joe had a cold and the president spent time sitting next to his bed, talking to him.

It finally came time for the president to leave. Autumn had already been hard upon the land, and it was far too cold for Joe to sit out on the porch to see the president off. His son came up to his bedroom and said good-bye. The president did not like to hug or kiss other men, but he kissed his crippled father, and then he hurried out of the room.

As soon as the president left, Joe motioned toward the balcony, insisting that his bed be moved there to watch the presidential helicopter depart. The world waited on his son now, the way it once had waited on him, but the helicopter did not lift off. Joe's face twisted up in disbelief. As he sat looking down on the grass where his sons had so often played, he felt a touch on his shoulder. "Look who's here, Dad," the president said.

Kennedy had come to say good-bye again. He felt he had to touch his father one more time. He wrapped his father in his arms and kissed him. Then he was gone, and within a moment the helicopter lifted off into the leaden skies.

On Tuesday, November 12, Cartha DeLoach, the new FBI liaison to the White House, walked into the Oval Office. That first visit to the presidential office inspired a moment of awe in the most sanguine of men. Kennedy had just started wearing glasses except when he was in public, and he did not quite have the youthful, forceful look by which most Americans knew him. DeLoach introduced himself and listened to Kennedy, but he kept staring at the president's hands. They were shaking. As the president continued talking, he put his hands under the desk as if he did not want DeLoach to see the uncontrollable tremor.

The following Saturday, DeLoach was in the auditorium when Kennedy gave a speech to the National Academy of the FBI. DeLoach tried to pay attention, but he kept looking at the president's hands. "His hand was shaking terribly, and I couldn't think that was from stress or the strain of the speech," DeLoach said. "I thought it must be a disease of some kind."

DeLoach was not the only one to notice that Kennedy's hands often shook. The *Boston Globe*'s Bob Healy observed the trembling hands as well, though it was not something that he would consider writing about in his paper. Since Kennedy had entered the White House, the only major publication to write about his health problems in an important way was the scandal sheet *Confidential*, in an almost scholarly "special report" on the potential effects of Addison's disease. Other than that July 1962 article titled "Medical Facts on President Kennedy's Mysterious Malady," the press had largely left the story alone.

Kennedy was heading into an election year in which he would have the punishing task of running for president while he continued to bear all the onerous burdens of office. The campaign would have tested the mettle and stamina of the healthiest of men, and it was a question whether Kennedy was up to such a challenge. It was not simply that his hands trembled, that he was corseted with a metal and cloth brace to protect his back, and that in privacy at Hyannis Port he often used crutches. The question was whether his regimen of drugs and all the endless tensions of his office had begun to break the man down. How much longer could he maintain the greatest of all his many illusions—that he was a vibrantly healthy young president?

Politics is always defined by endings, and as the political year concluded, Kennedy's legislative program was in trouble. His tax cut bill was spat back at him by a vote of 12–2, and the Senate Finance Committee decided to table the matter until 1964. His civil rights bill was pushed back again, its delay fostered primarily by southern Democrats who could win in procedure what they would lose in votes. At least his foreign aid bill was passed, but the recalcitrant Senate stripped it of $800 million of the $4.5 billion he had requested, and that was still $200 million beyond what the House was willing to give him.

On the very weekend that this disastrous news led the nation's newspapers, the White House could at least announce that Jackie would be going along with her husband on his trip to Texas on November 21. It was no small matter that Jackie had agreed to travel with him and become part of the still-unannounced reelection campaign. In the political scope of things, her involvement might weigh heavier than all his losses in Congress. In the first place, if Jackie backed away from politics much longer, she risked becoming a political albatross—a first lady who appeared to enjoy living everywhere but the White House.

Jackie's European sojourn had created headlines that might please a king, but not a democratic leader—"Mrs. Kennedy Aegean Island–Hopping," "Jackie Follows Script as Hollywood Wrote It," "Jackie Sails in Splendor." Betty Beale, a Washington social columnist, reported that Jackie's European trip had caused "complaints . . . to pour in from all quarters and it may hurt politically." Marianne Means, a Hearst columnist and reporter, wrote: "During her nearly three years in the White House, she has consistently refused all invitations to appear with the president at political functions and most public events, outside the realm of the arts. She did not once accompany him last fall as he campaigned for Democratic congressmen up for reelection. And she has never traveled with him on any of his trips around the country."

Jackie had a radiant popularity all her own that would help create the almost frenetic excitement that would translate into votes next November. In 1960 Kennedy's advisers had thought Jackie might be a liability; in 1964, in a close campaign, she might prove a crucial asset.

Joe Kennedy had long ago taught his sons that time was the most precious of commodities, and the president filled every cranny of his life as richly as he could, even the short helicopter ride from the White House lawn to Andrews Air Force Base as he left for Texas. "Where's John?" the president asked his wife's maid, Provi, as he moved through the second floor looking for his son.

"Well, I don't know," Provi replied defensively. "It's raining, and Miss Shaw does not want him to go."

"Go down the hall and make sure he's dressed," the president ordered. "I want to take him with me."

John Jr. trudged toward his father wearing a raincoat and a sou'wester rain hat. The president hated hats, but he donned one for a moment as he and little John ran out in the rain to the helipad where three helicopters sat waiting. John Jr. would not be going on the long trip west, but spending time with his son was one of the president's pure delights, even on the short hop to Andrews Air Force Base. The president could bid an especially happy good-bye to his son since, upon his return, on November 25, they would be celebrating his namesake's third birthday.

Kennedy was relieved that Jackie was going with him, but as he sat in the helicopter waiting for her to arrive, he was once again reminded of a woman's prerogative to be late even if her husband was the president of the United States. His son was with him but he beat out a tattoo with his fingers on his leg, while his aides scurried back into the White House to attempt to hurry his wife. As he sat waiting on her, he apparently knew that on his return he had the most onerous and difficult of duties: to deal with the question of O'Donnell and O'Brien's possible corruption. There were rumors in the White House that he was going to fire the two when he returned from Texas. That may not have been true, and while some former staff members are convinced that the men were fully culpable, others are not so sure. Still others who knew them outside the White House say that it was impossible that they would have sought financial benefits for themselves. Whether their still-loyal friends are correct, the matter had finally to be faced by a president who abhorred such confrontations and personal dealings.

Jackie finally arrived, and the craft rose up off the manicured lawn to fly southeast to Andrews Air Force Base. Below, the city appeared an exquisite rendering of geometric shapes. From the obelisk-shaped tower of the Washington Monument and the rectangles of the Mall to the dome of the Capitol, Washington seemed a city of perfect forms, patterns, and designs. Kennedy knew as well as any man that political Washington was nothing like that at all, but a city of dark labyrinths and twisted alleyways as well as grand streets and noble buildings. In the exercise of power, he traversed the whole political city, going places that few knew he went, leaving images of himself that sometimes had little to do with the man flying high above.

As *Air Force One* flew westward from the Maryland air force base, Kennedy was heading to a Texas that was a far different place from what it first appeared to be. From above, the state looked like a place of endless vistas and almost limitless horizons. As Kennedy well knew, political Texas was a narrow, convoluted, dangerous place rife with betrayals and mistrust. The Texas Democrats were feuding, wasting their energy fighting one another instead of their common Republican opponents. That was politics as it always was, an impossible brew of the sublime and the ugly, the passionate and the calculated, public idealism and private cynicism. Matters were so bad that the state's liberal Senator Ralph Yarborough and its conservative Governor John Connolly could hardly manage a civil conversation.

That was a matter that Kennedy would deal with, but he was traveling here primarily because a man who was running for president followed the scent of money, and Texas was big money. For months he had been pushing the Texas Democrats to set up a major fund-raiser, an event that would allow him to take home a million dollars or more, money that was fuel to power his campaign.

Every time Kennedy left the safe, confining presence of the White House, he had to know that someone might be out there looking at him with the eyes of a killer, if not the hands and the will. Of the thirty-four presidents who had preceded him, three had been assassinated, and attempts had been made on the lives of three others. During the first thirty-four months of his presidency, the Secret Service had noted twenty-five thousand threats against the president and listed one million names on its "security index" of those who represented possible threats.

Early in November, the president's trip to Chicago to attend the Army-Navy game had been canceled, possibly because the Secret Service heard that Cuban exiles were planning to kill Kennedy as he sat watching the game. His trip to Miami went on as planned, though an FBI informant had warned of a plan to assassinate Kennedy from a high building in a southern city and some of the former members of Brigade 2506 had boasted that they would kill him.

On the morning of November 22, 1963, Dallas was festooned with five thousand handbills headlined: WANTED FOR TREASON. The president's face stared out like a criminal's picture tacked to the post office wall. The papers declared that Kennedy was wanted for "treasonous activities" for such measures as giving up American sovereignty to the "communist controlled United Nations," betraying the forces of a Free Cuba, and approving the Nuclear Test Ban Treaty. There was a full-page advertisement in the *Dallas Morning News* that made similar charges, the page bordered in black like a mourning card.

That morning the president was sitting in his suite at the Hotel Texas in Fort Worth when he saw the advertisement. "Can you imagine a thing like that?" he said, looking at the page with distaste. Kennedy was frightened by the dreadful propensity of mass man to follow demagogues leading him to disaster with mindless slogans. "You know," the president said, turning to Jackie, "we're heading into nut country today." As Kennedy paced the floor, his thoughts turned not to the surly protesters who might line the streets, stimulated by such fare as the advertisement, but to the previous evening, when he had been among enthusiastic supporters at a massive rally at Houston Coliseum. It had been a splendid evening, though one of the Texas advance men, Jack Valenti, had noticed that Kennedy tried to keep his hands out of sight beneath the rostrum. They were "vibrating so violently at times that they seemed palsied," and the president had nearly dropped his five-by-seven cards.

"You know, last night would have been a hell of a night to assassinate a president," Kennedy said, almost as an aside. "I mean it," he continued. "There was the rain, and the night, and we were all getting jostled. Suppose a man had a pistol in a briefcase." He pantomimed a phantom killer, pulling out his gun and firing away. "Then he could have dropped the gun and the briefcase and melted away in the crowd."

The president and first lady set out in the motorcade to journey from Dilles' Love Field to his speech at the Texas Trade Mart. Kennedy had both a politician's savvy and a husband's pride in his wife. He knew nothing of dress designers other than the bills that he paid, but Jackie looked exquisite in her pink suit, pink pillbox hat, and white gloves.

The top was down on the presidential limousine so that the people could see their president and first lady. Jackie reached to put on sunglasses against the glaring Texas sun, but Kennedy asked her to put them back in her lap, telling his wife that the onlookers wanted to see her eyes. Kennedy liked to shield himself from the throngs that sought in him something they did not have and he could not give, but each time he drove in a motorcade through an American city he looked out into those restless faces as a gauge of his own political future. Whatever doubts he had when he set out this morning, as Kennedy scanned the faces this day, he had reason to believe that these people loved him, shouting his and Jackie's names, pleading with him to look their way, waving at him, celebrating not only the president but also the presidency.

For Jackie, it was often a blur of faceless humanity out there on the streets, but Kennedy seemed to look at each face, locking in his gaze so that thousands would walk away feeling that they had connected with their president. At Lemmon and Lomo Drive a group of little children stood holding a sign: MR. PRESIDENT. PLEASE STOP AND SHAKE OUR HANDS. "Let's stop here, Bill," Kennedy told Bill Greer, the driver. As Kennedy descended from the limousine, the crowd surged forward, pressing around him. Spectators wanted nothing more than a handshake, an autograph, a touch, but they would have drowned him in their adulation.

As the motorcade crawled through the downtown streets, the crowds were at times a dozen deep, pushing against barricades, surging into the streets, pressing for a glimpse of their president. Kennedy waved and smiled, and yet as always there was a dispassionate quality to him, as if he were watching himself watching the crowds watching him. Though the crowds could not possibly hear him, he kept saying, "Thank you, thank you, thank you." He was president, and he needed the votes of the Mexican-Americans shouting his name, the secretaries leaning out of the windows waving at him, the businessmen standing there clapping. If he was to win reelection, he needed Texas, and he rightfully saw these crowds as a good omen.

In the summer he had doodled on a sheet of paper, "To govern is to choose." He might as easily have written, "To live is to choose." That was the axiom of his life. As a boy lying in a hospital bed, he had chosen health over illness, refusing to live as a near-invalid. Today, as the motorcade moved through the downtown streets, Kennedy looked like a vital, youthful president. No one in these cheering crowds knew that to protect his back he wore a brace of canvas and steel that held him supernaturally erect. No one knew about the drugs that he took, the pain that he felt, and the price that he paid to maintain the illusion of heath.

Kennedy chose to enter politics and to stand for Congress, the Senate, and the presidency. He chose the issues he thought mattered and those that did not. In his press conferences as president, he tried to educate Americans about the hard choices they and he faced, to make them realize that men made war and peace by their conscious decisions. In his first 1,036 days in office, no issue had so defined his presidency as had Cuba. At the Bay of Pigs he had suffered his most pathetic defeat, and a year and a half later he had stood strong against the choice of war during the missile crisis. He then chose to continue the attempt to bring Communist Cuba down in the dangerous folly of a secret war. Even at this moment one of Castro's would-be assassins was meeting in Paris with several CIA officers and receiving a Papermate pen with a tiny hidden needle that he was to use to prick Castro's skin, killing him with poison. Cubela said that he preferred to do the deed with "two high-powered rifles with telescopic sights" that could be used to kill Castro from a distance.

Kennedy was riding through the clamorous streets because he chose to be there. Everything he had done led him to Dallas. He was a politician, and he needed Texas money. The top was down because he had made a covenant with the people, and in a democracy the people saw their leaders. Kennedy carried the seeds of his own death within his pained, weakened body, and if he had not let his maladies haunt his days, he surely would not let the fear of assassination haunt him either. All of his life he contemplated courage and its meaning. For him, riding on a sun-dappled day in an open limousine through the streets of Dallas was not restless disregard of the dangerous circumstances. It was the essence of his life.

The speech that he was about to give at the Texas Trade Mart evoked many of the themes of his life. What made it different was that in recent months he had grown distressed at the growing hysteria on the dark edges of American politics. There were critics who called strength weakness and slandered their enemies as traitors. Kennedy was a politician running for reelection, and it was natural that he would not look kindly at those who condemned his stewardship of America, but this was far deeper than a calculated ploy. Since his youth, he had seen that the strength of democracy was its people and the leaders they freely chose. "America's leadership must be guided by the lights of learning and reason, or else those who confuse rhetoric with reality and the plausible with the possible will gain the popular ascendancy with their seemingly swift and simple solutions to every world problem." He condemned the far Left and its apparent assumption "that words will suffice without weapons," and the warriors of the far Right who thought that "peace is a sign of weakness."

"We, in this country, in this generation, are—by destiny rather than choice—the watchmen on the walls of world freedom," he was to say. "We ask therefore that we may be worthy of our power and responsibility, that we may exercise our strength with wisdom and restraint, and that we may achieve in our time and for all time the ancient vision of 'peace on earth, goodwill toward men.' " It was an uneasy, difficult destiny, and Kennedy professed it passionately, even if he did not always listen to the words he spoke or act on the ideals he believed.

As the limousine moved past the cheering crowds, the Secret Service men looked out on the restless, happy faces, searching for one with a hard, purposeful look. They were trained to look away from the president, to look into the faces of the crowd, to look up at the buildings, to look for a malevolent glint of steel, but never to look into Kennedy's eyes.

Ahead stood the Texas Book Depository, and on the sixth floor crouched a man with a cheap rifle and a dream of immortality. The next moment has become so much a part of the American psyche that it is as if we are all riding beside the first lady in the back seat of that black Lincoln. We shiver soundlessly at a loss beyond loss. In that instant all the certitudes and easy optimism of American history were blown away.

32

Requiem for a President

The last of the flocks of geese had glided across the gray sky flying south. The Kennedys had already sent the *Honey Fitz* to Florida; if it had not been for the traditional family Thanksgiving, Joe would already have headed to the sun and warmth of Palm Beach. The president would be coming for the holiday, and Joe would have weathered even grimmer November days to spend another long weekend with his son.

As Joe sat slouched in his wheelchair, even the expressions on his face appeared diminished. At times the seventy-five-year-old man seemed little more than an inert form to be hauled from bed to wheelchair and back again. Those who took care of him tiptoed around his diminished life, but in the minute circumference of his days there was little he did not see, and few nuances of life at Hyannis Port that he did not grasp.

Joe had fallen asleep for his afternoon nap when Ann Gargan awakened him. "Uncle Joe, there's been a terrible accident," Ann said. Joe had hardly begun contemplating these words when Rita Dallas, the nurse, took Ann's arm and pushed her outside. A minute later she returned and stood beside her groggy charge. "Uncle Joe, it's Wilbur," she said, mentioning the head gardener. "He's been hurt, but he's all right, he's all right."

Joe went back to sleep. After Joe's nap, Frank Saunders came rushing into the room, his voice charged with enthusiasm. "Hey, chief, it's movie time!" the chauffeur exclaimed as he prepared to wheel Joe down to the private theater where he had once sat with mistresses and famous guests. Joe usually liked Elvis Presley movies, but after a few minutes of *Kid Galahad*, he became restless, and Saunders took him back to his room, where Joe was told his television set was not working.

As Joe lay there perusing a magazine, Eunice and Teddy burst into the room. Eunice moved toward her father, took his hand, and kissed him. "Daddy, Daddy, there's been an accident," she whispered, as if her words were a secret. "But Jack's okay, Daddy. Jack was in an accident."

Eunice did not want to say what had to be said, and then she said it. "Jack's dead. He's dead. But he's in heaven. He's in heaven. Oh God, Daddy."

"Jack's okay, isn't he, Daddy?" Her father always made everything right.

Teddy could no longer even pretend to pretend. He fell on his knees in front of his father's bed and covered his stricken face with his hands.

"Dad, Jack was shot," Teddy said.

"He's dead, Daddy," Eunice said. "He's dead."

Rose entered to see her husband lying there with a stony stare and a hand that beat against the sheet. "He was so angry, mad at the world for doing it again," she later told her niece, Kerry McCarthy. "It worried me more that he was so angry. He took everyone to task, including God. I would rather have him be in pain, but his anger at the Almighty, you can't deal with that. Only later came the pain."

Bobby had been struck his own grievous wound, but the assailant had hit his heart, not his head, and he stumbled onward as if he had not been felled. He had always sought solace in action, and so he did now, moving to protect his brother in death as ultimately he could not protect him in life. Even before he knew that the president was dead, he called Bundy at the White House and ordered that his brother's personal files be removed to the NSC offices in the Old Executive Office Building and kept under around-the-clock guard. He also ordered that the secret taping system be dismantled so that no one would know that the president had recorded their meetings.

Bobby set out to try to learn who had murdered his brother. His first instinct was not to seek out the assassin among America's enemies or hidden somewhere on a list of psychopaths and the deranged. He turned toward his own government and the agency he had helped to hone into a murderous machine. He had ordered the CIA to do whatever was necessary to destroy Cuba, and the agency was ready to garrote a watchman in the night and to poison a head of state.

Bobby telephoned McCone and asked the CIA director to come over to Hickory Hill. When McCone arrived, Bobby asked him whether the CIA had killed the president. The director swore that the intelligence agency had nothing to do with the president's death.

Next Bobby called Harry Williams, a Cuban exile leader close to the attorney general, who was at the Ebbit Hotel in Washington preparing a new series of attacks from secret bases in Central America. He knew then that a man linked to Castro protests in New Orleans, Lee Harvey Oswald, had been arrested in Dallas. "One of your boys did it," Bobby said. The attorney general had created this exile army, and now he feared that it had turned against him and expressed its deadly wrath against his brother.

Bobby called Walter Sheridan, who was down in Nashville prosecuting Jimmy Hoffa for jury tampering, and asked the Justice Department lawyer to check out Hoffa. The labor leader was another man of evil intent, his fury honed on Bobby's own obsessive, vengeful justice. Then he called Julius Draznin, a labor lawyer in Chicago with knowledge of the mob, and asked him about the Mafia, and in particular Sam Giancana.

One way or another all these calls led back to Cuba, though the one possibility Bobby did not consider was that Castro himself might feel he had a right to kill the man who had tried to kill him. "Castro could have made a very strong case that what he did was justified," said former Secretary of State Alexander Haig, then working on the Cuban Coordinating Committee. "We were attacking his country, and he was fighting an enemy. That's different from an assassination in a pure sense of the word." Everywhere Bobby looked, everywhere he turned, whatever he thought, he found more enemies, and every potential assassin he looked at this day was an enemy he himself had made, or had helped to make.

Bobby ran away from the deadly prospect of contemplation. He listened instead of talking, and talked instead of remembering, trying to give solace to the inconsolable. He was there when *Air Force One* flew into Andrews Air Force Base carrying his brother's body, the president's widow, and Lyndon Baines Johnson, the thirty-sixth president of the United States. He was there riding in the hearse, with Jackie still in her blood stained dress, hearing her tortured retelling of the murderous assault. He was there at Bethesda Naval Hospital that evening, talking endlessly on the phone, beginning to plan the details of the state funeral. He was there late in the evening at the White House trying to gently nudge the others off to bed.

Bobby finally lay down on the bed in the Lincoln Room, but his eyes would not close and sleep would not come, and he asked Spalding to join him there. "Listen, you ought to take a sleeping pill," the president's friend said, and left to seek the sedative. When he returned, the attorney general was still haunted with sleeplessness. "It's such an awful shame," Bobby said without a hint of emotion. "The country was going so well. We really had it going."

Spalding said good night, shut the door, and turned to walk down the corridor. It was then that he heard sobs of sorrow. He thought that Bobby was saying, "Why, God, why? What possible reason could there be in this?" Bobby cried until finally the pill took hold and he fell silent.

On the day John F. Kennedy was buried, the skies were dark. The leaders of America and the world walked in solemn procession the five blocks from the White House to St. Matthew's Cathedral. Along the route tens of thousands stood paying their homage, while millions watched on black-and-white television sets.

The president's flag-draped coffin sat on a caisson drawn by six gray horses. Then came a riderless horse. Behind walked Jackie, with Bobby on her right arm and Teddy on her left, and behind them world leaders who were nothing but a shuffling mass of mourners. There were mourners too throughout the world. Millions knew his name as a symbol of hope, not because of anything he had promised or could promise, but because he had become emblematic of a new spirit celebrating human aspiration and challenge.

Bobby walked beside Jackie to steady his brother's widow in her time of need. But who would steady him as he plodded onward? "May Joe find solace in the unexampled triumphs of his son in the assurance that Bobby will repeat Jack's career," Lord Beaverbrook had cabled Rose, as if life was a grand football game: when one player was hurt, another grabbed a helmet and loped onto the field to take his place. Bobby had hardly begun the terrible contemplation of what inadvertent role he might have played in his brother's death, and already his name was being called.

Bobby had incalculable burdens to bear, and one of them, Jackie, was walking beside him. He would be responsible for his brother's widow and her two children. The president's life had justified so much of all that the Kennedys had suffered. Joe Jr. had died a hero's death in a ball of flame, and the president had carried on with his brother's bold dreams. Bobby's father had taught him that he was supposed to pick up the burden, but how could any man lift what was set out before Bobby this day?

Teddy had decked himself out in his brothers' lives, and now as he marched forward to the funeral dirge, he wore the pants that the president had worn for his inauguration and a pair of his gloves. Teddy's rental suit of a morning coat and striped trousers had arrived incomplete, and Kennedy's valet had let out the president's pants and pressed a pair of the president's gloves into Teddy's oversized hands. There had been no hat to fit his large head, and so he wore none, and neither did Bobby. And when the kings and prime ministers and others heads of states and world and American leaders walked on that gray day, they too marched hatless in the thirty-degree weather, their heads slightly bowed as if in prayer.

As the pallbearers carried the body into the church, there was silence across the land. Joe sat watching on his television. In Times Square traffic stopped, and New Yorkers stood, heads bowed, while high from the marquee of the Astor Hotel "Taps" echoed across the stilled and empty square.

Jackie thought of her husband as being Greek in that "the Greeks fought the gods" and had a "desperate defiance of fate." Americans had never been imbued with a Greek sense of tragedy, but they had it now as they buried their martyred president with the most profound dignity and reverence.

Bishop Philip Hannon read passages from the Bible that Jackie had chosen. "Your old men shall dream dreams," he said. "Your young men shall see visions. And where there is no vision the people perish."

Then the bishop read what he called "the final expression of his ideals and aspirations, his inaugural address." The phrases that the prelate spoke did not sound like mere political perorations. In its power and depth, the passage had a biblical resonance. "Let the word go forth from this time and place, to friend and foe alike, that the torch has been passed to a new generation of Americans." Hannon intoned words that a little less than three years before a young president had spoken in a strong voice, standing coatless on a frigid day, looking out with anticipation at the challenges of a dangerous time. "And so, my fellow Americans, ask not what your country can do for you; ask what you can do for your country." That was the most famous sentence Kennedy had ever spoken. Whatever Kennedy's faults, he believed that Americans must reach out beyond their narrow self-interest to each other and to the world. "With a good conscience our own sure reward, with history the final judge of our deeds, let us go forth to lead the land we love, asking His blessing and His help, but knowing that here on earth God's work must truly be our own."

As the coffin was brought out after the low pontifical mass and placed back on the caisson, Jackie whispered to her son: "John, you can salute Daddy now and say good-bye to him." Her son raised his right hand in a salute to his father. As he held his hand up, the picture that tens of millions watched on television shimmied slightly.

The gun carriage carrying the coffin set out to Arlington National Cemetery, its way announced by the sound of muffled drums, followed by the dignitaries in a series of limousines. The procession moved slowly past the spire of the Washington Monument and the Lincoln Memorial.

And so they buried him that day. He was a hero of war if not yet of peace, and he was buried among rows upon rows of men who had, many of them, given up their lives for their country. He had followed his father's laws of manhood, and they had led him to the most exalted position in the land, and to Dallas on a November afternoon, and to his burial here among the brave and true.

As the mourners stood silent on this darkest of afternoons, a thunder of noise erupted in the midst of the stillness. Fifty low-flying Air Force F-105s roared across the sky, disappearing as soon as they arrived. Behind them flew *Air Force One*, so low that it was not like a plane in the sky, but a great bird that had come to pay its last homage. As the plane that had flown Kennedy around the world and carried his body back from Dallas moved across the cemetery plot, its wings dipped, and then the plane was gone. There were those at the gravesite who had stayed strong until this moment, but now they broke down.

As the burial ended, the president's widow walked to the grave holding a lit taper in her hands. She reached forward with the rod and a flame burst into life. She gave the taper to Bobby, and he touched it against the flame, as did Teddy. From now on, the Kennedys would come here and look at the eternal flame and seek the light they had lost.

Source Abbreviations

AAML: Archives of the Andrew Mellon Library, Choate

AKP: Arthur Krock Papers, Seeley G. Mudd Manuscript Library, Princeton University

ASP: Arthur Schlesinger Jr. Papers, JFKPL

ASPU: Adlai Stevenson Papers, Seeley G. Mudd Manuscript Library, Princeton University

ATD: Arthur Schlesinger Jr., *A Thousand Days* (Boston: Houghton Mifflin, 1965)

AWRH: Carl Sferrazza Anthony, *As We Remember Her: Jacqueline Kennedy Onassis in the Words of Her Friends and Family* (New York: HarperCollins, 1997)

AWRJ: John F. Kennedy, ed., *As We Remember Joe* (Cambridge, Mass.: privately printed, 1945)

BP: Joan and Clay Blair Jr. Papers, American Heritage Center, University of Wyoming

BPL: Boston Public Library

CUOH: Columbia University oral history, New York

CY: Michael R. Beschloss, *The Crisis Years: Kennedy and Khrushchev, 1960–1963* (New York: HarperCollins, 1991)

DHP: C. David Heymann Papers, State University of New York at Stonybrook

DP: William Manchester, *The Death of a President: November 20—November 25, 1963* (New York: Harper & Row, 1967)

DPP: David Powers Papers, JFKPL

FBIFOI: Federal Bureau of Investigation Freedom of Information Act request.

FMC: *Forbes* Magazine Collection

FRUS: *Foreign Relations of the United States, 1961–1963* (Washington, D.C.: U.S. Department of State, 1988), many of the documents also available on the Department of State web site, www.state.gov/www/about_state/history/ frusken.html.

HSCA: House Select Committee on Assassinations

HTF: Amanda Smith, ed., *Hostage to Fortune: The Letters of Joseph P. Kennedy* (New York: Viking, 2000); many of these documents also available at JFRPL and other archives.

HUA: Harvard University Archives, Cambridge, Massachusetts

IR: *Alleged Assassination Plots Involving Foreign Leaders: An Interim Report of the Select Committee to Study Governmental Operations with Respect to Intelligence Activities* (Washington, D.C.: U.S. Senate, 1975)

JEP: Judith Exner Papers accessed as part of lawsuit *Judith Exner* vs. *Random House*, et al.

JFKMM: Victor Lasky, *JFK: The Man and the Myth* (New York: Macmillan, 1963)

JFKOA: Richard D. Mahoney, *JFK: Ordeal in Africa* (New York, Oxford, 1983)

JFKPL: John F. Kennedy Presidential Library

JFKPP: John F. Kennedy Personal Papers, JFKPL

JMBP: James MacGregor Burns Papers, NHP. Courtesy James MacGregor Burns

JPKP: Joseph P. Kennedy Papers, JFKPL.

K: Theodore C. Sorensen, *Kennedy* (New York: Harper & Row, 1965)

KLOH: JFKPL oral history

KMPK: Kerry McCarthy, "P. J. Kennedy: The First Senator Kennedy," unpublished manuscript, courtesy Kerry McCarthy

KP: Koskoff Papers, JFKPL

KR: Michael R. Beschloss, *Kennedy and Roosevelt: The Uneasy Alliance* (New York: Norton, 1980)

LC: Library of Congress, Washington, D.C.

LL: Laurence Leamer

LM: David Cecil, *Lord M, or the Later Life of Lord Melbourne* (London: Constable and Co., 1954)

MP1960: Theodore H. White, *The Making of the President 1960* (New York: Atheneum, 1961)

NA: National Archives

NHP: Nigel Hamilton Papers,

NPSOH: National Park Service oral history

OTR: off the record

PC: Personal Collection

PFP: Paul Fay Papers, Stanford University

PIC: John F. Kennedy, *Profiles in Courage* (New York: Harper, 1955; memorial edition, 1964)

PJFK: Herbert S. Parmet, *JFK: The Presidency of John F. Kennedy* (New York: Dial, 1983)

PRIUM: Presidential Recordings, "Integration of the University of Mississippi," JFKPL

PS: Pierre Salinger, *P.S.: A Memoir* (New York: St. Martin's, 1995)

RCP: Robert Coughlin Papers, in author's, possession

RFK:Victor Lasky, *Robert F. Kennedy: The Myth and the Man* (New York: Trident Press, 1968)

RFKCB: C. David Heymann, *RFK: A Candid Biography of Robert F. Kennedy* (New York: Dutton, 1998)

RKHT: Arthur Schlesinger Jr., *Robert Kennedy and His Times* (Boston: Houghton Mifflin, 1978)

RKIHOW: Edwin O. Guthman and Jeffrey Shulman, eds., *Robert Kennedy: In His Own Words* (New York: Bantam, 1988)

RL: Franklin D. Roosevelt Presidential Library, Hyde Park, New York

RWC: Robert White Collection, Florida International Museum, St. Petersburg. Courtesy of Robert White

RWP: Richard Whalen Papers, JFKPL. Courtesy of Richard Whalen,

SB: Richard D. Mahoney, *Sons and Brothers: The Days of Jack and Bobby Kennedy* (New York: Arcade, 1999)

SJFK: Herbert S. Parmet, *Jack: The Struggles of John F. Kennedy* (New York: Dial, 1980)

TD: Robert F. Kennedy, *Thirteen Days* (New York: Norton, 1969)

TEEK: Burton Hersh, *The Education of Edward Kennedy* (New York: William Morrow, 1972)

TEW: Robert F. Kennedy, *The Enemy Within* (New York: Harper, 1960)

TFB: Edward M. Kennedy, ed., *The Fruitful Bough: A Tribute to Joseph P. Kennedy* (privately printed, 1965)

TKL: Theodore C. Sorensen, *The Kennedy Legacy* (New York: Macmillan, 1969)

TOB: Burton Hersh, *The Old Boys: The American Elite and the Origins of the CIA* (New York: Scribner's, 1992)

TR: Rose Fitzgerald Kennedy, *Times to Remember* (Garden City, N.Y.: Doubleday, 1974)

WES: John F. Kennedy, *Why England Slept* (New York: Wilfred Funk, 1940)

WK: Pierre Salinger, *With Kennedy* (New York: Doubleday, 1966)

WNJ: C. David Heymann, *A Woman Named Jackie* (New York: Lyle Stuart, 1989)

YM: David Cecil, *The Young Melbourne* (New York: Bobbs-Merrill, 1939)

Notes

1. A True Man

3 arranged for her only son: This account of Joe Kennedy's journey to Boston to deliver hats is based on a LL interview with Mary Lou McCarthy, and on KMPK.

3 piercing, dismissive eyes: LL interview with Mary Lou McCarthy, and KMPK.

3 Driven from their land: Andrew Buni and Alan Rogers, *Boston, City on a Hill* (1984), p. 76.

4 an estate of: 1860 census, ward 2, East Boston, June 1860, p. 203, Boston Vital Records.

4 As the driver guided: *East Boston Argus-Advocate*, souvenir edition, May 1897.

4 She sent one daughter: Laurence Leamer, *The Kennedy Women* (1994), p. 20.

4 shot glass filled: Interview, Joe Kane, KP.

4 "slick as grease": *East Boston Argus-Advocate*, August 20, 1892.

5 He and his business associates: Leamer, pp. 99–100.

5 the largest Jewish community: Sari Roboff, *East Boston: Boston 200 Neighborhood History Series* (1976), p. 6.

6 jammed together: Buni and Rogers, p. 92.

6 if an Italian: ibid., p. 93.

6 Susan Southworth and Michael Southworth, *The AIA Guide to Boston* (1984), p. 437.

7 quasi-apes, as looming, salivating simian wretches: L. Perry Curtis Jr., *Apes and Angels: The Irishman in Victorian Caricature* (1997), p. 58.

7 "simply an Americanized . . .": Stephen Halpert and Brenda Halpert, introduction and narrative, *Brahmins and Bullyboys: G. Frank Radway's Boston Album* (1973), p. 3.

8 a rented house: Tax Assessor's Records, BPL, 1886, p. 94.
9 had Joe photographed in a long dress: Joseph Kennedy sent a copy of the photo to his son, Edward Kennedy. "What I would particularly like you to observe is the sharp piercing eyes, the very set jaw and the clenched left fist," he wrote. "Maybe all of this meant something!" Quoted in TFB, p. 8.
10 "Any nation that cannot . . .": Mary Cable, *The Little Darlings: A History of Child Rearing in America* (1975), p. 172.
10 "a perfect gentleman . . .": E. Anthony Rotundo, *American Manhood: Transformations in Masculinity from the Revolution to the Modern Era* (1993), p. 269.
10 "An able-bodied young . . .": G. Stanley Hall, *Youth: Its Education, Regimen, and Hygiene* (1904), p. 94.
10 "Better even . . .": ibid., p. 100.
10 hitched a ride: Richard J. Whalen, *The Founding Father: The Story of Joseph P. Kennedy* (1964), p. 21.
10 playing with a toy pistol: P. J. Kennedy letter in Loretta Kennedy Connelly collection, courtesy Mary Lou McCarthy and Kerry McCarthy.
11 One Memorial Day: TFB, p. 7.
11 One summer Joe got together: Whalen, p. 21.
11 finest public school: Philip Marson, *Breeder of Democracy* (1963), p. 68.
12 His grades were pathetic: Boston Latin School transcript, HUA.
12 Joe took his friend: interview, Walter Elcock Jr., RWP.
12 prove their manhood: Hasia R. Diner, *Erin's Daughters in America: Irish Immigrant Women in the Nineteenth Century* (1983), pp. 22–23.
12 "homosexuality . . .": G. Stanley Hall, *Life and Confessions of a Psychologist* (1923), pp. 132–33.
14 "in a very roundabout way": Whalen, p. 24.

2. Gentlemen and Cads

16 "Compared to any . . .": M. M., "The Yard Dormitories," *Harvard Advocate*, 1909, p. 3, KUA.
16 "A hundred or so . . .": quoted in Charles Hawthorne Weston, "The Problem in Democracy at Harvard," *Harvard Advocate*, Spring 1912, KUA.
16 "Three Cs and . . .": A. M. Schlesinger Jr., "Harvard Today," *Harvard Advocate*, September 1936, KUA.
16 "Our friendships . . .": H. E .P., "The Importance of Being a Sport," *Harvard Advocate*, May 1908, HUA.
17 About two-thirds: Ronald Story, *Harvard and the Boston Upper Class: The Forging of an Aristocracy, 1800–1870* (1980), p. 173.
17 young men largely dominated: Morrison I. Swift wrote in the *Harvard Illustrated* in May 1911: "The irreducible fact [is] that the rich men's sons, whether confessedly or not, are the central figures of the college," KUA.

17 almost all Irish immigrant: James Joseph Kenneally, *The History of American Catholic Women* (1990), p. 113.
17 One of them: M. A. DeWolfe Howe, *Barrett Wendell and His Letters* (1924), p. 47.
17 "over-civilized man . . .": Theodore Roosevelt, *The Strenuous Life* (1904), p. 7.
18 "When the students entered . . .": *Harvard Crimson*, October 19, 1908, KUA.
18 "Our ancestors have bred . . .": William James, *The Moral Equivalent of War and Other Essays* (1971), pp. 5–7.
18 finance committee of the Freshman: *Harvard Crimson*, February 20, 1909, KUA.
18 one of the fifteen ushers: ibid., March 12, 1909, KUA.
19 "the private school . . .": ibid., April 10, 1912, KUA.
19 in one typical . . .: ibid., December 6, 1911, KUA.
19 "the most magnificent sight . . .": Thomas Goddard Bergin, *The Game: The Harvard-Yale Football Rivalry, 1875–1983* (1984), p. 100.
19 graduate of Worcester: *Harvard Crimson*, December 6, 1911, KUA.
20 Fisher was class: Harvard class alumni bulletin, 1912, p. 101, KUA.
20 "Important fall baseball . . .": *Harvard Crimson*, October 1, 1908, KUA.
20 "We're the two . . .": TFB, p. 283.
21 "For a short while . . .": Joseph F. Dinneen, *The Kennedy Family* (1959), p. 14, and Whalen, p. 30.
21 in a typical year: Weston, "The Problem in Democracy . . ."
22 The tap on Joe's door: Doris Kearns Goodwin, p. 215.
22 "Everywhere was to be seen . . .": *Harvard Crimson*, February 25, 1911, KUA.
23 "a peculiar kind . . .": Winfield Scott Hall, *A Manual of Sex Hygiene* (1913), pp. 75–76.
23 "He talked himself . . .": interview, Arthur Goldsmith, RWP.
23 first professional baseball coach: Joe Bertagna, *Crimson in Triumph* (1986), p. 151.
24 thrown out his arm: Whalen, p. 27.
24 he did as well: Joe Kennedy batted .285, getting two hits in seven times at bat. *Harvard Crimson*, September 26, 1911, KUA.
24 "No year and no . . .": ibid., May 13, 1909.
24 "carrying in his . . .": Henry James, *Charles W. Eliot: President of Harvard University*, vol. 2 (1930), p. 60.
24 On the football: ibid., p. 69.
25 "a rather hysterical . . .": *Harvard Crimson*, May 6, 1909, KUA.
25 "Baseball is on trial . . .": ibid., May 4, 1912, KUA.
25 At the beginning of : John A. Blanchard, ed., *The H Book of Harvard Athletics: 1852–1922* (1923), p. 148.
26 "rather crude material": *Harvard Alumni Bulletin*, 1911, KUA.

26 The game stayed close: *Boston Globe*, June 24, 1911.
27 "My father and I . . .": TR, p. 61.
27 did not even mention Joe: *Harvard Crimson*, June 22, 1911, KUA.
27 one of only 36: At the end of the football season, the *H* men at the university doubled to seventy. ibid., December 6, 1911.
27 "2,262 undergraduates . . .": ibid., December 15, 1911, KUA.
27 petitioned to graduate: Joseph P. Kennedy to P. J. Kennedy, September 30, 1911, Joseph P. Kennedy Records, HUA.
27 during his four years: Joseph P. Kennedy records, HUA.

3. Manly Pursuits

28 "If you're going . . .": interview, Joe Kane, KP.
28 With the barest hint: LL interviews with Mary Lou McCarthy and Kerry McCarthy.
29 "I had read . . .": interview, Rose Kennedy, RCP.
29 seventy-five guests: Gail Cameron, *Rose: A Biography of Rose Fitzgerald Kennedy* (1971), p. 69.
29 Honey Fitz began yelling: LL interview with Geraldine Hannon.
29 "is in a condition . . .": Hall, *Manual*, p. 69.
30 "the most desperate cases . . .": Barnarr A. Macfadden, *The Virile Powers of Superb Manhood: How Developed, How Lost, How Regained* (1900), p. 38.
30 Some experts: Hall, *Manual*, p. 83.
30 "Now listen, Rosie . . .": Doris Kearns Goodwin, p. 392.
30 11 percent: Bruce A. Phillips, *Brookline: The Evolution of an American Jewish Suburb* (1990), p. 28.
32 "by accepting the idea . . .": Doris Kearns Goodwin, p. 272.
32 "natural cynicism": Joseph P. Kennedy to Lord Max Beaverbrook, October 23, 1944, NHP.
34 "The strikers . . .": *Boston Globe*, November 1, 1917.
34 "probably no one . . .": David Palmer, "Organizing the Shipyards, Unionization at the New York, Federal Ship and Fore River, 1898–1945," Brandeis University Ph.D. diss., 1989, p. 19, NHP.
34 "The female sex . . . ": Charles G. Herbermann, et al., *The Catholic Encyclopedia* (1912), p. 687.
35 suffered from an ulcer: TR, p. 80.
35 "Tommy, it's so easy . . .": interview, Oscar Haussermann, RWP.
35 close to seven hundred thousand dollars: Ronald Kessler, *The Sins of the Father* (1996), p. 31.
35 high-stakes game: ibid.
36 In January 1920: Doris Kearns Goodwin, p. 305.
36 "If you need more . . .": quoted in ibid., p. 307.
36 "I don't know how . . .": Joseph P. Kennedy to Vera Murray, August 15, 1921, JPKP, HTF, p. 29.

37 "I hope you have . . .": Joseph P. Kennedy to Arthur Houghton, September 19, 1921, HTF, p. 31.
37 There were 125 beds: Doris Kearns Goodwin, p. 310.
37 half his wealth: *Boston Globe*, May 23, 1963.
38 forward-looking men: Lynn Dumenil, *The Modern Temper: American Culture and Society in the 1920s* (1995), p. 233.
38 In 1926 the Canadian: A&E documentary, *Prohibition*, based on the book by Edward Behr, *Prohibition: Thirteen Years That Changed America* (1996), an LL interview with Edward Behr.
39 "there was a . . .": interview with Cartha DeLoach.
39 "Joe brought the . . .": LL interview with Patty McGinty Gallagher.
39 He founded the famous: Hank Messick, *The Silent Syndicate* (1967), p. 163.
39 Fitzgerald says: LL interview with Benedict Fitzgerald.
39 confirmed by Q. Byrum Hurst: LL interview with Q. Byrum Hurst.
39 vowed he did not commit: Graham Nown, *The English Godfather* (1987), p. 47.
39 George Raft in his screen: ibid., p. 76.
39 greased with ample payoffs: ibid.
40 "everybody knew . . .": LL interview with George Smathers.
40 "Joe was having . . .": LL interview with Zel Davis.
40 "When I worked . . .": videotaped oral history, Mel Shoemaker, courtesy Gus Russo.
41 "About ten years . . .": LL interview with Christopher Kennedy.
41 $302: Joseph P. Kennedy to Robert Potter, August 17, 1920, HTF, p. 24.
42 "It's all right . . .": Doris Kearns Goodwin, pp. 324–25.
42 Were the good gentlemen: Joseph P. Kennedy was not the only observer of this hypocrisy. The late Thomas "Tip" O'Neill wrote in his autobiography of working as a groundskeeper at Harvard during the 1927 Harvard commencement: "They were . . . drinking champagne, which was illegal. . . . I remember that scene like it was yesterday, and I can still feel the anger I felt then. . . . Who the hell do these people think they are, I said to myself, that the law means nothing to them?" It was a question that Kennedy and O'Neill, the former speaker of the House of Representatives, answered in different ways. Tip O'Neill, with William Novak, *Man of the House* (1987), p. 6.

4. "Two Young 'Micks' Who Need Discipline"

43 "Gee, *you're* a great mother . . .": TR, p. 93.
43 four or five weeks: interview, Rose Kennedy, RCP.
43 "better take it in stride . . .": Doris Kearns Goodwin, p. 353.
44 "a more distant figure": interview, John F. Kennedy, JMBP.
44 eaten by lions: *Time* file, June 1, 1960, NHP.
44 "at exactly the same . . .": L. Emmett Holt, M.D., *The Care and Feeding of*

Children: A Catechism for the Use of Mothers and Children's Nurses (1923), p. 88.
44 "gentle shaking": quoted in Christina Hardyment, *Dream Babies: Three Centuries of Good Advice on Child Care* (1983), p. 127.
44 "twice as much food . . .": ibid., p. 126.
45 she purchased approved books: Rose Kennedy, NPSOH.
45 "I wouldn't have . . .": TR, p. 111.
45 "Lack of precision in the mother . . .": Mrs. Burton Chance, *The Care of the Child* (1909), p. 20–21.
45 Kikoo Convoy: TFB, p. 195.
46 "I didn't think . . .": interview, Rose Kennedy, *Time* file, 1960–61, RWP.
46 "He told me once . . .": interview, Henry Luce, RWP.
46 "The mood of . . .": Alfred Adler, *The Individual Psychology of Alfred Adler: A Systematic Presentation in Selections from His Writings*, edited and annotated by Heinz L. Ansbacher and Rowena R. Ansbacher (1956), pp. 380–81.
46 ". . . Physically we used to have . . .": interview, John F. Kennedy, JMBP.
46 the two boys kicked: interview, Rose Kennedy, RCP.
46 "Remember that Jack . . .": Joseph P. Kennedy to Joseph P. Kennedy Jr., July 28, 1926, HTF, p. 48.
47 "When his sister was . . .": interview, Rose Kennedy, RCP.
47 "man of broad hips . . .": Joseph Collins, *The Doctor Looks at Love and Life* (1929), pp. 68, 74.
47 "they are the most . . .": ibid., p. 74.
47 not allowed to play: LL interview with Robert Bunshaft.
47 "The boys have . . .": TR, p. 93.
48 English chauffeur: interview, Rose Kennedy, RCP.
48 "because the more . . .": Ralph LaRossa, *The Modernization of Fatherhood: A Social and Political History* (1997), p. 180.
48 "once children are . . .": Chester T. Cromwell, *American Mercury*, October 1924.
49 "The old man . . .": LL interview with Tom Finneran.
49 "Good Luck": Joseph P. Kennedy to John F. Kennedy, May 19, 1926, HTF, p. 7.
49 "This is a shrine!": Hank Searls, *The Lost Prince: Young Joe, the Forgotten Kennedy* (1969), p. 44.
49 "When they'd shoot . . .": LL interview with Holton Wood.
50 "very pugnacious . . .": LL interview with Augustus Soule Jr.
50 "Well, coach, you're . . .": Willard K. Rice, KLOH.
50 "a wonderful idea": ibid.
51 As a back, Joe Jr.'s greatest: Searls, p. 44.
51 Jack, a wiry: undated clipping, Dexter archives.
51 "At that time": interview, Rose Kennedy, RCP.
52 "As long as they": ibid.

53 "sexual affection for *men*": James M. O'Toole, *Militant and Triumphant: William Henry O'Connell and the Catholic Church in Boston, 1859–1944* (1992), p. 191.
53 "the screen's . . . leading family man": Terry Ramsaye, "Intimate Visits to the Homes of Famous Film Magnates," *Photoplay*, September 1927.
53 "spendthrift clause . . .": Kessler, p. 41.
53 "spit in his eye": ibid., p. 44.

5. Moving On

54 "I felt it was no place . . .": Joe McCarthy, *The Remarkable Kennedys* (1960), p. 42.
54 "It is not a pleasant . . .": *Boston Globe*, April 17, 1945.
55 her check bounced: LL interviews with Mary Lou McCarthy and Kerry McCarthy.
55 Eunice was suffering: Doris Kearns Goodwin, p. 368.
56 "He moved so quickly . . .": Gloria Swanson, *Swanson on Swanson* (1980), p. 369.
57 catered food: interview, Rose Kennedy, RCP.
57 "Well, he felt sorry . . .": ibid.
58 the next occupant of Gloria's: Betty Lasky, *RKO The Biggest Little Major of Them All* (1984), pp. 55–57. See also Roland Flamini, *Scarlett, Rhett, and a Cast of Thousands* (1975), p. 146.
58 "Would you please . . .": Kathleen Kennedy to "Dear Daddy," January 31, 1930, Eunice Kennedy Shriver letters.
58 "How is little Gloria?": ibid., March 23, 1930.
58 "he would have seen . . .": LL interview with Kerry McCarthy.
59 between $200,000 and: interview, James Landis, RWP.
59 "Who are you?": interview, Rose Kennedy, RCP.
59 "The story got around . . .": ibid.
60 "Forty is a . . .": LL interview with Harvey Klemmer.
60 "was always the first . . .": TFB, p. 194.
60 "Daddy did not . . .": Kathleen Kennedy to "Dear Mother," February 13, 1932, Eunice Kennedy Shriver letters.
60 "They were considered . . .": LL interview with David Wilson.
61 Some of the teenagers: LL interview with Paul Morgan.
61 "People said, 'Why do you . . .' ": interview, Rose Kennedy, RCP.
61 "After that we became . . .": LL interview with Manuel Angulo.
62 "Perhaps Joe . . .": ibid.
62 Jack was so shy: ibid.
62 "I'm going to be . . .": LL interview with Alan Gage.
62 "As a mother . . .": interview, Rose Kennedy, Laura Berquist papers, Boston University.
63 could not remember: K, p. 34.

63 jumping again and again: Paul Healy, "Investigator in a Hurry," *Sign*, August 1957.
63 into a glass door: RKHT, p. 22.
63 A half-hour later: *New York Herald Tribune*, September 13, 1961.
63 Jack won a commencement prize: Riverdale School archives.
63 "Bobby looked . . .": interview, Lem Billings, ASP.
64 "Bobby got along . . .": interview, Jean Kennedy Smith, ASP.
64 "I put an end . . .": interview, Rose Kennedy, RCP.
64 "might have a religious vocation": Sister M. Ambrose to Robert F. Kennedy, September 29, 1957, RFK preadministrative working files, JFKPL.
64 "Bishop Bernard from the Bahama . . .": ibid., August 24, 1957.
64 "where they'd meet . . .": interview, Rose Kennedy, RCP.
64 "He is a rare youth . . .": Frank S. Hackett to George St. John, May 29, 1929, AAML.
65 "My nose my leg . . .": quoted in Nigel Hamilton, *JFK: Restless Youth*, (1992), p. 86.
65 "began to get sick . . .": John F. Kennedy to Joseph P. Kennedy, 1930–31, JFKPP.
65 "Joe fainted twice . . .": ibid.
65 "I see things blury": "Jack" to "Dear Mother," undated letter on Canterbury School stationery, NHP.
65 "I smashed into . . .": ibid.
66 "I hope my marks . . .": John F. Kennedy to Joseph P. Kennedy, April 15, 1931, NHP.
67 Joe Jr. took Teddy: AWRJ, p. 59.
67 never played . . . again: interview, Rose Kennedy, RCP.
67 "My lord, this is . . .": LL interview with Harry Fowler.
67 "Don't come in second . . .": TR, p. 143.
68 "He always trusted . . .": LL interview with Eunice Kennedy Shriver.
68 "All children can be de-throned . . .": Adler, p. 381.
68 "I wasn't supposed to . . .": interview, Edward Kennedy, RCP.
69 "Probably three times . . .": ibid.
69 "I wanted him . . .": *Boston Globe,* December 11, 1937.
70 "I doubt that . . .": interview, Frank Waldrop, BP.
71 "He, himself . . .": C. H. Cramer to John F. Kennedy, undated quoting Roy Howard to Newton D. Baker, Baker Collection, Library of Congress, LC.
71 "Roosevelt was under . . .": ibid.
71 "You may rest . . .": Joseph P. Kennedy to William Randolph Hearst, October 19, 1932, William Hearst papers, Bancroft Library, University of California at Berkeley.
71 To them, he appeared: C. H. Cramer to John Kennedy, quoting a letter of July 12, 1932, Roy Howard to Newton D. Baker, found in Baker collection, LC.
71 "If we live . . .": Eddie Dowling, CUOH.
72 Joe helped Jimmy: Koskoff, p. 51; Whalen, p. 136.

72 "I didn't want anything . . .": James Warburg, CUOH.
72 Joe finagled: Koskoff, p. 53.
73 "Your taste in dumb cruise . . .": KR, p. 98.
73 Roosevelt laughed loudly: ibid.

6. "Most Likely to Succeed"

75 "in any part Hebraic": Joseph P. Kennedy Jr., application, May 1, 1929, AAML.
75 students voted: Choate yearbooks, AAML.
76 "friendliness": Clara St. John to Rose Kennedy, October 7, 1931, ibid.
76 One of the teachers: Sheldon Stern, JFKPL resident historian, interviewed Harold Taylor for the library. Taylor was afraid of a tape recorder, and this information is based on Stern's notes.
76 "some one send to . . .": George St. John to Joseph P. Kennedy, October 20, 1932, AAML.
76 "We'll try to show . . .": ibid., April 22, 1932.
76 "one of the 'big boys' . . .": Mrs. George C. St. John to Rose Kennedy, October 7, 1931, AAML.
77 kept at school: George St. John to Joseph P. Kennedy, December 3, 1929, AAML.
77 "too easily satisfied . . .": Hank Searls, *The Lost Prince* (1969), p. 64.
78 "When Joe came home . . .": John F. Kennedy to Joseph P. Kennedy, December 9, 1931, family correspondence, JFKPP.
78 "a lavender bathrobe . . .": Mrs. George St. John to Rose Kennedy, January 18, 1932, AAML.
79 "mumps, and the doctor . . .": Rose Kennedy to Mrs. George St. John, n.d., AAML.
79 "full of pop": Mrs. George St. John to Earl Leinbach, January 25, 1932, AAML.
79 Jack had to return: George St. John to John F. Kennedy, July 30, 1932, AAML.
79 special built-up shoes: Choate secretary to Rose Kennedy, March 10, 1933, AAML.
79 ten letters from Rose: Seymour St. John, "JFK Fiftieth Reunion of 1,000 Days at School," June 1985, AAML.
79 "pink eye": Choate secretary to Rose Kennedy, February 14, 1933, AAML.
79 "a little grippy cold": ibid., February 17, 1933.
79 Jack's most distinguished contribution: interview, Earl Leinbach, JMBP.
80 "The golf is going good . . .": John F. Kennedy to Joseph P. Kennedy, n.d. (Sunday), JFKPP.
80 "rather outstanding": Lem Billings, KLOH.
80 "My father did try . . .": ibid.
80 "I think that . . .": ibid.
81 "a very popular hero": *Boston Globe*, June 11, 1933.

81 "These boys, when . . .": interview, Rose Kennedy, RCP.
82 "He had set his heart . . .": AWRJ, p. 43.
82 Whitelaw recalled: Aubrey H. Whitelaw to John F. Kennedy, June 26, 1958, JFKPP.
82 "the heads of all": Joseph P. Kennedy Jr. to Joseph P. Kennedy, April 23, 1934, RFK papers, JFKPP.
83 "We were going . . .": Aubrey H. Whitelaw to John F. Kennedy, June 26, 1958, JFKPP.
83 "far beyond his . . .": Joseph P. Kennedy to Joseph P. Kennedy Jr., May 4, 1934, HTF, p. 133.
83 "Joe came back . . .": John F. Kennedy to Lem Billings, July 25, 1934, JFKPL.
83 "Joe, if you feel . . .": interview, Rose Kennedy, RCP.
84 "At first his [Jack's] attitude . . .": Hamilton, p. 106.
84 "I'm afraid it . . .": Seymour St. John, "JFK Fiftieth Reunion."
84 "There is . . . little . . .": St. John, "JFK Fiftieth Reunion."
84 "I can't tell you how . . .": JPKP, HTF, p. 120.
84 "I thought I would . . .": John F. Kennedy to Joseph P. Kennedy, December 2, 1934, JFKPP.
85 "I would be lacking . . .": Joseph P. Kennedy to John F. Kennedy, December 7, 1934, RCP.
85 "possibly contributed . . .": Joseph P. Kennedy to George Steele, January 5, 1935, AAML.
85 "I have no memory . . .": LL interview with Hugh Sidey.
86 One of Jack's: LL interview with Larry Baker.
86 "Maury Shea, Maury Shea . . .": Joan and Clay Blair Jr., *The Search for JFK* (1976), p. 35.
86 "We damn near . . .": interview, Ralph Horton, BP.
87 "They made fun of me": LL interview with Larry Baker.
87 spoiled snob: St. John recollections, AAML.
87 "It is only one . . .": Joseph P. Kennedy to Joseph P. Kennedy Jr., May 4, 1934, HTF, p. 134.
87 a prayer for Jack: Lem Billings, KLOH.
87–88 "Well, you know, Jack . . .": Mrs. George St. John to Rose Kennedy, February 6, 1934, AAML.
88 "If this had happened . . .": ibid.
88 "The reason . . .": John F. Kennedy to Lem Billings, June 27, 1934, JFKPL.
88 "God what a beating . . .": ibid., June 22, 1934.
88 "You've never smelt . . .": ibid., June 21, 1934.
89 "Yesterday I went through . . .": ibid., June 27, 1934.
91 "Will you please . . .": George St. John to Joseph P. Kennedy, February 11, 1935, AAML.
91 other faculty members: Sheldon Stern's notes of his JFKPL interview with Harold Taylor.
91 "My God, my son": Doris Kearns Goodwin, p. 488.

91 "We reduced Jack's conceit": Goddard Lieberson, *John F. Kennedy: As We Remember Him* (1965), p. 17.
92 "definitely in a . . .": St. John, "JFK Fiftieth Reunion."
92 "this silly episode . . .": TR, p. 183.
92 did not receive: Choate records for John F. Kennedy, RWC.
92 "Jack has rather . . .": John F. Kennedy, Princeton application, RWC.
93 "Finally, Pete Caesar came . . .": Lem Billings, KLOH.

7. The Harvard Game

94 $125 a month: Joseph P. Kennedy to Joseph P. Kennedy Jr., December 7, 1934, JPKP, HTF, p. 147.
94 He had his own valet: AWRJ, p. 31.
95 His picture was featured: ibid., p. 27.
95 He broke his arm: Joseph P. Kennedy Jr. to Joseph P. Kennedy, telegram, April 19, 1935, HTF, p. 153.
95 The police arrived: Searls, p. 93.
95 "the Life-saver": ibid., p. 94.
95 "Get out of this room!": AWRJ, p. 14.
96 "I always thought that": LL interview with Robert Purdy.
96 "My impression . . .": ibid.
96 he chaired the: AWRJ, p. 17.
97 Galbraith wanted to choose: John Kenneth Galbraith, *A Life in Our Times* (1981), p. 51.
97 He thanked her: interview, Mrs. Josephine Fulton, NHP.
98 When he went: AWRJ, p. 51.
98 "You didn't go . . .": undated 1963 clipping by Gene School and Mina Wetzig, HUA.
98 "I have had a very . . .": John F. Kennedy to Lem Billings, September 29, 1935, JFKPP.
99 "Today was most embarrassing . . .": ibid., October 1935.
99 "a very good looking . . .": ibid., n.d.
99 "turning yellow . . .": Hamilton, p. 144.
99 Once again it was suspected: JPKP, HTF, p. 166.
100 "the most harrowing . . .": John F. Kennedy to Lem Billings, January 3, 1936, JFKPP.
100 "I don't know . . .": ibid., January 1936.
100 "I'm writing Rip . . .": ibid., January 18, 1936.
100 "It seems to me . . .": ibid., January 1936.
100 "The next time . . .": ibid., January 3, 1936.
100 "I think he was making . . .": LL interview with James Rousmanière.
101 "getting rather fed up . . .": John F. Kennedy to Lem Billings, January 27, 1936, JFKPP.
101 "Eat drink and make . . .": ibid.
101 gave his father's: John F. Kennedy, Harvard application, May 8, 1935, NHP.

101 "I ended up . . .": John F. Kennedy to Lem Billings, May 9, 1936, JFKPL.
102 "Dr. Wild, I want . . .": Payson S. Wild, KLOH.
102 even sleeping: LL interview with Robert Purdy.
102 "Four of us had dates . . .": John F. Kennedy to Lem Billings, October 16, 1936, JFKPP.
102 "I am now known . . .": ibid.
103 "I swear I don't think . . .": interview, Ralph Horton Jr. BP.
103 "I was very young": Searls, p. 99.
103 "Gertrude Niesen was . . .": James Rousmanière, CUOH.
104 "I guess he [Macdonald] . . .": LL interview with James Rousmanière.
104 "It was obvious . . .": ibid.
105 "attached ear . . .": *Harvard Crimson*, March 24, 1939, KUA.
105 "One did not cultivate . . .": Galbraith, p. 53.
105 "Jack devoted himself . . .": interview, John Kenneth Galbraith.
105 "Did you see . . .": John F. Kennedy to Lem Billings, February 13, 1936, JFKPP.
106 "Jack, if you want . . .": Doris Kearns Goodwin, p. 505.
106 "Mind your own . . .": ibid.
106 "I suppose I knew . . .": AWRJ, p. 3.
108 donated $62,500: Thornwell Jacobs, president of Oglethorpe University, to Bernard Baruch, January 15, 1937, KP.
108 "I never heard . . .": Joseph P. Kennedy to Bernard Baruch, n.d., KP.
108 Two years later: Jay Pierrepoint Moffat, *The Moffat Papers: Selections from the Diplomatic Journals of Jay Pierrepoint Moffat*, 1919–1943 (1956), p. 156, KP.
108 "The old man hired . . .": interview, Charles Houghton Jr., BP.
109 Joe was no more disloyal: Frank R. Kent, "The Great Game of Politics," 1937, KP.
109 "a shining star . . .": *Boston Sunday Post*, August 16, 1936, and Joseph P. Kennedy to Father Charles Coughlin, August 18, 1936, HTF, p. 187.
109 "dissension and division . . .": Whalen, p. 183.
109 Weisl recalled: LL interview with Edwin Weisl Jr.
110 "You should think . . .": Doris Kearns Goodwin, p. 480.
110 "He wanted to know . . .": Searls, p. 102.
111 To add a modicum: ibid., p. 105.

8. Mr. Ambassador

112 "That neat little scheme . . .": Frank Kent to Joe Kennedy, April 18, 1938, Maryland Historical Society, Frank R. Kent papers, KP.
113 "thereby breaking . . .": Joseph P. Kennedy diary, March 8, 1938, JPKP, HTF, p. 239.
113 "An unemployed man . . .": Joseph P. Kennedy to Frank Kent, March 21, 1938, KP.

113 "I think it is not . . .": David E Koskoff, *Joseph P. Kennedy: A Life and Times* (1974), p. 127.
113 "I have shown . . .": Cordell Hull to Joseph P. Kennedy, March 13, 1938, KP.
114 "difficult to let . . .": Joseph P. Kennedy to Bernard Baruch, March 21, 1938, KP.
114 "parts of it fell . . .": Joseph P. Kennedy diary, March 18, 1938, JPKP, HTF, p. 245.
114 "but rather the loud . . .": *Boston Globe*, July 17, 1949.
114 "complete poppycock": ibid
114 "his hope of . . .": quoted in Carey McWilliams *A Mask for Privilege*, (1948), pp. 20–21.
115 In 1922, President: ibid, p. 38.
115 "Well, they brought . . .": interview, Harvey Klemmer.
115 "been to no . . .": Joseph P. Kennedy to Frank Kent, March 21, 1938, KP.
116 "come to some form of": Beschloss, *Kennedy and Roosevelt* p. 172.
116 "write a story . . .": LL interview with Walter Trohan.
116 "the chilling shadow . . .": *Chicago Tribune,* June 23, 1938.
117 "It was a true . . .": JPKP, HTF, p. 266.
117 For all Joe's: LL interview with Walter Trohan; KR, pp. 171–72; and Joseph P. Kennedy, unpublished diplomatic memoir, JPKP, HTF, p. 266.
118 Teddy took umbrage: interview, Edward Kennedy, RCP.
118 "Any problems . . .": LL interview with Luella Hennessey Donovan.
118 "His name was connected . . .": LL interview with Harvey Klemmer.
118 "Rose, this is a . . .": TR, p. 221.
119 "sticking up and retrieved it": Joseph P. Kennedy diary, April 9, 1938, JPKP, HTF, p. 250.
119 "for the life . . .": KR, p. 174.
120 "'slippery' . . .": John Martin Blum, *From the Morgenthau Diaries*, vol. 1 (1959) , p. 518.
120 "trying to keep up . . .": KR, p. 178.
120 "I don't have . . .": Joseph P. Kennedy to Cordell Hull, September 10, 1938, JPKP, HTF, p. 274.
120 strongest in the world: Joseph P. Kennedy to Cordell Hull, September 22, 1938, JPKP, HTF, p. 281.
120 "not lose our heads": KR, p. 175.
120 "I know what difficult days . . .": Franklin Roosevelt to Joseph P. Kennedy, August 25, 1938, RL.
121 "Everyone unutterably shocked . . .": TR, p. 238.
121 "I'm feeling very blue . . .": KR, p. 176.
121 "I hope this doesn't mean . . .": Joseph P. Kennedy to Cordell Hull, September 28, 1938, JPKP, HTF, p. 288.
121 "Isn't it wonderful?": KR, p. 177.
121 "after all . . .": Koskoff, p. 159.

122 "hardly prepared": ibid. p. 178.
122 "The Navy Day . . .": John F. Kennedy to parents, n.d. (1938), JFKPP.
122 "he would be a . . .": KR, p. 182.
122 "I have never made . . .": ibid., p. 183.
123 "Germany is still bustling": Searls, p. 123.
124 "Luella, I need a Band-Aid": LL interview with Luella Hennessey Donovan.
124 "Eddie Moore came in . . .": interview, Edward Kennedy, RCP.
124 "Joe Jr. used to tease . . .": LL interview with Luella Hennessey Donovan.
125 Among them were a: William Koren Katz and Marc Crawford, *The Lincoln Brigade: A Picture History* (1989), p. xi.
125 At Harvard, Joe Jr. wrote: Searls, pp. 108–9.
125 "The last few weeks . . .": Joseph P. Kennedy Jr., unpublished memoir, JFKPL.
126 "It made me sick": *Atlantic Monthly*, October 1939.
126 Stray dogs and cats: George Hill, *The Battle for Madrid* (1976), p. 174.
127 "We were touched . . .": Joseph P. Kennedy Jr. to Joseph P. Kennedy, JPKP, HTF, p. 323.
128 "amateur and temporary . . .": ibid., p. 301.
128 "the natural Jewish reaction": Robert F. Kennedy, "Answer to Lippmann Editorial About Dad, November 14th," RFK papers, JFKPL.
129 Young Bobby already understood: "Notes on Master Robert Kennedy's Cornerstone Ceremony," April 1, 1939, RFK papers, JFKPL, and *New York Herald*, April 1, 1939.
129 "Why?": interview, Rose Kennedy, RCP.
129 "there is only one . . .": unpublished memoir, Joseph P. Kennedy Jr., "Germany," RFK papers, JFKPL.
130 "I had always thought . . .": ibid.
130 "Don't get . . .": Gerald Walker and Donald A. Allan, "Jack Kennedy at Harvard," *Coronet*, May 1961, and Hamilton, p. 241.
130 "rather unpleasant contact . . .": John F. Kennedy to Lem Billings, NHP.
131 Jack left a faucet: interview, Charles Houghton, BP.
131 "Well, we went to . . .": Lem Billings, KLOH.
132 "Story of father . . .": John F. Kennedy, diary entry, July 25, [1937], JFKPP.
132 "In the afternoon . . .": ibid., July 26, [1937].
132 "Fascism is the thing . . .": quoted in SJFK, p. 53.
133 "Jack didn't discuss . . .": interview, James Rousmanière.
133 "went to court . . .": John F. Kennedy to Lem Billings, March 23, 1939, JFKPL.
134 "terrific diamond . . .": ibid, April 28, 1939.
134 "The whole thing . . .": ibid.
134 "talked with the . . .": ibid., May 1939.
134 "What Germany will . . .": ibid.

134 "All of the young . . .": ibid.
135 "evince[d] that capacity . . .": YM, p. 45.
135 "The ideal was the . . .": ibid., p. 8.
135 "a skeptic in . . .": ibid., p. 76.
135 "Nature had meant . . .": ibid., p. 78.

9. "It's the End of the World, the End of Everything"

136 he shared with Chamberlain: Moffat, December 16, 1938.
136 The British feared: British Foreign Office, preparations for Evian meeting, "International Assistance Refugees," signed by R. M. Marke, 23/5, KP.
137 "The baby is tossed . . .": JPKP, HTF, p. 303.
137 "I have a couple . . .": *Liverpool Post,* May 19, 1937.
137 "He is, after all . . .": Franklin Roosevelt to Joseph P. Kennedy, July 22, 1939, JPKP, HTF, p. 353.
138 "the chief thing . . .": Joseph P. Kennedy to Franklin Roosevelt, August 9, 1939, JPKP, HTF.
139 "I could barely . . .": interview, Edward Kennedy, RCP.
139 "Joe [Jr.] took Teddy . . .": interview, Rose Kennedy, RCP.
140 "It's the end . . .": KR, p. 190.
140 "The natural shock . . .": "The Athenia Affair," John F. Kennedy, JFKPL.
141 "there are signs of decay . . .": Joseph P. Kennedy to Franklin Roosevelt, September 30, 1939, RL.
141 "Maybe I do him . . .": Joseph P. Kennedy, unpublished diplomatic memoir, James M. Landis Papers, LC
142 "For Christ's sakes": Moffat, pp. 297–98.
143 "That is taking . . .": The account of the meeting is based on Joseph P. Kennedy, unpublished memoir, Landis Papers, LC.
143 "Supposing, as I do not . . .": KR, p. 197.
143 "Jack rushed madly . . .": Joseph P. Kennedy Jr. to Joseph P. Kennedy, March 17, 1940, JFKPL.
144 "The [British] nation . . .": John F. Kennedy, "Why England Slept," thesis, JFKPP.
144 "resort[ing] to . . .": J. E. Fuller, D. F. Parry, and A. W. Sulloway, "Sex by the Yard," *Harvard Advocate,* December 1937, KUA.
144 far less than 1 percent: ibid.
144 "gone too far . . .": Joseph P. Kennedy to John F. Kennedy, May 20, 1940, JFKPP.
145 Blair Clark: interview, Blair Clark.
145 "it is already a best-seller . . .": Joseph P. Kennedy to Winston Churchill, August 14, 1940, NHP.
145 the book sold: William Roulet, president, Wilfred Funk, Inc., to Senator John F. Kennedy, November 30, 1959, and Joel Satz to John F. Kennedy, August 23, 1940, NHP
145 "He was stunned . . .": LL interview with Martha Sweatt Reed.

145 "Well, if Bill . . .": John F. Kennedy to Mrs. Sweatt, n.d., courtesy Martha Sweatt Reed.
146 "At Choate . . .": ibid.
146 he was taking: Joseph P. Kennedy to Rose Kennedy, August 2, 1940, JPKP, HTF, p. 454.
147 "Actually the incident": Raimund von Hofmannsthal to Clare Boothe Luce, March 20, 1940, Clare Boothe Luce papers, LC.
147 "His behavior as ambassador . . .": interview, Henry Luce, RWP.
147 "that Mr. Kennedy . . .": British Foreign Office files, FOI 371/24251, January 18, 1940, KP.
147 To the British, who: ibid., February 9, 1940.
147 In reality, Joe: JPKP, HTF, p. 232.
147 "quite unpopular . . .": British Foreign Office files, FOI 371/24251, March 5, 1940, KP.
148 "playing off . . .": ibid., March 6, 1940.
148 "treat us rough": Jay Pierrepont Moffat diary, February 13, 1940, KP.
148 "It is always difficult . . .": *The Spectator*, March 8, 1940.
148 "as he [the American ambassador] . . .": Viscount Halifax to Lord Lothian, British Foreign Office, 3242/131/45, May 13, 1940, KP.
148 "running the government": Joseph P. Kennedy to Secretary of State, August 2, 1940, KP.
149 "How can we . . .": Joseph P. Kennedy to Cordell Hull, July 31, 1940, KP.
149 He told a story: Breckinridge Long, *The War Diary of Breckenridge Long* (1966), November 6, 1940, p. 146.
149 "I am depressed . . .": Joseph P. Kennedy to Arthur Krock, November 3, 1939, KP.
149 could "run the show": Joseph P. Kennedy to Joseph P. Kennedy Jr., June 6, 1940, JPKP, HTF, p. 436.
149 "I wanted to ring you . . .": Joseph P. Kennedy, "Memorandum of Telephone Conversation Between the President and Myself at 5:00 P.M., August 1, 1940," RCP.
150 "For the United States . . .": Joseph P. Kennedy to Secretary of State, September 11, 1940, KP.
150 "I certainly don't get . . .": Joseph P. Kennedy to Edward Kennedy, September 11, 1940, NHP.
151 "in bedroom all morning": Clare Boothe Luce diary, April 2, 1940, Clare Boothe Luce papers, LC.
151 "Yesterday a Messerschmitt . . .": Joseph P. Kennedy to Clare Boothe Luce, October 1, 1940, Clare Boothe Luce papers, LC.
152 "25 million Catholic votes": Kessler, p. 207.
152 "I urge you to . . .": *Boston Globe*, undated clipping, NHP.
153 "WHEN YOU LAND . . .": Mrs. Henry R. Luce, cable to Joseph P. Kennedy, October 21, 1940, Clare Boothe Luce papers, LC.
153 In early October: Viscount Halifax to Lord Lothian, British Foreign Office, FO371/2425, October 10, 1940, KP.

153 "not to make . . .": Franklin D. Roosevelt to Joseph P. Kennedy, October 17, 1940, RL.
153 The president insisted: Franklin D. Roosevelt to Joseph P. Kennedy, October 26, 1940, RL.
153–54 "The president sent . . .": Arthur Krock private memorandum, December 1, 1940, AKP.
154 Roosevelt had alerted: This account of the meeting is based on Joseph P. Kennedy's diary, JPKP, HTF, pp. 480–82; Krock's memo of the meeting, KP; and TR, p. 274.
155 According to Roosevelt's: KR, p. 218
155 "then he would support": ibid., p. 221.
155 "his father's greatest . . .": FBI director, February 23, 1942, FBIFOI.
156 "self-success": Torbert Macdonald to John F. Kennedy, n.d., JFKPP.
156 "In my years . . .": JPKP, HTF, pp 482–89, RWP.
156 "There was that radio . . .": interview, Henry Luce, RWP.
157 "more brains than . . .": *Boston Globe*, November 10, 1940.

10. Child of Fortune, Child of Fate

158 "seemed to . . .": Searls, p. 156.
159 "Still can't get used . . .": John F. Kennedy to Lem Billings, October 4, 1940, JFKPP.
159 "Because of his back . . .": WNJ, p. 149.
159 "I'm not interested . . .": Hamilton, p. 358.
160 "I think Jack knew . . .": interview, Harriet Price, BP.
160 Jack was chosen: *Stanford Daily*, October 18, 1940.
160 "This draft has caused . . .": John F. Kennedy to Lem Billings, November 14, 1940, JFKPP.
160 "When I hear these . . .": Joseph P. Kennedy to John F. Kennedy, September 10, 1940, JFKPP.
160 "I think his father . . .": quoted in Hamilton, p. 351.
161 "WHEN WILL OUTLINE . . .": Joseph P. Kennedy to John F. Kennedy, December 5, 1940, JFKPP.
161 "I have seen the English stand . . .": John F. Kennedy to Joseph P. Kennedy, December 6, 1940, JFKPP.
161 "The danger of our . . .": John F. Kennedy to Joseph P. Kennedy, "Flying the Mainliner," n.d., JFKPP.
162 "As I remember . . .": John F. Kennedy to Harriet Price, read in interview, BP.
162 "I think in that . . .": quoted in RKHT, p. 43.
163 "a personal favor . . .": Joseph P. Kennedy to Joseph P. Kennedy Jr., March 17, 1941, ASP.
163 private flying lessons: LL interview with Benedict Fitzgerald.
163 "Year after year . . .": Sen. Edward M. Kennedy, "We Want to Remember His Life, Not Relive His Death," *Parade*, June 30, 1988.

164 "Remember, too . . .": Rose Kennedy to Robert F. Kennedy, January 12, 1942, ASP.
164 Bobby began to tremble: LL interview with Pierce Kearney.
165 Rose sent Teddy to join: RKHT, p. 34.
165 arrived in short pants: interview, Pierce Kearney, KP.
165 ten in all: TEEK, p. 38.
165 "That was hard to take . . .": LL interview with Edward Kennedy.
165 "You'll just have to . . .": RKHT, p. 31.
165 "I had been . . .": LL interview with Edward Kennedy.
166 "she'd read a Peter . . .": ibid.
166 "The house for us . . .": ibid.
166 "Teddy went outside . . .": interview, Rose Kennedy, RCP.
167 "I saw a man and . . .": LL interview with Joseph Gargan.
167 "I don't know . . .": John F. Kennedy to Cam Newberry, July 8, 1941, NHP.
167–68 "The boy has taken . . .": A. G. Kirk to Captain C. W. Carr, Chelsea Naval Hospital, March 24, 1941, BP.
168 "I just had . . .": interview, Edward Kennedy, RCP.
169 "Mr. Kennedy was so afraid . . .": LL interview with Luella Hennessey Donovan.
169 In 1941: Elliot S. Valenstein, ed., *The Psychosurgery Debate: Scientific, Legal, and Ethical Perspectives* (1980), p. 25.
169 the medical kings of lobotomy: ibid., p. 22.
169 They had performed: see Walter Freeman, M.D., and James W. Watts, M.D., *Psychosurgery: Intelligence, Emotion, and Social Behavior Following Prefrontal Lobotomy for Mental Disorders* (1942).
170 Over half a century: This author was in a nearly empty restaurant in Nashville, the Pancake Palace, one afternoon in 1998. The only other customers were an elderly woman and a middle-aged woman in the next booth. The author overheard the older woman telling her daughter in mesmerizing detail about Rosemary Kennedy's lobotomy and how guilty she still felt. When the author introduced himself to the women, they were at first startled and worried that they had been followed. But when they understood, the elderly woman told her story in detail.
170 Rosemary was shipped away: According to the institution's records, Rosemary Kennedy did not arrive at her current home at what was then called the St. Coletta School for Exceptional Children in Jefferson, Wisconsin, until July 1949. Until then, according to Luella Hennessey Donovan, she was kept at Craig House. Rosemary's sister, Eunice Kennedy Shriver, says that she has no recollection of where Rosemary was kept.
171 "the Navy Department . . .": Rose Kennedy to "Dear Children," February 2, 1942, JFKPL.
171 "I am sorry Gunther's . . .": Rose Kennedy to "My Darlings," December 5, 1941, JFKPL.
171 "There is nothing comparable . . .": AWRJ, p. 35.

172 put in his hours: interview, S. A. D. Hunter, BP.
172 "He's coming . . .": Inga Arvad, unpublished memoir, NHP.
172 "the Royal Theatre . . .": Inga Arvad to John F. Kennedy, January 20, 1942, NHP.
173 "To an indefinite . . .": In the late 1990s the photo was in the possession of a collector, Ben Swearingen. Robert White showed the author a copy of the photo.
173 "the conversation slid . . .": FBI report, November 16, 1940, FBIFOI.
173 "lost control of myself . . .": W. H. Welch, FBI report, May 6, 1942, FBIFOI.
174 "a skirt chaser": interview, Ronald McCoy, BP.
174 "I've got another . . .": interview, Frank Waldrop, BP.
174 under FBI surveillance: FBI laboratory report, February 13, 1942, FBIFOI.
174 "He can help me . . .": W. H. Welch, FBI report, May 6, 1942, FBIFOI.
174 "We've got ten . . .": interview, Ronald McCoy, BP.
174 "I think he was . . .": LL interview with Betty Coxe Spalding.
175 "She said that he began . . .": LL interview with John White.
175 FBI inventoried her possessions: inventory of Inga Arvad possessions, n.d., FBIFOI.
175 "more possibilities than": D. M. Ladd to Mr. Kramp, January 9, 1942, FBIFOI.
176 Instead, Jack took what he: LL interview with John White.
176 "One of ex-Ambassador . . .": quoted in Hamilton, p. 438.
176 "He is full of enthusiasm . . .": Inga Arvad to John F. Kennedy, January 26, 1942, JFKPP.
177 "You are going away": Inga Arvad to John F. Kennedy, n.d., NHP.
177 "Maybe your gravest": Inga Arvad to John F. Kennedy, Wednesday (by textual analysis, probably dated January 28, 1942), JFKPL.
177 "stinking New Dealism": memo FBI director, "Re: Mrs. Paul Fejos, Espionage," February 1, 1942, FBIFOI.
177 "stinks of defeat": John F. Kennedy to Ralph Horton, n.d. (early 1942?), 1992, JFKPL.
177 "may call for us . . .": ibid.
178 "all around us . . .": John F. Kennedy to Lem Billings, February 12, 1942, JFKPL.
178 "Washington is begginning [*sic*]": John F. Kennedy to Rip Horton, n.d., NHP.
178 talked to the Church: FBI memo, February 17, 1942, FBIFOI.
179 "go to bed with him": FBI ARV translation, February 24, 1942, FBIFOI.
179 "If I were but . . .": Inga Arvad to John F. Kennedy, March 11, 1942, JFKPP.
179 "If you feel anything . . .": ibid.
180 "For God's sake!": Hamilton, p. 488.
180 "could be of real . . .": Franklin Roosevelt to Joseph P. Kennedy, March 7, 1942, RL.

180 He also had: James Landis to Joseph P. Kennedy, March 10, 1942, Landers Papers, LC.
180 "just be a hindrance . . .": Joseph P. Kennedy to Franklin Roosevelt, March 12, 1942, RL.
180 "unknown reliability": J. Edgar Hoover to Edwin M. Watson, April 20, 1942, FBIFOI.
181 "In the moving picture . . .": Joseph P. Kennedy, special service contact, Boston field office, October 11, 1945, FBIFOI.
181 "When I saw Mr. Roosevelt . . .": Joseph P. Kennedy to Lord Beaverbrook, August 12, 1942, Beaverbrook papers, KP.
181 "He has become disgusted . . .": Joseph P. Kennedy to Joseph P. Kennedy Jr., June 20, 1942, JFKPL.
181 "limping monkey . . .": ARV summary, July 24, 1942, FBIFOI.
182 an eighty-foot armed vessel: Dick Keresey, "Farthest Forward," *American Heritage*, July–August 1998.
183 "I swear to God . . .": Torbert Macdonald to John F. Kennedy, June 20, 1942, JFKPP.
183 "going into the Quartermaster Corps": Hamilton, p. 512.
183 "It was wintertime . . .": LL interview with Holton Wood.

11. A Brothers' War

185 told a new friend: interview, James Reed, BP.
185 his favorite book: interview, John F. Kennedy, JMBP.
185 "For the chosen few . . .": John Buchan, *Pilgrim's Way: An Essay in Recollection* (1940), p. 50.
186 "We made fun . . .": Ted Guthrie to John F. Kennedy, October 3, 1961, David Powers papers, JFKPL.
186 "I was only . . .": ibid.
186 "He suddenly threw . . .": John F. Kennedy to Lem Billings, May 6, 1943, JFKPP.
186 "lackadaisical way . . .": John F. Kennedy to "Dear Dad & Mother," May 14, 1943, JFKPP.
187 Jack willed himself: interview, John Iles, BP.
187 refusing to sign in: interview, Catherine Holway Kelley, BP.
187 That spring weekend Chuck Spalding: LL interview with Chuck Spalding.
187 "he will want to be back . . .": John F. Kennedy to "Dear Dad and Mother," May 14, 1943, JFKPL.
187 earning himself the nickname: interview, Charles Albert Harris, BP.
187 On the night of August 1, 1943: This account of the sinking of PT-109 is based primarily on interviews with the survivors conducted by Joan and Clay Blair Jr., BP; see also John Hersey, "Survival," *New Yorker*, June 17, 1944.
188 looked like a fishing bobber: interview, George Ross, BP.
190 He stopped swimming: Hersey, "Survival."

190 NAURO ISL NATIVE KNOWS POSIT: Robert J. Donovan, *PT-109: John F. Kennedy in World War II* (1961), p. 179.

190 suffering from fatigue: Bureau of Medicine and Surgery, Navy Medical Records, August 9, 1943, JFKPL.

191 "I can say in all honesty . . .": LL interview with Bryant Larson.

191 "There is no doubt . . .": Deputy Commander, South Pacific Force, to Vice Admiral R. S. Edwards, "Subject: Lieutenant (j.g.) Jack Kennedy, Commanding Officer, PT-109," January 11, 1944, DPP.

191 "Our reaction . . .": interview, George Ross, BP.

191 "He was furious!": interview, John Iles, BP.

191 "Americans can never . . .": John F. Kennedy to Inga Arvad, draft letter, RWC. Kennedy edited this phrase in his own hand and dropped it from the final letter.

192 "I received a letter . . .": John F. Kennedy to Inga Arvad, September 26, 1943, NHP.

192 "A number of my illusions . . .": ibid., n.d., NHP.

192 "Munda or any . . .": ibid.

193 "He never really . . .": John F. Kennedy to "Dear Mother and Dad," received September 12, 1943, JFKPL.

193 "I used to have . . .": John F. Kennedy to Inga Arvad, n.d., NHP.

193 "You said that . . .": ibid. and RWC.

194 "Previous to that . . ." John F. Kennedy to "Dear Mother and Dad," received September 12, 1943, JFKPL.

194 "When I read . . .": ibid.

195 "pride in . . .": quoted in Doris Kearns Goodwin, p. 647.

195 "Jack, you know . . .": Rose Kennedy to the family, October 9, 1942, JFKPL.

195 "When I returned home . . .": Joseph P. Kennedy to Joseph P. Kennedy Jr., August 31, 1943, JFKPP.

196 "With the great quantity . . .": Joseph P. Kennedy Jr. to "Dear Family," August 29, 1943. JFKPP.

196 sinking a ship and a half a day: information available at: Uboat.net/allies/documents/usaaf.

196 "In their long brotherly . . .": TR, p. 285.

196 almost started to cry: Doris Kearns Goodwin p. 623.

197 Joe Jr. snuck him onto: Searls, p. 196.

197 He talked about: LL interview with Mark Soden.

197 "He had a special . . .": ibid.

197 "Joe was what . . .": LL interview with Robert Duffy.

198 "They really don't . . .": Joseph P. Kennedy Jr. to "Dear Family," August 29, 1943, JFKPP.

198 "In my talks . . .": John Daly made these comments in his original statements for *As We Remember Joe*, the memorial tribute edited by John F. Kennedy. They were edited out of the final document. JFK personal correspondence, 1933–50, JFKPP.

198 five trips cross-country: Joseph P. Kennedy Jr. to "Dear Family," August 29, 1943, JFKPP.
198 Judge John Burns stood: Doris Kearns Goodwin, p. 663.
199 "To Ambassador . . .": Searls, pp. 202–3.
199 seven schools: RKHT, p. 42.
199 arrived in a checked coat: LL interview with David Hackett; and David Hackett, KLOH.
199 Bobby invited friends: LL interview with Sam Adams.
200 "I played . . .": quoted in RKHT, p. 46.
200 "the opportunity to meet . . .": Joseph P. Kennedy to Robert Kennedy, October 14, 1942, JPKP, HTF, pp. 549–50.
201 "if this *friend* of mine . . .": Robert F. Kennedy to David Hackett, April 3, 1944, ASP.
201 "He would turn away . . .": LL interview with Sam Adams.
201 "This boy will . . .": "Norton Says," newspaper clipping, September 10, 1943, KP.
201 Sam had a splendidly: LL interview with Sam Adams.
201 Bobby's sister Jean felt: interview, Jean Kennedy Smith, ASP.
201 "Of course they were . . .": Robert F. Kennedy to Dave Hackett, January 1945, ASP.
202 "I was paddled fifteen . . .": interview, Edward Kennedy, RCP.
202 One night at Fessenden: LL interview with Dan Burns.
202 "Your youngest brother . . .": Joseph P. Kennedy to Kathleen Kennedy, January 17, 1944, JFKPL.
202 "I succeed in dispersing . . .": Searls, p. 204.
204 "During the winter months . . .": Squadron diary, JFKPL.
206 "the wind . . .": Searls, pp. 221–22.
206 "My love life is still . . .": Joseph P. Kennedy Jr. to "Mother and Dad," January 31, 1944, JFKPL.
206 "Several people have . . .": Joseph P. Kennedy Jr. to "Mother and Dad," February 19, 1944, JFKPL.
206 "What you gonna . . .": LL interview with Robert Duffy.
206 he confided: LL interview with Lynn McTaggart; and Lynne McTaggart, *Kathleen Kennedy: Her Life and Times* (1983), p. 146.
207 "I'll still take . . .": *New York Sun,* dispatch dated May 1, 1944.
208 "I wonder if the next . . .": quoted in Doris Kearns Goodwin, p. 676.
208 "entitled to the . . .": Joseph P. Kennedy to Joseph P. Kennedy Jr., February 21, 1944, JFKPL.
208 "let all the rest . . .": Doris Kearns Goodwin, p. 676.
209 "Never did anyone . . .": AWRJ, p. 54.
209 "HEARTBROKEN. FEEL . . .": quoted in Doris Kearns Goodwin, p. 677.
209 "I have finished . . .": Joseph P. Kennedy Jr. to "Dear Mother and Dad," May 8, 1944, JFKPL.
210 On his last mission: Searls, p. 237.

210 "He seems to be . . .": Joseph P. Kennedy to Joseph P. Kennedy Jr., July 19, 1944, JFKPL.

210 "I think at this point . . .": Robert F. Kennedy to Joseph P. Kennedy Jr., n.d., ASP.

211 "pride—his sense of . . .": miscellaneous JFK correspondence, n.d., JFKPL.

212 "the whole squadron . . .": Joseph P. Kennedy Jr, to John F. Kennedy, August 12, 1944, JFKPL.

213 Timilty thought: interview, Joseph Timilty, BP.

213 "He . . . is obviously . . .": "Special Examination and Treatment Request," August 4, 1944, U.S. Navy Medical Records, JFKPL.

214 As Jack lay: interviews, Leonard and Kate Thom, BP.

214 On the evening before: Jack Olsen, *Aphrodite: Desperate Mission* (1970), p. 223.

214 Late that afternoon: Searls, pp. 265–66.

12. A New Generation Offers a Leader

219 "Children, your brother . . .": AWRJ, p. 207.

220 When Joe called: LL interview with Mary Lou McCarthy.

220 "Joe's death has . . .": Joseph P. Kennedy to James Forrestal, September 5, 1944, Mudd Manuscript Library, Princeton University.

220 "It came at a time . . .": interview, John F. Kennedy, JMBP.

220 "enjoyed great health": AWRJ, p. 5.

221 "You are the . . .": Mike Grace to John F. Kennedy, n.d., JFKPL.

221 "the best ones . . .": Mrs. John S. Pillsbury to John F. Kennedy, August 16, 1944, JFKPL.

221 "a completeness to . . .": AWRJ, p. 5.

221 "There must be . . .": Harriet Price to John F. Kennedy, n.d., JFKPL.

221 "You will have some . . .": Barbara Ellen Spencer to John F. Kennedy, August 15, 1944, JFKPL.

221 "natural cynicism": Joseph P. Kennedy to Lord Beaverbrook, October 23, 1944, NHP.

222 "Who'd you ever . . .": interview, Joe Kane, KP.

222 "the Hopkins . . .": diary notes on the 1944 political campaign, HTF, p. 608

222 "felt that Roosevelt": HTF, p. 608.

222 "The people one . . .": Kathleen Kennedy to "Dearest Family," September 15, 1946, RFK papers, JFKPL.

223 "Colitis chronic": JFK medical record, Bureau of Medicine and Surgery, U.S. Navy Department, December 15, 1944, JFKPL.

223 "Sometime in the . . .": John F. Kennedy to Red Fay, November 21, 1944, PFP.

223 "yellow as saffron . . .": interview, J. Patrick Lannan, BP.

223 "It just seems . . .": Kathleen Kennedy to John F. Kennedy, October 31, 1944, JFKPL.

223 patterned in part: interview, John F. Kennedy, JMBP.
223 "I am returning . . .": John F. Kennedy to Lem Billings, February 20, 1945, JFKPL.
224 "Oh, you mean . . .": interview, J. Patrick Lannan, BP.
224 "Frankly, don't think . . .": Joseph P. Kennedy to Red Fay, March 26, 1945, PFP.
224 "Everyone evidently . . .": Robert Kennedy to John F. Kennedy, envelope dated January 1, 1945, JFKPL.
224 "Anyways their . . .": John F. Kennedy to Lem Billings, February 20, 1945, JFKPL.
224–25 "I took a piece . . .": ibid.
225 "he had it . . .": interview, Chuck Spalding, BP.
226 "Kennedy will not . . .": Arthur Krock, KLOH.
226 "an international . . .": *New York Journal-American*, May 2, 1945, BP.
226 "throwing curves": ibid., May 14, 1945.
226 "juggled the ball": ibid., May 16, 1945.
226–27 "Americans can now . . .": ibid., May 4, 1945.
227 "the product of . . .": ibid., May 20, 1945.
227 "Our preoccupation . . .": ibid.
227 tired of nothing: Hearst Newspapers, July 27, 1945.
228 They played with: interview, J. Patrick Lannan, BP.
228 "He really didn't . . .": interview, Pat Stammers, BP.
228 "All the centers . . .": Deirdre Henderson, ed., *Prelude to Leadership: The European Diary of John F. Kennedy* (1995), p. 43.
228 "raping and looting": ibid., p. 45.
228 "easily won the": ibid., p. 15
228 "You can easily . . .": ibid., p. 74.
229 "the eventual . . .": ibid., p. 7.
229 "Pappy's eyes . . .": Paul B. Fay Jr., *The Pleasure of His Company* (1966), p. 132, and LL interview with Paul B. Fay.
229 Lem thought: Lem Billings to John F. Kennedy, January 1, 1946, JFKPP.
230 "was to convince . . .": *Time*, July 1, 1946.
230 "aggressively shy": interview, David Powers, BP.
230 "I think I know . . .": Kenneth P. O'Donnell and David F. Powers, *Johnny, We Hardly Knew Ye: Memories of John Fitzgerald Kennedy* (1972), p. 54.
231 They told Jack: LL interview with Mark Dalton.
231 Joe Kane, who was: Hamilton, p. 674.
232 Joe had Eddie Moore: interview, David Powers, BP.
232 When Powers offered: ibid.
232 "They gave me favors": Steve Buckley, "The Other Joe Russo," *Boston* magazine, June 1993.
232 "Naval hero of the South Pacific": news release for weekly papers from Kennedy for Congress headquarters, n.d., JFKPP.
233 Red was a talker: LL interview with Paul B. Fay.
233 "I am not sure . . .": Robert F. Kennedy to Dave Hackett, April 1945, ASP.

233 "I know that . . .": Robert F. Kennedy to Joseph P. Kennedy, n.d., on Harvard Law School stationery, ASP.
234 "from the lowest grade . . .": RKHT. p. 60.
234 "mentally the most . . .": John F. Kennedy speech to Veterans of Foreign Wars, July 2, 1946, DPP, JFKPL.
235 "I have noticed . . .": John F. Kennedy, speech dated 1945 or 1946, BP.
235 "If we turn our . . .": John F. Kennedy, VFW speech, 1946, DPP.
236 "You don't feel good . . .": Thomas Broderick, KLOH.
236 sat soaking: interview, Paul Fay, BP.
236 One afternoon: TEEK, p. 48.
236 "Sinatra! Sinatra!": untitled clipping, n.d., in JFK pre-presidential scrapbook from December 1945–46, JFKPL.
237 "He appeared to me . . .": SJFK, p. 161.
237 "I got the impression . . .": LL interview with Mark Dalton.

13. A Kind of Peace

238 any student who lived: *Harvard Crimson*, freshman issue, 1946, n.d., HUA.
238 Men found themselves: "War and Peace: Five Years, Two Classes in Retrospect," *Harvard Album*, DHP.
238 There were 659 Harvard: *Harvard Crimson*, January 6, 1947, HUA.
239 As a 165-pound: Harvard University 1947 roster, HUA.
239 "For Christ's sake, would . . .": LL interview with Wally Flynn.
240 "I think my leg's . . .": ibid.
240 "Oh, those guys were . . .": ibid.
240 Harvard had a predominantly: On April 6, 1948, the *Harvard Advocate* took a presidential straw poll in all the houses of Harvard College. The three leading candidates were Republicans. Harold Stassen, Dwight Eisenhower, and Arthur Vandenberg received 55.5 percent of the vote. Henry Wallace, a Democrat, was fourth with 12.5 percent, while President Truman received a pathetic 4.4 percent of the 1,464 votes cast. HUA.
240 Sam Adams, Bobby's: LL interview with Sam Adams.
241 "He was one of us": LL interview with Chuck Glynn, Wally Flynn, Sam Adams, Nick Rodis, and Paul Lazzaro.
241 "Nick, I think he . . .": LL interview with Wally Flynn.
241 When the team played: LL interview with Chuck Glynn.
241 Bobby's football teammate: ibid.
242 "people like the Jews . . ." : John Deedy, "Whatever Happened to Father Feeney," *The Critic*, May–June 1973, ASP.
242 "What am I gonna . . ." : LL interview with Chuck Glynn.
242 "tough and rough . . .": RKHT, p. 68.
242 Bobby's friends: LL interview with Chuck Glynn.
243 "Bobby and his mother . . .": LL interview with Wally Flynn.
243 In his postwar tenure: Harvard College Record 1948, ASP.

243 "How long would . . .": LL interview with Billy Sutton, Billy Sutton KLOH, and Chris Matthews, *Kennedy and Nixon: The Rivalry That Shaped Postwar America* (1996), p. 44.
244 unemployment stood: John Lewis Gaddis, *We Now Know: Rethinking Cold War History* (1997), p. 47.
244 "From Stettin in . . .": ibid., p. 41.
244 "Was it politically . . .": *Time*, March 24, 1947.
244 Joe was all for: ibid.
245 "free peoples . . .": Gaddis, p. 49.
245 "The greatest danger . . .": John F. Kennedy, "Aid for Greece and Turkey," Record of House of Representatives, April 1, 1947, BP.
246 Jack had Mark Dalton: LL interview with Mark Dalton.
246 "There is no need to . . .": John F. Kennedy, "Labor-Management Relations Act, 1947," Record of House of Representatives, April 16, 1947, BP.
246 "Could you tell . . .": Paul F. Healy, "Galahad in the House," *The Sign*, July 1950, BP.
247 "an effective anti-Communist . . .": ibid.
247 The union was: *Time*, March 31, 1947.
247 in 1941, at the time: SJFK, p. 178.
247 "The responsibility for . . .": quoted in Seymour Topping, *Journey Between Two Chinas* (1972), p. 141.
247 "Jack was fearless . . .": LL interview with Mark Dalton.
248 "Has anybody talked . . .": quoted in SJFK, p. 183.
248 "one field in which . . .": John F. Kennedy, speech, September 28, 1946, DPP.
248 "Everything you said . . .": JFK to Clare Boothe Luce, January 21, 1947, Clare Boothe Luce papers, LC.
249 The Associated Press: Leonora Ross, "Boston Girls Spurn Rich Bachelors, They Prefer Personality," JFK scrapbooks, 1946–48, JFKPL.
249 "Joe was using me . . .": LL interview with George Smathers.
249 On one occasion: interview, Ralph Horton, BP.
249 "Guess I just haven't . . .": Paul I. Murphy, "Unmarried Millionaires: Jack Kennedy," JFK scrapbooks, 1946–48, JFKPL.
250 "He was a guy . . .": LL interview with George Smathers.
250 "Jack was crippled . . .": ibid.
251 "I spent . . .": John F. Kennedy to Professor James Burns, August 25, 1959, JMBP.
251 "When we got home . . .": LL interview with Pamela Churchill Harriman.
252 He may well have: Dr. Elmer Bartels believed that Jack could have had the disease no more than a year before it was diagnosed. Dr. Dorothea E. Hellman, who knew about his illness as a faculty member at the Harvard Medical School and later observed his treatment in Washington, was convinced that Jack's illness had begun years before his attack in London. "At that time he was already pigmented and had obviously lost weight," Dr. Hellman stated to Joan and Clay Blair Jr. "Many subsequent photographs confirm this impres-

sion and can be added to the . . . multiple puzzling illnesses that required long stays in hospitals both in New England and in the United Kingdom." BP.

252 *Queen Elizabeth*, where: C. A. Buchanan to Secretary of the Navy James Forrestal, cable, October 7, 1947, James Forrestal papers, Mudd Library, Princeton University.

252 the photos of Jack: *Fitchburg Sentinel,* October 10, 1947, and *Boston Post,* October 17, 1947.

253 "He had to take . . .": interview, Dr. Elmer C. Bartels, BP.

253 "agitation, euphoria, insomnia . . .": "Steroids," *Mayo Clinic Health Letter*, September 1994, www.mayohealth.org.

253 As it was: This section, as well as all other parts of *The Kennedy Men* dealing with JFK's health, have been read by Dr. Mauro Di Pasquale, a world-renowned expert on steroids.

253 "Jack, the way . . .": Paul Fay, unedited manuscript, *The Pleasure of His Company*, Myrick E. Land papers, Boston University Special Collections.

254 "They actually didn't . . .": interview, Dick Clasby, RCP.

254 "You better stop that": interview, Joseph C. Kernell, BP.

254 "Excuse me . . .": LL interview with Bruce Sundlun.

255 At Catholic Cranwell: Adam Clymer, *Edward M. Kennedy: A Biography* (1999), p. 17.

255 In the first two summers: LL interview with Joe Gargan.

256 "He was a marvelous . . .": interview, Edward Kennedy, RCP.

256 "Finally it stalled . . .": Edward M. Kennedy to Joseph P. Kennedy, April 11, 1948, RCP.

256 "This morning I served . . .": ibid., April 25, 1948, RCP.

257 Without even one visit: TFB, p. 45, and *The Economist*, February, 4, 1977.

257 $35 million: Kessler, p. 254.

257 Joe gave one-quarter: *Washington Star*, March 22, 1947.

257 In 1946, Joe divested: HTF, p. 521.

258 Jack was there: *Holyoke Transcript*, August 13, 1946.

258 $2,609,000: press release, February 16, 1956. "A copy of my article, in its present form, is enclosed," Cushing wrote Joe. "If you have any changes to suggest I will be very happy to follow them." Cushing to Joseph P. Kennedy, December 15, 1955, PC.

258 "They [the gifts] are made . . .": *The Pilot*, February 11, 1956.

259 "If one has been . . .": Joseph P. Kennedy to Archbishop Richard Cushing, February 13, 1956. PC.

259 "But it's not just . . .": *Boston Herald*, May 20, 1948, HUA.

260 "Time is man's . . .": TFB, p. 122.

261 "As far as I remember . . .": Mrs. Christopher Bridge, KLOH.

261 He had left: TEEK, p. 78.

261 "He stood alone": quoted in RKHT, p. 78.

261 He thought to himself: Joseph P. Kennedy to the Duchess of Devonshire, September 1, 1948, HTF, p. 637.

262 "broke down like . . .": quoted in Jerry Oppenheimer, *The Other Mrs. Kennedy* (1994), p. 117.
262 Twice authorities picked: Robert F. Kennedy diary, March 29, 1948, Palestine, JFKPL.
262 "Met officers . . .": ibid.
263 "Many of the leading . . .": Israel, 4–8, ASP.
263 "I do not think . . .": *Boston Post*, June 6, 1948.
263 "all eyes will now . . .": *Boston Sunday Advertiser*, January 16, 1949.
263 Mindszenty was a great: Donald F. Crosby, *God, Church, and Flag: Senator Joseph R. McCarthy and the Catholic Church, 1950–1957* (1978), p. 12.
264 "The farmhouse he . . .": Robert F. Kennedy diary, June 1948, JFKPL.
264 "freckled face . . .": Joan Winmill Brown, *No Longer Alone* (1975), p. 47.
264–65 "He talked about . . .": LL interview with Joan Winmill Brown.
265 "He was so . . .": ibid.
265 When a handsome: Leamer, p. 390.
268 On one occasion: Oppenheimer, p. 113.
268 "It was very nice . . .": LL interview with Nick Rodis.
268 "I think Ethel was . . .": Gerald Tremblay, KLOH,
269 "Don't cry . . .": quoted in Oppenheimer, p. 119.
269 "about my coming . . .": ibid.
270 "My financee [*sic*] . . .": RKHT, p. 88.

14. The Grease of Politics

271 they had consumed: Paul E. Murphy to John F. Kennedy, June 30, 1950, and Alfred Martinez to John F. Kennedy, July 14, 1950, JFKPP.
272 kicking him full force: RKHT, p. 88.
272 Without Ethel's knowledge: Oppenheimer, p. 131.
272 his bride traveled: Robert F. Kennedy to Joseph P. and Rose Kennedy, Honolulu, Hawaii, HTF, p. 643.
273 "Wally, come in": LL interview with Wally Flynn.
273 "You're Mary": Mary Davis, KLOH,
274 "What is the rationalization": Robert F. Kennedy, "A Critical Analysis of the Conference at Yalta, February 4–11, 1945," University of Virginia Law Library, DHP.
274 "Mr. Syngman Rhee's . . .": *Virginia Law Weekly*, December 14, 1950, DHP.
275 The students were all: RKHT, p. 86.
276 Bobby's football teammate: LL interview with Wally Flynn.
276 the amenable friend sat up front: TEEK, p. 79.
277 "How are you . . .": Adam Clymer interview with Senator Edward Kennedy.
278 "Initially, my father . . .": interview, Edward Kennedy, RCP.
278 "The father was terribly . . .": TEEK, p. 81.
278 "If I had a . . .": interview, Edward Kennedy, RCP.
278 "They had three . . .": ibid.

279 "However, upon considering . . .": Edward Kennedy to "Dear Mother and Dad," n.d., RCP.

279 Scott Fitzgerald's wife, Zelda: Peter D. Kramer, "How Crazy Was Zelda?" *New York Times Magazine*, December 1, 1996.

280 Archbishop Richard Cushing had told him: Doris Kearns Goodwin, p. 642.

280 arranged for the building: This information on Rosemary Kennedy's treatment at Craig House and her arrival at St. Coletta's is based on the institution's records as stated by Alan K. Borsari, Alverno House director in 1993 when the author visited the institution.

280 The first visit: Bob Healy of the *Boston Globe* accompanied JFK on a campaign trip to Wisconsin in 1958 when he says JFK made the visit. LL interview with Bob Healy.

280 missing over one-quarter: *Time*, November 7, 1960.

280 as often as not with one woman: The pattern is discernible in JFK's many cables: Waldorf-Astoria Hotel, April 9, 1949 ("reserving single room for Mrs. Pamela Farrington"), to Miss Jane Blodgett in New York City, April 27, 1949 ("Dear, Could you go out Friday night"), to Waldorf Astoria Towers, May 11, 1949 (requesting "double room for Miss Ilse Bay"), and to Mrs. Adele O'Connor, June 9, 1949 (saying he was coming to New York and would she "be interested in going to Cape for few days"). JFKPP.

281 He was an internationalist: Clay Blair Jr., "The Evolution of Cabot Lodge," *Saturday Evening Post*, October 22, 1960.

281 "My God, man . . .": interview, George Smathers, BP.

281 "Florida will not . . .": quoted in Robert G. Sherill, "The Power Game: George Smathers, the Golden Senator from Florida," *The Nation*, December 7, 1964.

281 "I'm running": interview, George Smathers, BP.

282 "turned a strong . . .": Joseph Alsop, with Adam Platt, *I've Seen the Best of It* (1992), p. 411.

282 "He had leukemia . . .": interview, Rose Kennedy, RCP.

282 "intermittent slight . . .": Dr. Vernon S. Dick to Dr. William P. Herbst Jr., March 20, 1953, JFKPP.

283 needed an imprimatur: RKIHOW, p. 436.

283 "Yugoslavia—Belgrade—Stones . . .": John F. Kennedy's diary of his 1951 trips to Europe and Asia has been ably transcribed by Robert White, who has made sense of JFK's often almost illegible handwriting, RWC and JFKPL.

283 "the Italian economy . . .": ibid.

284 "Why should they . . .": "Statement of the Honorable John F. Kennedy," February 22, 1951, BP.

284 it was not without reason: "Kennedy Acquiring Title, 'America's Younger Statesman,' " *Boston's Political Times*. Quoted in SJFK, p. 220.

284 "a pain in the ass": Peter Collier and David Horowitz, *The Kennedys: An American Drama* (1984), p. 181.

284 "courage is the virtue": PIC, p. xi.
284 had been on crutches: SJFK, p. 225.
285 "Eisenhower looking very fit . . .": John F. Kennedy, 1951 diary, October 3, 1951, JFKPL.
286 "You can feel . . .": ibid., n.d.
286 "It was almost . . .": RKIHOW, p. 436.
286 "It depended on . . .": John F. Kennedy 1951 diary, RWC and JFKPL.
287 the young man had : LL interview with Wilson Gathings.
287 "*I met murder* . . .": quoted in John F. Kennedy diary, October 7, 1951?, JFKPL.
287 "Drove to Haifa . . .": Robert F. Kennedy diary, October 5, 1951?, JFKPL.
288 Four days after: JFK and RFK met with the Pakistani leader on October 12, 1951 (JFK diary). Prime Minister Liaquat Ali Khan was murdered on October 16, 1951.
288 "Bored by westerners . . .": John F. Kennedy 1951 diary, n.d., RWC and JFKPL.
288 "when [Indian] independence . . .": ibid.
289 "would not eat . . .": ibid.
289 "I would like to . . .": LL interview with Seymour Topping, p. 142.
289 "Two years ago . . .": ibid.
290 "One great reason . . .": John F. Kennedy 1951 diary, RWC and JFKPL.
290 "that French realize": ibid.
291 "And everybody there . . .": RKIHOW, p. 438.
291 "He'd be in the hospital . . .": Grace Burke, KLOH.
291 "Foreign policy today . . .": John F. Kennedy, "Report on the Trip to the Middle and Far East." Mutual Broadcasting Network, November 15, 1951, BP.
292 "Our resources . . .": address to the Boston Chamber of Commerce, November 19, 1951, AAML.
292 "unconscious of the fact . . .": ibid
292 "It is France . . .": John F. Kennedy, address to the Boston Chamber of Commerce, November 19, 1951, AAML.
293 Jack spoke as: *Meet the Press,* December 2, 1951, BP.
293 "young college graduates . . .": *Lowell Sun*, December 9, 1951, and *Springfield Union*, December 19, 1951, quoted in Ronald J. Nurse, "America Must Not Sleep: The Development of John F. Kennedy's Foreign Policy Attitudes, 1947–1960," thesis, Michigan State University, 1971, p. 89.
293 "the closest . . .": interview, David Powers, BP.
294 had such a strong Republican: David Halberstam, *The Fifties* (1993), p. 224.
294 "no one will ever know . . .": quoted in "Report of Meeting with Wickliffe W. Crider and John Elliott of the Staff of BBD&O in New York City, Tuesday, February 26, 1952, Some Observations and Recommendations" JFKPL.

294 "marked by informality . . .": Mark Dalton to Joseph P. Kennedy, February 20, 1952, and "Report of Meeting with Wickliffe W. Crider and John Elliott of the Staff of BBD&O in New York City, Tuesday, February 26, 1952, Some Observations and Recommendations," JFKPL.
295 "Oh, there's Kennedy!": LL interview with Mark Dalton.
295 "The father wanted me . . .": ibid.
295 "Mark didn't like . . .": interview, John Galvin, BP.
296 "Mark Dalton was . . .": oral history, Robert F. Kennedy, ASP.
296 "before the revolution . . .": John Droney, KLOH.
296 He got right to it: interview, Anthony Gallucio, BP.
296 "like he burned my bridges": LL interview with Sam Adams.
296 "How many people . . .": interview, Joe Gargan, HP.

15. The Golden Fleece

298 "He was all . . .": interview, Congressman John McCormack, BP.
298 Jack read the extraordinary: *Congressional Quarterly News Feature*, October 3, 1952, pp. 964–65, JFKPP.
298 "His theme was to . . .": Edward C. Berube, KLOH,
298 Lodge had started: interview, Henry Cabot Lodge, BP.
299 In August he was sick: Vernon S. Dick to Dr. William P. Herbst Jr., March 20, 1953, JFKPP, and Dr. T. A. Morrissey to John F. Kennedy, December 21, 1954, RWC.
299 From then on: SJFK, p. 239.
299 Indeed, when Lodge: interview, Henry Cabot Lodge, BP.
299 card-carrying Communists: There is a debate over the number of Communists McCarthy named. This is the smallest figure mentioned.
300 Michael S. Sherry, *In the Shadow of War: The United States Since the 1930s* (1995), p. 170.
300 February 1947 instituted: Truman's loyalty program provided a rude approximation of the problem of Communists in government. Of the 4,722,278 names checked, the FBI found information serious enough to refuse government positions to only 575 individuals. Most of them were considered "security risks," perhaps having had contacts with a questionable person, and only a few were named "loyalty risks," a far more serious accusation. Another 3,634 government employees quit before the government finished their dossiers.
300 Cincinnati Reds renamed: Sherry, p. 171.
301 "The theme of today . . .": John F. Kennedy, speech to 1952 Newton College of the Sacred Heart graduation, JFKPP.
301 "knew Joe pretty well . . . ": John P. Mallan, "Massachusetts: Liberal and Corrupt," *The New Republic*, October 13, 1952, BP.
301 750,000 Irish Catholics: SJFK, p. 245.
302 "the *Rashomon* drama . . . ": quoted in James Thomas Gay, "1948: The Alger Hiss Spy Case," thehistorynet.com.

302 "How dare you . . .": SJFK, p. 245.
302 "Oh, Bob, come . . .": LL interview with Sam Adams.
303 "I told you before . . .": interview, Phil David Fine, BP.
304 "Rabbi John": SJFK, p. 248.
304 The campaign: Jackson Holtz, KLOH.
304 Joe had offered: Joseph P. Kennedy to Arthur Krock, June 24, 1936, HTF, p. 186.
304 enough to ensure: interview, Henry Cabot Lodge, BP.
304 Lodge told the good news: interview, Joseph Timilty, BP.
305 "I don't remember . . .": RKIHOW, p. 444.
305 "You know, we . . .": SJFK, p. 511.
305 he reported: Ralph Coghlan to John F. Kennedy, July 15, 1952, BP.
305 "I will work out . . .": TFB, p. 127.
305 "At last the Fitzgeralds . . .": Collier and Horowitz, p. 189.
305 Jack sang "Sweet Adeline": John Droney, KLOH.
306 Thomas "Tip" O'Neill, insisted years: Congressman Thomas O'Neill, interview with Jerry Williams on WBZ-TV, JFKPP, and interview, Thomas O'Neill, BP.
306 "Paul is not going . . .":TFB, p. 246.
306 During the campaign: Anthony Gallucio's name has been spelled various ways in previous books. This is the spelling used at the JFKPL.
306 "I'm going to run . . .": interview, Anthony Gallucio, BP.
306 "very quickly and . . .": interview, Jean Kennedy Smith, ASP.
306 "What are you . . .": Jean Stein and George Plimpton, eds., *American Journey: The Times of Robert Kennedy* (1970), p. 45.
307 "aiming for the post . . .": *Cape Cod Times*, December 24, 1952.
307 "I'll talk to [Senator John] McClellan . . .": LL interview with Maurice Rosenblatt.
307 "political, not ideological": TKL, p. 41.
307 "I walk in . . .": LL interview with Maurice Rosenblatt.
309 "Teddy had not been . . .": L. B. Nichols to Clyde Tolson, May 11, 1954, "subject: Edward M. Kennedy," FBIFOI.
309 "apparently some of . . .": ibid.
310 "was angling his columns . . ." J. J. Kelly to J. Edgar Hoover, August 28, 1953, FBIFOI.
310 "the most vocal . . .": H. G. Foster to J. Edgar Hoover, July 2, 1954, FBIFOI.
310 "Knowing him from . . .": interview, David Powers, BP.
311 "Dad says don't . . .": interview, Charles Bartlett, BP.
311 "I've got no . . .": interview, Anthony Gallucio, BP.
311 "Well, I can't pay . . .": Mary Davis, KLOH.
311 "Mary, you wouldn't . . .": interview, Thomas P. O'Neill, BP.
312 When Sorensen was: SJFK, p. 262.
312 "far more interested . . .": interview, Robert F. Kennedy, ASP.

313 After this first: James M. Landis, to Joseph P. Kennedy, January 21, 1953, Landis papers, LC.
313 nearly empty Senate: interview, John F. Kennedy, BP.
313 "The Atlantic Monthly . . .": Theodore C. Sorensen to James M. Landis, November 1, 1953, James M. Landis papers, LC.
314 "chattel . . .": interview, James Reed, BP.
314 "I remember . . .": interview, Ben Smith, BP.
314 "You know, they're going . . .": interview, Charles Bartlett, BP.
315 "Excuse me, officer . . .": Hugh D. Auchincloss, "Growing Up with Jackie, My Memories 1941–1953," JFKPL.
315 decked out her stepbrother: ibid.
315 When she went to see: ibid.
315 "If I could be a . . .": quoted in AWRH, pp. 53–54.
316 "If you're so much . . .": Janet Lee Auchincloss, KLOH.
316 The prospective groom: John F. Kennedy to Paul Fay, PFP.
316 "Flo Pritchett's birthday! SEND DIAMONDS": Blair and Blair, p. 553.
316 "Florence Pritchett was a serious . . .": interview, Paul Fay, BP.
316 "I was very stuck . . .": interview, John F. Kennedy, JMBP.
317 "too young and too old": John F. Kennedy to Paul Fay, n.d., PFP.
317 "I am hoping . . .": Joseph P. Kennedy to Torbert Macdonald, July 22, 1953, HTF, p. 662.
317 "Jack! What are . . .": This section on the relationship between John F. Kennedy and Gunilla Von Post is based on LL interviews with Von Post and on her book, *Love, Jack* (1997).
319 "Their wealth is from. . . .": original manuscript of Fay, *The Pleasure of His Company*, PFP.
320 "Oh Mummy, you . . .": TFB, p. 232.
320 "When he turned . . .": ibid.
320 "I was seated next . . .": John Droney, KLOH.
320 "I was seated next . . .": LL interview with Paul Fay, CBS interview, HP, and Fay, p. 145.
321 "Wish you were here. Jack": AWRH, p. 80.
321 on to Pebble Beach: LL interview with Paul Fay.
322 "the kind of honeymoon . . .": Fay, p. 141.
322 He even may: LL interview with Jewel Reed.
322 "*He would find love* . . .": quoted in AWRH, p. 85.

16. Aristocratic Instincts

323 "I don't know why!": Lord Harlech, KLOH.
323 Jackie worked on a private: AWRH, p. 87.
323–24 a drawing of a: interview, Langdon Marvin Jr., HP.
324 "a sad look . . .": interview, Charles Bartlett, BP.
324 sell many of her wedding gifts: LL interview with Carey Fisher.

324 "Jack went crazy . . .": Collier and Horowitz, p. 196.
324 "Miss New Zealand isn't . . .": J. V. Bouvier III to Senator John F. Kennedy, June 13, 1956, JFKPP.
324 "Harvard's outstanding [1941] graduate": Edward B. Lockett, "Life's Reports: Progress of a Harvard Man," *Life*, February 7, 1944.
325 "Do you remember . . .": John F. Kennedy to Gunilla Von Post, March 2, 1954, courtesy Gunilla Von Post.
325 "I thought I might get . . .": ibid., June 28, 1954, courtesy Gunilla Von Post.
325 had to cable her: interview, Gunilla Von Post, and Von Post, pp. 39–49.
325 "We plopped him . . .": interview, Langdon Marvin Jr., HP.
325 "Renaissance ideal . . .": YM, p. 8.
325 Down in Palm Beach: There have been some suggestions that Kennedy was reading *Young Melbourne*. Cecil's second volume had just been published, and it is far more likely that he was reading volume 2.
326 "all salt and sunshine . . .": LM, p. 13.
326 "more interesting because . . .": handwritten notes by JFK while sitting by the pool late in 1955, JFKPP.
326 "Jack did have aristocratic . . .": LL interview with Charles Bartlett.
327 He knew—and this rankled: LL interview with Paul B. Fay.
327 "tend[s] to lack . . .": John F. Kennedy essay or speech, box 39, JFKPP.
327 "Life is real . . .": Henry Wadsworth Longfellow, "A Psalm of Life" (1839).
327 "Let us have wine . . .": Lord Byron, *Don Juan*, canto II.
327 "rather pompous": handwritten notes by JFK while sitting by the pool late in 1955, JFKPP.
328 "gratuitous insult": Chalmers M. Roberts, KLOH.
328 "I am frankly . . .": speech, U.S. Senate, April 6, 1954, www.geocities.com.
328 "a kind of disease . . .": quoted in Gaddis, p. 184.
329 "Vietnam represents . . .": quoted in Thomas C. Reeves, *A Question of Character: A Life of John F. Kennedy* (1991), p. 254.
329 "Fighting thousands of miles . . .": notes for speech, JFKPP.
329 embarrassed to be seen: Crosby, p. 35.
330 "I want to give you . . .": Sidney Zion, *The Autobiography of Roy Cohn* (1988), quoted in C. David Heymann, *RFK: A Candid Biography of Robert F. Kennedy* (1998) p. 83.
330 "whenever I said . . .": ibid., p. 84.
330 "As you were on . . .": memorandum for Senator Henry M. Jackson, June 2, 1954, RFK Papers JFKPL.
331 "I want you to tell . . .": New York *Daily News*, June 12, 1954.
331 Both the children and the dogs: LL interview with David Hackett.
332 "Bobby . . . could play . . .": Clymer, p. 20.
332 Not only was Holcombe: ibid.
332 "I think academic . . .": Lester David, *Ted Kennedy: Triumphs and Tragedies* (1972), p. 70.
333 "seems a little technical . . .": Robert F. Kennedy to Edward M. Kennedy, December 14, 1953, ASP.

333 During freshman year: LL interview with Claude Hooton Jr.
333 Teddy and Hackett: TEEK, p. 85.
333 He and Hooton went off: LL interview with Claude Hooton Jr.
334 "I talked to Dad . . .": Robert F. Kennedy to Edward M. Kennedy, January 11, 1955, RFK papers, JFKPL.
334 "Teddy had the . . .": interview, Rose Kennedy, RCP.
334 "So I ended up . . .": Adam Clymer, interview with Edward Kennedy.
335 "I think . . . [Jack's] most . . .": interview, Edward Kennedy, RCP.
336 shouted too loud: Clymer, p. 21.
336 "My brothers and I were . . .": Adam Clymer, interview with Edward Kennedy.
336 "Janet": LL interview with Janet Des Rosiers Fontaine.
337 "He'd say, 'Well, you know . . .' ": ibid.
338 "One night I wasn't home . . .": ibid.
338 "Have a long, happy . . .": Rose Kennedy to Janet Des Rosiers, m.d., courtesy Janet Des Rosiers Fontaine.
339 The Kennedys had been members: interview with Lamar Harmon, resident manager of Everglades Club, conducted by FBI, Joseph P. Kennedy, special inquiry, White House, February 17, 1956, FBIFOI. In February 2001, an Everglades member, Dick Prentice, confirmed the Kennedys' membership with the club's current secretary.
339 "You know, I would like . . .": LL interview with Janet Des Rosiers Fontaine.
340 "I don't know if . . .": interview, Dr. Elmer C. Bartels, BP.
340 "deemed dangerous . . .": James A. Nicholas, J.D, Charles L. Burstein, M.D., Charles J. Umberger, Ph.D., and Philip D. Wilson, M.D., "Management of Adrenocortical Insufficiency During Surgery," *Archives of Surgery*, November 1955.
340 Jackie was sitting beside: James MacGregor Burns, *John Kennedy: A Political Profile* (1960), p. 156.
341 "So I was just . . .": interview, John F. Kennedy, JMBP.
341 "acute attack of . . .": Dr. Thomas A. Morrissey to John F. Kennedy, December 21, 1954, RWC.
341 "Tell them you're . . .": LL interview with Priscilla Johnson.
341 "We stay in session . . .": John F. Kennedy to Gunilla Von Post, December 18, 1954, courtesy Gunilla Von Post.
341–42 "You could see . . .": Janet Lee Auchincloss, KLOH.
342 "I think your . . .": interview, John F. Kennedy, JMBP.
342 believed that the Kennedy: LL interview with Mark Dalton.
342 "it would be . . .": David M. Oshinsky, *A Conspiracy So Immense: The World of Joe McCarthy* (1983), p. 489.
343 "So I was rather in ill grace . . .": interview, John F. Kennedy, JMBP.
343 Jack was wrapped: New York *Daily News*, December 22, 1954.
344 "How is it now?": LL interview with Chuck Spalding.
344 "anybody we did a favor . . .": John F. Kennedy, memorandum, February 16, 1955, RWC.

344 "I don't know . . .": ibid., March 7, 1955.
345 "he needed someone . . .": LL interview with Gunilla Von Post.
345 "I am anxious . . .": John F. Kennedy to Gunilla Von Post, n.d., courtesy Gunilla Von Post.
345 his crutch collapsed: *Boston Globe*, May 22, 1955.
345 He looked: Dr. Janet Travell, KLOH.
346 "I don't want her . . .": AP, November 20, 1968, and *Washington Star* files, Martin Luther King Library.
346 she diagnosed that one leg: One of Kennedy's legs may have been shorter than the other, but it is curious that a few years later Dr. Travell, who was White House physician, also told the presidential aide Mike Feldman that one of his legs was shorter than the other. LL interview with Myer Feldman.
346 "She began to really . . .": interview, John F. Kennedy, JMBP.
346 "The use of novocaine . . .": Dr. Alexander Preston, M.D., to James MacGregor Burns, July 29, 1959, JMBP.
347 "Tender? Oh, he was tender": LL interview with Gunilla Von Post.
348 "I just got word . . .": John F. Kennedy to Gunilla Von Post, August 22, 1955, courtesy Gunilla Von Post.
348 "I wish *you* . . .": John F. Kennedy to Gunilla Von Post, September 1955, courtesy Gunilla Von Post.
348 "Jack has called me . . .": Lem Billings to Gunilla Von Post, n.d., courtesy Gunilla Von Post.
349 "I had a wonderful . . .": John F. Kennedy to Gunilla Von Post, date uncertain, courtesy Gunilla Von Post.
349 "I am looking forward . . .": John F. Kennedy to Gunilla Von Post, 1966, courtesy Gunilla Von Post.
349 scholarly tone of the first draft: draft dated February 8, 1955, RWC.
349 "When I'd finished . . .": *Boston Globe*, May 22, 1955.
349 "opening and closing . . .": SJFK, p. 327.
349 "lying on a board . . .": Arthur Krock, KLOH.
350 "the public's appreciation . . .": PC, p. 4.
350 "Our political life . . .": ibid., p. 19.
350 "And only the very courageous . . .": ibid., p. 20.
351 "The stories of past . . .": ibid., p. 258.
351 Jack used: "Kennedy's Magic Formula," *Saturday Evening Post*, August 13, 1960.
351 "One of his slogans . . .": interview, Rose Kennedy, RCP.
352 the members may well have been: SJFK, p. 397.

17. The Pursuit of Power

353 more than one hundred million: Sig Mickelson, "Two National Political Conventions Have Proved Television's News Role," *Quill*, December 1956.
353 "I am Senator . . .": Senator John Kennedy, final text of narration for Democratic keynote film, RWC.

353 sounded overwrought: "Look and Listen with Donald Kirkley," *Baltimore Sun*, August 15, 1956.
354 "Not much": *Berkshire Eagle*, undated clipping, 1956.
354 "his intelligence, farsightedness . . .": AWRH, p. 98.
355 "spend[ing] the whole . . .": John F. Kennedy to James Finnegan, May 2, 1956, KP.
355 "if someone who is as . . .": John F. Kennedy to Walter T. Burke, Esq., May 11, 1956, RWC.
355 "that for the first . . .": Stan Karson, interoffice memo, May 22, 1956, KP.
355 "Should be married . . .": Burns, p. 182.
356 "As long as you . . .": Jerry Williams, interview with Congressman Thomas P. O'Neill Jr., WBZ-TV, Boston, and O'Neill papers, John J. Burns Library, Boston College.
356 One citizen: RFK, p. 98.
356 "She had seen me . . .": interview, John F. Kennedy, JMBP.
356 "My point was that . . .": ibid.
357 "too much time": TEEK, p. 100.
357 "So many people . . .": RFK, p. 99.
357 "a forlorn figure": AWRH, p. 100.
358 "one of the ablest . . .": *Clearwater Sun*, August 24, 1956.
358 two-week visit: AP, August 18, 1956.
358 "I do not wish . . .": John F. Kennedy to H. W. Richardson, April 10, 1956, RWC.
358 Jack arranged: W. F. B. Morse and William Thompson, memorandum of agreement May 2, 1956, RWC.
358 Those who disliked: For descriptions of Thompson, see Edward Klein, *All Too Human* (1996), pp. 195–97, and Seymour M. Hersh, *The Dark Side of Camelot* (1997), pp. 388–90.
359 "Unique among them . . .": quoted in Collier and Horowitz, p. 209.
359 "exhaustion and nervous . . .": UPI, undated clipping, JFKPL.
360 "I told him, 'You . . .": LL interview with George Smathers.
360 Bobby attempted to ingratiate: A. Rosen to L. N. Conroy, February 9, 1955, FBIFOI.
360 visited the FBI chief again the next year: M. A. Jones to Mr. Nichols, July 1, 1955, FBIFOI.
360 "Kennedy was completely uncooperative . . .": J. Edgar Hoover to Mr. Nichols, July 20, 1955, FBIFOI.
360 Bobby grasped the Bible: William O. Douglas, *The Court Years, 1939–1975* (1980), p. 306.
360 "Russian doctors are . . .": ibid., p. 307.
361 about twenty pounds: *Boston Globe*, October 23, 1955.
361 "As a result . . .": Douglas, p. 307.
362 Bobby not only edited: Robert F. Kennedy to David Lawrence, October 5, 1955, RFK papers, JFKPL.
362 invited a reporter: *Boston Globe*, October 23, 1955.

362 Yet he agreed: Maggie Duckett to Robert F. Kennedy, September 27, 1955, RFK papers, JFKPL.
362 "The pictures were . . .": quoted in RKHT, p. 127.
362 He was still mining: *New York Times Magazine*, April 8, 1956.
362 "I hope the United States . . .": Robert F. Kennedy to J. Edgar Hoover, September 11, 1956, FBIFOI.
362 "cherish it even more": C. A. Evans to Mr. Rosen, March 5, 1957, FBIFOI.
363 Ethel's losses began: *New York Times*, October 5, 1955.
363 sometimes he had to: "Because of weather Ethel refuses to fly so we're delaying our trip until next week." Robert F. Kennedy to Murray Kempton, telegram, October 8, 1957, RFK papers, JFKPL.
363 "Bobby accompanied . . .": interview, Rose Kennedy, RCP.
363 "He's got no balls":WNJ, p. 115.
364 "an appearance of insincerity": RKHT, p. 135.
364 "Oh, my God . . .": quoted in John Bartlow Martin, *Adlai Stevenson and the World* (1977), p. 235.
364 women nearly fainted: "Dem Ladies Mob 'Elvis' Kennedy," clipping, JFKPP.
364 "I hate to impinge . . .": Robert F. Kennedy to Adlai Stevenson, Illinois State University at Normal, DHP.
365 "He had a tendency . . .": Clark R. Mollenhoff, *Tentacles of Power* (1965), p. 129.
365 "Occasionally, I . . .": ibid., p. 124.
365 He argued with: RKHT, p. 142.
366 "Unless you are . . .": ibid., p. 141.
367 "out of his investigation . . .": *Louisville Courier-Journal*, March 24, 1957.
367 walked out of a speech: Joseph E. Persico, *Edward R. Murrow: An American Original* (1997), p. 485.
367 "were charming and poised . . .": *New York Times*, September 14, 1957.
367 "We have been working . . .": Francis X. Morrissey to Robert F. Kennedy, February 10, 1958, RFK papers, JFKPL.
367 taken steps to ensure: ibid., March 28, 1958.
367 "I think some reporter . . .": Joseph P. Kennedy to Robert F. Kennedy, n.d. (probably 1957), family correspondence, RFK papers, JFKPL.
368 his Christmas gift list contained: 1957 gift list, RFK papers, JFKPL.
368 The next summer: James Brady, "How Conveniently We Forget What Hoffa Really Was," *Advertising Age*, December 2, 1996.
368 he even suggested: Murray Kempton to Robert F. Kennedy, telegram, n.d., RFK papers, JFKPL.
368 "ought to be cleared . . .": Robert F. Kennedy to Jack Anderson, April 18, 1957, RFK papers, JFKPL.
368 "My objective has always . . .": Jack Anderson to Robert F. Kennedy, April 26, 1957, RFK papers, JFKPL.
368 "if there are some steps . . .": Robert F. Kennedy to FCC Chairman George C. McConnaughey, May 30, 1957, RFK papers, JFKPL.

368 "Spoken bits of praise . . .": Clark R. Mollenhoff to Robert F. Kennedy, December 23, 1957, RFK papers, JFKPL.
368 Bobby called in: Patrick Munroe, KLOH.
369 "[Bobby's] 'a great . . .": *Boston Globe*, March 20, 1957.
369 "the look of a man . . .": TEW, p. 74.
370 "Nobody can describe . . .": quoted in Dan E. Moldea, *The Hoffa Wars: Teamsters, Rebels, Politicians, and the Mob* (1978), p. 26.
370 he wanted to be present when: Mr. Jones to Mr. DeLoach, September 19, 1959, FBIFOI.
370 suggesting that Cheasty was anti-black: Arthur A. Sloane, *Hoffa* (1991), p. 71.
371 "Come on now . . .": Stein and Plimpton, pp. 57–58.
371 "the work of Williams . . .": TEW, p. 58.
371 "Oh, I used to . . .": RFK, pp. 107–8.
371 "a young, dim-witted, curly-headed . . .": *Kansas City Star*, July 23, 1961, FBIFOI.
371 "Is there any . . .": RFK, p. 111.
372 "relentless, vindictive . . .": Alexander Bickel, "The Case Against Him for Attorney General," *The New Republic*, January 9, 1961.
372 "The sworn statements . . .": Clark Mollenhoff to Robert F. Kennedy, January 8, 1957 (but possibly 1958), HP.
373 Every day Ethel sat: *Washington Star*, March 24, 1957.
373 "mash notes": New York *Daily News*, August 24, 1957.
373 "It's ridiculous to wait . . .": *San Francisco News*, July 12, 1957.
373 The two men went: Stein and Plimpton, p. 56.
373–74 "I don't understand . . .": TEW, p. 144.
374 He took startling testimony: Walter Sheridan, *The Fall and Rise of Jimmy Hoffa* (1972), pp. 99–100.
374 interviewing 1,525 witnesses: Sloane, p. 159.
374 15 unions and more than 50 companies: Robert F. Kennedy, "An Urgent Reform Plan," *Life*, June 1, 1959.
374 "appalling public apathy": ibid.
374 He ate gourmet meals: Sloane, p. 139.
375 "At thirty-three . . .": notes on *The Jack Paar Show*, June 23, 1959, Museum of Broadcasting, DHP.
375 "My God, this is . . .": oral history, Howard E. Shuman, U.S. Senate Historical Office.
375 "Did that tragic episode . . .": quoted in JFKOA, p. 20.
376 Of the 138 editorials: ibid., p. 21.
376 "You lucky mush": Burns, p. 196.
377 "I've got a very important . . .": LL interview with Arnold Beichman, and *New York Post*, December 12, 1997.

18. The Rites of Ambition

378 "wary of being known . . .": SJFK, p. 407.
378 "It's awful . . .": interview, John F. Kennedy, JMBP.
379 Then the dog that Jackie: Dr. Janet Travell, KLOH.
379 In the middle of September: ibid.
379 "You know, that's . . .": ibid.
379 "a virus infection": Dr. Janet Travell, KLOH.
379 canceled a dinner: John F. Kennedy to Lord Beaverbrook, September 28, 1957, NHP.
379 "discouraged": Dr. Janet Travell, KLOH.
380 "You know, what you need . . .": ibid.
381 "as robust as a sumo wrestler": WNJ, p. 200.
381 "Now, Lem, which . . .": Janet Lee Auchincloss, KLOH.
381 accompanied . . . by Bill Thompson: George A. Smathers, oral history by Donald A. Ritchie, U.S. Senate Historical Office.
382 took great interest: CY, p. 99.
382 had his picture taken: The author has seen this picture in Patty McGinty Gallagher's home in Palm Beach, Florida.
382 Jack had been accompanied: Mr. Rosen to Mr. Boardman, March 4, 1958, FBIFOI.
382 "Batista had a . . .": LL interview with George Smathers.
383 Sarge came from: Robert A. Liston, *Sargent Shriver: A Candid Portrait* (1964), p. 25.
383 "I found a man . . .": Leamer, p. 426.
383 "White Slave activities . . .": FBI investigation, January 29, 1954, FBIFOI.
384 start Cleary Brothers: Dinneen (1959), p. 187.
384 Jean settled for a more: Collier and Horowitz, pp. 215–16.
384 "one of the great American fortunes": *Forbes*, October 21, 1991.
384 "I think they will . . .": Archbishop Cushing to Joseph P. Kennedy, May 17, 1956, PC.
385 "I didn't get . . .": Archbishop Cushing to Joseph P. Kennedy, June 8, 1956, PC.
385 "weak as a cat": quoted in TEEK, p. 104.
386 "darn good looking fellow": LL interview with Joan Kennedy.
387 "What do you . . .": LL interview with Joan Kennedy, and interview, Joan Kennedy, RCP.
387 "proposed in a . . .": ibid.
387 "When the interview . . .": ibid.
388 "I was young and naive . . .": interview, Joan Kennedy, RCP.
388 engagement ring that: ibid.
389 an entourage of pretty young women: Vincent J. Celeste, KLOH.
389 "customarily surrounded by . . .": quoted in SJFK, p. 452.
389 slapped as many bumper: TEEK, p. 108.
389 "house to house": interview, John F. Kennedy, JMBP.

389 Celeste thought that he: Vincent J. Celeste, KLOH.
390 "a professional tattletale": TEEK, p. 54.
390 "to a law firm . . .": Eddie O'Dowd to John F. Kennedy, June 20, 1955, RWC.
390 "have been running . . .": "Could not give my name for good reasons but am a resident of the block" to John F. Kennedy, June 20, 1955, RWC.
391 paying Feldman $15,000: LL interview with Myer Feldman.
391 "I think it's the . . .": ibid.
391 At each stop he introduced: Paul Fay, CBS interview, n.d., NHP.
391 "I really hate this": LL interview with Joseph P. Miller, and Joseph P. Miller, unpublished memoir.
392 "Be sure to keep . . .": John F. Kennedy to Steve Smith, "Saturday afternoon (en route to California)," n.d., RWC.
392 "probably as nearly perfect . . .": O'Donnell and Powers, p. 145.
392 "Should we get . . .": John F. Kennedy to Steve Smith, "Saturday afternoon . . ." RWC.
392 an estimated $1.5 million: SJFK, pp. 451–52.
392 "I wouldn't change . . .": *Boston Sunday Herald*, April 7, 1957, HUA.
393 "I have told . . .": Joseph P. Kennedy to Archbishop Cushing, May 1, 1958, PC.
393 "Among our civic . . .": Archbishop Cushing to Joseph P. Kennedy, n.d. (1958), PC.
393 "I do hope that sometime . . .": Bishop of Covington to Archbishop Cushing, May 9, 1958, PC.
393 "if you can get any . . .": Archbishop Cushing to Bishop Malloy, May 14, 1958, PC.
394 "a program was carried . . .": Lawrence F. O'Brien to Robert F. Kennedy, February 19, 1953, RFK miscellaneous correspondence, JFKPL.
394 normally reserved for bishops: Archbishop Cushing to Joseph P. Kennedy, December 4, 1958, PC.
394 "I am really . . .": Joseph P. Kennedy to Cushing, May 13, 1958, PC.
395 "I haven't been . . .": Joseph P. Kennedy to Lord Beaverbrook, June 27, 1958, NHP.
395 he assumed it was one of: interview with campaign aide, OTR.
395 "At the beginning . . .": ibid.
396 was outraged: LL interview with Marcus Raskin, and OTR interview with Kennedy's mistress.
396 "These men . . . died for . . .": "Handwritten Notes of Senator Kennedy to Be Used by Him in Making Memorial Day Speeches at Brookline and Dorchester, Massachusetts, on May 30, 1958," JFKPP.
398 changed his mind: oral history, Father Cavanaugh, Notre Dame University.
398 "We almost . . .": LL interview with Joan Kennedy.
398 "I remember at . . .": Edward M. Kennedy to Lord Beaverbrook, November 18, 1958, NHP papers.
399 "He stayed . . .": interview, John F. Kennedy, JMBP.

400 "We were dumped . . .": LL interview with Joan Kennedy.
400 "When Ted told . . .": interview, Joan Kennedy, RCP.

19. "A Sin Against God"

402 On the first day: K, p. 119.
403 "quiet confidence": ibid.
403 Non-Catholic staff members: ibid.
403 Jack's staff had already: David Hackett, KLOH, LL interview with David Hackett, and LL interview with Myer Feldman.
403 "I want you to . . .": interview, Robert Wallace, DHP.
404 "Now wait . . .": ibid.
404 "Where the hell . . .": Joe Miller, unpublished recollection, and LL interview with Joe Miller.
405 "Virtually everyone . . .": ibid.
405 "You have been . . .": ibid.
405 Already, in March: Mr. Jones to Mr. DeLoach, July 13, 1960, FBIFOI.
406 Florence Kater wrote: *The Thunderbolt*, November 1963.
406 She had in her possession: ibid.
406 "racing like a . . .": Florence Mary Kater to Robert Kennedy, April 2, 1963, Hedda Hopper collection, Margaret Herrick Library, Los Angeles.
406 "Here comes the agony . . .": LL interview with Chuck Spalding.
407 "I was taken into . . .": Dictated letter to Jacqueline Kennedy on weekend in Rhode Island, n.d., presidential recording, cassette L, JFKPL.
407 "Jenny Ryan . . .": ibid.
407 I Jacqueline . . . : Jacqueline Bouvier Kennedy to Evelyn Lincoln, January 11, 1960, RWC.
408 he was checking . . . : C. D. DeLoach to Mr. Mohr, April 19, 1960, FBIFOI.
408 "in Miami he had . . .": To Director, FBI, from Special Agent in Charge, New Orleans, March 23, 1960, FBIFOI.
409 "some bimbos and . . .": interview, Blair Clark, HP.
409 by the following evening: Judith Exner, *My Story* (1977), pp. 52–54.
409 "the main topic . . .": ibid, pp. 90–91. In a deposition in *Judith Exner* vs. *Random House*, et al., Exner stated that these were among the truthful statements in her autobiography.
410 "kind of spooky": Gerri Hirshey, "The Last Act of Judith Exner," *Vanity Fair*, April 1990.
410 alimony of $433.33: testimony of FBI informant at Bank of America, FBI file LA, 92–113, JEP.
410 "financially independent": deposition of Judith Exner, in *Judith Exner, Plaintiff, vs. Defendants Random House, Inc., Ballantine Books, Villard Books, Laurence Leamer, Milt Ebbins,* does 2–100, December 4, 1997, JEP.
410 "family money . . .": Hirshey, p. 221.
410 "without sufficient funds . . .": Exner deposition, December 3, 1997, JEP.
410 owed $2,784 . . . : Field Audit Company, March 4, 1959, JEP.

410 behind in her car payments: Utter Pontiac Company, July 16, 1959, JEP.

410 about $100 a week: Judith Exner deposition, December 3, 1997, JEP.

410 took out a number: Stephen Monka, Los Angeles, 92–113, August 8, 1962, FBIFOI, JEP.

411 Just two weeks before: Exner stayed at the Sands from January 23, 1960, to January 25, 1960. Ellwood signed the bill for $110.82. JEP.

411 "almost certainly . . .": Charles Rappleye and Ed Becker, *All-American Mafioso: The Johnny Rosselli Story* (1991), p. 208.

411 "shacking up with . . .": C. A. Evens to Mr. Belmont, "Subject: Judith E. Exner, Associate of Hoodlums," March 20, 1962, FBIFOI, JEP.

411 "It was common for Sinatra . . .": HSCA interview of Mr. Fred Otash, July 24, 1978, JEP.

411 she stayed at: Tropicana Hotel receipt. Exner's bill ran to $98.99, JEP.

411 apparently paying: On the few occasions when the mode of payment is listed, it is inevitably given as cash, JEP.

412 final settlement: deposition of Judith Exner, December 3, 1997, JEP.

412 The $6,000: two checks for three thousand dollars to Judith Campbell, from Executive Business Management, Trustee for William J. Campbell, March 4 and March 5, 1960, JEP.

412 "It was a wonderful night . . .": Kitty Kelley, "The Dark Side of Camelot," *People*, February 29, 1988.

412 "Johnnie Roselli who . . .": unpublished memoir of Jeanne Humphreys, courtesy, Gus Russo.

412 A friend who had: FBINY 92–793, November 6, 1960, HSCA records, NA.

412 Clarke made the mistake: SAC Chicago to Director, FBI, April 3, 1961, HSCA records, NA.

413 "quietly arrange a meeting . . .": Kelley, "The Dark Side of Camelot."

413 "I was the one person . . .": Seymour Hersh, p. 307.

414 "She came to Chicago . . .": William Brashler, *The Don: The Life and Death of Sam Giancana* (1977), p. 210.

414 "She found out about . . .": LL interview with Joe Shimon.

415 "Politics has . . .": This dictation was undated, but based on references made by Kennedy, the JFKPL has dated it 1959.

416 "I moved into . . .": John F. Kennedy, dictation concerning entrance into politics, presidential recording, cassette N, JFKPL.

416–17 "We planned . . .": Carroll Kilpatrick, KLOH.

417 "something positive . . .": interview, Rose Kennedy, RCP.

417 "very low estimate . . .": C. H. Cramer to John Kennedy, quoting a letter of July 12, 1932, Roy Howard to Newton D. Baker, found in Baker collection LC, NHP.

417 "It definitely shows . . .": John F. Kennedy to Joseph P. Kennedy, n.d. (spring 1959), NHP.

418 "He had real contempt . . .": Joseph Alsop, KLOH.

418 "I know how . . .": Robert Monagan, KLOH.

418 "I'd be very happy . . .": Paul Healy, "Senate's Gay Young Bachelor," *Saturday Evening Post*, June 13, 1953.
418 "I don't want . . .": Ken Gormley, *Archibald Cox: Conscience of a Nation* (1997), p. 118.
419 "wore a rather bemused . . .": Robert W. Merry, *Taking on the World: Joseph and Stewart Alsop—Guardians of the American Century* (1996), p. 345.
419 "I don't think that I could say . . .": Chuck Spalding, RFK papers, KLOH.
420 clamped his teeth: LL interview with Bob Healy.
420 "I went to the top . . .": quoted in TEEK, p. 129.
420 "I wanted to get . . .": ibid.
421 crowd of more than eight thousand: interview, Ted Kennedy, RCP.
421 "I have never seen . . .": Joan Kennedy to Lord Beaverbrook, April 10, 1960, Beaverbrook papers, NHP.
421 *National Review* charged: Lasky, p. 132. Edwin Guthman confirms that Corbin was behind the pamphlets. LL interview with Edwin Guthman.
422 "My poll . . . in W.V. . . .": These JFK notes are owned by the *Forbes* Magazine Collection and are used here with their permission and acknowledgment.
422 no memory of: LL interview with Pierre Salinger.
422 I talked today . . . : John F. Kennedy, memo, April 8, 1960, RWC.
423 "house of prostitution": Seymour Hersh, p. 113.
423 she had been jailed: ibid.
423 "public knowledge could have . . .": ibid., p. 117.
423 "He'd pick up . . .": O'Neill, p. 92.
424 two briefcases: Raymond Chafin and Topper Sherwood, *Just Good Politics: The Life of Raymond Chafin* (1994), p. 143.
424 Humphrey spent $25,000: MP1960, p. 120.
424 James McCahey Jr., a Chicago: LL interview with Mrs. James McCahey.
424 "They had a . . .": LL interview with Charles Peters.
425 "I'll be the first . . .": LL interview with Bruce Sundlun.
425 "Well, Senator . . .": ibid.
425 "When Frank came down . . .": interview, Rose Kennedy, RCP.
426 "I remember discussing . . .": Myer Feldman, KLOH, and LL interview with Myer Feldman.
426 "Bobby had been . . .": discussion between Franklin D. Roosevelt Jr. and Arthur Schlesinger Jr., February 12, 1975, ASP.
426 "Is FDR Jr. there tonight?": John F. Kennedy, n.d., FMC.
426 "There's another candidate . . .": JFKMM, p. 344.
426–27 "I don't have any . . .": quoted in Whalen, p. 452.
427 "And the Star . . .": John F. Kennedy, n.d., FMC.
427 "There was a lot . . .": discussion between Franklin D. Roosevelt Jr. and Arthur Schlesinger Jr., February 12, 1975, ASP.
428 "'simply because he . . .": memo to files, AKP.

428 "whatever one's . . .": Fletcher Knebel, "A Catholic in 1960," *Look*, March 3, 1959.
428 "the crash of . . .": February 24, 1959, religious issues file, JFKPP.
428 "Wherever I go . . .": Cardinal Cushing to Joseph P. Kennedy, May 20, 1960, PC.
428 "This letter really . . .": Joseph P. Kennedy to Richard Cardinal Cushing, March 23, 1959, PC.
429 "I hope that . . .": ibid., May 20, 1959, PC.
429 "the religious issue . . .": Cardinal Cushing to Joseph P. Kennedy, May 22, 1959.
429 "Nobody asked me . . .": Chafin and Topper, pp. 123–24.
429 And pity poor Humphrey: MP1960, p. 119.
430 "A sin against God . . .": ibid., p. 117.
430 "I suppose if I . . .": John F. Kennedy, n.d., FMC.
431 a soft-core porn item. . . . : Benjamin C. Bradlee, *Conversations with Kennedy* (1975), p. 27.
431 Jackie stood alone: ibid., p. 28.
431 trying to gauge: Hugh Sidey, *Time* file, 1960.

20. A Patriot's Song

433 "Yes, this candidate . . .": Norman Mailer, *The Presidential Papers* (1963), p. 39.
433 The Los Angeles police guarding: LL interview with former LA Police Detective Dan Stewart.
433 "These things happened . . .": O'Neill, p. 92.
433 In the Maryland primary: Torbert Macdonald, KLOH.
434 "I remember . . .": quoted in TEEK, pp. 122–23.
434 With the help of New York City's: MP1960, pp. 153–54.
434 He helped to add Illinois: ibid., p. 195.
434 "Senator Kennedy, who appears . . .": "Should Health of Presidential Candidates Be a Campaign Issue?," *Medical World News*, February 9, 1976.
435 "outstanding possibility": Myer Feldman, KLOH.
435 "We need Texas": interview, Joseph Timilty, BP.
435 "I knew he hated Jack": Jeff Shesol, *Mutual Contempt: Lyndon Johnson, Robert Kennedy, and the Feud that Defined a Decade* (1997), p. 59.
436 "Well, you know . . .": LL interview with Myer Feldman.
436 "In your first move after . . .": The author draws heavily on the best account of this convoluted process in Jeff Shesol's *Mutual Contempt.*
436 "Jack, I don't want you . . .": LL interview with Charles Bartlett.
437 originally written by Sorensen: LL interview with Myer Feldman.
438 "an example of . . .": interview, Chuck Spalding, BP.
438 "Time Inc. realizes . . .": interview, Henry Luce, RWP.
438–39 "I see Otto . . .": Pierre Salinger, *P.S.: A Memoir* (1995), p. 76.
439 "a remarkable political demonstration": Myer Feldman, KLOH.

439 "Our next president . . .": quoted in Matthews, p. 136.
439–40 "anxiety and discomfort": *Time*, November 7, 1960.
440 "I think that in the . . .": interview, John F. Kennedy, JMBP.
441 "didn't like publicity": Dr. Janet Travell, KLOH.
441 "With respect . . .": Dr. Janet Travell and Dr. Eugene J. Cohen to John F. Kennedy, June 11, 1960, RWC.
441 the two doctors traveled: Janet G. Travell, M.D., to John F. Kennedy, June 24, 1960, RWC.
441 "about the security . . .": Dr. Janet Travell to Dr. Elmer C. Bartels, June 24, 1960, RWC.
441 "Senator Kennedy does not . . .": Janet Travell, M.D., to Robert F. Kennedy, July 11, 1960, DHP.
441 "You do not have . . .": Eugene J. Cohen, M.D., and Janet Travell, M.D., to John F. Kennedy, July 29, 1960, DHP.
441 "I tracked down . . .": Dr. Janet Travell, KLOH.
442 "No, Lyndon . . .": LL interview with Myer Feldman.
442 "lift its bloody hand . . .": quoted in K, p. 194.
442 "In every Catholic-dominated . . .": brochure of Baptist Sunday School Committee of the American Baptist Association, DHP.
442–43 "American freedom grew . . .": Carol V. R. George, *God's Salesman: Norman Vincent Peale and the Power of Positive Thinking* (1993), p. 198.
443 "Post Protestant Pluralism": ibid, p. 196.
443 "Recently I spent . . .": ibid., p. 200.
443 "moral character of . . .": ibid., p. 210.
443 "the old, strong, narrow . . .": ibid., p. 197.
443 "People have often . . .": William Geoghegan, KLOH.
445 "the right and duty . . .": *New York Times*, May 17, 1960.
445 twice as many Protestants: Charles R. Morris, *American Catholic: The Saints and Sinners Who Built America's Most Powerful Church* (1997), p. 281.
445 Jack's speech that morning: LL interview with Bill Wilson.
445 he could never be . . . : George, p. 202.
446 "Protestant America . . .": ibid., p. 208.
446 "Henry, if you . . .": Henry Brandon, KLOH.
446 "anxiety about . . .": Adelaide King Eisenmann to John Seigenthaler and Mike Feldman, Democratic National Committee interoffice memorandum, October 3, 1960, PA/PF Corr 59–60 JFKPL.
446 "Americans are wondering . . .": *U.S. News & World Report*, August 22, 1960.
447 "You know I came . . .": LL interview with Bonnie Williams.
447 the candidate pushed: Gormley, p. 130.
447 "I was just aghast . . .": LL interview with Archibald Cox.
447 wrote a futile letter: Gormley, p. 130.
448 called Feldman: Myer Feldman, KLOH, and LL interview with Myer Feldman.
449 Jack told his two aides: LL interview with Myer Feldman.

449 Jack believed that this area: ibid.
449 at home started injecting: LL interview with Betty Coxe Spalding.
449 Truman Capote might be: *New York Times*, December 4, 1972.
450 "The demands of his . . .": Dr. Max Jacobson, unpublished memoir, courtesy Mrs. Max Jacobson.
450 hitting his fist: Matthews, p. 146.
450 "No, Senator . . .": K, p. 198.
451 "Kick him in the balls": LL interview with Bill Wilson.
451 hit his kneecap: MP1960, p. 298.
452 87 percent: Mary Watson, *The Expanding Vision: American Television in the Kennedy Years* (1990), p. 8.
452 " 'Party,' not 'pawty' ": Myer Feldman, KLOH.
452 Reinsch hurried: Watson, p. 13.
453 "we must attempt . . .": quoted in Matthews, p. 166.
454 "Senator Kennedy made . . .":
454 the action would have taken: The original CIA memo for "A Program of Covert Action Against the Castro Regime," dated March 16, 1960, stated that "it will not reach its culmination earlier than 6 to 8 months from now." *Foreign Relations* 6 (1958–1960), pp. 850–51.
454 "in the full hearing . . .": *Washington Post*, August 8, 2000. See also Gus Russo, *Live by the Sword: The Secret War Against Castro and the Death of JFK* (1998), p. 365.
454 "we wish to give . . .": Russo, p. 365.
455 Manuel Artime, the political leader: ibid., pp. 169–70.
456 "I want you to . . .": Gerald S. and Deborah H. Strober, *Let Us Begin Anew: An Oral History of the Kennedy Presidency* (1993), pp. 325–26.
456 "I think Kennedy . . .": ibid, p. 325.
456 "She . . . absolutely curled . . .": Lord Harlech, KLOH.
456 "that it [the pregnancy] was planned . . .": *Washington Star*, September 2, 1960.
457 "I'm sure I spend less . . .": AP, September 15, 1960.
457 "oddly humorous . . .": quoted in Merry, pp. 353–54.
457 60 percent had voted: Taylor Branch, *Parting the Waters: America in the King Years 1954–1963* (1988), p. 374.
458 "And I know those . . .": ibid., p. 342.
458 $300,000 cash to get out the: ibid., p. 343.
458 King had been looking: Harris Wofford, *Of Kennedys and Kings* (1980), p. 12.
459 "super idealistic": LL interview with Harris Wofford.
459 "Look, our real . . .": ibid.
460 "Governor, is there . . .": Jack Bass, *Taming the Storm* (1993), p. 170.
460 "The trouble with": Wofford, p. 18.
460 "You bomb-throwers . . .": Branch, p. 364.
461 "Do you know": Wofford, p. 19.
461 "screwing up my brother's . . .": Branch, p. 367.

461 "we would lose . . .": Bass, p. 171.
461 "Did you see": Wofford, p. 28.
461 two million copies: Bass, p. 171.
461 "deeply indebted . . .": Branch, p. 369,
462 Later in Honolulu: LL interview with Dick Livingston.
462 In that final week: Don Shannon papers, JFKPL.
463 The next to the last day: MP1960, p. 371.
463 "If the younger . . .": quoted in JFKMM, p. 433.
463 "Man, I'm tired . . .": RKHT, p. 219.
463 "How much is that?": John Richard Reilly, KLOH.
464 Joe shut: Arthur Krock, KLOH.
464 "You know, there . . .": LL interview with Joe Dolan.
464 "I believe in this man . . .": Hersh, p. 138.
465 A different story: ibid., pp. 135–36.
465 Another story has Joe: SB, p. 44.
465 In yet another scenario: Anthony Summers, *Official and Confidential: The Secret Life of J. Edgar Hoover* (1993), p. 269.
465 In June of that election: The author has in his possession a letter from a private collection addressed to Joseph P. Kennedy at Cal-Neva Lodge and dated June 14, 1960, PC.
465 "many gangsters with . . .": Mahoney, p. 165, and J. Edgar Hoover, personal memo to the attorney general, August 16, 1962, FBIFOI.
466 Sinatra had boasted: The FBI taps of the Armory Lounge in Chicago show Giancana's displeasure. In 1975 the FBI noted: "When Kennedy was successfully elected and when Sinatra's representations as passed on by Giancana did not materialize when the Kennedy administration, especially the Department of Justice under Robert Kennedy which intensified the investigation of organized crime, Giancana lost considerable face because of the reliance of the leadership of organized crime on Sinatra's representations and Giancana never forgave Sinatra for this situation." CG 137–349, p. 11, Chicago, HSCA records, NA.
466 ninety minutes to move: *New York Times*, November 8, 1960.
467 "You know . . .": LL interview with Bob Healy.
467 "The campaign is now . . .": *New York Times*, November 8, 1960.
468 Jack was no more: K, p. 212.
468 "If I were he . . .": Salinger, p. 49, and LL interview with Pierre Salinger.
468 bill of about $10,000: MP1960, p. 377.
468 even without: Jack W. Germond and Jules Witcover, "'Dark Side of Camelot' Takes Liberties with the Truth," *Baltimore Sun*, November 19, 1997.
468 "Mr. President": LL interview with Myer Feldman.
468 his hands were trembling: MP1960, p. 380.

21. The Torch Has Been Passed

471 stranding ten thousand: *Time,* January 27, 1961, p. 9.
471 One million other Americans: *Washington Post,* January 21, 1961.
472 did not even applaud: ibid.
473 Sorensen had added: LL interview with Harris Wofford, and Wofford, p. 99.
473 four million Americans: *New York Times,* January 21, 1961.
473 black ministers in the South: Branch, p. 384.
473 peoples of Europe: A Gallup poll showed that 72 percent of Germans had a favorable feeling toward Kennedy, as did 60 percent of the French, and 59 percent of the British. *Washington Post,* January 31, 1961.
473 "begin anew": *New York Times,* January 21, 1961.
473 "radical improvement": ibid.
474 inundated with telegrams: David Powers, handwritten note, DPP.
474 "Jack doesn't belong . . .": *Life,* December 19, 1960.
475 "I assure you . . .": *Newsweek,* September 12, 1960.
475 "I can't": *Life,* December 19, 1960.
475 "Goddamn it . . .": LL interview with George Smathers.
476 "That's it, general": LL interview with John Seigenthaler.
476 known as the Irish Mafia: Stewart Alsop, "The White House Insiders," *Saturday Evening Post,* June 10, 1961.
477 arrived by bicycle: Kai Bird, *The Color of Truth: McGeorge Bundy and William Bundy: Brothers in Arms* (1998), p. 185.
477 Kennedy could not run: William Attwood, CUOH.
477 "I want to help . . .": Steve Smith to Arthur Schlesinger Jr., March 5, 1976, ASP.
477 visited the White House only once: TR, p. 393.
477 "We had this confidence . . .": LL interview with Myer Feldman.
478 "Kennedy put the knife . . .": LL interview with Marcus Raskin.
478 "I AM SURE . . .": Claude Hooton Jr. to Robert F. Kennedy, December 16, 1960, DHP.
478 "There is some talk . . .": Robert F. Kennedy to Claude Hooten Jr., December 22, 1960, DHP.
478 "Why can't a . . .": LL interview with Sam Adams.
478 "decorative butterflies . . .": Kay Halle, CBS interview, NHP.
479 they worked all night: See David Bell on Sorensen's efforts. David Bell, KLOH.
479 the same salary: WK, p. 63.
479 He wanted no staff meetings: ibid., p. 74.
479 "Listen, you sons of bitches . . .": quoted in Ralph Martin, *A Hero for Our Time* (1984), p. 223.
479 "I never knew . . .": LL interview with Myer Feldman.
479 "was capable because of . . .": Walter Rostow, KLOH.
480 "By 'toughness' I meant . . .": Adam Yarmolinsky, "Camelot Revisited," *Virginia Quarterly,* Autumn 1996.

480 He began shortly: Myer Feldman KLOH, and LL interview with Myer Feldman.
480 "People, even if . . .": Lord Harlech, KLOH.
481 left by the golf shoes: Myer Feldman, KLOH, and LL interview with Myer Feldman.
481 The president had to: LL interview with Myer Feldman.
482 The mysterious coverings: Joseph Alsop, KLOH.
482 "I'd like to make this . . .": Kay Halle, CBS interview, n.d., NHP, confirmed by John Kenneth Galbraith in letter to the author, September 2000.
482 One of the few calls: White House, telephone memorandum, January 26, 1961, JFKPL.
482 Mrs. Oswald had come: *Atlanta Constitution*, December 7, 1963.
482 "Jack feels that . . .": Arthur Schlesinger Jr., memo dictated February 22, 1961, ASP.
483 Jackie's stepbrother: The *New York Times*, a stickler on such matters, writes that "Jackie Kennedy was Vidal's mother's second husband's third wife's daughter." *New York Times*, October 15, 1995.
483 "I can't remember . . .": LL interview with Harris Wofford.
483 Dulles told the men: LL interview with Myer Feldman.
483 an accomplished man who had come: TOB, p. 427.
483–84 "for practical purposes . . .": McGeorge Bundy, memorandum of discussion on Cuba, January 28, 1961, in Mark J. White, *The Kennedys and Cuba: The Declassified Documentary History* (1999), p. 15. Wherever possible, more readily accessible sources than the files at the JFKPL will be cited. When otherwise not noted, the FRUS documents are from the Kennedy administration Cuban documents found on the State Department's estimable web site at www.state.gov/www/about_state/history/frusX/index.html.
484 from the original $4.4 million: "Inspector General's Survey of the Cuban Operation October 1961," in *Bay of Pigs Declassified: The Secret CIA Report on the Invasion of Cuba*, edited by Peter Kornbluh (1998), p. 58.
484 Kennedy had met: Clark Clifford, memorandum, January 24, 1961, Department of State, Rusk files: Lot 72 D 192, White House correspondence, 1/61–11/63, FRUS.
484 "continuing civil war": Central Intelligence Agency, memorandum, Washington, D.C., January 26, 1961, CIA, DDO/DDP files, job 78–01450R, box 5, area activity—Cuba (top secret), drafted by Bissell for a presidential briefing, FRUS.
484–85 "My belief from . . .": LL interview with Colonel Jack Hawkins.
485 "the way [would] then . . .": Colonel J. Hawkins, "memorandum for: Chief, WH/4, "Subject: Policy Decisions Required for Conduct of Strike Operations Against Government of Cuba," January 4, 1961, enclosure in Colonel J. Hawkins, "Clandestine Services History: Record of Paramilitary Action Against the Castro Government of Cuba, 17 March 1960–May 1961." www.geocities.com.

485 "our presently planned . . .": Washington, D.C., January 22, 1961, 10:00 A.M., Department of State, INR/IL historical files, Cuba Program, January 21, 1961, FRUS, author emphasis.

485 "final planning . . .": Central Intelligence Agency, *memorandum for the record*, Washington, D.C., January 28, 1961, DDO/DDP files, C. T. Barnes Chrono, January–July 1961, author's emphasis.

485 "You don't even know . . .": Lord Harlech, KLOH.

485 "They were a strange . . .": Robert Amory, KLOH.

486 Out of the sixty-nine thousand: Kornbluh, p. 77.

486 These errors were: In his "Record of Paramilitary Action Against the Castro Government of Cuba, 17 March 1960–May 1961," Colonel Hawkins writes: "These operations were not successful. Of 27 missions attempted only 4 achieved desired results. The Cuban pilots demonstrated early that they did not have the required capabilities for this kind of operation."

486 "not more than . . .": Colonel J. Hawkins, "Anti-Castro Resistance in Cuba: Actual and Potential, March 16, 1961," www.geocities.com/Capitol Hill.

486 "at least ten times . . .": ibid., enclosure.

486 "nearly 7,000 insurgents": CIA, "Cuban Operation, Document, April 12, 1961," reprinted in Kornbluh, p. 131.

486 CIA Director Allen Dulles: TOB, p. 431, and LL interviews with Burton Hersh and Samuel Halpern.

486 "I was never . . .": Robert Amory, KLOH.

486 done for security reasons: Richard N. Goodwin, *Remembering America* (1988), p. 176.

487 Nor did the one hundred thousand: Juan Carlos Rodriguez, *The Bay of Pigs and the CIA* (1999), p. 18.

487 When the CIA's operatives: These CIA figures are for the period from October 1960 to April 15, 1961. Hawkins, "Anti-Castro Resistance . . . ," p. 9.

487 "prompt free elections . . .": NSC action memorandum 31, Washington, D.C., March 11, 1961, reprinted in *Politics of Illusion: The Bay of Pigs Invasion Reexamined*, edited by James G. Blight and Peter Kornbluh (1998), p. 226.

487 "the risk is too . . .": President's Special Assistant (Schlesinger) to President Kennedy, memorandum, Washington, D.C., March 15, 1961, NSC files, countries series, Cuba, general, 1/61–4/61, JFKPL. See also Blight and Kornbluh, p. 222.

487 "Castro has been able . . .": memorandum of conference with President Kennedy, January 25, 1961, FRUS.

488 "to intimidate so as to obtain . . .": "Propaganda Action Plan in Support of Military Forces: D-Day Until the Fall of the Castro Regime," n.d., National Security Archive web site, www.gwn.edu/~nhsachiv/

488 "Assassination was intended . . .": CY, p. 134.

488 In late August: "Alleged Assassination Plots Involving Foreign Leaders (1975)," report of the Senate Select Committee on Intelligence Activities, chaired by Senator Frank Church, pp. 74–76.

488 These agents were in: interview of James O'Connell by Jim McDonald, September 25, 1978, House Select Committee on Assassinations, HSCA 189–10140–10362.

489 included those led: Felix I. Rodriguez, *Shadow Warrior* (1989), pp. 65–66.

489 "Poles, Germans, and Americans": John Henry Stephens, Church Committee interview by Bob Kelley, May 30, 1975, NA.

489 On March 29, 1961: cables in Mason Cargill, Church Committee memo, March 21, 1975, CIA DPD file. NA. See also Russo, pp. 53–55.

489 "I was part of that . . .": Blight and Kornbluh, p. 87.

489 "disposing of Castro": *CIA Targets Fidel: Secret 1967 CIA Inspector General's Report on Plots to Assassinate Fidel Castro* (1996), p. 26.

490 "the details were . . .": ibid., p. 35.

490 "The White House has twice . . .": ibid., p. 48.

490 "someone was supposed . . .": CY, p. 139.

490 at least twenty people: *CIA Targets Fidel*, pp. 45–46.

491 "he had met . . .": CG 92–349, 1960, HSCA records, NA.

491 "There's a question . . .": LL interview with Cartha DeLoach.

491 "particularly desires that no hint . . .": McGeorge Bundy, memorandum of discussion on Cuba, January 28, 1961, reprinted in Mark White, p. 17.

492 Articles about the training: Peter Wyden, *Bay of Pigs: The Untold Story* (1979), p. 46.

492 124 members: conversation between JFK and Richard Helms, March 6, 1963, cassette D, telephone conversations, JFKPL.

492 Beyond that, the president: Arthur Schlesinger Jr., memorandum for the president, "Subject: Howard Handleman on Cuba," and memorandum for the president, "Subject: Joseph Newman on Cuba," March 31, 1961, FRUS.

493 "David will again . . .": Washington D.C., February 17, 1961, NSC files, countries series, Cuba, general, 1/61–4/61, FRUS.

493 "Don't forget . . .": President's Special Assistant for National Security Affairs (Bundy) to President Kennedy, memorandum, Washington, D.C., March 15, 1961.

493 "I would have . . .": *Miami Herald*, April 16, 2000.

22. The Road to Girón Beach

494 the CIA presented: NSC action memorandum 31, Washington, D.C., March 11, 1961, NSC files, FRUS.

495 "done a remarkable job . . .": President's Special Assistant for National Security Affairs (Bundy) to President Kennedy, memorandum, Washington, D.C., March 15, 1961, NSC files, countries series, Cuba, FRUS.

495 "suitable for guerrilla . . .": CIA, "Revised Cuban Operation," Washington, D.C., March 15, 1961, NSC files, countries series, Cuba, FRUS.

495 "We were standing in . . .": LL interview with Colonel Jack Hawkins.

495 "have the right . . .": General Gray, summary notes prepared on May 9, 1961, NSC files, countries series, Cuba, "Subject: Taylor Report," FRUS.

495 "the invasion force . . .": Joint Chiefs of Staff to Secretary of Defense McNamara, memorandum, JCSM–166–61, Washington, D.C., March 15, 1961, "Subject: Taylor Report," FRUS.

495 fifty-fifty chance of: "Review of Record of Proceedings Related to Cuban Situation, May 5," Naval Historical Center, area files, Bumpy Road materials, Taylor Report, FRUS.

496 "a divorce between . . .": Robert A. Hurwitch, KLOH.

496 "Since I think you lean . . .": President's Special Assistant for National Security Affairs (Bundy) to President Kennedy, memorandum, Washington, D.C., February 18, 1961, FRUS.

496 "proscribe[d] the use . . .": Assistant Secretary of State for Inter-American Affairs (Mann) to Secretary of State Rusk, memorandum, Washington, D.C., February 15, 1961, FRUS.

496 When the president asked: Commander in Chief, Atlantic (Dennison), to Chairman of the Joint Chiefs of Staff (Lemnitzer), memorandum, Norfolk, Va., March 28, 1961. An editorial note refers to the March 29, 1961, meeting. FRUS.

496 Kennedy insisted: Notes Relating to Instructions on Bay of Pigs invasion February 9, 1963: Washington National Records Center RG 330, McNamara Files; FRC 71A3470, Cuba, FRUS.

496 "morale reasons": One of the participants in Cuban Study Group, probably Dulles, admitted this after the invasion, memorandum for the record, Washington, D.C., May 1, 1961, "Subject: Taylor Report," FRUS.

497 "to give this activity . . .": Wyden, p. 122.

497 Rusk was particularly upset: ibid., p. 148.

498 "he still wished . . .": General Gray notes, March 15, 1961, "Subject: Taylor Report," FRUS.

498 "I hear you don't think . . .": ATD, p. 259. See also RKIHOW, p. 242.

498 "entrenched Cold War ways": RKHT, p. 443.

499 "to deny any such CIA activity": memorandum for the president, "Subject: Cuba: Political, Diplomatic, and Economic Problems," April 10, 1961, president's office files, box 65, JFKPL. This document also may be found at www.parascope.com under "Foreign Ops" and in National Security Files, Country Series, Cuba, Genoma, FRUS.

499 "When lies must be": ibid.

500 Over the weekend: Jack Hawkins, "Classified Disaster: The Bay of Pigs Operation Was Doomed by Presidential Indecisiveness and Lack of Commitment," *National Review*, December 31, 1996, and LL interview with Colonel Jack Hawkins.

500 "Bissell said he felt . . .": LL interview with Colonel Jack Hawkins.

501 "to finance anti-Castro . . .": *CIA Targets Fidel*, p. 42.

501 Varona had no more: ibid., pp. 27–28, 32.

501 "felt it was . . .": Theodore C. Sorensen, KLOH.

501 "The minute I land . . .": Richard Goodwin, p. 174.

502 Suspicious reporters: Kornbluh, p. 304.

503 Steve Smith, a weekend guest, thought: Wyden, p. 194.
503 vowed that if: LL interviews with Grayston Lynch and Roberto San Román.
503 The UN ambassador was full: CY, p. 115.
503 "If Cuba now proves . . .": from: New York to: Secretary of State, no. 2892, April 16, 1961, 6:00 P.M., NSC files, box 40A, JFKPL.
503 no longer be able: Deputy Director of the CIA (Cabell) to General Maxwell D. Taylor, memorandum, Washington, D.C., May 9, 1961, "Subject: Taylor Report," FRUS.
503 "regularly harassed . . .": Joseph Alsop, KLOH.
503–04 "He used to drive him . . .": John Bartlow Martin interview with Robert Kennedy, May 14, 1964, ASP.
504 "who never quite . . .": RKIHOW, p. 205.
504 "I'm going down . . .": Senator Wayne Morse, KLOH.
504 "overriding considerations": memorandum for the record, Washington, D.C., Subject: Taylor report, May 4, 1961, FRUS.
504 Kennedy told Feldman: LL interview with Myer Feldman.
505 Rusk agreed: Deputy Director of the CIA (Cabell) to General Maxwell D. Taylor, memorandum, Washington, D.C., May 9, 1961, "Subject: Taylor Report," FRUS.
505 During that first day: Kornbluh, p. 314.
506 "I don't think . . .": quoted in RKIHOW, p. 242.
506 involved with clandestine aspects: CIA, memorandum, Washington, D.C., January 19, 1961, Special Group meeting, Cuba, FRUS.
506 "The real question . . .": President's Special Assistant for National Security Affairs (Bundy) to President Kennedy, memorandum, Washington, D.C., April 18, 1961, FRUS.
506 Bobby cautioned all those: Wyden, p. 289.
507 "Can anti-Castro forces . . .": Chief of Naval Operations (Burke) to Commander in Chief, Atlantic (Dennison), telegram, Washington, D.C., April 18, 1961, 3:23 P.M., Naval Historical Center, area files, Bumpy Road materials, FRUS.
507 "nobody knew . . .": memorandum of conversation, Washington, D.C., April 18, 1961, Naval Historical Center, area files, Bumpy Road materials, FRUS.
507 "pressing for every . . .": quoted in Wyden, p. 267.
507 Cleveland thought: ibid.
507 the president had allowed: For testimony on the way the Joint Chiefs were kept out of the crucial policy loop, see General Earle Wheeler, KLOH.
507 "The president is going . . .": memorandum of conversation, Washington, D.C., April 18, 1961, Naval Historical Center, area files, Bumpy Road materials, FRUS.
507 had to take down: Captain William C. Chapman, USN retired, "The Bay of Pigs: The View from Prifly," paper presented at the Ninth Naval History Symposium, U.S. Naval Academy, October 29, 1989, and LL interview with William C. Chapman.
508 "There's no doubt . . .": LL interview with William C. Chapman.

508 "Authorities [the president] would . . .": Chief of Naval Operations (Burke) to Commander in Chief, Atlantic (Dennison), telegram, Washington, D.C., April 18, 1961, 8:37 P.M., Naval Historical Center, area files, Bumpy Road materials, FRUS.

508 "Wounded should . . .": ibid.

508 "Evacuation of . . .": Commander in Chief, Atlantic (Dennison), to Chief of Naval Operations (Burke), telegram, Norfolk, Va., April 19, 1961, 2:01 A.M., Naval Historical Center, area files, Bumpy Road materials, FRUS.

508 "Burke, I don't want . . .": Wyden, p. 270.

508 as close to crying: CY, p. 123.

508 "We've got to do . . .": *Time* file for RFK cover, Hugh Sidey material, LL interview with Hugh Sidey, CY, p. 123, and Evan Thomas, *Robert Kennedy* (2000), p. 122.

509 "We are sincerely . . .": Embassy in the Soviet Union to Department of State, Moscow, telegram, April 18, 1961, 2:00 P.M., Department of State, FRUS.

509 most crucial creations: Kennedy wrote in *Why England Slept*: "The great advantage a democracy is presumed to have over a dictatorship is that ability and not brute force is the qualification for leadership. Therefore, if a democracy cannot produce able leaders, its chance for survival is slight." WES, p. 182.

509 "to put the guerrillas . . .": memorandum for the record, Washington, D.C., May 16, 1961, Naval Historical Center, Bumpy Road materials, FRUS.

510 "God damn it": "After Action Report on Operation Pluto, May 4, 1961," "Report on Activities on Barbara J.," p. 7, quoted in Chapman, "The Bay of Pigs," p. 29.

510 "Am destroying all . . .": CIA, memorandum prepared for the Cuban Study Group, Washington, D.C., May 3, 1961, NSC files, FRUS.

510 "to act or be judged . . .": CY, p. 124.

511 looked at the *Washington News*: Wyden, p. 290.

511 "Soviet Cuban": handwritten notes of President Kennedy during a meeting with José Miro Cardona, et al., re: Cuban situation, April 19, 1961, JKPPP.

511 "a landing of supplies . . .": Kornbluh, p. 319.

511 "present situation in Cuba . . .": Attorney General (Kennedy) to President Kennedy, memorandum, Washington, D.C., April 19, 1961, FRUS.

512 "I hope . . .": RKIHOW, p. 11.

512 "I understand that you advised . . .": *New York Times*, May 28, 1961.

513 "When I took exception . . .": notes on cabinet meeting, Washington, D.C., April 20, 1961, Yale University, Chester Bowles papers, FRUS.

513 plan to invade Cuba: Joint Chiefs of Staff to Secretary of Defense McNamara, memorandum, JCSM–278–61, Washington, D.C., April 26, 1961, FRUS.

513 "You people are so . . .": quoted in Richard Goodwin, p. 187.

513 "Jackie walked . . .": TR, p. 400.
514 As the pilots skimmed: LL interview with Stanley Montunnas, and Chapman, "The Bay of Pigs," pp. 20–34.
514 "One of the intelligence men . . .": ibid.
514 "We were all . . .": LL interview with Stanley Montunnas.
515 "fucking brass hats": SB, p. 114. For Kennedy's view of the Joint Chiefs, see also ATD, p. 295.
515 putting his arm around him: RKHT, p. 295.
515 "Let's go in . . .": LL interview with Edward Kennedy.
515 "he'd rather be called . . .": JFK discussion with General Lemnitzer at seventeenth meeting of Cuban inquiry, in Luis Aguilar, introduction, *Operation Zapata: The "Ultrasensitive" Report and Testimony of the Board of Inquiry on the Bay of Pigs* (1981), p. 331.
515 "Fuck!": quoted in SB, p. 115.
516 He believed: RKIHOW, p. 245. There is no evidence that this took place.
516 Bissell, as the attorney: CIA, memorandum for the record prepared by McCone, "Discussion with Attorney General Robert Kennedy," Washington, D.C., November 29, 1961, 9:00–10:20 A.M., FRUS.
517 "probability of being able . . .": Dulles testimony to Cuban Study Group, Secret Eyes Only, PC.
517 Most of the brigade: Grayston L. Lynch, *Betrayal at the Bay of Pigs* (1998), p. 138.
517 "aircraft were probably . . .": memorandum for the record, Washington, D.C., April 26, 1961, NSC files, FRUS.
518 "Men of all ages": Juan Carlos Rodriguez, p. 194.
518 except into the swamps: Juan Carlos Rodriguez, p. 195.
518 When they turned back: ibid., and LL interview with Grayston Lynch, and Cuban Study Group to President Kennedy, memorandum 1, Washington, D.C., June 13, 1961, FRUS.
519 "I told 'em . . .": LL interview with Grayston Lynch.
519 "75 percent of": memorandum for the record, Washington, D.C., April 24, 1961, National Defense University, Maxwell Taylor papers, FRUS.
519 "I don't specifically . . .": LL interview with Colonel Jack Hawkins.
520 "How could you . . .": LL interview with Roberto San Román, and Thomas, pp. 124–25.
520 "He [Kennedy] was taking . . .": RKHT, p. 447.
520 "a positive course . . .": recommendation 5, in Aguilar, p. 51.
521 other crucial sources: There were several extraordinarily perceptive memorandums, including a paper, with five attached annexes, prepared for the National Security Council by an interagency task force on Cuba, Washington, D.C., May 4, 1961, FRUS. The interagency task force on Cuba was composed of representatives of the Departments of State, Defense, and Justice, as well as the CIA and USIA. See also President's

Deputy Special Assistant for National Security Affairs (Rostow) to President Kennedy, memorandum, Washington, D.C., April 21, 1961, FRUS.

521 no longer fully trusted: LL interviews with Myer Feldman and John Kenneth Galbraith.

23. A Cold Winter

523 often stayed: Kay Halle, CBS interview, n.d., NHP.

523 For Kennedy, it was: Dr. Janet Travell, KLOH.

523 checked into the Mayflower Hotel: Mayflower Hotel receipt, May 6, 1961, JEP.

523 "This was the first . . .": Exner, p. 221.

523 The president strained: Dr. Janet Travell, KLOH.

523 "You could see . . .": LL interview with Ben Bradlee.

523 Because he often woke up: Dr. Janet Travell, KLOH.

523 He was so allergic: ibid.

523 To help immunize: ibid.

524 There were six pills: Dr. Janet Travell, KLOH.

524 He also took: James A. Nicolas, M.D., Charles L. Burstein, M.D., Charles J. Umberger, Ph.D., and Philip D. Wilson, M.D., "Management of Adrenocortical Insufficiency During Surgery," *AMA Archives of Surgery*, November 1955. The technical term is desoxycorticosterone acetate trimethylacetate, commonly known as BOCA.

524 "He's all hopped. . . .": Arthur Krock, "Memorandum: (Aspects of John F. Kennedy)," February 9, 1972, AKP.

524 If he took too much: H. J. Sturenburg, U. Fries, and K. Kunze, "Glucocorticoids and Anabolic/Androgenic Steroids Inhibit the Synthesis of GABAergic Steroids in Rat Cortex, *Neuropsychobiology* 35, no. 3 (1997): 143–46; S. S. Sharfstein, D. S. Sack, and A. S. Fauci, "Relationship Between Alternate-Day Corticosteroid Therapy and Behavioral Abnormalities," *JAMA* 248, no. 22 (December 10, 1982): 2987–89; Department of Neurology, University Hospital Hamburg-Eppendorf, Germany.

525 Dr. Cohen put a syringe: LL interview with Dr. David V. Becker.

525 two months before: Admiral George G. Burkley, KLOH.

525 Dr. Burkley and his colleagues: PJFK, p. 121.

526 "I had the opportunity . . .": Dr. Dorothea E. Hellman to Joan and Clay Blair Jr., March 31, 1977, BP.

526 "Dr. Cohen got her . . .": OTR interview with a physician long associated with Dr. Eugene Cohen.

526 "a deceiving, incompetent . . .": Eugene J. Cohen, M.D., to Admiral George Burkley, February 20, 1964, RWP.

526 a passionate doctor: LL interview with Lorraine Silberthau, and paid death notices, *New York Times*, July 17 and July 18, 1999.

526 "He was totally . . .": LL interview with Dr. David V. Becker.

526 in 1957 had drained: Janet Travell, *Office Hours: Day and Night* (1968), p. 320.
526 "Then, as you know . . .": Eugene J. Cohen, M.D., to Admiral George Burkley, February 20, 1964, RWP.
527 "Does JFK . . .": New York *Daily News*, May 12, 1961.
527 Then Dr. Jacobson: Dr. Max Jacobson, unpublished memoir, courtesy Mrs. Max Jacobson.
527 The treatment varied: ibid.
528 "That dosage would have . . .": LL interview with Dr. Mauro G. Di Pasquale.
528 He was becoming: LL interview with Joseph Paolella.
529 Bobby had set up the meeting: Aleksandr Fursenko and Timothy Naftali, "*One Hell of a Gamble": Khrushchev, Castro, and Kennedy, 1958–1964* (1997), p. 112.
529 "We were appalled . . .": LL interview with Cartha DeLoach.
529 "If this underestimation . . .": Fursenko and Naftali, p. 113.
529 "new progressive policy . . .": ibid., p. 114. The authors of this important book benefited from unprecedented access to Russian archives. The comparable documents from the American side may still not have been released by the JFKPL.
530 "woman of loose morals": ibid., p. 115.
530 "almost a taboo subject": Edward Lansdale, KLOH.
531 "understand what Robert Kennedy . . .": Fursenko and Naftali, p. 119.
531 "Goddamn it!": Frank Saunders, with James Southwood, *Torn Lace Curtain* (1982), p. 38.
531 He had gone: TFB, p. 264.
531 "The weather be . . .": Saunders, p. 38.
532 "Send in the broads": ibid., p. 43.
532 Lem had seen: interview, Ralph Horton, BP.
533 "opportunism . . .": "National Intelligence Estimate," November 9, 1960, Washington, D.C., December 1, 1960, FRUS.
533 "From the particular vantage point . . .": Embassy in Yugoslavia (Kennan) to Secretary of State (Rusk), telegram, at Paris/1/Belgrade, June 2, 1961, 3:00 P.M., FRUS.
533–34 "In an exchange . . .": Department of State, Washington, D.C., May 23, 1961, FRUS.
534 "First, and most important . . .": Washington, D.C., February 11, 1961, NSC files, countries series, USSR (top secret), drafted by McGeorge Bundy on February 13, 1961, "The Thinking of the Soviet Leadership, Cabinet Room," February 11, 1961, FRUS.
534 "when the push . . .": Martin J. Medhurst," "Reconceptualizing Rhetorical History: Eisenhower's Farewell Address," *Quarterly Journal of Speech* 80, 1994.
535 Instead, she perused: LL interview with Joseph Boccehir. Robert White has visited Mr. Boccehir and seen many of Jacqueline Kennedy's sketches.

535 "so identical": Sarah Bradford, *America's Queen: The Life of Jacqueline Kennedy Onassis* (2000), p. 196.
535 "This evening, Madame . . .": ibid.
535 "without mixing in politics . . .": AWRH, p. 150.
536 he turned away a glass: Don Shannon papers, JFKPL.
536 the doctor attended: Dr. Max Jacobson, unpublished memoir.
536 Travell shot him up: CY, p. 189.
536 "Once an idea . . .": memorandum of conversation, Vienna, June 3, 1961, drafted by Akalovsky, approved by the White House on June 23, 1961, FRUS.
536 "without affecting . . .": ibid.
538 "The calamities of a war . . .": memorandum of conversation, Vienna, June 4, 1961, drafted by Akalovsky and approved by the White House on June 23, 1961, FRUS.
538 "I had no . . .": Joseph Alsop, KLOH.
538 When Kennedy arrived: LL interviews with Marcus Raskin and Myer Feldman.
539 a myriad of advice: All of these letters are found in Dr. Travell's papers, file correspondence, box 1, JFKPL.
539 The doctor diagnosed: Dr. Janet Travell, KLOH.
539 "mild virus infection": "Kennedy Authorized Briefings on Illness," n.d., *Washington Star* collection, Martin Luther King Library, Washington, D.C.
539 "Were any of the exhibits . . .": memorandum for Secretary of Defense (McNamara), July 10, 1961, RWC.
540–41 "With this weekend's . . .": ibid., August 14, 1961.
541 "The fact is that . . .": Joseph Alsop KLOH.
541 "Ever since the crossbow . . .": Henderson, p. xxviii.
541 Decades later: LL interview with Hugh Sidey.
542 "The H-Bomb! The H-Bomb! . . .": "The Complacent Americans," available at: www.conelrad.com/media/atomicmusic/complacent.html.
542 "The every-family-for-itself approach . . .": David Arnold, "Blast from the Past: The Bombs Never Fell, but Underground Shelters Still Dot the Suburban Landscape," *Boston Globe Magazine,* December 12, 1999.
542 "Some of us had been . . .": Yarmolinsky, *Virginia Quarterly,* Autumn, 1996.
542 "There's no problem here": RKHT, p. 429.
543 Joe and the president: Dr. Janet Travell, KLOH.
543 In the six months: The author has in his possession the bills that Dr. Jacobson submitted for the period from May 13 to October 17, 1961. In that six-month period he charged $25 daily for incidental expenses for thirty-six days, plus travel expenses. This does not include the days he was with President Kennedy on his European trip. Expenses May 12, 1961, to October 17, 1961, Max Jacobson, M.D., October 20, 1961, RWC.
543 "He wasn't a real": LL interview with Joseph Paolella.
543 "I wasn't sure . . .": LL interview with George Smathers.

543–44 "Get it away from here!": Chuck Spalding, KLOH.
544 "That's out of the question": Dr. Max Jacobson, unpublished memoir.
544 "He felt strongly . . .": LL interview with Dr. David V. Becker.
544 "I am sorry . . .": Eugene J. Cohen, M.D., to John F. Kennedy, November 12, 1961, RWC.
544 Dr. Cohen told: This story was confirmed in an LL interview with Milton Gwirtzman and by two other sources.
544 "I further told him that . . .": Eugene J. Cohen, M.D., to Admiral George Burkley, February 20, 1964, RWC.
545 "Humbly and respectfully . . .": Eugene Cohen, M.D., to John F. Kennedy, n.d. (late 1961?), RWC.
546 "obvious intense personal interest": C. A. Evans to Mr. Belmont, June 4, 1962, FBIFOI.
546 "any barbiturates or narcotics . . .": R. H. Jevons to L. W. Conrad, "Subject: Examination of Medicines for the Attorney General," June 7, 1962, FBIFOI.
546 "there was no trace . . .": Dr. Max Jacobson, unpublished memoir.
546 when John Jr. rushed forward: LL interview with Larry Newman.
547 "I will not treat this patient . . .": Richard Reeves, p. 273. This was based on an interview with Dr. Kraus shortly before his death.
547 "Dr. Travell Quitting . . .": Travell, p. 396.
547 "I hate to use . . .": Eugene J. Cohen, M.D., to Admiral George Burkley, February 20, 1964, RWC. In her autobiography Dr. Travell asserts that the president told her that he did not want her resignation, and life went on as before. Travell, p. 397.
547 Dr. Burkley became in effect: Dr. George G. Burkley, KLOH. It was not until July 1963, when the new *Government Organization Manual* listed Dr. Burkley's title as "White House physician," that it became public knowledge that the navy doctor was Kennedy's personal physician.
547 many members of the staff: LL interview with Myer Feldman. Deputy Press Secretary Malcolm Kilduff says that when he asked what had happened with Travell, he was told not to ask any questions. LL interview with Malcolm Kilduff.
547 "I asked him many . . .": LL interview with Dr. David V. Becker.
548 frolicking in the White House pool: LL interview with Malcolm Kilduff.
548 "to give him up . . .": Cartha D. DeLoach, *Hoover's FBI: The Inside Story of Hoover's Trusted Lieutenant* (1997), p. 38.
548 "You must always . . .": LL interview with John Kenneth Galbraith.
549 "I would say I . . .": LL interview with Joseph Paolella.
549 "hit the White House . . .": oral history, Jacqueline Kennedy Onassis, University of Kentucky Library.
549 At the first state dinner: interview, Esther Van Wagoner Tufty, DHP, and Maxine Cheshire, with John Greenya, *Maxine Cheshire, Reporter* (1978), p. 44.
549 "There was a tremendous . . .": OTR interview.

550 "It all happened . . .": LL interview with John White.
551 "A moment later . . .": Exner, p. 205.
551 "I, Anthony Summers . . .": LL interview with Anthony Summers.
551 $5,000 in "expenses": deposition of Judith Exner, December 4, 1997, *Judith Exner, Plaintiff, vs. Random House, Inc., Ballantine Books, Villard Books, Laurence Leamer, Milt Ebbins,* does 2–100, and interview, Gerri Hirshey.
551 "I did try . . .": LL interview with Liz Smith.
551 this is denied by both: LL interviews with Mark Obenhaus and Seymour Hersh.
552 "from extensive bone . . .": transcript of proceedings, *Judith Exner vs. Random House, Inc., et al.,* December 17, 1998, JEP.
552 "Whenever she'd come . . .": LL interview with Robert J. McDonnell.

24. Bobby's Game

553 he hid on the plane: Thomas, p. 110.
553 "on the basis . . .": C. A. Evans to Mr. Belmont, "Subject: Prowler at the Home of the Attorney General," November 25, 1961, FBIFOI.
553 "The attorney general . . .": interview, Rose Kennedy, RCP.
553–54 "My father, all of us . . .": LL interview with Christopher Kennedy.
554 "I can understand . . .": LL interview with Harris Wofford.
554 about 70 percent: Branch, p. 374.
554 "in some areas . . .": RKIHOW, p. 79.
554 "in the interest . . .": ibid.
554–55 spent five weeks: Edwin Guthman, *We Band of Brothers: A Memoir of Robert F. Kennedy* (1971), p. 160.
555 There had been: ibid., p. 159.
555 "Yankee go home!": from: Special Agent in Charge, Atlanta to: Director, May 5, 1961, FBIFOI.
555 "Well, we're not going . . .": oral history, Scott I. Peek, administrative assistant to Senator George A. Smathers, 1952–63, U.S. Senate Historical Office.
556 His hands shook: Branch, p. 414.
556 "In the worldwide . . .": Honorable Robert F. Kennedy, address, May 6, 1961, University of Virginia Library, DHP.
557 "Stop them!": Wofford, p. 125.
557 "The meeting": LL Interview with Harris Wofford.
558 "It lasted": ibid.
558 "In the election": Wofford, pp. 128–29.
558 A few school districts: William Henry Kellar, *Make Haste Slowly: Moderates, Conservatives, and School Desegregation in Houston* (1999).
558 about 214,000 out of : *Meet the Press,* NBC, September 24, 1961, DHP.
559 The Democrats had lost: MP1960, p. 393.
559 "I think you should— . . .": RKHT, p. 296.
560 Floyd Mann, the chief: Branch, p. 448.

560 "a turning point . . .": quoted in Robert E. Thompson and Hortense Myers, *Robert F. Kennedy: The Brother Within* (1962), p. 167.
561 "You shouldn't have . . .": Branch, p. 464.
561 "John, it's more important . . .": ibid., p. 465.
562 "But when all has . . .": Thompson and Myers, p. 171.
562 "Lansdale (the Ugly American)": Attorney General Robert F. Kennedy handwritten notes, November 7, 1961, RFK papers, JFKPL.
563 "My idea is to stir . . .": ibid.
563 "the people themselves . . .": IR, p. 140.
563 "arouse premature actions . . .": ibid., p. 141.
563 "a dead Castro . . .": Chairman of the Board of National Estimates (Kent) to Director of Central Intelligence (Dulles), memorandum, Washington, D.C., November 3, 1961, FRUS.
564 "special intelligence estimate . . .": General Edward Lansdale to Attorney General Kennedy, memorandum, November 30, 1961, FRUS.
564 "no authoritarian regime . . .": memorandum for the record, Washington, D.C., December 14, 1961, CIA, DCI (McCone) files, FRUS.
564 "we may be heading . . .": Director of Intelligence and Research (Hilsman) to Deputy Undersecretary of State for Political Affairs (Johnson), memorandum, Washington, D.C., February 20, 1962, FRUS.
564 "This effort may . . .": memorandum from Chief of Operations (Mongoose) Lansdale, Washington, D.C., December 7, 1961, FRUS.
564 One of the stories": Sterling Seagrave, *The Marcos Dynasty* (1988), p. 145.
565 "with outside help . . .": Chief of Operations (Mongoose) Lansdale, program review, Washington, D.C., February 20, 1962, FRUS.
565 "to get off your ass . . .": IR, p. 141.
566 One scheme was: ibid., pp. 142–43.
566 "a major psychological . . .": notes on an Operations Group meeting, Washington, D.C., June 7, 1962, Department of State, FRUS.
566 "The top priority . . .": Chief of Operations in the Deputy Directorate for Plans (Helms) to Director of Central Intelligence (McCone), memorandum, Washington, D.C., January 19, 1962, CIA, FRUS.
566 "monolithic and ruthless . . .": quoted in Lawrence Freedman, *Kennedy's Wars: Berlin, Cuba, Laos, and Vietnam* (2000), p. 147.
567 "Life for him . . .": quoted in Maxwell Taylor Kennedy, ed., *Make Gentle the Life of the World: The Vision of Robert F. Kennedy* (1998), p. 133.
567 "Let me knock you down": Warren Rogers, *When I Think of Bobby: A Personal Memoir of the Kennedy Years* (1993), p. 81.
567 Shriver felt uncomfortable: Wofford, p. 386.
567 her children watched: New York *Daily News*, December 9, 1961.
568 "trash barrel assault": ibid.
568 In Rome at the end: George Dixon, "Gift to 'Crash Kennedy,' " clipping dated March 22, 1962, JFKPL.

568 Mrs. Nicholas Katzenbach was so bold: notes on Robert Kennedy, Mrs. Nicholas Katzenbach, ASP.
568 "What do you think . . .": John Seigenthaler, KLOH.
569 "When I go to Hickory Hill . . .": interview, Joan Kennedy, Laura Bergquist papers, Boston University.
570 "I never once . . .": Rowland Evans, KLOH.
570 "I say out!": LL interview with Charles Bartlett. Mr. Bartlett was present on this occasion at Hickory Hill.

25. Lives in Full Summer

572 "You know, the fact . . .": Letitia Baldrige Hollensteiner, KLOH.
572 For a ceremony at the: ibid.
573 "for the first time . . .": *Washington Star*, December 11, 1961.
574–75 "Well, I think . . .": Salinger, p. 115.
575 watched by forty-six: *Newsweek*, February 26, 1962.
576 "Where did you get . . .": LL interview with Hugh Sidey.
577 "Kennedy used . . .": LL interview with John Kenneth Galbraith.
577 Alsop noted that: Alsop, with Platt, p. 440.
578 "Do keep this . . .": Theodore H. White to Robert F. Kennedy, March 14, 1961, ASP.
578 "would be delighted . . .": Robert F. Kennedy to Theodore H. White, March 23, 1961, ASP.
578 "Kennedy's notion . . .": LL interview with Bob Healy.
578 "Corbin was abrasive . . .": LL interview with Larry Newman.
578 "Kenny just didn't . . .": LL interview with Bob Healy.
579 "whereas I think . . .": quoted in a review of Jules Witcover, *The Dark Side of Camelot, Columbia Journalism Review*, January–February 1998.
579 "I may have . . .": LL interview with Ben Bradlee.
579 Seigenthaler, moreover: LL interview with John Seigenthaler. See also John Seigenthaler, KLOH.
579 "Jolly Paul Corbin . . .": *Newsweek*, September 4, 1961.
579 "Fire him!": John Seigenthaler, KLOH.
580 "Good Lord, why do . . .": Teno Roncalio, KLOH.
580 "I was there . . .": interview, Evelyn Jones, RCP.
580 "We wanted to": LL interview with Joan Kennedy.
581 "Don't lose a day": Clymer, p. 31.
581 That was a lesson: *Boston Traveler*, September 25, 1961.
581–82 "Nobody forced me to run": TEEK, p. 147.
582 "wealthy personable lightweight": Edward M. Kennedy to Robert F. Kennedy, "Subject: Redbook Profile," n.d., RFK papers, JFKPL.
582 "the article should be fair . . .": ibid.
583 "little reason for . . .": ibid.

583 "there are those in . . .": William Peters, "Teddy Kennedy," *Redbook*, June 1962.
583 "the entire community": ibid.
583–84 "I feel that the Virginia . . .": Edward M. Kennedy to Robert F. Kennedy, "Subject: Redbook Profile," n.d., RFK papers, AG correspondence, JFKPL.
584 "The vital thing . . .": Robert F. Kennedy to Edward M. Kennedy, December 14, 1961, RFK papers, AG correspondence, JFKPL.
584 "Saddle up . . .": Betty Hannah Hoffman, "What It's Like to Marry a Kennedy," *Ladies' Home Journal*, October 1962.
585 Joan, unlike Jackie: *Louisville Times*, June 16, 1962.
586 "You rich boys . . .": Gerald Tremblay, KLOH.
586 "the naturalness of a newspaperman . . .": quoted in Laura Bergquist, "Life on the New Frontier," *Look*, January 2, 1962.
586 On one occasion: *New York Times*, June 12, 1977.
587 "My father and mother . . .": LL interview with Edward Kennedy.
587 One day he fell: Saunders, pp. 76, 94.
587 "Why, look at that . . .": LL interview with Luella Hennessey Donovan.
587 "looked like whores": Saunders, pp. 83–84.
588 "everyone in the room . . .": John Jay Hooker Jr., KLOH.
588 High up in the: Saunders, p. 100.
588 "All right, everybody . . .": Paul Fay, unedited manuscript PFP.
588 "He is so big . . .": Rose Kennedy, "Thanksgiving '61," HTF, p. 699.
589 Rose thought: ibid.
589 "It's so incredible . . .": LL interview with Frank Waldrop.
589 "I had to stay . . .": TFB, p. 227.
591 "He'll be fine . . .": Saunders, p. 121.
591 "nothing more to live . . .": interview, Steve Smith, March 5, 1976, ASP.
591 when she entered: Saunders, p. 134.
592 One of them covered: Rita Dallas, with Jeanira Ratcliffe, *The Kennedy Case* (1973), p. 87.
592 Bobby and Ethel hurried: ibid, p. 36.
592 "For a group of people . . .": interview, Dr. Henry Betts, CP.
593 "He was very . . .": ibid.
594 "I don't know, I don't . . .": ibid.
594 "Uncle Joe, the family . . .": Dallas, p. 125.
594 "I think she . . .": interview, Dr. Henry Betts, CP.

26. Dangerous Games

597 could have been fined: William J. Eaton, "Rule Notwithstanding, Brumus Joins RFK on Job," UPI, n.d., FBIFOI.
597 Bobby had scarcely: R. C. Renneberger to Mr. Callahan, "Subject: Attorney General Visit to Mechanical Section," February 1, 1961, FBIFOI.
597 That same evening: H. L. Edwards to Mr. Mohr, "Subject: Attorney Gen-

eral's Efforts to Get into the Bureau Gymnasium Tuesday Evening, 1/31/61," February 1, 1961, FBIFOI.

597 push-ups he performed: J. F. Malone to Mr. Mohr, "Subject: Visit of Attorney General to Basement Gymnasium," February 8, 1961, FBIFOI.

597 "is controlled and to a . . .": M. A. Jones to Mr. DeLoach, "Subject: Interview of Attorney General Robert F. Kennedy," March 20, 1961, FBIFOI.

598 "They are not important . . .": *Los Angeles Times*, November 5, 1961, FBI-FOI.

598 "that in regard . . .": John Edgar Hoover to Mr. Tolson, Mr. Belmont, Mr. Mohr, and Mr. DeLoach, memorandum, October 9, 1961, FBIFOI.

598 "Hoover was a . . .": LL interview with William Hundley.

599 Johnson spoke the language: William F. Roemer Jr., *Man Against the Mob* (1989), p. 206.

599 Bobby listened to the: ibid.

599 according to Cartha DeLoach: Thomas, p. 423.

599–600 "In one particular instance . . .": LL interview with Cartha DeLoach.

600 "You look old to me": Thomas, p. 118.

600 "ridiculous on the face of it": RKIHOW, p. 129.

600 "a Polish-Jewish refugee": To: Director, FBI, from Legat, Rome, January 30, 1961, FBIFOI.

600 FBI records: Seymour Hersh, p. 113.

600–01 more than six hundred wiretaps: James W. Hilty, *Robert Kennedy: Brother Protector* (1997), p. 233.

601 nearly eight hundred bugs: Thomas, p. 117.

601 Hoover, however, could produce: DeLoach, p. 52, and RKHT, p. 274.

601 "perhaps . . . he did not . . .": RKHT p. 278.

601 In the White House, Mike: LL interview with Myer Feldman.

601 He also boasted: C. A. Evans to Mr. Parsons, "Subject: Samuel M. Giancana Anti-racketeering," March 6, 1961, HSCA, NA.

601 "I hope Sinatra . . .": Paul Fay, unedited manuscript.

602 "The attorney general indicated . . .": J. Edgar Hoover to Mr. Tolson, Mr. Parsons, Mr. Rosen, and Mr. DeLoach, April 17, 1961, Giancana files, HSCA, NA.

602 "I wrote Sam's name . . .": ibid.

602 "& promptly": Special Agent in Charge, Chicago, to FBI Director Hoover, December 9, 1961, HSCA, NA. The memo refers to "John (LNU) possibly John Drew." Evan Thomas has concluded that this was Johnny Formosa, Thomas, p. 163. The memos themselves are courtesy of Evan Thomas.

603 "As the Bureau is aware . . .": Special Agent in Charge to FBI Director Hoover, January 18, 1962, HSCA, NA.

604 "The relationship between . . .": J. Edgar Hoover to Kenneth O'Donnell, February 27, 1962, HSCA, NA.

604 "So what do you . . .": LL interview with Joe Dolan.

605 "This man is going . . .": quoted in Seymour Hersh, p. 104.
605 "Well, she loved him . . .": LL interview with George Smathers.
606 "believe[d] his . . .": R. W. Smith to W. C. Sullivan, "Subject: Frank A. Capell," July 14, 1964, FBIFOI.
606 In the years since: Donald Spoto, *Marilyn Monroe* (1993), pp. 594–611.
606 Bobby's host and: ibid., pp. 562–63.
606 Beyond that: Special Agent in Charge, San Francisco, to FBI Director, August 6, 1962, FBIFOI.
606 On one of these: LL interview with Edwin Guthman.
606 "I'm dancing . . .": LL interview with Joe Naar.
606 "Eunice kidded . . .": LL interview with Charles Bartlett.
606 "The man in charge . . .": LL interview with Milton Gould.
607 "If there's . . .": LL interview with Milt Ebbins.
607 a decision that haunted: LL interview with Patricia Lawford Stewart.
607 "Bobby's three younger . . .": *Time*, August 3, 1962, p. 10.
607 The Forest Service, at: *U.S. News & World Report*, August 27, 1962, and Special Agent in Charge Milnes to Mr. Hoover, "Subject: Visit of Attorney General to Seattle and the Northwest," August 7, 1962.
608 That was October 31, 1960: IR, p. 125.
608 "If you have seen . . .": ibid., p. 133.
608 "He cautioned Larry Houston . . .": interview, Colonel Sheffield Edwards, Rockefeller Commission on CIA Activities Within the United States, April 9, 1975, Gerald R. Ford Library.
608 "I told the attorney general . . .": Memorandum for Messrs. Tolson, Belmond, Evans, Sullivan, DeLoach, and Malone, May 10, 1961, HSCA, NA.
609 "Mr. Bissell . . . in connection . . .": J. Edgar Hoover to Robert F. Kennedy, May 22, 1962, FBIFOI. See also IR, pp. 125–27.
609 "It must be understood . . .": LL interview with Myer Feldman.
609 "contingency plans": Brig. Gen. Lansdale, memorandum for the record, "Subject: Meeting with President," March 16, 1962, Courtesy Gus Russo. See also David Corn and Gus Russo, "The Old Man and the CIA," *Nation*, March 26, 2001.
610 "I commented that this was . . .": ibid.
610 "exploited": memorandum for the record, Caribbean Survey Group, March 21, 1962. Courtesy Gus Russo.
610 "His job was . . .": LL interview with Samuel Halpern. See also Thomas, p. 178.
611 Operation Mongoose was not: CIA Operations Officer for Operation Mongoose (Harvey) to Chief of Operations (Mongoose) Lansdale, memorandum, Washington, D.C., July 24, 1962, FRUS.
611 Castro was so successful: memorandum for the file, Washington, D.C., December 27, 1961, CIA, "Subject: Discussion with Attorney General Robert Kennedy," FRUS.

611 "expressed skepticism . . .": guidelines for Operation Mongoose, Washington, D.C., March 14, 1962, Department of State, S/S files: Lot 65 D 438, FRUS.
611 "Bobby came over almost . . .": LL interview with Dino A. Brugioni.
612 "I don't think . . .": Dino A. Brugioni, *Eyeball to Eyeball: The Inside Story of the Cuban Missile Crisis* (1991), p. 69.
612 "If you're going to . . .": LL interview with Samuel Halpern.
613 By the end of July 1962: CIA Operations Officer for Operation Mongoose (Harvey) to Chief of Operations (Mongoose) Lansdale, memorandum, Washington, D.C., July 24, 1962, NSC files, FRUS.
613 "Which one is Meredith?": Branch, p. 648.
614 "Thousand Said Ready to Fight for Mississippi": ibid., p. 653.
614 "I love our customs!": ibid., p. 659.
614 "And now—Governor . . .": RKHT, p. 320.
615 "I appreciate your interest . . .": PRIUM, belt 4-A.
615 "should be Mandrake the Magician": RKIHOW, p. 167.
616 "Mr. President, let me say . . .": PRIUM, belt 4-A.
616 "I say I'm going to . . .": PRIUM, belt 4-C.
616 Bobby gave him: Branch, p. 660.
616 "the orders of the court . . .": ibid., p. 665.
617 "The attorney general announced today . . .": PRIUM, audiotape 26.
617 "I knew that he . . .": RKIHOW, p. 165.
618 "They better fire . . .": PRIUM, audiotape 26.
618 "was the avoidance . . .": RKIHOW, p. 160.
618 "Well, I think we're gonna have . . .": PRIUM, audiotape 26A.

27. "A Hell of a Burden to Carry"

621 Kennedy had hurried: Saunders, p. 182.
621 "The South Korean army should have . . .": conversation with General Douglas MacArthur, August 16, 1962, presidential recordings, tape 12, JFKPL.
621 "assess accurately . . .": President's Military Representative (Taylor) to President Kennedy, memorandum, Washington, D.C., August 17, 1962, NSC files, FRUS.
621 "noise level": ibid.
622 "wondered whether . . .": memorandum of conversation, Washington, D.C., January 30, 1962, FRUS.
622 "wondered whether . . .": ibid.
622–23 During a meeting: Press Secretary (Salinger) to President Kennedy, memorandum for the president, February 1, 1962, FRUS.
623 "At the time I called . . .": Aleksei Adzhubei, report to Central Committee of the Communist Party of the Soviet Union, March 12, 1962, quoted in Fursenko and Naftali, p. 153.

623 new $133 million: ibid., p. 154.

623 Although historians debate: For varying analyses of the reasons, see Tony Judt, "On the Brink," *New York Review of Books*, January 15, 1998, and Ernest R. May and Philip D. Zelikow, eds., *The Kennedy Tapes: Inside the White House During the Cuban Missile Crisis*, (1997), pp. 666–70.

623 "Tell Fidel . . .": Fursenko and Naftali, p. 182.

623 Khrushchev would arrive: ibid.

624 "these developments . . .": intelligence memorandum, OCI 3047/62, "Subject: Recent Soviet Military Aid to Cuba," August 22, 1962, FRUS.

624 "the instantaneous commitment . . .": memorandum of meeting with President Kennedy, August 23, 1962, CIA, DCI (McCone) files, FRUS.

625 remained deeply suspicious: interview with Myer Feldman.

625 "ominous reports": Fursenko and Naftali, p. 205.

625 "Put it back . . .": Marshall Carter, memorandum, September 7, 1962, CIA, DCI (Dulles) files, FRUS.

625 "Goddamn it!": Thomas, pp. 207–8. See also Fursenko and Naftali, pp. 193–94.

626 "suspect[ed] the presence . . .": notes prepared by Acting Director of Central Intelligence Carter, September 6, 1962, CIA, NSC meeting, FRUS.

626 "Out of respect . . .": Fursenko and Naftali, p. 209.

626 "These congressmen . . .": telegram 616 from Moscow, September 7, 1962, FRUS.

627 Soviet plans were: Fursenko and Naftali, pp. 188, 217.

627 each probably carrying: *Newsweek*, May 14, 2001.

627 50,874 Soviet troops: ibid., p. 188.

627 double the number: ibid.

627 He wanted American pilots: ibid., p. 217.

628 "general impression . . .": memorandum of Operation Mongoose meeting, October 4, 1962, FRUS.

628 "nothing was moving . . .": ibid.

629 "over twenty times . . .": Fursenko and Naftali, p. 217.

629 "The weapons that . . .": ibid., p. 219.

629 "a quiet evening . . .": McGeorge Bundy, *Danger and Survival: Choices About the Bomb in the First Fifty Years* (1990), pp. 684–85.

630 "Oh shit!": Brugioni, p. 223.

630 "general dissatisfaction . . .": memorandum for the record, CIA, "Mongoose Meeting with the Attorney General," October 16, 1962, FRUS.

630 There were two meetings: The participants in the Ex Comm varied somewhat from meeting to meeting. They included John and Robert Kennedy, and from the State Department, not only Secretary Rusk but Charles Bohlen, a Russia expert (who in the midst of the crisis left to become ambassador to France), Undersecretary of State George Ball, Deputy Secretary U. Alexis Johnson, Assistant Secretary of State for Latin America

Edwin M. Martin, and, newly returned from the Soviet Union, Ambassador at Large Llewellyn Thompson. The Defense Department contingent, headed by Secretary McNamara, included his deputy Roswell Gilpatric, Assistant Secretary Paul Nitze, and General Maxwell Taylor, the chairman of the Joint Chiefs. The CIA group, led by McCone, included his deputy Marshall Carter and the head of NPIC, Arthur Lundahl. The other members included Vice President Johnson, UN Ambassador Adlai Stevenson, McGeorge Bundy, Ted Sorensen, Secretary of the Treasury C. Douglas Dillon, USIA Deputy Director Don Wilson, and Kenneth O'Donnell. Others were called in at various times.

631 "What is the strategic . . .": Ex Comm meeting, October 16, 1962, tapes 28 and 28A, JFKPL. The Ex Comm meetings were secretly tape-recorded by President Kennedy. They have been transcribed and annotated in May and Zelikow, *The Kennedy Tapes.* Sheldon M. Stern, the longtime historian at the JFKPL, argued in the *Atlantic* in May 2000 that there are so many errors in the early editions that "*The Kennedy Tapes* cannot be relied on as an accurate historical document." Stern expands his argument in "Source Material: The 1997 Published Transcripts of the JFK Cuban Missile Crisis Tapes: Too Good to Be True?" *Presidential Studies Quarterly*, September 2000, pp. 586–93. The author has listened to the tapes and, like Stern, found differences at various places from the May and Zelikow transcription. Students of this period are strongly advised to listen to the tapes themselves and to read a new edition of *The Kennedy Tapes* that promises to deal with the pattern of errors that Stern so assiduously discovered. They are doubly urged to do so since the author has radically truncated the dialogue at the Ex Comm meeting (while attempting to remain true to the spirit of these exchanges) in order to write a narrative that the average reader will find of tolerable length. Individual exchanges are cited from the tapes themselves. Dialogue from meetings that were not taped is cited from other sources.

632 Kennedy and Bohlen sauntered: Merry, p. 386.

633 By the time: May and Zelikow, p. 122.

633 "If we wanted to . . .": Ex Comm meeting, October 18, 1962, 11:00 A.M., tapes 30 and 30A, JFKPL.

634 the top White House: LL interviews with Malcolm Kilduff and Myer Feldman.

634 "We figured . . .": LL interview with Myer Feldman.

634 "I think it's the whole . . .": The secret tape recordings of this and other Ex Comm meetings for the most part parallel other contemporaneously narrated accounts of the events, but in this instance there are major differences. McCone was a meticulous man, but his account of this meeting does not even mention Robert Kennedy's passionate doubts as recorded. It may be that McCone considered them unimportant or did not record them because they went against his own considered judgment. Memo-

randum for the file, Washington, D.C., October 19, 1962, CIA, DCI/ McCone file, Job 80–B01285A, "Meetings with the President (top secret)," drafted by McCone, FRUS.

634–35 "Cuba belonged . . .": memorandum of conversation, Washington, D.C., October 18, 1962, 5:00 P.M., NSC files, drafted by Akalovsky on October 21, 1962, approved by the White House on October 23, 1962, FRUS.

635 "The president of the . . .": TD, p. 33.

635 Only now: May and Zelikow believe that the president was "possibly accompanied by his brother," but there is no evidence that Bobby remained in the Oval Office. May and Zelikow, p. 171.

635 several secretaries, and almost certainly Bobby: It is clear that Robert Kennedy knew about the taping system because one of his first actions after he learned of President Kennedy's shooting in Dallas on November 22, 1963, was to order the system dismantled. Thomas, p. 276.

636 "Secretary McNamara, Deputy Secretary Gilpatric . . .": monologue of John F. Kennedy, October 18, 1962, tape 31, JFKPL.

637 "I think the benefit this morning . . .": Ex Comm meeting, October 19, 1962, tape 31, JFKPL.

638 "useful play . . .": The author has listened to this passage of the tape many times and remains unsure whether Kennedy says "play" or "ploy."

640 "I don't think I . . .": Lord Harlech, KLOH.

640 "This thing . . .": Thomas, p. 217.

640 "That's not what . . .": ibid.

641 "strange flip-flops": Bird, p. 233.

641 "Well, I'm having . . .": ibid., p. 234.

641 "a bit disgusted": ibid., p. 232.

641 "spoken with the president . . .": record of meeting, October 19, 1962, FRUS.

28. "The Knot of War"

643 eight hundred sorties: May and Zelikow, p. 197.

643 Bobby sat next: TD, p. 38.

643 "if we used . . .": minutes of the 505th meeting of the National Security Council, Washington, D.C., October 20, 1962, 2:30–5:10 P.M., FRUS.

644 He scribbled: In his note to himself, Kennedy said that Dillon called the missiles "flops," but the NSC account of the meeting states that "Dillon recalled that we sent United States missiles to Europe because we had so many of them we did not know where to put them." Minutes of the 506th meeting of the National Security Council, Washington, D.C., October 21, 1962, 2:30–4:50 P.M. FRUS.

644 "We don't want to . . .": Ex Comm meeting, October 22, 1962, 3:00 P.M., tapes 33 and 33A, JFKPL.

644 "that in the days . . .": May and Zelikow, p. 201.

645 The president was constantly: Bradford, p. 239.
645 "Who's that? Eddie?": Edward Berube, KLOH.
646 "It's a very difficult choice . . .": meeting with members of Congress in the Cabinet Room, October 22, 1962, tape 33A, JFKPL.
647 "They're gonna keep . . .": Ex Comm meeting, October 23, 1962, 6:00 P.M., tape 35, JFKPL.
647 "whatever was done . . .": memorandum for the files, October 23, 1962, CIA, DCI/McCone files, "Meetings with the President," FRUS.
648 "It looks really mean . . .": Ex Comm meeting, October 23, 1962, 6:00 P.M., tape 35, JFKPL.
648 "Robert Kennedy and his circle . . .": Fursenko and Naftali, pp. 249–50.
649 "I called Bolshakov . . .": LL interview with Charles Bartlett.
649 A few minutes later: Thomas, p. 222.
649 "had almost daily conversations": Anatoly Dobrynin, *In Confidence* (1995), p. 76.
649 "was far from being . . .": ibid., pp. 82–83.
649 "he had a very helpful . . .": RKHT, p. 514.
650 "an idea of the . . .": Dobrynin, p. 82.
650 That same evening: Bradford, p. 240.
650 He told the president: TD, pp. 51–52.
650 "He [Kennedy] took. . . .": Bradford, p. 240.
650 The air force's massive: Brugioni, pp. 366, 398.
651 "the danger and concern . . .": TD, p. 53.
651 "Here is the exact situation": Ex Comm meeting, October 24, 1962, 10:00 A.M, Cabinet Room, tapes 34 and 35, JFKPL.
651 "for a few fleeting . . .": This account of RFK's mindset is based on his own recollections in *Thirteen Days*, pp. 53–54, a handwritten note in RFK's personal papers, JFKPL, seen by two authors who have had access to the document: Thomas, p. 225, and RKHT, p. 514; and the tape transcripts of the Ex Comm meeting. There are subtle differences in these three accounts that the author has attempted to reconcile in a reasonable way.
652 The next afternoon: Bradford, p. 240.
652 And an ironic: ibid., p. 241.
653 "Lansdale feels badly . . .": memorandum for the director, "Subject: Mongoose Operations and General Lansdale's Problems," October 25, 1962.
653 "stated that he understood . . .": memorandum of Mongoose meeting, October 26, 1962, FRUS.
653 "that would leave . . .": memorandum of telephone conversation between President Kennedy and Prime Minister Macmillan, October 26, 1962, NSC files, FRUS.
653 "the primary Soviet . . .": paper prepared by the Planning Subcommittee of the Executive Committee of the National Security Council, NSC files, JFKPL, October 25, 1962, FRUS.
653 "it was not incoherent . . .": TD, p. 66.

653 "not transport armaments . . .": embassy in the Soviet Union to the Department of State, Moscow, telegram, October 26, 1962, 7:00 P.M., president's office files, FRUS, JFKPL.
654 "as not an unreasonable solution": Ex Comm meeting, October 27, 1962, 10:00 A.M., JFKPL.
654 "The missiles [in Turkey] were . . .": transcript of a discussion about the Cuban Missile Crisis, 1983, NSC archives 03307, DHP.
655 Joint Chiefs were preparing: Ex Comm meeting, October 27, 1962, FRUS.
655 In an invasion: Fursenko and Naftali, p. 276.
655 anyone within half a mile: ibid., p. 242.
655 "We all know. . . .": Ex Comm meeting, October 27, 1962, 4:00 P.M., Cabinet Room, tapes 42 and 43, JFKPL.
656 "Well, the only . . .": ibid.
657 "spoiling for a fight": Dobrynin, p. 87.
658 Bobby was almost crying: Thomas, p. 228.
658 "that some day . . .": Dobrynin, p. 90.
658 Castro himself had admonished: Fursenko and Naftali, pp. 272–73.
658 "In order to save . . .": ibid., p. 284.
659 "has given a . . .": Chairman Khrushchev to President Kennedy, Moscow, October 28, 1962, Department of State, FRUS.
659 "the greatest danger . . .": John F. Kennedy, speech on aid for Greece and Turkey, record of House of Representatives, April 1, 1947.
660 "Once we've got these . . .": National Security Council meeting, November 7, 1962, tape 53A. The dialogue here and in the rest of this chapter are from tapes at the JFKPL that are here transcribed for the first time.
661 "conducting surveillance . . .": NSC meeting, November 12, 1962, tape 56, JFKPL.
662 "Bobby's notion is . . .": NSC meeting, either November 14 or 15, 1962, tape 58, JFKPL.
662 "personal opinion": Fursenko and Naftali, p. 300.
662 "definite schedule . . . let's say . . .": ibid., p. 303.
662 "We have the firm impression . . .": ibid, p. 310.

29. The Bells of Liberty

664 "Ted Kennedy in Italy": Joseph A. Page, "The Precocious Ted Kennedy," *The Nation,* March 10, 1962.
665 When he left Panama: Walter Trohan, *Political Animals: Memoirs of a Sentimental Cynic* (1975), p. 327.
665 "some 200 million . . .": quoted in Clymer, p. 35.
665–66 "Bobby was opposed . . .": John Sharon, KLOH.
666 "The only argument . . .": LL interview with Chuck Spalding.
666 "Can you put it in . . .": LL interview with Bob Healy. See also Clymer, p. 36, and TEEK, p. 158.

667 "Teddy and his brothers . . .": LL interview with Milton Gwirtzman.
668 The count was: Clymer, p. 39.
669 ". . . it was wrong . . .": ibid, p. 38.
669 "Now listen, Eddie": Fay, p. 226.
670 talked to Trohan: LL interview with Walter Trohan.
670 "In Israel he almost . . .": quoted in TEEK, p. 181.
670 "Now, Mr. President . . .": LL interview with Milton Gwirtzman.
670–71 "his candidacy has already . . .": Joe McCarthy, "One Election JFK Can't Win," *Look*, November 6, 1962.
671 "widely regarded here. . . .": quoted in Clymer, p. 38.
671 "You want something . . .": ibid., p. 47.
671 "Teddy, these are. . . .": ibid.
672 "The major wheels. . . .": *Washington Star*, April 27, 1962.
672 "In the field of party . . .": Sidney Hyman, "Why There's Trouble in the New Frontier," *Look*, July 2, 1963.
672 "In rough issues . . .": ibid.
672 There were 15,600: David Kaiser, *American Tragedy: Kennedy, Johnson, and the Origins of the Vietnam War* (2000), p. 201.
673 "They're not certain . . .": presidential recordings, telephone conversations, cassette E, JFKPL.
673 The Americans placed: LL interviews with John Nolan and Barrett E. Prettyman Jr.
674 In a phrase: Thomas, p. 238.
674 "The time will probably . . .": notes of President Kennedy's remarks at the 508th NSC meeting, January 22, 1963, NSC files, FRUS.
674 "putting glass . . .": ibid.
674 "He felt a very . . .": LL interview with John Nolan.
674 On April 3: Thomas, p. 239.
674 "contingencies such as the . . .": summary record of the second meeting of the NSC Cuba Standing Group, NSC files, April 23, 1963, FRUS, JFKPL.
675 Neither Bobby: LL interview with John Nolan.
675 they brought a wet suit: John Nolan has a picture of Castro in the wet suit hanging in his Washington law office.
675 the CIA had prepared: IR, p. 86.
675 "very interested": Gordon Chase, memorandum for the record, "Subject: Mr. Donovan's Trip to Cuba," March 3, 1963, NSC archives, www.gwu.edu/~nsarchiv/.
676 "always the possibility": President's Special Assistant for National Security Affairs (Bundy) to the Standing Group of the National Security Council memorandum, April 21, 1963, NSC files, "The Cuban Problem," FRUS.
676 "desire for some noise level . . .": Gordon Chase, NSC staff, to President's Special Assistant for National Security Affairs (Bundy), memorandum, April 11, 1963, FRUS.
676 "a railway bridge . . .": ibid.

676 "might initially intensify . . .": memorandum for the NSC Cuba Standing Group, NSC files, countries series, Cuba, general, May 1–15, 1963, FRUS.
677 Manuel Artime, their leader, received: Russo, p. 172.
677 "I had many chances . . .": Blight and Kornbluh, p. 121.
678 One Sunday morning: Thomas, p. 239.
678 "in the dark-of-the-moon . . .": memorandum for the record, June 19, 1963, CIA, "Subject, Meeting at the White House Concerning Proposed Covert Policy and Integrated Program of Action Towards Cuba," FRUS.
678 "sabotage of Cuban . . .": paper prepared by the CIA for the NSC Cuba Standing Group, Washington, D.C., June 8, 1963, NSC files, FROUS JFKPL.
678 when they had burned: LL interview with Samuel Halpern.
678 "We fixed that": LL interview with Bradley Earl Ayers.
679 "set Cuba aflame": LL interview with Grayston Lynch, and Lynch, p. 171.
679 "We could float . . .": summary record of the tenth meeting of the NSC Cuba Standing Group, July 16, 1963, NSC files, FRUS, JFKPL.
680 "I need a phrase . . .": LL interview with Myer Feldman.
681 "if he had said . . .": Lord Harlech, KLOH.
681 "These were the three happiest . . .": Dorothy Tubirdy, KLOH, and LL interview with Dorothy Tubirdy.
682 "more British than Irish": Dr. Thomas J. Kiernan, KLOH.
682 "You'll have to move it": LL interview with Malcolm Kilduff.
683 "I think he felt back home": LL interview with Mary Ryan.
683 "He had always . . .": Jacqueline Kennedy, in a worldwide broadcast on what would have been JFK's forty-seventh birthday, clipping, May 30, 1964, JFKPL.
685 "Bells mark . . .": JFK notes written sometime on European trip, June 1963, JFKPP.
685 "Heroes of the past . . .": President John F. Kennedy, "Heroes Watch Us!: Some Words to Live by for Independence Day," ibid.
686 "Admiral, I want to express . . .": audiotape of 1963 reunion. Courtesy R. F. Duffy.

30. The Adrenaline of Action

688 At the first dance of the winter of 1962–63: Nina Burleigh, *A Very Private Woman: The Life and Unsolved Murder of Presidential Mistress Mary Meyer* (1998), p. 217.
688 Once, at Hickory Hill: ibid., p. 177.
688 For a year: ibid., pp. 208–16.
688 "She was a very interesting . . .": LL interview with Ben Bradlee.
689 "Mary, where have . . .": Burleigh, p. 217.
689 "What is the scandal?": presidential recordings, telephone conversations, John F. Kennedy and Arthur Schlesinger Jr., March 22, 1963, cassette E, JFKPL.

690 "How serious do . . .": interview, George Smathers, BP.
690 "Have we learned . . .": Evans to Belmont, July 2, 1963, FBIFOI.
691 Baker's assertion: LL interview with Bobby Baker. See also Thomas, p. 255.
691 "the best oral sex I ever had": Thomas, p. 444.
691 LaVern Duffy, one of Bobby's: ibid., p. 257.
691 "The White House was . . .": LL interview with Marcus Raskin.
692 "I don't know . . .": LL interview with Joseph Paolella.
692 "debaucher of a girl . . .": memorandum to Mr. Mohr, October 27, 1961, FBIFOI.
692–93 "I think that all men . . .": *The Thunderbolt*, no. 54, November 1963.
693 In July, the FBI: M. Jones to Cartha DeLoach, July 9, 1963, FBIFOI.
693 "would have been . . .": LL interview with Edwin Guthman.
693 Kennedy's aide Mike Feldman: LL interviews with Myer Feldman and Cartha DeLoach.
693 "I know how you dislike . . .": Charles Bartlett to John F. Kennedy, July 19, 1963, PC.
694 "the committee would have . . .": Charles Bartlett, memo of conversation with Kenny O'Donnell, February 1, 1963, PC.
694 "the buffer and the string cutter": LL interview with John Seigenthaler.
694 "O'Donnell's remark . . .": Charles Bartlett, untitled memo, July 19, 1963, PC.
694 Bartlett admitted years: LL interview with Charles Bartlett.
694 Newman had not: Larry Newman has verified this statement by showing the author his work records at the Secret Service.
694 He was bewildered: When Newman was transferred to the Washington field office shortly before the end of the standard two-year tour, he suspected even more that something was wrong. In September just before he left, he went into the Oval Office to have his picture taken with the president. The Secret Service agent might have spoken about the matter, but he knew that was not done. "I was afraid to say something," Newman said three decades later, when the matter still profoundly rankled him and he continued to feel that his honor had been betrayed. "I thought if I said something my whole career might blow up in my face." LL interview with Larry Newman.
695 "Charley, there are . . .": Charles Bartlett to Larry Newman, April 21, 1997, PC.
695 "The president never gave . . .": LL interview with Charles Bartlett.
695 It was supposed to cost: *Newsweek*, March 25, 1963.
695 "It's the only house . . .": AWRH, p. 190.
696 "I just hope . . .": James Reed, KLOH.
697 The thirty-one-foot-long device: *Atlanta Constitution*, August 9, 1963.
697 "to keep her courage up": O'Donnell and Powers, p. 377.
698 "My dear Jack, let's go . . .": *Look*, November 17, 1964.
698 "Did you ever think . . .": *Cape Cod Times*, November 20, 1983.
698 "Jack was one of these . . .": LL interview, with Myer Feldman.

698 "I think that Jack . . .": LL interview with Betty Coxe Spalding.
699 "Take that out . . .": LL interview with Evelyn Lincoln.
699 "Jackie—Onassis.": presidential doodles, September 20, 1963, JFKPP.
699 "The yacht has been secured . . .": President Kennedy, notes written during his stay at the Bing Crosby residence in Palm Beach, presidential doodles, JFKPP, and press release, September 30, 1963, JFKPL.
699 "twisting in the historic . . .": *Newsweek*, October 28, 1963.
699 "Well, why did you let . . .": LL interview with Marie Rider.
699 "What's the helicopter coming . . .": LL interview with Malcolm Kilduff.
700 "I'm a great big bear": Kay Halle, CBS interview, n.d., NHP.
701 "I loved you from . . .": Jacqueline B. Kennedy to John F. Kennedy, October 5, 1963. Robert White shared excerpts from this letter with the author in late 1997.
701 it could have been written: LL interview with Nina Burleigh.
701 "Why don't you . . .": JFK letter, n.d., PC.
701 Kennedy scribbled: notes for toast, JFKPP.
702 When it came time: James Reed, KLOH, and LL interview with Joan Kennedy.
702 When Bartlett gave: Fursenko and Naftali, pp. 321–22, and LL interview with Charles Bartlett.
702 "Everyone will suppose . . .": ATD, p. 835.
702–03 "through the motions . . .": CY, p. 570.
703 "It was personal . . .": interview, Robert F. Kennedy, ASP.
704 King sent: Branch, p. 757.
704 "assurance that a Negro": James Baldwin, *The Fire Next Time* (1963), p. 4.
705 "You don't have no . . .": quoted in Branch, p. 810.
706 "You've got . . .": ibid, p. 811.
706 "*the* most dramatic . . .": ibid.
707 "Pretty good job . . .": Presidential Recordings, tape 88, JFKPL.
707 "I'd like to bet . . .": Strober and Strober, p. 287.
707 "My impression of . . .": ibid.
708 "the back of an envelope . . .": RKIHOW, p. 200.
709 "Do you think we . . .": RKHT, p. 348.
709 "this was going . . .": RKIHOW, p. 176.
709 "They're going to come . . .": Thomas, p. 443.
709 "a financial pillar . . .": Branch, p. 209.
710 Hoover and his colleagues: David J. Garrow, *The FBI and Martin Luther King Jr.* (1981), p. 91.
710 "I assume you know . . .": Branch, p. 837.
710 "You've read . . .": ibid., pp. 835–37.
711 "there was considerable doubt . . .": Courtney Evans to Belmont, June 25, 1963, FBIFOI.
711 "grab one little brother": Thomas, p. 263.
711 "So you're down here . . .": ibid.
712 "This doesn't have anything . . .": Branch, p. 884.

713 "a secret member of . . .": RKIHOW, p. 141.
713 Bobby had gone: *New York Post*, March 10, 1961.
713 Instead, he might suddenly: LL interview with David Hackett.
713 He learned that there were: Robert F. Kennedy, "Buying It Back from the Indians," *Life*, March 23, 1962.
714 "I'll talk to Ethel . . .": William V. Shannon, "Bobby's Day," *New York Post*, May 6, 1963.

31. To Live Is to Choose

715 "long, hard fight . . .": *Boston Globe*, November 17, 1963.
715 "The New Frontier . . .": AP, August 5, 1963.
715 "Of course, they are . . .": John F. Kennedy and Harold Hughes, telephone conversation March 9, 1963, presidential recordings, tape E, JFKPL.
716 "I don't know . . .": John F. Kennedy and C. Douglas Dillon, IRS rules on expense accounts, March 12, 1963, ibid.
716 "We might just as well . . .": John F. Kennedy and Paul Fay, naval base closings, March 12, 1963, ibid.
716 "to keep them . . .": John F. Kennedy and Arthur Schlesinger Jr., prospective posting to Latin America for Samuel Beer, March 22, 1963, ibid.
716 "withdrawal from Vietnam . . .": presidential doodles, April 2, 1963, JFKPP.
717 "Military aid . . . aid etc . . .": notes of President Kennedy written during a NSC meeting, April 20, 1963, ibid.
717 In January 1963, General: Kaiser, p. 188.
717 "they didn't know . . .": quoted in ibid., p. 245.
717 Diem's own elder brother: Ellen J. Hammer, *A Death in November: America in Vietnam, 1963* (1987), p. 62.
718 almost been killed in 1962: Kaiser, p. 275.
718 "an authoritarian organization . . .": ibid., p. 61.
718 "To govern is to choose": John F. Kennedy, notes, August 26, 1963, JFKPP.
718 That summer: *Cape Cod Times*, November 20, 1983, and O'Donnell and Powers, p. 13.
719 "felt that he had . . .": RKIHOW, p. 394.
719 "plans to withdraw 1,000 U.S. . . .": national security action memorandum no. 263, "Subject: South Vietnam," Department of State, S/S-NSC files: Lot 70 D 265, NSC meetings (secret), printed in *Public Papers of the Presidents of the United States: John F. Kennedy, 1963,* pp. 759–60.
720 "Viet-Nam is not a . . .": from Saigon to secretary of State, October 16, 1963 (top secret), Roger Hilsman papers, JFKPL.
720 "The United States can . . .": Henry Cabot Lodge Jr. papers, vol. 2, Massachusetts Historical Society.
720 "McCone hated . . .": RKIHOW, p. 397.

720 "Diem and Nhu were undoubtedly . . .": Department of State, memorandum of conversation, "Subject: Viet-Nam," August 28, 1963 (no distribution), Roger Hilsman papers, JFKPL.

720–21 "You'd better get . . .": quoted in Kaiser, p. 264.

721 "The difficulty is, I'm sure . . .": president's office files, October 29, 1963, meetings recordings, tape 118/A54, JFKPL.

722 A top-secret October 25: "Top Secret Check-List of Possible U.S. Actions in Case of Coup," October 25, 1963, Roger Hilsman papers, JFKPL.

723 "taken alive . . .": memorandum of August 30, 1963, quoted in Hammer, p. 295.

723 "It's hard to believe he'd commit . . .": assessment of the coup in Vietnam, Ex Comm meeting, November 2, 1963, president's office files, meetings recordings, tape A55, JFKPL.

726 "One two three . . .": president's office files, telephone recordings addendum, cassette M, JFKPL.

728 bases in Costa Rica and Nicaragua: memorandum for the record, Special Group meeting, November 12, 1963, FRUS.

728 "Bob Kennedy, it seems . . .": Blight and Kornbluh, p. 122.

728 trained for an operation: Jack Anderson, "Oil Raid Story Told," *St. Paul Dispatch*, April 22, 1971; LL interviews with Bradley Ayers and Samuel Halpern; and Bradley Earl Ayers, *The War that Never Was* (1976), pp. 210–36.

729 "He [Fitzgerald] offered me . . .": Rolando Cubela Secades, HSCA interview, August 28, 1978, p. 10.

729 "I have looked . . .": quoted in Russo, p. 433.

729 the administration began: Peter Kornbluh, "JFK and Castro: The Secret Quest for Accommodation," *Cigar Aficionado*, September–October 1999.

729 "the U.S. must require . . .": memorandum for the record, minutes of the special meeting of the Special Group, November 5, 1963, FRUS.

730 FitzGerald was convinced: Russo, p. 390.

730 The president spent: appointments book, president's office files, JFKPL.

730 "Look who's here, Dad": Dallas, p. 11.

731 At least his foreign aid: *Boston Globe*, November 16, 1963.

732 "complaints . . . to pour . . .": *Boston Globe*, November 3, 1963.

732 "During her nearly . . .": *Advertiser*, November 3, 1963.

734 "vibrating so violently . . .": DP, p. 85.

734 "You know, last night . . .": ibid., p. 121.

736 "two high-powered rifles . . .": *CIA Targets Fidel: Secret 1967 Inspector General's Report on Plots to Assassinate Fidel Castro* (1996), p. 90.

32. Requiem for a President

738 already sent the *Honey Fitz:* LL interview with Ham Brown, Secret Service agent attached to Joseph P. Kennedy.

738 "Uncle Joe, there's . . .": Dallas, p. 14.

738 "Hey, chief, it's . . .": Saunders, p. 227.
739 "He was so . . .": LL interview with Kerry McCarthy.
739 Bobby asked him: Thomas, p. 277.
740 "One of your boys did it": Thomas, p. 277, and Russo, p. 303.
740 Then he called Julius Draznin: Thomas, p. 277.
740 "Castro could have made . . .": interview, Alexander Haig.
741 Teddy's rental suit: DP, p. 576.
741 New Yorkers: *Atlanta Constitution*, November 26, 1963.
741 the picture that tens of millions: *TV Guide*, January 25, 1964.
742 now they broke down: interview, Malcolm Kilduff, and DP, p. 597.
743 seek the light: Kerry McCarthy recalls Rose Kennedy talking literally about the light they had lost.

Bibliography

Abbott, Edith. *Historical Aspects of the Immigration Problem.* 1926. Reprint, New York: Arno Press, 1969.

Adams, George. *East Boston Directory*. Boston: Adams, 1849.

Adams, William Forbes. *Ireland and Irish Emigration to the New World from 1815 to the Famine.* New Haven, Conn.: Yale University Press, 1932.

Adler, Alfred. *The Individual Psychology of Alfred Adler: A Systematic Presentation in Selections from His Writings.* Edited and annotated by Heinz L. Ansbacher and Rowena R. Ansbacher. New York: Basic Books, 1956.

Aguilar, Luis. *Operation Zapata: The "Ultrasensitive" Report and Testimony of the Board of Inquiry on the Bay of Pigs.* Frederick, Md.: Aletheia Books, 1981.

Ainey, Leslie G. *Boston Mahatma: A Biography of Martin Lomasney.* Boston: Bruce Humphries, 1949.

Aldrich, Nelson W., Jr. *Old Money: The Mythology of America's Upper Class.* New York: Knopf, 1988.

Alsop, Joseph. *FDR, 1882–1945: A Centenary Remembrance.* New York: Viking, 1982.

Alsop, Joseph, and Robert Kintner. *American White Paper: The Story of American Diplomacy and the Second World War.* New York: Simon and Schuster, 1940.

Alsop, Joseph, with Adam Platt. *I've Seen the Best of It.* New York: Norton, 1992.

Amory, Cleveland. *The Proper Bostonians.* New York: E. P. Dutton & Co., 1947.

Anthony, Carl Sferrazza. *First Ladies: The Saga of the President's Wives and Their Power,* vol. 2. New York: William Morrow, 1990.

———. *As We Remember Her: Jacqueline Kennedy Onassis in the Words of Her Friends and Family.* New York: HarperCollins, 1997 (AWRH).

Arensberg, Conrad. *The Irish Countryman: An Anthropological Study.* New York: Macmillan, 1937.

Arensberg, Conrad, and Solon Kimball. *Family and Community in Ireland.* Cambridge, Mass.: Harvard University Press, 1940.

Ariès, Philippe. *Centuries of Childhood: A Social History of Family Life.* Translated from the French by Robert Baldick. New York: Vintage Books, 1962.

Associated Press. *Triumph and Tragedy: The Story of the Kennedys.* New York: Morrow, 1968.

Ayers, Bradley Earl. *The War that Never Was.* Indianapolis: Bobbs-Merrill, 1976.

Bailyn, Bernard, Donald Fleming, Oscar Handlin, and Stephan Thernstrom. *Glimpses of the Harvard Past.* Cambridge, Mass.: Harvard University Press, 1986.

Ballowe, James, ed. *George Santayana's America.* Urbana: University of Illinois Press, 1967.

Baltzell, E. Digby. *Puritan Boston and Quaker Philadelphia.* New York: Free Press, 1979.

Bass, Jack. *Taming the Storm: The Life and Times of Judge Frank M. Johnson Jr. and the South's Fight over Civil Rights.* New York: Doubleday, 1993.

Bayley, Edwin R. *Joe McCarthy and the Press.* Madison: University of Wisconsin Press, 1981.

Beebe, Lucius. *Boston and the Boston Legend.* New York: Appleton-Century, 1935.

Behr, Edward. *Prohibition: Thirteen Years that Changed America.* New York: Arcade, 1996.

Beran, Michael Knox. *The Last Patrician: Bobby Kennedy and the End of American Aristocracy.* New York: St. Martin's Press, 1998.

Bergin, Thomas Goddard. *The Game: The Harvard-Yale Football Rivalry, 1875–1983.* New Haven, Conn.: Yale University Press, 1984.

Bergquist, Laura, and Stanley Tretick. *A Very Special President.* New York: McGraw-Hill, 1965.

Bertagna, Joe. *Crimson in Triumph: A Pictorial History of Harvard Athletics, 1852–1985.* Lexington, Mass.: Stephen Greene Press, 1986.

Beschloss, Michael R. *Kennedy and Roosevelt: The Uneasy Alliance.* New York: Norton, 1980 (KR).

———. *The Crisis Years: Kennedy and Khrushchev, 1960–1963.* New York: HarperCollins, 1991 (CY).

Bird, Kai. *The Color of Truth: McGeorge Bundy and William Bundy: Brothers in Arms.* New York: Simon and Schuster, 1998.

Birmingham, Stephen. *The Right People.* Boston: Little, Brown, 1958.

———. *Real Lace: America's Irish Rich.* New York: Harper & Row, 1973.

Blair, Clay, Jr., and Joan Blair. *The Search for JFK.* New York: Berkley, 1976.

Blanchard, John A. *The H Book of Harvard Athletics, 1852–1922.* 1923.

Blight, James G., and Peter Kornbluh, eds. *Politics of Illusion: The Bay of Pigs Invasion Reexamined.* Boulder, Colo.: Lynne Rienner, 1998.

Blocker, Jack S., Jr. *American Temperance Movements: Cycles of Reform.* Boston: Twayne Publishers, 1989.

Blum, John Martin. *From the Morgenthau Diaries.* 3 vols. Boston, Houghton Mifflin, 1959–67.

Bradford, Sarah. *America's Queen: The Life of Jacqueline Kennedy Onassis.* New York: Viking, 2000.

Bradlee, Benjamin C. *Conversations with Kennedy.* New York: Norton, 1975.

Branch, Taylor. *Parting the Waters: America in the King Years, 1954–1963.* New York: Simon and Schuster, 1988.

Brashler, William. *The Don: The Life and Death of Sam Giancana.* New York: Harper & Row, 1977.

Brinkley, David. *Washington Goes to War.* New York: Random House, 1998.

Brooks, Van Wyck. *The Flowering of New England.* New York: E. P. Dutton & Co., 1936.

———. *Scenes and Portraits: Memories of Childhood and Youth.* New York: E. P. Dutton, 1954.

Brown, Joan Winmill. *No Longer Alone.* Old Tappan, N.J.: F. H. Revell Co., 1975.

Brown, Thomas. *JFK: History of an Image.* Bloomington: Indiana University Press, 1988.

Brugioni, Dino A. *Eyeball to Eyeball: The Inside Story of the Cuban Missile Crisis.* New York: Random House, 1991.

Buchan, John. *Pilgrim's Way: An Essay in Recollection.* Boston: Houghton Mifflin, 1940.

Bullitt, Orville H., ed. *For the President: Personal and Secret Correspondence Between Franklin D. Roosevelt and William C. Bullitt.* Boston: Houghton Mifflin, 1972.

Bundy, McGeorge. *Danger and Survival: Choices About the Bomb in the First Fifty Years.* New York: Vintage Books, 1990.

Buni, Andrew, and Alan Rogers. *Boston, City on a Hill: An Illustrated History.* Boston: Windsor Publications, 1984.

Burleigh, Nina. *A Very Private Woman: The Life and Unsolved Murder of Presidential Mistress Mary Meyer.* New York: Bantam, 1998.

Burner, David, and Thomas R. West. *The Torch Is Passed: The Kennedy Brothers and American Liberalism.* New York: Atheneum, 1984.

Burns, James MacGregor. *John Kennedy: A Political Profile.* New York: Harcourt, 1960.

Byrnes, James F. *All in My Lifetime.* New York: Harper and Brothers, 1958.

Cable, Mary. *The Little Darlings: A History of Child Rearing.* New York: Scribner's Sons, 1975.

Cameron, Gail. *Rose: A Biography of Rose Fitzgerald Kennedy.* New York: Putnam, 1971.

Carmody, Denise L., and John T. Carmody. *Roman Catholicism: An Introduction.* New York: Macmillan, 1990.

Cassini, Oleg. *In My Own Fashion: An Autobiography.* New York: Simon and Schuster, 1987.

Cecil, David. *The Young Melbourne.* New York: Bobbs-Merrill, 1939 (YM).

———. *Lord M., or the Later Life of Lord Melbourne.* London: Constable and Co., 1954 (LM).

Cecil, Russell L., ed. *A Textbook of Medicine*. Philadelphia and London: W. B. Saunders Co., 1947.

Chafin, Raymond, and Topper Sherwood. *Just Good Politics: The Life of Raymond Chafin*. Pittsburgh: University of Pittsburgh Press, 1994.

Chance, Mrs. Burton. *The Care of the Child*. Philadelphia: Penn, 1909.

Cheshire, Maxine, with John Greenya. *Maxine Cheshire, Reporter*. Boston: Houghton Mifflin, 1978.

Chomsky, Noam. *Rethinking Camelot: JFK, the Vietnam War, and U.S. Political Culture*. Boston: South End Press, 1993.

Churchill, Winston S. *Step by Step: 1936–1939*. New York: Putnam's, 1939.

———. *The Unrelenting Struggle*. Boston: Little, Brown, 1942.

CIA Targets Fidel: Secret 1967 CIA Inspector General's Report on Plots to Assassinate Fidel Castro. Melbourne: Ocean Press, 1996.

Clymer, Adam. *Edward M. Kennedy: A Biography*. New York: William Morrow, 1999.

Cohn, Roy. *McCarthy*. New York: New American Library, 1968.

Cole, Wayne S. *America First: The Battle Against Intervention, 1940–1941*. Madison: University of Wisconsin Press, 1953.

———. *Charles A. Lindbergh and the Battle Against American Intervention in World War II*. New York: Harcourt Brace Jovanovich, 1974.

Collier, Peter, and David Horowitz. *The Kennedys: An American Drama*. New York: Summit Books, 1984.

Collins, Joseph. *The Doctor Looks at Literature: Psychological Studies of Life and Letters*. New York: George H. Doran Co., 1923.

———. *The Doctor Looks at Love and Life*. Garden City, N.Y.: Garden City Publishing Co., 1929.

Corn, David. *Blond Ghost: Ted Shackley and the CIA's Crusades*. New York: Simon and Schuster, 1994.

Corry, John. *Golden Clan: The Murrays, the McDonnells, and the Irish American Aristocracy*. Boston: Houghton Mifflin, 1977.

Craig, Gordon A., and Felix Gilbert, eds. *The Diplomats, 1919–1939*. Princeton, N.J.: Princeton University Press, 1953.

Crosby, Donald F. *God, Church, and Flag: Senator Joseph R. McCarthy and the Catholic Church, 1950–1957*. Chapel Hill: University of North Carolina Press, 1978.

Cross, Robert D. *The Emergence of Liberal Catholicism in America*. Cambridge, Mass.: Harvard University Press, 1958.

Curtis, L. Perry, Jr. *Apes and Angels: The Irishman in Victorian Caricature*. Washington, D.C.: Smithsonian Institution Press, 1997.

Cutler, John Henry. *"Honey Fitz": Three Steps to the White House*. Indianapolis and New York: Bobbs-Merrill Co., 1962.

———. *Cardinal Cushing of Boston*. New York: Hawthorn, 1970.

Dallas, Rita, with Jeanira Ratcliffe. *The Kennedy Case*. New York: Putnam, 1973.

Damore, Leo. *The Cape Cod Years of John Fitzgerald Kennedy*. Englewood Cliffs, N.J.: Prentice-Hall, 1967.

David, Lester. *Ted Kennedy: Triumphs and Tragedies.* New York: Grosset & Dunlap, 1972.

Davis, John. *The Bouviers: Portrait of an American Family.* New York: Farrar, Straus, 1969.

———. *The Kennedys: Dynasty and Disaster.* New York: McGraw-Hill, 1984.

De Bedts, Ralph F. *Ambassador Joseph Kennedy, 1938–1940: An Anatomy of Appeasement.* New York: Peter Lang, 1985.

DeLoach, Cartha D. *Hoover's FBI: The Inside Story of Hoover's Trusted Lieutenant.* Washington, D.C.: Regnery, 1997.

Demarco, William. *Enclaves: Boston's Italian North End.* Ann Arbor, Mich.: UMI Research Press, 1981.

Diner, Hasia R. *Erin's Daughters in America: Irish Immigrant Women in the Nineteenth Century.* Baltimore: Johns Hopkins University Press, 1983.

Dinneen, Joseph F. *The Purple Shamrock: The Honorable James Michael Curley of Boston.* New York: Norton, 1949.

———. *The Kennedy Family.* Boston: Little, Brown, 1959.

Ditzion, Sidney. *Marriage, Morals, and Sex in America: A History of Ideas.* New York: Octagon Books, 1975.

Dobrynin, Anatoly. *In Confidence: Moscow's Ambassador to America's Six Cold War Presidents.* New York: Times Books, 1995.

Donovan, Robert J. *PT-109: John F. Kennedy in World War II.* New York: McGraw-Hill, 1961.

Douglas, William O. *The Court Years, 1939–1975.* 1980. Reprint, New York: Vintage, 1981.

Duis, Perry R. *The Saloon: Public Drinking in Chicago and Boston, 1880–1920.* Urbana and Chicago: University of Illinois Press, 1983.

Dumenil, Lynn. *The Modern Temper: American Culture and Society in the 1920s.* New York: Hill and Wang, 1995.

Dunn, Judy, and Carol Kendrick. *Siblings: Love, Envy, and Understanding.* Cambridge, Mass.: Harvard University Press, 1982.

Erie, Steven P. *Rainbow's End: Irish-Americans and the Dilemmas of Urban Machine Politics, 1840–1985.* Berkeley: University of California Press, 1988.

Espaillat, Arturo. *Trujillo: The Last Caesar.* Chicago: Henry Regnery, 1963.

Evans, E. Estyn. *Irish Folk Ways.* London: Routledge, 1957.

Exner, Judith, as told to Ovid Demaris. *My Story.* New York: Grove Press, 1977.

Fay, Paul B., Jr. *The Pleasure of His Company.* New York: Harper & Row, 1966.

Fields, Suzanne. *Like Father, Like Daughter: How Father Shapes the Woman His Daughter Becomes.* Boston: Little, Brown, 1983.

Flamini, Roland. *Scarlett, Rhett, and a Cast of Thousands.* New York: Macmillan, 1975.

Foster, R. F. *Modern Ireland, 1600–1972.* New York: Viking Penguin, 1988.

Frank, Richard B. *Guadalcanal: The Definitive Account of the Landmark Battle.* New York: Random House, 1990.

Frederick, Mrs. Christine. *The New Housekeeping: Efficiency Studies in Home Management.* Garden City, N.Y.: Doubleday, 1918.

———. *Household Engineering: Scientific Management in the Home*. Chicago: American School of Home Economics, 1920.

Freedman, Lawrence. *Kennedy's Wars: Berlin, Cuba, Laos, and Vietnam*. New York: Oxford, 2000.

Freeman, Walter, M.D., and James W. Watts, M.D. *Psychosurgery: Intelligence, Emotion, and Social Behavior Following Prefrontal Lobotomy for Mental Disorders*. Springfield, Mass.: Charles C. Thomas, 1942.

Furlong, Nicholas. *County Wexford in the Rare Oul' Times*. vol. 1. Wexford: Old Distillery Press, 1985.

Fursenko, Aleksandr, and Timothy Naftali. *"One Hell of a Gamble": Khrushchev, Castro, and Kennedy, 1958–1964*. New York: Norton, 1997.

Fussell, Paul. *The Great War and Modern Memory*. New York: Oxford, 1975.

Gaddis, John Lewis. *We Now Know: Rethinking Cold War History*. Oxford: Clarendon Press, 1997.

Galbraith, John Kenneth. *A Life in Our Times*. Boston: Houghton Mifflin, 1981.

———. *Letters to Kennedy*. Edited by James Goodman. Cambridge, Mass.: Harvard University Press, 1998.

Galbreth, Lillian. *The Homemaker and Her Job*. New York: Appleton-Century, 1927.

Gallagher, Mary Barell. *My Life with Jacqueline Kennedy*. New York: McKay, 1969.

Garrow, David J. *The FBI and Martin Luther King Jr.* New York: Penguin, 1981.

Gay, Peter. *The Education of the Senses: The Bourgeois Experience: Victoria to Freud*. Vol. 1. New York: Oxford, 1984.

———. *The Tender Passion: The Bourgeois Experience: Victoria to Freud*. vol. 2. New York: Oxford, 1986.

Gentry, Curt. *J. Edgar Hoover: The Man and the Secrets*. New York: Norton, 1991.

George, Carol V. R. *God's Salesman: Norman Vincent Peale and the Power of Positive Thinking*. New York: Oxford, 1993.

Gilbert, Martin, and Richard Gott. *The Appeasers*. Boston: Houghton Mifflin, 1963.

Goddard, Henry Herbert, Ph.D. *Feeble-mindedness: Its Causes and Consequences*. New York: Macmillan, 1914.

Goodwin, Doris Kearns. *The Fitzgeralds and the Kennedys*. New York: Simon and Schuster, 1987.

Goodwin, Richard. *Remembering America*. Boston: Little, Brown, 1988.

Gormley, Ken. *Archibald Cox: Conscience of a Nation*. Foreword by Elliot L. Richardson. Reading, Mass.: Addison-Wesley, 1997.

Gossett, Thomas F. *Race: The History of an Idea in America*. 1963. Reprint, New York: Oxford, 1997.

Graham, Katharine. *Personal History*. New York: Knopf, 1997.

Graves, Robert. *Good-bye to All That*. New York: Doubleday, 1929.

Graves, Robert, and Alan Hodges. *The Long Week End: A Social History of Great Britain, 1918–1939.* New York: Macmillan, 1941.

Green, Martin. *The Problem of Boston: Some Readings in Cultural History.* New York: Norton, 1966.

———. *The Great American Adventure.* Boston: Beacon Press, 1984.

Griswold, Robert L. *Fatherhood in America: A History.* New York: Basic Books, 1993.

Groves, Ernest. *The Drifting Home.* Boston: Houghton Mifflin, 1926.

Groves, Ernest T., and Gladys H. Groves. *Parents and Children.* Philadelphia: Lippincott, 1928.

Guthman, Edwin. *We Band of Brothers: A Memoir of Robert F. Kennedy.* New York: Harper & Row, 1971.

Guthman, Edwin, and Jeffrey Shulman, eds. *Robert Kennedy: In His Own Words.* New York: Bantam, 1988.

Halberstam, David. *The Fifties.* New York: Villard, 1993.

Hall, G. Stanley. *Adolescence.* Vol. 1. New York: D. Appleton and Co., 1904.

———. *Youth: Its Education, Regimen, and Hygiene.* 1904. Reprint, New York: Arno Press, 1972.

———. *Morale: The Supreme Standard of Life and Conduct.* New York: D. Appleton and Co., 1920.

———. *Life and Confessions of a Psychologist.* 1923. Reprint, New York: Arno Press, 1977.

Hall, Winfield Scott. *A Manual of Sex Hygiene.* Chicago: Howard-Severence Co., 1913.

Halpert, Stephen, and Brenda Halpert, eds. *Brahmins and Bullyboys: G. Frank Radway's Boston Album.* Boston: Houghton Mifflin, 1973.

Halsey, William. *The Survival of American Innocence: Catholicism in an Era of Disillusionment, 1920–1940.* Notre Dame, Ind.: University of Notre Dame, 1980.

Hamilton, Edith. *The Greek Way.* New York: Norton, 1930.

Hamilton, Nigel. *JFK: Restless Youth.* New York: Random House, 1992.

Hammer, Ellen J. *A Death in November: America in Vietnam, 1963.* New York: Dutton, 1987.

Handlin, Oscar. *Boston's Immigrants: A Study in Acculturation.* 1941. Reprint, Cambridge, Mass: Harvard University Press/Belknap Press, 1959.

———. *The Uprooted.* Boston: Little, Brown, 1951.

Hansen, Marcus Lee. *The Atlantic Migration, 1607–1860.* Cambridge, Mass.: Harvard University Press, 1940. Reprint, New York: Harper Torchstone, 1961.

Hardyment, Christina. *Dream Babies: Three Centuries of Good Advice on Child Care.* New York: Harper & Row, 1983.

Haynes, Roy A. *Prohibition Inside Out.* New York: Doubleday, 1926.

Hellman, John. *The Kennedy Obsession: The American Myth of JFK.* New York: Columbia University Press, 1997.

Henderson, Deirdre, ed. *Prelude to Leadership: The European Diary of John F. Kennedy.* Washington, D.C.: Regnery, 1995.

Herbermann, Charles G., et al. *The Catholic Encyclopedia.* New York: Appleton Co., 1912.

Hersh, Burton. *The Education of Edward Kennedy.* New York: William Morrow, 1972 (TEEK).

———. *The Old Boys: The American Elite and the Origins of the CIA.* New York: Scribner's, 1992 (TOB).

Hersh, Seymour. *The Dark Side of Camelot.* Boston: Little, Brown, 1997.

Heymann, C. David. *A Woman Named Jackie.* New York: Lyle Stuart, 1989 (WNJ).

———. *RFK: A Candid Biography of Robert F. Kennedy.* New York: Dutton, 1998 (RFKCB).

Higgins, Trumbull. *The Perfect Failure: Kennedy, Eisenhower, and the CIA at the Bay of Pigs.* New York: Norton, 1987.

Hill, George. *The Battle for Madrid.* New York: St. Martin's Press, 1976.

Hilty, James W. *Robert Kennedy: Brother Protector.* Philadelphia: Temple University Press, 1997.

Hitchcock, James. *Catholicism and Modernity.* New York: Seabury Press, 1979.

Hoffman, Elizabeth Cobbs. *All You Need Is Love: The Peace Corps and the Spirit of the 1960s.* Cambridge, Mass.: Harvard University Press, 1998.

Holmes, Oliver Wendell. *Dr. Holmes's Boston.* Edited by Caroline Ticknor. Boston: Houghton Mifflin, 1915.

Holt, L. Emmett. *The Care and Feeding of Children: A Catechism for the Use of Mothers and Children's Nurses.* 1894. Reprint, New York: Appleton, 1923.

Hooker, N. H., ed. *Jay Pierrepont Moffat, 1919–1943.* Cambridge, Mass.: Harvard University Press, 1956.

Horsman, Reginald. *Race and Manifest Destiny: The Origins of American Racial Anglo-Saxonism.* Cambridge, Mass.: Harvard University Press, 1981.

Howe, M. A. De Wolfe. *Barrett Wendell and His Letters.* Boston: Atlantic Monthly Press, 1924.

Hunt, C. W. *Booze, Boats, and Billions.* Toronto: McClelland and Stewart, 1988.

———. *Whisky and Ice: The Saga of Ben Kerr, Canada's Most Daring Rum-runner.* Toronto: Dundrun Press, 1995.

Ickes, Harold Le Claire. *The Secret Diary of Harold L. Ickes: The First Thousand Days, 1933–1936.* New York: Simon and Schuster, 1953.

———. *The Secret Diary of Harold L. Ickes,* vol. 2, *The Inside Struggle, 1936–1939.* New York: Simon and Schuster, 1954.

———. *The Secret Diary of Harold L. Ickes,* vol. 3, *The Lowering Clouds, 1939–1941.* New York: Simon and Schuster, 1954.

James, Henry. *Charles W. Eliot: President of Harvard University.* Boston: Houghton Mifflin, 1930.

James, William. *The Moral Equivalent of War and Other Essays.* Edited, and with an introduction, by John K. Roth. New York: Harper & Row, 1971.

Kaiser, David. *American Tragedy: Kennedy, Johnson, and the Origins of the Vietnam War.* Cambridge, Mass.: Harvard University Press/Belknap Press, 2000.

Katz, William Koren, and Marc Crawford. *The Lincoln Brigade: A Picture History*. New York: Atheneum, 1989.

Kay, Jane Holtz. *Lost Boston*. Boston: Houghton Mifflin, 1980.

Kellar, William Henry. *Make Haste Slowly: Moderates, Conservatives, and School Desegregation in Houston*. College Station: Texas A&M University Press, 1999.

Kenneally, James Joseph. *The History of American Catholic Women*. New York: Crossroad, 1990.

Kennedy, Edward M., ed. *The Fruitful Bough: A Tribute to Joseph P. Kennedy*. Privately printed, 1965 (TFB).

Kennedy, John F. *Why England Slept*. New York: Wilfred Funk, 1940 (WES).

———, ed. *As We Remember Joe*. Cambridge, Mass: privately printed at the university press, 1945 (AWRJ).

———. *Profiles in Courage*. New York: Harper, 1955. Memorial edition, HarperPerennial, 1964 (PIC).

Kennedy, Joseph P. *I'm for Roosevelt*. New York: Reynal and Hitchcock, 1936.

Kennedy, Maxwell Taylor. *Make Gentle the Life of the World: The Vision of Robert F. Kennedy*. New York: Harcourt Brace, 1998.

Kennedy, Robert F. *The Enemy Within*. New York: Harper, 1960 (TEW).

———. *Thirteen Days*. New York: Norton, 1969 (TD).

Kennedy, Rose. *Times to Remember*. Garden City, N.Y.: Doubleday, 1974 (TR).

Kessler, Ronald. *The Sins of the Father*. New York: Warner, 1996.

Kett, Joseph. *Rites of Passage: Adolescence in America, 1790 to the Present*. New York: Basic Books, 1977.

Kevles, Daniel. *In the Name of Eugenics*. New York: Knopf, 1985.

Kimmel, Michael. *Manhood in America: A Cultural History*. New York: Free Press, 1996.

Klein, Edward. *All Too Human: The Love Story of Jack and Jackie Kennedy*. New York: Pocket Books, 1996.

Kobler, John. *Ardent Spirits: The Rise and Fall of Prohibition*. New York: Da Capo, 1993.

Koestenbaum, Wayne. *Jackie Under My Skin: Interpreting an Icon*. New York: HarperCollins, 1995.

Kornbluh, Peter, ed. *Bay of Pigs Declassified: The Secret CIA Report on the Invasion of Cuba*. New York: New Press, 1998.

Koskoff, David E. *Joseph P. Kennedy: A Life and Times*. Englewood Cliffs, N.J.: Prentice-Hall, 1974.

Krock, Arthur. *Memoirs: Sixty Years on the Firing Line*. New York: Funk and Wagnalls, 1968.

———. *In the Nation: 1932–1966*. New York: Paperback Library, 1969.

Lambert, Angela. *1939: The Last Season of Peace*. New York: Weidenfeld & Nicolson, 1989.

LaRossa, Ralph. *The Modernization of Fatherhood: A Social and Political History*. Chicago: University of Chicago Press, 1997.

Lasky, Betty. *RKO: The Biggest Little Major of Them All*. Englewood Cliffs, N.J.: Prentice-Hall, 1984.

Lasky, Victor. *JFK: The Man and the Myth*. New York: Macmillan, 1963 (JFKMM).

———. *Robert F. Kennedy: The Myth and the Man*. New York: Trident Press, 1968 (RFK).

Lawford, Patricia Seaton. *The Peter Lawford Story*. New York: Carroll & Graf, 1988.

Leamer, Laurence. *The Kennedy Women*. New York: Villard, 1994.

Leighton, Isabel, ed. *The Aspirin Age, 1919–1941*. New York: Simon and Schuster, 1949.

Lender, Mark Edward, and James Kirby Martin. *Drinking in America: A History*. New York: Free Press, 1987.

Leuchtenburg, William E. *Perils of Prosperity, 1914–1932*. 2d ed. Chicago: University of Chicago Press, 1993.

Leutze, James, ed. *The London Journal of General Raymond E. Lee: 1940–1941*. Boston: Little, Brown, 1971.

Levin, Murray B. *Kennedy Campaigning: The System and the Style as Practiced by Senator Edward Kennedy*. Boston: Beacon Press, 1966.

Lieberson, Goddard. *John F. Kennedy: As We Remember Him*. Philadelphia: Courage Books, 1965.

Lippman, Theo, Jr., *Senator Ted Kennedy: The Career Behind the Image*. New York: Norton, 1976.

Liston, Robert A. *Sargent Shriver: A Candid Portrait*. New York: Farrar, Straus, 1964.

Logevall, Fredrik. *Choosing War: The Lost Chance for Peace and the Escalation of War in Vietnam*. Berkeley: University of California Press, 1999.

Long, Breckinridge. *The War Diary of Breckenridge Long: Selections from the Years 1939–1944*. Edited by Fred L. Israel. Lincoln: University of Nebraska Press, 1966.

Lynch, Grayston L. *Betrayal at the Bay of Pigs*. Washington, D.C.: Brassey's, 1998.

Macfadden, Barnarr A. *The Virile Powers of Superb Manhood: How Developed, How Lost, How Regained*. 1900. Reprint, New York: Physical Culture Publishing Co.

Madsen, Alex. *Gloria and Joe*. New York: Morrow, 1988.

Mahoney, Richard D. *JFK: Ordeal in Africa*. New York: Oxford, 1983 (JFKOA).

———. *Sons and Brothers: The Days of Jack and Bobby Kennedy*. New York: Arcade, 1999 (SB).

Mailer, Norman. *The Presidential Papers*. New York: Putnam, 1963.

Manchester, William. *Portrait of a President: John F. Kennedy in Profile*. Boston: Little, Brown, 1962.

———. *The Death of a President: November 20–November 25, 1963*. New York: Harper & Row, 1967 (DP).

———. *The Glory and the Dream: A Narrative History of America, 1932–1972*. New York: Bantam, 1980.

———. *The Last Lion: Alone: 1932–1940*. Boston: Little, Brown, 1988.

Mangan, J. A., and James Walvin, eds. *Manliness and Morality: Middle-Class Masculinity in Britain and America, 1800–1940*. New York: St. Martin's, 1987.

Marson, Philip. *Breeder of Democracy*. Cambridge: Schenkman, 1963.

Martin, John Bartlow. *Adlai Stevenson and the World: The Life of Adlai E. Stevenson*. Garden City, N.Y.: Doubleday, 1977.

Martin, Ralph. *A Hero for Our Time*. New York: Ballantine, 1984.

———. *Henry and Clare: An Intimate Portrait of the Luces*. New York: Putnam's, 1991.

Matthews, Chris. *Kennedy and Nixon: The Rivalry that Shaped Postwar America*. New York: Simon and Schuster, 1996.

May, Ernest R., and Philip D. Zelikow, eds. *The Kennedy Tapes: Inside the White House During the Cuban Missile Crisis*. Cambridge, Mass.: Harvard University Press/Belknap Press, 1997.

May, Henry F. *The End of American Innocence: A Study of the First Years of Our Own Time, 1912–1917*. New York: Knopf, 1959.

McCarthy, Joe. *The Remarkable Kennedys*. New York: Dial Press, 1960.

McNamara, Robert S. *In Retrospect: The Tragedy and Lessons of Vietnam*. New York: Times Books, 1995.

McTaggart, Lynne. *Kathleen Kennedy: Her Life and Times*. New York: Dial Press, 1983.

McWilliams, Carey. *A Mask for Privilege: Anti-Semitism in America*. Boston: Little, Brown, 1948.

Mead, Frederick S., ed. *Harvard's Military Record in the World War*. Boston: Harvard Alumni Association, 1921.

Merry, Robert W. *Taking on the World: Joseph and Stewart Alsop—Guardians of the American Century*. New York: Viking, 1996.

Merwick, Donna. *Boston Priests, 1848–1910: A Study of Social and Intellectual Change*. Cambridge, Mass.: Harvard University Press, 1973.

Messick, Hank. *The Silent Syndicate*. New York: Macmillan, 1967.

Michaelis, David. *The Best of Friends*. New York: William Morrow, 1983.

Miller, Alice. *The Drama of the Gifted Child*. New York: Basic Books, 1981.

Mintz, Steven, and Susan Kellogg. *Domestic Revolutions: A Social History of American Family Life*. New York: Free Press, 1988.

Moffat, Jay Pierrepoint. *The Moffat Papers: Selections from the Diplomatic Journals of Jay Pierrepoint Moffat, 1919–1943*. Edited by Nancy Harvison Hooker. Cambridge, Mass.: Harvard University Press, 1956.

Moldea, Dan E. *The Hoffa Wars: Teamsters, Rebels, Politicians, and the Mob*. New York: Paddington Press, 1978.

Mollenhoff, Clark R. *Tentacles of Power*. Cleveland: World, 1965.

Montgomery, Frances Trego. *Billy Whiskers: The Autobiography of a Goat*. Akron, Ohio: Saalfield, 1902.

Morison, Samuel Eliot. *Three Centuries of Harvard, 1636–1936*. Cambridge, Mass.: Harvard University Press, 1936.

———. *One Boy's Boston, 1887–1901*. Boston: Houghton Mifflin, 1962.

Morris, Charles R. *American Catholic: The Saints and Sinners Who Built America's Most Powerful Church*. New York: Random House, 1997.

Mosley, Leonard. *Dulles: A Biography of Eleanor, Allen, and John Foster Dulles and Their Family Network*. London: Hodder and Stoughton, 1978.

Mosley, Sir Oswald. *My Life*. New Rochelle, N.Y.: Arlington House, 1968.

Muggeridge, Malcolm. *The Thirties*. London: Hamish Hamilton, 1940.

Nasaw, David. *Children of the City*. Garden City, N.Y.: Anchor Press, 1985.

Neustadt, Richard E. *Report to JFK: The Skybolt Crisis in Perspective*. Ithaca, N.Y.: Cornell University Press, 1999.

Newfield, Jack. *Robert Kennedy: A Memoir*. New York: Dutton, 1969.

New Yorker Book of War Pieces, The. New York: Reynal & Hitchcock, 1947.

Nicolson, Harold. *Diaries and Letters, 1930–1964*. London: Penguin, 1984.

November 22: The Day Remembered as Reported by the Dallas Morning News. Dallas: Taylor, 1990.

Nown, Graham. *The English Godfather*. London: Ward Lock, 1987.

O'Brien, David J. *The Renewal of American Catholicism*. New York: Oxford, 1972.

O'Connor, Thomas H. *The Boston Irish*. Boston: Back Bay Books, 1995.

O'Donnell, Helen. *A Common Good: The Friendship of Robert F. Kennedy and Kenneth P. O'Donnell*. New York: William Morrow, 1998.

O'Donnell, Kenneth P., and David F. Powers. *Johnny, We Hardly Knew Ye: Memories of John Fitzgerald Kennedy*. Boston: Little, Brown, 1972.

Olsen, Jack. *Aphrodite: Desperate Mission*. New York: G. P. Putnam's Sons, 1970.

O'Neill, Thomas, with William Novak. *Man of the House*. New York: Random House, 1987.

Oppenheimer, Jerry. *The Other Mrs. Kennedy*. New York: St. Martin's Press, 1994.

Osherson, Samuel. *Finding Our Fathers: How a Man's Life Is Shaped by His Relationship with His Father*. New York: Fawcett Columbine, 1986.

Oshinksy, David M. *A Conspiracy So Immense: The World of Joe McCarthy*. New York: Free Press, 1983.

O'Toole, James M. *Militant and Triumphant: William Henry O'Connell and the Catholic Church in Boston, 1859–1944*. Notre Dame, Ind.: University of Notre Dame Press, 1992.

Parmet, Herbert S. *Jack: The Struggles of John F. Kennedy*. New York: Dial, 1980 (SJFK).

———. *JFK: The Presidency of John F. Kennedy*. New York: Dial, 1983 (PJFK).

Peiss, Kathy. *Cheap Amusements: Working Women and Leisure in Turn-of-the-Century New York*. Philadelphia: Temple University Press, 1986.

Pepper, Claude Denson, with Hays Gorey. *Pepper: Eyewitness to a Century*. New York: Harcourt Brace Jovanovich, 1987.

Persico, Joseph E. *Edward R. Murrow: An American Original*. New York: Da Capo Press, 1997.

Phillips, Bruce A. *Brookline: The Evolution of an American Jewish Suburb*. New York: Garland, 1990.

Phillips, Cabell. *From the Crash to the Blitz, 1929–1939*. New York: Macmillan, 1975.

———. *The 1940s: Decade of Triumph and Trouble*. New York: Macmillan, 1975.

Pilat, Oliver. *Drew Pearson: An Unauthorized Biography*. New York: Warner/Pocket Books, 1973.

Rachlin, Harvey. *The Kennedys: A Chronological History, 1823 to the Present*. New York: World Almanac, 1986.

Rappleye, Charles, and Ed Becker. *All-American Mafioso: The Johnny Rosselli Story*. New York: Doubleday, 1991.

Redmon, Coates. *Come as You Are: The Peace Corps Story*. San Diego: Harcourt Brace Jovanovich, 1986.

Reeves, Richard. *President Kennedy: Profile of Power*. New York: Simon and Schuster, 1993.

Reeves, Thomas C. *A Question of Character: A Life of John F. Kennedy*. New York: Macmillan, 1991.

Renehan, Edward J., Jr. *The Lion's Pride: Theodore Roosevelt and His Family in Peace and War*. New York: Oxford, 1998.

Rhodes, James Robert, ed. *Chips: The Diaries of Sir Henry Channon*. London: Weidenfeld & Nicolson, 1967.

Roboff, Sari. *East Boston: Boston 200 Neighborhood History Series*. Boston: The Boston 200 Corporation, 1976.

Rodriguez, Felix I., and John Weisman. *Shadow Warrior*. New York: Simon and Schuster, 1989.

Rodriguez, Juan Carlos. *The Bay of Pigs and the CIA*. New York: Ocean Press, 1999.

Roemer, William F., Jr. *Man Against the Mob*. New York: D. I. Fine, 1989.

Rogers, Warren. *When I Think of Bobby: A Personal Memoir of the Kennedy Years*. New York: HarperCollins, 1993.

Roosevelt, James, with Bill Libby. *My Parents: A Differing View*. Chicago: Playboy Press, 1976.

Roosevelt, Theodore. *The Strenuous Life: Essays and Addresses*. New York: Century, 1902. Reprint, St. Clair Shores, Mich.: Scholarly Press, 1970.

Rosovsky, Nitza. *The Jewish Experience at Harvard and Radcliffe*. Cambridge, Mass.: Harvard University Press, 1986.

Rotundo, E. Anthony. *American Manhood: Transformations in Masculinity from the Revolution to the Modern Era*. New York: Basic Books, 1993.

Rumbarger, John J. *Profits, Power, and Prohibition*. Albany: State University of New York Press, 1989.

Russell, Francis. *The President Makers: From Mark Hanna to Joseph P. Kennedy*. Boston: Little, Brown, 1976.

Russo, Gus. *Live by the Sword: The Secret War Against Castro and the Death of JFK*. Baltimore: Bancroft Press, 1998.

Ryan, Dennis. *Beyond the Ballot Box: A Social History of the Boston Irish, 1845–1917.* East Brunswick, N.J.: Associated University Presses, 1983.

Ryan, Dorothy, and Louis J. Ryan. *The Kennedy Family of Massachusetts: A Bibliography.* Westport, Conn.: Greenwood Press, 1981.

St. John, George. *Forty Years at School.* New York: Henry Holt, 1959.

Salinger, Pierre. *With Kennedy.* New York: Doubleday, 1966 (WK).

———. *P.S.: A Memoir.* New York: St. Martin's, 1995 (PS).

Saunders, Frank, with James Southwood. *Torn Lace Curtain.* New York: Holt, Rinehart and Winston, 1982.

Schlesinger, Arthur, Jr. *A Thousand Days.* Boston: Houghton Mifflin, 1965 (ATD).

———. *Robert Kennedy and His Times.* Boston: Houghton Mifflin, 1978 (RKHT).

Schoor, Gene. *Young John Kennedy.* New York: Harcourt, Brace & World, 1963.

Seagrave, Sterling. *The Marcos Dynasty.* New York: Harper & Row, 1988.

Searls, Hank. *The Lost Prince: Young Joe, the Forgotten Kennedy.* New York: World Publishing Co., 1969.

Shackleton, Robert. *The Book of Boston.* Philadelphia: Penn Publishing Co., 1916.

Shannon, William V. *The American Irish.* New York: Macmillan, 1963.

Shepard, Tazewell J. *John F. Kennedy: Man of the Sea.* New York: William Morrow, 1965.

Sheridan, Walter. *The Fall and Rise of Jimmy Hoffa.* New York: Saturday Review Press, 1972.

Sherry, Michael S. *In the Shadow of War: The United States Since the 1930s.* New Haven, Conn.: Yale University Press, 1995.

Sherwood, Robert E. *Roosevelt and Hopkins: An Intimate History.* New York: Harper, 1948.

Shesol, Jeff. *Mutual Contempt: Lyndon Johnson, Robert Kennedy, and the Feud that Defined a Decade.* New York: Norton, 1997.

Sloane, Arthur A. *Hoffa.* Cambridge, Mass.: MIT Press, 1991.

Smith, Amanda, ed. *Hostage to Fortune: The Letters of Joseph P. Kennedy.* New York: Viking, 2000 (HTF).

Smith, J. David. *Minds Made Feeble: The Myth and Legacy of the Kallikaks.* Rockville, Md.: Aspen Systems Corp., 1985.

Smith, Gene. *Dark Summer: An Intimate History of the Events that Led to World War II.* New York: Macmillan, 1987.

Smith, Richard Norton. *The Harvard Century: The Making of a University to a Nation.* New York: Simon and Schuster, 1986.

Smith-Rosenberg, Carroll. *Disorderly Conduct: Visions of Gender in Victorian America.* New York: Knopf, 1985.

Solomon, Barbara Miller. *Ancestors and Immigrants: A Changing New England Tradition.* 1956. Reprint, Chicago: University of Chicago Press, 1972.

Sorensen, Theodore C. *Kennedy*. New York: Harper & Row, 1965 (K).

———. *The Kennedy Legacy*. New York: Macmillan, 1969 (TKL).

Southworth, Susan, and Michael Southworth. *The AIA Guide to Boston*. Chester, Conn: Globe Pequot Press, 1984.

Spada, James. *Peter Lawford: The Man Who Knew Too Much*. New York: Bantam, 1991.

Spoto, Donald. *Marilyn Monroe: The Biography*. New York: HarperCollins, 1993.

Steel, Ronald. *Walter Lippmann and the American Century*. New York: Vintage Books, 1980.

———. *In Love with Night: The American Romance with Robert Kennedy*. New York: Simon and Schuster, 2000.

Stein, Jean, and George Plimpton, eds. *American Journey: The Times of Robert Kennedy*. New York: Harcourt Brace Jovanovich, 1970.

Stivers, Richard. *The Hair of the Dog: Irish Drinking and American Stereotype*. University Park: Pennsylvania State University Press, 1976.

Story, Ronald. *Harvard and the Boston Upper Class: The Forging of an Aristocracy, 1800–1870*. 1980. Reprint, Middletown, Conn.: Wesleyan University Press, 1985.

Strasser, Susan. *Never Done: A History of American Housework*. New York: Pantheon, 1982.

Strober, Gerald S., and Deborah H. Strober. *Let Us Begin Anew: An Oral History of the Kennedy Presidency*. New York: HarperCollins, 1993.

Sullivan, Robert E., and James M. O'Toole, eds. *Catholic Boston: Studies in Religion and Community, 1870–1970*. Boston: Roman Catholic Archdiocese of Boston, 1985.

Sulloway, Frank J. *Born to Rebel: Birth Order, Family Dynamics, and Creative Lives*. New York: Pantheon, 1996.

Summers, Anthony. *Official and Confidential: The Secret Life of J. Edgar Hoover*. New York: Putnam, 1993.

Sumner. William H. *A History of East Boston*. Boston: J. E. Tilton and Co., 1858.

Sutherland, Daniel. *Americans and Their Servants*. Baton Rouge: Louisiana State University Press, 1981.

Sutton-Smith, Brian, and B. G. Rosenberg. *The Siblings*. New York: Holt, Rinehart and Winston, 1970.

Swanson, Gloria. *Swanson on Swanson*. New York: Random House, 1980.

Sykes, Christopher. *Nancy: The Life of Lady Astor*. London: Collins, 1972.

Synnott, Marcia Graham. *The Half-Opened Door: Discrimination and Admissions at Harvard, Yale, and Princeton, 1900–1970*. Westport, Conn.: Greenwood Press, 1979.

Szulc, Tad. *Fidel: A Critical Biography*. New York: William Morrow, 1986.

Tager, Jack, and John W. Ifkovic, eds. *Massachusetts in the Gilded Age: Selected Essays*. Amherst: University of Massachusetts Press, 1985.

Tannen, Deborah. *You Don't Understand: Women and Men in Conversation*. New York: William Morrow, 1990.

Taylor, Telford. *The Breaking Wave: The Second World War in the Summer of 1940.* New York: Simon and Schuster, 1967.
———. *Munich: The Price of Peace.* Garden City, N.Y.: Doubleday, 1979.
Thayer, Mary Van Rensselaer. *Jacqueline Bouvier Kennedy.* New York: Doubleday, 1961.
Theoharis, Athan. *Seeds of Repression: Harry S Truman and the Origins of McCarthyism.* Chicago: Quadrangle Books, 1971.
Thernstrom, Stephan. *The Other Bostonians: Poverty and Progress in the American Metropolis, 1860–1870.* Cambridge, Mass.: Harvard University Press, 1964.
Thomas, Evan. *Robert Kennedy.* New York: Simon and Schuster, 2000.
Thompson, Robert E., and Hortense Myers. *Robert F. Kennedy: The Brother Within.* New York: Macmillan, 1962.
Topping, Seymour. *Journey Between Two Chinas.* New York: Harper & Row, 1972.
Townsend, Kim. *Manhood at Harvard: William James and Others.* Cambridge, Mass.: Harvard University Press, 1996.
Travell, Janet. *Office Hours: Day and Night.* New York: World Publishing Co., 1968.
Trohan, Walter. *Political Animals: Memoirs of a Sentimental Cynic.* Garden City, N.Y.: Doubleday, 1975.
Trout, Charles H. *Boston: The Great Depression and the New Deal.* New York: Oxford, 1977.
Tull, Charles. *Father Coughlin and the New Deal.* Syracuse, N.Y.: Syracuse University Press, 1965.
Valenstein, Elliot S., ed. *The Psychosurgery Debate: Scientific, Legal, and Ethical Perspectives.* San Francisco: W. H. Freeman, 1980.
———. *Great and Desperate Cures: The Rise and Decline of Psychosurgery and Other Radical Treatments for Mental Illness.* New York: Basic Books, 1986.
Vogel, Morris J. *The Invention of the Modern Hospital: Boston 1870–1930.* Chicago: University of Chicago Press, 1980.
Von Bremscheid, Rev. Matthias. *The Christian Maiden.* Translated from the German by Members of the Young Ladies Sodality Holy Trinity Church, Boston. 2d ed. Boston: Angel Guardian Press, 1905.
Von Post, Gunilla. *Love, Jack.* New York: Crown, 1997.
Warner, Sam B., Jr. *Streetcar Suburbs: The Process of Growth in Boston, 1870–1900.* Cambridge, Mass.: Harvard University Press/MIT Press, 1962.
Watson, Mary. *The Expanding Vision: American Television in the Kennedy Years.* New York: Oxford, 1990.
Watt, Donald Cameron. *Why War Came: The Immediate Origins of the Second World War, 1938–1939.* New York: Pantheon, 1989.
Wayman, Dorothy G. *Cardinal O'Connell of Boston: A Biography of William Henry O'Connell, 1859–1944.* New York: Farrar, Straus and Young, 1954.
Wecter, Dixon. *The Saga of American Society.* New York: Scribner's, 1937.
West, J. B., with Mary Lynn Kotz. *Upstairs at the White House: My Life with the First Ladies.* New York: Coward, McCann and Geoghegan, 1973.

Whalen, Richard J. *The Founding Father: The Story of Joseph P. Kennedy*. New York: New American Library, 1964.

White, Mark J. *The Kennedys and Cuba: The Declassified Documentary History*. Chicago: Ivan R. Dee, 1999.

White, Theodore H. *The Making of the President 1960*. New York: Atheneum, 1961 (MP1960).

———. *In Search of History: A Personal Adventure*. New York: Warner Books, 1978.

———. *America in Search of Itself: The Making of the President 1956–1980*. New York: Harper & Row, 1982.

Whitehall, Walter Muir. *Boston in the Age of John Fitzgerald Kennedy*. Norman: University of Oklahoma Press, 1966.

Wicker, Tom. *Kennedy Without Tears*. New York: William Morrow, 1964.

Wills, Garry. *The Kennedy Imprisonment: A Meditation on Power*. Boston: Little, Brown, 1981.

Winsor, Justin, ed. *The Memorial History of Boston, 1630–1880*. 5 vols. Boston: Ticknor & Co., 1880.

Wittke, Carl. *The Irish in America*. Baton Rouge: Louisiana State University Press, 1956.

Wofford, Harris. *Of Kennedys and Kings*. New York: Farrar, Straus & Giroux, 1980.

Woodham-Smith, Cecil. *The Great Hunger: Ireland 1845–1849*. London: Hamish Hamilton, 1962.

Woods, Robert A. *The City Wilderness: A Settlement Study*. 1898. Reprint, New York: Arno Press, 1979.

———. ed. *Americans in Process: A Settlement Study*. Boston: Houghton Mifflin, 1902.

Woods, Robert A., and Albert J. Kennedy. *The Zone of Emergence*. 2d ed. Cambridge, Mass.: MIT Press, 1962.

Workers of the Writers' Program of the Work Projects Administration in the State of Massachusetts. *Boston Looks Seaward: The Story of the Port, 1630–1940*. Boston: Bruce Humphries, 1941.

Wyden, Peter. *The Bay of Pigs: The Untold Story*. New York: Simon and Schuster, 1979.

Zelizer, Barbie. *Covering the Body: The Kennedy Assassination, the Media, and the Shaping of Collective Memory*. Chicago: University of Chicago Press, 1992.

Zion, Sidney. *The Autobiography of Roy Cohn*. Secaucus, N.J.: Lyle Stuart, 1988.

Acknowledgments

The manuscript of *The Kennedy Men* ran 1,100 pages, and it was a major imposition to ask busy people to read it looking for errors. No one read with a more knowing eye than did Myer "Mike" Feldman, President Kennedy's deputy counsel. Sheldon Stern, the longtime historian at the John F. Kennedy Presidential Library, carefully read these many pages. Fellow Kennedy authors Burton Hersh, Nigel Hamilton, Gus Russo, and Dan Moldea read the book, each with his own unique expertise. Kerry McCarthy, who is Joseph and Rose Kennedy's grand-niece and a historian of the family, gave the manuscript a judicious reading. Sam Halpern, a former high-ranking CIA official, read the Cuban material. Professor Barton Bernstein of Stanford University read the White House chapters. So did former Senator Harris Wofford, who served in the Kennedy administration. Dr. Mauro Di Pasquale read the material on President Kennedy's health, as did Dr. David V. Becker, professor of radiology and medicine, Weill Medical College of Cornell University, New York Presbyterian Hospital. Others who read all or part of the manuscript included Vesna Leamer, Lee Tomic, Count Alexandre De Bouthri Báthory, Joy Harris, Herb Gray, Jim Morrissey, Howard Reed, Raleigh Robinson, Patrick Flynn, and Diane Leslie. In an act of great generosity, Kristina Rebelo Anderson, the prominent investigative journalist, took weeks away from her important book on mercury poisoning to copyedit and fact-check the entire manuscript before I submitted it to my publisher. Of course, none of these people are responsible for any errors that may remain in the book or interpretations that are mine alone.

The Kennedy Men has a great deal of new material in it largely because many people put great trust in its author. For the eleven years that Evelyn Lincoln served as John F. Kennedy's secretary, she kept a secret archive, including many documents that were to have been destroyed. She willed these papers to Robert White. Interspersed through much of this book, they are by far the most important private trove of Kennedy documents in the world. White, a prominent collector, has generously given me first-time use of all these documents.

Patricia Coughlan has given me the material that Robert Coughlan, her husband, accumulated while he was writing Rose Kennedy's autobiography, *Times to Remember*. These materials include many hours of taped interviews with Kennedy family members that are among the most intimate, candid interviews they have ever given. I would like to thank the late Jim Connor and his widow, Pat, for their friendship and support. Mary Lou and Kerry McCarthy gave me the first-time use of many unique photos from the Loretta Kennedy Connelly Collection.

I must also especially thank Gunilla Von Post for allowing me the first-time use of the letters that John F. Kennedy wrote her during their love affair. I had these letters verified by an expert, but it is common knowledge among those close to the Kennedys that Von Post's story is true. Janet Fontaine, who was Joseph P. Kennedy's mistress for a decade, also cooperated with me. Her only request was that I treat her story with dignity, a promise that I hope I have kept. During the West Virginia primaries, when candidate Kennedy was losing his voice, Fontaine was the stewardess on his plane. Kennedy communicated by writing messages on sheets of paper, many of which Fontaine saved. I thank Mrs. Fontaine and the *Forbes* Magazine Collection, the current owner of the documents, for permission to use the notes. Profesor James MacGregor Burns has graciously given me permission to quote from his revealing 1959 interview with the then Senator John F. Kennedy.

Although we live in a culture of mistrust, I was extremely blessed by the people who talked to me. Some of these sources spoke with me for the first time, and most of them spoke with compelling candor and depth. This is not in any way an authorized book, and I am especially grateful to the various Kennedys I have interviewed during the writing of my books on their family: Senator Edward M. Kennedy, Joan Kennedy, Eunice Kennedy Shriver, Patricia Kennedy Lawford, Jean Kennedy Smith, Kathleen Kennedy Townsend, Robert Kennedy Jr., Christopher Kennedy, Douglas Kennedy, Timothy Shriver, Mark Shriver, Anthony Shriver, Kerry Kennedy Cuomo, and Rory Kennedy.

I am not including in these acknowledgments the names of people I interviewed for volume 2 of this work, *The Kennedy Men, 1963–2003*. Those people I have interviewed about the Kennedys over the years about the period through 1963, or who have helped me in other ways, include: Sam

Adams, Kristina Rebelo Anderson, Andrew, Duke of Devonshire, Manuel Angulo, Carl Anthony, Bradley Earl Ayers, Bobby Baker, Larry Baker, Charles Bartlett, Dr. David Becker, Ed Becker, Edward Behr, Arnold Beichman, Kai Bird, Brad Blank, Marvin Blank, Joseph Boccehir, Ben Bradlee, Ham Brown, Joan Winmill Brown, Dino Brugioni, Robert Bunshaft, Nina Burleigh, Dan Burns, John Burns, Fox Butterfield, William C. Chapman, Blair Clark, Adam Clymer, Fred Cohen, Jeanne Conway, Alan K. Corsair, Pat Coughlin, Archibald Cox, John Henry Cutler, Mark Dalton, Zel Davis, Cartha DeLoach, Ann Denove, Ahmed Desouky, Frank Dillow, Dr. Mauro G. Di Pasquale, Joe Dolan, Janet Donovan, Luella Hennessey Donovan, William Douglas-Home, Robert Duffy, Milt Ebbins, Luis Estevez, Paul B. Fay Jr., Myer Feldman, Bob Filardi, Tom Finneran, Carey Fisher, Benedict Fitzgerald, Patrick Flynn, Wally Flynn, Janet Des Rosiers Fontaine, Harry Fowler, Alan Gage, John Kenneth Galbraith, Patty McGinty Gallagher, Barbara Gamarekian, Nancy Gardiner, Joe Gargan, Wilson Gathings, Chuck Glynn, Fred Good, Milton Gould, Arthur Grace, John Greenya, Edwin Guthman, Milton Gwirtzman, David Hackett, Alexander Haig, Sam Halpern, Geraldine Hannon, Pamela Harriman, Colonel Jack Hawkins, Bob Healy, Deirdre Henderson, Burton Hersh, Seymour Hersh, Sally Roche Higgins, Gerri Hirshey, Claude Hooton Jr., Maria Hooton, David Horowitz, Ron Howard, William Hundley, Q. Byrum Hurst, Mrs. Max Jacobson, Terry Kahn, Professor William Kaufman, Pierce Kierney, Malcolm Kilduff, Tom Killefer, Harvey Klemmer, Peter Kornbluh, Bryant Larson, Paul Lazarro, Helen Leamer, Evelyn Lincoln, Dick Livingston, Grayston Lynch, Scott Malone, Phil Manuel, Francis McAdoo, Mrs. James McCahey, Senator Eugene McCarthy, Kerry McCarthy, Mary Lou McCarthy, Lisa McCormick, Robert J. McDonnell, Priscilla Johnson McMillan, Lynn McTaggart, Elizabeth Mehren, Joe Miller, Melody Miller, Stanley Montunnas, Paul Morgan, Joe Naar, Timothy Naftali, Bob Neal, Pat Newcomb, Larry Newman, John Nolan, Ken Norwick, Mark Obenhaus, Mirko Obradovic, Ivanka Ostojic, Joe Paolella, June Payne, Charles Peters, Barrett E. Prettyman Jr., Dick Prentice, Robert Purdy, Charles Rappleye, Marcus Raskin, Coates Redmon, Martha Sweatt Reed, Howard Reed, Jewel Reed, Shafica Reed, Marie Rider, Marilyn Riesman, Bill Robinson, Terri Robinson, Nick Rodis, Teno Roncalio, Maurice Rosenblatt, John Rosenthal, James Rousmanière, Mary Ryan, Pierre Salinger, Roberto San Román, Arthur Schlesinger Jr., John Seigenthaler, Joe Shimon, Ed Shorter, Hugh Sidey, Lorraine Silberthau, George Smathers, Liz Smith, Mark Soden, Augustus Soule Jr., Betty Coxe Spalding, Chuck Spalding, Dan Stewart, Pat Lawford Stewart, Anthony Summers, Bruce Sundlun, Bill Sutton, Kingsley Swan, Lee Tomic, Michael Tomic, Seymour Topping, Walter Trohan, Dorothy Tubirdy,

William Vanden Heuvel, Sue Vogelsinger, Gunilla Von Post, Frank Waldrop, Edwin Weisl Jr., Richard Whalen, John White, Bonnie Williams, David Wilson, Bill Wilson, Nadine Witkin, and Holton Wood.

I must also acknowledge Melody Miller in Senator Edward Kennedy's office. I have disagreed with her at times over matters concerning the family she loves so much and serves so well, but I've always known that I was dealing with a person of the highest integrity and honor. I am also deeply indebted to the many other authors whose work I have found insightful.

I will not list here the various libraries and research institutions I visited since they are listed in the endnotes. I would, however, like especially to thank Harriet Young of Orlando, Florida, for giving me her wonderful collection of Kennedy materials that she has collected over a lifetime of interest in the family. I would like to make special note of the foreign relations series on the U.S. State Department web site. The documents about the Kennedy administration have been wonderfully annotated, and they are accessible to everyone. This speaks eloquently of a government agency that is not afraid of the truth.

The Kennedy Men employs dialogue and long narrative scenes in many places. These passages are neither invented nor imaginative reconstructions, but are based on sources cited in the endnotes including interviews, oral histories, diaries, letters, and tapes. The most important of these sources are the wide range of telephone conversations and meetings secretly recorded by President Kennedy from the summer of 1962 until his death. *The Kennedy Men* is the first book to make full use of these recordings. Kennedy intended these tapes for his own personal purposes. There is no indication that he thought that historians would ever use them. They are hardly the calculated utterances of a president aware that he is being recorded for posterity. The Kennedys destroyed some of these tapes before they were given to the Kennedy Presidential Library; others probably would have been destroyed if Kennedy's secretary, Evelyn Lincoln, had not secreted them away. I transcribed many of these tapes. In a few places where I was not sure of the speaker, I asked Myer Feldman and Sheldon Stern to listen and to give their best judgments.

At the John F. Kennedy Presidential Library, I would especially like to thank Megan F. Desnoyers, who, during her tenure as the acting archivist, performed splendid service to the American public. Allan Goodrich and his associate in the audiovisual division, James Hill, do an exemplary job. In the library itself, Maura Porter, June Payne, and Steve Plotkin have unfailingly helped to answer my endless questions. My stays in Boston were always a pleasure thanks in part to Skip Brandt, the security director at the Boston Park Plaza Hotel. As always, Kenneth Norwick, my literary attorney, offered astute advice.

My agent, Joy Harris, is legendary for her commitment to her authors. In taking this book to William Morrow, she brought me into contact with the best team of professionals with whom I have ever worked. Paul Bresnick originally acquired the book before his early retirement. Meaghan Dowling gave the manuscript a deeply nuanced, sensitive edit, only one of the many contributions she made to this project. Her assistant, Kelli Martin, is a backup catcher whom the New York Mets should consider signing. Sharyn Rosenblum is a sterling publicist who daily performs at least one miracle for a William Morrow author. Other members of the William Morrow team that I would like to thank include its publisher, Michael Morrison, Lisa Gallagher and Libby Jordan in marketing, Brenda Segel and Michele Corallo in sub rights, Cindy Buck and Andrea Molitor in copyediting, and Jim Fox in legal, with whom I am working for the third time.

Mike Foster is now a teacher and librarian and did not work for me full-time, as he did on *The Kennedy Women*, but he was still highly helpful. Don Spencer is now working in television, but he has continued to transcribe all my interviews. I could not have written this book in the time I did without Zootsoftware, and I want to thank Tom Davis, the program's inventor, for his support.

I must mention my daughter, Daniela, whose valuable work as a teacher helps to inspire me, and my two brothers, Edward and Robert. Many times these past years I have thought that you do not have to be a Kennedy man to have great and true brothers. I want to thank Vesna Obraduvic Leamer, my wife, last of all. Vesna's contributions are immeasurable. She helps me in the research and is forever running out to the library, tracking down some fact or another. She saves me from innumerable errors, the least of which would have appeared in the pages of my books. By rights, I should dedicate every word that I write to her, and in a way, I do.

INDEX

For audio excerpts of John F. Kennedy's
secret tapes, speeches, and other archive materials
go to www.kennedymen.com.

(((LISTEN TO)))

THE KENNEDY MEN

WRITTEN AND READ BY
LAURENCE LEAMER

"A brilliant narrative without parallel in its depth of thought, faithfulness to fact, and pure pleasure."
—Myer Feldman, Deputy Special Counsel to President Kennedy

Includes excerpts from President Kennedy's secret White House recordings of phone conversations and private meetings.

$29.95 ($44.95 Can.)
9 Hours • 6 Cassettes
ISBN 0-694-52648-7
ABRIDGED

Available wherever books are sold, or call 1-800-331-3761 to order.